SYMMETRY

International Series in
MODERN APPLIED MATHEMATICS AND COMPUTER SCIENCE

Volume 10
Series Editor: Ervin Y. Rodin, Washington University
(Volumes in the Series have also been published as Special
Issues of the journal *Computers and Mathematics with Applications)*

RELATED TITLES
Other Volumes in the Series

COOPER & COOPER	Introduction to Dynamic Programming
SAATY & ALEXANDER	Thinking with Models
SURI	Resources Management Concepts for Large Systems
BELLMAN et al	Mathematical Aspects of Scheduling and Applications
CERCONE	Computational Linguistics
WITTEN	Hyperbolic Partial Differential Equations
SAATY & KEARNS	Analytical Planning
CHOI	Statistical Methods of Discrimination and Classification

Other Books of Interest

AVULA et al	Mathematical Modelling in Science and Technology
KRISHNA	Crystal Growth and Characterization of Polytype Structures
MARCHUK	Ocean Tides: Mathematical Models and Numerical Experiments
SINAI	Theory of Phase Transitions: Rigorous Results
STAHL	Operational Gaming

Related Journals*

Bulletin of Mathematical Biology
Computers and Graphics
Computers and Mathematics with Applications
Mathematical Modelling
Pattern Recognition
Progress in Crystal Growth and Characterization
System
Technology in Society

* Sample copies available on request

SYMMETRY
Unifying Human Understanding

Edited by
ISTVÁN HARGITTAI

Hungarian Academy of Sciences and
University of Connecticut, Storrs

International Series in
MODERN APPLIED MATHEMATICS AND COMPUTER SCIENCE,
Volume 10

Series Editor
Ervin Y. Rodin
Washington University

PERGAMON PRESS

New York • Oxford • Toronto • Sydney • Frankfurt

U.S.A.	Pergamon Press Inc., Maxwell House, Fairview Park, Elmsford, New York 10523, U.S.A.
U.K.	Pergamon Press Ltd., Headington Hill Hall, Oxford OX3 OBW, England
CANADA	Pergamon Press Canada Ltd., Suite 104, 150 Consumers Road, Willowdale, Ontario, M2J 1P9, Canada
AUSTRALIA	Pergamon Press (Aust.) Pty. Ltd., P.O. Box 544, Potts Point, N.S.W. 2011, Australia
FEDERAL REPUBLIC OF GERMANY	Pergamon Press GmbH, Hammerweg 6, D-6242 Kronberg, Federal Republic of Germany
BRAZIL	Pergamon Editora Ltda., Rua Eça de Queiros, 346, CEP 04011, São Paulo, Brazil
JAPAN	Pergamon Press Ltd., 8th Floor, Matsuoka Central Building, 1-7-1 Nishishinjuku, Shinjuku, Tokyo 160, Japan
PEOPLE'S REPUBLIC OF CHINA	Pergamon Press, Qianmen Hotel, Beijing, People's Republic of China

ISBN 0-08-033986-7

Cover art: Marko Spalatin.—Marko Spalatin, a painter, sculptor, and graphic artist, was born in Zagreb, Yugoslavia. He emigrated to the United States in 1963 and received recognition for his works at an early age. Today his works are exhibited in permanent collections internationally, in museums such as the Museum of Modern Art, New York City; Tate Gallery, London; Musee d'Art Moderne, Paris; Muzej Moderne Umijetnosti, Belgrade; as well as many other museums, institutions, galleries, and private collections.

Mr. Spalatin is a chronicler of nature in his works, interweaving nature's colors and symmetries. In his words:

"If anything, nature must be my greatest source of inspiration. There is no place more colorful and dramatic in form than a coral reef submerged in a crystal sea. I grew up near the Adriatic coast in Yugoslavia, spending many hours skindiving. However, now, for most of the year, I live where seasonal changes are very intense, with incredible transitions in color and form. I have learned to recognize and appreciate natural symmetries, whether they occur in sea life or in the fields and woods surrounding my home. . ."

The editors wish to express their deep appreciation to Mr. Spalatin for the beautiful representation of symmetry that he provided for the cover of this volume.

Published as a special issue of the journal *Computers and Mathematics with Applications*, Volume 12B, Numbers 1–4 and supplied to subscribers as part of their normal subscription.

Printed in Great Britain by A. Wheaton & Co. Ltd., Exeter

CONTENTS

v

vi   Contents

FOREWORD

The project "SYMMETRY" presented to our readers here was a very ambitious one. Its scope is tremendous, with subjects ranging from fractals through court dances to crystallography and literature. The 70 + authors hail from all corners of the world; it was not only an interdisciplinary, but also an international undertaking.

But why does it belong to *Computers and Mathematics with Applications*?

To answer this question, we must remind ourselves what were and are the basic aims of our publication. The genesis of our endeavor was, of course, the long held conviction of many on our Editorial Board that there is a one-to-one relationship between an amazingly large number of physical, biological, and other phenomena on the one hand and aspects of mathematics on the other; together with the realization that the advent of computers is making not only a quantitative, but also a qualitative difference in the understanding, and unravelling of these relationships.

The past few years have continued to confirm rapidly the validity of such views. However, it also became clear that the mere utilization of computers as machines churning out data or figures is, in all but very simple situations, an endeavor leading to blind alleys. By now it is strongly felt by most large-scale users of computers that the superimposition of certain principles is absolutely essential.

What might such principles be?

That Symmetry is one of the leading candidates as such a principle is a clear conclusion. In the words of C. N. Yang:

> . . . the general structure of the periodic table is essentially a direct and beautiful consequence of symmetry . . . , the isotropy of the Coulomb force; the existence of the antiparticles . . . were anticipated in Dirac's theory, which was built on the principle of relativistic symmetry. In both cases, as in other examples, nature seems to take advantage of the simple mathematical representation of the symmetry laws. The intrinsic elegance and beautiful perfection of the mathematical reasoning involved and the complexity and depth of the physical consequences are great sources of encouragement . . . One learns to hope that nature possesses an order that one may aspire to comprehend. [Chen Ning Yang, *Elementary Particles*. Princeton University Press (1962).]

It seems therefore that the notion of symmetry may be one which plays an ephemeral, but yet absolutely essential role in our understanding of the physical universe and our role in it. To pursue the Kantian view, perhaps symmetry is playing a role complementary to space and time. We do not do anything with space and time; but all of our understanding is couched in terms of them. They are a framework within which all else has to be placed.

It must be true that whatever is placed in this space-time framework, has to be put there according to certain rules.

Symmetry could possibly be one of the axiomatic rules of that portion of space time in which we exist.

But, as we said, it is an ephemeral rule—it makes its appearance here and there—once we notice and once we do not. Having noticed, we may remember the event—or, it may become so commonplace in our minds that we may forget. We do not particularly think of the symmetries of flowers or other things that we are in day to day contact with—we just admire their beauty. But upon looking at a picture of Escher's, its very unusual nature impells us to ask: why is it so interesting? And we will definitely remember it as an object having the property of symmetry.

It is unlikely that court dances or Moorish decorations will be of much direct use to the atomic physicists or to crystallographers (although see Feynman's latest book . . .). But if the scientist will read about them, they may occupy a special niche in his mind, because of their very unusual nature. Then, perhaps these associations may indeed bear even direct fruit in his own discipline . . .

Regardless of whether direct contacts and/or interactions are likely, it seems that a certain conclusion is still inevitable. If indeed symmetry plays such an important role in so many human endeavors and phenomena, and if it may be one of the clues to a better understanding of where, what and who we are, we ought to speak more about this subject. And we should not be like the three blind men with the elephant. We ought to talk about this to each other—in cross-disciplinary ways.

We are doing so in this volume.

It is a work of love, dedication and purpose—all exhibited by its initiator and Editor, Professor István Hargittai; and by the many authors represented here.

So, in conclusion of this Foreword, I wish to record the thanks and admiration of our entire Editorial Board to all of the authors and to Professor Hargittai for having compiled a work of a scope and quality which are the hallmarks of truly lasting reputations.

ERVIN Y. RODIN
General Editor

PREFACE

During the past year as this special issue was being organized, many people have asked me the question: how did it happen? Some 14 months ago, while sitting in my office in the Physics Department of the University of Connecticut, I received a phone call from Professor Ervin Y. Rodin of Washington University in St. Louis. He introduced himself as editor of an international journal *Computers and Mathematics with Applications* and asked me whether I would be willing to organize a special issue on Symmetry for his journal. After some hesitation, I said yes.

There was, of course, some prehistory to this phone conversation. I developed the symmetry passion at the beginning of my molecular structure studies, as I was rediscovering how symmetry can aid the determination of molecular geometry. Soon my fascination for symmetry extended beyond the molecular realm, and I started collecting examples wherever I saw them. During the mid-seventies, when our children were small, I renewed my interest in photography and started taking color slides. Then once in the late seventies, I was giving a seminar at the Chemistry Department of the Moscow State University; at the end of the talk there remained some time, and I had a few "symmetry" slides to show. The favorable response induced me to give a symmetry talk two years later when I was visiting Moscow again. Thus started my popular symmetry talks, which have been given during the eighties in such diverse places as Steinfurt (FRG); Atnasjø (Norway); Paks and Pécs (Hungary); Baku (Azerbaidzhan); Rome (Italy); Storrs, CT; Washington, D.C.; Newark, NJ; and Gainesville, FL. These lectures have generated a lot of interest and much interaction with other people from various fields. At about the same time that my lectures began, I wrote a paperback in Hungarian, *Symmetry through the eyes of a chemist*[1], for a popular science series of the Hungarian Academy of Sciences. It appeared in Budapest during the fall of 1983 just as my visiting professorship was beginning in Storrs, CT. I received a lot of copies of my book and I was giving them away liberally. Through some mutual (non-scientific) friends, a copy reached Ervin Rodin who apparently became interested . . .

The present volume is no mere accident, however. Throughout my years of meeting numerous people—mainly scientists, but some artists as well—a desire to see more information on symmetry has been expressed repeatedly. It is of course widely recognized that symmetry is a general unifying concept, and specialists of different fields see it also as a means to relate their areas. Among a considerable number of artists there is the desire to reach out to scientists, and symmetry is viewed as a convenient vehicle for communication. Besides, there is some nostalgia for earlier projects, like the symmetry enterprise of Studium Generale in 1949[2] and the later volumes organized by Gyorgy Kepes[3].

As our project got under way, the response and interest was overwhelming. Even the few people who turned down my invitation declined in most cases on the ground that such a worthy project should be done in a more central journal. I was given only one negative comment, which I interpreted as positive anyway: I was warned that such a project would definitely get out of hand and I was opening a Pandora's box.

In order to explain the motivation behind and the goals of our project, I would like to cite here from my letter of invitation originally sent to prospective contributors:

> It is, of course, commonplace today that not only is symmetry one of the fundamental concepts in science, but it is also possibly the best bridging idea crossing various branches of sciences, the arts, and many other human activities. Whereas symmetry has been considered important for centuries primarily for its aesthetic appeal, this century has witnessed a dramatic enhancement in its recognition as a cornerstone scientific principle. In addition to traditionally "symmetry-oriented" fields such as spectroscopy or crystallography, the concept has made headway in fields as varied as reaction chemistry, nuclear physics or the study of the origin of the Universe. At the same time, in its traditional fields its meaning and utility have greatly expanded. It may be sufficient to refer to antisymmetry, dynamical symmetry, generalized crystallography, or the symmetry analysis of music and of artistic design.

This special issue is envisioned exactly in the spirit of making it primarily a contribution to this bridging role of symmetry. The plan is to bring together the most diverse fields from mathematics to psychology, space research, musicology, chemistry, the folk arts, literature, crystallography, aesthetics, modern physics, philosophy and others.

Outstanding representatives from the various fields are being asked each to prepare a chapter of quasi-monographic character. It is left to the author to what extent a general survey of a particular area will be attempted or a more personal approach will be emphasized.

Complete coverage of the topic is intrinsically impossible. The objective is to have the special issue present the contemporary status of symmetry studies and man's views of its role and importance. Further, it is hoped that the special issue will open new avenues in symmetry studies, by utilizing existing scientific and artistic achievement, and also by re-juvenating interest in fields where symmetry studies have not recently advanced.

The title of the journal should not mislead anyone into narrowing the scope and depth of the contributions. The journal considers it a service to both the Two Cultures to house this special issue. As an outgrowth it would not be surprising, however, to see fruitful consequences of this undertaking appear as widened interest in the applications of computers and mathematics in fields concerned with symmetry aspects.

As the scope of the special issue will be interdisciplinary to the extreme, its readership will similarly be anticipated as such. It would thus be appreciated to have the contributions be prepared in such a way as to appeal to a relatively wide audience. It is clear, however, that this may be achieved in different fields to different extents.

Obviously my intentions were somewhat overambitious, but in some unmodest moments I feel pleased at how much of it could be accomplished during less than a one-year period. The experience we are sharing with Dr. Rodin and the much hoped-for success of this initial project may open the way for continuation. The original objective may prove too elusive ever to reach completely, but it is worthy of aiming for anyway.

The interdisciplinary character of this series of special issues on symmetry is expressed also by the loose arrangement of the contributions. The papers are not grouped by specialization, except in isolated instances where explicit interaction between two papers made such an arrangement useful. We believe that even a glance through papers remote from one's own field will stimulate further interest. Be aware, I am not attempting an "umbrella" introduction of what follows; the symmetry concept is the most potent linkage of the papers in these issues.

There is one more bridging feature of these special issues worth mentioning. It is a truly international undertaking. Contributions from 18 nations appear in these issues, and I hope that representatives of many other countries will join in in the future. Symmetry is making a symbolic contribution to human communication on this broader scale.

These thoughts are rather personal, but this symmetry volume is the result of the collective effort of many people. The most important part is, of course, the contributions themselves, and I thank every one of the authors for it. Many authors, and others, have made their imprint on this project in other kinds of participation as well. From providing encouragement, advice, suggestions, up to refereeing the papers, we have benefited from an enormous amount of assistance. I would like to list here at least some of the names of the people to whom I personally, and the project as a whole, owe a lot: Leonid V. Azaroff (Storrs, CT), Ralph H. Bartram (Storrs, CT), Donald W. Crowe (Madison, WI), Julius A. Elias (Storrs, CT), Myriam Giambiagi (Rio de Janeiro), Vladislav V. Goldberg (Newark, NJ), Anding Jin (Nanjing), Jay Kappraff (Newark, NJ), Gyorgy Kepes (Cambridge, MA), Kozo Kuchitsu (Tokyo), Alfred H. Lowrey (Washington, D.C.), Alan L. Mackay (London), David Markowitz (Storrs, CT), Saul Patai (Jerusalem), Vladimir Prelog (Zürich), Wolfgang Schirmer (Berlin), Eugene P. Wigner (Princeton, NJ).

I would also like to record my appreciation to the secretarial staff of the Institute of Materials Science of the University of Connecticut, Ms. Lane Witherell, Ms. Carol Blow, and Ms. Doris Rook, for efficient assistance. The conception, organization, and preparation of the special issue occurred during my visiting professorship (in physics, 1983/84, and in chemistry, 1984/85) at the University of Connecticut; the pleasant and helpful atmosphere of this School greatly helped to carry out this rather unique undertaking.

As I look back on this period of enthusiasm and anguish, hard work, and pleasure and

excitement at bringing out these issues, I find it most natural to dedicate my part in this project
to my wife Magdi, who shares it all.

February, 1985

Institute of Materials Science and
Departments of Physics and Chemistry,   ISTVÁN HARGITTAI†
University of Connecticut,
Storrs, CT 06268, U.S.A.

REFERENCES

1. I. Hargittai, *Szimmetria egy kémikus szemével*. Akadémiai Kiadó, Budapest (1983). Revised and expanded English
   version will be published in 1986: I. Hargittai and M. Hargittai, *Symmetry through the eyes of a chemist*. VCH
   Verlagsgesellschaft, Weinheim and VCH Publishers, Deerfield Beach, FL.
2. *Studium Generale*, Zweiter Jahrgang, Viertes und fünftes Heft, pp. 203–278 (1949).
3. G. Kepes (Ed.), *Vision + Value Series*. G. Braziller, New York (1965, 1966).

†Visiting Professor (1983/85). On leave from the Hungarian Academy of Sciences, Budapest, P.O. Box 117,
H-1431 Hungary.

SYMMETRY
Unifying Human Understanding
(Part 1)

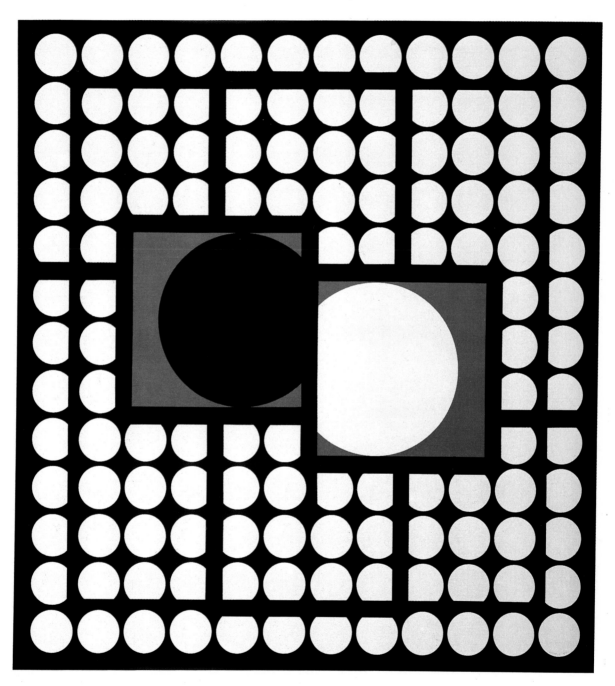

Victor Vasarely, "BELLA-NEG" (1957), courtesy of the artist

Comp. & Maths. with Appls. Vol. 12B, Nos. 1/2, pp. 1–17, 1986
Printed in Great Britain.

0886–9561/86 $3.00 + .00

LIMITS OF PERFECTION

Ist-ván Hargittai†

Institute of Materials Science and Departments of Chemistry and Physics, University of Connecticut,
Storrs, CT 06268, U.S.A.

Abstract—This personal narrative is an introduction to a collective effort by a number of scientists and artists to examine the role and significance of symmetry in the most diverse domains of nature and human activity. Material symmetry, devoid of the rigor of geometrical symmetry, is viewed applicable to material objects as well as abstractions with limitless implications.

To mark the 350th anniversary of Johannes Kepler's death, the Hungarian Post Office issued a beautiful memorial stamp (Fig. 1). Next to Kepler's portrait his famous model of the planetary system is shown. This was fitting since of all of Kepler's discoveries this is the best known to the general public, although it is viewed by some as his most spectacular failure[1].

A closer look at Kepler's activities, however, justifies the selection of the Hungarian Post Office. Although he is most famous for his three laws of heavenly mechanics, there is another piece of work that was also a milestone in a different branch of science, crystallography. If one is astonished by the depth of his understanding of the physics of the sky with the then available data, it is not less astonishing that Kepler could discuss the "atomic" arrangement in crystals two hundred years before Dalton and three hundred years before X-ray crystallography began. In his new year's gift of the hexagonal snowflake Kepler[2] not only examines the hexagonal symmetry of the snow crystals but lays the foundation of the principle of densest packing in crystal structures. Densest packing is then, of course, the key to the symmetry of crystal habit. The planetary model from the regular solids is also a densest packing model. Kepler's search for harmony was the bridge between his two lines of activities. Although the snowflake paper seems to be almost an accident on the background of his astronomy, the Hungarian stamp gave more credit to the *complete* Kepler than was probably envisioned by the planners of the stamp themselves.

According to the regular solids model, taking the *six* planets, known to Kepler, in order, the greatest distance of one planet from the Sun stands in a fixed ratio to the least distance of the next outer planet from the Sun. There are, of course, conveniently five ratios for the six planets. A regular solid can be interposed between two adjacent planets so that the inner planet, when at its greatest distance from the Sun, lies on the inscribed sphere of the solid, while the outer planet, when at its least distance, lays on the circumscribed sphere.

There are molecular structures which can be best described by polyhedra enveloping other polyhedra. The structure of $[Co_6(CO)_{14}]^{4-}$ is shown is Fig. 2: an omnicapped cube of carbonyl oxygens envelopes an octahedron formed by cobalt atoms[3].

One of today's most successful models in structural chemistry is based on extremely simple considerations of space distribution. The valence shell electron pair repulsion (VSEPR) model[4] postulates that the geometry of the molecule is determined by the space requirements of the electron pairs in the valence shell of its central atom. The bond configuration around atom A in an AX_n molecule is such that the electron pairs in the valence shell be at maximum distances from each other. Thus the arrangement may be visualized so that the electron pairs occupy well-defined parts of the space about the central atom. In a different concept, these space segments are called localized molecular orbitals.

It is easy to demonstrate the three-dimensional consequences of the VSEPR model. Only a few balloons have to be blown up and connected at their narrowing ends in groups of two, three or four[5]. The linear, equilateral triangular, and tetrahedral arrangements of these assemblies are what the VSEPR model predicts for the electron pairs. Another beautiful analogy is found on walnut trees[6]. As two, three, or four walnuts grow sometimes together, the above

†Visiting Professor (1983/85). Permanent address: Hungarian Academy of Sciences, P.O. Box 117 Budapest, H-1431, Hungary.

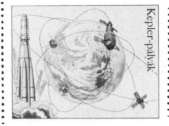

Fig. 1. Kepler memorial stamp. (Hungarian Post Office, 1980.)

arrangements occur unfailingly (see Fig. 3). The soft balloons and hard walnuts may even be viewed as representing weaker and stronger interactions and thereby represent even more subtle analogies for molecular structure. Can molecular geometry be so simple? There is obvious oversimplification in the model. Of the many effects determining molecular structure, one is taken into account and all the others are ignored. The model is applicable where this particular effect, to wit the repulsions of the electron pairs due to their space requirements, is dominant. The VSEPR model is successful because this effect is important enough in extensive classes of compounds.

In comparing the complexity of planetary motion and molecular structure, the real analogy is in the possibility of selecting a dominant effect and ignoring the others. This approach works much better for the planetary motion where the dominant effect is the gravitational attraction of the Sun while the others are perturbations. In the world of molecules, the dominant effect may change from one compound to another. In spite of the tremendous amount of accumulated knowledge about molecular structure, its basic principles are still being clarified. One of the characteristics of the forces keeping the molecule together is that they are very strong, whereas the gravitational forces are very weak. To discover the law of gravitational interactions, the observations had to be made on a large scale. The laboratory was the planetary system itself.

As we compare the symmetries of molecules and crystals, a striking difference is that there are no limitations for molecules and there are well-defined limitations for crystals. A consequence is the finite number (32) of symmetry classes for crystal habit with no such limits for molecules.

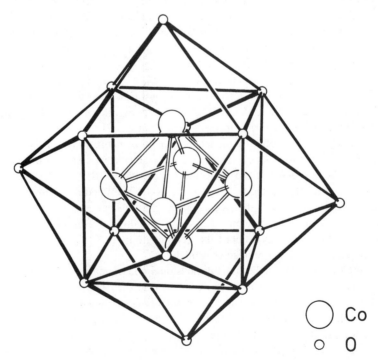

Fig. 2. The structure of $[Co_6(CO)_{14}]^{4-}$, the omnicapped cube of the carbonyl oxygens envelopes the cobalt octahedron. (After[3].)

Fig. 3. Walnut clusters with two, three, and four walnuts with linear, equilateral triangular, and tetrahedral arrangements. (Photographs by the author following the idea of Niac and Florea[6].)

The molecules are a more fundamental building unit in the hierarchy of structures than the crystals and many crystals themselves are built from molecules. Unfortunately, molecules are usually not to be seen by the naked eye whereas the crystals are. They are so appealingly symmetrical that they have become a sort of idol for symmetry.

For the Béla Bartók centenary a couple of years ago (1981), Victor Vasarely produced a limited edition of ten serigraphs created for ten of Bartók's musical pieces. Each serigraph was accompanied by a poem, each written by a different living Hungarian poet. More than once the word *crystal* was related to Bartók's music. If one considers their symmetry properties in a strictly technical sense, however, they could not be farther from each other. Bartók's music is interweaved throughout by the Fibonacci numbers and the golden section. These symmetries characterize for example the scattered leaf arrangements of many plants and are omnipresent in other domains of the animate world as well. On the other hand, "crystallization is death", as crystallographers themselves like to point out. Incidentally, pentagonal symmetry is conspicuously present in primitive organisms and crystallographer Nikolai Belov[7] suggested that it was their means of self-defense against crystallization.

The restrictions on crystal symmetry start with the fact that, strictly speaking, it exists only theoretically. Its main characteristic is translational symmetry, i.e. *infinite* periodic repetition, and in reality, of course, crystals always end somewhere; but apart from this, there is no five-fold symmetry axis, nor are there axes with higher order than six. Considering the more easily visualized two-dimensional surface of the regular polygons, only the equilateral triangle, the square and the regular hexagon can be used to completely cover an area without gaps, whereas the regular pentagon, heptagon, etc., can not. Similar limitations extend to the third dimension.

In spite of the limitations in crystal symmetry, the crystals have a unique appeal and are widely used also in analogies. Quote the Czech writer Karel Čapek[8] on his visit to the mineral collection of the British Museum

> . . .But I must speak again about crystals, shapes, colors. There are crystals as huge as the collonade of a cathedral, soft as mould, prickly as thorns; pure, azure, green, like nothing else in the world, fiery, black; mathematically exact, complete, like constructions by crazy, capricious scientists, or reminiscent of the liver, the heart. . . There are crystal grottos, monstrous bubbles of mineral mass, there is fermentation, fusion, growth of minerals, architecture and engineering art. . . Egypt crystallizes in pyramids and obelisks, Greece in

Fig. 4. Illustration for phyllotaxis. (Photograph by the author.)

columns; the middle ages in gilly-flowers; London in grimy cubes. . . Like secret mathematical flashes of lightning the countless laws of construction penetrate the matter. To equal nature it is necessary to be mathematically and geometrically exact. Number and phantasy, law and abundance—these are the living, creative strengths of nature; not to sit under a green tree but to create crystals and to form ideas, that is what it means to be at one with nature!

This is then exactly the point where Bartók and the crystals meet. While the composer invariably refused to discuss the technicalities of his work, he liked to state "We create after Nature" and he meant it literally.

As a crystal is being built from molecules, and the energetically most favorable arrangement is being achieved, the molecules come into close, touching range with each other. Compared with the free molecules, their interactions may have perturbing effects on their structures. One of the simplest consequences may be the lowering of their original symmetry. Discussing the structural consequences of densest packing in molecular crystals, the following explanation is attributed to crystallographer Aleksandr Kitaigorodskii, "The molecule also has a body. When this body is hit, the molecule feels hurt all over." This analogy emphasizes the importance of spatial requirements of the molecules in building the molecular crystal rather than the peculiarities of its electronic structure which would be of greater importance in a more chemical behavior as in a chemical reaction. Personifying the molecule obviously had an appeal to the scientist, and it also has an appeal for children who generally like to do so with various objects. When some years ago the author's daughter was asked in the kindergarten about her father's occupation, she said he cured sick molecules. The tendency to make metaphors, however, seems to diminish with coming of age. Some valuable things may be lost in the educational process. It is much easier to get through to children the notion that there is much more to symmetry than what we call bilateral symmetry. Adults seem to be more narrow-minded and more indoctrinated.

Not everything is perfect, however, even with the most symmetrical molecules. Concerning their symmetry, that is. When the symmetry of a molecule is described, it is usually the motionless, frozen molecule that is meant. This structure would correspond to the minimum energy. The molecules are never motionless, however. Even if they could be cooled to the absolute coldest temperature of 0 K or $-273.16°C$, they would not come to a standstill.

The molecular vibrations often lead to some instantaneous distortion or lowering of the molecular symmetry. This is true for the relatively rigid molecules and even more so for the very flexible molecules. Intramolecular motion may even lead to a continuous permutation of the atoms in the molecule.

Imagine a ring molecule of five atoms with four being in one plane and the fifth atom sticking out of the plane. This arrangement lowers the symmetry of the five-member ring to a

Fig. 5. Henri Matisse: Dance. (The Hermitage, Leningrad.) Reproduced by kind permission from The Hermitage.

mere symmetry plane. The point is that this molecule may be performing such intramolecular motion that during every second a million times or more the sticking out position switches from one atom to the next and to the next and so on.

Consider now five dancers in a circle instead of the five atoms (see Fig. 5). Let them make a jump one after the other in a quick succession. If we take photographs with very short exposures, we can catch various configurations, including even the most symmetrical one in which all the five dancers are on the ground. On the other hand, a longer exposure leads to a blurred picture all around the circle. The apparent symmetry of the dancing group obviously depends on the length of the exposure used, and also, of course on the speed of their movement.

The molecular structures and crystal structures represent two well-separated cases from the point of view of their symmetry properties. The molecule is characterized by point-group symmetry as it has at least one unique point in the whole structure. Crystal structures are characterized by space-group symmetry or translational symmetry as they have no unique point in their structure. When point-group and space-group symmetries are compared, it is not obvious how to distinguish between higher and lower symmetries. Within each domain, however, and the molecules and crystals are merely examples, there is a hierarchy of symmetries.

Increasing the symmetry beyond some limits may lead, however, to sterility and certainly diminishes the information content. Scientific instruments with ever increasing perfection may filter out important peculiarities which do not conform with the general pattern. Perfect symmetry may be aimless, and it irritates many. Perhaps symmetry considerations could facilitate relating the perception of structures and the world of emotions? On the level of analogies this seems to be possible as is illustrated by a poem by Ann Wickham[9]:

GIFT TO A JADE

For love he offered me his perfect world.
This world was so constricted and so small
It had no loveliness at all,
And I flung back the little silly ball.
At that cold moralist I hotly hurled
His perfect, pure, symmetrical, small world.

Geometrical symmetry is strict: it allows for no "degrees" of symmetry. Something is either symmetrical or not. What may be called *material symmetry*, on the other hand, implies a continuous spectrum of the degrees of symmetry. The term material symmetry here refers generally to non-geometrical symmetry and may be applied to real material objects as much as to any abstraction.

The human face is an obvious example of bilateral symmetry. However, none of us has a perfectly symmetrical face (Fig. 6). It may be a matter of flattery on the part of the painter to show more perfect symmetry than there is, or this may even be demanded by the paintee. Old religious paintings or contemporary political portraits may show more facial symmetry than there is. Any personality cult produces very symmetrical face images. However, minor asymmetries may have a strong appeal and the notion about the beauty of a face may also be changing. There was a minor uproar some time ago when a Budapest theater critic praised the acting of an actress and remarked also on her pretty, *modern* face. In fact, hers was conspicuously asymmetric, so people were wondering whether their faces were modern or old-fashioned. There is considerable interest currently in the origin and meaning of facial asymmetry.

Another fascinating symmetry/asymmetry relationship exists between our hands (Fig. 7). Distinguishing between left and right has definite connotations in almost all fields of human activity. The importance of chirality is ever growing in the sciences. Even human attitude toward handedness is evolving. Figure 8 shows two pictures from classrooms of the University of Connecticut. The older classrooms of the Chemistry Department have homochiral chairs only, designed for right-handed students. The more contemporary classrooms of the Mathematics Department are furnished with heterochiral chairs to accommodate both the right-handed and the left-handed students.

Fig. 6. Eszter Hargittai in front of a shop-window, 1984. (Photograph by the author.)

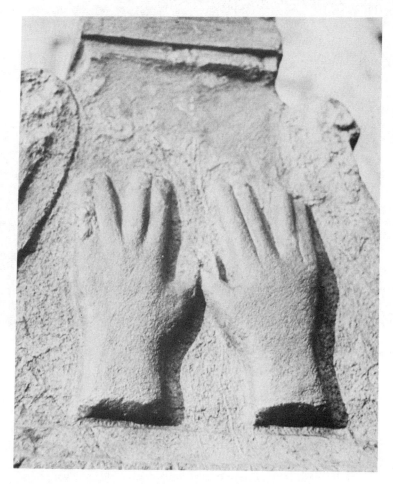

Fig. 7. Tomb in the Jewish cemetery, Prague. (Photograph by the author.)

Fig. 8(a). Classroom in the Chemistry Department, University of Connecticut, 1984.

Fig. 8(b). Classroom in the Mathematics Department, University of Connecticut, 1984. (Photographs by the author.)

Artistic creation as reflection may vary in a limitless range. The two photographs in Fig. 9 illustrate this point. Construction machinery on the bank of the Danube is shown in one of the photographs. Its reflection in the Danube is captured in the other. Imagine a mirror-perfect water surface with much more likeness, or gale conditions destroying any trace of resemblance. The origin of reflection itself as seen by Jean Effel is depicted in Fig. 10.

The sciences of the 20th century have opened up a new world, one which had previously been inaccessible to man's instruments, let alone his senses. To parallel this development, artistic expression is coping, or is at least attempting to do so, in reflecting our world on an

Fig. 9(a). Building construction machinery on the bank of the Danube.

Fig. 9(b). Reflection of the building construction machinery in the Danube. (Photographs by the author.)

entirely new level. It is not that the artist is expected to bend over the screen of an electron microscope and paint ''after nature'', but the newly discovered domains and phenomena must and do find their reflections in artistic expression. It may be considered symbolic that X-ray crystallography and the Black Square of Kazimir Malevich were born about the same time. The picture from 1915 is reproduced in Fig. 11 along with the title page of a later Malevich work.

Current progress mandates the expansion of the well-established frameworks of the symmetry concepts. One of the cradles of modern symmetrology, crystallography is transforming itself to embrace all structural science on the atomic level[10]. Liquids, colloids, amorphous

Institution du reflet

— Les riverains auront, gracieusement, leur portrait à l'aquarelle...

Fig. 10. The creation of reflection, by Jean Effel. [La Creation du Monde]. (''Those who dwell by the river will have their portrait, gracefully, in water colors. . .''). (Reproduced by kind permission from Mme. Jean Effel.)

solids cannot be put into the existing ''perfect'' systems even though they are not without structure. We are witnessing their emancipation. We quote Ann Wickman[9] again:

THE WOMAN AND HER INITIATIVE

Give me a deed, and I will give a quality.
Compel this colloid with your crystalline.
Show clear the difference between you and me
By some plane symmetry, some clear stated line.
These bubblings, these half-actions, my revolt from unity.
Give me a deed, and I will show my quality.

John Bernal was the pioneer of generalized crystallography, and Belov noted in his obituary ''. . .his last enthusiasm was for the laws of lawlessness.'' Did Ann Wickham's metaphors parallel Bernal's discoveries? Although several disciplines apart, geographically and chronologically they operated in close range (London).

In any case the non-classical, irregular, unstable, unusual, unexpected are gaining impor-

Fig. 11(a). Kazimir Malevich: Black square (1915?).

Fig. 11(b). Kazimir Malevich: Suprematism, 34 drawings, UNOVIS, Vitebsk, 1920, Title page.

tance in the sciences. The "morphology of the amorphous" is being investigated and "mapped"[11]. The 1978 Nobel laureate physicist Philip Anderson stated that "the next decade is very likely to be the most 'disordered' decade in theoretical physics."[12]

The increasingly recognized importance of non-periodic structures makes even more conspicuous the absence or near-absence of the three-dimensional space groups outside the world of crystals. The honeycomb built by the bees is a notable exception. Its regular hexagonal structure is then partially copied by the construction of the concrete base of off-shore oil platforms. There is unmistakable similarity to the honeycomb in the hexagonal joints of basaltic sheets resulting from contraction during cooling. These are examples that meet the eye and there may be others hidden.

How much visible and how much less visible symmetry is there in the arts? There is a lot of visible symmetry in Bach's music and there is also a lot in Bartók's—not so visible but well-established by research. There is a wealth of symmetry phenomena in other music as well as in other arts, not only in paintings and sculptures whose symmetry properties are most commonly considered.

One of the less frequently perceived symmetries is inversion, a combination of applying a two-fold rotation axis and a symmetry plane. Here is an example from Hungarian author Frigyes Karinthy's short story entitled "The same in man"[13] represented by some edited fragments. There are three characters: Bella the beloved lady, Fox the employee and Bella's suitor Sándor who is also Fox's boss. The editing means to present the two meeting in a parallel way rather than consecutively.

BELLA	FOX
Sometimes I just gaze before me without thinking of anything. . .	Sometimes I just gaze before me without thinking of anything. . .

SÁNDOR/BOSS

| Bella! If only you knew how beautifully you expressed yourself. . . | On my money? Then you'd better go to a lunatic asylum, that's where cases like you are treated. . . |

BELLA	FOX
Sometimes I have the feeling that I'd like to be somewhere else than I am. I can't say where, somewhere I haven't been before.	I often have the feeling that I'd like to be somewhere else than I am. I don't know where, anywhere, somewhere, I haven't been before.

SÁNDOR/BOSS

| Bella, how true, how wonderful. . . How did you put it? Let me engrave it in the records of my mind. . . | The nuthouse, man, the nuthouse. That's where you belong. |

There is inversion in James Reston's description of New Zealand in his Letter from Wellington, Search for End of the Rainbow[14]:

> Nothing is quite the same here. Summer is from December to March. It is warmer in the North Island and colder in the South Island. The people drive on the left rather than on the right. Even the sky is different—dark blue velvet with stars of the Southern Cross—and the fish love the hooks.

Wellington and Madrid are approximately connected by a straight line going through the center of the Earth which is then the inversion center for them. It is too bad that the journalist did not date his letter from Madrid.

The black-and-white variation is the simplest case of color symmetry and it is also the simplest example for antisymmetry (see Fig. 12). The relationship between matter and antimatter is another example. "Operations of antisymmetry transform objects possessing two possible values of a given property from one value to the other"[15]. According to this general definition antisymmetry can be given broad interpretation and application. Geometrically less strict but in their atmosphere truly black and white antisymmetries are presented in Fig. 13.

Another literary example is taken from Karinthy's writing, this time to illustrate antisymmetry. It is edited from a short story entitled "Two Diagnoses"[16]. The same person Dr. Same goes to see a physician at two different places. At the recruiting station he would obviously

(a)

(b)

Fig. 12. Antisymmetries. (a) Hungarian motif after [17]. (b) Zagorsk. (Photograph courtesy of Dr. A. A. Ivanov, Moscow.)

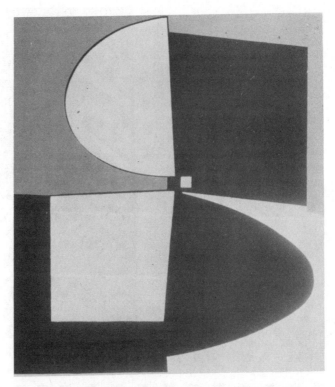

Fig. 13(a). Victor Vasarely: ''P62-Basilan'', 1951. (Reproduced by kind permission from the artist.)

Fig. 13(b). Neizvestnii: N. S. Khrushchev's tomb, Novodevichi cemetery, Moscow.

Fig. 14. Two restaurants in downtown Washington, D. C.: *the* Sans Souci and *a* McDonald's. (Photograph by the author.)

Fig. 15. Logo of sporting goods store in downtown Boston, MA. (Photograph by the author.)

Fig. 16. A breed apart. Poster. (Reproduced by kind permission from Merrill Lynch Company.)

Fig. 17. Military jets and a sea gull, off Bodø, Norway. (Photograph by the author.)

like to avoid being drafted, while at the insurance society he would like to acquire the best possible terms for his policy. His answers to the identical questions of the two physicians are related by antisymmetry.

DR. SAME	PHYSICIAN	DR. SAME
At the recruiting station		At the insurance society
Broken-looking, sad, ruined human wreckage, feeble masculinity, haggard eyes, wavering movement.		*Young athlete with straightened back, flashing eyes.*

How old are you?

Old. . . very old, indeed.	*Coyly,* On, my gosh, I'm almost ashamed of it. . . I'm so silly. . .

Your I.D. says you're thirty two.

With pain. To be old is not to be far from the cradle—but near the coffin.	To be young is not to be near the cradle, but far from the coffin.

Are you ever dizzy?

Don't mention dizziness, please, Doctor, or else I'll collapse at once. I always have to walk in the middle of the street, because if I look down from the curb, I become dizzy at once.	Quite often, sorry to say. Every time I'm aboard an airplane and it's up-side-down, and breaking to pieces. Otherwise, not. . .

Two restaurants stand side by side in downtown Washington, D. C. One is the one-of-a-kind exclusive *San Souci*, the other is a *McDonald's* of the famous fast-food chain. The antisymmetry plane appears physically as a vertical wall between the two restaurants (Fig. 14).

Antisymmetry may be powerful in focusing attention, showing contrast, emphasizing a point. Figure 15 shows the logo of a sporting goods store in downtown Boston. The antisymmetry plane emphasizes that both winter *and* summer sport fans are welcome. A poster of the Merrill Lynch investment company is reproduced in Fig. 16. There is a horizontal antisymmetry plane in "a breed apart". Military jets and a sea gull fly on the two sides of an imaginary antisymmetry plane in our ultimate example (Fig. 17).

REFERENCES

1. A. Koestler, *The Sleepwalkers*, The University Library. Grosset and Dunlap, New York (1963).
2. J. Kepler, *Strena, seu De Nive Sexangula*, 1611. English translation, *The Six-cornered Snowflake*. Clarendon Press, Oxford (1966).
3. R. E. Benfield and B. F. G. Johnson, The structures and fluxional behaviour of the binary carbonyls; A new approach. Part 2. Cluster carbonyls $M_m(CO)_n$ (n = 12, 13, 14, 15, or 16). *J. Chem. Soc. Dalton Trans.*, 1743–1767 (1980).
4. R. J. Gillespie, *Molecular Geometry*. Van Nostrand Reinhold Co., London (1972).
5. H. R. Jones and R. B. Bentley, Electron-pair repulsions, a mechanical analogy. *Proc. Chem. Soc.*, 438–440 (1961).
6. G. Niac and C. Florea, Walnut models of simple molecules. *J. Chem. Educ.* **57**, 429–429 (1980).
7. N. V. Belov, *Ocherki po strukturnoi mineralogii*. Nedra, Moskva (1976).
8. K. Čapek, *Anglické Listý*. Československý Spisovatel, Praha (1970). The English version cited here was kindly prepared by Dr. Alan L. Mackay, Birkbeck College (University of London), 1982.
9. A. Wickham, *Selected Poems*. Chatto and Windus, London (1971).
10. A. L. Mackay, Crystallography—the continuous re-definition of the subject. *Indian J. of Pure and Appl. Phys.* **19**, 765–768 (1981).
11. B. B. Mandelbrot, *The Fractal Geometry of Nature*. Freeman, New York (1983).
12. V. F. Weisskopf, Contemporary Frontiers in Physics. *Science* **203**, 240–244 (1979).
13. F. Karinthy, *Grave and Gray*, Selections from his Work. Corvina Press, Budapest (1973).
14. J. Reston, Letter from Wellington. Search for end of the rainbow. *International Herald Tribune*, Thursday, May 7, 4-4 (1981).
15. A. L. Mackay, Expansion of Space-Group Theory. *Acta Cryst.* **10**, 543–548 (1957).
16. F. Karinthy, *Selected Works* (in Hungarian). Szépirodalmi, Budapest (1962). The English translation was kindly checked by Dr. R. B. Wilkenfeld, Professor of English, University of Connecticut, 1984.
17. Gy. Lengyel, *Handiwork. New Techniques—New Solutions* (in Hungarian). Kossuth, Budapest (1975).

Comp. & Maths. with Appls. Vol. 12B. Nos. 1/2, pp. 19–20, 1986
Printed in Great Britain.

BUT WHAT IS SYMMETRY?

ALAN L. MACKAY
Department of Crystallography, Birkbeck College (University of London). Malet Street,
London WCIE 7HX U.K.

Abstract—An attempt is made to show that "symmetry", by prescribing equalities between objects at a particular level, implies that there are higher levels of organisation and thus a hierarchy of ordering.

"Symmetry" is the classical Greek word ΣYM-METPIA, "the same measure", due proportion. Proportion means equal division and "due" implies that there is some higher moral criterion. In Greek culture due proportion in everything was the ideal. The word and the usage have been taken over as a technical term into most European languages. The Chinese word, also embedded deeply in Chinese culture, indicates reciprocity.

To say that an object or a situation is symmetrical in space-time coordinates x, y, z, t, means that part of the object (etc.) has the same measure as another part. Measure implies identity to within the limits of the measuring device employed. Philosophically we may, legitimately or otherwise, extrapolate to "absolutely symmetrical", but this may lead us into paradox or contradiction. Assuming absolute indistinguishability leads us to the strange world of wave mechanics, so markedly different from the world at our everyday scale.

Symmetry is a working concept. If all the object is symmetrical, then the parts must be halves (or some other rational fraction) and the amount of information necessary to describe the object is halved (etc.).

Thus, once having identified an object as "symmetrical", we can divide it into *motif* and *rule of repetition*. That is, we can see it on two levels at once; the level of physical structure and the level of informational or organisational structure. Paradoxically, by becoming simpler (through being composed of two equal parts) it becomes more complex (because we see that there is a higher general law relating the two parts). Seeing the general law enables us to infer probable properties (through the principle of Pierre Curie and of Neumann) since the properties cannot be less symmetrical than the structure. [Neumann's Principle is: "The symmetry elements of any physical property of a crystal must include the symmetry elements of the point group of the crystal". Thus, a centro-symmetric crystal cannot be pyroelectric, which would require the two symmetrically related ends to behave differently towards a change of temperature.]

If an object (etc.) such as a large crystal with 10^8 repetitions in each direction, were extremely symmetrical, then it would be less interesting: the motif and the rules for its repetition would be too small. If an object is completely without rules of structure, as molecules of water in a mist, then its structure is all on one level and again it is not so interesting. Thus, in between there must be some maximum of interest. On the basis of information theory, we might guess this to be when the necessary information was distributed roughly equally on all levels.

The total information should have some preset value, matched to the observer or participant. It is ideal that a masterpiece should exactly occupy the observer's processing capacity: too small and the observer is bored (his channel capacity or bandwidth is pre-empted but is not used): too much and he is overwhelmed.

In a structure where information is to be put into a number of categories at the same level the most probable distribution is found by maximising

$$\Sigma \left(-p_i \log_i \right).$$

If nothing is known *a priori*, then the probabilities are expected to be equal. Given the total information in bits, how should it be distributed into a hierarchy? Usually the pre-knowledge which we bring to a situation is that it can be so distributed. It is the essence of the human perceptive machine that it should be organised as a hierarchy and our minds are geared to doing this. Probably this is a question of education, culture, habit and capacity, but let us say that we may be able to handle seven levels of conceptualisation (to choose a number out of the air without much justification—eventually a better understanding of the brain should be able to supply a figure). Roughly speaking N bits at one level will be represented by $\log_2 N$ bits at the next higher level.

The army is a model hierarchy. In the Turkish army the ranks are *Onbashi* (head of ten—Corporal), *Yuzbashi* (head of a hundred—Captain), *Binbashi* (head of a thousand—Major). Here the numbers are logarithms to base 10 in the corresponding level. An army with a different "fan-out ratio", having for example an officer for every two men, would have quite different properties. We may note that the Corporal and the Captain are in locally symmetrical situations, where each commands ten below and reports as one of ten to the level above.

To see symmetry is to see regularity and redundancy but also to see *regularity in the rules* and so on hierarchically. We approach Goedel's paradox in that in looking at an apparently closed symmetrical system we see that it is not, after all, closed.

Undoubtedly the brain has a very powerful drive to relate everything to everything else in some internal network. This may be continued until we reach a final state, sometimes described as the "oceanic state" a feeling where the self and the universe are perceived as harmoniously related.

Recapitulating our view, we can see that the highest approbation of a work of art (a painting by Bosch, a novel by Mann, a play by Shakespeare) is that it is true on several levels at once: it is technically excellent; it is in the style of a certain time and place: it is an expression of the sentiments of certain people: it is representative of the "*Zeitgeist*": all together, at each level, it has the right blend of expectedness (symmetry) and unexpectedness (information).

The measure of mankind is man, but equally the measure of man is mankind. Symmetry not only ties objects together at the same level, it shows that objects at different levels have the same measure—συμμετρια.

Comp. & Maths. with Appls. Vol. 12B, Nos. 1/2, pp. 21–37, 1986
Printed in Great Britain.

0886-9561/86 $3.00 + .00
© 1986 Pergamon Press Ltd.

GENERALISED CRYSTALLOGRAPHY

ALAN L. MACKAY

Department of Crystallography, Birkbeck College (University of London), Malet Street,
London WC1E 7HX, U.K.

Abstract—Classical crystallography consists of the study of those atomic assemblies which can be generated by the function GROUP (MOTIF) ⇒ PATTERN. We are now concerned with the more general pattern-generating function: PROGRAM (MOTIF) ⇒ STRUCTURE, which requires us to look into the concept of INFORMATION in which the PROGRAM is written. Some structures may contain their own PROGRAMS where the information is represented as a material structure. Inorganic and living matter are part of a continuous spectrum of organisation. We present the outlines of a philosophy of structuration or structuralism based on the scientific rather than on the literary culture.

> A crystal lacks rhythm from excess of pattern, while a fog is unrhythmic in that it exhibits a patternless confusion of detail

—Alfred North Whitehead (1861–1947). [*The Principles of Natural Knowledge*. Cambridge (1925) p. 198]

> The difference between a piece of stone and an atom is that an atom is highly organised, whereas a stone is not. The atom is a pattern, and the molecule is a pattern, and the crystal is a pattern; but the stone, although it is made up of these patterns, is just a mere confusion. It's only when life appears that you begin to get organisation on a larger scale. Life takes the atoms and molecules and crystals, but, instead of making a mess of them like the stone, it combines them into new and more elaborate patterns of its own.

—Aldous Leonard Huxley (1894–1963). [*Time Must Have a Stop*. London, (1945)].

> Those reductionists who try to reduce life to physics usually try to reduce it to primitive physics— not to good physics. Good physics is broad enough to contain life, to encompass life in its description since good physics allows a vast field of possible descriptions.

—Carl Friedrich von Weissaecker (1912–   ). [*Theoria to Theory*. (1968)].

LITERARY STRUCTURE

A book is normally a linear stream of words. If it is a biography, this often corresponds to the unidirectional passage of time in the life of the subject. The introduction of "flashbacks", where the time variable jumps, is one technique for elegant variation. In a scholarly book— *On the Shoulders of Giants* by Robert K. Merton comes to mind—there is often a mass of cross-references, footnotes, footnotes on footnotes and appendices, which gives a much more net-like structure, corresponding to the ramifying thoughts. An index may provide threads (clue = thread), tying each item to a central nervous system for retrieval. However, not all topics (from *topos* = a place) have time as the principal variable. To describe a static scene in two or more dimensions (which thus cannot be projected into a linear sequence without disruption), other techniques are necessary.

These one-dimensional scans may be analogues to the raster scan of the television screen or to the vectorial movements of the beam in a computer graphics system. Indeed, when an author finds great difficulty in describing a scene, he may resort to an actual picture. Further, when the three-dimensional reality is inadequately described even by a picture (as in the case of protein molecules), solid stereoscopic representations may be necessary. When we are concerned with the changes in material structures in time, further dimensions are needed. This outlook is admirably caught by A. T. Winfree with his stimulating title: *The Geometry of Biological Time*, which is a study of spatial and temporal rhythms in biology and chemistry. It should be as significant a book for scientists as Proust's *A la recherche du temps perdus* (1912) was for the literary culture.

The problem is to construct a non-lineal book or article. [Here, "lineal" alludes to lineage. "Non-linear" is used for phenomena where the response is not proportional to the stimulus]. I first found such a book in 1942 in *Meet Yourself as You Really Are* by Prince Leopold

Loewenstein and William Gerhardi (1936, Penguin, 1942) which had a much more complicated structure. In it, the unit items of text, each perhaps half a page in length, led from one to another through addresses which followed the items. Each item might have several alternative addresses for the next module. Thus, the information was structured as in a modern computer data base and could be read in a vast number of different sequences. The actual geometry might be that of a three- [or greater] dimensional network of high coordination number. This is what is needed for describing the association paths in the brain, which is just such a network. A network of this type is the basis of Table 1, but it can only be a vestige of a real mental network.

INTRODUCTION

We have a Pythagorean strain in our culture which has continually made congenial the idea that somehow the symmetrical geometrical figures—the Five Platonic Solids in particular—are at the bottom of things. This attitude was caricatured in Swift's *Gulliver's Travels*, where the philosophers of Laputa carried about actual solid models of the concepts which they wished to discuss. If we wish to discuss spatial structure then we have effectively to do the same. The great power of computer graphics is that it gives us the possibility of conjuring up shapes from a data base and combining them in new ways (as Prospero does in Shakespeare's play "The Tempest"). Discourse about solid structures is impossible without effectively being able to call up pre-fabricated concepts, level upon level, the simplest being the Platonic solids, as we will. Literary labels, such as the words "rhombic triacontahedron" or "para di-chloro-benzene" have precise meanings. If we do not know enough of them, then we cannot even begin to use the hierarchically structured tree of concepts which is modern science.

In the *Timaeus*, Plato put forward the idea that everything was made of exactly those units which are to be found today in every schoolboy's geometry set—45, 45, 90 degree and 30, 60, 90 degree setsquares. These were then built into more complex objects and gave rise eventually to the properties of materials.

> Plutarch explained: Is their opinion true who think that he ascribed a dodecahedron to the globe, when he says that God made use of its bases and the obtuseness of its angles, avoiding all rectitude, it is flexible, and by circumtension, like globes made of twelve skins, it becomes circular and comprehensive. For it has twenty solid angles, each of which is contained by three obtuse planes, and each of these contains one fifth part of a right angle. Now it is made up of twelve equilateral and equiangular quinquangles (or pentagons), each of which consists of thirty of the first scalene triangles. Therefore it seems to resemble both the Zodiac and the year being divided into the same number of parts as these." [Quaestiones Platonicae 5.1, 1003C. (R. Brown trans.) in *Plutarch's Morals*, (ed. W. W. Goodwin) (1870), vol. 5, p. 433].

Visiting our present day society as a cultural anthropologist, what would Plato or Plutarch make of the World (Football) Cup ceremonies in which just such an object is kicked around in scenes of mass hysteria—clearly rites of cosmological significance?

In fact Plato put forward three important concepts:

(1) the idea that everything was made of an infinite number of a few units of pattern,
(2) the idea of groups (abstract patterns without particular motif) and
(3) the idea that the structure of a material determined its properties.

The important analogy with the alphabet, where with a small number of letters one can write an infinite number of different sentences, has also been familiar since antiquity. The invention of the alphabet, one of the greatest steps for mankind, giving a physical representation of an abstract structure, built up hierarchically from 20 or 30 signs to the most complex intellectual concepts, clearly prompted the ideas that the manifold forms of matter were similarly built up from atoms. Epicurus (341–270 B.C.) put it quite simply: "The atoms come together in different order and position like the letters, which, though they are few, by being placed together in different ways, produce innumerable words".

ATOMS AND REDUCTIONISM

There have always been two schools, one the Democritean, based on the view that the complexities of living material were due to the possibilities latent in the combinations of simple inanimate units and the vitalist, theistic view that life was a reflection of some more perfect life pre-existing elsewhere. The Democritean view was characterised as "reductionist" which became a term of opprobrium. (Von Weissaecker's comment at the head of this paper tries to counter this view). Although everything is "just atoms" everything is intrinsically an inseparable whole. David Bohm, who has long worked towards this goal, claims that "there are no things, only processes". There is no absolute separation and thus no absolute symmetry. Many paradoxes arise if we claim that something is "absolutely symmetrical", or "absolutely isolated". However, we can make working rules for everyday use, although these may let us down in certain special cases.

The conflict between materialism and spiritualism or religion has continued to be fought out in many theatres. The struggle can be followed in the literary as well as the scientific world. One cannot but look round the present-day world and repeat the often quoted phrase from Lucretius: "tantum religio potuit suadere malorum". Crystallography cannot help taking sides with Lucretius, Boscovich (Jesuit though he was), Maxwell and others who tried to work out the consequences of the combinations of atoms. Karl Marx even, who wrote his doctoral thesis on Epicurus and Democritus, summed up the position:

> Lucretius is the truly Roman heroic poet; his heroes are the atoms, indestructible, impenetrable, well-armed, lacking all qualities but these; a war of all against all, the stubborn form of eternal substance. Nature without gods, gods without a world.

Seeing in this a description of the processes of molecular dynamics we may agree with J. L. Borges that, in the scientific field as in the literary, "Every writer creates his own precursors", a phenomenon quantified in the Science Citation Index.

GROUP THEORY

Groups are groups of operations, particularly symmetry operations, and they are distinctive because not all arbitrary combinations of operations are possible. Only certain combinations of operations can occur and these are called groups. The five Platonic solids are examples of pure structure. A motif, one of the triangles mentioned by Plato, is repeated by the symmetry operations to give the solid polyhedron. Abstract groups are nothing but relationships between operations—they imply no specific operations and no arguments. Similarly, perhaps in the wave equation

$$\nabla\psi = \lambda\psi$$

the function ∇ is understood, but what is the argument ψ?

For a century or so the science of crystallography has been dominated by this beautiful edifice of group theory which found a close expression in the actual structure of crystals. At first this was speculative but, from 1912, the X-ray diffraction analysis of crystals, essentially the X-ray microscope, has shown that the symmetrical patterns predicted (by Kepler, Barlow, Pope, Dalton and others) actually exist at the atomic level.

Ptolemy (the Hellenistic astronomer) was captivated by the Platonic solids and constructed a universe of crystal spheres or polyhedra, concentric about the Earth, carrying the celestial bodies. Kepler tried to quantify this by showing that the radii of the planetary orbits about the sun corresponded to a nesting of the five Platonic solids. However, Kepler's greatness lay in his eventual discarding of this model in favour of a dynamic model embodying his three laws of planetary motion. The pattern of radii is generated by the internal dynamics and not by the imposition of external static structures. The music of the spheres model was unsuitable for the structure of the planetary system but it appeared again as the solution of Schroedinger's equation for the eigenstates of the hydrogen atom. The vibrating atom has indeed the symmetries of the regular polyhedra postulated by Kepler. Just as Kepler escaped from too rigid a pre-conceived framework of symmetrical structure for the phenomena of Nature so also modern crystallographers have gradually generalised their concepts of symmetry, escaping from the rigid formalism of the 230 space groups. In the present period we must acknowledge the great influence of J. D. Bernal in promoting ideas of a generalised crystallography which would gradually encompass our growing knowledge of the properties of living and inanimate matter.

We will not deal with the symmetries of the atom nor with the more remarkable symmetries now revealed in the structures of the nuclear world, but remain with the symmetries of the structures made of many atoms, which arise from their interactions.

The idea of symmetry has been with us for two thousand years and is still extremely important although thought has been greatly enriched recently by ideas of "symmetry-breaking". This latter trend was first introduced by Jean Buridan (1297–1358) who described the plight of a strictly philosophical ass who found himself symmetrically placed between two identical bundles of hay but who, since he could see no reason for going to one rather than the other, starved to death. This is one of the paradoxes which arise when we try to make absolute statements. How an apparently symmetrical regime breaks up, even under an isotropic agent like change in temperature, to give fresh structure, is now an extremely active area of research. We see now how symmetry is not imposed on a system but results from the way it works.

The study of form is again progressing and, after the successes in genetics, the whole question of morphogenesis is in motion again. The great goal of Ernst Haeckel (1866) to establish a "crystallography of organic form" has come round again in new guises, most notably that of "cellular automata".

The fabric of the universe is undoubtedly hierarchical. All levels of organisation of matter overlap a little, but some separations are clearer than others. We realise now that all parts of the universe are coupled together, some tightly and some loosely. If there were any regions not coupled, then they could not be part of this universe, but would be reachable only, perhaps, through a black hole. We may distinguish: . . . quarks/nucleons/nuclei/atoms/molecules// cells/organisms/societies//planets/stellar systems/galaxies/ . . . Accordingly, we also run into paradoxes of Zeno's type if we discuss symmetry in absolute terms. The significant question, with respect to a physical rather than a mathematical system is: "how symmetrical?" When working with physical instruments you have to know their limits of accuracy and now, with computers, you have to begin by defining the number of significant digits in a representation in order to define whether two numbers are equal or not. How many repetitions is effectively infinite?

Crystals are perhaps distinguished in that their span of existence is in some cases very large. Crystals of silicon can now be produced which run for a metre or 3×10^9 layers almost without dislocation. If we consider a very small crystal of 1000 units, ten repetitions in each direction, then half of the units will be on the outside of the crystal and thus in environments very different from those enclosed within. Biological crystals show more frequently only a few repetitions in any direction before some new feature supervenes.

CRYSTALLOGRAPHY

At first, in the nineteenth century, crystallography was pre-eminently the science of spatial symmetry but, as a subject, it has been spreading and moving, by the accretion of new material on one side and by erosion of the other side like the Island of San Serriffe (Tucker, 1977), so that, although the name remains the same and no sudden change has taken place, the domain of study is nowadays quite different from what it once was.

Until 1912, crystallography consisted mainly in the classification of crystals of mineral substances into one or other of the 32 crystallographic point groups, according to their idealised external forms. This was an economically important activity because crystals are only formed of identical units and their formation is thus a test for purity. The hand-picking of ores, recognising crystals of a particular mineral, is the foundation of all metallurgy and thus of civilisation itself.

One of the earlier questions to be answered was: what symmetries are possible for an infinite lattice of points? This was answered by several people with the number 32, the number of the crystal classes, proved with varying degrees of rigour. Given that, as Hauy had postulated, a crystal was a regular assembly of unit cells (molecules integrantes) what symmetries were possible? Speculation had taken place as to their internal structure and this had resulted in the remarkable formulation of the 230 crystallographic space groups.

The question was: in what different ways can an infinite number of identical asymmetric units be arranged in space so that all have identical surroundings. Here "identical" may include also the mirror image of a configuration being reckoned as identical.

This question was answered independently by E. S. Fedorov, a mineralogist in St. Petersburg, A. M. Schoenflies, a professor of rational mechanics in Goettingen and by William Barlow, a businessman and amateur chemist in Cambridge and London.

Crystallography is the science of symmetrical structure, primarily at the level of the arrangement of atoms. Atoms of the same element are practically and theoretically indistinguishable. Symmetry means that the structural motif of one part of the design is repeated elsewhere. Unit plus recurrence rules or operations gives overall pattern. But, the paradox is that the group in fact appears as a function of the motif. Different motives generate different groups, and in biology different motives give different programs.

We may take "crystallography" to be the science studying the structures and properties of assemblies made of large numbers of copies of a single type of unit, or of a few types of unit. The word "leptonics" was suggested by E. Haeckel for the science of fine structure but it did not become popular. But after all, "physics" and "chemistry" are hardly self-defining names. The name has the meaning which becomes attached to it. So also with "crystallography".

FROM SYMMETRY GROUPS TO PROGRAMS

Many kinds of operations of recurrence appear as operators in our computer programs generating diversity from identity. These include: *Rotations*, crystallographic, rational, irrational, infinitesimal. *Translations*: identical or similar: *Centre of inversion*: (and inversion axes); *Reflection*; *Similarity*: *Statistical* or stochastic operations. Except for the last three items, these are the subjects of classical crystallography and, since there are many excellent manuals, they will not be examined in detail here. We are concerned here with going beyond them.

Atoms, electrons, molecules are exactly equal (with molecules, isotopic incoherence and different excited states begin to creep in) (although, in the particle physics world there may be some difficulty in distinguishing two different particles from two excited states of the same type of particle) higher up, organisms begin to be only quasi-equivalent. (Viruses, ants, . . . human beings). (Abraham Lincoln, following Euclid's style, began with the axiom: "All men are created equal"!)

Mathematical symmetry depends on decoupled systems, easier to achieve conceptually than physically. Partially decoupled systems require the quantification of symmetry.

There are really three ways of looking at the symmetry groups:

(1) Taking the structure as a whole and asking: Under what groups of operations is this arrangement invariant? This leads us to the traditional 230 space groups.

(2) Setting up a framework or scaffolding of interlocking symmetry elements and then putting in a single motif of pattern and seeing how it is repeated infinitely by the symmetry operations. Each particular space group has a set of operations defining how the motif is to be repeated.

(3) Starting from the more recent idea of *cellular automata* and asking how the large-scale pattern emerges by the interactions of neighbouring elements. This new attitude is enabling theory both to escape from the classical constraints and also to connect with physical, biological and computer science. The system GROUP(MOTIF) = PATTERN has been thoroughly studied: interest is now in the more general function PROGRAM(MOTIF) = STRUCTURE.

INFORMATION

The new understanding of the genetic code has shown us a pair of structural systems, the nucleic acid (DNA/RNA system of the genetic code) and the protein structures which operate as functional and structural components and which are related to each other as language is to referent. Thus we have found another natural language in which a symbolic system and its material substrate exist side by side. In the DNA/Protein system we have the severe constraint that the description, the language, must also be a material structure of atoms and must obey also the laws of chemistry and physics which apply to all such systems. The symbolic systems of thoughts in the brain and of writing on paper or magnetic tape must also obey the relevant physical laws.

Having seen that genes are primarily material structures specialised to carry information, we can ask about the information content of various physical systems. In other structures the information is not localised as it is in genes, but the medium is the message. Seeing a protein molecule with a biochemical function we may ask what information corresponds to it. The information contained in the corresponding gene describes how to make the protein molecule given the appropriate backing machinery. The information also says something about the evolutionary history of the protein. Step by step protein and gene evolved dialectically by the passage of information from one to the other under the pressure of the environment. We can see that the protein structure corresponds to a certain amount of information.

If we consider a very complex inorganic structure, for example the mineral Paulingite, which contains several thousand atoms in its unit cell, we may ask where are the genes which code for it. Where is the information which makes this structure in the appropriate environmental conditions. The answer must be that the structure is its own information carrier. The information it carries is not arbitrary but relates only to the mineral itself. As we move from non-living to living matter we see a progressive segregation of the information-carrying function.

Gregory Chaitin has developed a useful definition of randomness. Given a sequence of binary digits, can this sequence be described by a description which is much shorter than the sequence itself? For example, the sequence (of 30 bits)

$$000110001100011000110001100011$$

is simply the sequence 00011 repeated six (110) times. Thus 30 bits can be reduced to about 8 (in the framework of a particular system of logic). Thus, the sequence is not random but can be described by about 8, instead of 30, bits of information.

It is possible, however, to find inorganic structures which do carry arbitrary information. Polytypes such as silicon carbide perhaps do so. We can now make, by molecular beam evaporation, arbitrary sequences of particular compounds. Indeed, Graham Cairns–Smith has been pressing the view that kaolin, clay, has many of the properties of a gene and can carry arbitrary information (in the distribution of the Al atoms among the silicon atoms in the sheets. Armin Weiss has evidence that this distribution can be replicated and that particular distributions can be selected for by particular environmental conditions. Thus, there is a *prima facie* case that inorganic precursors of genes may in fact exist.

We may define, in the most general terms, a crystal as being a structure, the description of which is much smaller than the structure itself.

In a typical inorganic crystal the information about the structure of the crystal is carried in a non-localised form by the rules of interaction between the particular atoms present. In a protein molecule the information about the sequence from which its crystal structure flows is carried in the associated DNA sequence but this assumes also the lower level regularities in the interactions between the constituent atoms.

CELLULAR AUTOMATA

With the ready availability of computers with graphical displays operating in real time, a new and powerful metaphor has entered crystallographic theory. This is the concept of *cellular automata*. It was begun, about 1950, by Stanislaw Ulam, and achieved wide popularity with the "Life Game" developed by J. H. Conway. A cellular automaton is essentially a pattern in space generated by the interaction of individual units, usually thought of as "cells" in space but being also perhaps material objects, with their neighbours. Each unit may have several states, such as being occupied by an atom or not, and the state of each unit is determined by the states of a certain number of neighbouring units. The rules of interaction may be simple, but could be substantial stretches of computer program. The whole pattern continually re-arranges itself on these local criteria. S. Wolfram has recently classified the behaviour of cellular automata into four classes. It is very tempting to identify these with different categories of material structure. These categories are:

(I) Patterns which settle down to the same result whatever the starting configuration. (We might think of a sodium chloride crystal, which appears inevitably whenever the appropriate ions are mixed and allowed to settle. There is no alternative structure.)

(II) Patterns where there are a few definite outcomes which depend on the starting positions. (This may be the production of various phases of a material, for example the various cement compounds in the complex phase diagram of $CaO/SiO_2/H_2O$ according to the paths taken in the temperature (pressure) composition configuration space.)

(III) Patterns which are extremely sensitive to the exact values of the starting positions. (Here we may have complex structures which can store arbitrary information, like DNA sequences producing particular proteins with very different properties which are extremely sensitive to certain mutations.)

(IV) Patterns which are extremely unpredictable. For these there is no method of predicting their behaviour which is simpler than the system itself. Certain of these automata appear to be general computers in the sense of Turing. (Here we see some of the phenomena of living systems and life itself.)

We may expect that this computational analogy will bring in a new era in crystallography by establishing a secure link between the inorganic and living biological systems brought about by an understanding of the role of information in material structure.

CONCLUSIONS

Today, crystallography is no longer the science dealing only with the symmetry of the external forms of mineral crystals but has exfoliated to become a much more general science dealing with structure at the atomic level—the way in which the aggregations of atoms combine to give the manifold properties of matter, living as well as inanimate. The connections between topics are so numerous that they can only be described by the multi-dimensional network which I have endeavoured to provide in Table 1. The numbers in brackets after each entry indicate a few of the connections with other entries.

Francis Galton invented the "word association game" and believed that words or ideas were stored in the brain such that some were nearer together and some were further apart. He showed that this network was strongly characteristic of an individual. If we move from the central core of classical crystallography, then we can see how all kinds of other questions group round it. Since the brain is a three-dimensional structure, the ideas in the brain should be representable by a three-dimensional network. H. G. Wells commented: "Queen Victoria was like a great paper-weight that for half a century sat upon men's minds, and when she was removed their ideas began to blow about all over the place haphazardly." Once we see that the classical theory of the crystallographic space groups is not the only structuring principle, then we can lift off from our minds the incubus of the "International Tables for X-ray Crystallography" and begin to wander about and to realise that there is a whole complex of connected concepts which also form the hunting-grounds of crystallographers.

Notes, expanding the catch-words of the diagram follow, so as to enable the mental associations to be understood. To present this mental topography would require three-dimensional paper, but items can be examined individually. The numbers in square brackets given after each item refer to closely associated concepts in the table but in fact the network is much more richly interconnected than is indicated.

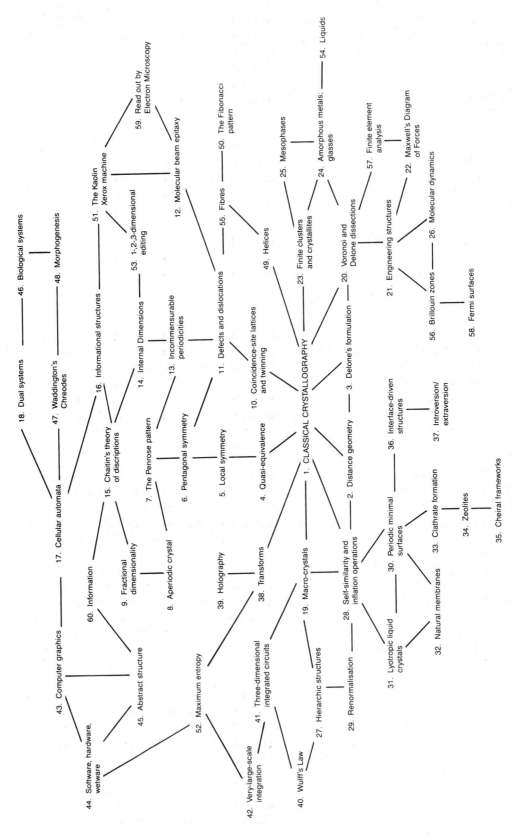

Table 1. Topics connected with the generalisation of crystallography—a concept-association network

1. *Classical crystallography*

The theory of perfect infinite crystals where every unit cell is the same and every atom is in its proper place. Every unit of pattern (asymmetric unit or fundamental region) has surroundings which are identical with those of every other. The crystal as a whole belongs to one or other of the 230 space groups. Since the crystal is infinite it can have no faces. The external forms of crystals are dealt with by considering the possible combinations of point symmetry elements (having no translational components) which may transform an infinite point lattice into itself. This gives the 32 crystallographic point groups or crystal classes. The formalism describes the arrangement of the atoms with respect to axes defined by the symmetry elements.

The concepts of classical crystallography may be set down as follows:

(a) a crystal consists of an infinite repetition by translation in three non-coplanar directions of a unit cell. Thus all unit cells in the crystal (deemed infinite) are identically situated.

(b) the motif of pattern or asymmetric unit of structure is repeated inside the unit cell by the elements of symmetry so that each motif of pattern has surroundings identical with those of every other. Thus every crystal structure belongs to one or other of the 230 crystallographic space groups.

(c) the unit cell contains an integral number of formula units or molecules.

(d) the external form of a crystal (here not infinite but consisting of a very large number of unit cells) belongs to one or other of the 32 crystallographic point groups of symmetry.(2,4,10,19,20,21,39,49)

2. *Distance geometry*

The coordinates of the atoms in a crystal structure are conventionally given with respect to axes defined by the symmetry. These x,y,z coordinates have individually no physical significance. It is, however, possible to describe the whole structure in terms of internal coordinates, distances, bond angles and torsion angles, which are of physical significance. Since a bond angle can be represented by three distances and a torsion angle by six, distances between all pairs of atoms describe a molecule (or complete structure). Interconversion between systems is possible.(1,3)

3. *Delone's formulation*

Delone, Shtogrin, Galiulin and later P. Engel have recast the formalism of space group theory to show that if the local surroundings of each lattice point are identical with those of every other, out to a certain finite distance (2R to 6R, where R is the radius of the largest empty ball (not containing a lattice point) then the usual space groups are obtained.(1)

4. *Quasi-equivalence*

The classical theory of crystals requires every unit of pattern to have surroundings identical to that of every other (out to an infinite range). In many cases it is sufficient that surroundings should be only quasi-equivalent. In correspondence with physical reality there may only be limited equivalence and thus symmetry elements of limited range. The theory of crystallography on a spherical surface (as is the circumstance for many viruses), was developed by Caspar and Klug on this basis.(1,5)

5. *Local symmetry*

In an infinite crystal there may be extra elements of symmetry which operate only over a limited range. These may be seen by non-space-group extinctions in the diffraction pattern. For example in crystals of units of tobacco mosaic virus there are local 17-fold axes corresponding to the symmetry of what are recognisable as units made of 2×17 subunits. As here the local operations need not be "crystallographic".(4,6)

6. *Pentagonal symmetry*

There may be a conflict between the symmetry possible for an isolated unit and that possible for a unit in an infinite lattice. Thus, a five-fold axis is not one which an infinite lattice may exhibit and is thus a non-crystallographic symmetry element. However, it may nevertheless occur locally.(5,7)

7. *The Penrose pattern*

This appears to be a unique quasi-lattice which has local five-fold axes of symmetry. It may occur in two-dimensional or three-dimensional form, the plane pattern being a section through the three-dimensional. The pattern has recently been observed experimentally in splat-cooled Al_6Mn. It is hierarchic and non-periodic, although made out of copies of two types of tiles infinitely repeated. It is possible that a similar tiling with seven-fold symmetry exists, if negative tiles are allowed in order to remove the overlapping of closely packed heptagons.(6,8,13)

8. *Aperiodic crystal*

It was suggested by Schroedinger ("What is Life?", 1945) that arbitrary information could be transmitted by an "aperiodic crystal", something like a regularly punched tape, that is, by frequency or amplitude modulation of a regular structure. This has been found to be realised in the sequences of nucleic acids in the "double helix".(7,9)

9. *Fractional dimensionality*

Geometrical structures, which have perimeters and areas which are fractional powers of their linear dimensions, instead of being proportional to the first and second powers respectively, have been known for some time and have been, as "fractals", extensively discussed by B. Mandelbrot. Looking into natural structures examples can be seen. The Penrose pattern and snowflake-like patterns, produced by recursive operations, have such properties.(8,15)

10. *Coincidence-site lattices and twinning*

The regular structure of a crystal may be interrupted by a sudden change of orientation (or by a translation jump—giving an anti-phase domain) across a boundary where the two domains retain some symmetry elements in common. This twinning may happen once or it may be repeated. If it is repeated regularly then we recognise a superstructure.(1,11)

11. *Defects and dislocations*

The perfection of classical crystals may be marred by defects of a wide variety of types. There may be additional atoms or absences or displacements of atoms or larger lines, planes or volumes of the crystal may be disturbed. A large sector of materials science is concerned with the mechanics of such defects.(6,10,12,13)

12. *Molecular beam epitaxy*

Using ionic beams for depositing layers of atoms or for implanting guest atoms in specific regions, together with electron beam etching for removing material, it is now possible to construct materials of almost arbitrary structure and composition on a nearly atomic scale. Such materials form the basis for coming generations of electronic circuits. Conversely, details of arbitrary structure at the atomic level can be read our by high resolution electron microscopy and other techniques. The whole question of crystal growth is reaching the level of the individual atom.(11,51,59)

13. *Incommensurable periodicities*

Perfect crystal structures may be modulated by displacements of atoms in static waves which are not related to the periodicities of the crystal lattice. These modulations show up as extra spots in the diffraction pattern which may be as strong as the normal spots. Thus, not all extra lines in a powder diffraction pattern must be due to the presence of an extra phase or impurity. That is, the regular structure spontaneously adopts a symmetry-breaking arrangement.(7,11,14)

14. *Internal dimensions*

By holding energy in rotational modes, a gas may carry more than the energy $(3/2)KT$ belonging to the translational motion of the molecules. The molecules appear to have internal dimensions and the adiabatic gas constant gamma for nitrogen, for example, may be greater than expected. A structure being microporous, may appear to have 'internal area' and to have a fractal area.(13,15,61)

15. *Chaitin's theory of descriptions*

G. Chaitin characterises a structure, the simplest structure being a sequence of N binary digits (0s and 1s), as non-random if it can be described by many fewer than N binary digits, for example if it repeats periodically. If about N digits are needed and the description cannot be made shorter than the structure itself, then the sequence is random. What structure you can see in a sequence may depend on how intelligent you are and what you know already.(14,16,60)

16. *Informational structures*

An abacus is a structure with movable beads which can be used for the storage of arbitrary information such that the information should be stable against random thermal disturbances. The DNA chain is such an abacus and similar devices can be imagined as made of layer sequences of SiC or as patterns of substitution of Al for Si in kaolin sheets.(15,17,51)

17. *Cellular automata*

There exists a very large branch of mathematics and computer science, known as ''automata theory'' which deals with the generation and description of patterns in one or more dimensions. One class of such patterns is ''cellular automata''. Examples of these are ''the life game'' of J. H. Conway and ''The Ising model'' of a lattice of magnetic spins. Some automata are reversible and some not.(16,19,43,47)

18. *Dual systems*

One structure may carry arbitrary information from which another structure can be generated. There may be dual systems (such as that of protein sequence/DNA sequence) where equivalent information may exist in two forms and be translated from one to another, much as in a dynamic system energy may exist in two forms (kinetic or potential) and may pass from one form to another.(17,58)

19. *Macro-crystals*

Crystals need not have atoms as the units which form periodic arrays. In, for example, opal, microspheres of silica (ca. 2500 A in diameter) pack under colloidal forces to form crystals. beta-FeOOH, polystyrene and tungstic oxide particles in monodisperse suspension also form macro-crystals. If material is transported from convex surfaces to concave in such a macro-crystal, then a periodic surface of minimal curvature may result.(1,27,28,41)

20. *Voronoi and Delone dissections*

It is only conventional that crystal structures should be described by parallelepipedal unit cells. An equivalent description can be given in terms of space-filling convex polyhedra which are obtained by taking the planes perpendicularly bisecting all lines between lattice points. Dissecting further, it is possible to form a Voronoi polyhedron around each atom. As a modification, instead of the perpendicularly bisecting plane, a number of other rules can be followed. The most useful is to take the radical plane. Thus, a structure can be treated as a packing (without interstices) of convex polyhedra. The Delone dissection is a unique dissection into tetrahedra which follows from the Voronoi or other dissection.(1,3,21,24,57)

21. *Engineering structures*

It is possible to regard a crystal of atoms held together by interatomic forces as an engineering structure and to apply the methods of computational engineering for computing its properties. Finite element analysis is one of the techniques which can be carried over.(20,22,61)

22. *Maxwell's diagram of forces*

J. C. Maxwell showed that for reasonable physical conditions a network, reciprocal to a given three-dimensional framework could be drawn, such that the areas of the faces of the reciprocal polyhedra represented in magnitude and direction the forces between pairs of nodes in the primary framework. This can be applied to atomic arrangements.(21,57)

23. *Finite clusters and crystallites*
As crystals get smaller the importance of the surface gets greater and, for very small particles, may be a dominating concern in determining the stable structure. Small clusters can be modelled by computation. The basic characteristic of crystals that all units of pattern must be in identical surroundings has to break down from sheer topology. In a crystal $10 \times 10 \times 10$ units cells, half the unit cells are on the outside. Thus there is an absolute scale of size in the number of repeated units.(1,24,25)

24. *Amorphous metals; glasses*
The unit cell formalism of crystals cannot be carried over to amorphous materials where there is usually a random network arrangement of atoms (or other units of structure) but the Voronoi space-filling dissection methods of description can be applied to both. Similarly, liquid structures can be brought within the ambit of ''crystallography'' with appropriate extensions of concepts, such as that of coordination polyhedra.(23,25,54)

25. *Mesophases*
Between liquids and crystals there is a range of intermediate states of order usually called ''liquid crystals'' or ''meso-phases''. The principal types of liquid crystal are ''nematic'', ''smectic'' and ''cholesteric'' but many sub-classes are to be distinguished.(23,24,55)

26. *Molecular dynamics*
A large technology of simulating the behaviour of atomic/molecular systems, roughly following the kinetic theory begun by Maxwell, but using large computing facilities, has developed and enables global parameters of materials, (heat capacity, dielectric properties, etc) to be calculated from interatomic forces. The obtaining of global parameters in agreement with experiment confirms that the interatomic forces were correctly chosen. The calculation resembles that of making a ''cellular automaton'' but the interacting components do not lie on a lattice.(21,61)

27. *Hierarchic structures*
In most complex structures, particularly those of living systems, larger units consisting of smaller units, which themselves consist of still smaller units, can be distinguished. That is, most structures are hierarchic. The spans of the different levels in the hierarchy may differ. Some units may be composed of only 10 sub-units. Crystals are rather exceptional in that some crystals may consist of 10^{24} sub-units (atoms or units of chemical formula). H. A. Simon's parable of the watchmakers clearly shows the greater probability of the formation of hierarchic structures.(19,29,40)

28. *Self-similarity and inflation operations*
Some structures, well above the level of atoms, appear unchanged when magnified by a certain factor. This is a self-similarity operation, more easily implemented in mathematical than in physical system. Spirals typically have this property. The Penrose pattern (q.v.) has this property. A. V. Shubnikov and later others have classified the similarity space groups.(1,29,30,31)

29. *Renormalisation*
Renormalisation is a technique for handling critical point transitions, for example the Curie point, below which temperature a lattice of randomly oriented magentic moments may spontaneously orient itself. Each small group of spins is replaced by a single larger unit of suitably averaged properties, and so on hierarchically until the unit considered comprises the whole volume of the specimen. It parallels the physical process by which correlation waves get longer as the critical point is approached.(27,28)

30. *Periodic minimal surfaces*

These are geometrical invariants which actually occur in a variety of physical systems. A "soap-film" surface of zero mean curvature (Gaussian curvature everywhere negative, (hyperbolic) or zero) can repeat in three dimensions according to one or other of the 230 space groups to divide all space into two congruent regions. A. H. Schoen (1969) has found some 17 such types of surfaces only 5 of which were known before 1900 to H. A. Schwarz and E. R. Neovius.(28,31,33,36)

31. *Lyotropic liquid crystals*

The clearest occurrence of the periodic minimal surfaces of H. A. Schwarz is in phases of glycerol mono-oleate where the lipid bilayer takes up the shape of the F-surface (Longley and McIntosh, 1981). Other surfaces may be expected in this type of system.(28,30,32)

32. *Natural membranes*

There are a number of natural systems in which the building elements are membranes (just as there are many where the elements are fibres). These include etioplasts, silicate sheets, lipid bilayers as in myelin as well as notional membranes such as grain boundaries. In these, the geometry of the system may be controlled by bulk properties such as surface tension or curvature [Fontell, 1981].(30,31)

33. *Clathrate formation*

Clathrate compounds are those where a network in three-dimensions forms around isolated and inactive guest molecules, such as gas molecules. The guest molecules may or may not be capable of removal without breaking the network. A series of the gas hydrates and certain cage silicates have isomorphous structures. Besides water and silica the cage may be formed from urea, quinone and a variety of other molecules. Clathrate formation on templates is thus an important building principle. The matter is not only of academic importance since much of the Russian natural gas is found as solid clathrate crystals.(30,34)

34. *Zeolites*

Zeolites represent a large and important class of cage silicates, "molecular sieves", where arrays of cavities and tubes of various sizes can be constructed [Wells, 1977].(33,35)

35. *Cheiral frameworks*

It is still uncertain whether a system of cheiral cavities can be constructed as a silicate framework for use in filtering one hand of molecules from their mirror images. One of the theoretically constructed periodic minimal surfaces, the gyroid, (Schoen, 1969) divides all space into two regions one of which is the mirror image of the other. A cheiral filter would be of value in separating optical isomers for biological purposes. It has been suggested that silica gel, precipitated around cheiral molecules which are then removed, may have such properties.(34)

36. *Interface-driven structures*

In a number of structures, for example a foam of soap films or grains of aluminium, the surface tension may be a controlling factor. Aluminium grains grow on annealing minimising the interfacial area.(30,37)

37. *Introversion/extraversion*

We might use these terms as synonyms for hydrophobic/hydrophilic forces which summarise the individual forces between a large number of atoms. Protein molecules are seen, for example, as micelles, where a minimum energy configuration is achieved by concentrating hydrophobic ("oily") groups on the inside of the globule which is surrounded with hydrophilic groups which then interact with the aqueous medium. Milk, for example, consists of such micelles and by churning, these can be inverted into butter, so that the fat becomes the continuous phase and includes globules of water.(36)

38. *Transforms*

A density distribution in three-dimensions can undergo a Fourier transformation (and reversely). This weighted reciprocal lattice is what can be observed in the "X-ray or electron microscopy" of crystals on which our knowledge of their atomic arrangements depend. The description of structures in real and in reciprocal (transform) spaces are dual to each other.(39)

39. *Holography*

The process of the "X-ray microscopy" of crystal structure is formally very similar to that of holography.(1,38)

40. *Wulff's Law*

G. V. Vul'f pointed out that the equilibrium shape of a crystal, the faces of which had different surface tensions, was the one which minimised the total surface energy. Conversely, the relative surface energies of the different faces can be calculated from the equilibrium proportions of the crystal.(4,27)

41. *Three-dimensional integrated circuits*

Integrated circuits and structures of molecular dimensions are approaching each other in scale. Molecular electronics is a developing technology since speed and memory requirements push towards three-dimensional arrays of circuit elements. It is possible that we might arrange circuit elements, molecules or actual two dimensional chips, on the minimal surfaces which occur in lyotropic colloids, in silicate frameworks, and in a variety of other structures. Here we have crystallography appearing at a new level, where the units are not atoms or molecules but much larger units.(19,31,40,42)

42. *Very large scale integration (V.L.S.I.)*

Already arrays of microprocessors resemble two-dimensional crystals and techniques of crystallography are used for making them.(41,56)

43. *Computer graphics*

A large technology has developed for handling three-dimensional images or models of molecular and crystal structures of all kinds. Experiments in the interactions between molecules can now be made "in machina" (as well as *in vivo* and *in vitro*) since reliable value of the interatomic interactions are available. Further, anything which can be imagined can be drawn. Experiments can be done to see the results of all kinds of pattern-generating relationships in which the description of a structure and the structure itself may be linked.(17,48,56)

44. *Software, hardware, wetware*

Software is the programme or description of a process. Hardware is the computer system on which the programme is executed and wetware now represents the molecular sensors which interface the computer to the real chemical and biological world.(45,56)

45. *Abstract structure*

Geometry, group theory, network theory, information theory and automata theory are examples of abstract mathematical areas which bear directly on the needs of generalised crystallography. Practical questions arising from crystallography have stimulated many areas of pure mathematics. A computer program is an abstract structure.(44,60)

46. *Biological systems*

Real biological systems have structures which are specialised for the storage and transcription of information. The information-carrying elements have also to obey the normal laws of physics just as much as the other structural and operational material.(18,48)

47. *Waddington's Chreodes*

C. H. Waddington used the idea of a multi-dimensional landscape to visualise the development of an organism. The route of development ran down valleys meeting branch points from time to time at which critical choices were made. In a much more restricted sense the topography of the reaction paths in molecules has been mapped in N-dimensions. J. D. Dunitz has applied the classical space group theory to the description of symmetries in this configuration space.(17,48)

48. *Morphogenesis*

The whole field of the genesis of shape and form in living organisms, beginning with the gastrulation of a cell, is now actively developing with the stimulation of automata theory and a number of other branches of mathematics. The aims of Ernst Haeckel, articulated in his *Generelle Morphologie der Organismen* (1866), to establish a "crystallography of organic forms" are now being approached in a more sophisticated way.(46,47)

49. *Helices*

The most general relationship in space between two units is that one can be brought to superposition with the other by a rotation and a translation which moves it along a helix. The first great break in classical crystallography (about 1950, with the discovery of the alpha-helix of protein chains) was the realisation that irrational helices actually occur in materials, and thus that not all structures fall into one or other of the 230 space groups. The double helix of DNA (1953) emphasised this situation and it is now realised that between crystals and "amorphous" structures there is a great variety of different kinds of organisation.(1,55)

50. *The Fibonacci pattern*

The seeds in the head of a sunflower grow in a spiral pattern which involves the Fibonacci series and results from the local interactions between the seeds which press on each other and are unable to move far. There is evidence that when fibres pack in a dense bunch, twisted like a rope, they produce this kind of pattern. It may be observed as the lines of Retzius in elephant ivory and perhaps in collagen as well as in commercially produced ropes.(55)

51. *The kaolin xerox machine*

G. Cairns–Smith has evidence that kaolin plates are able to retain, in the pattern of Al and Si atoms, a certain amount of arbitrary information and that this may be replicated by the growth of further crystallites from initial seeds. Thus, we may have an alternative to the DNA/Protein system as a naturally-occurring self-replicating system. It may be another example of the self-replicating system postulated by J. von Neumann.(16,59,61)

52. *Maximum entropy*

A powerful mathematical principle enabling the most information to be extracted from a diffraction pattern or from an image of a structure and employing to the best advantage any *a priori* information about the structure which may be available (Bricogne, 1984).(38)

53. *1-,2-,3-dimensional editing*

Regarded as a physical process, editing, the alteration of the neighbourhood of a particular item in a structure, is clearly easiest for a one dimensional structure (a linear sequence of text) than for a two-dimensional one (a picture or, for example, a newspaper page). A three-dimensional structure is hard to edit since there is no further dimension into which items can be removed to transport them to somewhere else. Editing has to be done by shuffling vacancies. Thus, it is not surprising that the main natural information structure, the genetic code, is a one-dimensional sequence which can be readily chopped up and re-assembled in the course of evolution or of artificial bio-engineering. Two-dimensional editing is just possible, but the advantages of sheet storage of information may be that it is readily duplicable, like a natural Xerox machine. Information locked in a three-dimensional structure is secure against alteration.(14,51)

54. *Liquids*

Liquids represent a particular challenge as regards the specification of their rules of ordering which must be determinate at close-range but random at longer-range. Under the influence of J. D. Bernal, who developed a geometrical (random network) theory of liquid structure, which forced thermodynamic and other theories of liquids to become more material and realistic, the structure of liquids has come to be included in the subject matter of generalised crystallography.(24,25)

55. *Fibres*

From the early 1930s, when W. T. Astbury took up the examination of all kinds of fibres by the X-ray diffraction methods then hitherto used for crystals, the study of materials ordered in less than three dimensions has been part of crystallography. This kind of study culminated, in 1953, with the discovery of the structure of the double helix of DNA by J. F. C. Crick and J. D. Watson.(25,49,50)

56. *Brillouin zones*

In the study of waves in crystals the reciprocal lattice has played a large part and this was developed, initially by L. Brillouin, to give an elaborate geometrical structure of surfaces in reciprocal space which explain well the electrical properties of metals. The Brillouin zones into which this imaginary space is divided resemble the minimal surface found in lyotropic colloids and elsewhere.(20,56,58)

57. *Finite element analysis*

With the advent of computer power theoretical engineering has developed to the study of larger and more complex structures. It can be seen that there are strong similarities between the theoretical chemical studies of the mechanics of protein molecules and of the modes of vibration of a complex engineering structure. Finite element analysis consists in the subdivision of a continuous structure, such as a curved steel container, into a network of discrete elements with interactions which are mathematically describable. This procedure enables powerful network algorithms to be applied by computer.(20,22)

58. *Fermi surfaces*

The Fermi surface is a surface in reciprocal space and relates to the energies of waves propagated in particular directions in a crystalline metal. It treats the propagation of electron waves as a particular case of Brillouin's treatment of the motion of waves in a periodic structure.(56)

59. *Read out by Electron Microscopy*

The resolution of the electron microscope has improved to the point where, under good conditions, the positions of single atoms can be seen. This affords the possibility of being able to read out such positions from a structure as arbitrary information. The complementary writing process might be implantation of ions by a similar device.(12,51)

60. *Information*

REFERENCES

(In a global reconnaissance, such as we have provided above, we cannot give full references for every statement, but will confine our documentation to general accounts with only a few specific items. Thus, this is more a reading list than a list of references keyed to specific citations.)

1. S. Andersson, On the description of complex inorganic crystal structures. *Angew. Chem. Int. Ed. Engl.* **22**, 69–81 (1983). (A modern but classical account of complex structure.)
2. J. D. Bernal, (1958). The importance of symmetry in the solids and liquids. *Acta Physica* (Acad. Sci. Hung.) **8**, 269–276 (1958).
3. J. D. Bernal and C. H. Carlisle, The range of generalised crystallography. *Soviet Physics—Crystallography* **13**, 811–831 (1969). (This is the first and clearest articulation of the program of "generalised crystallography".)
4. L. Blumenthal, *Distance Geometry*. Oxford (1951).
5. W. L. Bragg, *The Crystalline State*, Vol. I. Bell, London (1933). (The oldest and clearest account of the range of classical crystallography after the discovery of diffraction.)
6. G. Bricogne, Maximum entropy and the foundations of direct methods. *Acta Cryst.*, **A40**, 410–444 (1984). (A very difficult but authoritative summary of the principles of "generalised microscopy".)
7. G. Cairns–Smith, *Genetic Takeover*. Cambridge University Press (1980). (The examination of alternative information-carrying systems.)

8. G. J. Chaitin, Randomness and mathematical proof. *Scientific American* May, 47 (1975).
9. B. N. Delone, N. P. Dobilin, M. I. Shtogrin and R. V. Galiulin, A local criterion for regularity of a system of points. *Sov. Math. Dokl.* **17**, 319–322 (1976). (Mathematical theorems of the "new" classical crystallography.)
10. G. Donnay and D. L. Pawson, X-ray studies of echinoderm plates. *Science*, **166**, 1147–1150 (1969), H-U. Nissen, *Science* **166**, 1150–1152 (1969).
11. R. C. Evans, *Crystal Chemistry*. Cambridge University Press (1939). (A comprehensive statement of the range of crystal chemistry as developed by the Cambridge school.)
12. K. Fontell, Liquid crystallinity in lipid—water systems. *Mole. Cryst. Liq. Cryst.* **63**, 59–82 (1981).
13. E. Haeckel, *Generelle Morphologie*. Berlin (1866). (An early program for the study of biological morphogenesis.)
14. B. Hayes, (1984). Cellular automata—computer recreations. *Sci. Amer.* **250**, 10–16, (1984).
15. *International Tables for X-ray Crystallography*. International Union of Crystallography. (But previous editions from 1935. The authoritative statement of the space groups of crystallography and everything to do with the notations of classical crystallography.)
16. W. Longley and T. J. MacIntosh, A bicontinuous tetrahedral structure in a liquid-crystalline lipid. *Nature* **303**, 612–614 (1983).
17. A. L. Mackay, Generalised crystallography. *Izvestija Jugoslav. Centr. za Kristallogr.* **10**, 25–36 (1975). (A more recent statement of the scope of generalised crystallography.)
18. A. L. Mackay, Crystal symmetry. *Physics Bulletin*, 495–7 (1976). (An early introduction of the idea of cellular automata into crystallography.)
19. A. L. Mackay, (1981). De Nive Quinquangula. *Soviet Physics (Crystallography)*, (*Kristallografiya*, **26**, 909–918). (A discussion of the role of the "forbidden" five-fold symmetry in crystal systems.)
20. B. Mandelbrot, *Fractals*, Freeman, San Francisco (1977). (A statement of hierarchic structure in self-similar systems. A key extension of classical ideas.)
21. M. O'Keefe and A. Navrotsky, (eds.), *Structure and Bonding in Crystals*. Academic Press (1981). (The solid chemical foundations of the structure of crystals.)
22. F. C. Phillips, *An Introduction to Crystallography*.
23. Longmans, London (1949). (and later editions—a clear statement of the classical science of the external forms of crystals.)
24. D. Schechtman, I. Blech, D. Gratias and J. W. Cahn (1984), "Metallic phase with long-range orientational order and no translational symmetry," *Phys. Rev. Letters*, **53**, No. 2, 1951–1953 (12 Nov. 1984).
25. H. A. Simon, The Architecture of Complexity. *Proc. Amer. Phil. Soc.* **106**, 467–482 (1962). (H. A. Simon played a key role in the development of the association between information and structure.)
26. A. Tucker, *The Guardian*, p. 21, 1 April (1977).
27. A. F. Wells, (1962). *Inorganic Crystal Chemistry*, 3rd edn. (but also subsequent editions) Oxford (1962). (A standard manual of inorganic crystal structure.)
28. A. F. Wells, *Three-dimensional Nets and Polyhedra*. Wiley, New York (1977).
29. A. T. Winfree, *The Geometry of Biological Time*. Springer Verlag (1980). (A pioneering introduction to the complexities of structures which change in time as well as space introducing a host of new vistas.)
30. S. Wolfram, J. D. Farmer and T. Toffoli, (1984). Cellular automata. *Physica D* **10D**, Nos. 1 and 2 (1984). (The most substantial manifesto of the new school working out the structuring properties of cellular automata.)

Comp. & Maths. with Appls. Vol. 12B, Nos. 1/2, pp. 39–62, 1986
Printed in Great Britain.

0886–9561/86 $3.00 + .00

SEEN AND UNSEEN SYMMETRIES: A PICTURE ESSAY

GYÖRGY DOCZI
6837 47th N. E., Seattle, Washington 98115, U.S.A.

Abstract—Nature's and art's symmetrical patterns can be traced to basic pattern-forming processes, which also shape the invisible patterns of harmonious human life. Two of these are identified as the "sharing of limits", and the "union of complementary opposites" or in one word: **dinergy**. Jointly they create a third one: **dinergic symmetry**, an archetypal patterning process, that has appeared throughout the ages in countless forms such as "squaring the circle", "uniting heaven and earth", or the "oneness of life and death". **Dinergic symmetry** is paradoxical, numinous, yet not necessarily religious, even though religions have tried to make it so by using and abusing it. It is the source of **creativity**, which may yet bring to bloom the wonder of the world: the thousand-petaled golden flower of **peace**.

To what extent are symmetrical patterns in art and in nature matched by similarly symmetrical— but invisible—patterns of human values? Before I attempt to answer this question let me briefly stop at the word "**symmetry**", which has a broader and a narrower meaning. (1)

The narrower, newer, meaning indicates equality between the two opposite sides of a plane or a line, while the broader, older meaning refers to "beauty or harmony of all parts". It is in this wider and older meaning that I shall use it here, following Blake: "Tiger, tiger burning bright, in the forest of the night, what immortal hand or eye could frame thy fearful **symmetry**".

Leonardo da Vinci's illustration to Vitruvius' thesis that the human body fits into a circle centered on the navel and touching the outstretched extremities (Fig. 1), while the height and the width of the outstretched arms correspond to a square, can also be extended to include a pentagram or a pentagonal star. By drawing the square **concentric** with the circle (not resting on the ground as Leonardo did, to illustrate Vitruvius' point) the circle cuts out of the upper line of the square a **chord** that is exactly the length of a side of the **pentagon**, or the distance between the **pentagram's** tips (Fig. 2). Thus, the pentagon and pentagram have the same significance in the structure of the human body as the circle and the square.

The pentagonial star contains the three basic musical root harmonies: the diapason or 1 to $2 = 0.5$ or octave a string pulled 1/2 its length produces the same tone an octave higher. When pulled at $2:3$, it equals 0.666 and gives a diapente or quint. When pulled at $3:4$, it equals 0.75 or a diatessaron or quart (Fig. 3). These are *limits* in the length of strings that produce pleasant sounds, when pulled together with the full string, or one fundamental note. They also constitute the first overtones, which make the difference between musical sound and noise. As Pythagoras noted, $1:2$, $2:3$ and $3:4$ express these limits in numerical terms, and they can just as well be expressed in spacial terms by placing 2, 3 and 4 squares upon each other.

In the human body arms, legs and torso constitute a whole, by themselves, yet the hand shares this relationship to the arm, paralelling the relationship of foot to leg and skull to torso (Fig. 4). The hand, foot and skull again represent a whole in themselves and they again share, within themselves, the same proportions. Thus, the toes relate to the sole, or metatarsus, the same way as the wrist relates to the palm, or metacarpus, and the mandible relates to face (from upper teeth to nasal bone) in a relationship of $2:3$ or quint (Fig. 5). In the same sense, the heel relates to the sole and toes, as the fingers relate to palmbones and wrist; and the cranial box to the face and mandible in a relationship of $3:4$ or quart. These relationships have been examined in nine different skeletons, (2) and are true to the extent to which simple measurements can be relied upon. The graph (Fig. 6) shows to what extent the measurements approximate the $2:3$ and $3:4$ proportions. They are seldom exact, but close enough for a general tendency to be observed.

The pentagram's $2:3$ or 0.666 relationship is very close to the Golden Section's, 0.618 ratio (see Fig. 3), where the *minor* part relates to the *major* one in the same way as the *major* relates to the *whole* (Fig. 7). I call this a **dinergic** relationship because it reconciles opposites, minor and major, thereby generating a new whole. Professor C. G. Waddington[3] proposed

G. DOCZI

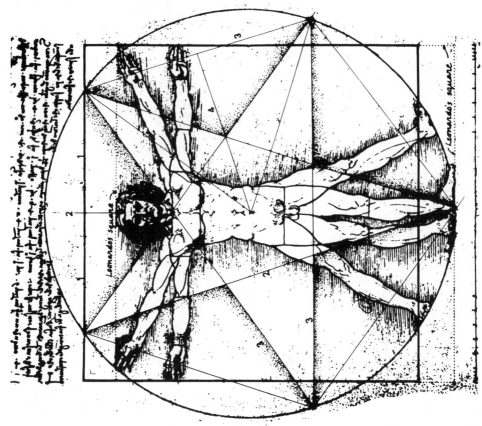

Fig. 2. Illustration to *Vitruvius* by Leonardo da Vinci. (1452–1519) with pentagonal star added.

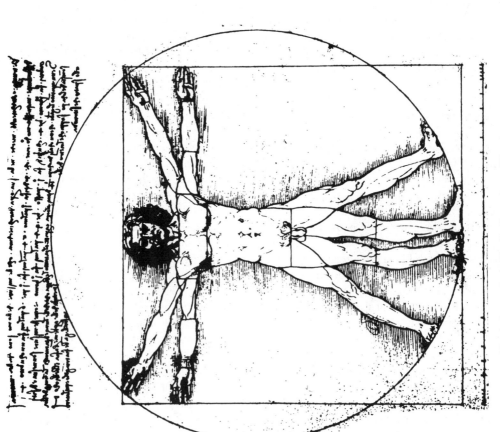

Fig. 1. Leonardo de Vinci's illustration to *Vitruvius*.

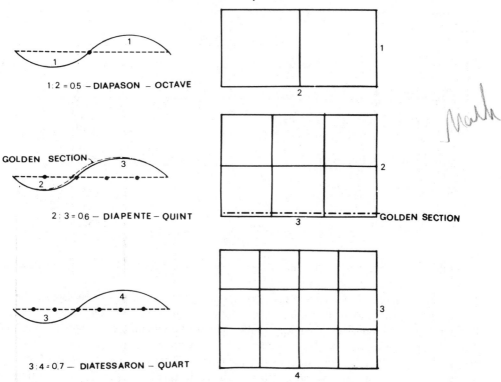

Fig. 3. Limits: basic musical Root-Harmonies—overtones—and their spatial equivalents.

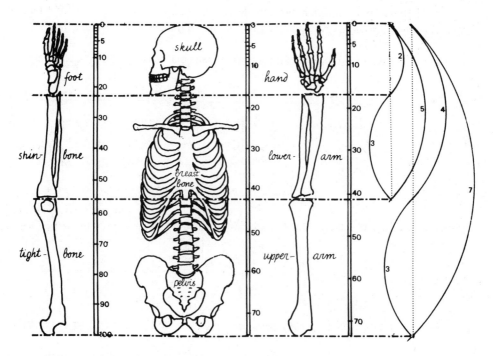

Fig. 4. Relationships of leg, torso and arm.

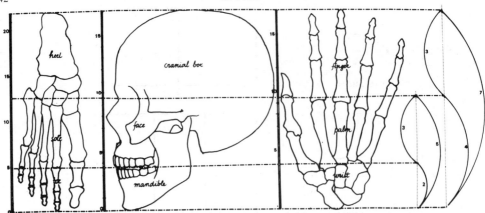

Fig. 5. Relationships of foot, skull and hand.

to call this "the relatedness of neighbors", since in organic growth this relationship frequently appears between old and new parts.

In the renaissance the name "divine proportion" was coined for the Golden Section probably because its ratio can never be reached with whole numbers. It can only be approximated. This ratio is an irrational number like π, the ratio of the circumference of the circle to its diameter. The Golden Section's construction, a square within a semicircle, creates a rectangle that is $\sqrt{5} = 2.236$ long, consisting of two reciprocal golden rectangles. The ratio of neighboring members in the summation series of Fibonacci numbers is 0.618, moving from the smaller to the larger numbers and 1.618 vice-versa.

The square, the circle and the pentagram united in the human body is a paradigm of supreme importance. The square represents the earth, the circle, heaven. Combined in the human body, they add up to "squaring the circle", which means that we unite in ourselves heavenly as well as earthly qualities, a belief firmly held throughout the ages. The discovery that the pentagram is also among our constituents suggests that it is through the musical root harmonies that we accomplish this. This is the paradigm of **dinergic symmetry**.

The Lutf Allah mosque is one of the most famous buildings in Iran, built on the principle of a round dome over a square base (Fig. 8). As far as we can tell, this construction method goes back to the ancient Zoroastrian fire temples in the Iranian desert (Fig. 9) and has the advantage of being much more open than the fully supported dome (as for instance the Pantheon in Rome).

From Iran, the idea spread, probably through Armenian masons to St. Sophia in Byzantium,

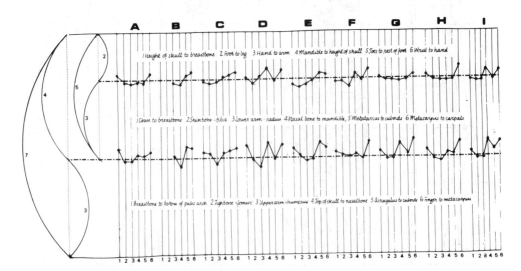

Fig. 6. Relationships of skull to torso 1; foot to leg 2; hand to arm 3; cranial box to skull 4; toes to foot 5; and fingers to hand 6.

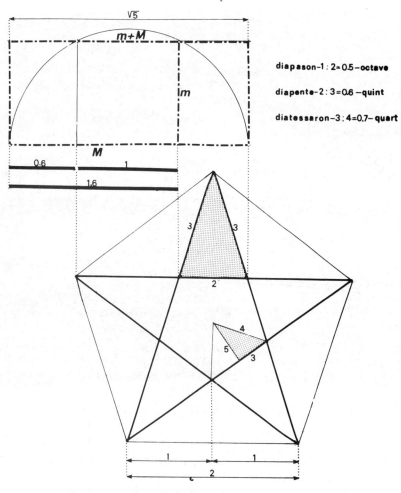

diapason-1 : 2=0.5-octave

diapente-2 : 3=0.6 -quint

diatessaron-3 : 4=0.7- quart

Fibonacci series · 1 : 2 : 3 : 5 : 8 : 13 : 21 : 34 : 55... ➤0.618... ◄—1.618...

Fig. 7. Dinergy. The Golden Section: a paradigm of dinergy.

to the Dome of the Rock in Jerusalem, to the Dome of Firenze, St. Marks in Venice, St. Peters in Rome, St. Pauls in London, and the Pantheon in Paris. From here it took off over the Atlantic to become the scheme of the "sanctuary" of democracy; the Capitol of Washington and a whole series of state houses from Texas to Washington to Colorado. It moved east to the Taj Mahal of India, to the great churches of Kiev and Moskva, and finally to the parliament of Budapest.

That the curve of this dome is indeed an example of **dinergic symmetry** uniting minor parts with major parts, can be seen from the fact that it happens to be identical with the China clam's cross section (Fig. 10), which is illustrated here by a series of radiating straight lines and circles that are at logarithmic distances. The same dinergic construction method (dinergic because it unites circles and straight lines, using their meeting points) can be used to reconstruct other shells, for instance the beautiful heart-shaped cockle (Fig. 11), which also fits into two $1 \times \sqrt{5}$ rectangles.

Even some vegetable shells like the walnut follow the same pattern of dinergic symmetry (Fig. 12), which I call *archetypal*. This following C. G. Jung, who gave this term a new important meaning, very generally: ancient, universal, unconscious pattern.

The Dome of Firenze's Santa Maria del Fiore church, one of the first domes over a square base in Europe, is a good example of incorporating dinergic symmetry (Fig. 13). Plan and elevation fit exactly in a golden section rectangle. Nave height (k) and dome height (m) again share the golden proportion or quint. The height of the dome C and the length of the floor plan B are in 3:4 relationship and the central axis divides the building into two equal parts; an octave.

Fig. 8. Lutf Allah mosque, Isphahan, Iran.

That heaven and earth are indeed united in human beings—which is an archetypal idea, dinergic as well as symmetrical, which has never relinquished its hold over the imagination—can be seen most dramatically in one of the masterpieces of modern architecture: the Sidney Opera House (Fig. 14). The architect, Jørn Utzon, was inspired by the clouds. He wanted to bring them down to earth to cover his buildings, as he himself says[4]. The great white arches indeed rest like clouds over the buildings, lightly, as if they could lift off at any moment.

This is not a domed structure but consists of a series of arches which rest upon only a few points. There is an intricate relationship between most of these points and the height of arches A, B, and C over the water's surface, respectively the various platform levels upon which they rest (Fig. 15). These relations follow the golden section as can be seen in the elevation. In addition, two reciprocal golden rectangles cover the entire structure. The larger one contains the three arches of the stage and the auditorium while the smaller one covers the vestibule.

The entire plan is contained in a golden rectangle (dashdotted line) within which the two main buildings are each contained in a $\sqrt{5}$ rectangle with a major and minor components following the respective reciprocal golden rectangles, as in the elevation (Fig. 16). The smallest building is a single golden rectangle. Perhaps it is not surprising that these relationships exist, since the architect started with *freehand* sketches of the curved lines.

That the dinergic symmetry of heavenly and earthly qualities is the source of all creative activity is mythologically expressed by the Creation myth of the Bible, "And God said: 'let us make man in our own image, after our likeness. . .' And God created man in his own image, in the image of God created He him"[5].

In Michelangelo's fresco on the ceiling of the Sistine chapel, we see the first man but there is not yet any life in him. (Fig. 17). We catch the very last moment of creation when he stretches out his finger to be touched by the Creator. And the point where the two fingers meet is the golden section point on the entire length. Both the Creator and Adam share the golden section's quint proportions. Each is contained within his own golden section rectangle within the overall shape, which is $\sqrt{5}$ long.

As a matter of fact, the whole ceiling shares the **dinergic symmetry** (Fig. 18): it is composed

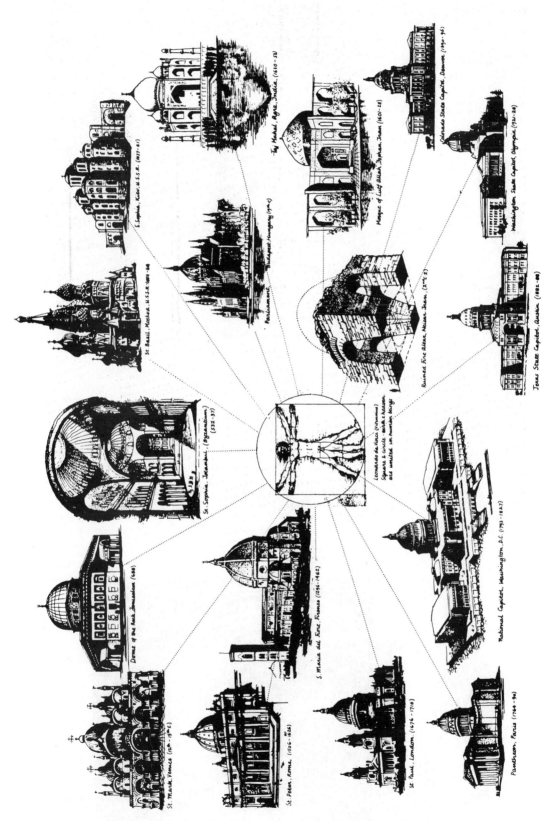

Fig. 9. Dome resting on a square base. Historical development.

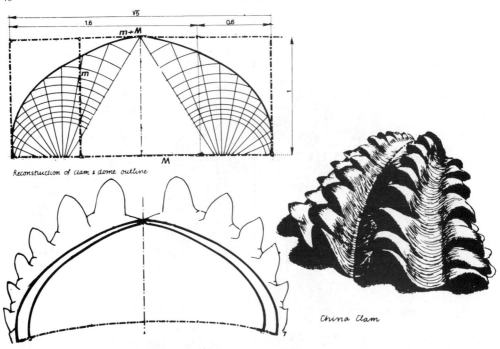

Reconstruction of clam & dome outline

China Clam

Fig. 10. Two halfs of China Clam & Lutf Allah dome outline.

of three golden rectangles; a large horizontal one and two small vertical ones on either side. The width of the subframe shares this same relationship with the overall width, and the pictures of the sun and moon's and Adam's creation, the fall and the deluge follow each other and the other scenes of creation in 3:4 harmony. Michelangelo must have been familiar with theories of the golden section, well known among artists and architects of the renaissance. However it

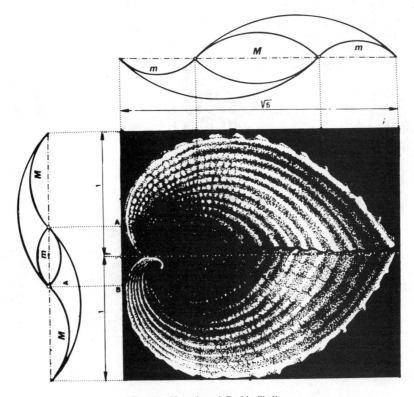

Fig. 11. Heartshaped Cockle Shell.

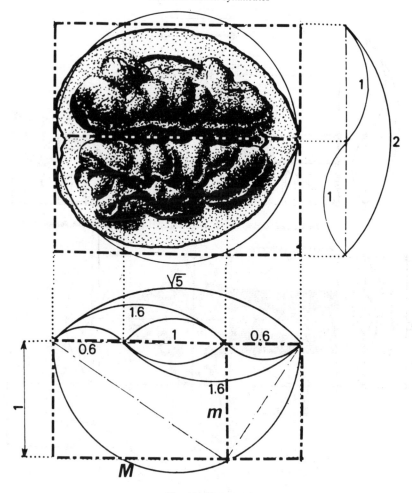

Fig. 12. Walnut.

is not necessary to be a Michelangelo in order to create patterns of dinergic symmetry. After all, it is a pattern-forming process of nature; and he was a child of nature. His greatness was in using the processes of nature consistently. "Not I, not I, but the wind that blows through me"[6].

It is now time, after having seen examples of **dinergic symmetry** in nature and art, to see to what extent it also exists in folk art and in the unsophisticated habits of preliterary and prehistoric people.

The "squaring of the circle" and "the unity of heaven and earth" was not "invented" by the Zoroastrian priests in the Iranian desert. It has been used since time immemorial by many people, including for instance the Pawnee Indians of Nebraska (Fig. 19). Similar ideas also appear in many other places and among peoples, for example in the purification lodges of the Sioux[8]. The earth lodge was round, symbolizing *heaven*, with four poles representing the four directions of the *earth*. These poles held up the roofbeams, leaving an opening at the center for the sacred fire which represented Wakan Tanka or the Great Spirit who lives in heaven and on earth, in the center of all beings. Beds were arranged around the perimeter with two reserved for the virgins on each side of the altar, and very narrow they were indeed. The earth lodge was a sanctuary as well as a home for two families and every single feature in it had sacred significance. It was heaven upon earth, manifesting itself in **dinergic symmetry**.

The hand that does the carving is built according to the principles of **dinergic symmetry** and the carving that emerges simply reflects this. Evidence of this can be seen in Figs. 20–23.

We may go even a step further and find **dinergic symmetry** in the earliest period of prehistoric carvings (Fig. 24). The frontal view of the Venus of Willensdorf is twice as high as it is wide: a clear example of octave harmony. The navel is exactly at the golden section

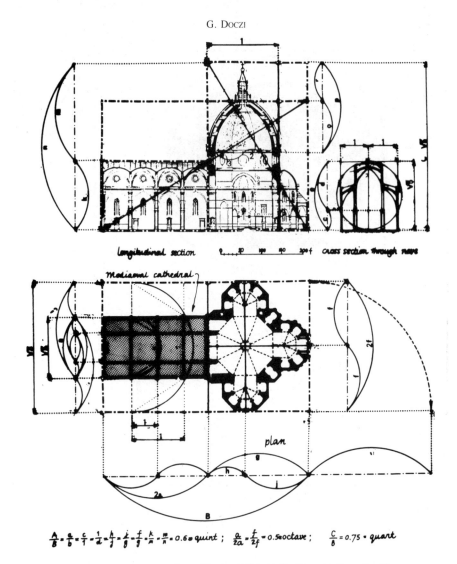

longitudinal section   0   30   60   90   120 f   cross section through nave

Mediaeval cathedral

plan

$\frac{A}{B} = \frac{a}{b} = \frac{c}{f} = \frac{1}{d} = \frac{h}{j} = \frac{i}{g} = \frac{f}{g} = \frac{k}{g} = \frac{m}{n} = 0.6 =$ quint ; $\frac{a}{2a} = \frac{f}{2f} = 0.5 =$ octave ; $\frac{C}{B} = 0.75 =$ quart

Fig. 13. Santa Maria del Fiore, Firenze (1296–1462). Dome by Brunelleschi.

Fig. 14. Jørn Utson: Sidney Opera House.

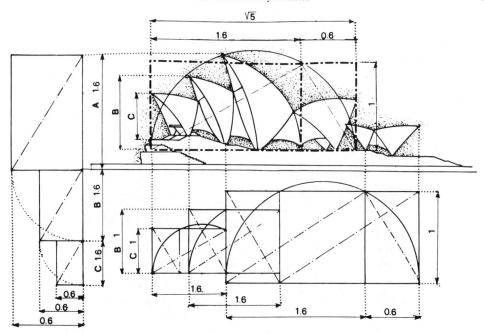

Fig. 15. Sidney Opera. Elevation.

point of the height. In profile she fits neatly into a $\sqrt{5}$ rectangle. The female statuette, Lespugue, of France, dating from the late Ice Age is carved in ivory and fits also into a $\sqrt{5}$ rectangle in frontal view, while in profile, additional golden rectangles are added for head and feet.

In a prehistoric burial cave in New Grange, Ireland, double spirals have been found engraved in the stone (Fig. 25). Similar patterns have been found elsewhere in connection with burial grounds. Many believe—Robert Graves among them[9]—that if one follows the lines with the finger, one spiral moves inward as the other moves outward: a symbol of the **dinergic symmetry** of life and death.

There is something awe inspiring in the fact that the same pattern can also be found written into the dactyloscopic patterns of our fingerprints, the double helix of DNA and also in the microtubules of axonemes (which are tiny intracellular structures only recently discovered and seen here 90,000 times enlarged)[10]. A power makes itself known that is greater than we.

This power chooses to express itself in patterns of **dinergic symmetry** in nature as well as in art, and by doing so opens our eyes to that which we cannot see: the **limitless**. Wendell Berry speaking of an ancient sycamore tree says it is "a fact, sublime, mystical and unassailable"[11].

The ancient Indians sensed this power and worshiped it with their arms raised to **heaven** and their heads bent down to **earth**. They instituted sacred rites to be fully aware of it. The initiation rites served this purpose (as do our baptismal and confirmation rites); they taught the young that beyond their fleshly father and mother, they have a cosmic parentage: mother Earth and father Heaven[12].

The realization of this amounted to a *rebirth*. The Nootka Indians of the Northwest carved it in wood, and it follows the $\sqrt{5}$ rectangle, head and legs being in golden proportions to the body. The Aztec of Teotihuacan modeled it in clay and this image fits into a 3:4 rectangle while significant parts, for instance the opening of rebirth, are between the quint and octave points of its full height (Figs. 26, 27). It is significant in this connection that Socrates, whose mother was a midwife, refers to himself not as a teacher, but as a **midwife**, helping his fellow beings in their rebirth[13].

Let us now turn to an issue which is vital for all of us: can we find a **dynergic symmetry** between **war and peace**? Are there any signs that underneath the staggering surface differences that separate people from people today, they actually **do agree** on basic archetypal issues? It is reassuring to note that true to ancient tradition, which always considered the pentagon a

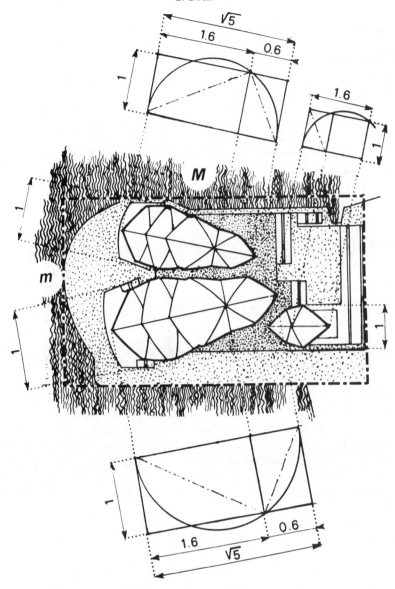

Fig. 16. Sidney Opera Plan.

sacred portent of good fortune (and even though this was very likely not a determining issue of design), our own defense establishment is housed in a building bearing **this** name and shaped accordingly (Fig. 28).

In addition, the pentagonal star is repeated in our national flag, for each of the states of the Union, and it is likewise proudly displayed in the flags of Soviet Russia, China and Cuba, to mention only those nations who most vehemently disagree with us politically. We all swear our most solemn allegiance to our flag, which we cherish above all, yet it carries the same pentagram.

Money we cherish in yet a different way. While, for instance, we disagree with the Hungarians on so many important issues, the Hungarian 10 forint note which bears the great poet Petöfi's picture, and the one dollar note with George Washington's picture, are practically the same shape (Fig. 29). Both are contained in a $\sqrt{5}$ rectangle and the circle which establishes the golden section might just as well be the symbol of the **one world** which they share.

The Moebius strip is perhaps the best symbol of **dinergic symmetry** (Fig. 30). The legendary two people who lived on its two sides could never climb over its edges to meet until

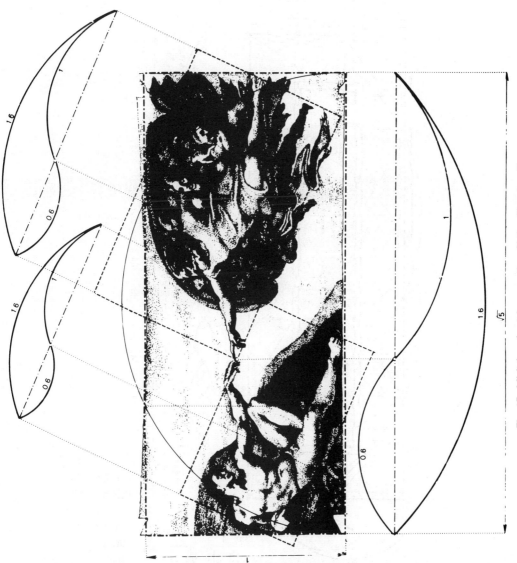

Fig. 17. Michelangelo: Adams creation. Sistine Chapel, ceiling.

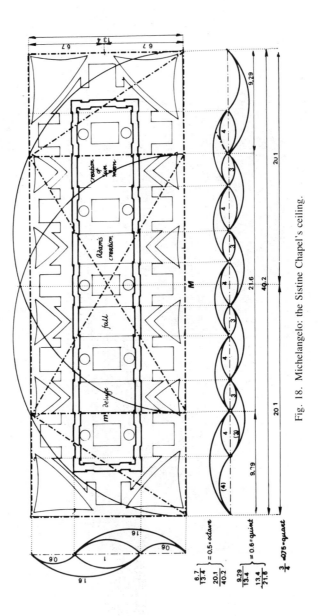

Fig. 18. Michelangelo: the Sistine Chapel's ceiling.

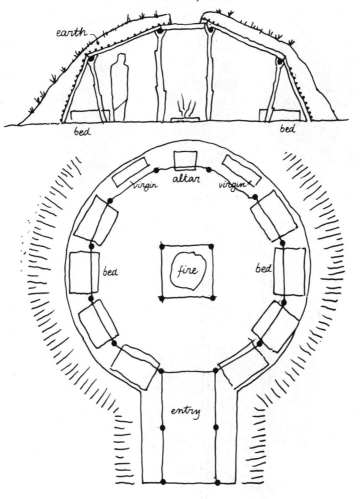

Fig. 19. Pawnee earth lodge.

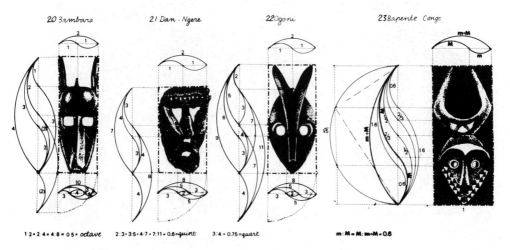

Figs. 20–23. African dance masks.

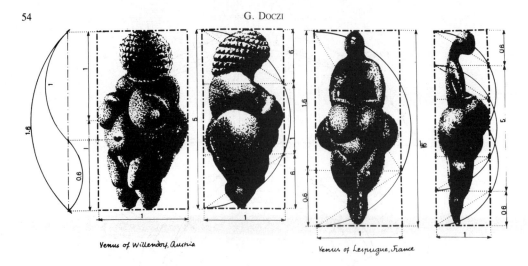

Venus of Willendorf, Austria

Venus of Lespugue, France

Fig. 24. Venus of Willendorf, Austria. Venus of Lespugue, France.

Prehistoric spiral mazes at New Grange, Ireland.

The Whorl finger pattern,

Axoneme, the core of a single
axopod, shown in cross-section. Enlarged 90,000 diameters.

Fig. 25. Power: dinergic symmetry in prehistoric carvings and in our body.

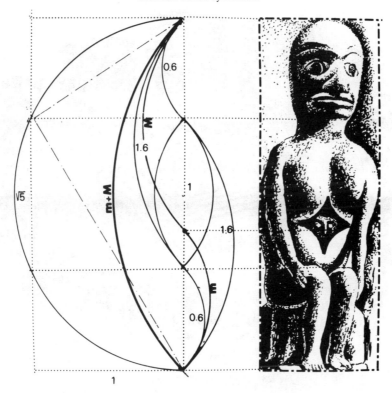

Fig. 26. Nootka carving.

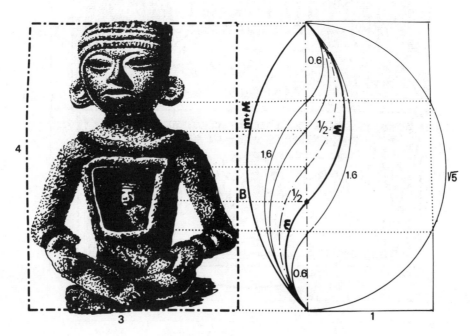

Fig. 27. Teotihuacan clay figurine.

Fig. 28. Universality of the pentagonal star.

Fig. 29. Identity of Hungarian and American banknote proportions.

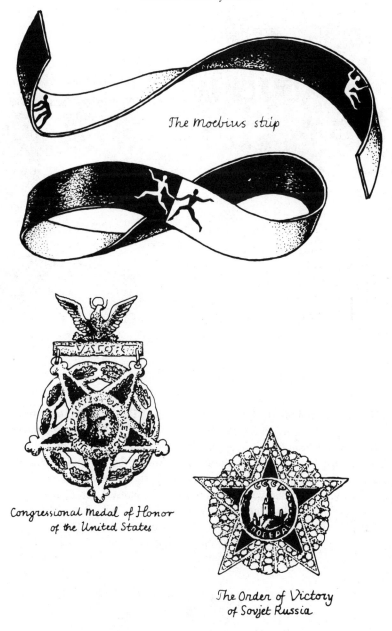

The Moebius strip

Congressional Medal of Honor
of the United States

The Order of Victory
of Soviet Russia

Fig. 30.

Dr. Moebius came along and with a twist, connected the *opposite* ends. The world waits with baited breath for the Dr. Moebius of statesmanship to arrive. We do not need to climb over the edges of our strip, we need only to realize that our world **is one** and that we **share** it with all the other "twolegged, fourlegged and winged ones"[14]. Anyway, the United States and the Soviet Union have in their highest military orders the same pentagonal star.

Meanwhile, we build ever taller skyscrapers which are the marvel of the world. Their sheer audacity takes the breath away. We are proud of the luxuries they provide us. Some of us hardly notice the dark alleys between them with garbage cans and cockroaches (Fig. 31).

We are about as tall in relation to our skyscrapers as the termites are to their heap, reaching about 20 feet in the Australian desert. We do not know what path evolution took to bring them about from their cockroach-like ancestors. But since they are seemingly impossible to exterminate, they very likely would survive a nuclear holocaust. Death may not be as horrible as survival under such circumstances.

It might well be as Socrates said, "No one knows whether death is not the greatest of

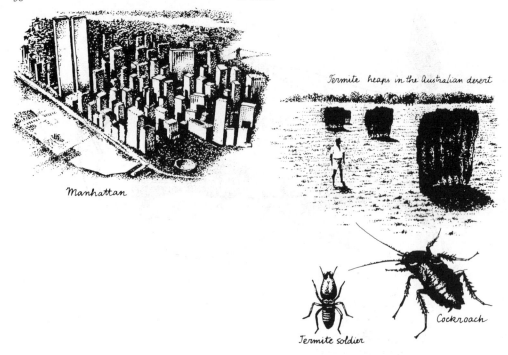

Fig. 31. Alternatives.

blessings''[15]. Indeed, looking at this deathmask (Fig. 32), taken from a woman who was pulled out of the Seine, it is hard to escape the notion that she is happy like Simon ben Jochai, the ancient rabbi, about whom his disciples said, ''When he died he celebrated his wedding''.

Peter the Great rests in the city which he built (now called Leningrad) and the reassured and reassuring smile around his lips seems to say that he **now** knows the answer to the riddle of death (Fig. 33). When one really **ceases** to think about **oneself** one is **all-sharing**.

Philippo Brunelleschi, the master builder who constructed the dome of Firenze also smiles in his death (Fig. 34). Bach's cantata 161 entreats, ''Come sweet death''.

The cemeteries in Transsylvania are among the happiest places on earth; they are planted

Fig. 32. L'Inconnue de la Seine. Deathmask. Paris morgue XIXc.

Fig. 33. Peter the Great's deathmask, Museum of Anthropology, Leningrad.

with fruit trees. Indeed, there is a **dinergic symmetry** between life and death. The limits of life cease to exist, and we become one with the limitless. But do not hasten it. Life is **also** one of the great blessings or rather it is a **gift** and a **responsibility**.

Brunelleschi left us a crucifix in the Santa Maria Novella church[16]. This drawing (Fig. 35) shows how the pentagram as well as the circle and square fits it just as it fit Leonardo's man of which I spoke earlier. Vasari[17] describes how Brunelleschi inadvertently hurt his friend Donatello's feelings when he remarked about his friends crucifix, "My friend, you modeled it from a peasant!" (Meaning that there was nothing divine about it.) Angrily, Donatello

Fig. 34. Filippo Brunelleschi, deathmask, made by his fosterson Cavalcanti (Buggiano) 1446.

Fig. 35. Crucifix, Santa Maria Novella, Firenze, by Filippo Brunelleschi. (1377–1446).

Fig. 36. Lifemask of John Keats, National Portrait Gallery, London.

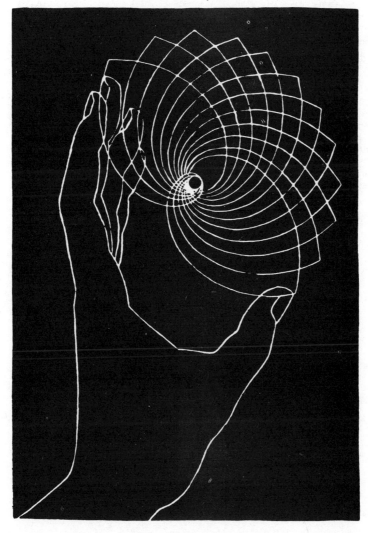

Fig. 37. The creative hand.

replied, "You do it better!" Brunelleschi went home and did. When he was finished he invited Donatello for dinner and rather slyly sent him ahead with the groceries pretending an errand. When he returned home he found the groceries scattered on the floor and Donatello on his knees in front of the crucifix praying.

There are invisible, numinous presences, archetypal patterns of compelling power, just as there are unheard melodies (Fig. 36). John Keats says, "Heard melodies are sweet, but those unheard are sweeter. Beauty is truth, truth beauty"(18).

The human hand is at present balled into a fist. It cannot create. But any fist can open up. As it does so all the fingers describe logarithmic spirals, jointly creating a stroboscopic image (Fig. 37), the wonder of the world, the thousand petaled lotus of enlightenment, the archetypal pattern of **Peace**.

Acknowledgement—My gratitude is due to my wife Aina, without whose devoted practical support and constructive criticism this work could not have been accomplished.

ENDNOTES

1. I am grateful to Istvan Hargittai who has called my attention to this fact.
2. For permission to work at the Biological Structures Laboratory of the University of Washington I express my gratitude to Dr. Daniel O. Garney.
3. C. H. Waddington, "The Modular Principle in Biological Form" Module, Proportion, Symmetry, Rhythm (ed. György Kepes).

4. GA 54 Jørn Utzon, Sidney Opera House.
5. *Genesis*, 1: 27, 28.
6. D. H. Lawrence, *Song of a Man Who Has Come Through.*
7. *Pawnee Life*, Field Museum of Natural History, Chicago.
8. Joseph Eppes Brown, *The Sacred Pipe.*
9. Robert Graves, *The White Goddess.*
10. Drawn after *Ultrastructure Research* **30**, 7–37 (1970).
11. Wendell Berry, *Openings. The Return.*
12. Mircea Elisde, *Cosmos and History & The Sacred and the Profane.*
13. Plato *Theateteus*, 149.
14. *The Sacred Pipe*, (ed. Joeseph Eppes Brown).
15. Plato *The Last Days of Socrates, The Apology.*
16. Eugenio Battisti, *Phillipo Brunelleschi, The Complete Works.*
17. Giorgio Vasari, *The Life of the Painters, Sculptors and Architects.*
18. John Keats, *Ode on a Grecian Urn.*

Comp. & Maths. with Appls. Vol. 12B, Nos. 1/2, pp. 63–75, 1986
Printed in Great Britain.

0886–9561/86 $3.00 + .00
© 1986 Pergamon Press Ltd.

SYMMETRY AND MODULARITY

A. L. LOEB

Department of Visual and Environmental Studies, Carpenter Center for the Visual Arts,
Harvard University, Cambridge MA 02138, U.S.A.

Abstract—Many complex structures may be generated from simple modules with the aid of an algorithm or generating rule. The properties of space generally impose restrictions on the shape of the module. As examples we consider tilings of the euclidean plane and of the sphere, lattices and lattice complexes, Dirichlet Domains and space-filling, and the decomposition and recombining of some polyhedra into space fillers.

INTRODUCTION

We tend to look at the perfect symmetry of an organism such as a sea urchin with astonishment and admiration, comparing it to the less than perfect symmetry of the human species. Yet it is precisely the simple organisms which display more perfect symmetry, whereas complex systems suffer perturbations of their symmetries. The indistinguishability of electrons accounts for the symmetries of molecular bonding, and the symmetry of single crystals can be accounted for by the assumption that identical atoms or ions tend as much as possible to have identical environments[1]. In more complex systems perturbations will introduce imperfections, resulting in a loss of symmetry. Complex systems can frequently be considered as hierarchical superpositions of several levels of symmetry[2], much as functions may be expressed through Fourier analysis as superpositions of sine and cosine functions. We must therefore look upon perfectly symmetrical structures as basic, elementary building blocks for our real world, in the same way that musical tones are analyzed in terms of 'pure' sine and cosine waves.

Symmetrical structures occur as a result of the juxtaposition of multitudes of identical modules, each of which interacts in such a manner with all others that all have as identical environments as possible. Such structures may be described in terms of the module and the algorithm used to generate them. Consider, for instance, the simplest module, a point, together with the following algorithm: "Every point is in the vicinity of only one other point." The structure so generated consists of a pair of points. On the other hand, changing the algorithm to: "Every point is equidistant from *two* other points." generates either an infinite array of equally spaced points on a line, or an array of equally spaced points along the circumference of a circle. In the latter case the question arises whether closure can be achieved, that is to say, whether a full 360-degree traversal around the circle's circumference will produce a point which coincides with the initial point. If that is the case, then a polygon will have been generated; if not, then one or more additional traversals of the circle's circumference may eventually produce closure, or the circle will be traversed indefinitely, with the result that points will be generated which come arbitrarily close to each other.

Which of these possibilities occurs, depends on the angle between the line segments joining any point to its nearest two neighbors. If this angle equals $[(k - 2)/k] \times 180$ degrees, then a k-gon is generated after a single rotation. Any other rational fraction of 180 degrees will produce closure after several rotations, with the result that a stellated polygon is generated: an angle of $[(k - 2j)/k] \times 180$ degrees will produce k-fold rotational symmetry after j traversals of the circle. If, however, the angle joining each point to its two nearest neighbors equals an irrational fraction of 180 degrees, then a continuum of points will be generated along the circumference of the circle.

The conclusion to be drawn is that the algorithm does not uniquely specify the pattern generated: one needs to refine either the algorithm or the module. In this instance it is convenient to redefine the module by adding to the point two semi-edges representing a valence bond from the point to each of its neighbors. The angle between these valence bonds determines the pattern generated by the algorithm.

REGULARLY SPACED POINTS

The patterns just considered are made up of divalent points: every point is joined to two neighbors. One can analogously use an algorithm for trivalent points, each being equivalent to its three neighbors. If all three bonds are coplanar, a hexagonal tessellation is generated. If the bonds are not coplanar, but the angles between the bonds are all equal, three regular solids are generated:

$$
\text{If the angle equals} \left\{ \begin{array}{c} 60 \\ 90 \\ 108 \end{array} \right\} \text{degrees, the} \left\{ \begin{array}{c} \text{tetrahedron} \\ \text{cube} \\ \text{pentagonal-} \\ \text{dodecahedron} \end{array} \right\} \text{is generated.}
$$

The three angles need not all be equal; when they are not, the semi-regular solids are generated. For instance, when one angle equals 60 degrees and the other two 120 degrees, a truncated tetrahedron results (see Fig. 1). The mutually equivalent points or vertices lie on the surface of a common sphere; the edges joining them may be projected onto the sphere so that they become parts of geodesic circles which tessellate the surface of the sphere. In general, the valence r may be greater than three, but its value depends on the total number of edges joining the vertices E and on the number of edges surrounding each polygonal tile on the sphere surface, n. Although all vertices must be equivalent, all faces are not necessarily identical, for the number of edges surrounding each face may vary, as exemplified by the truncated tetrahedron (Fig. 1). The average number of edges of all faces is denoted by n_{av}. The relation between these variables is derived from Euler's relation between the number of vertices, edges and faces of a polyhedron; if all vertices lie on a common singly connected surface, it is[3]:

$$
\frac{1}{r} = \frac{1}{2} + \frac{1}{E} - \frac{1}{n_{av}}. \tag{1}
$$

This is a Diophantine equation: all its variables are rational; in point of fact, only n_{av} is not an integer. The equation has a finite number of solutions, each corresponding to one of the regular or semi-regular structures[4]. The tessellations of the euclidean plane are included among these solutions; they correspond to the solutions having E equal to infinity. (The truncated tetrahedron, for example, has r equal to 3, E equal to 18. It has four triangular and four hexagonal faces, with the result that n_{av} equals 4.5; these values fit Eq. 1.)

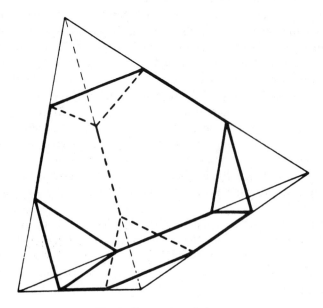

Fig. 1. Truncated tetrahedron.

We have demonstrated therefore that the module consisting of an r-valent vertex together with the algorithm that each vertex be joined to r equidistant vertices on a singly connected surface (including the euclidean plane), will generate all regular and semi-regular structures. Their duals, structures having all faces equivalent, represent all tilings of the sphere and the euclidean plane with only one type of tile. There is a corresponding Diophantine equation for these structures, which is derived from Equation 1 by interchanging n and r:

$$\frac{1}{n} = \frac{1}{2} + \frac{1}{E} - \frac{1}{r_{av}}. \tag{2}$$

Physically, Eq. 2 represents the restriction placed on the polygons which will singly tile the sphere and euclidean plane. It is notable that only Euler's relation and the requirement of equivalence of vertices (resp. tiles [faces]) were used here; no use was made of the magnitude of angles or distances, and symmetry was implied, not used explicitly. The reasoning can be summarized in the following steps:

1. Define a module.
2. Define the algorithm by which these modules are to be combined.
3. The module and the algorithm may or may not be compatible; in our example they are compatible only under certain conditions, expressed in terms of a Diophantine equation, which has but a finite number of solutions.
4. These solutions each correspond to one of the structures which may be generated from the given module, which are thus exhaustively enumerated and annotated.

The mutually equivalent vertices may not lie on a single common surface. Rather, they may be distributed in three-dimensional space, forming a structure called a *lattice* if their environments are not only identical, but also identically oriented. If their environments are merely identical, but not necessarily identically oriented, the structure is called a *lattice complex*[5]. Whereas structures comprising mutually equivalent vertices that lie on a single common surface properly must be considered two-dimensional, and are characterized by two connection parameters, r (the valency of each vertex) and n_{av} (the average value of the number of edges surrounding each face), the lattices and lattice complexes are three-dimensional. In three dimensions the number of parameters increases to eight[6], but only five diophantine equations relate these eight parameters, with the result that there is much greater latitude in the formation of three-dimensional structures comprising mutually equivalent vertices than there is for corresponding two-dimensional structures. Fischer, Burzlaff, Hellner and Donnay have found 402 lattice complexes (including lattices).

In two dimensions the highest possible valence for an array of mutually equivalent vertices is *six*. The corresponding structure is the triangular tessellation, in which each vertex is surrounded by six neighbors equivalent to each other and to the central vertex. In three dimensions the corresponding maximum valence is *twelve*; the twelve mutually equivalent neighbors are located at the vertices of a cuboctahedron surrounding the central vertex, which in turn is equivalent to each of the twelve neighbors (Fig. 2).

TILINGS

The tilings of the euclidean plane correspond to those solutions of Eq. 2 having E equal to infinity. As that equation is derived directly from Euler's relation, and hence deals only with numbers of connections, not with actual distances or magnitudes of angles, the edges of the tiles need not be straight lines, but may be curved. The plane may be tiled with any straight-edged triangle; this becomes evident when one realizes that any pair of such triangles may be juxtaposed to form a parallelogram, and a parallelogram can tile the plane by translation. To generalize to curved edges, the symmetry implications of the last statement are helpful. The translational symmetry resulting from tessellating with a parallelogram requires the opposite sides of the parallelogram to be pair-wise congruent, whereas the pairing of the triangles is

A. L. LOEB

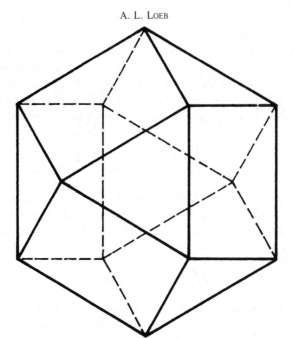

Fig. 2. Cuboctahedron.

achieved by rotation, hence requiring the diagonal of the parallelogram to contain a center of two-fold rotational symmetry. These constraints are satisfied if each side of the tessellating triangle contains a two-fold center (see Fig. 3).

The plane may also be tessellated by *any* quadrilateral. This is achieved by rotating the quadrilateral 180 degrees around the centers of each edge. As a result, curved-edge quadrilateral tessellations, like the analogous triangular ones, need two-fold centers on all edges (Fig. 4). Both the triangular and the quadrilateral tessellations contain four distinct sets of two-fold centers of rotational symmetry[7], the latter at the centers of edges, the former at the centers of edges as well as at the vertices. Although any straight-edged triangle or quadrilateral will tessellate the plane, only special pentagonal and hexagonal tessellations are possible[8]. Moreover, that no polygons having more than six edges can tessellate the plane follows from Eq.

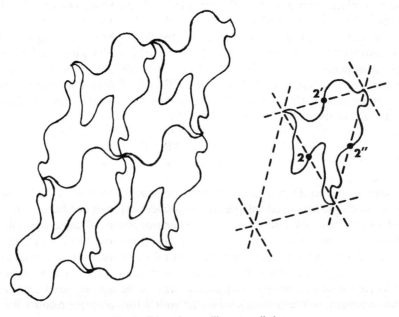

Fig. 3. Triangular curvilinear tessellation.

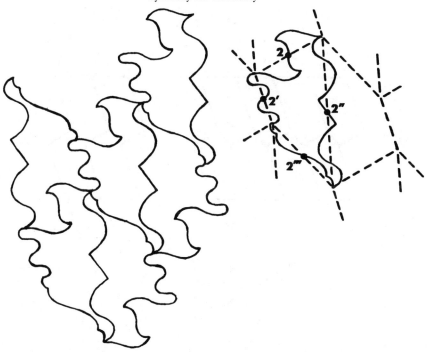

Fig. 4. Quadrilateral curvilinear tessellation.

2: if we set E equal to infinity, we find:

$$\frac{1}{n} = \frac{1}{2} - \frac{1}{r_{av}}.$$

Since r_{av} may not be less than three, n may not exceed six, q.e.d.

Three of the special pentagonal tessellations are shown in their straight-edged appearance in Figs. 5 and 6. Figure 5(a) shows only a strip; these strips may be repeated either congruently or enantiomorphically, adjacent strips being each others' mirror images. The requirement imposed on the pentagon in these instances is that two non-adjacent sides must be parallel and equal in length. Whichever combination of strips is used, there are four distinct sets of rotational symmetry; when all strips are congruent, two sets lie on the straight edge joining the strips, namely on the vertices and halfway in between. The other two lie on the centers of the edges forming the zigzag pattern in the center of the strips. When alternate strips are enantiomorphs, the "zigzag" sets and their mirror images constitute the four distinct sets of two-fold rotocenters.

More interesting is the pentagonal tessellation in Fig. 5(b). The pentagon must have right angles at two non-adjacent vertices; the edges joining at either vertex must be equal in length, but all four need not be equal. Symmetry theory is helpful for understanding why this tessellation works. The pentagons are arranged in 244' symmetry[9], the right angles fitting around the two distinct centers of fourfold rotational symmetry. Since these two centers are distinct, the edges meeting at one of them need not be equal in length to those meeting at the other. The location of the two-fold centers of rotational symmetry is determined by those of the two distinct sets of fourfold centers: the three form the vertices of an isosceles right triangle. In order for the pentagon to tessellate, the two-fold center must lie on the center of that pentagon edge which does not terminate at one of the right-angled vertices. The question which remains to be answered is whether it is generally true that the center of this fifth edge is indeed equidistant from the two right-angled vertices and subtends a right angle with them. The author has proven this to be the case; the detailed proof would take too much space here, but Fig. 6 gives the coordinates of the points in question, plotted in a suitably chosen cartesian coordinate system from which the desired result is easily derived.

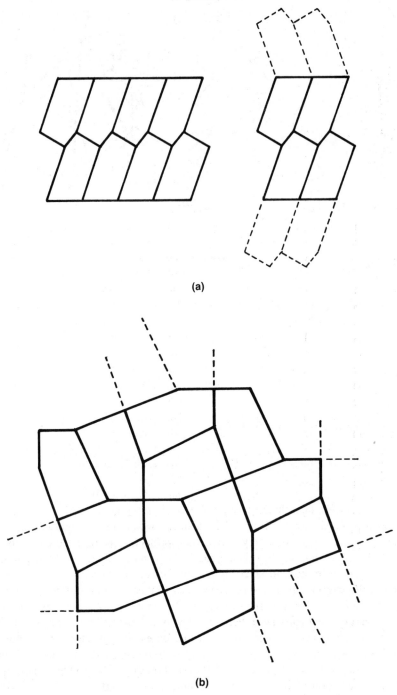

(a)

(b)

Fig. 5. Pentagonal tessellation.

Hexagonal tiles will tessellate the euclidean plane if one pair of opposite sides are parallel and the other four sides contain centers of twofold centers of rotational symmetry[10] (see Fig. 7). As not more than four distinct centers of rotational symmetry may coexist in a plane[11], we exhaust with these the tessellations of the euclidean plane; the hyperbolic plane has infinitely many tessellations[12].

SPACE FILLERS

To every lattice or lattice complex there corresponds a *Dirichlet Domain*. This domain is a region which contains a single point of the lattice (complex) and which has the property that every location within it is closer to that particular point than to any other point of the lattice

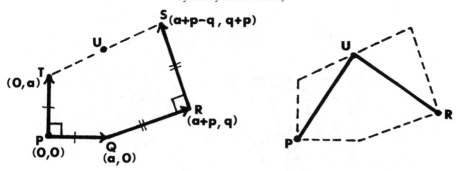

Fig. 6. Proof of the tessellability of this pentagon.

(complex). A Dirichlet Domain is constructed by perpendicularly bisecting every line joining a particular lattice (complex) point to all others; the innermost polyhedron formed by these bisecting planes is the Dirichlet Domain. From this definition it follows that Dirichlet Domains are space fillers, for if Dirichlet Domains are constructed around all the points of a lattice (complex), then every location must be closer to at least one of these points than to all others, hence there can be no location that does not belong to the Dirichlet Domain of one of the lattice (complex) points.

Those points which share a common face on the surface of their Dirichlet Domains are called *neighbors*. The polyhedron whose vertices are the locations of all the neighbors of a given lattice (point) is called the *coordination polyhedron* of that point. In order to constitute a three-dimensional lattice complex, each point must have at least four near neighbors. In the Diamond lattice complex each point is at the center of a regular tetrahedron, and connected to four other points at the vertices of that tetrahedron. The Dirichlet Domain of this lattice complex is a stellated truncated tetrahedron: the truncated tetrahedron (Fig. 1) is stellated on its four triangular faces by triangular pyramids whose apex angle is arccos ($-1/3$) (see Fig. 8). Thus there are four nearest neighbors sharing the large hexagonal faces, and twelve more distant neighbors sharing the small triangular faces[13].

Six nearest neighbors at the vertices of a regular octagon define a lattice whose Dirichlet Domain is a cube; this is the primitive cubic lattice. Eight nearest neighbors at the vertices of a cube might at first glance appear to have a regular octahedron as Dirichlet Domain. A regular octahedron, however, is not a space-filler: the Dirichlet Domain is actually a truncated octa-

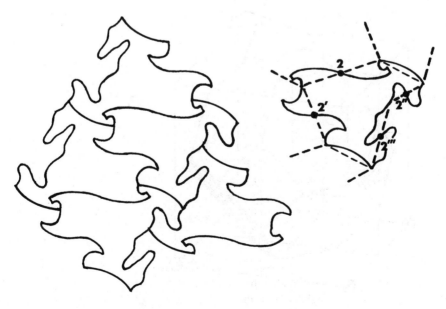

Fig. 7. Hexagonal tessellation.

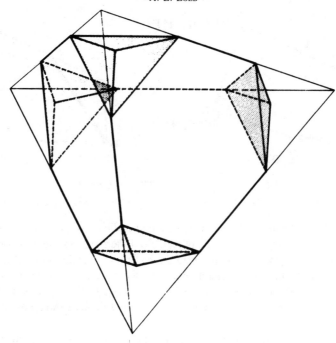

Fig. 8. Dirichlet domain of the diamond lattice complex.

hedron, with the eight nearest neighbors sharing the eight hexagonal faces, and six slightly more distant neighbors sharing the six square faces (see Fig. 9). These fourteen neighbors constitute the vertices of a rhombic dodecahedron (Fig. 10). This lattice is called *body-centered cubic*.

As stated above, twelve is the maximum number of equidistant neighbors permitted in three-dimensional lattices (see Fig. 2). The lattice characterized by this configuration of neighbors is called *cubically close-packed*. Its Dirichlet Domain is the rhombic dodecahedron (Fig. 10), which therefore fills a dual function, namely as coordination polyhedron for the body-centered cubic lattice, as well as the Dirichlet Domain for the cubically close-packed lattice[14, 15].

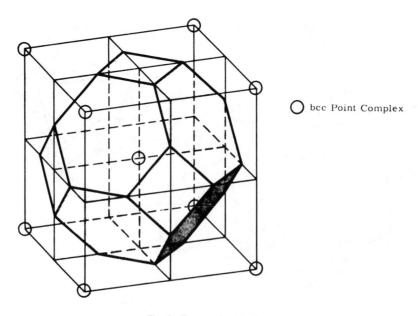

◯ bcc Point Complex

Fig. 9. Truncated octahedron.

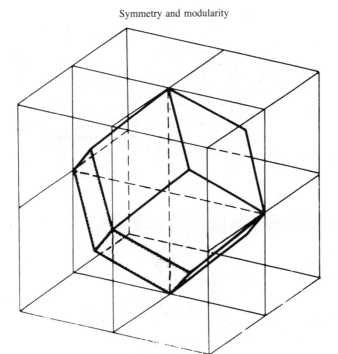

Fig. 10. Rhombic dodecahedron.

THE RHOMBIC DODECAHEDRON

This polyhedron constitutes one of the principal modules for generating space frames. When packed to fill space, the centers of these dodecahedra constitute a cubically close-packed lattice, whereas the centers together with the vertices constitute a body-centered cubic lattice. It may be subdivided into sub-modules in two distinct manners. As shown in Fig. 11, six square pyramids may be lopped off, leaving a cube. The six pyramids may in turn be combined to constitute a second cube congruent with the first one. On the other hand, eight triangular pyramids may be lopped off to leave a regular octahedron; these pyramids may be combined to form two regular tetrahedra, as in Fig. 12. Since the dodecahedra are space fillers, it follows that regular octahedra and tetrahedra having the same edgelength will together fill space, two tetrahedra being needed per octahedron.

The apex angles of the square pyramids which were combined to form a cube, equal arccos 1/3. Two such pyramids may be paired into a square bipyramid by joining their square bases. The result is an irregular, but space-filling octahedron having one diagonal shorter than the other two. Six of these octahedra, in three distinct orientations, will reconstitute the dodecahedron, hence are themselves space fillers.

The apex angles of the triangular pyramids, four of which were combined to constitute a regular tetrahedron, equal arccos (− 1/3). These triangular pyramids are identical with those mentioned above in connection with the Diamond lattice's Dirichlet Domain; that Domain was made up of a truncated tetrahedron combined with four of the triangular pyramids. Since we now see that these four will together just constitute a regular tetrahedron, we may conclude that a truncated regular tetrahedron together with a regular tetrahedron having the same edgelength will fill space in equal proportions. It must be noted in passing that the angles arccos 1/3 and arccos (− 1/3), which are slightly smaller than 71 degrees and slightly larger than 109 degrees respectively, figure prominently in modular structures; they equal as well the angles between the body diagonals of the cube.

THE TRUNCATED OCTAHEDRON

The truncated octahedron having all edgelengths equal to each other is the Dirichlet Domain of the body-centered lattice. It may combine with eight octants of itself to form a cube as in

A. L. LOEB

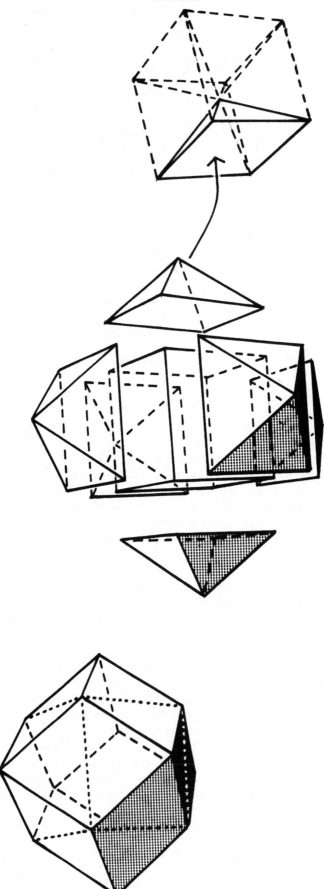

Fig. 11. 1 rhombic dodecahedron = 2 cubes.

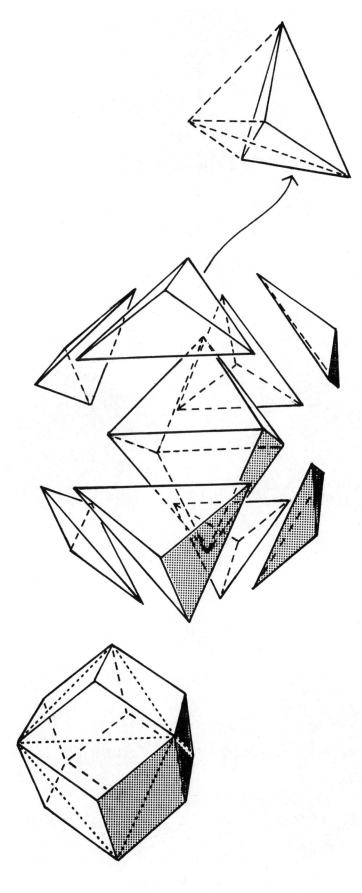

Fig. 12. 1 rhombic dodecahedron = 1 octahedron + 2 tetrahedra.

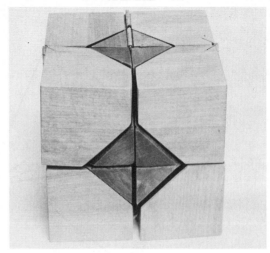

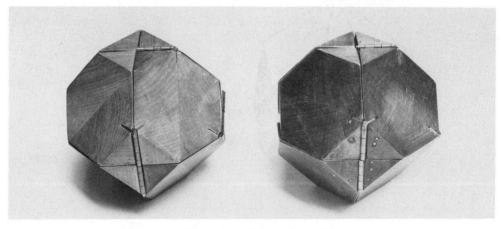

Fig. 13. 1 cube = 2 truncated octahedra.

Fig. 13. Whereas the rhombic dodecahedron may be considered twice a cube, this polyhedron would be a half cube, as illustrated by some of its transformations in Fig. 13.

CONCLUSIONS

By starting with a collection of mutually equivalent points, confined to a singly connected surface or distributed through three-dimensional euclidean space, we have noted that space is not a passive vacuum, but has distinct properties which determine the shapes of the modules which can fit in. Symmetry is the result of the restrictions placed by the properties of space on the possible juxtapositions of these modules. These restrictions may be expressed in terms of a few remarkably simple Diophantine equations.

Acknowledgement—The author is pleased to acknowledge the assistance of William Lyon Hall in the drawing of Figs. 3–7.

REFERENCES

1. A. L. Loeb, A systematic survey of cubic crystal structures. *J. Solid State Chem.* **1**, 237–267 (1970).
2. Cyril S. Smith, in *Hierarchical Structures* (edited by Lancelot L. Whyte, A. C. Wilson and D. Wilson). Elsevier (1969).
3. A. L. Loeb, *Space Structures, their Harmony and Counterpoint*, Ch. 4 and 5. Addison-Wesley, Reading, Massachusetts (1976).
4. A. L. Loeb, *Space Structures*, p. 56 and Chapter 12.
5. W. Fisher, H. Burzlaff, E. Hellner and J. D. H. Donnay, *Space Groups and Lattice Complexes*. U.S. Department of Commerce, National Bureau of Standards Monograph 134 (1973).
6. A. L. Loeb, *Space Structures, their Harmony and Counterpoint*, pp. 10 and 22. Addison Wesley (1976).
7. A. L. Loeb, *Color and Symmetry*, p. 31. Wiley, New York (1971), reprinted by Krieger.
8. Marjorie Rice and Doris Schattschneider, The incredible pentagonal versatile. *Math. Teaching* **93**, 52–53 (1980); Doris Schattschneider, Tiling the plane with congruent pentagons. Math. Magazine, **51**, 29–44 (1978).
9. A. L. Loeb, *Color and Symmetry*, p. 19. Wiley, New York, (1971).
10. Doris Schattschneider, Will it tile? Try the Conway Criterion! *Math. Magazine* **53**, 224–233 (1980).
11. A. L. Loeb, *Color and Symmetry*, pp. 27 and 28.
12. D. Dunham, J. Lindgren and D. Witte, Creating repeating hyperbolic patterns, *Computer Graphics* **15**, 215–223 (1981).
13. A. L. Loeb, A systematic survey of cubic crystal structures. *J. Solid State Chem.* **1**, 245.
14. A. L. Loeb, *Vector equilibrium in crystals*, in *Synergetics* by R. Buchminster Fuller, p. 860–876. Macmillan, New York (1975).
15. A. L. Loeb, *Space Structures, their Harmony and Counterpoint*, p. 147. Addison-Wesley, Reading, MA (1976).

Comp. & Maths. with Appls. Vol. 12B, Nos. 1/2, pp. 77–82, 1986
Printed in Great Britain.

SYMMETRY AS AN AESTHETIC FACTOR

HAROLD OSBORNE[†]
Kreutzstrasse 12, 8640 Rappersvill SG, Switzerland

Abstract—In classical antiquity symmetry meant commensurability and was believed to constitute a canon of beauty in nature as in art. This intellectualist conception of beauty persisted through the Middle Ages with the addition doctrine that the phenomenal world manifests an imperfect replica of the ideal symmetry of divine Creation. The concept of the Golden Section came to the fore at the Renaissance and has continued as a minority interest both for organic nature and for fine art. The modern idea of symmetry is based more loosely upon the balance of shapes or magnitudes and corresponds to a change from an intellectual to a perceptual attitude towards aesthetic experience. None of these theories of symmetry has turned out to be a principle by following which aesthetically satisfying works of art can be mechanically constructed. In contemporary theory the vaguer notion of organic unity has usurped the prominence formerly enjoyed by that of balanced symmetry.

From classical antiquity the idea of symmetry in close conjunction with that of proportion dominated the studio practice of artists and the thinking of theorists. Symmetry was asserted to be the key to perfection in nature as in art. But the traditional concept was radically different from what we understand by symmetry today—so different that "symmetry" can no longer be regarded as a correct translation of the Greek word *symmetria* from which it derives—and some acquaintance with the historical background of these ideas is essential in order to escape from the imbroglio of confusion which has resulted from the widespread conflation of the two.

To the Greeks symmetry meant commensurability, and two magnitudes were said to be commensurable if there exists a third magnitude which divides into both without remainder. As applied to works of art symmetry meant the visible commensurability of all the parts of a work to one another and to the whole. A statue, for example, or a building was symmetrical if it has some easily discernible part such that all other parts were exact multiples of it so that it served as a visible and apprehensible module. This idea of beauty or perfection was assumed to apply in nature as well as in art. It was thought that every species or natural kind has an ideal set of proportions, an ideal symmetry, from which all individuals deviate in a greater or less degree but by their approximation to which their beauty can be assessed. The ideal of the artist was to reproduce, not the imperfect natural objects which were his models, but the ideal symmetry characteristic of their species and his success was judged by the extent to which he was thought to have embodied this ideal symmetry in his work more perfectly than it was manifested in his models. The many lost "canons" of antiquity, the most renowned of which was the *Canon* of the sculptor Polyclitus (active circa 450–420 B.C.) were attempts to discover practical rules for the realisation of these ideal proportions. Similar principles were current for architecture and a rather confused account of them was given by the Roman architect Vitruvius in his treatise *De Architectura* (written shortly before 27 B.C.), which, rediscovered by Pollio and printed in 1486, remained for centuries the aesthetic bible of Europe for architecture. Vitruvius wrote:

> The design of a temple depends on symmetry, the principles of which must be carefully observed by the architect. They are due to proportion. Proportion is a correspondence among the measures of the members of an entire work, and of the whole to a certain part selected as a standard. From this result the principles of symmetry. Without symmetry and proportion there can be no principles in the design of any temple; that is, there is no precise relation between its members, as in the case of those of a well-shaped man. (111,i)

This preference for intellectually apprehensible rather than sensuous beauty corresponded to the basic cast of the Greek temperament, their demand for order and comprehensibility everywhere and their abhorrence of all that is vague and formless. It was an outlook on the

[†]President of The British Society of Aesthetics.

world which found characteristic expression in the numerological mysticism of Pythagoras and explains why the philosopher Plato could declare that the five regular solids are the most beautiful of things. It lies at the root of what has come to be known as the "idealism" of classical art. As Professor Rhys Carpenter has said in his study *Greek Art* (1962):

> The sculptor who adheres to a simple integral canon, however narrowly its prescriptions may conform to anything observable in his race and period, will never produce an individually characterized face, whether true portrait or plausible possible likeness, but only a geometrically ideated abstraction. The so-called "idealism" of classical art is not simply due to a generalisation and regularisation of the contours, but more than all to this deliberate substitution of integral ratio, intact and entire, for the chance approaches to it discoverable in nature's own structural symmetry. The difference in meaning between our word "symmetry" and the Greek "*symmetria*," from which it is descended, defines very exactly the difference between *ideal* beauty as we conceive it and *ideated* beauty as Greek sculpture created it by applying a modular canon. (p. 160)

This intellectual conception of beauty persisted through the Middle Ages. Thus St. Augustine could say:

> I am delighted by the highest equality, which I apprehend not with the eyes of my body, but with those of my mind. I, therefore, believe that the more what I see with my eyes draws near to what I apprehend with my spirit, the better it is. [*De vera rel.*, XXXI, 57]

And towards the close of the period St. Thomas Aquinas wrote:

> Beauty . . . is the object of cognitive power, for we call beautiful things which give pleasure when they are seen; thus beauty rests on proper proportion, because the senses delight in things with proper proportion as being similar to themselves; for the sense and all cognitive power is a kind of reason. . . . [*Summa theologiae*, 1, q. 5]

The two fundamental assumptions of classical aesthetics continued virtually unquestioned at the time of the Renaissance: first, the belief that the same principles of perfection apply in nature and in art and, second, contrary as it very often was to practice, that the representational realism asserted to be the goal of the artist should depict not the imperfect models available in nature but an ideal symmetry of the type, which can be discovered mathematically. It was this which led to the belief, which inspired the Academies and culminated in Le Brun (1619–90), that the principles of taste and expression can be formalised and imposed as disciplinary doctrine. To their basic aesthetic assumptions the artists and connoisseurs of the Renaissance also added the belief that the ideal symmetry which is imperfectly manifested in nature was concretely exemplified in the works of the ancient sculptors—although at that time these were known mainly in Roman copies or in fragments. It was this belief which was the central inspiration of the Neo-Classical movement, whose chief voice was Johann Joachim Winckelmann (1717–68).

Renaissance artists and scholars were intoxicated with mathematics and it was at this time that the doctrine of the Golden Section or, as it was then generally known, the *Divine Proportion* came to prominence. This sacrifices exact commensurability but imposes a single ratio throughout. Called by Kepler "one of the two great treasures of geometry" and "a precious jewel," it was held to be the key to aesthetic configuration in all the arts. Fra Luca Pacioli published his influential *Divina Proportione* in 1509 with plates by Leonardo da Vinci and incorporating an Italian translation of the *De Quinque Corporibus Regularibus* of Piero della Francesca. This Divine Proportion arises when a line is divided in extreme and mean ratio, i.e. so that the ratio of the whole line to the greater section is equal to the ratio of the larger to the smaller section, or more generally the proportion which arises between two relations which have one term in common, when one of the three terms is the sum of the other two terms. Interest in this proportion was revived by A. Zeising in the middle of the nineteenth century† and for long

---

†A. Zeising, *Neue Lehre von den Proportionen des menschlichen Körpers* (1854) and *Aesthetische Forschung* (1855).

was believed at least by a minority of interested persons to be the universal key to aesthetic symmetry in nature and in all the arts. Gustav Theodor Fechner, a pioneer of experimental psychology, published in *Vorschule der Aesthetik* (1876) the results of experiments which he claimed to have shown that rectangles formed in accordance with the Golden Ratio are aesthetically preferred. But later experiment has thrown considerable doubt upon these results. In *The Curves of Life* (1900) Theodore Cook claimed to have shown that the spiral curves typical of organic growth display a symmetry equivalent to that embodied in the Divine Proportion. A still more detailed study by d'Arcy Wentworth Thompson, *Growth and Form* (1917), showed that the spirals in some organic structures correspond with the famous Fibonacci series. These books have been influential but not convincing. There are interesting but more cautious papers by C. H. Waddington, F. G. Gregory and Joseph Needham in *Aspects of Form* (1951), edited by Lancelot Law Whyte. Here Waddington says:

> We come then to conceive of organic form as something which is produced by the interaction of numerous forces which are balanced against one another in a near-equilibrium that has the character not of a precisely definable pattern but rather of a slightly fluid one, a rhythm.

Comparing this with the products of fine art, he adds:

> There is, in a human work of sculpture, no actual multitude of internal growth-forces which are balanced so as to issue in a near-equilibrium of a rhythmic character. We should therefore not expect that works of art will often arrive at the same type of form as we commonly find in the structures of living matter. Much more can we anticipate an influence of man's intellectualising, pattern-making habit of simplification, diluted perhaps by an intrusion of unresolved detail.

There has in addition to all this been a very considerable volume of work done in the attempt to prove the importance and prevalence of Golden Section symmetry by the direct analysis of acknowledged masterpieces and works of outstanding repute. The many studies are too numerous to enumerate individually but mention should be made of *Ad Quadratum* (1921), a detailed study of classical and medieval church architecture by F. M. Lund, who purports to show "the aesthetic superiority over all other proportions" of the Golden Section. A typical example of the application of this method to famous paintings is *Geometry in Pictorial Composition* (1969) by Brian Thomas. The theory of the classical idea of symmetry has been admirably expounded in several books by Matila C. Ghyka, including *Esthétique des Proportions dans la Nature and dans les Arts* (1927) and *Essai sur le Rythme* (1938).

If it were possible to discover complex sets of "ideal" proportions which could serve as a standard by their approximation to which the beauty of natural objects might be assessed, it should not be difficult to produce statues conforming to the "classical" or "ideal" tradition of art even more successfully by means of computers than by taking life-casts from models chosen for their near-approximation to the ideal symmetry. (Painting introduces other complications arising from the demands of composition within an area. One would have to assume the universality of the Golden Section or some other norm of symmetry, apply it to the whole area and drop the notion of approximating individual figures to different sets of proportions.) But despite the enormous mass of data that has been accumulated, no positive conclusion has been reached about the key position of Golden Section symmetry. The effect of optical illusion was noticed as early as the sculptor Lysippos of Sikyon (active in the latter part of the 4th century B.C.), of whom Pliny records from an earlier source: "There is no Latin word for the term *symmetria*, a quality which he preserved with the utmost precision by a new and previously unattempted system which involved altering the "square" figures of earlier sculptors; and he often used to say that by them men were represented as they really were, but by him they were represented as they appeared." [*Nat. Hist.* xxxiv, 65] (Perhaps what he meant was that instead of making his figures to embody the ideal norm of symmetry, he made them so that they *appeared* to do so. A work which actually and measurably conforms to a given pattern of symmetry will not usually *look* as if it does so.)

An attempt was made to allow for optical illusion by some of those who repeated Fechner's experiments on aesthetic preference with simple rectangles of different proportions. But in pictures and other more complex objects its effects are complicated and far-reaching. They can be studied in the standard works on optical illusion or in such books as M. H. Pirenne's *Optics, Painting and Photography* (1970) and they have been exploited by exponents of Op Art for their own purposes. Because they are so many and complex exact measurement in relation to the proportion indicated by the Golden Section has not proved a feasible project. One can say that other things being equal an upright line looks shorter than a horizontal line of the same length, that figures of equal size in receding perspective seem to be of different sizes. But the visual action of one shape on another in three-dimensional painting, the impact of psychological prominence when shapes have representational meanings, the influences of colour contrasts, and so on, differ with every example and cannot be exactly calculated. Yet the application of geometrical analysis to paintings without taking these factors into account must lead to unreal results. Moreover, in the language of Phenomenology, what one should be studying is the visual object, that which we see, not the physical substrate of what we see. (This may have been what Lysippos was getting at). For these reasons among others it has not proved possible to verify the ubiquity of Golden Section symmetry in art that has been acknowledged to be great. And certainly no modern artist has ever produced great art by the deliberate application of the Golden Section. Whether the ancient artists did so or whether they followed some more rudimentary rules of thumb is an open question.

As has been said, the modern conception of symmetry is very different from that inherited from classical times and given new prominence at the Renaissance. The change is consistent with radical though not always deliberate changes in fundamental aesthetic assumptions. Our idea of beauty in art has become emotional and expressive; intellectual perspicuity counts for less than it did in the past. Modern thinking is less than convinced that the same aesthetic principles apply in nature and in art. We repudiate the basic assumptions of Renaissance aesthetics, the belief that there is an "ideal" symmetry of natural objects, whether based on the Golden Section, on "dynamic symmetry"† or on any other mathematical formula, and we do not accept the belief that artistic excellence can be achieved by realistic representation of a naturally beautiful object. It is no longer necessary to argue, as that too little remembered critic R. H. Wilenski so ably argued in his book *The Meaning of Modern Sculpture* in 1932, that a life-cast of a fine figure of a man or a beautiful woman does not automatically produce a work of art. And we accept without a qualm what artists have long enough known, that the representation of what is ugly, mean or repulsive in nature can have its proper place in pictorial art. We happily accept that representation itself is neither a necessary nor a sufficient condition of fine art. If Golden Section symmetry retains an interest for us, it must be against the background of these changed aesthetic attitudes. In fact when people speak of symmetry today we naturally think of some sort of geometrical regularity and repetition of forms rather than of any specific mathematical ratio. In his *Principles of Symmetry* (1917) Professor F. M. Jaeger defines it as follows: "Symmetrical figures are such as are similar to themselves in more than one way." This is rather more general than what is commonly understood by symmetry—that constructions are symmetrical if they repeat either their main forms or a mirror image of their main forms on either side of a medial axis.

In the first place it is at once obvious that symmetry involving the repetition of elementary or easily distinguishable elements is characteristic of the sort of patterns which properly belong to wall-papers, textiles, carpets, etc., and perhaps finds its finest expression in the decorative aspects of Islamic architecture. Even so, it is noticeable that in fine decorative art produced by craftsmen rather than by mass production methods in a factory the pattern elements, though remaining recognizably the same through repetition, are deliberately subjected to variations in colour or design or orientation and conscious irregularity of a minor character is often introduced. This is the case in old oriental carpets, pre-Inca textiles of Peru, Byzantine mosaics, and so on. Symmetry is also sometimes attributed to art objects as a quality of the composition as a

---

†The concept of *dynamic symmetry* was introduced by Jay Hambidge in his book of that name in 1920. It depends on the idea of commensurability among rectilinear areas instead of commensurability among linear magnitudes. Hambidge reduced all design to rectilinear design and called those designs symmetrical which are constructed from commensurable areas. He spoke of "static symmetry" when designs constructed in commensurate areas display linear commensurability among the lines by which the areas are bounded and "dynamic symmetry" of designs composed in commensurate areas whose bounding lines are not commensurable. The evidence that "dynamic symmetry" so conceived was known and used by the ancient Greek artists is inconclusive.

whole when there is no exact duplication of forms but a certain balance, or equality of "weighting," about an imaginary axis. In this less precise sense Baroque or Rococo art may sometimes be called symmetrical despite its deliberate asymmetry. In this vaguer sense symmetry is often described by the still more vague term "harmony" and its relevance for aesthetic quality has long been recognised. In the light of his general world-view it may well be that Plotinus had something very like this in mind when he wrote:

> Now by almost all persons it is maintained that it is the symmetry of the different parts with respect to each other, and the beautiful colour, which produce beauty for visual observation— beauty is identical with symmetry and being shaped after fixed proportions.

If he did mean something like this and was not merely repeating the classical view of symmetry current in his day, he may have anticipated what Dürer had in mind when he said: "The accord of one thing with another is beautiful, therefore want of harmony is not beautiful. A real harmony linketh together things unlike." (I used these two quotations in an earlier work, *Theory of Beauty*, 1952.)

It remains now to consider the relevance of this modern concept of symmetry for the furtherance of aesthetic experience. For this purpose it is necessary to consider briefly that thorny topic about which more nonsense has been written than most—the nature of aesthetic observation and aesthetic satisfaction.

Aesthetic perception—I must be dogmatic for the sake of brevity—is a skill which needs to be cultivated and developed. It is the more difficult to acquire because it runs counter to our ordinary habits of perception. In everyday life we look out on the world with the practical purpose of discriminating things and the relations in which things stand to one another and we automatically cease to attend to our visual impressions when they have served this practical purpose. Aesthetic perception is the contrary of this: we look at selected objects for the purpose only of seeing and in the endeavour to perceive the whole of the presentational content without practical inhibitions. In the words of Professor Monroe C. Beardsley, the primary marks of aesthetic character are "the presence in the object of some notable degree of unity and/or the presence of some notable intensity of regional quality."† It is upon these that we direct attention, upon what has been called the *organic unity* of the object (it offers itself to perception as a unified whole not only as a bundle of analytical parts) and upon the expressive character of its contained qualities. Beardsley rightly indicates that this is the source of what he calls aesthetic "gratification" or aesthetic "satisfaction." I have gone a step further than this in my own writing on aesthetics and have advanced the view that the much vaunted pleasure which many people experience in aesthetic perception derives from the successful exercise of enhanced perceptual activity upon an object adequate to arouse and sustain it at more than ordinary intensity. It is therefore analogous to the pleasure we derive from the enhanced exercise of any other human faculty for its own sake—to the pleasure some people take, for example, in the enhanced exercise of the reasoning power demanded by higher mathematics, logic, ontology or aesthetics. It is to things which are suitable to arouse and competent to sustain enhanced percipience for its own sake that we attribute aesthetic quality or call works of art.

Let us now consider the aesthetic relevance of these various modes of symmetry. First, it is clear that the symmetry of repeating pattern provides a very elementary aesthetic stimulus. It may serve to arouse attention, particularly if the repeating elements are unfamiliar or if they carry personal associations. But it cannot hold or enhance perceptual attention. At most such patterns provide a congenial perceptual background, unobtrusive and undemanding—which, indeed, is their purpose. The deliberate variations introduced by craftsmen augment the visual interest, but do not alter the basic character of the genre. Much the same thing can be said about the symmetry of individual elements and of individual objects such as crystals or snowflakes. They can arouse interest, curiosity and admiration. But visual interest in them is shortlived and superficial: In contrast to the impact of an artistic masterpiece, perceptual attention soon wanders, never goes deep. There is no enhancement of perception.

---

† See the essay "The Aesthetic Point of View" reprinted in the volume of selected essays under that name, edited by Michael J. Wreen and Donald M. Callen, 1982.

The more complex kind of symmetry which consists not so much of multiple repeated elements as of balance and weighting about a medial axis has closer affinities with the aesthetic. It is the sort of construction which often inspires non-iconic abstract art. But to be effective it must not be too obvious but remain subordinate to other sources of perceptual interest. Certainly such symmetry of balance can occur in objects—advertisements, documentary photographs, maps, etc.—which no one would dream of using for aesthetic ends. Furthermore, it is subject to optical conditions. For example, horizontal symmetry of this sort is much more obtrusive than vertical, as can easily be proved in the following way. Take a photograph of scenery which is fairly evenly weighted about a medial axis, giving horizontal symmetry. Turn it on end so that the weighting becomes vertical above and below a horizontal axis, and the sense of symmetry virtually disappears. It seems that for aesthetic efficiency there must be some repetitive regularity which acts as a perceptual background without obtruding too openly into awareness. If it attracts attention too openly on itself, our awareness becomes intellectual rather than perceptual, and the modern concept of aesthetic experience is perceptual rather than intellectual. Finally, the more recondite forms of balance and interplay of parts often described as harmony come close to what is known as organic unity, although "organic unity" is the wider term and includes qualities which could not sensibly be brought within the category of symmetry. Even so, symmetry in this extended sense cannot be subjected to formula or rule but is directly detected by those who have developed the skill of artistic appreciation.

It is as certain as anything in this field can well be that in none of these senses can symmetry be regarded as a necessary condition of successful art. Much American abstract expressionism and much European gestural art has not only neglected symmetry but has deliberately eschewed it. Some Chinese painting "balances" large expanses of empty space against small areas of motif in a way which contradicts Western ideas of symmetry. And these are but a few examples from many. Nor is symmetry a sufficient condition. Too much or too obvious symmetry defeats its own purpose. When it is unobtrusively subordinated to other perceptual stimuli symmetry may enhance the overall aesthetic potentiality of a work; otherwise the aesthetic appeal is annulled.

The foregoing article distinguishes two separate ideas which are commingled in the current concepts of symmetry. They are *commensurability*, which from the time of the Renaissance tended to be interpreted in the form of a single pervading ratio such as the Golden Section, and *balanced repetition*. The former aims at great precision, the latter is more vaguely conceived. The aesthetic implications of these ideas are briefly touched upon.

Comp. & Maths. with Appls. Vol. 12B, Nos. 1/2, pp. 83–91, 1986
Printed in Great Britain.

0886-9561/86 $3.00 + .00
© 1986 Pergamon Press Ltd.

SYMMETRY INDUCED BY ECONOMY

L. Fejes Tóth
Mathematical Institute, Hungarian Academy of Sciences, Budapest, P.O.B. 127, H-1364, Hungary

Abstract—Various extremum problems are presented which lead to highly symmetric geometrical configurations.

1. INTRODUCTION

The word "symmetry" evokes the amazing structure of the bees' honeycomb, the wall ornaments of ancient Egyptians and Moors, the Platonic solids, and other regular figures produced by man or nature which all incarnate perfection of symmetry. Due to their intrinsic beauty, and to their close connection with natural science and mathematics, regular figures attracted attention through the ages.

An intelligent schoolboy, trying to construct cardboard models of solids with equal regular faces would hardly fail to rediscover the regular solids. He also would come across the trigonal and pentagonal dipyramids and the few further convex solids with regular triangular faces. But he certainly would realize the higher degree of symmetry of the five solids with equivalent vertices. This may have been the way the Greeks discovered the regular solids.

Generally, a configuration is said to be regular if it consists of equivalent components. The trigonal dipyramid consists of equivalent faces but is not regular according to the traditional definition of regular solids. In the classical theory of regular figures we start with a definition of regularity, and try to give a complete enumeration of the respective figures. This theory, which may be called the *systematology* of regular figures, is contrasted by the *genetics* of regular figures which is based on the perception that certain economy postulates, in a sufficiently wide sense, imply regularity. Here regular figures are not defined but they come into being from irregular figures, and unordered chaotic sets in virtue of the ordering effect of an extremum requirement. In what follows we try to illustrate this theory by some examples.

We shall use the Schläfli symbol $\{p, q\}$ both for a regular polyhedron and a regular tiling with p-gonal faces and q-valent vertices. Polyhedra and tilings with regular faces and equivalent vertices are called uniform. They are denoted by a symbol $(l, m, \ldots)$ giving the number of sides of the faces around one vertex in their cyclic order.

2. POLYHEDRA AND SPHERICAL CONFIGURATIONS

The ancient legend about surrounding the site of Carthage by straps cut out of the skin of a steer suggests the problem of maximizing the area of an n-gon of given perimeter. The solution to this problem is known to be the regular n-gon. We phrase two analogous problems in space: Among the convex polyhedra of given surface-area having (i) a given number f of faces, (ii) a given number v of vertices find that one of maximal volume. It is known[9,11,15,21] that for $f = 4$, 6, and 12 the solutions are given by the respective regular solids ($\{3, 3\}$, $\{4, 3\}$, $\{5, 3\}$), and the same is conjectured to be true for $v = 6$, and 12 ($\{3, 4\}$, $\{3, 5\}$).

J. Steiner (1796–1863) proved a weaker statement for the octahedron: Among all polyhedra of the topologic type of the regular octahedron the regular one is the best. All attempts to prove Steiner's conjecture about the same extremum property of the regular icosahedron failed so far. This shows the difficulties often involved in similar problems.

However several inequalities are known which express extremum properties of all five Platonic solids[11, 15, 20]. We recall the simplest one: Let R be the circumradius and r the inradius of a convex polyhedron with f faces, v vertices, and $e(= f + v - 2)$ edges. Let $p = 2e/f$ be the average number of sides of the faces, and $q = 2e/v$ the average number of edges meeting at the vertices. Then

$$\frac{R}{r} \geq \tan \frac{\pi}{p} \tan \frac{\pi}{q}$$

and equality holds only for the five regular solids. This implies that among the convex polyhedra with either 4, 6 or 12 faces or 4, 6, or 12 vertices the respective regular solid approximates best the shape of the sphere in the sense of minimizing R/r.

There are some extremum requirements posed on all convex polyhedra irrespective of the number of faces and vertices which is fulfilled by one or another Platonic solid. As an example we mention the following theorem: Among all convex polyhedra containing a ball, the circumscribed cube has the least total edge-length[2].

On the pollen-grains of flowers there are small orifices. When the pollen-grain sticks to the stigma from an orifice near the point of adhesion a tube outgrows to the female nucleus enabling the process of fertilization. Some flowers have spherical pollen-grains on which the orifices are rather uniformly distributed. Trying to explain the peculiar arrangements of the orifices, the Dutch biologist Tammes made the hypothesis that on each grain nature tries to produce the maximal number n of orifices under the condition that no two orifices are allowed to get nearer to one another than a certain distance depending only on the species. Now the question arises about the smallest sphere which accommodates n orifices under the above condition. This problem is equivalent with the following: On the unit sphere distribute n points so as to maximize the least distance between pairs of them. This problem is often referred to as the problem of Tammes.

The problem was investigated by several authors[5,8,26,29], and completely solved for $n \leq 12$, and $n = 24$. We emphasize the cases of $n = 3, 4, 5$, and 12 points when the best arrangements are given by the vertices of $\{3, 2\}$, $\{3, 3\}$, $\{3, 4\}$ and $\{3, 5\}$, and the cases of $n = 8$, and 24 points which lead as solutions to the vertices of the uniform polyhedra $(3, 3, 3, 4)$ and $(3, 3, 3, 3, 4)$ (see Fig. 1).

Let p_n be the convex hull of the extremal set of n points. The sequence $P_4, P_5, \ldots$ can be considered as a natural extension of the set of the trigonal Platonic solids. Polarity with respect to the sphere yields a similar extension of the set of the trihedral Platonic solids. However, there is a great variety of other extremum problems which provide essentially different ''generalisations'' of the notion of regular solids.

The problem of Tammes can be reformulated as follows: On the sphere find the densest packing of n equal circles (spherical caps). The *density* of a set of domains lying on the sphere is defined by the total area of the domains divided by the surface-area of the sphere. If no two domains overlap then they are said to form a *packing*. Thus the problem is to find the biggest circle whose n congruent replicas can be placed on the sphere without overlapping each other.

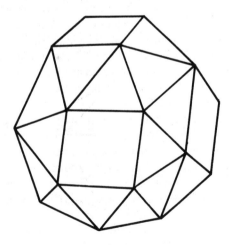

Fig. 1. The uniform solid $(3, 3, 3, 3, 4)$.

The following formulation of the problem suggests its connection with stereochemistry: What is the maximal number of non-overlapping equal balls of prescribed radius which can be brought in contact with a unit ball?

Analogous problems arise in higher dimensions in information theory. For the efficient transmission of informations through a noisy channel we have to design a code. Suppose that we need n code words each consisting of a sequence of d pulses of discrete voltage levels. The code words can be represented in the d-dimensional space as points whose coordinates are the d voltage levels. The energy needed to transmit a pulse is proportional to the square of the voltage level. So the total power required to transmit one code word is proportional to the square of the distance between the origin and the point representing the code word. It is convenient to choose code words whose transmission requires the same energy. This means that we have to choose the n points on the boundary of a d-dimensional ball. Another requirement is to choose the code words so that they could be well distinguished from one another. Under the supposition of a background noise of constant intensity this requirement turns out to be equivalent with distributing the points so that no two should get closer to each other than a prescribed distance. Adding the last condition of minimizing the total energy which is needed to transmit an information, we have the problem of finding the smallest d-dimensional ball whose boundary can hold n points under the above condition.

It is interesting to note that the problem of Tammes, as well as its d-dimensional analogue were brought up independently of their applications by pure geometric considerations, and were solved in some highly interesting cases by geometrical methods. Among others it turned out[3] that the extremal distribution of 120 points on a four-dimensional ball is given by the vertices of the regular 600-cell, one of the four-dimensional analogues of the Platonic solids discovered by L. Schläfli in the middle of the last century. But recently even more efficient methods were developed based on sophisticated considerations in analysis. We will return to some results obtained by this method later.

We still recall a jocular interpretation of the problem of Tammes[23]: Over a planet n inimical dictators bear rule. How should the residences of these gentlemen be distributed so as to get as far from one another as possible? Our next problem is the problem of the allied dictators, who want to set up their residences so as to control the planet as well as possible. More exactly, on a sphere mark n points so as to minimize the greatest distance between a point of the sphere and the mark nearest to it.

The solution is known for $n \leq 7$ and $n = 10, 12$ and 14. For $n = 3, 4, 6$ and 12 we have the vertices of $\{3, 2\}, \{3, 3\}, \{3, 4\}$ and $\{3, 6\}$, similarly as in the problem of the inimical dictators. The solution of the remaining cases can be summarized along with the cases when $n = 6$ and 12 as follows. For $n = 5, 6$ and 7 we have the vertices of a dipyramid, and for $n = 10, 12$ and 14 the vertices of an "antiprismatic dipyramid" (Fig. 2) with the respective number of vertices[6, 28].

It is not difficult to show that there are only finitely many numbers n such that the problems of n inimical and n allied dictators have identical solutions, and it is conjectured that the only such numbers are 2, 3, 4, 5, 6 and 12[7].

Some dome structures and geometrical sculptures consist of spherical circle packings[30]. Here the problem arises of minimizing the total material of the circles simultaneously requiring

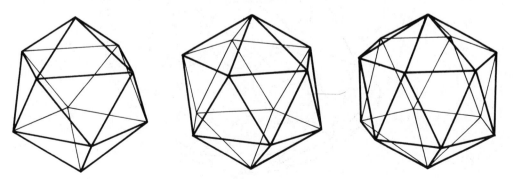

Fig. 2. Antiprismatic dipyramids with 10, 12, and 14 vertices.

a prescribed stability of the framework. The circles need not to be congruent but for practical reasons their size must be bounded from below. In order to formulate an exact problem we introduce a notion.

On the unit sphere let $P = \{c_1, \ldots, c_n\}$ be a *locally stable packing* of circles defined by the property that each circle is fixed by the others so that no circle of P can be moved alone without overlapping another circle of P. On the boundary of c_i let a be a greatest arc which is not touched by a circle c_j, $j \neq i$. Let $2\lambda_i$ be the angle subtended by a at the center of c_i. We call λ_i the *lability* of c_i, and define the lability λ of P by $\lambda = \max_{1 \leq i \leq n} \lambda_i$. Since the smallest circle is touched by at most five, and at least two circles, we have $\pi/5 \leq \lambda \leq \pi/2$.

The problem we are interested in is to find among all locally stable packings of circles with lability not exceeding a prescribed bound λ, and radii not less than a given value r that one of minimal density. For various particular values of λ and r this problem leads to many regular configurations. We consider the locally stable packing consisting of equal circles centered at the vertices of any of the following tilings: $\{4, 3\}$, $\{5, 3\}$, $(3, 4, 4)$, $(3, 6, 6)$, $(3, 8, 8)$, $(3, 10, 10)$, $(3, 4, 4, 4)$, $(4, 6, 6)$, $(5, 6, 6)$. Let λ_0 and r_0 be the lability and radius of the circles. Then for $\lambda = \lambda_0$ and $r = r_0$ the solution to the problem is the respective packing under consideration[18]. The packing generated by the "football tiling" $(5, 6, 6)$ is exhibited in Fig. 3.

Before discussing our last problem about spherical arrangements we make a digression. In an Euclidean or non-Euclidean space let B be a ball. I called the maximal number of congruent non-overlapping copies of B which can touch B the *Newton number* of B[16]. This name refers to the controversy between Newton and D. Gregory about the Newton number of an ordinary ball which, 180 years later, was proved to be twelve as claimed by Newton. In a non-Euclidean space the Newton number depends besides the dimension on the size of the ball.

In a packing of circles $c_1, \ldots, c_n$ of radius r let c_i be touched by k_i circles. Our problem is to find the maximum of the average number of points of contact:

$$M(r) = \max \frac{1}{n}(k_1 + \ldots + k_n)$$

extended over all packings of circles of radius r.

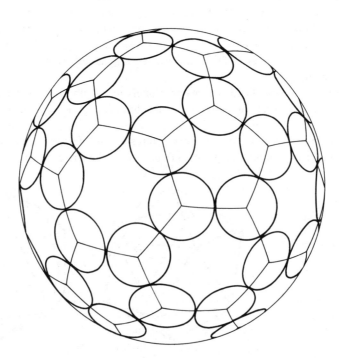

Fig. 3. A locally stable packing of circles centered at the vertices of the tiling $(5, 6, 6)$.

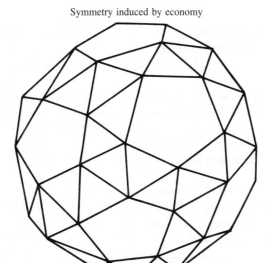

Fig. 4. The uniform solid (3, 3, 3, 3, 5).

Among the varied extremal packings special attention is due to those in which each circle is touched by as many circles as its Newton number. We call such a packing maximal neighbour packing, in short *maximal packing*.

A maximal packing of n equal circles exists only if $n = 2, 3, 4, 6, 8, 9, 12, 24, 48, 60,$ or 120[16,27]. For $4 \leq n \leq 24$ the circles constitute a densest packing, and the same is conjectured to be true for $n = 48$ and 120. For $n = 60$ the circles are centered at the vertices of (3, 3, 3, 3, 5) (Fig. 4). The maximal packings of 48 and 120 circles have the same symmetry groups as the maximal packings of 24 and 60 circles, notably the rotation groups of {3, 4} and {3, 5}, respectively. Accordingly, they exist in two enantiomorphous varieties.

We still emphasize a further particular case: we have $M(\pi/6) = 4$. The extremal packing is not unique. The centers of six circles are equally spaced on the equator, leaving room for three further circles on both hemispheres in two different ways: the packing is either symmetric with respect to the equator or with respect to the center of the sphere. The two configurations of unit balls which touch the central unit ball at the centers of these circles occur in nature in the crystal structure of some metals.

3. ARRANGEMENTS IN THE PLANE

We start with some fundamental concepts. Let $\mathbf{e}_1, \ldots, \mathbf{e}_d$ be vectors which span the d-dimensional Euclidean space. Applying the translations $k_1\mathbf{e}_1 + \ldots + k_d\mathbf{e}_d$ with all possible d-tuples of integers $k_1, \ldots, k_d$ to a body B, we obtain a lattice of translates of B. Lattice-translations are fundamental symmetry operations of all regular arrangements which extend through the whole space.

The *density* of an infinite set of bodies scattered through the whole space is defined by a limiting value. Instead of the exact definition we confine ourselves to the vivid interpretation of the density as the total volume of the bodies divided by the volume of the whole space.

After these general remarks we consider arrangements in the ordinary plane.

Let D be an arbitrary centro-symmetric convex disc. Among all possible regular or completely irregular packings of congruent copies of D we want to find a packing of maximal density. The answer is given by the following theorem[10,11,15]: The density of a packing of equal centro-symmetric convex discs never exceeds the density of the densest lattice-packing. Vividly expressed this means that at the command to fill the greatest possible part of the plane the disorderly lying discs will get into parallel position, and align in a lattice.

It must be noted that this interpretation is rough because the requirement of maximal density does not determine the packing uniquely. The regularity of the packing can be disturbed by "breaking lines" and other kinds of irregularities without changing the density in the whole plane. In addition there are special discs which allow a regular non-lattice-packing having the same density as the densest lattice-packing.

For general convex discs a similar theorem holds only for packings of translates of the disc[25].

In the 1930s German scientists studied the problem of drawing up the plan of economic human settlements laying the foundation of the so called location theory. Among others they raised the following problem. In a uniformly populated big country we want to plant a certain number n of factories which produce the same kinds of goods. Each point of the country is provided by the factory nearest to it. How should the factories be distributed so as to minimize the total haulage?

We try to formulate the problem exactly. Let D be a domain, P a point, and $f(x)$ a strictly increasing function defined for $x \geq 0$. We consider the moment of D with respect to P defined by $M(D, P) = \int_D f(PA) da$, where da is the area element at the point A. Let $P_1, \ldots , P_n$ be n points. Let D_i be the Dirichlet cell of P_i consisting of those points of D which are nearer to P_i than to any other point P_j. The problem is to distribute the points $P_1, \ldots , P_n$ so as to minimize the sum $\sum_{i=1}^{n} M(D_i, p_i)$.

It was conjectured that for great values of n we obtain the best distribution by putting the points in the vertices of a tiling $\{3, 6\}$. Prompted by purely geometrical considerations, the problem was raised and studied again confirming the correctness of the above conjecture[9,11].

J. Nigli gave a complete survey over the infinite connected regular circle-packings enumerating 31 types of such packings. Four of these packings are solutions to the problem of minimizing the density of an artibrary packing of circles under the condition that the lability of the packing (defined as on the sphere) should not exceed a prescribed value[15]. The respective packings consist of equal circles centered at the vertices of the tilings (3, 12, 12), (4, 8, 8), $\{6, 3\}$ and $\{4, 4\}$ (Fig. 5).

We call a circle-packing in which each circle is touched by at least k circles k-neighbour packing. Two consecutive rows in the densest lattice-packing of circles form a 4-neighbour packing with zero density. But any 5-neighbour packing of equal circles has positive density. What is the thinnest 5-neighbour packing of equal circles? The solution is another of the packings enumerated by Nigli[17] in which the circles are centered at the vertices of the tiling (3, 3, 3, 3, 6) (Fig. 6).

The regular shape of the honeycomb fascinated man ever and again. According to a widely spread (but questionable[14]) hypothesis, the bees aim at using the minimum amount of wax per cell. Since the bee-cells are deep as compared with the diameter of their openings, the problem of constructing not necessarily congruent ''bee-cells'' of given volume so as to minimize the total area of the cell-walls can be approximated by the problem of decomposing a plane region into a great but given number of convex polygons of equal area having minimal total perimeter. This problem leads to the tiling $\{6, 3\}$, in accordance with the shape of the honeycomb[13].

Similar problems are suggested by succulent vegetable tissues in which the cells are crammed tightly together in a part of space without filling it completely. Suppose that: (1) under the condition of equal constant surface-area the cells try to expand so as to maximize their total volume and (2) under the condition of equal constant volume the cell-walls try to contract so as to minimize their total surface-area. What shape and arrangement will the cells assume under these conditions?

In the stem of plants the cells are largely elongated in the axial direction. This propounds the two-dimensional analogues of the above problems in which the part of volume and surface-area are played by area and perimeter. In both problems we have an array of extremal packings depending on the prescribed perimeter p and area a of the cells which we consider as plastic convex discs packed into a domain. For a great number of cells the asymptotic behaviour of the two arrays are similar[12,15,19]. For small values of p and a the cells are equal small circles packed anyhow. Along with p and a the circles will increase, and at certain values of p and a will get into the densest packing forming the incircles of the faces of a tiling $\{6, 3\}$. Increasing forth p and a, the discs will turn into smooth hexagons which arise from the faces of $\{6, 3\}$ by rounding off their corners with equal circular arcs. Finally these arcs will shrink to points so that the cells will fill the whole available room.

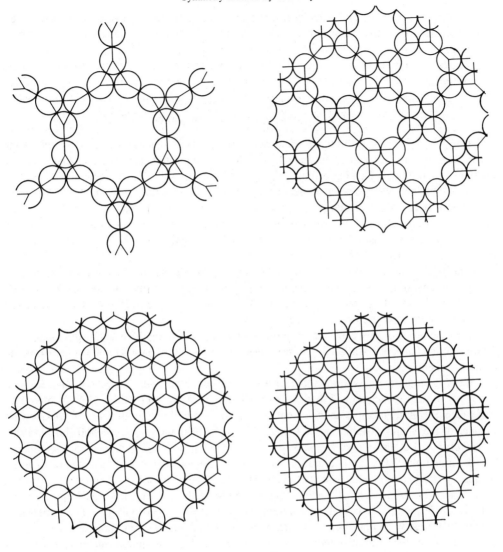

Fig. 5. Locally stable packings of circles centered at the vertices of (3, 12, 12), (4, 8, 8), {6, 3}, and {4, 4}.

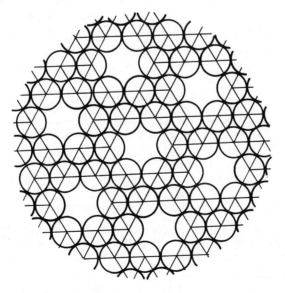

Fig. 6. Thinnest 5-neighbour packing of equal circles.

Observe the general conditions in the above problems: The congruences of the discs, their regular shape and regular arrangement are all induced by one simple and natural extremum requirement.

The configuration of smooth hexagons can be observed in microscopic sections of the stem of some plants.

4. ARRANGEMENTS IN THE SPACE

Little is known about the three-dimensional analogues of the problems considered in the plane. The difficulties inherent in these problems are demonstrated by the fact that even the problem of the densest packing of unit balls is a long-standing unsolved problem.

According to a well-founded conjecture among the solutions there are two regular packings of different types. Both are built up of hexagonal layers consisting of balls centered at the vertices of a tiling {3, 6} of edge-length two. The regular packings in question arise by putting the layers together so as to form a packing in which each ball is touched by twelve others in one of the two configurations described at the end of Sec. 2.

The layers can be put together also by letting the two configurations alternate from layer to layer in any order. Since the Newton number of a ball is twelve all these packings are maximal packings. It is conjectured that in three-dimensional Euclidean space all maximal packings consist of hexagonal layers[16].

Still the proof of this conjecture seems to be difficult. On the other hand, the method we referred to in connection with higher dimensional analogues of the problem of Tammes brought some surprising results.

Let N_d denote the Newton number of a d-dimensional ball. Obviously, we have $N_1 = 2$ and $N_2 = 6$, and we mentioned that $N_3 = 12$. These were the only values of N_d known until quite recently. A remarkable achievement of late years was the determination of the Newton number of the 8- and 24-dimensional ball[22,24]: $N_8 = 240$ and $N_{24} = 196560$. For no other values of $d > 3$ is the value of N_d known.

It turned out that the configurations of 240 and 196560 balls touching a central ball are unique[1]. Each one occurs in a particular lattice-packing which in 8-dimensional space is proved, and in 24-dimensional space is conjectured to be the densest lattice-packing. So these lattices are the unique maximal packings in the respective dimensions. In other dimensions higher than three we do not know whether a maximal packing exists at all or not.

In spite of the difficulties there is a hope of elaborating the genetics of three- and more-dimensional regular distributions. To conclude we present an encouraging result.

In our ordinary space let P be a packing of unit balls. Let r be the least upper bound of the radii of all balls disjoint to the balls of P. We call $1/r$ closeness of P. We want to find among all possible packings of unit balls the closest one, i.e. that one for which r assumes the least possible value. The solution to this problem is a lattice-packing. The centers of the balls form a so called body-centered cubic lattice which consists of the centers and the vertices of cubes constituting a face to face tiling of the space[4].

REFERENCES

1. E. Bannai and N. J. A. Sloane, Uniqueness of certain spherical codes. *Canad. J. Math.* **33**, 437–449 (1981).
2. A. S. Besicovitch and H. G. Eggleston, The total length of the edges of a polyhedron. *Quart. J. Math. Oxford Ser.* (2) **8**, 172–190 (1957).
3. K. Böröczky, Packing of spheres in spaces of constant curvature. *Acta Math. Acad. Sci. Hungar.* **32**, 243–261 (1978).
4. K. Böröczky, Closest packing and loosest covering of the space with balls. *Studia Sci. Math. Hungar.* **17**, (to appear).
5. L. Danzer, Endliche Punkmengen auf der 2-Sphäre mit möglichst grossem Minimalabstand. Habilitationsschrift. Göttingen (1963).
6. G. Fejes Tóth, Kreisüberdeckungen der Sphäre. *Studia Sci. Math. Hungar.* **4**, 225–247 (1969).
7. G. Fejes Tóth and L. Fejes Tóth, Dictators on a planet. *Studia Sci. Math. Hungar.* **15**, 313–316 (1980).
8. L. Fejes Tóth, Über die Abschätzung des kürzesten Abstandes zweier Punkte eines auf einer Kugelfläche liegenden Punkt-systems. *Jber. dtsch. Math.-Ver.* **53**, 66–68 (1943).
9. L. Fejes Tóth, The isepiphan problem for *n*-hedra. *Amer. J. Math.* **70**, 174–180 (1948).
10. L. Fejes Tóth, Some packing and covering theorems. *Acta Sci. Math. Szeged* **12A**, 62–67 (1950).

11. L. Fejes Tóth, *Lagerungen in der Ebene, auf der Kugel und im Raum*, Zweite Aufl. Springer Verlag, Berlin (1972).
12. L. Fejes Tóth, Filling of a domain by isoperimetric discs. *Publ. Math. Debrecen* **5**, 119–127 (1957).
13. L. Fejes Tóth, On shortest nets with meshes of equal area. *Acta Math, Acad. Sci. Hungar.* **11**, 363–370 (1960).
14. L. Fejes Tóth, What the bees know and what they do not know. *Bull. Amer. Math. Soc.* **70**, 468–481 (1964).
15. L. Fejes Tóth, *Regular Figures*. Pergamon Press, Oxford (1964).
16. L. Fejes Tóth, Remarks on a theorem of R. M. Robinson. *Studia Sci. Math. Hungar.* **4**, 441–445 (1969).
17. L. Fejes Tóth, Five-neighbour packing of convex discs. *Periodica Math. Hungar.* **4**, 221–229 (1973).
18. L. Fejes Tóth, Stable packing of circles on the sphere. *Structural Topology* **10** (1985).
19. L. Fejes Tóth and A. Heppes, Filling of a domain by equiareal discs. *Publ. Math. Debrecen* **7**, 198–203 (1960).
20. A. Florian, Ungleichungen über konvexe Polyeder. *Monatsh. Math.* **60**, 288–297 (1956).
21. M. Goldberg, The isoperimetric problem for polyhedra. *Tohoku Math. J.* **40**, 226–236 (1935).
22. V. I. Levenstein, Bounds for packing in n-dimensional Euclidean space (in Russian). *Dokl. Akad. Nauk SSSR* **245**, 1299–1303 (1979) [*Soviet Math. Dokl.* **20**, 417–421 (1979)].
23. H. Meschkowski, *Ungelöste und unlösbare Problems der Geometrie*. Viewing & Sohn, Braunschweig (1960).
24. A. M. Odlyzko and N. J. A. Sloane, New bounds on the number of unit spheres that can touch a unit sphere in n-dimensions. *J. Combin. Theory Ser. A* **26**, 210–214 (1979).
25. C. A. Rogers, The closest packing of convex two-dimensional domains. *Acta Math.* **86**, 309–321 (1951).
26. R. M. Robinson, Arrangements of 24 points on a sphere. *Math. Ann.* **144**, 17–48 (1961).
27. R. M. Robinson, Finite sets on a sphere with each nearest to five others. *Math. Ann.* **179**, 296–318 (1969).
28. K. Schütte, Überdeckung der Kugel mit höchstens acht Kreisen. *Math. Ann.* **129**, 181–186 (1955).
29. K. Schütte and B. L. van der Waerden, Auf welcher Kugel haben 5, 6, 7, 8 oder 9 Punkte mit Mindestabstand Eins Platz? *Math. Ann.* **123**, 96–124 (1951).
30. T. Tarnai, Spherical circle-packing in nature, practice and theory. *Structural Topology* **9**, 39–58 (1984).

Comp. & Maths. with Appls. Vol. 12B, Nos. 1/2, p. 93–96, 1986
Printed in Great Britain.

0886–9561/86 $3.00 + .00
© 1986 Pergamon Press Ltd.

MAGIC RECTANGLES IN THE TABLE OF THE 32 CLASSES OF SYMMETRY

I. I. SHAFRANOVSKII

Mining Institute, Leningrad, U.S.S.R.

Abstract—A reexamination of the classical table of the 32 crystallographic classes of symmetry is presented. Interesting features emerge as pair-wise products of the symmetry orders are formed. The Table of the 32 classes reveals additional information concerning simple and composite twins as well as epitaxy growth.

The 32 classes of finite crystallographic symmetry is generally recognized as the basis of classical geometrical crystallography. The symmetry classification of natural and artificial crystals is fully covered by them and the system is complete. The table of the 32 classes is associated with the names of M. L. Frankenheim (1826)[1], J. F. H. Hessel (1830), and A. V. Gadolin (1867).

The present paper shows that there is a wealth of information in the Table of the 32 classes hitherto unrecognized in spite of the long history of applications of this Table in crystallography and mineralogy. In addition to its value in classification, the Table contains a number of interesting mathematical regularities as will be indicated.

A variant of the Table of the 32 classes was communicated by Boldyrev and Dolivo–Dobrovolskii in 1934[2]. This variation is used for further consideration in the present work. A simplified version is given in Table 1 which is similar to those found in some text books on elementary crystallography. The classes are denoted (in italics) according to C. Hermann and C. Mauguin. The triclinic and monoclinic systems are combined in one according to Fedorov[3]. The main advantage of this table is in its rigor and simple visualization at the same time. The format also facilitates the understanding of the regularities which are the focus of this paper.

Notice now the numbers parenthesized in each cell of Table 1 under the Hermann–Mauguin notations. These numbers give the order of symmetry for each class. This order is determined by the number of homogeneous points connected by the symmetry elements of a given class. It corresponds also to the number of faces of a simple form belonging to this class.

Multiplying the orders of column I from top to bottom by the orders of column V from bottom to top, in sequence, we always get the same number, 48:

$$
\begin{array}{rcl}
\mathrm{I} & & \\
\hline
1 \times 48 &=& 48 \\
2 \times 24 &=& 48 \\
3 \times 16 &=& 48 \\
4 \times 12 &=& 48 \\
6 \times 8 &=& 48 \\
12 \times 4 &=& 48 \\
\hline
& \mathrm{V} &
\end{array}
$$

The same result is obtained when multiplying the orders of column II from top to bottom by the orders of column IV from bottom to top, in sequence. The products of the orders in column III yield the same number if going gradually from the top and bottom towards the center, to wit

$$
\begin{array}{l}
\mathrm{III} \ \text{line 1 \& line 6,} \ 2 \times 24 = 48 \\
\text{line 2 \& line 5,} \ 4 \times 12 = 48 \\
\text{line 3 \& line 4,} \ 6 \times 8 = 48.
\end{array}
$$

Table 1. The 32 Symmetry Classes†

Types / Systems	I Primitive	II Central	III Planar	IV Axial	V Plane-axial	VI Inverse-primitive	VII Inverse-planar
1. Triclinic-monoclinic	1 (1)	$\bar{1}$ (2)	m (2)	2 (2)	$2/m$ (4)		
2. Ortho-rhombic	$\lvert 2 \rvert$ (2)	$\lvert 2/m \rvert$ (4)	mm (4)	222 (4)	mmm (8)		
3. Trigonal	3 (3)	$\bar{3}$ (6)	$3m$ (6)	32 (6)	$\bar{3}m$ (12)		
4. Tetragonal	4 (4)	$4/m$ (8)	$4mm$ (8)	422 (8)	$4/mmm$ (16)	$\bar{4}$ (4)	$\bar{4}2m$ (8)
5. Hexagonal	6 (6)	$6/m$ (12)	$6mm$ (12)	622 (12)	$6/mmm$ (24)	$\bar{6}$ (6)	$\bar{6}2m$ (12)
6. Cubic	23 (12)	$m3$ (24)	$\bar{4}3m$ (24)	432 (24)	$m3m$ (48)		

†The parenthesized numbers under the Herrmann–Mauguin notations stand for the order of symmetry.

Table 2. Compilation of pair-wise products

Labels	Symmetry Classes	Products	Examples
I/1 & V/6	1 & $m3m$	$m3m$	
I/2 & V/5	2 & $6/mmm$	$12/mmm$	Graphite
I/3 & V/4	3 & $4/mmm$	$m3m$ and $12/mmm$	Pseudo-boléite, rutile
I/4 & V/3	4 & $3m$	$m3m$ and $12/mmm$	Salmiak
I/5 & V/2	6 & mmm	$12/mmm$	
I/6 & V/1	23 & $2/m$	$m3m$	Harmotone, phillipsite
II/1 & IV/6	$\bar{1}$ & 432	$m3m$	Plagioclase
II/2 & IV/5	$2/m$ & 622	$12/mmm$	
II/3 & IV/4	$\bar{3}$ & 422	$m3m$ and $12/mmm$	
II/4 & IV/3	$4/m$ & 32	$m3m$ and $12/mmm$	Quartz
II/5 & IV/2	$6/m$ & 222	$12/mmm$	
II/6 & IV/1	$m3$ & 2	$m3m$	Pyrite
III/1 & III/6	m & $\bar{4}3m$	$m3m$	Eulytine, helvine
III/2 & III/5	mm & $6mm$	$12/mmm$	
III/3 & III/4	$3m$ & $4mm$	$m3m$ and $12/mmm$	
VI/4 & VII/5	$\bar{4}$ & $\bar{6}2m$	$12/mmm$	
VI/5 & VII/4	$\bar{6}$ & $\bar{4}2m$	$12/mmm$	

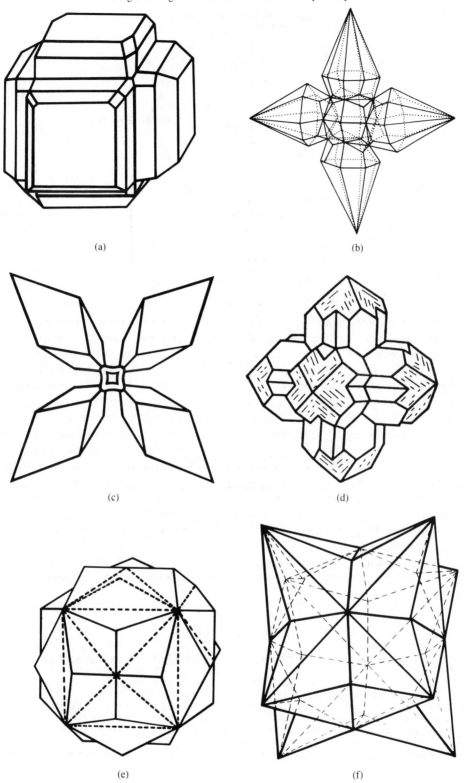

Fig. 1. Complex twins with generalized symmetry *m3m:* (a) Pseudo-boleite (*4/mmm*), (b) Salmiak (*''4/mmm''*), (c) Salmiak/aggregate (*''3m''*), (d) Phillipsite (*2/m*), (e) Pyrite (*m3*), (f) Eulytine ($\overline{4}3m$).

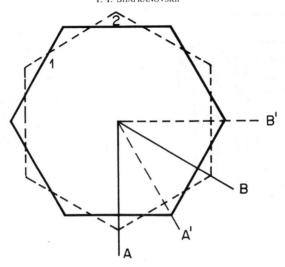

Fig. 2. Graphite twin with common symmetry *12/mmm*.

Incidentally, this can be done with the orders of either column II or column IV. Finally, multiplying the orders of columns VI and VII along the diagonal, again 48 is obtained.

Examine now the practical implications of the above relationships. All the products yielding 48 satisfy either *m3m* or *12/mmm*, or both as shown in Table 2. From here the conclusion suggests itself: the enumerated pairs of symmetry classes correspond to possible combinations of symmetry for crystal growth[4]; mainly complex twins, with generalized six-fold symmetries *m3m* (Fig. 1) and *12/mmm* (Fig. 2). The latter noncrystallographic class occurs, for example, in graphite twin with growth face (0001) and rotation angle of 30° around the rotation axis L_6[5]. A simple form with *m3m* symmetry is a 48-hedron—hexoctahedron. The symmetry *12/mmm* has not been found in separate crystals and may be considered as a generalized form for twins as a 48-hedron bis-dodecahedral bipyramid.

It is left to the reader to find further regularities in Table 1 which is in fact analogous to the "Magic Quadrangle". It is not our task here to interpret the mathematical nature of the above regularities. Our goal was simply to call attention to the crystallographic consequences emerging from them[6].

To summarize, the system of the 32 symmetry classes in the form as given in Table 1 provides ground for discussing regular crystal growth, both simple and composite, as well as epitaxy growth.

REFERENCES

1. J. J. Burckhardt, Die Entdeckung der 32 Klassen durch M. L. Frankenheim im Jahre 1826. *N. Jb. Miner. Mh.* **11**, 481–482 (1984).
2. A. K. Boldyrev, *Crystallography* (in Russian). ONTI, Leningrad–Moscow (1934).
3. E. S. Fedorov, *A course in crystallography* (in Russian). St. Petersburg (1901).
4. E. P. Makagonov, *The symmetry of growth* (in Russian). The Academy of Sciences of the U.S.S.R., The Urals Scientific Center, Sverdlovsk (1979).
5. G. I. Shafranovskii, Classical and nonclassical twins of graphite (in Russian). *Zapiski Vses. Mineralog. Obshchestva* **N4**, 577–581 (1983).
6. I. I. Shafranovskii, Essays on Mineralogical Crystallography (in Russian). *Mineralogicheskii Sbornik* **38**, N2 (1984).

Comp. & Maths. with Appls. Vol. 12B, Nos. 1/2, pp. 97–100, 1986
Printed in Great Britain.

0886-9561/86 $3.00 + .00

AN EXTENSION OF THE
NEUMANN–MINNIGERODE–CURIE PRINCIPLE[†]

J. Brandmüller
Sektion Physik der LM Universität München, D-8000 München 40, F.R.G.

Abstract—The Neumann–Minnigerode–Curie Principle (NMC Principle) which enables one to derive the selection rules for the physical properties from the symmetry of the object in question is given for the totally symmetric representation of the point group and thus for the static properties of the object. Its dynamical properties, however, are described by the non-totally symmetric irreducible representations of the point group to which the NMC Principle can be extended.

1. THE HISTORY OF THE NEUMANN–MINNIGERODE–CURIE PRINCIPLE

In about the middle of the last century some mineralogists and physicists began to put symmetry considerations concerning crystals and their physical properties in a more concrete form, i.e. they started to apply group theory. It would be an interesting task to examine this long and arduous road but I do not intend to do this here. F. E. Neumann (1798–1895), professor of physics and mineralogy at the University of Königsberg (today: Kaliningrad), had a clear idea that there was an essential relation between the geometrical structure of a crystal and its physical properties. This fact was mentioned by his student W. Voigt[1]. B. Minnigerode (1837–1896)[2], who was acquainted with Neumann, formulated the following "empirical principle" in 1884— just a hundred years ago—which appeared to be quite modern at that time: "The group of the structure of a crystal is contained in the group of each of its physical properties". On the 13th of November, 1884, P. Curie (1859–1906)[3] gave a fundamental lecture "Sur la Symétrie" at the French Mineralogical Society in Paris. Ten years later Curie stated his ideas about symmetry more precisely[4].

> The characteristic symmetry of a phenomenon is that symmetry which is best compatible with the existence of the phenomenon. A phenomenon can exist in surroundings which possess its characteristic symmetry or at least one subgroup of its characteristic symmetry. In other words, certain symmetry elements can exist together with certain phenomena but they are not necessary. But it is necessary that certain symmetry elements do not exist.

Then he added the sentence that has become classical in the meantime: "C'est la dissymétrie qui crée le phénomène". It is totally correct that Shubnikov and Koptsik[5] use the expression Neumann–Minnigerode–Curie Principle (NMC Principle) in this context. They write it in the following form:

$$G_{\text{object}} \subseteq G_{\text{property}} \tag{1}$$

and this is just what Minnigerode had already expressed in words (see above). In order that a physical property is allowed to exist within an object it is a necessary but not sufficient condition that the group of the symmetry operations of the object G_{object} be at least a subgroup of the group of the symmetry operations of the physical property G_{property}.

2. THE NMC-PRINCIPLE FOR THE TOTALLY SYMMETRIC REPRESENTATION

The above statement can be expressed equally as follows: The tensor describing a physical property (the so-called "property tensor") has to be invariant against all symmetry operations

[†]Dedicated to Dr. G. B. Semerano, Istituto Veneto di Scienze, Lettere ed Arti, Venezia, Italy.

of the object, e.g. of a crystal (crystal physics is often used as a model in this case). For polar *i*-tensors of rank v being *i*nvariant against time reversal one can write explicitly[6,7]:

$$d_{p,\ \underbrace{\rho\sigma\tau v...}_{v\ \text{indices}}} = \hat{O}_R d_{p,\ \underbrace{abcd...}_{v\ \text{indices}}} \tag{2a}$$

and for axial *i*-tensors of rank v:

$$d_{ax,\ \underbrace{\rho\sigma\tau v...}_{v\ \text{indices}}} = \det R \cdot \hat{O}_R d_{ax,\ \underbrace{abcd...}_{v\ \text{indices}}}. \tag{2b}$$

The indices ρ, σ, τ, v . . . and a, b, c, d . . . run over all three space coordinates x, y, z.

The linear operator $\hat{O}_R$, sometimes described as the Wigner Operator[8,9], is given by the relation

$$\hat{O}_R = \underbrace{R_{a\rho}R_{b\sigma}R_{c\tau}R_{dv}...}_{v\ \text{factors}}. \tag{3}$$

According to Einstein's convention one has to sum over indices which appear twice in Eqs. (2).

$R_{a\rho}$. . . etc. are the elements of the matrices which describe the individual symmetry operations R. It is possible to obtain enough relations for the calculation of the components of the property tensors $d_{\rho\sigma\tau v...}$ if one restricts oneself to the generators of a group. The definition of the property tensor goes back to Curie[4]. For the case of weak interaction the following linear tensor equation is valid describing the relation between the field tensor U of the cause and the field tensor E of the effect

$$E = dU. \tag{4}$$

For more than one cause and for strong interactions we have the relation

$$E = d^{(1)}U^{(1)} + d^{(2)}U^{(1)}U^{(2)} + d^{(3)}U^{(1)}U^{(2)}U^{(3)} + \ldots. \tag{5}$$

The formulations (2a, b) of the NMC Principle only refer to the totally symmetric representation of each point group. According to this procedure one obtains the property tensors for the static properties of the objects[6].

3. THE EXTENDED NMC-PRINCIPLE FOR THE NON-TOTALLY SYMMETRIC REPRESENTATIONS

The non-totally symmetric irreducible representations describe the dynamical effects of an object. Examples for these are infrared absorption and Raman effect. According to Bross[10] the NMC Principle in its extended version can also be applied to the non-totally symmetric irreducible representations which yields the following equations:

For *polar* i-*tensors* of rank v (invariant against time reversal)

$$\Delta_{ij}(R)\ ^j d_{p(v)} = \hat{O}_R\ ^i d_{p(v)}, \tag{6a}$$

for *axial* i-*tensors* of rank v

$$\Delta_{ij}(R)\ ^j d_{ax(v)} = \det R \cdot \hat{O}_R\ ^i d_{ax(v)}, \tag{6b}$$

$\Delta_{ij}(R)$ being the elements of the representation matrices for the symmetry operations R. i and j each run from 1 to the degree of degeneracy of the representation. To each many-dimensional representation belong just as many property tensors of a certain rank as the representation has dimensions. These different tensors of each individual representation are distinguished by the expressions ^1d, ^2d, . . etc. Again one needs only the generators of the group for the calculation.

4. APPLICATION TO THE CALCULATION OF THE TENSORS FOR THE DYNAMICAL PROPERTIES WITH RESPECT TO THE 32 CLASSICAL CRYSTALLOGRAPHIC POINT GROUPS

The relations (6a, b) can be used in order to calculate the property tensors in the non-totally symmetric irreducible representations of the 32 classical crystallographic point groups. Considering the irreducible representations of the third kind[11] which are one-dimensional and have complex conjugate characters one obtains tensors with complex components—which is not very sensible in physics. Following a suggestion of Birman[12] one can combine two one-dimensional irreducible representations having complex conjugate characters in one two-dimensional but now reducible representation with real matrices.

An alternative method to (6a, b) would be the application of the projection operator[13,14]. Both methods are used in order to calculate polar and axial tensors of rank 0 to 4 for all representations of the 32 classical crystallographic point groups[15].

5. APPLICATION TO THE CLASSICAL CURIE LIMITING GROUPS

In 1884, P. Curie introduced groups with infinity-fold rotation axes in order to be able to describe the symmetry of physical field quantities[3]. As an example the electric field has the symmetry ∞m in the international notation of Hermann–Mauguin. Generally this is the symmetry of a polar i-vector. One knows that there exist seven of these Curie limiting groups[5]. It is possible to calculate the property tensors for all of these Curie limiting groups, too, by means of (6a, b). In this case the extension of the NMC Principle by Bross (Eq. 6a, b) is more convenient than the projection operator method because of the appearance of $\cos\phi$ in the elements of the matrices describing the generators.

These Curie limiting groups are of quite great physical importance considering the fact that all linear molecules belong either to the point group ∞m or $\infty/m\ 2/m$. The symmetry of cellulose and poly-γ-benzyl-D-glutamate is given by the group $\infty 2$ and the symmetry of collagene, poly-caprolactane fibre, silk, tendons and bones is the same as the symmetry of the two-dimensional pure rotational group $O^+(2) \equiv \infty[16]$. Atoms and their different states are described by the rotational group $O(3) \equiv \infty/m\ \infty/m$ and its irreducible representations (taking into account the spin, however, makes it necessary to consider the group $SU(2)$). The elementary particle multiplet of the pseudo-scalar mesons $M(0^-)$, i.e. the mesons π, K, η and η' which all have spin zero and odd parity, belongs to the special three-dimensional rotational group $SO(3) \equiv \infty\ \infty$, the elementary particle multiplet of the vector mesons $M(1^-)$ with spin one and odd parity is described by the group ∞m with respect to its symmetry. The non-totally symmetric representations of these Curie limiting groups determine the dynamical properties of those objects.

6. DYNAMICAL PROPERTIES AND MAGNETIC GROUPS

The extension of the NMC Principle by Bross (Eq. 6a, b) is also applicable to the magnetic point groups. The unitary conjugate subgroup of a magnetic group is one of the 32 classical crystallographic groups. In order to complete the magnetic group one has to add a contribution containing all the anti-unitary symmetry operations[7]. For this contribution there is again a generating or—as it is sometimes said—"colouring" element for which one has to use the extended NMC Principle in addition to the calculations for the generators of the unitary conjugate subgroup. But it is now necessary to distinguish between i-tensors being invariant against time reversal and c-tensors not being invariant against time reversal when dealing with equilibrium (static) properties. With respect to the i-tensors the unchanged equations (2a, b) are still valid but in the totally symmetric representation of the group[6,7] we have

for *polar* c-*tensors* of rank v

$$d_{p,p\sigma\tau v\ldots} = -\hat{O}_R d_{p,abcd\ldots} \tag{7a}$$

and for *axial* c-*tensors* of rank v

$$d_{ax,p\sigma\tau v\ldots} = -\det R \cdot \hat{O}_R d_{ax,abcd\ldots}. \tag{7b}$$

Instead of Eqs. (7a, b), cf.[7, p. 132], the following equations have to be used for tensors describing transport properties (e.g. the electric conductivity) which contain an explicit time dependence and for tensors describing properties which can only be treated adequately by means of a microscopic description containing an implicit time dependence (e.g. Raman scattering by magnons). Instead of Eqs. (7a, b) we have

for *polar tensors* of rank v

$$D_{ij}(R) \, ^j d_{p(v)} = \hat{O}_R \, ^i d_{p(v)}^* \tag{8a}$$

and for *axial tensors* of rank v

$$D_{ij}(R) \, ^j d_{ax(v)} = \det R \cdot \hat{O}_R \, ^i d_{ax(v)}^*. \tag{8b}$$

The matrices of the co-representations $D_{ij}(R)$ can be calculated from the matrices $\Delta_{ij}(R)$ according to a procedure mentioned by Wigner[8].

This procedure can also be extended to the magnetic Curie limiting groups. One of these seven groups is e.g. $\infty/m \; 2'/m'$ which describes the symmetry of a magnetic field. On the other hand the symmetry of the magneto-electric tensor is $\infty/m' \; 2/m'$[6].

The calculations according to Eqs. (6) and (8) can be done by hand fairly conveniently up to tensors of rank two. However, starting with tensors of third rank the calculation of the property tensors is a typical task for a computer.

Acknowledgements—The author thanks his colleague H. Bross for many helpful discussions and Mr. Humer-Hager for the translation into English.

REFERENCES

1. W. Voigt, *Lehrbuch der Kristallphysik*. Leipzig, Berlin (1910).
2. B. Minnigerode, Untersuchungen über die Symmetrieverhältnisse und die Elasticität der Krystalle. *Nachr. Akad. Wiss. Göttingen, Math-phys./Klasse II a* **184**, 195–226 (1884).
3. P. Curie, Sur la symétrie. *Soc. minéralog. France Bull., Paris* **7**, 418–457 (1884).
4. P. Curie, Sur la symétrie dans les phénomènes physiques, symétrie d'un champ électrique et d'un champ magnétique. *J. de Phys., 3ᵉ série* **III**, 393–415 (1894).
5. A. V. Shubnikov and V. A. Koptsik, *Symmetry in Science and Art*. New York (1974).
6. R. R. Birss, *Symmetry and Magnetism*. Amsterdam (1966).
7. A. P. Cracknell, *Magnetism in Crystalline Materials*. Oxford (1975).
8. E. P. Wigner, *Group Theory and its Application to the Quantum Mechanics of Atomic Spectra*. New York (1959).
9. P. Rudra, On irreducible corepresentations of finite magnetic groups. *J. Mathem. Phys.* **15**, 2031–2035 (1974).
10. H. Bross, private communication.
11. C. J. Bradley and A. P. Cracknell, *The Mathematical Theory of Symmetry in Solids*. Oxford (1972).
12. J. F. Birman, Handbuch der Physik (S. Flügge) Band XXV/2b, *Licht und Materie I b*. Berlin (1974).
13. M. Hamermesh, *Group Theory and its Application to Physical Problems*. London (1962).
14. J. P. Elliot and P. G. Dawber, *Symmetry in Physics*, 2 volumes. London (1979).
15. J. Brandmüller and F. X. Winter, Die kartesischen irreduziblen Tensoren und ihre Bedeutung für die physikalischen Eigenschaften, Teil I: Die klassischen kristallographischen Punktgruppen. To be published in *Zeitschrift für Kristallographie*.
16. Landolt-Börnstein, Numerical data and functional relationships in science and technology, *News Series III*, **11** (1979).

Comp. & Maths. with Appls. Vol. 12B, Nos. 1/2, pp. 101–112, 1986
Printed in Great Britain.

0886-9561/86 $3.00 + .00
© 1986 Pergamon Press Ltd.

SYMMETRY AND ITS "LOVE–HATE" ROLE IN MUSIC

DANA WILSON
Ithaca College School of Music, Ithaca, NY, U.S.A.

Abstract—Symmetry plays a major role in Western music. The first portion of this discussion reveals how the symmetry operations of translation, reflection, and rotation manifest themselves in melodic material. The second portion discusses how symmetry operations are often used by a composer to set up expectation in the listener, only to prevent fulfillment of that expectation through denial of larger-scale symmetrical relations. The possible reasons for the presence of symmetry in music, and of the expectation-denial ("love–hate") relationship, are considered in light of their crucial roles.

Illustration above is the "Musical Heart" by Baude Cordier from the Chantilly Codex, c.1400.

To say that symmetry plays a major role in Western music may be to state the obvious. The very act of composition demands the exploitation of symmetry operations to the degree that most material found in a work can be traced to some basic pattern or idea stated near the work's beginning. Much musical terminology and most traditional formal structures are generated from the principles of symmetry.

Why this is so has been cause for speculation for centuries. A common approach has been music's relation to human bodily activities and emotional states. Symmetrical events such as breathing, the heartbeat, walking and love-making are all two-part operations involving a certain increase in activity followed by a decrease or temporary stasis, said to be captured musically in rhythmic and melodic gestures. Furthermore, the quickening and later slackening of the rate of breathing, heartbeat, and ambulation as a result of the onset and then decline of fear or excitement, are considered by many to influence directly larger-scale, tension-and-release musical patterns: dramatic pitch ascent followed by gentle descent, acceleration and then "calming" of rhythms or tempos, the increase-decrease of textural density or dynamic levels, and the like.†

Music has also long been associated with other arts, which may themselves be an expression of physiological patterning. The task of motivating two similar appendages to glide across the ballroom floor has required repetitive patterns. At least since the time of the ancient Greeks, music has been closely related to poetry and has often served to emphasize the metric and rhyming patterns of the verse. The degree to which this pulse—this sense of ebb and flow—

---

†Mozart, for example, used each of these devices to explicitly underscore such bodily activities in his operas.

provides the intellectual satisfaction of predictable order as much as physical intensification and relaxation probably varies from listener to listener, but it is clear that, in music's operating on many levels and serving different functions in many styles, its symmetrical relationships can be both crucial and complex.

The irony of symmetry in Western music is that, as much as patterning and a certain balance are desirable (the degree of each depending on the musical style as well as listener), total predictability leads to the listener's complacency, which can result in boredom. The task of the composer then is threefold: (1) to state an idea; (2) to corroborate it through symmetrical operations, thereby generating certain expectations in the listener; (3) then to deny the predicted larger-scale symmetry in the hopes of lifting the listener out of old "patterns" and generating a pang of uncertainty, tension, or the joy of surprise. Often this denial is coupled with an attempt to convince the listener of a new sense of balance, proportion, or release—providing, if you will, a new perspective.

In this discussion, we will first examine how symmetry operations serve to unify, balance, and direct a work's melodic material.† We will then use that insight to investigate how melodic symmetry, once established, generates listener expectation of symmetry on a larger scale, only to later deny it in order to elicit an emotional response. (The final illustration will demonstrate how, despite this "love–hate" of symmetry, a larger sense of balance is ultimately achieved.) Because of music's rather abstract nature, song will be the primary genre examined in the latter discussion, in the hope that the text, serving as a middle ground, will provide cues to specific melodic shapes and compositional decisions.

SYMMETRICAL OPERATIONS IN MUSIC

Before we enter the ambivalent realm of "love–hate," let us consider how basic symmetrical operations are at work in Western music in general. For the reader who does not read music notation, a graphic representation of melodic shape will accompany each example.

Translation

The most common and essential operation to be found in music is translation. The melodic patterns that are stated and then repeated, either exactly or on other pitch levels, provide the unity so necessary to this abstract medium. The opening measures of Mozart's Symphony No. 40 serve to illustrate simple, but effective, translation (Fig. 1). Note in Fig. 1 that this Classicist, so sensitive to balance and proportion, also employs symmetry in that broader sense. Not only does *a* (the entire first line of the example) translate down to *b* (the second line), but the short, repeated, downward gestures of *c* are balanced by the upward leap of *d*, and these two work

Fig. 1. Opening melody of Mozart's Symphony No. 40, in G minor.

---

†While melody may be the most crucial to the projection of symmetry, other elements, such as harmony and rhythmic figures, can also be important. However, to include them would demand technical discussion beyond the scope of this investigation.

together as a larger unit (that begins low and ascends) to be balanced by a descending pattern *e*.†

As gestures turn into phrases and phrases into sections, we express other musical devices and procedures that express translation:

> ostinato, passacaglia (simple repetition of a melodic gesture)
> canon, fugue (translation onto other pitch levels or at different time intervals)
> variation (translation with slight alteration)
> serialism (translation—as well as reflection—of a melodic/harmonic pattern)
> transposition (translation of a large musical unit to another pitch level).

Even the standard large-scale musical structures express translation. In all of the following common musical forms, each letter represents an entire section of a movement or work. Its reiteration suggests a substantial portion of that section. Taken together, the letters express the entire formal design:

> Ternary—A B A,
> Rondo—A B A C A or A B A C A B A,
> Sonata-Allegro—A (A); development of materials from A; A.

Finally, a massive work can connect movements through translation (and sometimes reflection) to help unify and balance the whole. Many nineteenth-century symphonies are referred to as cyclic because of this feature. Bach's *Musical Offering* is symmetrical on this largest level, as well as on several others:

Ricercar   5 Canons   Trio Sonata   5 Canons   Ricercar

Reflection

Reflection occurs in music in two ways. The most common might be called horizontal reflection, exemplified by the opening of Bach's Two-Part Invention No. 6, in E Major (Fig. 2). In this example, the ascending line *a* is reflected half a beat later by the descending line *b*. The plodding and quickly-predictable nature of this composite is then balanced by the dramatic textural and rhythmic change at *c*. The entire gesture is then repeated at *d*, but the roles of each hand are exactly reversed. Needless to say, the entire piece is an exploration of these symmetrical properties.

A marvelous example of reflection which undergoes a certain transformation in the process is found in Rachmaninoff's *Rhapsody on a Theme of Paganini*. The form of the work is generated by translation, with varying degrees of alteration, of the fast opening theme shown in Fig. 3(a). Later in the work, a quintessential Romantic theme forms the basis for a slow, contrasting section. It sounds "just right," yet is so contrasting in tempo, key, and articulation that a

Fig. 2. Opening of Bach's Invention No. 6, in E Major.

†The fact that these patterns later disintegrate in a turbulent frenzy exemplifies the "love–hate" thesis to be considered later.

Fig. 3. Rachmaninoff's *Rhapsody on a Theme of Paganini*. (a) opening, fast theme. (b) mid-section, slow theme. (© 1973 Belwin–Mills Publishing Co., used by permission.).

listener may not be consciously aware that the balance achieved is created by pitch reflection of the opening theme, as demonstrated in Fig. 3(b).

The other type of reflection—here referred to as vertical—looks effective in the score and certainly provides convincing symmetrical properties. However, since music is heard "from left to right," so to speak, and not from the center outward, this type of reflection is more difficult to perceive except on the smallest scale, as in the opening two phrases of George Gershwin's "I Got Rhythm."

On a larger scale, reflection is more often "felt" than heard and provides an "arch" form. Palindromes are found in music not containing conventional harmonic progressions (and thus not concerned with the problems posed by "reflecting" the progression), such as that of Webern and Hindemith in our century. One of the finest examples of vertical reflection is found in a work by Machaut (c.1300), as suggested in its title: "My End Is My Beginning."

While literal vertical reflection is not common, the idea of balance through the "what goes up must come down" principle has provided the aesthetic basis of melody throughout the history of Western music. The opening of Bach's *Two-Part Invention* No. 8, in F Major, illustrates this clearly (Fig. 4). Despite the apparent differences between the ascent and descent, the metric pulse underlying the music emphasizes the pitches circled so that the symmetry is clear. Furthermore, the abrupt and angular ascent is "balanced" by the faster but more gradual descent. Shown in the example is the right-hand part; as the descent begins, the left hand enters with the ascending line to balance that descent.

Rotation

Conventional musical notation suggests only one type of rotation that is audible as such, shown in Fig. 5(a). The relationship between the melodic gesture (pitches sounding one at a time) on the left and corresponding harmonic gesture (chord comprised of the same pitches) on the right has been exploited, of course, in all styles of music. The possibilities of rotation become even more apparent when we depart from conventional notation and arrange the same pitches on paper in a circular manner:

In this way we realize the rotational basis of much arpeggiated music (linear projection of chords), as demonstrated in the opening of the third movement to Beethoven's "Moonlight" Sonata, Op. 27, No. 2 (Fig. 5(b)). Although the notation requires linear deployment, comparing the pitch ordering left to right with the circle above will reveal the role of rotation in this

Fig. 4. Opening (right-hand part) of Bach's Invention No. 8, in F Major.

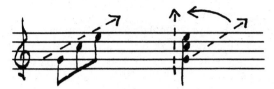

Fig. 5(a).

Fig. 5(b). Beginning (right-hand part) of Beethoven's "Moonlight" Sonata, Op. 27, No. 2, third movement.

passage. (The "sharps" are omitted for clarity.) Furthermore, the final chords (at the arrow) are a rotational summation of the previous arpeggiation, expressing the rotational relationship shown in Fig. 5(a).

SYMMETRY CREATED, SYMMETRY DENIED: LOVE–HATE

What separates music from several of the other arts is the fact that it happens "in time." It is largely because of this that an expectation of symmetrical shape or operation can be created in music, but not "fulfilled." Due possibly to *Gestalt* tendencies, the design in Fig. 6 will be described by most readers as involving partial translation, but lacking overall reflective symmetry due to the unique shape at the right. Now, however, the reader should place a piece of paper over this example and remove it by slowly sliding the paper to the right. In this manner, the example is gradually revealed "in time," as a musical structure would be. During the performance, the viewer/listener is exposed to new events, but tries to relate them to, and integrate them with, previous events while, simultaneously, forming predictions about future events. A good composition will allow a certain amount of this to happen easily so that dramatic alterations will have the desired impact. These alterations will not be perceived as merely contributing to asymmetry, but more accurately as a denial of the fulfillment of symmetry expectations.

In order to come even closer to the listening experience, the reader should have another piece of paper immediately following the first, to demonstrate that the comprehension of a temporal art is also dependent upon the precarious resources of human memory. The composer must work hard to implant in the listener certain aural images, but once assimilated, an entire movement can be "recalled" by a single gesture or phrase. *Perceived* symmetry in music, then, may be radically different from the amount of time actually devoted to each event. The fact that most climaxes occur approximately two-thirds of the way through a piece (often close to the "golden ratio") may have more to do with memory's sense of proportion than with a desire for asymmetry. A composer must adjust events so that the climb up the mountain takes longer (and is usually more arduous) than the run down the other side, even though upon reaching the base, the "hiker" perceives the mountain itself as basically symmetrical.

With that in mind, we will now examine symmetry expectation and denial in various contexts.

Fig. 6.

Symmetry denial through phrase extension

We begin this portion of the discussion with Robert Schumann's "Auf Einer Burg," from *Liederkreis,* Op. 39 (Fig. 7). Eichendorff's text is constructed of four verses, each of four lines in trochaic tetrameter (English translation by Sir Robert Randolph Garran.):

In a Castle	Auf Einer Burg
Fast asleep in ancient tower,	Eingeschlafen auf der Lauer
Still the aged knight is waiting,	Oben ist der alte Ritter,
Passes many a sudden shower,	drüben gehen Regenschauer
And the wood sighs through the grating.	und der Walt rauscht durch das Gitter.
Overgrown are beard and hair,	Eingewachsen Bart und Haare
Breast and ruffles turn'd to stone;	und versteinert Brust und Krause,
Many hundred years up there	sitzt er viele hundert Jahre
Sits he silent and alone.	oben in der stillen Klause.
Outside, it is still and lonely,	Draussen ist es still und friedlich,
All are gone into the valley;	alle sind in's Thal gezogen,
Forest birds are singing only	Waldesvögel einsam singen
In the ruins musically.	in den leeren Fensterbogen.
Down below a wedding passes,	Eine Hochzeit fährt da unten
In the sun the Rhine is sleeping;	auf dem Rhein im Sonnenscheine,
Blithe the violins and basses,	Musikanten spielen munter,
And the pretty bride is weeping.	und die schöne Braut, die weinet.

Schumann basically adheres to this structure in his simple folk-like setting. The *a* phrase is balanced within itself in that its second half is essentially a horizontal reflection of the first, but higher in pitch to create slight tension. The *a'* phrase is constructed similarly, but its inner reflection is lower in pitch, thus releasing the tension. Together the two phrases reinforce the gentle meter and phraseology (4 + 4 measures in the music) of the text's first verse.

The setting of the second verse relies on different symmetrical operations for construction, but upon listeners' expectations of symmetry—based upon their experience with the first verse setting—for effect. The *b* phrase maintains the expected meter and phraseology, but is generated from a translated rather than reflected pattern (constructed of elided sections of the *a* phrase). This translation upward generates so much energy that it continues over into the *b'* phrase, thus weakening the 4 + 4 phraseology by suggesting a 6-measure unit. To try to balance this ascent, a tortuous (and torturous) 4-measure descent completes the section, requiring an extension of 2 measures *beyond* the 2 "required" to satisfy the sense of 8-measure proportioning. Furthermore, that extended section, while based on the principle of descent suggested at the end of both *a* phrases, never descends low enough (nor do the actual pitches ever "resolve"† as they must according to tonal principles of Schumann's time) to dispel the tension or balance the song overall. By establishing symmetrical predictability and then gradually denying its fulfillment, the composer has structurally reinforced the pain of the old knight's loneliness (end of verse 2) and the irony of the bride's weeping amidst the gaiety (end of verse 4, not shown in Fig. 7, but the same as the ending shown), even though the 6-measure coupling and the final extension are not suggested by the meter of the text itself.

Denial of symmetry through truncation

In contrast to extension as created in the above examples, symmetry can be denied through truncation, or the shortening of the final section. The "asymmetry" of Schubert's "Die Ne-

---

†Cleverly, the pitches themselves finally do resolve with the beginning of the next song in the cycle, but this hardly provides closure.

Ein - ge-schla-fen auf der Lau-er o - ben ist der al - te Rit-ter;

drü - ben ge - hen Re - gen-schau-er, und der Wald rauscht durch das Git - ter.

Ein - ge-wach-sen Bart und Haa - re, und ver-stei-nert Brust und Krau-se,

sitzt er vie - le hun-dert Jah - re o - ben in der stil - len Klau - - se.

(Shown above is the setting of verses 1 and 2; the setting of verses
3 and 4 is melodically identical.)

Fig. 7. Melodic line of Schumann's "Auf Einer Burg."

bensonnen'' (''The Phantom Suns'') from *Winterreise*, reflects the verse structure of the text as follows:

Text	Music
$\left.\begin{array}{c}a\\a\end{array}\right\}$	$\left.\begin{array}{c}a\\\\a'\end{array}\right\}A$
$\left.\begin{array}{c}b\\b\end{array}\right\}$	
$\begin{array}{c}c\\c\\d\\d\end{array}$	$\left.\begin{array}{c}\\\\\\\end{array}\right\}B$
$\left.\begin{array}{c}e\\e\end{array}\right\}$	$a^2\ (A)$

The first verse describes the physical relationship between the singer and the suns, while the second verse deals with rejection. The third verse is an extension of the second yet states the poet's gloomy preference in only two lines (as though to discuss it further would be too painful), putting the reader off balance.

Schubert sets the first verse as two melodically identical four-measure phrases (though the harmony is different for the second) with simple, almost ''sing-song'' shapes (*A* in the music of Fig. 8). The second verse is set in a manner that reinforces the text's change of mood: with more disjunct lines, in a shifting minor mode, and higher in the singer's range for greater intensity. But it, too, is essentially two phrases of 4 measures each (*B* in the music).

The shorter last verse is set in the same manner as the first half of the first verse (*a*), but complete translation of *A* is not forthcoming. Nor is the short return of the opening material enough to comfortably balance the intense *B* section. Through the return to *A* material and the need to balance the *B* section, the music sets up expectation for symmetry beyond the metric and rhyming scheme of the text itself. The painfulness of betrayal and loneliness, and the tentative state of the person on the winter journey, are expressed through the text structure and the denial of large-scale formal symmetry in the music.

Denial of symmetry through imbalance
''The Man I Love,'' by George and Ira Gershwin, has always struck this writer as one of the saddest love songs ever written. The text describes waiting for love and, taken at face value, is full of optimism. It is the musical structure—specifically the relationship between the song's overall and internal shape—that infuses the text with the pain and futility of the wait.

The *A* section of the music (Fig. 9) is repeated almost exactly (*A'*), with both sections including three nearly-translational phrases of 2 measures each (1, 2, and 3 in the music). Each of these phrases has an upward, ''striving'' shape—hence, the optimism. But taken together as a complete section, there is a downward pull on the phrases, as shown by the dotted lines above the music. In almost all arrangements of this song, this downward pull is reinforced by a descending counterline (not shown), usually played by an instrument. The *B* section breaks out of this gloomy pattern at first, with a dramatic surge up to the highest pitch yet heard, but it soon drops in a manner similar to the *A* section. The same attempt is made again (translation, beginning at *b'*), but it drops even further.

And so we return to the opening material (at *A'* on the music) with its powerful descent. The ending (returning to the highest pitch) is a feeble attempt to balance all of the previous descent.

As illustrated by the shapes above the music, there is no question that symmetrical operations and balance are at work in this song. The ascending leaps in the *B* section are a welcome relief from the narrow, taunting figures in the *A* sections. However, each section's fundamental

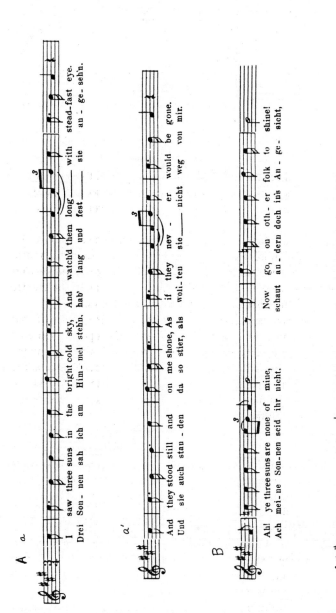
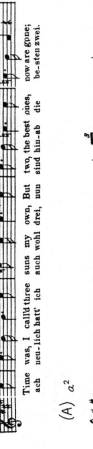

Fig. 8. Vocal line of Schubert's ''Nebensonnen.''

Fig. 9. Vocal line of Gershwin's "The Man I Love." (© Warner Bros. Music, used by permission.)

descent denies overall balance, creates a certain pathos, and, ironically, lends a desperate tone
to the text.

Symmetry and denial on different structural levels

''Dirge,'' a movement from Benjamin Britten's *Serenade for Tenor, Horn, and Strings,*
Op. 31, is a fascinating example of large-scale symmetrical shape generated from the antisymmetrical juxtaposition of small-scale structures which themselves are organized in terms of
symmetry operations. Having made this convoluted claim, let us begin with a study of the
small-scale structures and work toward a consideration of the large-scale form.

Perhaps the most unusual aspect of this movement is that it is a passacaglia—the singer
repeating the exact same melody nine times to express the nine verses of the fifteenth century,
anonymous text. The resulting effect is an intonation of sorts, as that of a religious chant, which
elucidates the symmetrical inevitability of the progression of death trials which commence on
Judgment ''Night.'' The melody itself (Fig. 10) is shaped through translation, while at the same
time suggesting the certain stages of descent from life to Whinnymuir, Brig o' Dread, and
finally Purgatory.

Taken alone, each verse demonstrates a clear, four-line rhyming scheme. Britten begins
the song in keeping with the text's meter, as shown at ''a'' in Fig. 10, but then sets the second
line of verse so that it is sung twice nearly as fast as the first. This is repeated for the third
(thus linking lines 2 and 3 aurally in juxtaposition to the ''abab'' rhyming scheme of the text).
The fourth line setting is linked to the second line setting in terms of rhyme and contour, but
to the first line setting in pitch and its longer duration of notes. The entire verse setting is thus
six measures in length. Of course, because this entire gesture is sung nine times, it provides
the work as a whole with a source of symmetry and a sense of uneasy stasis.

Supporting, coloring, and interpreting this prominent vocal line is the orchestra, whose
role is represented diagrammatically in Example 11. As shown in the example, the string sections
enter one by one in an ascending canon—literal translation from one line to the next, each time
on a higher pitch level. A certain predictability emerges and is reinforced as successive entrances
within the canon occur, but it has little to do with the phrase structure of the vocal line. This
is also true of the imitative gestures accompanying the fifth verse, except that the overall ascent
intensifies as canonic entrances get closer together (see Fig. 11).

Finally, with the entrance into Purgatory declared in the sixth verse, the orchestral climax
is achieved by the horn's entrance (in canonic imitation with the strings) as the strings play
unified gestures for the first time in the movement. But the climax is short-lived, and with the
seventh verse begins a gradual thinning and quieting despite the strings' persistent ''fire.''

The large-scale shape of the work, then, provides the balance of an arch with the high

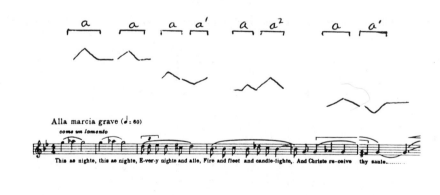

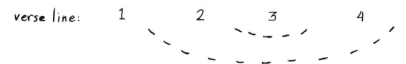

Fig. 10. Vocal melody, opening verse of Britten's ''Dirge.'' (© 1944 Hawkes and Son Ltd. used by permission
of Boosey and Hawkes, Inc.)

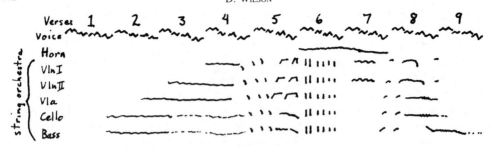

Fig. 11. Schema: Britten's "Dirge."

point corresponding to approximately that of the golden section. As discussed above, symmetry operations are obviously at work in both the vocal line and orchestral accompaniment, yet the anxiety associated with the steps to Purgatory is underscored by the lack of confluence of those elements while the horror is created by the inevitability of the large-scale thrust of the orchestra juxtaposed with the repetitive chant.

Love–hate—why?

The human need for symmetrical events in music was discussed earlier, but the need for symmetry-denial, regardless of whether a new sense of proportion is achieved, is even more curious. As T. S. Eliot wrote, "It is this contrast between fixity and flux, this unperceived evasion of monotony, which is the very life of verse."[1] The fact that music in the mainstream of Western development over the past millenium relies much more dramatically upon the assertion-denial of such patterns than is the case in many other cultures, may reflect our traditional compulsion to change from the norm, to progress. Or is the symmetry expectation-denial irony an expression of the Westerner's sense of self as being made "in God's image," but not capable of perfection in life or art? Is the discomfort/pleasure we feel in experiencing the irony part of a larger emotion, a larger expression?

However, the fact that overt repetition is becoming a mainstay in the music of several contemporary composers may be signalling a shift in our perception of the value of, and need for, change within the context of the acceptance and appreciation of ritual and tradition.† Does this "new" aesthetic also suggest that we are becoming more content with the human condition and the possibilities of the human spirit, or is this merely a phase of exorcism which will cause us to seek change, to deny symmetry in our lives—and our reflective arts—in the future?

REFERENCES

1. T. S. Eliot, Reflections on Vers Libre, *To Criticize the Critic*, NY Farrar, Straus, and Giroux, (1965).

†This is true, for example, in the music of Philip Glass and Steve Reich—two very popular, contemporary composers of so-called "art music." This is also true, of course, for a good deal of current "popular" music.

Comp. & Maths. with Appls. Vol. 12B, Nos. 1/2, pp. 113–121, 1986
Printed in Great Britain.

0886-9561/86 $3.00 + .00

THE COMPUTERIZED SOMA CUBE

Jon Brunvoll, Bjørg Cyvin, Einar Cyvin, Sven Cyvin, Aage Paus,
Martin Stølevik and Reidar Stølevik
The University of Trondheim, NLHT (N-7000) and NTH (N-7034), Trondheim, Norway

Abstract—The seven "Soma pieces" are produced as irregular combinations of 27 identical cubes, and may be joined to form a "Soma cube". The position of each Soma piece is expressed mathematically, and all possible positions are studied by computer programming along with geometrical considerations. It is found that the Soma cube may be constructed in 480 non-congruent ways; that excludes solutions which may be generated by rotations of the whole Soma cube. The solutions may be divided into 240 symmetrical pairs.

1. INTRODUCTION

The "Soma cube" was invented by the Danish engineer and poet Piet Hein. It is described along with some of its mathematical aspects by Gardner[1].

Seven "Soma pieces" are produced as irregular combinations of three (piece No. 1) or four (pieces Nos. 2–7) identical cubes as shown in Fig. 1. The pieces No. 5 and 6 are mirror images of each other. As Gardner[1] says, "it is an unexpected fact that these elementary combinations of identical cubes can be joined to form a cube again". Sets of Soma pieces have been traded as a popular toy; they have, however, considerable mathematical interest beyond that. The seven Soma pieces may in fact be arranged into one cube in many ways, but—to quote Gardner[1] again—"the exact number of such [essentially different] solutions has not yet been determined". In the present work we give the solution to this problem, which was solved with the aid of a computer program, combined with some geometrical considerations. This is an excellent example of the computerization of a mathematical problem.

2. MATHEMATICAL REPRESENTATION

In the Soma cube the $3 \times 3 \times 3 = 27$ positions of smaller cubes are identified in a coordinate system as shown in Fig. 2. Consequently any position of a Soma piece within the Soma cube is identified by three (for piece No. 1) or four (for pieces Nos. 2–7) triplets of the integers 0, 1 and 2.

3. TOTAL NUMBER OF POSITIONS

3.1 Definition

Consider the Soma cube with a given orientation. The first problem will be to derive all the ways in which each Soma piece may be placed within the cube. More precisely, we are looking for the number of positions for a given piece; two positions are defined as different when they have different coordinates in the mathematical representation (see above). Most of the pieces (Nos. 1, 3, 4, 5, 6) may be placed in a given position in two ways, and one of them (no. 7) in three, but it is not reasonable to distinguish between such possibilities. In other words we do not distinguish between the smaller cubes which build up a Soma piece.

3.2 Piece No. 1

Four ground positions as shown in Fig. 3 have the coordinate representations:

$$(1) \quad A_{xy}: \quad 000 \quad 010 \quad 100$$

$$(1) \quad B_{xy}: \quad 000 \quad 100 \quad 110$$

$$(1) \quad C_{xy}: \quad 010 \quad 100 \quad 110$$

$$(1) \quad D_{xy}: \quad 000 \quad 010 \quad 110$$

J. Brunvoll *et al.*

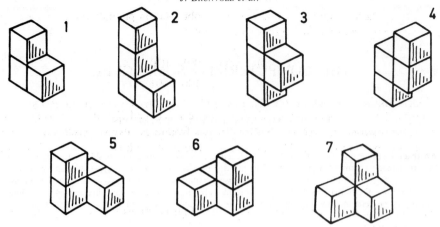

Fig. 1. The seven "Soma pieces".

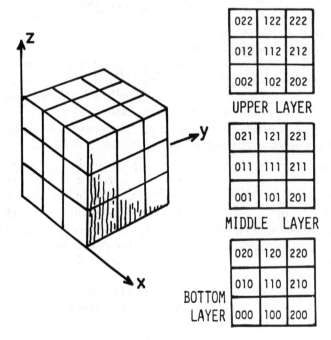

Fig. 2. Coordinates for the positions in an oriented "Soma cube".

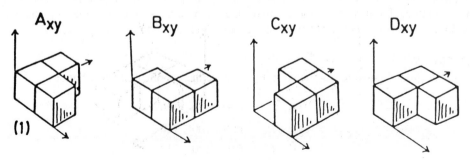

Fig. 3. Four ground positions of the Soma piece No. 1.

For each one of these positions two new ones are obtained on permuting the x, y and z axes. Thus, for instance, from the first position one derives

$$(1) \quad A_{yz}: \quad 000 \quad 001 \quad 010$$

$$(1) \quad A_{zx}: \quad 000 \quad 100 \quad 001$$

Altogether twelve positions emerge, where the piece is situated closely to the origin. All the other possible positions are obtained by translations from these twelve positions; in each case there are twelve positions covered by translations. The final result is $12 \times 12 = 144$ positions for piece No. 1.

3.3 *Piece No. 2*

The two positions shown in Fig. 4 give rise to four positions each by the rotation of axes parallel to the z-axis, *viz.*

$$(2) \quad A_{xy}: \quad 000 \quad 010 \quad 100 \quad 200$$

$$(2) \quad B_{xy}: \quad 000 \quad 100 \quad 110 \quad 120$$

$$(2) \quad C_{xy}: \quad 010 \quad 110 \quad 200 \quad 210$$

$$(2) \quad D_{xy}: \quad 000 \quad 010 \quad 020 \quad 120$$

and

$$(2) \quad A'_{xy}: \quad 000 \quad 010 \quad 110 \quad 210$$

$$(2) \quad B'_{xy}: \quad 000 \quad 010 \quad 020 \quad 100$$

$$(2) \quad C'_{xy}: \quad 000 \quad 100 \quad 200 \quad 210$$

$$(2) \quad D'_{xy}: \quad 020 \quad 100 \quad 110 \quad 120$$

Again we may derive two new positions from every one of the eight ones given above by permuting the x, y and z axes. Twenty-four positions with the piece closely to the origin emerge. In this case every one of these positions gives rise to six positions covered by translations. Consequently the total number of positions for piece No. 2 is $24 \times 6 = 144$.

3.4 *Pieces Nos. 3 and 4*

Fig. 5 shows one ground position each of pieces No. 3 and 4., *viz.*

$$(3) \quad A_{xy}: \quad 000 \quad 100 \quad 110 \quad 200$$

$$(4) \quad A_{xy}: \quad 000 \quad 100 \quad 110 \quad 210$$

Rotations without leaving the contact with the xy-plane give rise to three new positions in both cases. Again we may perform the permutations of the x, y and z axes, and finally the translations: six positions covered by translations are obtained from each position of the pieces near the origin. One arrives at the total number of positions as $12 \times 6 = 72$, the same for the pieces No. 3 and 4.

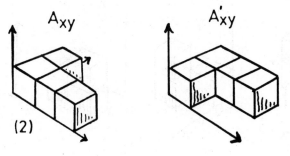

Fig. 4. Two ground positions of the Soma piece No. 2.

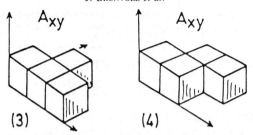

Fig. 5. Ground positions (one each) of the Soma pieces Nos. 3 and 4.

3.5 *Pieces Nos. 5 and 6*

It is clear that the pieces Nos. 5 and 6 have the same number of positions because of symmetry. For each of these pieces we define four ground positions, which are shown in Fig. 6. They have the following coordinates.

(5)	A_{xy}:	000	001	010	110	(6)	A_{xy}:	000	010	011	100
(5)	B_{xy}:	010	011	100	110	(6)	B_{xy}:	000	010	110	111
(5)	C_{xy}:	000	100	110	111	(6)	C_{xy}:	010	100	101	110
(5)	D_{xy}:	010	101	110	111	(6)	D_{xy}:	011	100	110	111

Two new positions are again derived in each case on permuting x, y and z. The resulting twelve positions for each of the two pieces are listed in Table 1, where the number of contacts of the small-cube faces with the coordinate planes are indicated. The twelve positions represent all the possibilities clustered around the origin. Each of them gives rise to eight positions covered by translations. Hence the total number of possibilities is $12 \times 8 = 96$ for each of the pieces No. 5 and 6.

3.6 *Piece No. 7*

Fig. 7 shows the two ground positions

(7)	A :	000	001	010	100
(7)	A':	011	101	110	111

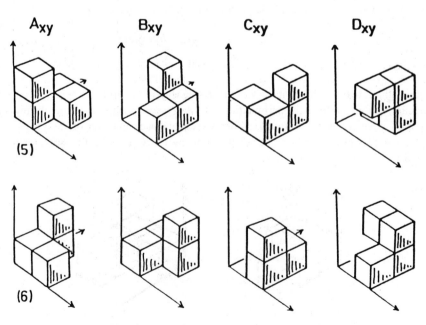

Fig. 6. Four ground positions each of the Soma pieces Nos. 5 and 6.

Table 1. Number of contacts of small-cube faces with the coordinate planes for the Soma pieces 5 and 6 in different positions

	Planes		
Position	xy	yz	zx
A_{xy}	3	3	2
A_{yz}	2	3	3
A_{zx}	3	2	3
B_{xy}	3	2	1
B_{yz}	1	3	2
B_{zx}	2	1	3
C_{xy}	3	1	2
C_{yz}	2	3	1
C_{zx}	1	2	3
D_{xy}	2	1	1
D_{yz}	1	2	1
D_{zx}	1	1	2

Each of them generate four positions, say $\{A, B, C, D\}$ and $\{A', B', C', D'\}$, respectively, on rotating around an axis parallel with z in the same way as was done with piece No. 1 (cf. Fig. 3). Each of these eight positions near the origin may be translated into seven new ones. Thus one arrives at the total number of positions as $8 \times 8 = 64$.

3.7 Summary

Table 2 summarizes the total number of positions derived for the different Soma pieces. The symmetry of each piece is also indicated in terms of the appropriate point group. All these point groups[2] are subgroups of O_h, the symmetry of a cube.

4. CONGRUENT SOLUTIONS

Two solutions for the Soma cube are defined as congruent when they may be transformed into each other by rotations of the whole cube. We will also speak about congruent positions of a single Soma piece when they are connected by rotations of the whole cube.

5. SOLUTION OF THE PROBLEM

5.1 Starting with piece no. 7

The different positions of piece No. 7 are treated in Sec. 3.6, where the eight positions near origin, viz. $\{A, B, C, D\}$ and $\{A', B', C', D'\}$ are described. Here we consider the solutions for the Soma cube where piece No. 7 has one of these eight positions. Thus we start with the most symmetrical one of the pieces; it is the only one with trigonal symmetry.

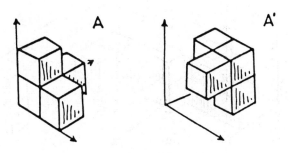

Fig. 7. Two ground positions of the Soma piece No. 7.

Table 2. Symmetry and number of positions of the Soma pieces

Piece No.	Symmetry	Total number of positions
1	C_{2v}	144
2	C_s	144
3	C_{2v}	72
4	C_{2h}	72
5	C_2	96
6	C_2	96
7	C_{3v}	64

Assume first position A for piece No. 7 (cf. Fig. 7). A computer program was designed and used to deduce all the solutions in this case; the number of such solutions appeared to be 1218. One of them is given below in the mathematical representation.

$$
\begin{array}{llllll}
(1) & : & 022 & 112 & 122 & \\
(2) & : & 201 & 202 & 212 & 222 \\
(3) & : & 200 & 210 & 211 & 220 \\
(4) & : & 011 & 111 & 121 & 221 \\
(5) & : & 020 & 021 & 110 & 120 \\
(6) & : & 002 & 012 & 101 & 102 \\
(7) & A: & 000 & 001 & 010 & 100
\end{array}
\qquad (i)
$$

Because of the trigonal symmetry of piece No. 7 it is clear that the 1218 solutions consist of congruent triplets. For instance, the solution (i) forms a triplet with the two following ones.

$$
\begin{array}{llllll}
(1) & : & 121 & 220 & 221 & \\
(2) & : & 012 & 022 & 122 & 222 \\
(3) & : & 002 & 102 & 112 & 202 \\
(4) & : & 110 & 111 & 211 & 212 \\
(5) & : & 101 & 200 & 201 & 210 \\
(6) & : & 011 & 020 & 021 & 120 \\
(7) & A: & 000 & 001 & 010 & 100
\end{array}
\qquad (ii)
$$

$$
\begin{array}{llllll}
(1) & : & 202 & 211 & 212 & \\
(2) & : & 120 & 220 & 221 & 222 \\
(3) & : & 020 & 021 & 022 & 121 \\
(4) & : & 101 & 111 & 112 & 122 \\
(5) & : & 002 & 011 & 012 & 102 \\
(6) & : & 110 & 200 & 201 & 210 \\
(7) & A: & 000 & 001 & 010 & 100
\end{array}
\qquad (iii)
$$

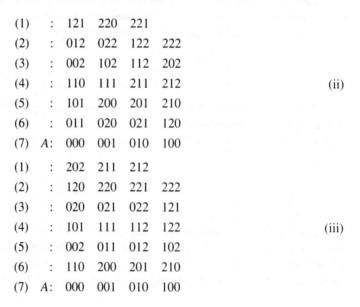

Fig. 8. Three positions of the Soma piece No. 7.

Hence the number of non-congruent solutions with piece No. 7 in position A is $1218:3 = 406$.

In the positions B, D and C' of piece No. 7 the small-cube faces have 3, 3 and 1 contacts with the three respective coordinate planes. These three positions are congruent. It was found by the computer program that no solutions are compatible with position B (and hence neither D or C') of piece No. 7.

In the congruent positions C, B' and D' (see Fig. 8) the piece No. 7 has 3, 1 and 1 contacts for the small-cube faces with the respective coordinate planes. Hence a solution with the piece No. 7 in position C gives immediately rise to two congruent solutions with the same piece in B' and D', respectively. All the C solutions were derived by the computer program. One of them is given below.

(1)	:	000	001	101	
(2)	:	200	201	202	210
(3)	:	211	220	221	222
(4)	:	002	102	112	212
(5)	:	012	022	121	122
(6)	:	011	020	021	120
(7)	C:	010	100	110	111

(iv)

The number of C solutions (which is the same for B' and D') was found to be 74.

It remains to investigate the possibilities with the piece No. 7 in position A' (cf. Fig. 7). It was found that no such solutions exist.

In conclusion, the Soma cube may be constructed in $406 + 74 = 480$ non-congruent ways.

5.2 *Starting with piece no. 2*

Additional information on the nature of the solutions for the Soma cube is obtained when one starts with one of the most unsymmetrical pieces, *viz.* No. 2. It has only a symmetry plane. The possible positions of this piece are discussed in Sec. 3.3.

The position A_{xy} (cf. Fig. 4) belongs to 24 congruent positions. They are derived on transforming this position by the 24 proper rotations of the O_h group[2], *viz.* $\{E, 8C_3, 3C_2, 6C_4, 6C_2'\}$. The position D_{xy} belongs to this set. Another set of 24 positions are derived on transforming A_{xy} by the improper rotations of O_h, *viz.* $\{i, 8S_6, 3\sigma_h, 6S_4, 6\sigma_d\}$. They are mutually congruent. Furthermore, they are symmetrical, but not congruent with the former set. The positions B'_{xy} and C'_{xy} belong to the latter set. In this way $24 \times 2 = 48$ positions are derived. In a similar way two sets of 24 positions each are derived from A'_{xy} (cf. Fig. 4). There remain two additional types of positions, which may be derived from (see also Fig. 9)

(2)	E:	010	011	110	210
(2)	F:	010	011	111	211

On transforming one of these positions by all the 48 symmetry operations of O_h only 24 congruent

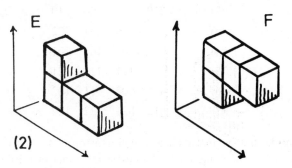

Fig. 9. Two positions of the Soma piece No. 2.

positions emerge; each one occurs twice. Altogether we arrive at $48 \times 2 + 24 \times 2 = 144$ positions, the same as were derived in another way in Sec. 3.3.

With piece No. 2 in the position A_{xy} all the 219 solutions were derived. One of them is:

(1)	:	101	201	202	
(2)	A_{xy}:	000	010	100	200
(3)	:	020	110	120	220
(4)	:	111	112	210	211
(5)	:	121	212	221	222
(6)	:	001	002	011	102
(7)	:	012	021	022	122

(v)

This solution (v) is congruent with (i), (ii) and (iii). The position A_{xy} of piece No. 2 is symmetrical, but not congruent with for instance C'_{xy}. Correspondingly there are 219 solutions with C'_{xy}; they are symmetrical, but not congruent with the former set. We may, as an example, choose the symmetry plane which brings A_{xy} into C'_{xy}. The following solution is symmetrical with (v) with respect to the chosen symmetry plane:

(1)	:	001	002	101	
(2)	C'_{xy}:	000	100	200	210
(3)	:	020	110	120	220
(4)	:	010	011	111	112
(5)	:	102	201	202	211
(6)	:	012	021	022	121
(7)	:	122	212	221	222

(vi)

Here the pieces Nos. 1, 2, 3, 4 and 7 have been reflected in the chosen symmetry plane. The pieces 5 and 6 have both been reflected and have changed place.

Assume now the position A'_{xy} for piece No. 2. The 21 possible solutions were found. One of them is (vii) below.

(1)	:	002	012	112	
(2)	A'_{xy}:	000	010	110	210
(3)	:	011	020	021	022
(4)	:	001	100	101	200
(5)	:	102	201	202	211
(6)	:	111	120	121	220
(7)	:	122	212	221	222

(vii)

(1)	:	112	202	212	
(2)	C_{xy}:	010	110	200	210
(3)	:	211	220	221	222
(4)	:	000	100	101	201
(5)	:	020	111	120	121
(6)	:	001	002	011	102
(7)	:	012	021	022	122

(viii)

We may again consider the symmetry plane which transforms A'_{xy} into C_{xy}. It produces a new set of 21 symmetrical, but not congruent solutions. The partner of (vii) is given above (viii).

Finally it was found that no solutions exist with piece No. 2 in any of the positions E or F (cf. Fig. 9).

In conclusion, there are 219 + 21 = 240 symmetrical pairs of solutions. The number of non-congruent solutions is 240 × 2 = 480, in agreement with the conclusion of Sec. 5.1.

REFERENCES

1. M. Gardner, *More Mathematical Puzzles and Diversions*. Penguin Books, New York (1961).
2. E. P. Wigner, *Group Theory and its Application to the Quantum Mechanics of Atomic Spectra*. Academic Press, New York and London (1959). Additional applications of group theory in several areas, especially in molecular symmetry, are included in many text-books; see, e.g.: F. A. Cotton, *Chemical Applications of Group Theory*, 2nd Edn. Wiley-Interscience, New York (1971); J. M. Hollas, *Symmetry in Molecules*. Chapman and Hall, London (1972); A. Vincent, *Molecular Symmetry and Group Theory*. Wiley, London (1977).

Comp. & Maths. with Appls. Vol. 12B, Nos. 1/2, pp. 123–138, 1986
Printed in Great Britain.

0886-9561/86 $3.00 + .00
© 1986 Pergamon Press Ltd.

THE SYMMETRY OF M. C. ESCHER'S "IMPOSSIBLE" IMAGES

Caroline H. MacGillavry

Vijverhoef 40, NL-1081 AX Amsterdam, Netherlands

Abstract—Escher's preoccupation with symmetry is well known. His periodic plane-filling patterns, his Circle limits, have been fully analyzed by several authors. Less attention has been given to the symmetry—much less evident—in his graphic work. In some of his prints, it is directly visible: then it is two-dimensional symmetry. In other cases one has to visualize or construct the three-dimensional image that is evoked by the print. Examples of both cases are given.

1. WHAT IS IMPOSSIBLE?

This is the first question. Not Escher's prints and drawings: They exist! He not only thought them out but also executed them. However, among the images that they evoke in the spectator's mind, and were meant by the author to do so, there are many that we feel to be impossible, inconceivable to exist in our "objective" three-dimensional reality.

This is the case with the imagery of many modern artists. What distinguishes Escher's work is, first, that every print or drawing is composed of items (animals, plants, fragments of buildings, etc.) which are reproduced with academic, "life-like", precision, so that the spectator can immediately identify each separate object as something known from everyday experience. These details tend to distract the mind from Escher's intentions with the print as a whole, when he combines the conceivable fragments into an "impossible" whole. These intentions which he wants to convey to the onlooker, are mostly quite abstract: the properties of space itself, such as symmetry; the geometric relations between an object in three-dimensional space with its representation on the flat plane of the drawing-board; the effect of visual illusions; approaches to infinity within the contours of the print or drawing. Thus, and this is the second point to make, Escher's prints appeal to our intellect rather than to our aesthetic or emotional receptiveness. Emotions, although composed of different and conflicting tendencies, overwhelm us instantaneously, whereas the intellect works step by step. Picasso's "Guernica", also composed of many individually recognizable details, evokes an immediate image of the horrors of war. Escher's "Concave and convex" (Fig. 1), full of details like "Guernica", at first sight only evokes bewilderment. Something does not fit, but what? The worst thing happens in the lower middle of the print: a floor on which a man sits dozing (left side) changes towards the right into a ceiling, from which a lamp hangs. At this point a spectator either turns away in disgust, or tries to understand what has been in Escher's mind.

2. DIFFERENT VIEWPOINTS COMBINED IN ONE PRINT

Let us follow the second course, step by step. First, strip the building from all its movable objects, such as human figures, lizards, flower pots. Except the two little pavilions, left and right, the building then appears to have the bi-lateral symmetry that most monumental buildings in all cultures have. What then goes wrong when the animals, ladders, etc. are put back in their places? The key to the picture is the flag, hanging from the wall and the balcony on the right. The pattern embroidered on it shows a well-known visual trick: a set of three adjoining parallelograms, repeated in two directions, is interpreted by the eye and the brain as a stack of blocks in two different ways: Seen from below, the small grey face is the bottom of each block, the left edges of the white faces seem to stick out of the drawing. But one can also "see" the small grey faces as the tops of blocks; then the right edges of the white faces are protruding. It is impossible to see the two representations simultaneously. However, Escher has separated them in the print. In nearly the whole right half we are forced by lamp, flagpole, and especially

Fig. 1. "Concave and Convex". (Copyright M. C. Escher heirs c/a Cordon Art, Baarn, Holland. Used by permission.)

by the bottom of the little pavilion, to view this part of the print from below upwards ("grey is bottom"). In the left part of the picture our view is forced downwards ("grey face is top"): the woman walks on a bridge which by symmetry corresponds to a vault on the right. The roof of the pavilion on the left becomes the ceiling of the one on the right, etc. What about the central strip? It is truly ambidextrous. The lizards below do not seem to mind, but the man sitting above them, slightly towards the left, will fall off the ceiling when he ventures across the border line between left and right.

In other prints where different directions of view are combined, this may lead to different kinds of symmetry. Figure 2, "Another world", shows the interior of a cube-shaped building. In the faces of this cube there are large "windows", separated by pillars. The axes of all these pillars converge towards one point, exactly in the center of the print. Consequently, the building is clearly drawn in perspective, but how? Divide the print in three segments: a) upper wall with upper half of the right-hand side wall, (b) left side with central square, and (c) lower wall with lower half of right-hand side wall. Make three masks, each covering two of the segments, leaving a view on the third one. Covering (b) and (c) one sees a moon-like landscape through the two windows of segment (a); a bird is sitting on the sill of the upper window and a horn hangs from the arched top of the right one. The direction of view is downward, practically in the plane of the print. The seemingly parallel pillars converge towards the center of the print, which is the so-called *nadir* of segment (a). If only segment (b) is left uncovered, the direction of view is perpendicular to the plane of the print. One sees again the bird, now from the side, the horn on the left, the moon landscape viewed horizontally, and the sky above it. The *vanishing point* is again in the center of the picture and now lies on the horizon of the central panel. The third view (c) is from below, directly opposite to the first view, converging towards the *zenith*, showing the sky. These three directions of view, at angles of about 90°, induce an apparent fourfold rotational symmetry about an axis across the middle of the print, parallel to its lower and upper edges. This symmetry is seen in the three representations of the same bird in the

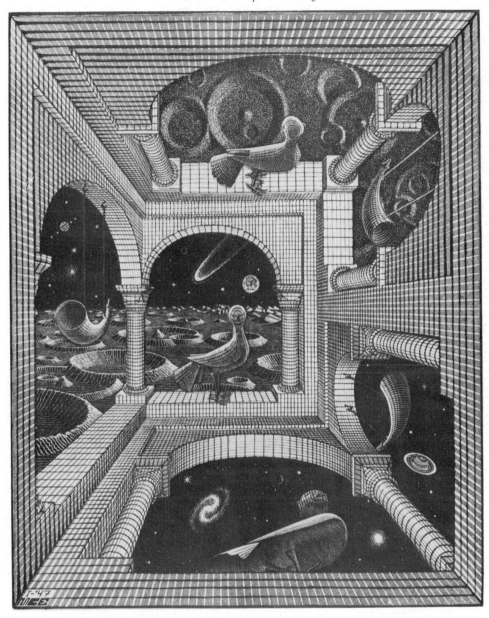

Fig. 2. "Another World". (Copyright M. C. Escher heirs c/a Cordon Arts, Baarn, Holland. Used by permission.)

three central panels. As to the side walls: the right one follows the rotation over 180° from view (a) to view (c). During view (b) this wall is covered. The left wall is only visible in view (b) so it is not bound by the symmetry.

Another print revealing three directions of view is "Relativity" (Fig. 3(a)). By eliminating all the niches, balconies, gardens and doors, we obtain the schematic Fig. 3(b). It shows that basically there are three walls, which appear to be perpendicular to each other, intersecting along the lines marked OA, OB and OC. By following in Fig. 3(a) the sets of lines supposedly parallel to these three lines, we find that they converge to three different vanishing points. Thus there are, as in "Another world", three directions of view, which in space would be perpendicular to the three walls. However, in this case the three corresponding vanishing points lie far outside the print, as indicated by the three arrows in Fig. 3(b). Next, take a look at the people who live in this strange building. The figures which I have colored blue are clearly subjected to a downward force, like we ourselves are attracted by gravity towards the center of the earth. These people are seen from below, in the direction AO towards the *zenith A*, as

Fig. 3(a). "Relativity". (Copyright M. C. Escher heirs c/a Cordon Art, Baarn, Holland. Used by permission.)

indicated in Fig. 3(b). A second group of people, the red ones, seem to be attracted by a force in the direction BO towards the left side wall, marked red. This wall is their floor; they do not feel the gravity that pulls the blue persons down. Neither do the green people who walk with their feet towards the right wall. The red and the green people are viewed from head to feet, so the vanishing points of their worlds are towards the *nadirs B* and *C* respectively.

Finally, look at the set of staircases in Fig. 3(a). There are three major ones which appear to be arranged as an equilateral triangle. In fact, if the print represents a view into a cubic room, then the direction of view on the print as a whole would be close to that of a space diagonal of the cube, which does have trigonal symmetry. These three stairs are fixed sideways to the three "walls", BOC, COA, and AOB respectively, as shown in Fig. 3(b). The one perpendicular to the blue people's ceiling—and of course also to their floor, not visible in the print—has steps which are parallel to the red and the green wall respectively, as shown in Fig. 3(b). So the red and the green people can walk on these stairs. When they both go from left to right, one is going down the stairs, the other is going up, as shown in Fig. 3(a). Why this difference? The red man climbs up from his floor, the green one descends towards his. The banister of this staircase is marked blue in Fig. 3(b), because it is parallel to the blue ceiling. The banisters of the other two big staircases are marked red and green respectively because their stairs are fixed sideways against the red (left) and green (right) wall. Again, on these two staircases the two sets of "other"-colored people can walk: the blue and green ones on the "red" staircase; the blue and red people on the green. Remarkably, in these latter two cases the people of one color walk on one side of the stairs, the other group on what would appear to the first group to be the reverse of the stairs. Why this difference with the top staircase A? That one goes between the *floor B* and the *floor C*. The crystallographic denomination of the plane of that staircase would be (011). The other big staircases, marked B and C run between the *floor C* (resp. *B*) and the *ceiling A*. Crystallographic planes ($\bar{1}01$) and ($\bar{1}10$). Thus, although these three staircases seem to be equivalent, in "reality" they are not. The same holds for the three vanishing points: zenith A, but nadirs B and C, as shown in Fig. 3(b). There are several other stairs in Fig. 3(a); some of them are indicated in Fig. 3(b). It is left to the reader to find out to which of the two types of stairs these belong. Crystallographers can also denominate the threefold axis, mentioned above, running perpendicular to the triangle of the three big staircases.

3. PLANE AND SPACE

Another quite frequent approach in Escher's work is the following. He starts from one of his truly periodic plane-filling patterns[1], constructed from adjoining, more or less schematically drawn animals (including angels and devils). By giving some of them an appropriate sort of relief in the drawing, these become more "life-like", they give a spatial suggestion and Escher then makes them walk or fly in seemingly three-dimensional space. Of course, they are still drawn on his plane drawing board! (See for example Fig. 4 "Reptiles".) His notebook is lying open, showing a trigonal pattern of reptiles in three different colors (in the litho: three different shades). The motifs are closely interlocked, and although recognizable as reptiles, they function simply as motifs in the pattern. However, two of them, in the lower left hand corner of the drawing, do come to life, one stretching a paw over the lower edge of the notebook and raising its head from the paper. The other has freed its whole right side, the left paws are still integrated in the repeating pattern. The print then shows the subsequent adventures of the creature: it climbs on a zoology book, and along a drawing-triangle on to the top of a regular polyhedron. From there it descends over an ashtray down into the pattern again, where it loses its individual identity.

Fig. 4. "Reptiles". (Copyright M. C. Escher heirs c/a Cordon Art, Baarn, Holland. Used by permission.)

It is the polyhedron to which I wish to draw attention. It represents Plato's regular pentagondodecahedron, a very symmetrical body, having *inter alia* five-fold rotation axes normal to each of its twelve faces and three-fold symmetry about each of its twenty corners. It seems that Escher wanted to say: "Look, in the plane pattern the creatures are arranged in close-packed hexagons. Three hexagons meet at each corner. But as long as you stay in the plane of the drawing, you cannot fill it in this way with regular pentagons. However, when you allow your paper to fold in three-dimensional space, and you fit the edges of adjoining pentagons together, you get the object on which the little alligator blows out its blast of triumph."

A possible sequel to this story is the following. Imagine that suddenly each face of the dodecahedron starts expanding in the five directions beyond its edges until it meets the extensions from the five adjoining pentagons. On the top of each face there is now formed a pentagonal pyramid. The originally convex body has become star-shaped with twelve fivefold spikes sticking out in the directions of the fivefold axes. If a little monster now tries to climb this stellated dodecahedron, it is lucky to find that the five side-faces of each spike have openings through which it can stick out its head and its four feet. (Its tail, if it has one, has to be tucked inside the pyramid.) Fig. 5 shows six of these monsters: one sitting on a central face and the others on its five adjoining faces; each under its own pyramid. Since they do not seem to fall off in any direction, they are apparently attracted by a gravitational force directed towards the center of the whole polyhedron. This is why Escher called the print "Gravitation". Note that the pyramids which should be at the back of the figure are also inhabited by monsters: some of their heads and feet are partly visible. The symmetry of the stellated polyhedron is still that of the original pentagondodecahedron. The creatures are not symmetrically arranged: Escher left them some freedom and individual rights within their prisons.

Among Escher's prints there are several more examples of creatures in his repeating patterns "coming to life" and seemingly going off into three-dimensional space, but in fact always staying in the plane of the picture. This is what Escher means to show: "Drawing is Fake!" In this group there is one print which from the point of view of symmetry needs further analysis. This is the well-known wood-cut "Day and Night" (Fig. 6). The landscape, seen in bird's eye view, appears quite symmetrical. Practically all the elements in the left-hand, "day" side of the print are found in mirror image on the right-hand side, as seen during the night: little town, river, etc. Remarkably, this most famous print is based on a typically Dutch landscape, in contrast with the many Italian ones of Escher's earlier period.

How do the birds come into the picture? Concentrate on the lower middle of the print, near the edge. Forgetting the rest of the picture, this part looks like the corner of a chess board. Going upwards and fanning out towards left and right, two changes occur. The fields of the chess board become fields in the landscape, receding in perspective towards the horizon. But at the same time, the plane of the chess board appears to fold upwards and to change gradually into a pattern of interlocking white and black birds. In the middle this pattern extends to the top of the print, blocking out the view on the landscape. Going to the right, the bird pattern resolves into a flock of white birds flying into the night, while the black birds fade away so the landscape is again visible. Towards the left, the complementary change occurs.

Why do I call this an "impossible" image? Because it is the superposition of two images: the vertical plane in which the birds move and the horizontal plane of the landscape. These two planes are welded together at the bottom of the print. By projecting the landscape on the plane of the birds, the complete picture is regenerated! Note how the contours of a vanishing bird cover more and more fields as you move up. The last point to make is about symmetry. The black–white mirror symmetry of the landscape seems at first sight also to be present in the bird pattern, although here coupled with a shift upwards over half the distance between two birds of the same color. However, the black birds are not the mirror images of the white ones: look at their tails!

4. CHANGING THE SCALE

Escher was much preoccupied by the problem of how to suggest infinity. Drawing in perspective is one solution, as we saw in "Day and night". Escher explored other, less conventional ways of the principle of perspective. His remarkable results have been fully described

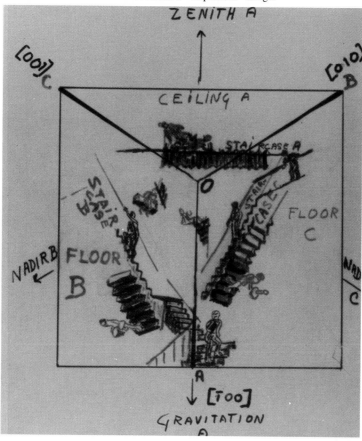

Fig. 3(b). Schematic drawing of "Relativity".

Fig. 5. "Gravitation". (Copyright M. C. Escher heirs c/a Cordon Art, Baarn, Holland. Used by permission.)

Fig. 6. "Day and Night". (Copyright M. C. Escher heirs c/a Cordon Art, Baarn, Holland. Used by permission.)

by Bruno Ernst in his book *The Magic Mirror of M. C. Escher*[2]. Another approach can be seen in Escher's numerous plane-filling repeating patterns: in principle these could be extended, like wall-paper or block printed material, to any desired length. But this did not satisfy him. He wanted to construct repeating patterns from motifs with ever smaller size, so that he could confine the design within the contours of his printing block or drawing paper. His first attempts had very small motifs in the middle of the picture, which grew larger and larger towards the edges. Again he did not find this satisfactory, because also here ever larger motifs could be added, so enlarging the size of the drawing. Inspired by a construction devised by the mathematician H. S. M. Coxeter, he used this frame of curves to construct his later "Circle-Limits I to IV", in which the largest motifs are in the center, becoming ever smaller towards the circumference of the circle. In principle, the construction would indeed allow a convergence towards infinitely small items. The very complicated, non-Euclidean symmetry of these circle limits has been fully analyzed by Coxeter[3], so I will not go into this. Instead I wish to draw attention to Fig. 7, "Fishes and scales". As always with Escher's work, it is essential to explore the print step by step. Below we see alternating rows of black and white fishes interlocking, but seeming to swim in opposite directions. Whereas the black fishes increase in size as they swim to the right, the white ones become smaller in their swimming direction. In the fourth and fifth rows from below, this effect gets enhanced. While the white fish in the lower right-hand side still stays in its row, the black one in the next row grows so large, both sideways and upwards, that it tends to occupy nearly the whole upper half of the figure. Something very odd happens: with the change of scale, its scales change into fishes! From here on the whole

Fig. 7. "Fishes and Scales". (Copyright M. C. Escher heirs c/a Cordon Art, Baarn, Holland. Used by permission.)

story repeats itself from right to left and then from top to bottom *ad infinitum*. What happens to the white fishes in the sixth row from below? As you follow them counter to their swimming direction, they get larger, as before, until they reach the center of the print. There they curve towards the right, changing their swimming direction and changing into scales of the big black fish. They become smaller and smaller until they reach the head of this fish. The only fixed point in this merry-go-round is the center of the print. You can check that it is a twofold rotation point: draw the contours of the fishes in the lower half of the print on a transparent sheet, rotate this sheet one half turn about the center of the print and you will see that it fits exactly the contours of the upper half. The only deviation from this twofold symmetry is the position of mouths and eyes: all the fishes look upward. Could this arrangement be extended beyond the limits of the print? One feels it could, but it would have required the geometric genius of Escher to do it. On the other hand, his sense of aesthetics may well have restrained him.

In Fig. 8, "Print gallery", the change of scale is an even more intriguing feature. You enter this gallery in the right lower corner of the figure and admire the prints (Escher's prints!). You pass a visitor and then you see a boy with somewhat abnormal body proportions. Comparing his head to his hand you see that he becomes larger from foot to crown. Also the framework of the gallery has extended abnormally. What at first, on the lower right, was a rather low arch, is now growing out of the top of the picture, just like the big fish in Fig. 7. The young man is looking at a print which undergoes the same transformation as he himself and as the gallery. Exploring it further, upwards and then towards the right this print expands more and more until its frame disappears out of the picture. Then, going down on the right, you find to your surprise that the gallery in which you started your exploration is part of the print. Going further around

Fig. 8. "Print Gallery". (Copyright M. C. Escher heirs c/a Cordon Art, Baarn, Holland. Used by permission.)

clockwise you find again the boy. So he is looking at a print in which he himself must be present somewhere. But then, where *is* he in the print, apart from being outside it? Start again from him, and this time go around withershins as witches do (after all, this *is* magic!). Now you find yourself rotating towards the white spot in the middle. Since, during this counterclockwise turn, everything gets smaller, then boy must be inside the white spot, somewhere to the right, under the slanting roof of the gallery, of which you see the left end. But pursuing this gallery towards the right, is it not the same gallery from which you started? The riddle is solved when we try to imagine what is going on inside the white spot. Still moving counterclockwise the print should become smaller and smaller until it dwindles in the center. Then, now turning again clockwise, the infinitesimally small print starts expanding spiralwise until it becomes the print as pictured on the left. Expanding further, the next layer of the spirally extending sheet begins to overlap the foregoing turn. The line where this happens starts just beyond the right of Escher's signature. It runs along the second rib of the slanting roof. Towards the right and down is the new layer of the spiral plane, towards the left, slightly upwards, the old one, as is seen by the two small arches. The second one, the top of which is just visible beyond the white spot, corresponds to the missing top of the large arch under which the boy in the left foreground stands! The whole litho is a sort of analogy in two dimensions of the Birth of the Universe. The "Big Bang" would take place in the center of the white spot; the overlapping sheets of the expansion spiral represent the time-scale of the expanding Universe.

Symmetry in this print is only a hint of a repetition along a spiral on an ever larger scale.

5. IMPOSSIBLE BUILDINGS

Under this heading I will discuss those buildings to which Escher himself gave this name. Of course, the structures in Figs. 1, 2, and perhaps 3(a) could have been included in this group.

The three prints to be considered now are all based on optical illusions. In Fig. 9(a), "Belvedere", the essential clues are the piece of paper lying on the floor and the object that the man sitting on a bench nearby holds in his hands. The sketch on the paper shows the conventional way of drawing a cube in elementary stereometry lessons. It can be interpreted either as a three-dimensional transparent cube seen from above, or alternately, the same cube seen from below (see the flag in Fig. 1!). In either case, the two pairs of lines that in the drawing intersect at points marked by a small circle, in space do not intersect but cross. One of the vertical edges of the cube is in front, the other is in the back of the cube. The corresponding horizontal lines represent edges in the back and in the front respectively. Just as in the flag of Fig. 1, when one changes the interpretation "seen from above" into "seen from below", those edges that first seemed to be in front, now appear to be in the back of the cube, *vice versa*. What happens if one starts with the two horizontal squares, the top and bottom faces of the cube, and makes the wrong connections between the two squares, so that the corners in front of the upper square are connected to the ones in the back of the lower square, and *vice versa*? When one does this, and it can be done with some distortion, one gets the object that the man on the bench holds in his hand. It combines the two views of the drawing on the paper: from above and from below. The man seems amazed to see that the bottom square appears to be rotated with respect to the top one. Above his head, this is actually the case: the man on the floor of the loggia in the Belvedere looks away to the right, whereas the lady upstairs turns half-face towards the onlooker. Evidently the distortion becomes less when the vertical edges, the pillars, are much longer than the edges of the base. If one follows the course of the pillars carefully, one sees that the four in the front row of the loggia floor are fixed to the back row of the ceiling, and *vice versa*. Escher has very cleverly camouflaged this crossing of the pillars by two tricks: firstly, the balustrade of the loggia floor is viewed nearly horizontally so that it is not immediately evident which pillars have their base in the front row, and which in the back. Secondly, the crossing over of the pillars is seen against a background of rugged mountain ridges, which distracts the onlooker's attention. Fig. 9(b) shows a very simple model of the relevant middle section of the building, the "loggia". In the view direction of Fig. 9(b) the pillars appear to be parallel, just as in Escher's print. Seen from the shorter side, Fig. 9(c), the crossing-over is evident. The symmetry of the loggia is that of the first stage of a fourfold screw-axis: rotation of 90° coupled with a shift from bottom to ceiling. This screw movement

Fig. 9(a). "Belvedere". (Copyright M. C. Escher heirs c/a Cordon Art, Baarn, Holland. Used by permission.)

is evidently caused by the torsion induced by alternately bending the pillars forwards and backwards. Because the rectangular floor and the ceiling of the loggia are askew, it is possible for a ladder standing on the floor to lean against the outside of the balustrade of the top story of the Belvedere. This story is completely "normal". The same holds for the ground floor of the Belvedere: inside is inside, so there is no escape possible for the raving prisoner.

Figure 10, "Ascending and descending", has been fully discussed by Bruno Ernst[2]. This complicated building has a courtyard; the roof of the central part of the building surrounding it seems to be a staircase on which two groups of monks walk. Those on the inner side seem to climb down the stairs, while those on the outer side climb up. However, after a full turn around the courtyard, every monk ends up in exactly the same place and at exactly the same height as before! How is this possible? The answer must be: if the stairs go neither up nor

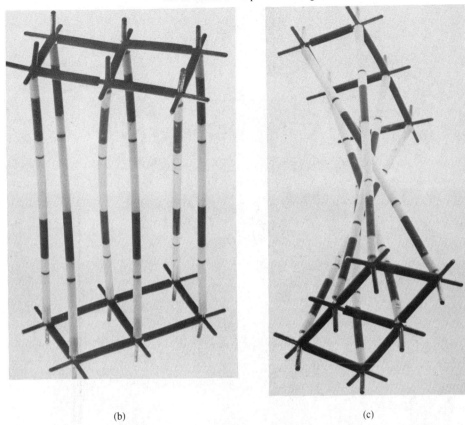

(b)   (c)

Fig. 9(b). Simplified model of the loggia (middle story) in "Belvedere", viewed as in Fig. 9(a); (c). Same model seen at an angle of about 90° from the view in (b).

down, then the plane of the stairs must be horizontal: but then, what about the floors of the building? The ledge of the floor of the top story is seen to be at an angle with the plane of the roof; the floors on the sides of the building that are visible in the print appear to be parallel to the top floor, so all are at the same angle towards the roof. The problem is solved when you imagine the floors at the back of the building to go up from right to left. This is actually seen to be the case in the inner side of the courtyard! In other words: the building is a "spiral ramp", just like a modern car garage. You get in on street level, drive round and round until you find a parking place. Then on leaving your car, you discover that you have to take an elevator which brings you from the n-th floor down to street level again. How does Escher trick you into a false sense of what is horizontal? Note that this is the only Escher building which stands as it were in an empty world: no foreground, no background, no horizon! Only a vague shadow at the bottom of the building. The illusion is further enhanced, as pointed out by Bruno Ernst, by a false perspective suggested by the shape of the courtyard. Escher[4] describes it as "rectangular". However, by counting the sections of the banisters, you see that this is not the case. The four sides of the courtyard are all of different length! In this print Escher does not give you any clue, but fools you by false information. Thus, the apparent orthogonal symmetry of the helicoidal central part is broken by this deviation from orthogonality. It is further impaired by changes in the pitch of the helix: the stories are of different height.

Thus, whereas neither the Belvedere nor the Monastery are really impossible buildings, they are certainly uncomfortable to live in, just like the house in Fig. 3, "Relativity". On the other hand, the building of Fig. 1 "Concave and convex", with its pavillions added, is truly impossible. Equally impossible is the building in the last print I will discuss, "Waterfall" (Fig. 11). The water tumbling down from the left-hand tower ("Tower I") turns the wheel of a mill. Then the water runs down a duct zigzagging between the two towers until it reaches the *high* level of the left tower, from where it falls down again. The basic error in this apparent *perpetuum mobile* is again an optical illusion, the so-called Penrose triangle[5]. This is a set of three

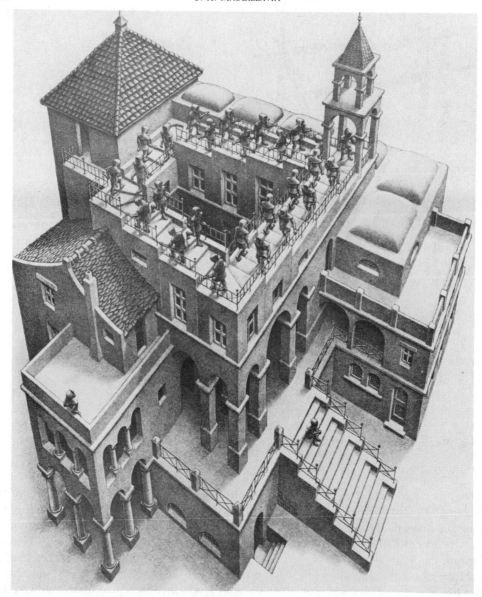

Fig. 10. "Ascending and Descending". (Copyright M. C. Escher heirs c/a Cordon Art, Baarn, Holland. Used by permission.)

beams, two fixed together at an angle of 90°, the third vertically linked to the end of one of the first two horizontal beams, so that the three are arranged like three consecutive edges of a rectangular parallelepiped. Viewed, with one eye closed, from the top of the vertical beam towards the end of the farthest horizontal beam, you are tricked into the illusion of "seeing" a triangle with three angles of 90°. This trick has been applied three times in Fig. 11. From the bottom of the left-side Tower I, the water-duct appears to run down towards the right, and *away* from you. Then, meeting the pillars of Tower II, it makes a right turn, still supposedly downward and away from you, until it appears to meet the second level pillars of Tower I: the first Penrose triangle! From there, running further again at right angles, it seems to meet Tower II at the bottom of the second set of pillars: the second Penrose triangle is formed. Then, along the last zigzag back in the direction of Tower I suggests the third Penrose triangle. What tricks you? First, the apparently very solid attachment of the ends of the three mutually orthogonal sides of the Penrose triangle. In reality these ends would be at a distance of the body-diagonal of the orthogonal parallelepiped with edges a and b of the zigzag of the duct and c, the height of a pillar. If the duct would really run away towards the horizon, then the zigzags should become shorter and narrower as seen in perspective. Moreover, the low parapets of the duct suggest a downward course. But seen against the background of a terraced hillside, they seem to run slightly up. Thus the whole representation of the duct is full of inconsistencies. They puzzle you, because—as Escher knew very well—it is impossible for the human brain to register momentaneously such a complicated print as a whole. The only really symmetrical objects in this print are the two blocks on top of the towers: on Tower I three interlocking cubes, on Tower II a stellated regular rhombo-dodecahedron. Both blocks still have full cubic symmetry!

6. FINAL REMARKS

In this article I have discussed only a limited number of Escher's "impossible" prints which show more or less symmetry, in total or in details. An analysis of all of them would fill a book which to a large extent would overlap Bruno Ernst's *Magic Mirror*[2].

Since my point of view was to be "symmetry", my analysis covers only the "cerebral"

Fig. 11. "Waterfall". (Copyright M. C. Escher heirs c/a Cordon Art, Baarn, Holland. Used by permission.)

side of Escher's work. This does not imply that his work only impresses me from this point of view. In the introduction I mentioned other reactions to works of art: aesthetic and emotional. These are of course highly individualistic. Some prints which I know to be much appreciated by many people, do not appeal to me personally. Some others I admire and love. Perhaps Escher's own opinion may be quoted (Bruno Ernst, l.c. Chapter 3).: ''I consider my own work as the most beautiful, and also as the ugliest.'' I would not go so far either way in my appreciation. But I always admire the profoundness of Escher's thoughts, the intensity of his concentration, and the excellency of his craftmanship, in which he put his pride.

Acknowledgements—I am indebted first and foremost to Escher himself: to what I learned from his writings, his lectures, but above all to what he himself told me about his work. I gained much insight into the mathematical background by studying Bruno Ernst's *Magic Mirror* and several articles by H. S. M. Coxeter. I hope there is still enough personal interpretation in this paper to make it a worthy tribute to the memory of a remarkable man and valued friend, Maurits Cornelis Escher.

REFERENCES

1. Caroline H. MacGillavry, *Symmetry Aspects of M. C. Escher's Periodic Drawings*. Oosthoek, Utrecht (1965); Second Edition under the Title: *Fantasy and Symmetry*. Harry N. Abrams, New York (1976).
2. Bruno Ernst, *The Magic Mirror of M. C. Escher*, Random House, New York (1976).
3. H. S. M. Coxeter, The non-Euclidean Symmetry of Escher's Picture ''Circle Limit III''. *Leonardo* **12**, 19–25, 32 (1979).
4. M. C. Escher, *Grafiek en Tekeningen*. Tijl, Zwolle (1959).
5. L. S. Penrose and R. Penrose, Impossible Objects, a special type of visual Illusion. *Brit. J. of Psychology* **49**, 31 (1958).

Comp. & Maths. with Appls. Vol. 12B, Nos. 1/2, pp. 139–153, 1986
Printed in Great Britain.

0886–9561/86 $3.00 + .00
© 1986 Pergamon Press Ltd.

HYPERBOLIC SYMMETRY

Douglas Dunham

Department of Mathematical Sciences, University of Minnesota—Duluth, Duluth,
Minnesota 55812, U.S.A.

Abstract—The use of computer graphics to create repeating patterns of the hyperbolic plane is a recent and natural development in the history of hyperbolic patterns. A program is described which generates such patterns in the Poincaré model of hyperbolic geometry. The program can easily be extended to create patterns with color symmetry.

1. INTRODUCTION

Few hyperbolic patterns have been created compared with the number of Euclidean plane patterns [1] and [2]. The main reason for this is the requirement for many precise geometric constructions or numerical calculations. A natural solution to this problem is to use a computer to perform the computations and a computer-driven output device to display the pattern. Several programs have been written to generate hyperbolic patterns and a new one will be described below. A sample of the output from this program is shown in Fig. 1. This new program allows for the creation of patterns with color symmetry, an advance over previous programs. Figure 1 provides an example of 2-color (black-white) symmetry; 3-color and 4-color symmetry are exhibited in Figs. 5(a), 13 and 14.

Most hyperbolic patterns are represented in the Poincaré model of hyperbolic geometry. This model and regular hyperbolic tessellations are described in Secs. 2 and 3. The symmetry group of a pattern is defined in Sec. 4. The computer-generation of hyperbolic patterns can be broken down into two steps: (1) creation of the basic subpattern and (2) replication of the subpattern. These steps are detailed in Secs. 5 and 6. Finally it is shown how the replication algorithm may be modified to incorporate color symmetry.

2. HYPERBOLIC GEOMETRY

Unlike the Euclidean plane and the sphere, the entire (i.e. complete) hyperbolic plane cannot be isometrically embedded in 3-dimensional Euclidean space. Thus, any model of hyperbolic geometry in Euclidean 3-space must distort distance. However, there are conformal models (i.e. where the hyperbolic measure of an angle is just its Euclidean measure).

The *Poincaré circle model* is conformal and has the additional property that it is represented in a bounded region of the Euclidean plane—this is useful when we desire to show an entire pattern. The points of this model are the interior points of the *bounding circle*. The hyperbolic lines are circular arcs orthogonal to the bounding circle, including diameters. For example, the backbones of the fish in Fig. 1 lie on hyperbolic lines. Also, all the fish in Fig. 1 are the same hyperbolic size, showing that equal hyperbolic distances are represented by decreasing Euclidean distances as one approaches the bounding circle.

3. REPEATING HYPERBOLIC PATTERNS

We will define a *repeating hyperbolic pattern* to be a pattern composed of hyperbolically congruent copies of a basic subpattern or *motif*. Either the right side or the left side of any one of the fish of Fig. 1 serves as a motif for that pattern if color is disregarded. (If color is taken into account, a motif may be formed from half of a black fish together with half of an adjoining white fish.)

Other important examples of repeating hyperbolic patterns include the regular tessellations $\{p, q\}$ of the hyperbolic plane by regular p-sided polygons, or p-gons, meeting q at a vertex (see Coxeter and Moser[3, Chapter 5]). It is necessary that $(p - 2)(q - 2) > 4$ to obtain a

Fig. 1. The top view of flounder-like fish arranged in a repeating hyperbolic pattern in the style of M. C. Escher's picture "Circle Limit I."

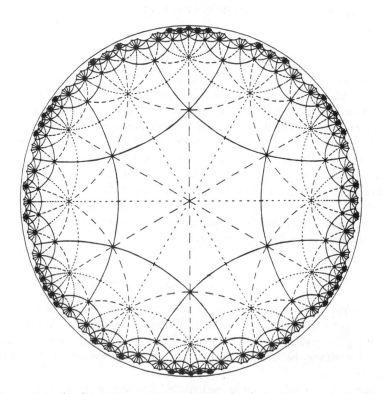

Fig. 2. The tessellation {6, 4} (solid lines), the dual tessellation {4, 6} (dotted lines), and other lines (dashed) of reflective symmetry of the two tessellations.

hyperbolic tessellation. Figure 2 shows the tessellation {6, 4} in solid lines and its dual tesselation, {4, 6}, in dotted lines.

If $(p - 2)(q - 2) = 4$ or $(p - 2)(q - 2) < 4$, one obtains regular tessellations of the Euclidean plane or sphere, respectively.

4. SYMMETRY GROUPS

A *symmetry* of a pattern is a congruence or isometry (hyperbolic distance-preserving transformation) of the hyperbolic plane which transforms the pattern onto itself. Reflections across the backbone lines are symmetries of Fig. 1. All the solid, dotted, and dashed lines of Fig. 2 are lines of reflective symmetry of that pattern. Hyperbolic reflections are either Euclidean reflections across diameters of the bounding circle or inversions with respect to orthogonal circular arcs. A line of reflective symmetry, i.e. the fixed line of a reflection, is called a *mirror*. Disregarding color, other symmetries of Fig. 1 include quarter-turns (rotations by $\pi/2$) about points where trailing edges of fin-tips meet, and translations by one fish-length along backbone lines. The *symmetry group* of a pattern is the set of all symmetries of that pattern.

The inradii and circumradii of $\{p, q\}$ lie on mirrors and divide each p-gon into $2p$ right triangles with acute angles of π/p and π/q. The symmetry group of $\{p, q\}$, denoted $[p, q]$, can be generated by reflections across the three sides of any such triangle, i.e. all symmetries of $\{p, q\}$ can be obtained by successively applying a finite number of those reflections (see Coxeter and Moser[3, page 54]). Thus, [6, 4] is the symmetry group of the pattern of Fig. 2. Note that the groups $[p, q]$ and $[q, p]$ are isomorphic.

As mentioned above, we may have $(p - 2)(q - 2) = 4$, in which case we obtain the Euclidean groups $[4, 4] = p4m$ and $[3, 6] = p6m$. As pointed out in [3, Sec. 4.6 and Table 4], these groups contain all the 17 plane crystallographic groups as subgroups. Similarly, the groups $[2, q]$, $[3, 3]$, $[3, 4]$ and $[3, 5]$ contain all the discrete spherical groups as subgroups. Thus, the groups $[p, q]$ are quite general and have significance beyond hyperbolic geometry.

The orientation-preserving subgroup of index 2 in $[p, q]$ is denoted by $[p, q]^+$ and consists of all symmetries which can be obtained by successively applying an even number of the reflections which generate $[p, q]$. The group $[p, q]^+$ can be generated by any two of the rotations by $2\pi/p$, $2\pi/q$, or π about the vertices of the right triangle formed by the mirrors of the reflections which generate $[p, q]$. In fact, those rotations can be produced by successive applications of pairs of the reflections which generate $[p, q]$. This is because reflections across intersecting lines produce a rotation about the intersection point equal to twice the angle of intersection, just as in the Euclidean case. Figure 3 shows a pattern with symmetry group $[5, 5]^+$—all symmetries of that pattern are orientation-preserving since only the left sides of the fish show. In Fig. 4(a), we see a pattern with symmetry group $[5, 4]^+$.

There is another subgroup, $[p^+, q]$ (q must be even), of index 2 in $[p, q]$, which contains rotations by $2\pi/p$ about the centers of the p-gons of $\{p, q\}$ and reflections across the sides of the p-gons (this is why q must be even). In fact, $[p^+, q]$ can be generated by a p-fold rotation about the center of any p-gon in $\{p, q\}$ and a reflection across the side of that p-gon. Figure 4(b) shows a pattern with symmetry group $[5^+, 4]$—note that adjacent pinwheels spin in opposite directions. These groups have been studied by Coxeter[4] and Sinkov[5].

The Dutch artist M. C. Escher used two instances of these groups, $[3^+, 8]$ and $[4^+, 6]$, for his patterns "Circle Limit II" (if the differences in shading are disregarded) and "Circle Limit IV;" these patterns are reproduced in Figs. 5(a) and 5(b). If color is disregarded, Fig. 1 also has symmetry group $[4^+, 6]$. Notice that mirrors pass through the center of the bounding circle in Figs. 1 and 5, whereas Fig. 4(b) has a 5-fold center of rotation at the center of the bounding circle. For more about the groups $[p, q]$ and their subgroups, see Coxeter and Moser[3, Secs. 4.3 and 4.4].

5. CREATING A MOTIF

Having laid the mathematical foundation, we now turn to the algorithmic structure of the pattern-creating program. The design of the motif is the first step in the pattern-creation process;

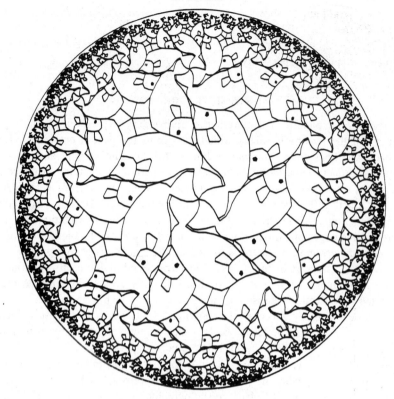

Fig. 3. A pattern with symmetry group $[5, 5]^+$.

the second step, replication, involves transforming copies of the motif about the hyperbolic plane and will be discussed in the next section.

In order to design the motif, we must choose a symmetry group. If the hyperbolic plane is covered without overlap by transformed copies of a connected set under elements of a symmetry group, that set is called a *fundamental region* for the symmetry group (copies of the fundamental region may overlap along boundaries). If there are reflections in the symmetry group, at least part of the boundary of the fundamental region must lie along mirrors of those reflections (the boundary certainly cannot cross a mirror). Thus, the only choice for a fundamental region for [6, 4] is one of the right triangles of Fig. 2, for if the fundamental region did not fill out such a triangle, it would violate the condition that copies of it cover the hyperbolic plane.

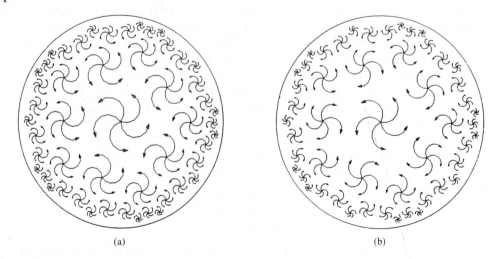

(a)   (b)

Fig. 4. (a) A pattern with symmetry group $[5, 4]^+$. (b) A pattern with symmetry group $[5^+, 4]$.

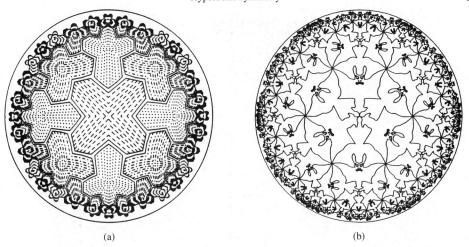

(a)   (b)

Fig. 5. (a) After M. C. Escher's pattern "Circle Limit II." (b) After M. C. Escher's pattern "Circle Limit IV."

There is more flexibility in choosing a fundamental region for $[p^+, q]$. The boundary may be formed by (1) a curve from the center of a p-gon to the p-gon's boundary, (2) a copy of that curve rotated by $2\pi/p$ about the p-gon center (if the copy does not cross the original curve) and (3) that part of the p-gon's boundary between the two curves. A sample fundamental region for $[4^+, 6]$ with a bold boundary is shown in Fig. 6. There is even more flexibility in choosing a fundamental region for $[p, q]^+$: the bounding curve must pass through one p-fold, one q-fold, and one 2-fold center of rotation (without crossing itself; it may pass through two p-fold or two q-fold centers of rotation).

Fig. 6. A pattern with symmetry group $[4^+, 6]$ showing a motif which fills a fundamental region with a bold boundary, and adjacent copies of the motif numbered 1, 2, 3, and 4.

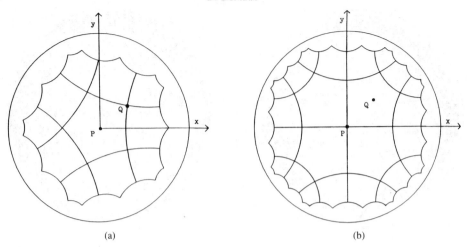

Fig. 7. (a) A diagram showing the positions of the points P and Q on the centered tessellation $\{p, q\}$. (b) A diagram showing the positions of the points P and Q on the tessellation $\{p, q\}$ having a vertex at the origin and an edge along the positive x-axis.

In order to facilitate the replication process, coordinates are chosen so that a p-gon of the underlying $\{p, q\}$ is centered at the origin and the positive x-axis bisects a side of that p-gon. To fix notation, we will label the origin P and the upper vertex of the side bisected by the positive x-axis will be labelled Q, as in Fig. 7(a). Thus, a fundamental region may always be chosen so that its boundary contains P and Q. If it is desired to generate a pattern with symmetry group $[p^+, q]$ with its mirrors passing through the center of the bounding circle, coordinates are chosen so that a p-gon of the underlying $\{p, q\}$ has a vertex at the origin P, one of its edges along the positive x-axis and its center Q above the x-axis (see Fig. 7(b)).

In order that copies of the motif not cross each other in the final pattern, the motif must be drawn within a fundamental region. In fact, the motif may fill out the fundamental region, thus creating a pattern of interlocking motifs—a characteristic feature of M. C. Escher's repeating patterns. In all the patterns of Figs. 1–6 except Figs. 5(a) and 5(b), the motif coincides with a fundamental region; in Figs. 5(a) and 5(b) the curved arrow motif lies strictly inside a fundamental region in each case.

The motif can be entered as a sequence of points by using an input device such as a cursor, thumbwheel-controlled crosshairs, or light pen. As each new point is entered, a line segment may be drawn from the previous point, or the line segment may be omitted if it is desired to start a new series of line segments. After a line segment has been entered, the computer program shows transformed copies of the segment in adjacent copies of the fundamental region. Thus, as the motif is being entered in a fundamental region, transformed copies of the motif are built up in adjacent copies of the fundamental region. This is the feature of the program that facilitates the creation of interlocking motifs. In Fig. 6, the original motif is outlined in bold lines and adjacent copies are numbered 1, 2, 3 and 4 (the motifs numbered 3 and 4 are not used in creating the interlocking motif since they are separated from the original motif by mirrors).

6. REPLICATING THE PATTERN

After the motif has been entered, copies of it may be transformed about the hyperbolic plane; this replication step generates the final pattern. The replication step takes place in two stages. In the first stage, all motifs within fundamental regions having the center of the bounding circle P as a boundary point are combined to form a larger subpattern called the p-gon pattern. The p-gon pattern is formed by reflecting or rotating (depending on the symmetry group) the original motif about P, i.e. these are ordinary Euclidean reflections and rotations. The p-gon pattern of Figs. 3, 4(a) (and 4(b)), 5(a) and 5(b) are the group of five centered fish, the centered pinwheel, the central cross, and the three central devils, respectively.

The advantages of forming the p-gon pattern are simplification of the replication algorithm and reduction of the number of transformations needed for the second stage of the replication process. For patterns with symmetry groups $[p, q]$, $[p, q]^+$, or $[p^+, q]$, the reduction factors are $2p$, p and p respectively. If the motif forms an interlocking pattern, so will the p-gon pattern.

The algorithm for accomplishing the second stage of the replication step depends on the observation that the p-gons of the tessellation $\{p, q\}$ form layers, and therefore so do copies of the p-gon pattern. The first layer of p-gons is just the central p-gon. The $(k + 1)$-st layer is defined inductively as the set of those p-gons not in any previous layer, but which share an edge or a vertex with the k-th layer. The first three layers of $\{6, 4\}$ are shown in Fig. 2 (solid lines), and the first five layers of $\{7, 3\}$ are shown in Fig. 8. We will build up the entire repeating pattern with layers of p-gon patterns in the same way that $\{p, q\}$ may be built up with layers of p-gons.

The first layer of the repeating pattern is formed by simply drawing the p-gon pattern. In order to extend the pattern from the k-th layer to the $(k + 1)$-st layer, first recall that there is a vertex Q of the central p-gon of $\{p, q\}$ on the boundary of the fundamental region containing the original motif. Next, suppose that the p-gon pattern has been transformed to the k-th layer so that Q is transformed to a vertex Q' common to the k-th and $(k + 1)$-st layers. Then the pattern may be extended (locally) to the $(k + 1)$-st layer by successively rotating or reflecting the transformed p-gon pattern about the vertex Q' to all p-gon positions in the $(k + 1)$-st layer (except that the last position will be covered by a transformation about the next vertex). Thus, all p-gon patterns of the $(k + 1)$-st layer may be obtained by transforming suitable p-gon patterns of the k-th layer about vertices common to the two layers.

We now describe in more detail the algorithm for obtaining the entire repeating pattern from the p-gon pattern for the groups $[p, q]$ and $[p, q]^+$ when $p > 3$ and $q > 3$. The algorithm must be modified slightly to handle the group $[p^+, q]$ or the cases $p = 3$ or $q = 3$. The algorithm will be described using the Pascal programming language; the data structures describing the p-gon pattern and the transformations will be given after the algorithm. Following the description of the data structures, we will define a procedure DrawPgonPattern which draws a transformed p-gon pattern, given the transformation.

The first layer, i.e. the p-gon pattern itself, is drawn by:

DrawPgonPattern(Identity)

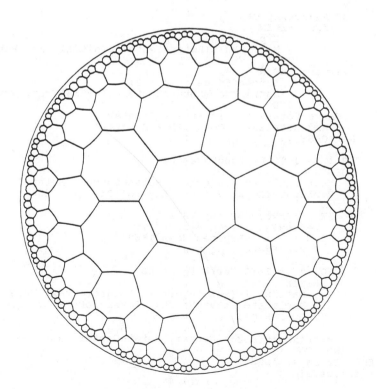

Fig. 8. Five layers of 7-gons of the tessellation $\{7, 3\}$.

If the number of desired layers, *nLayers*, is greater than one, then a recursive procedure, *Replicate* (Fig. 9), may be used, as described above, to extend the pattern from one layer to the next. Replicate has three parameters: (1) the transformation which takes the *p*-gon pattern to the present position in the *k*-th layer, (2) the number of additional layers to be drawn after the *k*-th layer and (3) the manner in which the *p*-gon in the present position lies adjacent to the $(k - 1)$-st layer of *p*-gons—it either shares an edge or a vertex with the previous layer (in the case $q = 3$—not considered in this discussion—*p*-gons in one layer share either one or two edges with the previous layer of *p*-gons). In order to be precise when dealing with this third parameter, we make the following type definition:

$$\text{AdjacencyType} = (\text{Edge, Vertex})$$

The way Replicate works is that the *p*-gon pattern is first drawn in the present position (in the *k*-th layer). Then, if there are additional layers to be drawn, Replicate generates recursive calls to itself, after computing the transformations to those *p*-gon positions in the next layer which share an edge or vertex with the *p*-gon in the present position.

Replicate uses the constant counter-clockwise rotations, *RotateP* by $2\pi/p$ about P (the center of the bounding circle), and *RotateQ* by $2\pi/q$ about the *p*-gon vertex Q (see Fig. 7(a)). Replicate also uses the variable rotations RotateCenter about *p*-gon centers, and RotateVertex about *p*-gon vertices. All these transformations are represented by matrices.

```
PROCEDURE Replicate(InitialTran: Transformation;
                    LayersToDo: INTEGER;
                    Adjacency: AdjacencyType);

VAR i, j, ExposedEdges, VertexPgons: INTEGER;
        RotateCenter, RotateVertex: Transformation;

BEGIN
  DrawPgonPattern(InitialTran);

  {If there are more layers to be drawn, compute the
   transformations to appropriate positions in the next
   layer and call Replicate with those transformations}
  IF LayersToDo > 0 THEN
    BEGIN
      CASE Adjacency OF
         Edge: BEGIN
                 ExposedEdges := p - 3;
                 MatrixMult(RotateCenter, InitialTran, Rotate3P)
               END;
        Vertex: BEGIN
                 ExposedEdges := p - 2;
                 MatrixMult(RotateCenter, InitialTran, Rotate2P)
               END
        {Where Rotate2P = RotateP * RotateP
           and Rotate3P = Rotate2P * RotateP}
      END {CASE};

      FOR i := 1 TO ExposedEdges DO
        BEGIN
          MatrixMult(RotateVertex, RotateCenter, RotateQ);
          Replicate(RotateVertex, LayersToDo - 1, Edge);

          IF i < ExposedEdges THEN
            VertexPgons := q - 3
          ELSE {IF i = ExposedEdges THEN}
            VertexPgons := q - 4;

          FOR j := 1 TO VertexPgons DO
            BEGIN
              MatrixMult(RotateVertex, RotateVertex, RotateQ);
              Replicate(RotateVertex, LayersToDo - 1, Vertex)
            END {FOR j};

          MatrixMult(RotateCenter, RotateCenter, RotateP)
        END {FOR i}
    END {IF LayersToDo > 0}
END {Replicate}
```

Fig. 9. The recursive procedure *Replicate* which extends the pattern from one layer of *p*-gon patterns to the next.

The following Pascal code will generate the second and subsequent layers:

```
RotateCenter := Identity;
FOR i := 1 TO p DO
  BEGIN
    MatrixMult(RotateVertex, RotateCenter, RotateQ);
    Replicate(RotateVertex, nLayers − 2, Edge);
    FOR J := 1 TO q − 3 DO
      BEGIN
        MatrixMult(RotateVertex, RotateVertex, RotateQ);
        Replicate(RotateVertex, nLayers − 2, Vertex)
      END;
    MatrixMult(RotateCenter, RotateCenter, RotateP)
  END
```

There are several observations to be made about Replicate (Fig. 9) and the Pascal code above. First, MatrixMult(C, A, B) is just a procedure to perform the matrix multiplication $C \leftarrow AB$. Eventually we will represent the position of points by column vectors. Thus, $T\mathbf{X}$ will denote the transformed vector obtained by multiplying the original vector $\mathbf{X}$ by the matrix T.

Next, suppose that T is a transformation taking the original fundamental region to a new position. Then the rotations T RotateP T^{-1} and T RotateQ T^{-1} about the transformed points TP and TQ are the analogs of RotateP and RotateQ. So, if we apply the analog of RotateP to a transformed point $T\mathbf{X}$, we obtain

$$(T \text{ RotateP } T^{-1}) \, (T\mathbf{X}) = T \text{ RotateP } \mathbf{X},$$

i.e. we obtain the same effect by applying RotateP and T in reverse order to the original point (which may be assumed to be in the original motif). Of course, the same observation holds for RotateQ and its analog. These observations are used to simplify Replicate (and the Pascal code above) by avoiding the computation of conjugates of RotateP and RotateQ. Thus, we may think of RotateP and RotateQ as being their conjugated analogs if we read the effect of the matrix products from left to right (e.g. the effect of applying T RotateP to a point is the same as first applying T and then applying the analog of RotateP).

Finally, note that once the p-gon pattern for $[p, q]$ is formed (using reflections), the entire pattern can be built up from the p-gon pattern using RotateP and RotateQ alone.

In the replication algorithm it is convenient to use the *Weierstrass model* of hyperbolic geometry whose points lie on the upper sheet of the hyperboloid $x^2 + y^2 - z^2 = -1$ (and $z > 0$). The efficacy of this model arises from the fact that isometries can be represented by 3×3 Lorenz matrices (Lorenz matrices preserve the quadratic form $x^2 + y^2 - z^2$). Consequently, the composition of two symmetries is represented by the product of the matrices representing those two symmetries. The hyperbolic lines of this model are the branches of the hyperbolas formed by the intersections of the upper sheet of the hyperboloid with planes passing through the origin. For more on the Weierstrass model, see Faber[6, Chapter 7].

The Weierstrass model is related to the Poincaré circle model, embedded as the unit disc in the xy-plane, by stereographic projection toward the point $(0, 0, -1)^t$ (the "South Pole" of the hyperboloid). Specifically, the projection is given by:

$$\begin{bmatrix} x \\ y \\ z \end{bmatrix} \longrightarrow 1/(1 + z) \begin{bmatrix} x \\ y \\ 0 \end{bmatrix}.$$

The inverse projection is given by:

$$\begin{bmatrix} x \\ y \\ 0 \end{bmatrix} \longrightarrow 1/(1 - x^2 - y^2) \begin{bmatrix} 2x \\ 2y \\ 1 + x^2 + y^2 \end{bmatrix}.$$

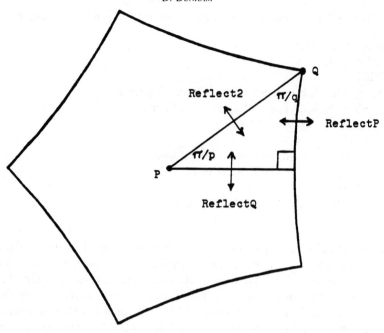

Fig. 10. A diagram showing a right triangle whose sides are the mirrors for the reflections ReflectP, ReflectQ
and Reflect2.

The transformations RotateP and RotateQ are defined relative to the centered tessellation
$\{p, q\}$ described in Sec. 5 and shown in Fig. 7(a). Let *ReflectP*, *ReflectQ* and *Reflect2* denote
the reflections across (1) the side of the central p-gon bisected by the positive x-axis, (2) the
x-axis, and (3) the circumradius PQ, respectively, as shown in Fig. 10. The matrices representing
ReflectP, ReflectQ, and Reflect2 are:

$$\begin{bmatrix} -\cosh(2b) & 0 & \sinh(2b) \\ 0 & 1 & 0 \\ -\sinh(2b) & 0 & \cosh(2b) \end{bmatrix}, \begin{bmatrix} 1 & 0 & 0 \\ 0 & -1 & 0 \\ 0 & 0 & 1 \end{bmatrix},$$

and

$$\begin{bmatrix} \cos(2\pi/p) & \sin(2\pi/p) & 0 \\ \sin(2\pi/p) & -\cos(2\pi/p) & 0 \\ 0 & 0 & 1 \end{bmatrix},$$

respectively, where

$$\cosh(b) = \cos(\pi/q)/\sin(\pi/p)$$
$$\cosh(2b) = 2\cosh^2(b) - 1$$
$$\sinh(2b) = \sqrt{\cosh^2(2b) - 1}.$$

The counterclockwise rotations, RotateP by $2\pi/p$ about P and RotateQ by $2\pi/q$ about Q are
represented by the matrix products Reflect2 ReflectQ and ReflectP Reflect2, respectively.

A convenient data structure to represent the points of the hyperboloid version of the p-gon
pattern is a record composed of a 3-vector and a pen attribute which is of the following type:

PenAttribute = (Up, Down, Red, Yellow, Blue)

Then the *p*-gon pattern may be represented by a record consisting of the number of points in the *p*-gon pattern and an array of points. Pascal type declarations for these data structures and the procedure DrawPgonPattern are shown in Fig. 11 (in which VectorMult($\mathbf{Y}$, M, $\mathbf{X}$) multiplies the vector $\mathbf{X}$ by the matrix M and places the result in $\mathbf{Y}$: $\mathbf{Y} \leftarrow M\mathbf{X}$). Note that PgonPattern is a global variable to DrawPgonPattern.

In summary, the pattern may be replicated by (1) using the inverse of the stereographic projection to map the *p*-gon pattern up onto the hyperboloid, (2) transforming the *p*-gon pattern to a new position by a Lorenz matrix representing a symmetry of the pattern, (3) stereographically projecting the transformed *p*-gon down onto the unit disc and (4) repeating steps (2) and (3) until the desired number of layers have been completed. Figure 12 shows the pattern of Fig. 5(b) on the hyperboloid.

7. COLOR SYMMETRY

Suppose that each of the motifs of an uncolored pattern receives one of *k* colors. That colored pattern is said to have *k-color symmetry*, or simply *color symmetry*, if each symmetry in the symmetry group of the uncolored pattern maps all motifs of one color to motifs of another (possibly the same) color, i.e. the symmetries of the uncolored pattern cause permutations of the colors of the colored pattern. See Loeb[7], Senechal[8] and [9], Shubnikov and Koptsik[10], and Wieting[11] for more on color symmetry.

```
TYPE PenAttribute = (Up, Down, Red, Yellow, Blue);

     PointType   = RECORD
                      X: ARRAY [1..3] OF REAL;
                      Pen: PenAttribute
                   END;


VAR PgonPattern: RECORD
                    Npoints: 1..MaxPoints;
                    Points: ARRAY [1..MaxPoints] OF PointType
                 END


PROCEDURE DrawPgonPattern(T: Transformation);

VAR     i: INTEGER;
    Xtrans: ARRAY [1..3] OF REAL;
      u, v: REAL;

BEGIN
  WITH PgonPattern DO
    FOR i := 1 TO nPoints DO
      WITH Points[i] DO
        BEGIN
          {Compute position of transformed point}
          VectorMult(Xtrans, T, X);

          {Project Xtrans to the xy-plane}
          u := Xtrans[1] / (1 + Xtrans[3]);
          v := Xtrans[2] / (1 + Xtrans[3]);

          {Take appropriate pen action}
          CASE Pen OF
               Up: MoveTo(u, v);
             Down: DrawTo(u, v);
              Red: Color := Red;
           Yellow: Color := Yellow;
             Blue: Color := Blue
          END {CASE}
        END {WITH Points[i]}
  END {DrawPgonPattern}
```

Fig. 11. Relevant type declarations and the procedure DrawPgonPattern which draws a transformed *p*-gon pattern, the transformation being a parameter of DrawPgonPattern.

Fig. 12. The pattern of Fig. 5(b) on the hyperboloid of the Weierstrass model of hyperbolic geometry.

Figure 1 and the checkerboard pattern are examples of 2-color symmetry. In Fig. 1, the colors black and white are interchanged by a quarter-turn about the meeting point of fin-tips or by a translation by one fish-length along a line of fish. However, reflection across a backbone line preserves the colors.

Figures 5(a) and 13 exhibit 3-color symmetry. The colors are represented by three different hatchings: dotted, dashed, and dot-dashed. In Fig. 5(a), rotations of $2\pi/3$ about meeting points of three crosses cyclically permute the three colors, whereas all reflections in the symmetry group preserve colors. On the other hand, some of the reflections in the symmetry group of Fig. 13 interchange two of the three colors.

In Fig. 14, we see an example of 4-color symmetry, the fourth color being represented by solid line hatching. Figures 13 and 14 are interesting in that they are examples of the only two kinds of hyperbolic color symmetry not having Euclidean or spherical analogs among patterns with 2-, 3-, and 4-color symmetry and (uncolored) symmetry groups $[p, q]$, $[p, q]^+$, and $[p^+, q]$.

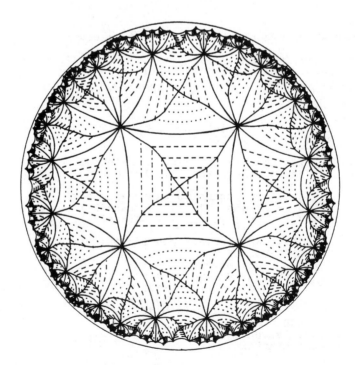

Fig. 13. A pattern with 3-color symmetry (the three colors are represented by dotted, dashed, and dot-dashed hatchings).

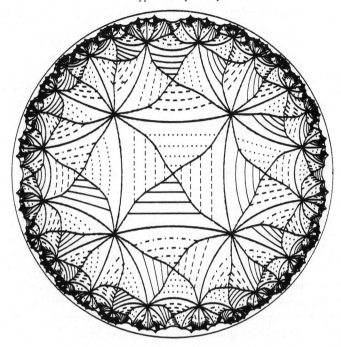

Fig. 14. A pattern with 4-color symmetry (the four colors are represented by solid, dotted, dashed, and dot-dashed hatchings).

The procedure DrawPgonPattern of Fig. 11 allows for the drawing of the original motif in three different colors, but all motifs will be colored exactly the same way—i.e. there will be no (non-trivial) color symmetry. To simplify the following discussion, we will assume that the original motif has been drawn in a single color from now on.

In order to produce a pattern with color symmetry, we must do the following: when a symmetry is applied to the original motif to bring it to a new position, the color permutation induced by that symmetry must be applied to the color of the original motif to obtain the color of the motif in the new position. We can achieve this by including the color permutation induced by a symmetry in the representation of that symmetry as a transformation.

Permutations are conveniently represented by arrays. For example, if Perm is the array representing the permutation Red → Blue, Yellow → Red, and Blue → Yellow, then Perm[Red] is Blue, Perm[Yellow] is Red, and Perm[Blue] is Yellow. Thus, we may represent symmetries as records containing a matrix part and a permutation part:

```
ColorType     = Red..Blue;
Transformation = RECORD
                    Matrix: ARRAY [1..3, 1..3] OF REAL;
                    Perm:   ARRAY [ColorType] OF ColorType
                 END
```

The only necessary modification to Replicate (Fig. 11) is to replace the calls to MatrixMult by calls to TranMult:

```
PROCEDURE TranMult(VAR C: Transformation; A, B: Transformation);
VAR Color: ColorType;
BEGIN
   MatrixMult(C.Matrix, A.Matrix, B.Matrix);
   FOR Color := Red TO Blue DO
      C.Perm[Color] := A.Perm[B.Perm[Color]]
END
```

In DrawPgonPattern, we must replace T by its matrix part in the call to VectorMult:

$$\text{VectorMult}(\textbf{Xtrans}, T.\text{Matrix}, \textbf{X})$$

and we must modify the color changes in the CASE statement to:

Red: Color := T.Perm[Red];
Yellow: Color := T.Perm[Yellow];
Blue: Color := T.Perm[Blue]

Also, when the p-gon pattern is being built up from the original motif, each copy of the motif must begin with a "point" indicating a pen action to select the appropriate (permuted) color.

It is easy to see how the above techniques could be extended to more than three colors. In the remainder of the discussion of color symmetry, we will use positive integers to represent colors (instead of their actual names) and the standard notation for permutations.

We need only specify color permutations for the generators of a symmetry group, since all other symmetries can be expressed in terms of the generators. However, the permutations must satisfy the same relations as the generators. For instance, the permutation part of a reflection must consist of disjoint transpositions: (1 2), (1 2)(3 4), (1 6)(2 3), etc. Similarly, the permutation part of a rotation must consist of disjoint cycles whose lengths divide the period of the rotation (the rotations of interest will all have finite periods). There may be additional restrictions imposed by other relations.

As an example, the generators for the group $[p^+, q]$ satisfy the relations:

$$(\text{RotateP})^p = \text{ReflectP}^2 =$$
$$(\text{RotateP}^{-1}\ \text{ReflectP}\ \text{RotateP}\ \text{ReflectP})^{q/2} = \text{Identity}.$$

Thus, if $p = 4$ and $q = 8$, the permutations (1 2 3 4) and (1 2)(3 4) may be used to represent the permutation parts of RotateP and RotateQ, respectively. However, if we replace (1 2)(3 4) by (1 2), the third relation above is no longer satisfied.

8. CONCLUSION

The hyperbolic pattern-creating program has been used to draw many patterns, including the ones in this paper. An earlier, more specialized version of the program[12] was used to generate all four of M. C. Escher's hyperbolic Circle Limit patterns. The present program has also been used in the classification of the types of 2-, 3-, and 4-color symmetry associated with the groups $[p, q]$, $[p, q]^+$ and $[p^+, q]$.

Possible directions of further research include extending the program to draw patterns whose symmetry groups are not subgroups of $[p, q]$ or are other subgroups of $[p, q]$ besides $[p, q]^+$ and $[p^+, q]$ (Escher's "Circle Limit I" and "Circle Limit III" fall into this latter category; see Coxeter[13]). Another direction would be the investigation of hyperbolic k-color symmetry for $k > 4$.

REFERENCES

1. J. Bourgoin, *Arabic Geometrical Pattern and Design*. Dover, New York (1973).
2. D. S. Dye, *Chinese Lattice Designs*. Dover, New York (1974).
3. H. S. M. Coxeter and W. O. J. Moser, *Generators and Relations for Discrete Groups*, 4th Edn. Springer-Verlag, New York (1980).
4. H. S. M. Coxeter, The groups determined by the relations $S^l = T^m = (S^{-1}T^{-1}ST)^p = 1$. *Duke Math. J.* **2**, 61–73 (1936).
5. A. Sinkov, The groups determined by the relations $S^l = T^m = (S^{-1}T^{-1}ST)^p = 1$. *Duke Math. J.* **2**, 74–83 (1936).
6. R. L. Faber, *Foundations of Euclidean and Non-Euclidean Geometry*. Marcel Dekker, New York (1983).
7. A. L. Loeb, *Color and Symmetry*. Wiley, New York (1971).
8. M. Senechal, Coloring symmetrical objects symmetrically. *Math. Mag.* **56**, 3–16 (1983).
9. M. Senechal, Color symmetry and colored polyhedra. *Acta Cryst.* A**39**, 505–511 (1983).

10. A. V. Shubnikov and V. A. Koptsik, *Symmetry in Science and Art*. Plenum Press, New York (1974).
11. T. W. Wieting, *The Mathematical Theory of Chromatic Plane Ornaments*. Marcel Dekker, New York (1982).
12. D. J. Dunham, J. E. Lindgren and D. Witte, Creating repeating hyperbolic patterns. *Computer Graphics* **15**(3), 215–223 (1981).
13. H. S. M. Coxeter, The non-Euclidean symmetry of Escher's picture "Circle Limit III." *Leonardo* **12**, 19–25 (1979).

Comp. & Maths. with Appls. Vol. 12B, Nos. 1/2, pp. 155–167, 1986
Printed in Great Britain.

0886–9561/86 $3.00 + .00
© 1986 Pergamon Press Ltd.

ASYMMETRY AND SYMMETRY IN CELLULAR ORGANIZATION

HEINZ HERRMANN

U–125, Molecular Genetics and Cell Biology, University of Connecticut, Storrs, CT 06268, U.S.A.

Abstract—It is proposed that the basic cell functions are associated with asymmetric structures. This is exemplified by three cell systems: The transduction of extracellular, molecular signals into intracellular responses by the cellsurface membrane; the two main steps in gene expression, the synthesis of messenger RNA and of cell proteins; the conversion of metabolic energy into mechanical work in muscle contraction. Asymmetry may be a requirement for the dynamics of functional activity. In some instances, e.g. muscle contraction, the asymmetry of individual functional units is compatible with an organization of multiple cooperative units into symmetric patterns.

Symmetry is one of the concepts that satisfy the search for a simplified representation of nature. During some periods, symmetry has been a principle not only of rational understanding but also of aesthetic beauty[2]. The value of the concept of symmetry for the analysis of inanimate systems is demonstrated in the articles in this volume.

Attempts to introduce symmetry into biology as an explanatory concept are not lacking. One of the outstanding examples is the treatment of symmetry by D'Arcy Thompson [13, pp. 357–443]. In Thompson's thought, symmetry indicates a thermodynamic state of equilibrium in living systems. In this sense, the concept of symmetry may be useful in the description of the end products of biological processes such as protein molecules, the organization of structural filaments, pigment patterns, or the forms of whole organs and organisms[9].

As a thesis that we will develop here: it is proposed that the processes that maintain the structure and function of cells, the basic units of living matter, depend on an asymmetric organization of cellular macromolecules. This is suggested by the following description of, (1) cell surface membranes, (2) the main phases of gene expression and (3) the biological conversion of metabolic energy into mechanical work.

Before discussing these separate cellular systems, it may be appropriate to survey briefly the overall organization of a generalized cell type as indicated in Fig. 1. All cells are enveloped by a very thin cell surface membrane that is described in more detail in the next section of this paper. The cell surface of some cells shows multiple protrusions, microvilli, for more efficient uptake of nutrients. Various forms of cell surface invaginations indicate processes for internalization (endocytosis, pinocytosis, phagocytosis) of surface bound molecules or engulfed liquid or solid material. Much internalized material is degraded in special organelles, the lysosomes. The degradation products can be used for resynthesis of cell specific macromolecules or for metabolic energy production.

One type of the internal membranes, the endoplasmic reticulum, translocates newly synthesized proteins to another membrane compartment, the Golgi complex. From the Golgi complex, the proteins are carried to the cell surface and become integral membrane constituents interacting with hormone type substances and with cellular and non-cellular surfaces.

Alternatively, the proteins are secreted into the extracellular space including the circulation and the intestines of animals and function there as hormones and digestive enzymes. In both membrane compartments some proteins are modified by addition of carbohydrate groups that are determinants of protein specificity, e.g., in the establishment of specific blood groups.

Other membranous organelles are the main energy producing units of the cell. Particles known as mitochondria produce from metabolic sources the reactive adenosine triphosphate molecule that is a general energy donor for cellular function. In plant cells, chloroplast particles use light for conversion of carbon dioxide and water into starch that is used as a source of metabolic energy. A network of protein filaments (not indicated in Fig. 1) is associated with the cell surface membrane and internal membrane organelles aiding the translocation of the latter.

The genetic information system is located in the cell nucleus. In bacteria and related cell forms, known as prokaryotes, the genetic material is not segregated from the rest of the cell.

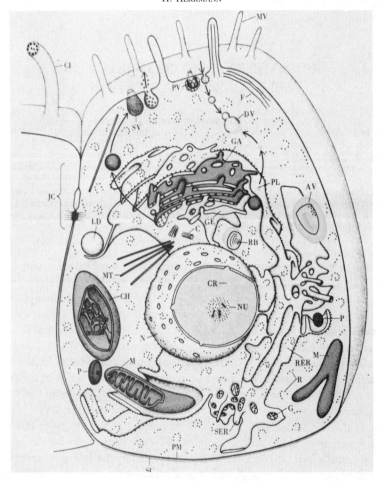

Fig. 1. A three dimensional schematic representation of the distribution of the main components of a cell of higher animals. A network of filaments, the cytoskeleton, found throughout the cytoplasm and in particular near the surface membrane and the nuclear envelope are omitted in the drawing. *AV*, autophagic vacuole; *C*, centriole; *CH*, chloroplast; *CI*, cilium; *CR*, chromatin; *DV*, digestion vacuole; *F*, filaments; *G*, glycogen; *GA*, Golgi apparatus; *GE*, GERL; *JC*, junctional complex; *LD*, lipid droplet; *M*, mitochondrion; *MT*, microtubules; *MV*, microvillus; *N*, nucleus; *NU*, nucleolus; *P*, peroxisome; *PL*, primary lysosome; *PM*, plasma membrane; *PV*, pinocytic vesicle; *R*, ribosomes and polysomes; *RB*, residual body; *RER*, rough endoplasmic reticulum; *SC*, extracellular coat (as drawn, ‘‘basal lamina’’); *SER*, smooth endoplasmic reticulum; *SC*, extracellular coat (as drawn, ‘‘basal lamina’’); *SER*, smooth endoplasmic reticulum; *SV*, secretion vacuole. The organelles have been drawn only roughly to scale. Also, the sizes and relative amounts of different organelles can vary considerably from one cell type to another. For example, only plant cells show chloroplasts. (From *Cells and Organelles*, Third Edition by Eric Holtzman and Alex Novikoff. Copyright © 1984 by CBS College Publishing. Reprinted by permission of CBS College Publishing.)

In other cells, those of unicellular protists and of plants and animals, the genetic material is separated from the rest of the cytoplasm by a membrane, the nuclear envelope.

The genetic information is stored in the double helical strands of DNA. By association with proteins, the DNA forms more complex structures, the chromosomes, and is supported by a network of protein filaments, the nuclear matrix.

The distribution of the cell structures represented in Fig. 1 changes continuously in response to the stimuli that reach the cell and require functional adaptation. Thus, the organization of a cell is in a highly dynamic state. Fixation of the cell structure shows no indication of a distinctive symmetry in the distribution of the cellular elements. However, most of the cell components interact with each other in a highly specific manner. Therefore, it is not appropriate to define cell organization as a form of random distribution. We will now pursue the topic of cellular asymmetry by the description of the three particular cell systems mentioned earlier.

1. ASYMMETRY OF THE CELL SURFACE ORGANIZATION

A cell of average size (20 microm. dia.) is separated from its surroundings by a membrane of less than 10 nm, that is 1/2000 of the cell diameter. If an imaginary cell were the size of a classroom 20 m long, the cell surface membrane would be only 1 cm thick. The membrane must have sufficient cohesive strength to maintain its continuity as a surface cover and boundary. At the same time, the membrane fulfills the role of mediator between the extracellular and intracellular compartments and it is perhaps the most multifunctional cell organelle.

A sequestration of small volumes from the bulk of the primeval ocean was probably one of the main conditions for the evolution of life. The confinement of macromolecules in a limited space must have facilitated the interactions of serial enzymatic processes for production of metabolic energy (ATP) and the use of some primitive genetic information system for the synthesis of macromolecules and their assembly into cellular structures. A cell surface membrane averts the influx of harmful material into the cell, prevents the loss of macromolecules from the cell and maintains favorable concentrations of ions and essential nutrients. It becomes the site of receptors for extracellular agents such as hormones, neurotransmitters, immune proteins and for the conversion of molecular signals conveyed by these agents into activation of intra-cellular targets. Finally, the surface membrane mediates the adhesion of many unicellular organisms to solid substrates. The surface membranes of the same or of different types of cells interact with each other in forming tissues and organs in the course of embryonic development and in rejecting immunologically incompatible material.

The results of numerous and diverse investigations indicated that the cell surface membrane consists of a lipid and a protein component and several models were suggested to account for the experimental findings. Eventually a model was proposed that accomodated theoretical con-siderations, analytical data and electronmicroscopi observations. This model became known as the "Fluid Mosaic Model of the Structure of Membranes" (Fig. 2)[10]. The basic theoretical requirement for the macromolecular organization in this model is the state of minimal free energy. The main structural component is a lipid double layer in which the polar hydrophilic groups of the lipid molecules are oriented toward the outside and the apolar hydrophobic groups are turned to the inside of the layer. A reduction of free energy is achieved in three phases of lipid assembly. The first reduction of free energy, and hence, the driving force in the formation of the lipid double layer is the exclusion of the hydrophobic portion (fatty acid chains) from the aqueous phase by the strong interaction of water molecules. Secondly, the weak interactions between the hydrophobic non-polar groups lead to the closer or looser alignment of these extended portions of the lipid molecules, depending upon the degree of saturation or unsaturation of fatty acid chains. As a third component in the reduction of free energy, the polar heads of the lipid molecules interact with the water phase and contribute to the orientation of the lipids in the formation of the bilayer.

The fluid state of the lipid layer has been inferred in studies which have determined the transition temperature of the lipid phase from the paracrystalline non-fluid condition to a viscous state. Temperatures which are compatible with the life of the cell are above these transition temperatures; therefore, the lipids can be expected to be at least in part in a fluid state. Direct evidence for the fluid state of the membrane lipids is mainly derived from direct measurements of the lateral translocation of the membrane macromolecules. Rotatory and oscillatory motion of membrane molecules is determined by interpretations of nuclear magnetic resonance spectra, electron spin resonance spectra or fluorescence polarization.

Although most schematic drawings of surface membrane models give the appearance of a symmetric distribution of lipid molecules, an actual enzymatic analysis shows an asymmetric distribution of the different lipid types in the outer and inner leaflet of the membranes[15]. For example, in the erythrocyte membrane, the outer layer is greatly enriched in phosphatidylcholine and sphingomyelin and contains only a small amount of phosphatidylethanolamine. In contrast, phophatidylethanolamine and phosphatidylserine are the prevalent lipids in the inner layer. This shows a far reaching, though not absolute asymmetry in the distribution of the different classes of lipids in the membrane. Also, we have to consider that the lipid distribution is the result of a dynamic state. The lipid molecules move within the plane of their respective leaflet and there is some exchange, greatly enhanced by certain membrane proteins, of lipid molecules between leaflets. Parts of the membrane, including the lipids, are internalized. This shows that the asymmetry is not the result of a static equilibrium but of a dynamic state. Possible shifts in the proportions of the lipid classes in the two lipid layers may be related to changes in cell shape.

According to the Fluid Mosaic Membrane model, the major portion of membrane proteins are intercalated into the lipid layer. By definition these are proteins which are removable from

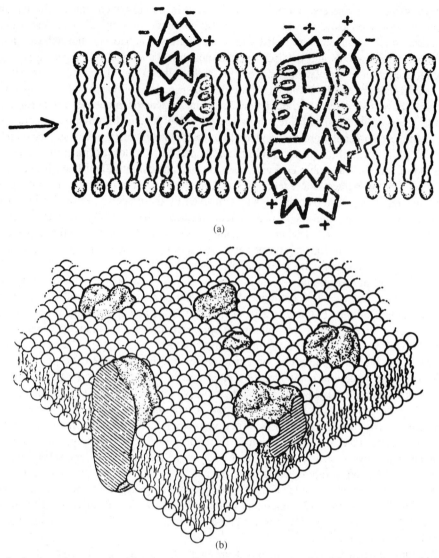

(a)

(b)

Fig. 2. A schematic crossection (a) and a three dimensional representation (b) of the molecular organization of a cell surface membrane according to the Fluid Mosaic Model. The drawings show the phospholipid molecules with their polar, hydrophilic heads oriented outward and the hydrophobic apolar tails oriented towards the interior of the membrane. The two layers are asymmetrical because they contain widely differing quantities of the main classes of phospholipids. Also, the drawings indicate the asymmetric disposition of the protein molecules in the membrane. Some proteins occupy only one half of the membrane. Of these proteins each class occupies only the outer or the inner half of the membrane. Therefore, the two halves of the membrane have qualitatively different protein populations. Another group of proteins, the transmembrane proteins, spans the entire width of the membrane. These proteins are asymmetrical because their polar groups that protrude on the inside or the outside of the membrane differ in amino acid composition and in the three dimensional conformation and functional role. (Reproduced from *Science* with permission from Professor S. J. Singer and the American Association for the Advancement of Science.)

membrane preparations only by drastic procedures such as treatment of the membrane with detergents. In attempting an isolation of such proteins, it is difficult to remove associated lipids. Subsequent analyses of the amino acid sequences in these proteins shows clusters of apolar amino acids which form hydrophobic regions in the peptide chain and interact with the apolar portion of the lipid molecules. The proteins which are so firmly embedded into the membrane have been designated as integral proteins. The integral proteins can be intercalated in the outer or the inner lipid layer with the polar groups exposed either on the outside or the cytoplasmic side of the lipid layers, respectively. The classes of integral proteins are asymmetrically distributed and each class is found only in the external or the internal portion of the cell surface membrane[11]. Other integral proteins span the entire width of the double lipid layer. The polar groups of this class of proteins are exposed on both sides of the membrane. Many external integral proteins carry chains of carbohydrates (e.g. acetylglucosamine, galactosamine, fucose, sialic acid) which contribute to the specificity of cell surface receptors in the binding of molecules from the cell environment.

In addition to the asymmetric distribution of cell surface constituents on the inside and outside of the membrane, most membrane macromolecules are highly asymmetric structures. The lipid molecules form, with the apolar part of their structures, the hydrophobic membrane layer. With their other polar portion, these molecules extend into the hydrophilic layers of the membrane. The same type of asymmetry holds for most proteins that are firmly embedded into the lipid layer of the membrane and contribute to the organization of the membrane layers. As pointed out earlier, these proteins function as intracellular enzymes or mediate specific interactions with extracellular proteins such as immunologically active substances, or induce aggregation with other cells.

Equally distinctive asymmetries are observed in transmembrane proteins that have two polar hydrophilic portions extending into both the outer and inner hydrophilic membrane layers. These transmembrane proteins have two important functions. They form structures that transfer, through the lipid layer, ions and other small hydrophilic molecules such as sugars and amino acids and they bind hormone-type substances. The binding reactions induce a conformational change in the transmembrane molecules that are directly or indirectly (e.g., by activation of phosphorylations) transmitted to the cell interior and elicit there a wide variety of responses. In all these transmembrane proteins, the external binding sites that react with ions or hormones differ from the internal binding sites with which they interact with other cell constituents. These differences in the functional role of the external and internal portions of the transmembrane molecules indicate the marked asymmetry in their structure. This is illustrated here for two transmembrane receptors that have been extensively analyzed (Fig. 3(a–b))[3,12].

As a last but important example, we should realize that the system of enzyme proteins that convert metabolic or light energy into adenosine triphosphate (ATP), the principal energy donor for cellular function, are asymmetrically located on either the outside or the inside of certain cellular membranes and includes highly asymmetric transmembrane proteins (ATPases).

2. ASYMMETRIES IN THE GENETIC INFORMATION SYSTEM

The genetic information that specifies the cellular properties is contained in the well known DNA double helix. This is a structure in which two long DNA molecules (up to a few cm/single DNA strand) are wound around each other. Each of the two molecules consists of purine and pyrimidine bases, each covalently linked to deoxyribose and a phosphate group. These base-sugar-phosphate complexes are the nucleotides. In one DNA molecule many nucleotide molecules are tied to one another by ester links between the phosphate group of one nucleotide and the deoxyribose of the next nucleotide. The two long strands of nucleotides are held together by hydrogen bonds between the purine and pyrimidine bases of two separate DNA strands. Different combinations of triplets of successive nucleotides within one of the two DNA molecules specify the use of a particular amino acid in the synthesis of proteins. However, this information is not directly used. The sequence of nucleotides encoding a certain protein in the DNA functions initially as a template for the enzymatic synthesis of a ribose nucleic acid molecule (RNA) with a nucleotide composition that is complementary to that of the DNA sequence. The RNA is modified in the cell nucleus and is eventually transferred to the cytoplasm where it becomes the template for alignment of amino acids in the synthesis of specific protein molecules. This latter process, known as translation, is described in more detail in one of the following sections (Asymmetry in the translational phase of gene expression).

Asymmetry of the linear nucleotide alignment

In the genetic information system asymmetries arise on several levels of organization. First, the genetic information is asymmetrically distributed within a single DNA strand. We can think of a long thread or ribbon with single lines of words from a novel printed on it in linear succession. Obviously, there is no symmetry in such printed material itself. In addition, regions expressing genetic information (exons) alternate asymmetrically with regions that are not expressed (introns) and may have regulatory functions. It has to be mentioned at this point that the distribution of genetic information is not necessarily throughout as heterogeneous as the

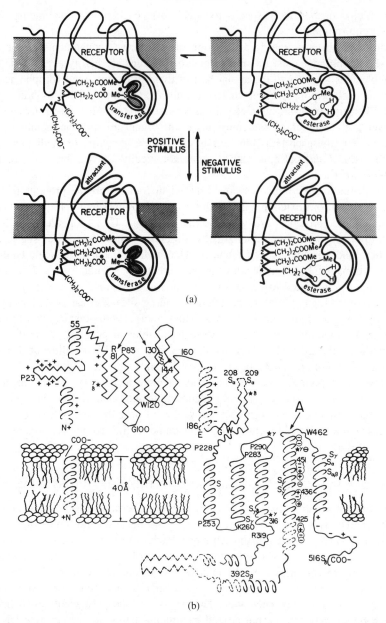

Fig. 3. The two dimensional appearance and the approximate location of two protein receptor molecules in the surface membrane of a bacterial (a) and an animal (b) cell. The bacterial receptor binds chemotactic attractants and repellents at its external portion. This interaction induces a conformational change of the molecule that is transmitted to the interior portion of the molecule and initiates there a methylation of the inner portion of the molecule. This modification triggers, in turn, a mechanism that controls the flagellar beat and with it the swimming response of the cell to the initial stimulus. The animal receptor molecule is a subunit of a complex that binds acetylcholine. This substance is release at the terminals of stimulated neurons. The binding to the receptor induces again a conformational change that is transmitted to another subunit of the complex and increases the ion flux through the membrane. The structural and functional differences in the outer and inner portions of the two transmembrane receptor molecules gives them a highly asymmetric character. (Fig. 3(a) reproduced from *J. Biol. Chem.* with permission from Professor D. E. Koshland and the Am. Soc. Biol. Chem., Fig. 3(b) reproduced from *Proc. Nat. Acad. Sci.* with permission from Professors J. Finer-Moore and R. M. Stroud and the Nat. Acad. of Sci.)

print in a book. A portion of the genetic information system consists of multiple repeats of the same nucleotide sequences (Repetitive DNA). In this case, certain symmetries exist between identical portions of the DNA molecule.

An additional source of asymmetry is the association, at one end of each of the DNA segment that encode one protein type, with a nucleotide sequence that is the signal for the beginning of the enzymatic process that produces an RNA molecule as the template used in the synthesis or proteins in the cytoplasm. At its other end, this DNA segment has a different nucleotide sequence that signals the termination of this template production. The genetic information for the production of a single protein is thus asymmetric not only by virtue of its heterogeneous content but also because each coding DNA has different regulatory nucleotide sequences at the initiating and terminating sites in the transcriptive phase of gene expression[6,8].

Asymmetries at the level of chromosomal organization

In the cell, the DNA is combined with several types of proteins, primarily histones. This organization is known as a chromosome. The cells of eukaryotic multicellular organisms can have from two to about eighty such chromosomes that are separated from the rest of the cytoplasm by a nuclear envelope. Within this envelope, the chromosomes are folded in highly irregular fashion without indication of symmetry[1,5]. The entire set of chromosomes is known as chromatin.

The replication process itself is asymmetric with the two DNA strands in one double helix being processed in different ways. One of the strands is replicated by an enzyme complex in a continuous fashion; the other DNA strand is replicated by another set of enzymes in a discontinuous fashion.

The replicated chromosomes are separated from each other in a process known as mitosis. As a first step in this process, the extended chromatin threads of the non-dividing cell are transformed into compact metaphase chromosomes that are readily visible in the light microscope. During this period there exists symmetry of the two sets of chromosomes. However, frequently the result of such a replication is again asymmetric. One of the two daughter cells may continue to replicate. The other daughter cell leaves the division cycle and transforms into a non-dividing specialized cell type.

Asymmetry in the translational phase of gene expression

The last phase in the conversion of genetic information into cell constituents occurs in the cytoplasm of the cell. The messenger RNA that contains the code of nucleotide triplets for the alignment of amino acids attaches to specific sites on particles, that are the structural center for the covalent linking of amino acids into proteins. These particles are known as ribosomes.

In an initial phase of ribosome analysis, the prokaryotic and eukaryotic ribosomes were defined as particles sedimenting at 70–80S respectively, each consisting of two subunits. The protein components of the subunits were identified and amino acid sequences were established. The ribosomal RNAs are among the most intensively studied cell components. Not only sequence and structural characteristics but transcription and processing analyses provide a remarkably complete picture of these molecules.

The difference in ribosomal composition is often regarded as one of the characteristics that separate prokaryotes and eukaryotes. However, if we consider the complexity of the ribosomal particle and the numerous steps in the translational process that depend on the particular structural properties of the ribosome, the similarities between the translational processes in the two types of organisms seem great and the dissimilarities less significant. Therefore, our account does not give a separate treatment of prokaryotic and eukaryotic translation.

The surface of these particles has sites for attachment of messenger RNA and of transfer RNA. There is a specific transfer RNA for each amino acid. The amino acids are translocated from the extracellular space to the cytoplasmic side of the cell surface by special protein carriers. An enzyme complex on the inside of the cell surface membrane covalently links the amino acids to transfer RNA. In the presence of additional protein factors and enzymes, one portion of the transfer RNA-amino acid complex attaches to a special site on the ribosome. Another portion of the complex attaches to the messenger RNA. The nucleotide triplet in the messenger RNA specifies the transfer RNA and at the same time the amino acid that is to be inserted at this point into the nascent protein molecule. In the jargon of the molecular biologist, we are speaking of the triplet on the transfer RNA as the anticodon.

Each ribosome has two sites for the binding of transfer RNA-amino acids. After the second complex has been bound, the amino acid at the first site is removed from its RNA carrier and is linked to the amino acid in the second site. The amino acid free transfer RNA at the first site is released and the complex at the second site is moved to the first site. This entire process is repeated, linking again and again single amino acids to the growing amino acid chain that eventually completes a protein molecule.

We are dealing here with a complex process that effects the actual synthesis of the protein molecule with its specific properties from a great number of small building blocks, the amino acids. For us it is important to see whether this process requires a symmetric or asymmetric organization of the ribosomal particle as the main mediator of this process. In describing here the ribosomal particle we will follow recent reviews of the overall structures of the molecular organization of the ribosome[7,14].

Each ribosomal particle consists of proteins and RNAs. The RNAs in the ribosomes differ from messenger and transfer RNAs. They are practically identical in all higher organisms and diverge only slightly from the corresponding molecules in the ribosomes of bacteria. Such a constancy of composition usually indicates that the respective molecules have a specific and important role in evolution that does not permit divergence from the established molecular specification. For the same reason, the ribosome proteins are also very similar.

Electronmicroscopic observation presents a direct representation of the ribosome structure, although preparative procedures can introduce considerable structural distortions. Despite such shortcomings, electronmicroscopy reveals certain basic features of the ribosome particle. It consists of two subunits. The smaller subunit is described as a prolate asymmetric particle with approximate dimensions of 23×11 nm. The larger part of this subunit is a flattened oval body that is connected by a narrowing portion to a more globular head.

The larger subunit is a hemispherical body of about 23 nm diameter with a flattened bottom. Its circumference extends into three processes from a hollowed plane surface. The central process (the "nose") is flanked by a 8–12 nm rod-shaped structure ("stalk") and a more rounded ("shoulder") process. The spatial relationship of the two ribosomal subunits has not been definitely determined but in the majority of ribosome models the head of the small subunit is located between the shoulder and the nose process of the larger subunit (Fig. 4)[14].

In the two subunits the RNA molecules assume a more central and the proteins a more peripheral distribution. The RNA in the smaller subunit has a V-shaped conformation. Immunelectronmicroscopy using antibodies against a RNA with methylated adenosines shows the distribution of the two RNAs in the subunits. Crosslinking of the different portions of the RNAs and between RNAs and proteins has contributed to the mapping of the spatial organization of each RNA within the subunits, also indicated in Fig. 4[7].

Information about the actual distribution of ribosomal proteins is obtained after electronmicroscopically discernible antibodies to the respective proteins are bound to the ribosomal particles.

Reassociation tests with the components of the small subunit revealed that one set of proteins binds directly to RNA. A second set of proteins binds to the particles after the initial RNA protein core has been formed. Binding of a third set of proteins depends on the presence of the second set of proteins. The reassociation of all components produces a particle that is functionally fully active.

From this description, we can see that the structural organization of the last phase in the expression of the genetic information system involves again a highly asymmetric structural organization. Neither the shape of the ribosome itself nor the distribution of the protein and RNA components within the ribosomal particles show detectable symmetry.

3. THE ASYMMETRY OF CELLULAR MECHANICAL WORK

As a last example of asymmetric organization, we will consider the conversion of metabolic energy into mechanical work by the functional unit in muscle contraction. Skeletal muscle very clearly shows several levels of organization. The molecular components of its contractile system are the proteins myosin, actin, tropomyosin and troponin. The myosin molecules form thick filaments and the actin complex forms thin filaments that are arranged in the muscle fiber with extraordinary regularity.

The basic functional unit in this system is asymmetric and consists of two components, the myosin molecule and the actin-tropomyosin-troponin complex, respectively. In the presence of Ca^{++} ions and supply of metabolic energy in the form of reactive adenosinetriphosphate molecule, the actin binds to one site on the globular head of the myosin molecule. A conformational

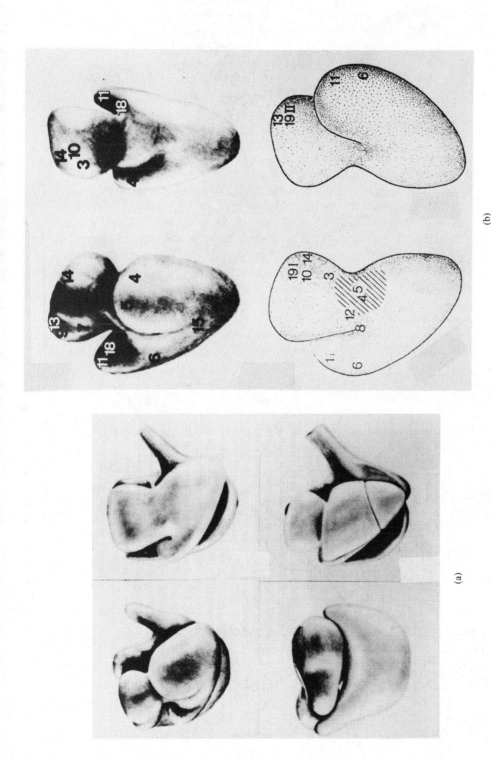

Fig. 4. Different types of asymmetries in the ribosomal particles of the protein synthesizing system. (a) different representations of the alignment of the small and large subunits of the ribosome. (b) the conformation of the small ribosomal subunit and the distribution on its surface of immune reactive groups of the proteins that constitute the small subunit. (c) a more comprehensive view of the distribution of protein groups within the small ribosomal particles. (d) the distribution of functional sites on the surface of the subunits. The sites indicate the areas for binding on the subunit surface of factors that activate or inhibit protein synthesis. (e) the conformation of the large ribosomal subunit and the distribution on its surface of the immune reactive groups of the proteins that constitute the large subunit. (f) a schematic representation of the conformation of the 16S rRNA molecules that are the backbone of the small ribosomal subunit. (Figs. 4 reproduced from *Annu. Rev. Biochem.* with permission from Annual Reviews, Inc., Figs. 4(a–e) with permission from Professor H. G. Whittmann, Fig. 4(f) with permission from Professor H. F. Noller.)

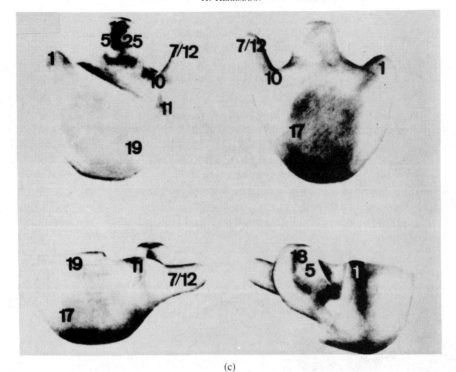

(c)

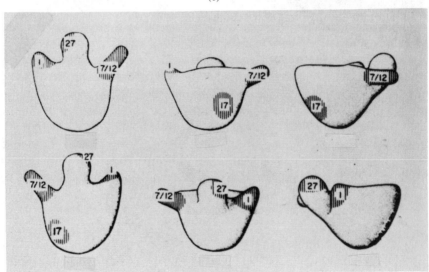

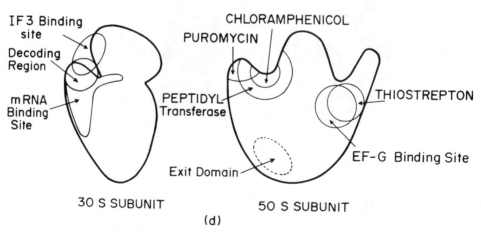

(d)

Fig. 4 (Continued).

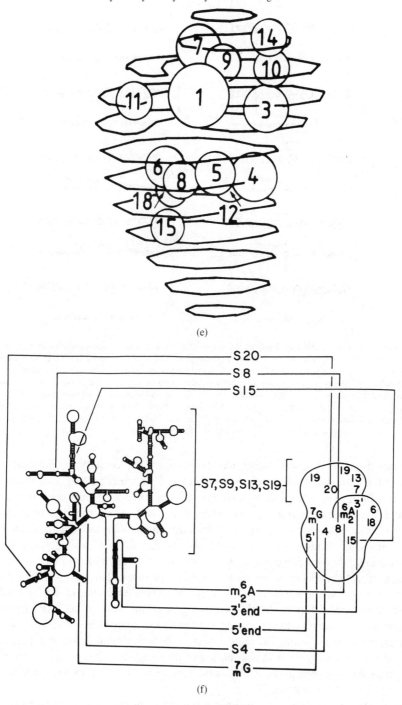

(e)

(f)

Fig. 4 (Continued).

change, presumably affecting the position of the myosin head, produces a translocation of the actin complex (Fig. 5)[4]. Both the myosin molecules and the actin complex are organized into filaments in parallel arrays so that the translocation is repeated many times. By this amplification of the molecular translocation, the actin filaments are pulled into the array of spaces between the myosin filaments and the entire muscle shortens on a macroscopic scale.

We can see in this example that the basic functional unit is again essentially asymmetrical. When such units are assembled into higher levels of organization, structures can arise of extraordinary regularity with several forms of symmetry. This is quite evident in longitudinal and transversal views of muscle structure.

This discussion of asymmetries in cellular organization does not imply that symmetry has no place in the cellular or organismic forms of life. We can hardly imagine symmetries more

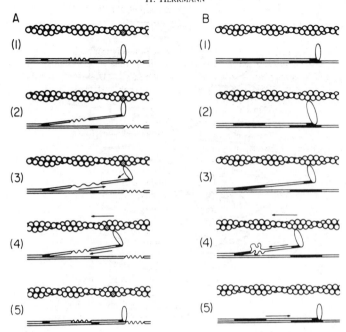

Fig. 5. Two models of the functional unit of muscle contraction showing the actin complex and the functional portion of the myosin molecule in the course of contraction. The models demonstrate that the change in the conformation of the myosin molecule leads to the translocation of the actin complex. The actin complex is represented by the double helix of polymerized single globular actin molecules and does not show the tropomyosin and troponin components of the actin complex. The single actin-myosin unit is asymmetrical. The alignment of multiple asymmetric units produces a highly regular repeat pattern of muscle structures on several dimensions that show many forms of symmetry. (Reproduced from *Cell and Muscle Motility*. Vol. 2, 1982 with permission from Professors S. C. Harvey, H. C. Cheung and Plenum Press.)

exquisite than those of the silicate skeleton of a diatom, the paracrystalline regularities of muscle filaments or the components of flagellae of ciliates and the many patterns and shapes in plants and animals. Both symmetry and asymmetry have their place in the organization of living matter. At any level of organization, functioning living systems are found sometimes in a state of symmetry and at other times they are in a state of asymmetry. The trotting horse, the individual smoking a pipe and reading the newspaper, the dancer, do not exhibit symmetry in their momentary poses. On the other hand, flying birds will exhibit symmetry most of the time. We are thinking of vertebrate organisms as symmetric structures. This symmetry is limited and refers only to the gross surface structure. It is general knowledge that the left and right halves of the vertebrate body differ in detail, as in the case of the human face. The interior of the vertebrate body has many asymmetric aspects such as the distribution of the digestive system and the liver, the heart and aorta. The brain appears to be highly symmetrical but it is functionally highly asymmetric. The symmetric distribution of the eyes of a larva of the flounder move to highly asymmetric positions in the adult animal that spends most of its existence lying on the mud of the bottom of the sea bed.

In living systems neither symmetry nor asymmetry are determined by some fundamental logical requirement. Both forms of organization are the result of evolutionary or developmental adaptations. In our world as a whole, symmetries may more readily attract attention and represent states of quasi equilibria that are amenable to theoretical treatment. They are in this sense useful artifacts of the human mind imposed on some forms of nature.

In conclusion, we propose as a rational that the asymmetric form of cellular organization is related to transformations of energy. Therefore, asymmetry may be associated with dynamic states of cell systems. Symmetry is perhaps primarily expressed in the more static end products of the energy utilizing process. Presumably, a cell with a perfectly symmetric structure would be dead and asymmetry is a requirement for the maintenance of the living state.

Parts of this paper are taken from a manuscript to be published at a future date by Harper & Row Publishers, Inc., N.Y. under the title, ''An Approach to Cell Biology''.

Acknowledgements—I am indebted to Dr. Judith Kelly for her valuable suggestions in preparing the manuscript.

REFERENCES

1. D. A. Agard and J. W. Sedat, Three-dimensional architecture of a polytene nucleus. *Nature* **302**, 676–681 (1983).
2. I. Bernal, W. C. Hamilton and J. S. Ricci, *Symmetry*. W. H. Freeman and Company, San Francisco (1972).
3. J. Finer-Moore and R. M. Stroud, Amphipathic analysis and possible formation of the ion channel in an acetylcholine receptor. *PNAS* **81**, 155–159 (1984).
4. C. H. Harvey and H. C. Cheung, *Myosin Flexibility. Cell and Muscle Motility*, Vol. 2, pp. 279–302 (Edited by R. M. Dowben and J. W. Shay). Plenum Press, New York (1982).
5. D. Mathog, M. Hochstrasser, Y. Gruenbaum, H. Saumweber and J. Sedat, Characteristic folding pattern of polytene chromosomes in drosophila salivary gland nuclei. *Nature* **308**, 414–421 (1984).
6. J. R. Nevins. The pathway of eukaryotic mRNA formation, *Annu. Rev. Biochem.* **52**, 441–466 (1983).
7. H. F. Noller. Structure of ribosomal RNA, *Annu. Rev. Biochem.* **53**, 119–162 (1984).
8. D. Pribnow. *Genetic Control Signals in DNA. Biological Regulation and Development*, Vol. 1: *Gene Expression*, pp. 219–277 (Edited by R. F. Goldberg). Plenum Press, New York (1979).
9. J. S. Richardson, The anatomy and taxonomy of protein structure, *Advances in Protein Chem.* **34**, 167–330 (1981).
10. S. J. Singer and G. L. Nicolson, The fluid mosaic model of the structure of cell membranes. *Science* **175**, 720–731 (1972).
11. T. L. Steck, The organization of proteins in the human red blood cell membrane. *J. Cell Biol.* **62**, 1–19 (1974).
12. J. B. Stock and D. E. Koshland, Jr, Changing reactivity of receptor carboxyl groups during bacterial sensing. *J. Biol. Chem.* **256**, 10826–10833 (1981).
13. D. W. Thompson, *On Growth and Form*. Cambridge University Press, London (1942).
14. H. G. Wittmann, Architecture of prokaryotic ribosomes. *Annu. Rev. Biochem.* **52**, 35–65 (1983).
15. R. F. A. Zwaal and B. Roelofsen, *Applications of Pure Phospholipases in Membrane Studies. Biochemical Analysis of Membranes*. pp. 352–377 (Edited by A. H. Maddy). John Wiley & Sons, Inc. New York. (1976).

Comp. & Maths. with Appls. Vol. 12B, Nos. 1/2, pp. 169–183, 1986
Printed in Great Britain.

0886-9561/86 $3.00 + .00
© 1986 Pergamon Press Ltd.

SYMMETRY AND SPONTANEOUSLY BROKEN SYMMETRY IN THE PHYSICS OF ELEMENTARY PARTICLES

PHILIP D. MANNHEIM

Department of Physics, University of Connecticut, Storrs, CT 06268, U.S.A.

Abstract—In this quasi-monographical article we review the role and the evolution in that role that the symmetry principle has played and is continuing to play in the development of the theory of the physics of elementary particles. We review first its use as an invariance principle which provides a group-theoretical classification for particles and for forces. In this classical, historical realm of the use of symmetry, symmetry acts only as a kinematic constraint which restricts and categorizes possible structures for the theory. With the development of local gauge theories it was found that rather than merely be a constraint on the dynamics the symmetry of local gauge invariance could also provide an origin for the existence of the forces in the first place, with the forces being described by the mediation of a set of associated vector gauge bosons. Further, following developments in the theory of phase transitions, it was found that these local gauge theories could then themselves be spontaneously broken by the dynamical effects of the forces to produce an asymmetric observable world in a theory with completely symmetric forces. We review all these developments emphasizing in particular the close analogy with many-body theory and solid state physics, and present some recent applications of these ideas, specifically the unified theory of electromagnetic and weak interactions due to Weinberg, Salam, and Glashow, and the grand-unified theories of all the fundamental forces which are currently being considered in the literature.

1. INTRODUCTION

The symmetry principle has proven to be one of the most powerful and durable principles in elementary particle physics, so that today it is one of its most fundamental cornerstones. And yet, perhaps even more remarkable than the degree to which symmetry has shown itself to be relevant has been the changing evolution in the role that symmetry has played in the development of elementary particle theory. Historically, symmetry first developed in elementary particle physics as it did in other areas of physics, namely it extended the general geometrical invariance concept of classical mechanics to invariances in quantum physics, so that applications could then be made to systems which were either exactly or approximately invariant under some symmetry transformation. In such cases the symmetry of the solutions to the equations of motion would be the same (i.e. either exact or approximate) as that of the equations themselves. Following fundamental developments made in the fifties in the many-body problem aspects of phase transitions and cooperative phenomena in solid state physics, it was gradually realized that in the presence of such collective phenomena it was possible for a situation to occur in which the solutions to the equations of motion could have less symmetry than the equations themselves. Such a situation is known as spontaneously broken symmetry wherein an exact invariance of a system of equations of motion is broken by its solutions. (The terminology though now standard is a little confusing since the symmetry is not broken at all at the level of the equations of motion. Nonetheless, we shall use it here while stressing that it is important to distinguish spontaneously broken symmetry from approximate symmetry, since an approximate symmetry in the sense introduced earlier is one which is in fact broken (approximately) in the equations of motion). The great virtue of spontaneously broken symmetry is that it allows an observable asymmetric world (the solutions to the equations of motion) to have associated with it a completely symmetric underpinning with completely symmetric interaction forces (which are governed by the equations of motion). It is this attractive possibility which first led to the adaptation of the spontaneously broken symmetry idea to elementary particle physics, and which eventually led to the recent dramatic experimental discovery of the weak interaction W and Z intermediate vector bosons. In this quasi-monographical review we shall endeavor to trace all these developments.

Philosophical discussion about symmetry and about the basic structure of matter dates back as far as the ancient Greeks, if not even further. The question "what is matter made of" drew the attention of Greek philosophers such as Anaxagoras and Democritos, who introduced the concept of the atom, and Empedocles who introduced the concept of elements, with four such elements (earth, fire, water and air) serving as the ultimate building blocks for all other matter. Additionally, the ancient Greeks found the concept of symmetry to be aesthetically attractive, particularly when applied to works of art or to Euclidean geometry. However, for the Greeks the structure of matter and the elegance of the symmetry concept were essentially separate philosophical notions to be studied independently, and then only abstractly. The modern theory of elementary particles and of fundamental processes which has now emerged through the repeated application of symmetry principles would thus surely have delighted the ancient Greeks. In a sense then the modern theory of matter represents the successful marriage of some of the most aesthetic of the Greek philosophical speculations.

2. HISTORICAL BACKGROUND

Before discussing symmetry in elementary particle physics it is useful to review briefly its earlier applications in physics since many of the essential ingredients are already there. One of the earliest significant applications of symmetry was made by Maxwell in the mid-nineteenth century when he constructed a theory of electromagnetism which was manifestly symmetric between the electric and magnetic fields. The theory predicted the existence of radio waves which would travel at the speed of light c. For our purposes here we note two things. Firstly, Maxwell's theory achieved a unification (another Greek dream) of some of the basic forces in nature, namely electricity, magnetism and light, showing that they were not associated with three independent and unrelated branches of physics at all but, rather, that they were intimately and profoundly connected. This was then the first unification in physics and it was the forerunner of the current unification program of modern elementary particle physics. Secondly, Maxwell's theory solved one problem only to raise another, since though his equations showed that radio waves travelled with the velocity of light the equations did not state in which frame this velocity was to be measured. There then followed a totally unsuccessful experimental search for a privileged aether frame in which to measure the velocity; and the problem was only resolved by Einstein in 1905 with his relativity postulate that the velocity of light is the same in all frames. With Einstein's special theory of relativity a new symmetry was introduced into physics one which treats space and time equivalently and is known as Lorentz invariance. Lorentz invariance augments the three ordinary spatial rotations (xy, yz, zx) with three others which mix space with time (xt, yt, zt) to give altogether the six parameter $SO(3, 1)$ Lorentz group of transformations which leave invariant the space-time interval $x^2 + y^2 + z^2 - c^2t^2$. Thus the symmetry between space and time grew out of the symmetry between electric and magnetic fields.

Even more profound than relativity was the development in 1925 of (non-relativistic) quantum mechanics by Heisenberg and Schrodinger. This new theory renounced classical mechanics with its real coordinate space and instead asserted that, rather, physics occurs in a complex Hilbert space and is described by wave functions and non-commuting operators. This change in the space in which physics is realized had far reaching implications for symmetry. Specifically, a wave function $\psi(x)$ could acquire a complex $\exp(i\alpha)$ phase under a symmetry transformation and further, a set of N such wave functions $\psi_i(x)$ $(i = 1, \ldots N)$ could also mix under a symmetry transformation, with the phase α then being an N-dimensional matrix. The invariance of the quantum-mechanical probability $\Sigma|\psi_i(x)|^2$ under symmetry transformations then restricted the $\exp(i\alpha)$ matrices to be unitary, so that for a set of N wave functions the relevant symmetry group would be an N-dimensional non-Abelian (i.e. non-commuting) unitary group $SU(N)$ with $N^2 - 1$ parameters. Thus quantum mechanics brought unitary groups (in contrast to the real orthogonal groups of classical physics) to the forefront in physics and opened the door to spectroscopy.

In spectroscopy the energy levels of a quantum-mechanical Hamiltonian are classified according to the irreducible representations of a symmetry group under whose transformations the Hamiltonian is either left invariant (exact symmetry) or transforms in a specific manner (approximate symmetry). This then leads to a classification scheme for energy levels and for transitions between them which was applied first to atomic energy levels, subsequently to the nuclear shell model, and eventually to elementary particles. Central to the spectroscopic use of symmetry is the requirement that the symmetry analysis be applied to systems with a finite, fixed number of degrees of freedom (such as a few electrons in an atom, a few nucleons, i.e. neutrons and protons, in a nucleus, or the three quarks in a nucleon). This then is the "classic" realm of the use of symmetry.

Shortly after the development of the non-relativistic quantum mechanics of Heisenberg and Schrodinger and concurrent with its spectroscopic applications to atoms with a fixed number of electrons, Dirac generalized Schrodinger's work and wrote down an equation (the Dirac equation) which satisfied both quantum mechanics and the Lorentz invariance of special relativity. The Dirac equation was a remarkable equation which gave the correct gyromagnetic ratio for the coupling of an electron to an external electromagnetic field (again a triumph for a symmetry, *viz.* Lorentz invariance). Additionally, the Dirac equation possessed a further symmetry, this one between positively and negatively charged particles (known as charge conjugation) which predicted the existence of the subsequently found positron. Thus the interplay of quantum mechanics with special relativity yielded the existence of anti-matter. Apart from this, the Dirac equation also brought about a major conceptual change in our view of the world. Specifically, since energy and mass are equivalent according to Einstein's special relativity (i.e. $E = mc^2$), the new implication of Dirac's work is that the mass of a given electron is equivalent to energy which in turn is (heuristically) equivalent to the mass of an electron plus the mass of an electron-positron pair. Hence a single electron is equivalent to two electrons plus a positron, so that it no longer has a fixed number of degrees of freedom. Repeating the argument ad infinitum then implies that there is no distinction between a single electron and an electron accompanied by an infinite number of electron-position pairs (the precise mathematical formulation of this heuristic description of pair creation is known as quantum field theory and is beyond the scope of this article). Thus, because of relativity we transit to a theory with an infinite number of degrees of freedom. While we now enter a new and far more complicated world we note that precisely because we have transited to an infinite number of degrees of freedom we now have the possibility that many-body cooperative phenomena can occur and thus provide a mechanism for the spontaneous breakdown of the original quantum-mechanical symmetries.

Thus to sum up, from the physics of the great breakthroughs of quantum mechanics and relativity elementary particle physics received two key notions, namely symmetry as a group-theoretical spectroscopic classification scheme whenever relativistic pair creation may be ignored, with the theory then being converted into a many-body problem whenever pair creation is relevant. The earlier developments of elementary particle physics concentrated on the spectroscopic aspects of the theory, while the more recent developments have emphasized the many-body aspects of spontaneous breakdown. We thus now review both of these types of developments just as they have unfolded in elementary particle physics.

3. SYMMETRY AS A GROUP-THEORETICAL CLASSIFICATION FOR THE ELEMENTARY PARTICLES

With the advent of big high energy accelerators in the fifties and sixties it became possible to explore the structure of matter down to nuclear and subnuclear distances of the order of 10^{-13} cm. After it became established that the nucleus was composed of neutrons and protons the quest began to see what the neutrons and protons themselves were made of. On bombarding neutrons and protons with other protons, with electrons, and with photons many new elementary particle states were discovered which fell into two broad classes; the baryonic resonances ("baryon" here means proton-like) with fermionic (i.e. half-integer) instrinsic spin angular momentum, and the meson ("meson" means intermediary in Yukawa's sense of mediating the nuclear force) resonances with bosonic (i.e. integer) spin angular momentum. As of today well over one hundred baryonic resonances and approaching one hundred mesonic ones have been established.

With such a proliferation of elementary particles it became necessary to organize them in some way, and in 1961 Gell–Mann[1] and independently Ne'eman[2] recognized that the particles nicely fitted into multiplets of a symmetry group, namely $SU(3)$. Thus the spectroscopy of atomic and nuclear physics was able to repeat itself again at a new level of matter. While the applications of spectroscopy proved to be extraordinarily successful these were some novelties and also some problems. The group $SU(3)$ was initially suggested by Gell–Mann and Ne'eman since it possessed eight-dimensional octet multiplets ($3^2 - 1 = 8$) which could accommodate nicely the eight low-lying (in mass) spin one-half baryons (the proton family), the eight low-lying spin zero pseudoscalar mesons (the pion family) and the eight low-lying spin one vector mesons (the rho family). However, the group $SU(3)$ also possessed a smaller multiplet than the octet, namely a complex three-dimensional triplet, the fundamental representation, out of which all the other representations of the group could be built. Hence just as the six parameter orthogonal $SO(3, 1)$ Lorentz group is built on the four real coordinates x, y, z and t, the eight parameter unitary $SU(3)$ group is built out of sets of three complex objects. Thus it was natural to inquire into the physical significance of this triplet, and so the quark model was born.

In 1964 Gell–Mann[3] and Zweig[4] introduced the quark model in which the objects in the triplet were called quarks, with the hadrons (the baryons and mesons) then being built out of quarks just as nuclei are built out of neutrons and protons, to give a shell model at the quark level as it were. In the $SU(3)$ triplet there are three types of quarks which are conventionally labelled up (u), down (d) and strange (s) (to denote the up and down isospin and the strangeness quantum numbers) to give three different varieties (or flavors, as we now say) of quarks. The quarks have novel fractional electric charge assignments (the up quark has charge $2e/3$ where e denotes the charge carried by a proton, while the down and strange quarks each have charge $-e/3$); also the quarks have spin one-half. In the quark model then the proton is built out of two up quarks and a down quark (to give it charge e), the neutron is built out of one up quark and two down quarks (to give it charge zero), while the other members of the proton family octet (see the appendix for a more detailed tabulation of this family octet and of all the other main elementary particles) are built out of various other combinations of three quarks with integer electric charge (specifically uus, uds, dds, a second uds, uss and dss). As well as a triplet of quarks, according to the Dirac equation there should also be an antimatter spin one-half triplet of antiquarks, $\bar{u}$, $\bar{d}$ and $\bar{s}$, with electric charges $-2e/3$, $e/3$ and $e/3$ respectively. Then the charge one, zero and minus one pi mesons of the pion family octet are built out of $\bar{d}u$, $\bar{u}u - \bar{d}d$, and $\bar{u}d$ respectively, with the rest of the octet containing mesons which are built out of $\bar{u}u + \bar{d}d - 2\bar{s}s$, $\bar{u}s$, $\bar{s}d$, $\bar{d}s$ and $\bar{s}u$. An analogous situation obtains for the rho family octet, and it can also be built out of quarks and antiquarks.

In this manner then all the known elementary particles can be classified according to the quark model, and as new particles were discovered the model was able to incorporate them either as higher u, d, s quark and/or antiquark excitations, or by increasing the number of flavors of quarks. As of today three additional flavors have been identified and they have been associated with three new quarks labelled charm (c), bottom (b), and top (t). Thus the $SU(3)$ symmetry scheme of Gell–Mann and Ne'eman has evolved into an $SU(6)$ of six flavors, and gives, all in all, an extraordinarily good and economical classification scheme for the elementary particles.

Now, as we noted above, there are some problems with the quark model. Firstly, no fractionally charged particles have actually been observed so far in accelerator collisions specifically designed to break up the proton; and secondly, the quark model failed completely to specify the nature of the forces which act between quarks so as to bind them together into hadrons in the first place. In the sixties there was a vigorous experimental search for quarks, but, reminiscent of the search for Maxwell's aether, it proved to be unsuccessful. Thus the quark model fell flat on its face with the much sought and much desired building blocks of matter simply not being observed experimentally. Worse, high energy electron proton scattering experiments were performed at the Stanford Linear Accelerator Center in the late sixties which were a further disaster for the quark model even while being a triumph. Specifically, the data were found to be consistent with each of the three quarks in the proton scattering off the electron independently and freely of the other quarks in the proton. Thus a free quark model explained the observed high energy data remarkably well[5]. But, if the quarks were effectively free at high energies they should then be able to escape completely from the interior of the proton and be observed in the very same experiments. And yet somehow they were not in fact observed. Seemingly, just as they were about to escape from the proton once and for all something prevented them from doing so at the very last moment as it were.

In light of this non-observation of quarks elementary particle theory started to move in a new direction, namely that of thinking of quarks as non-observable, permanently confined objects, with the forces between the quarks always pulling them back in and never letting them escape. The theoretical problem posed by the quark model then was to understand how they could be effectively free and thus weakly coupled to each other at high energies while at the same time binding into protons which themselves interact with very strong nuclear forces, and in such a way that the quarks are never released no matter how hard the protons are bombarded in an accelerator. Though no complete answer to the above problem has so far been given the use of symmetry and many-body ideas has provided us with much insight culminating in the development of non-Abelian gauge theories of the fundamental interactions.

4. LOCAL SYMMETRY

As we noted earlier, in quantum mechanics a wave function $\psi(x)$ would acquire a complex exp $(i\alpha)$ phase under a symmetry transformation. A natural extension of this idea is to allow the phase α to depend locally on the point x where the wave function is located. Then wave functions at different points acquire different phases. Since a derivative of $\psi(x)$ with respect to x would depend on the values of $\psi(x)$ at different points the differential equations of motion of the wave function would not be invariant under such local $\alpha(x)$ phase changes. It is possible to make them invariant under these so-called gauge transformations, however, by introducing additional particles, the so-called vector gauge bosons which, like derivatives, have spin one. Indeed, for the wave function of the electron the associated gauge particle is the photon. Thus the theory of interacting electrons and photons (quantum electrodynamics) may be developed as a local gauge theory of phase transformations, with the forces between the matter fields (the electron and positron) being mediated by the gauge bosons needed to maintain the local gauge invariance in the first place. With the advent of gauge theories we identify a completely new and essentially revolutionary role for symmetry. Previously symmetry was used kinematically as an invariance principle whose primary role was to constrain the forces between particles. Now, with local gauge invariance symmetry becomes a dynamical principle for actually producing the forces in the first place with those forces then being mediated by none other than the gauge bosons associated with the local symmetry. With local gauge invariance then there is now an intimate dynamical connection between symmetry and forces, and as we shall see below, this is the essential ingredient for the current unification program of elementary particle physics. Finally, we note that when the same gauge invariance principle is applied to the photon wave function it forces the photon to be massless, in exact agreement with experiment. Thus local gauge invariance is a very powerful concept.

The extension of gauge invariance to non-Abelian $SU(N)$ groups was developed by Yang and Mills[6] in the fifties, and they found that this time in order to maintain the local gauge invariance at the level of the equations of motion it is necessary to introduce N^2-1 spin one vector bosons. These would again be massless and would again mediate the forces between the $SU(N)$ matter fields.

While the Yang–Mills idea was conceptually elegant not much attention was given to it until the seventies when the phenomenon of asymptotic freedom was discovered[7,8]. It was noted that non-Abelian gauge theories possessed a peculiar self-quenching mechanism with the forces between particles essentially cancelling out at high energies. Thus the interactions between quarks would become free at asymptotic energies just as required for the interpretation of the high energy electron-proton scattering experiments to which we referred earlier. We therefore introduce a new symmetry classification for quarks, known as color, with the quarks then carrying color (just like they carry flavor). For phenomenological reasons it turns out that there are three colors (denoted by red, green and blue) of quarks, so that the local theory of color, known as quantum chromodynamics, would be an $SU(3)$ (another $SU(3)$, not the flavor $SU(3)$ of Gell–Mann and Ne'eman) non-Abelian gauge theory mediated by a set of eight massless colored gauge bosons known as gluons. It is generally believed today that the local $SU(3)$ theory of interacting quarks and gluons is the fundamental theory of the nuclear force, with the colored gauge bosons being for the nuclear force exactly what the photon is for the electromagnetic force.

The reason why quantum chromodynamics is asymptotically free may be understood heuristically. According to quantum field theory the ground state of a field theory is a many particle mixture of particle antiparticle pairs in exactly the same way as, as noted earlier, the electron is equivalent to an electron plus an infinite number of electron positron pairs. Thus even though the ground state of a quantum field theory is known as the vacuum it is far from empty, as may be anticipated since the physical creation of matter and antimatter which occurs in high energy accelerator experiments is due to the excitation of particles out of the vacuum. The vacuum thus has structure. The exploration of this structure is known as quantum field theory. Without needing to delve into the details of quantum field theory it is possible to use the above picture to understand some simple properties of the vacuum.

We consider first the vacuum polarization of quantum electrodynamics. A negatively charged electron is surrounded by a vacuum of electron positron pairs, and thus polarizes them causing the positrons to move nearer to the initial electron and the electrons to move further away, thereby inducing a dielectric field pointing away from the initial electron. This induced field tends to repel a positive test charge, which can then only feel the full attraction due to the initial electron when it gets very close to it. Hence in quantum electrodynamics the effective charge increases as objects get closer, and decreases as objects get further away.

The above situation is reversed in quantum chromodynamics. Now the vacuum is filled with the colored gauge bosons. Since they carry spin the vacuum acts as a magnetic medium rather than an electric one. Now for a vacuum filled with magnetic moments the effect of an applied magnetic field is paramagnetic and produces an induced field in the same direction as the applied field. Thus a quark polarizes its surrounding gluon vacuum so that the effective charge decreases as objects get closer and increases as objects get further away. This is then asymptotic freedom with the polarization of the gluon vacuum dynamically cancelling the forces between quarks at short distances deep inside the interior of the proton, just as required experimentally.

While this asymptotic freedom phenomenon was a considerable triumph for quantum chromodynamics the theory can go even further. Specifically, we note that the forces between quarks grow at large distances. Hence if they grow enough they may become so strong so as to confine the quarks altogether, so that they then could not escape from the interior of the proton. Though this is an attractive and widely anticipated possibility it has yet to be realized fully and is at the center of current research efforts. While the ultimate fate of quantum chromodynamics awaits a successful resolution of the confinement question (and of the related issue of showing that the hadrons are in fact dynamical bound states in quantum chromodynamics, and then calculating their masses) its success in high energy electron proton scattering is so encouraging that there is now a great measure of confidence in the theory, so much so in fact that it has stimulated and extended interest in non-Abelian gauge theories elsewhere in elementary particle physics, a topic to which we now turn.

5. SPONTANEOUSLY BROKEN SYMMETRY

While the discovery of asymptotic freedom greatly enhanced the status and utility of non-Abelian gauge theories, interest in them actually preceded the development of quantum chromodynamics. Specifically, already in the sixties Weinberg[9], Salam[10], and Glashow[11] had, independently, developed an earlier non-Abelian gauge theory, this one for the weak interactions (the interactions which are responsible for radioactive beta decay). Unlike quantum chromodynamics this particular gauge theory also incorporated the then relatively new many body concept of spontaneously broken symmetry. As well as using a new and untested concept the theory also had some other severe initial difficulties. For instance, it was not clear how to incorporate quarks into the theory correctly, and it was not at all clear whether the contribution due to the infinite number of particle antiparticle pairs in the vacuum would lead to finite (as opposed to infinite) values for the observable predictions of the theory (the so-called renormalization problem). So initially the idea could not be taken too seriously, and it was only in the seventies that all the outstanding theoretical issues were successfully resolved, and that experimental support for the whole idea was obtained.

The central problem for theories of the weak interactions has been around since Fermi's work in the thirties. Basically, Fermi suggested a phenomenological model of the weak inter-actions in which four spin one-half fermions interacted locally at the same space-time point. This model, together with subsequent modifications necessitated by the experimental discovery of non-conservation of parity in the fifties, gives an extremely good description of low energy weak interaction phenomenology. However, at higher energies of the order of 10^{11} electron Volts (i.e. 100 times the rest energy of the proton), because of the local nature of the four fermion interaction the theory implied that scattering processes which involved the four fermions would violate unitarity (i.e. conservation of probability). The local pointlike four fermion interaction was thus wrong in principle, and had to have some additional structure at high energy. The natural way to give the weak interaction some extra structure would be to have it mediated by an analog of the photon, thus suggesting the existence of a weak interaction intermediate vector boson which would then be for weak interactions exactly what the photon is for electromagnetism, and what the gluon now is for the strong interaction. Now the weak interaction is known to be a very short range interaction (typically of nuclear size or less), and since the range of an interaction is given in quantum field theory by the Compton wavelength ($\hbar/mc$) of the exchanged or mediating particle, the intermediate vector boson would have to be very heavy. (For electromagnetism the photon is massless and hence electromagnetic forces are long range). However, if, by analogy with electromagnetism, the intermediate vector boson is to be associated with a gauge principle (this anyway being the most elegant and restrictive possibility—and the only one with any strong predictive power) then, as we have seen, the same gauge principle would require the intermediate vector boson to be massless, and hence not yield the desired short range weak interaction after all. This is then the central problem of the weak interactions and it was ultimately resolved by the development of the Weinberg–Salam–Glashow theory.

The resolution of the problem stems from the fact that it is possible for a vector boson to actually acquire a mass in a guage invariant theory if the gauge symmetry is spontaneously broken, so that the solutions to the equations of motion (which yield the masses of the particles) have lower symmetry than that (exact gauge invariance) of the equations of motion themselves. The mass is then generated by the dynamics between the particles and is output to the theory rather than input. Being output the mass is then also in principle calculable.

In discussing the explicit way in which spontaneous breakdown works in field theory we have to distinguish between whether the underlying symmetry is global (phase exp $(i\alpha)$ constant in space-time) or local (phase exp $(i\alpha(x))$ space-time dependent). For the case of global symmetry the essential features are exhibited in the Heisenberg ferromagnet. Here the forces between the spins of the atoms at the lattice sites of a ferromagnetic material are short range (nearest neighbor typically) and also rotationally invariant. However, in the ground state of the system it is found that all the spins are aligned at absolute zero of temperature, so that the total spin magnetization of the system is non-zero and has a preferred, rotational non-invariant, direction. Thus though the forces are only short range the particles all conspire to lock in a macroscopic solution to the equations of motion with the associated locked in ground state then actually not being rotationally invariant. With all the spins locked into the same direction we thus have long range order (i.e. a correlation between the spins of atoms located on opposite sides of the crystal) even though the forces are only short range. This locking in is a many-body cooperative effect due to the presence of many particles, with the sheer fact of there being many particles over-whelming the fact that the individual interactions are only short range. Thus if the dynamics conspires to produce long range order at all then at the same time it will also conspire to lock in a preferred direction and yield a ground state which has lower symmetry than that of the underlying interparticle forces. This, then, is spontaneous symmetry breaking. Finally, whenever this happens we note additionally that it is found that the spontaneous breakdown also produces one further feature, namely a set of dynamical bound state collective modes. For the ferromagnet these modes are the spin waves.

For the case of local symmetry the essential features are exhibited in the Bardeen–Cooper–Schrieffer[12] theory of superconductivity. In a superconductor electrons form Cooper pairs which are correlated over rather long distances (i.e. long range order again) and flow through the superconductor in supercurrents which persist for very long times. When a magnetic field is applied along the axis of a cylindrical superconductor these supercurrents will typically flow in circles around the axis of the cylinder near to its surface (the skin effect) in such a direction so that, by Lenz's Law, they induce a magnetic field which will oppose the applied field. The magnetic field is thus cancelled in the interior of the superconductor (the Meissner effect) and mainly resides at the surface of the material. Thus as we go deeper into the superconductor the electromagnetic potential falls off very fast, typically as $\exp(-r/\lambda)/r$ where λ is of the order of the Cooper pair correlation length, in contrast to its pure $1/r$ behaviour in free space. Thus within the interior of a superconductor the magnetic field only penetrates the medium to a penetration depth of order λ. Within the medium of the superconductor then magnetic fields behave very differently than in free space[13,14].

As we noted earlier the vacuum of a quantum field theory is a many-body medium, and thus the above discussion of spontaneous breakdown in solid state physics carries over to elementary particle physics essentially in its entirety. However, because of constraints due to the Lorentz symmetry of special relativity, there are some significant new features in both the global and the local symmetry cases. In both the cases it turns out that the collective modes associated with the spontaneous breakdown of a symmetry are constrained to have two specific properties: they have to be massless, and they have to carry zero intrinsic spin angular momentum. Such modes are known as Nambu–Goldstone bosons[15,16], and they are characteristic of spontaneous symmetry breakdown in relativistic quantum field theory. The emergence of such strictly massless, spinless bosons is actually initially something of an embarrassment for elementary particle theory since none have ever been observed experimentally. Nonetheless, the whole idea has in fact been successfully applied in elementary particle physics with different physical realizations occuring in the global and the local cases.

We discuss first the global case. In the early sixties it became apparent that the spin zero pi mesons (the lightest of the known mesons) interacted with matter in a manner similar to that expected of Nambu–Goldstone bosons, suggesting that they were close to being such bosons. Specifically, it was found that the strong interactions in general, and quantum chromodynamics in particular, possessed a particular global symmetry (specifically a rather technical symmetry known as chiral flavor). However, this same global symmetry was only an approximate invariance of the weak interactions. (The weaker the interaction, the lower, in general, is the symmetry of the forces between particles). As we noted earlier in our discussion of quantum chromodynamics, the pi mesons are quark-antiquark bound states which are held together by the gluons. If, then, (and this has yet to be demonstrated conclusively) the polarization due to the many-body gluon vacuum brings about a spontaneous breakdown of the global chiral flavor symmetry (to thus provide an extremely delicate interplay between symmetry and dynamics) the pi mesons will then become exactly massless bound states and would thus be true Nambu–Goldstone bosons as far as the strong interactions are concerned. Then, finally, the weak interaction would break the global chiral flavor symmetry explicitly (i.e. in the equations of motion) but weakly, and thus partially destroy the Nambu–Goldstone phenomenon and give the pi mesons a small mass. Thus the pi mesons, whose observable masses are small but nonzero, are not exact Goldstone bosons, but are close enough that the theory still has predictive power. The theory thus explains why the pi mesons are so light compared to all the other known hadrons, with the Nambu–Goldstone picture of the pi mesons being in excellent accord with experiment.

We turn now to a discussion of the Nambu–Goldstone phenomenon in the case of spontaneous breakdown of a local symmetry. In this case a relativistic analog of the Meissner effect in a superconductor occurs with the massless vector gauge boson associated with the local gauge invariance acquiring a mass (the relativistic analog of the mass, $\hbar/\lambda c$, associated with a Compton wavelength of order the penetration depth of a superconductor) in the presence of an ordering (i.e. a spontaneous breakdown) of the quantum field theory vacuum. Thus the dynamics of the vacuum gives the gauge boson a mass despite the fact that the forces are gauge invariant, so that the gauge boson then mediates short range interactions. In counting the number of states of spin polarization of the vector boson (i.e. the number of different projections of its spin angular momentum vector along an axis of spin quantization), we note that while a massless vector boson possesses two spin projections (*viz.* left and right circularly polarized, just like the photon, the massless light quantum), a massive vector boson possesses three (left and right circularly polarized and also a state of zero spin projection, i.e. a state of longitudinal polarization). There is thus a question of where this third, longitudinal, state came from. The answer to this question was originally given by Higgs[17] who noted that the extra state came from none other than the spin zero (and hence spin projection zero) Nambu-Goldstone boson which accompanies the spontaneous breakdown. (i.e. the supercurrent makes the magnetic field short range in the interior of a superconductor). We thus recognize the Higgs phenomenon in which the would-be massless Nambu–Goldstone boson combines with the massless vector boson into one all-embracing massive vector boson so that no massless particle (spinless or vector) remains in the spectrum. Thus not only do we avoid the potential existence of massless vector gauge bosons, we get rid of the massless Nambu-Goldstone bosons as well. With the Higgs mechanism then we thus identify an explicit mechanism by which the many-body dynamics can conspire to give a mass to a vector gauge boson in a theory whose forces are strictly gauge invariant. Finally, then, this is precisely what we want for a theory of the weak interactions.

6. THE WEINBERG–SALAM–GLASHOW THEORY OF THE WEAK INTERACTIONS

At the time of the development of the Weinberg–Salam–Glashow theory the following properties of weak interactions were known. Fermi's four fermion interaction could be written in the form of a product of two local currents (each one containing a fermion and an antifermion, just like the electromagnetic current), one current carrying charge $+e$ and the other one carrying charge $-e$. At that time the only other established current was the electromagnetic current itself which carries zero charge. The weak interaction currents had both a leptonic (i.e. electron-like) and a hadronic piece. The leptons (*viz.* the electron (e), the muon (μ), and their respective neutral counterparts the neutrinos (ν_e and ν_μ)) contributed currents behaving as $\bar{e}\nu_e + \bar{\mu}\nu_\mu$ (charge $+e$) and $\bar{\nu}_e e + \bar{\nu}_\mu \mu$ (charge $-e$). The hadronic piece with charge $+e$ behaved like $(\bar{d}\cos\theta + \bar{s}\sin\theta)u$, while the piece with charge $-e$ behaved like $\bar{u}(d\cos\theta + s\sin\theta)$. (The elucidation of this particular form with d, s mixing angle θ is due to Cabibbo[18]). The four fermion interaction could thus describe radioactive beta decay processes such as the decay of a neutron into a proton plus an electron and the antiparticle of the electron neutrino, a decay which proceeds at the quark level via $d \rightarrow u + e^- + \bar{\nu}_e$.

In order to introduce vector bosons and give them masses, we now note as follows. The essential ingredient for implementing the Higgs mechanism is first to have a local gauge invariance for the exp ($i\,\alpha(x)$) phases based on some unitary $SU(N)$ group. Given the existence of the charge $+e$ and $-e$ weak interaction currents the smallest $SU(N)$ group we could use that would include these currents would be the non-Abelian $SU(2)$ group with $N^2 - 1 = 3$ spin one vector bosons. Hence in order to associate the charge plus and minus one intermediate vector bosons, W^+ and W^-, with the charged weak interaction currents the $SU(2)$ group theory necessitates that there actually be a third current and one further vector gauge boson. From the phenomenological structure of the weak interaction currents it turns out that this third current and its associated gauge boson should be electrically neutral. It would initially be very tempting to have this third current be the electromagnetic current, but its explicit structure does not fit in to the $SU(2)$ group theory. Thus there has to be another neutral current in nature. This is then a very deep and powerful symmetry based prediction, because it says that in order to give the charged W^+ and W^- gauge bosons a mass by spontaneous symmetry breaking there has to be a new neutral current with its own associated neutral gauge boson.

For the purposes of simply implementing the Higgs mechanism it would have been sufficient to stop with a weak $SU(2)$ group. However, because the needed third current was electrically neutral just like the electromagnetic current, Weinberg, Salam, and Glashow instead built not a three current model but a four current model, based on an $SU(2) \times U(1)$ group which contains altogether four gauge bosons. With the use of four currents the weak interaction currents could then be augmented with the electromagnetic current into a model which thereby unifies the weak interactions with electromagnetism. Even prior to the development of the $SU(2) \times U(1)$ theory there had been some speculation on a possible unification of the weak and electromagnetic interactions motivated primarily by the fact that they both had a current structure and by the fact that the hadrons and also the leptons participated in both the interactions. Standing in the way of such a unification was the fact that the weak interaction was short range while the electromagnetic interaction was long range. This difference could now be generated entirely by the Higgs mechanism. Thus with the advent of non-Abelian gauge theories we can have an observable asymmetric world (the intermediate vector bosons get a mass, the photon does not) emerge in a theory which treats all forces in a completely symmetric manner. Since the weak and electromagnetic interactions are thus treated symmetrically (i.e. at the level of the equations of motion) they are thereby unified. Merely generating a mass for the W bosons alone would already have been a significant development, but in fact the $SU(2) \times U(1)$ model goes a lot further since it provides a unification of two of the fundamental forces in nature, forces that on the face of it look highly disimilar, with all of this disimilarity being due to the dynamics of the vacuum.

In order to determine the explicit experimental predictions of the $SU(2) \times U(1)$ model it is necessary to specify the group theoretical content of the would-be Goldstone bosons of the model, and of the concommitant direction in which the vacuum breaks (i.e. the analog of the direction in which the rotational non-invariant total spin magnetization of a ferromagnet actually points). The simplest such specification is to break $SU(2) \times U(1)$ according to a doublet which contains two complex scalar fields (the doublet is the building block of $SU(2)$ just like the complex triplet builds $SU(3)$). The two complex fields themselves can be rewritten as four real fields. With one of these four real fields being needed to specify the direction of vacuum breaking there is room for only three would-be Goldstone bosons and thus only three of the four gauge bosons of $SU(2) \times U(1)$ can acquire masses by the Higgs mechanism. The one gauge boson which stays massless is identified with the photon while the three that acquire masses are the W^+ and W^- and a third electrically neutral gauge boson known as the Z. Additionally, since the real field which specifies the direction of vacuum breaking does not itself participate in the Higgs mechanism it must remain in the physical spectrum as an observable scalar, spinless field. Because it is the group-theoretic partner of the would-be Goldstone bosons this observable field (which turns out to be massive) is known as a Higgs boson. Since the model is unified the strength of the weak interaction is set by the known strength of the electromagnetic interaction, i.e. by the value of the electric charge. Using the above doublet pattern of breaking the masses of the W and Z bosons can then be completely determined, and it is found that the two charged W bosons acquire a mass of 83×10^9 electron Volts while the Z boson acquires a mass of 94×10^9 electron Volts. (Interestingly, this is just nicely in the energy range where, as we noted before, the pointlike four fermion interaction would have to start showing some structure).

As well as classify the gauge bosons, the $SU(2) \times U(1)$ model can also classify the leptons and the quarks. Specifically, the electron and its neutrino are found to form an $SU(2)$ doublet, and likewise the muon and its neutrino. Similarly, the up and down quarks form an $SU(2)$ doublet so that the up and down quarks which were originally classified according to the strong interaction up and down isospin (an $SU(2)$ subgroup of the flavor $SU(3)$ of Gell–Mann and Ne'eman) are now classified according to the weak interaction $SU(2)$ as well. However, not all was well initially with this weak interaction classification scheme for quarks, since the only other quark known at the time of the development of the $SU(2) \times U(1)$ model was the strange quark, which could not by itself fill up an entire doublet. To place the strange quark in a doublet it was necessary to introduce yet another quark, the charm quark[11], for which there was then no experimental support. Thus in the late sixties the status of the Weinberg–Salam–Glashow theory was that it was based on an untested Higgs mechanism idea, that it was a highly speculative attempt at unification, and that it made a whole host of unverified predictions—neutral currents, W bosons, Z bosons, Higgs bosons, and charmed particles. Finally, it was not even clear whether the model was even internally consistent (the renormalization problem).

This last, theoretical, difficulty was resolved by t'Hooft[19] who showed that non-Abelian gauge theories were in fact renormalizable, just like electromagnetism, so that the contributions due to the infinite number of particle antiparticle pairs in the vacuum would lead to finite observable predictions after all. Moreover, it turns out that theories in which non-Abelian gauge bosons already possess a mass at the level of the equations of motion are themselves not renormalizable. Thus the only renormalizable theory of massive non-Abelian gauge bosons is then one in which the gauge bosons acquire their masses by spontaneous symmetry breaking. This is then the hidden power of broken symmetry.

After this theoretical hurdle was disposed of people started to take non-Abelian gauge theories more seriously and a vigorous theoretical and experimental effort commenced. Neutral currents[20] and charmed particles[21] were soon found experimentally, and then, most spectacularly, the W[22] and Z[23] bosons themselves with just the right experimental masses. Additionally, a third charge $-e$ lepton (the τ lepton) and its associated neutrino were found, with these two particles then being nicely accommodated into another $SU(2)$ doublet; and then a third pair of quarks (b and t) which also fitted into an $SU(2)$ doublet were found. As of today the only unverified prediction of the Weinberg–Salam–Glashow theory is the existence of the Higgs boson, and it is currently the subject of a vigorous experimental search. Apart from this last (minor) question the Weinberg–Salam–Glashow theory is in extraordinary, even phenomenal, agreement with experiment, and today represents one of the best physical theories.

7. GRANDUNIFICATION

Armed with the success of the Weinberg–Salam–Glashow unification program (in fact, characteristically of elementary particle theory, prior to its actual experimental verification) the next natural step to take was to unify the electromagnetic and the weak interactions with the strong interactions[24,25]. Until the seventies this was a totally remote and highly whimsical idea, devoid of any serious underlying theoretical framework. What changed this situation was the advent of quantum chromodynamics and the discovery that the previously intractable strong interactions could be described by none other than a non-Abelian gauge theory. Thus the strong, the electromagnetic and the weak interactions were all found to be local gauge theories, and could thus all be unified into one all embracing framework to give a grandunification of these three fundamental forces.

To unify explicitly we note that the direct product group of quantum chromodynamics with the Weinberg–Salam–Glashow group is $SU(3) \times SU(2) \times U(1)$ with 12 ($= 8 + 3 + 1$) associated vector gauge bosons. We may therefore unify these three interactions group theoretically by embedding $SU(3) \times SU(2) \times U(1)$ into one single all-embracing larger grandunifying group. All of the three interactions are then completely equivalent and describable by one and the same gauge group and thus their interactions are identical at the level of the equations of motion. Of course their observable interactions are very different, and all of this difference must thus be attributable to the Higgs mechanism. In order to unify at all we must use some larger grandunifying gauge group, and this larger group will, consequently, necessarily generate some new and previously unanticipated interactions.

To study the experimental implications of these new interactions and also the explicit Higgs breaking pattern it is convenient to discuss the simplest and most popular of the various grandunifying gauge groups that have been proposed in the literature, namely $SU(5)$[25]. The local $SU(5)$ gauge group possesses a total of 24($= 5^2 - 1$) vector gauge bosons. Since 12 of them have already been identified with the $SU(3) \times SU(2) \times U(1)$ subgroup of $SU(5)$, the model therefore requires the existence of 12 other, new, vector gauge bosons altogether. With grandunification we unify not only the interactions, but also the matter fields by putting the quarks and leptons into common multiplets. In $SU(5)$, for instance, we put three down quarks, one of each of the three colors, a positron, and an electron antineutrino together in the five-dimensional representation of $SU(5)$ (viz. its building block). Thus, at the level of the equations of motion, quarks and leptons are also completely identical. The 12 new gauge bosons of the $SU(5)$ theory mediate transitions between the various leptons and quarks (precisely because the leptons and the quarks are in common representations) and are known as leptoquark gauge bosons. Such leptoquark gauge bosons thus mediate transitions in which quarks are converted into leptons, and thus cause the proton to decay into leptons (with typical decay processes such as $p \rightarrow e^+ + \pi^0$). Thus the extraordinary implication of grandunification is that matter is simply no longer stable, and that it will all ultimately decay.

Now of course the proton has never been seen to decay and thus these quark lepton transitions must be highly inhibited. Now, as we noted earlier for a superconductor, within the medium the potential behaves as $\exp(-r/\lambda)/r$. Therefore, if λ is extremely small (i.e. if $\lambda/\lambda c$ is extremely big) the potential will be very weak. Analogously then, if the 12 leptoquark gauge bosons acquire huge masses they will only mediate proton decay very weakly causing it to be highly inhibited and occur very slowly with a huge lifetime. Hence a Higgs mechanism which gives such huge masses to the leptoquark gauge bosons can nicely suppress proton decay.

While we see that we will need huge masses the great appeal of grandunified models is that we can actually calculate what the masses should be. Specifically, we recall that in asymptotically free theories (such as the non-Abelian $SU(5)$ theory) the effective charge due to the polarization of the gauge boson vacuum decreases as the distance decreases (i.e. as the mass or the energy increases). But we already know the effective charge in the 10^{11}-electron Volts region because that is where the Higgs mechanism is operative in the Weinberg–Salam–Glashow sector of the theory. Thus we can extrapolate back[26] to determine the effective charge in the leptoquark gauge boson mass region. For $SU(5)$ this extrapolation leads us to a leptoquark gauge boson mass scale of the order of 10^{23} electron Volts, yielding a proton lifetime of the order of 10^{30} years. (To get a feeling for the size of such a lifetime we note that the age of the universe is only 10^{10} years). We thus note the versatility and flexibility of spontaneous symmetry breaking. By generating all mass scales in the vacuum we are able to unify scales all the way from 10^6 electron Volts (the mass of the electron) to 10^{23} electron Volts, i.e. we can unify scales which differ by as much as 17 orders of magnitude, scales which would not otherwise appear to be even remotely related.

As for the experimental situation regarding the proton lifetime we note that it is not necessary to wait 10^{30} years to see a decay. We can simply put 10^{30} protons in a container and then, according to quantum mechanics, we should see one decay per year. Consequently, such experiments are currently being vigorously performed with sensitivities now ranging up to about 10^{33} years. As of today the experimental situation is a little discouraging since no proton decay has yet been seen despite the tremendous effort that has so far been invested. Moreover, because of their high sensitivity, the current generation of experiments have already been able to set a lower bound for the proton lifetime of 10^{32} years[27], thus ruling out simple grandunification schemes such as the $SU(5)$ model. Thus as of today there is absolutely no evidence at all for grandunification, and whether proton decay will actually be seen in the near future and whether some other of the many grandunification models (or a modified $SU(5)$) will correctly determine its lifetime remains to be seen and studied further. Nonetheless, what is important at the moment, is that with the advent of spontaneously broken non-Abelian gauge theories we have, for the first time, a framework in which realistic questions about a grandunification of the fundamental forces can be asked theoretically and be actually tested experimentally.

8. SUPERUNIFICATION

Though a grandunification of the strong, electromagnetic and weak interactions still awaits experimental confirmation (or denial), the theoretical developments that it has engendered invite an even more ambitious unification, namely that of the three grandunified forces with the only other known fundamental force, namely gravity. Such a unification of all of the four fundamental interactions is known as a superunification, and its achievement would yield the unified field theory that many physicists have dreamed of, and which no less a physicist than Einstein himself worked on for most of the latter part of his life. Indeed, Einstein's earlier and most celebrated work, namely his general theory of relativity, already contained some elements of unification in it since it was not just a theory of gravitation, but rather it was a unification of space and time with gravity. Moreover, because it led to a classical field theory of gravity it motivated Einstein to seek a unification of the gravitational field with the only other known macroscopic field, namely the electromagnetic one. For philosophical reasons, Einstein resisted quantum mechanics and thus sought an essentially geometric unification which was to be made at the level of the classical theory only. However, as we have already seen with the Weinberg–Salam–Glashow theory, it was only with the explicit inclusion of the quantum-mechanical effects associated with the spontaneous breakdown of non-Abelian gauge theories that unification became possible. Thus modern attempts at superunification have focussed on a unification with gravity at the quantum-mechanical level involving a quantized rather than a classical theory of gravity.

Such a possible superunification program has received encouragement from various sources. First, the unification of space, time and gravity embodied in the general theory of relativity is achieved through the imposition of a principle of generalized coordinate covariance (symmetry yet again) which requires that the predictions of the theory not depend on arbitrary choices and changes in the space-time coordinate frame of reference (the Lorentz transformations of special relativity then form a particular subset of these more general coordinate transformations). Since such arbitrary changes of frame of reference allow different changes to be made at different space-time points such changes are local, and hence in many ways analogous to the local gauge tranformations of non-Abelian gauge theories. Thus superunification may have a local gauge underpinning.

Secondly, from Newton's gravitational constant G, Planck's quantum of action $\hbar$ and the velocity of light c it is possible to form one unique combination with the dimension of a mass, viz. $(\hbar c/G)^{1/2}$, the so-called Planck mass. Numerically this Planck mass is of the order of 10^{-5} grams, i.e. of the order of 10^{28} electron Volts in energy units. This mass scale is remarkably close to the mass scale of grandunification, viz. 10^{23} electron Volts, suggesting that we only have to continue the grandunification program a little further to the higher Planck mass scale to obtain an overall superunification with the effects of quantized gravity then becoming important.

Our final motivation for superunification comes from an at first unlikely source. The division of the world into separate fermion and boson sectors that is present in elementary particle theory is somewhat ugly and artificial. To connect the two sectors it has been proposed (see e.g.[28] for a recent review) that there exist specific transformations which mix integer spin particles with half-integer ones. Such transformations are known as supersymmetry transformations. They are interesting transformations in their own right, but do not appear to be exact or even approximate invariances in nature. For our purposes here we note that since fermionic and bosonic fields have different dimensions, in order to connect them at all we have to have some additional operators in the supersymmetry group of transformations which can carry off the dimension balance. Such operators turn out to be none other than the familiar momentum operators associated with the space and time translation transformations of special relativity. Moreover, we can even go further by requiring that the supersymmetry transformations be local rather than global (just like we did with electromagnetism). Because of the above connection with space-time this leads to a local theory which turns out to be none other than a theory of gravity, called supergravity (see e.g.[29] for a recent review). Thus the road to superunification may be through supergravity. This exciting possibility and other potential superunification mechanisms are currently being explored in the literature with great vigor, and only time will tell their ultimate fate.

Acknowledgement—The writing of this article has been supported in part by the U.S. Department of Energy under grant No. DE-AC02-79ER10336.A.

REFERENCES

1. M. Gell-Mann, Symmetries of baryons and mesons. *Phys. Rev.* **125**, 1067–1084 (1962).
2. Y. Ne'eman, Derivation of strong interactions from a gauge invariance. *Nucl. Phys.* **26**, 222–229 (1961).
3. M. Gell-Mann, A schematic model of baryons and mesons. *Phys. Letts.* **8**, 214–215 (1964).
4. G. Zweig, Cern reports TH401 (1964) and TH412 (1964), unpublished; Fractionally charged particles and $SU(6)$, in *Symmetries in Elementary Particle Physics*, Proceedings of the 1964 International School of Physics "Ettore Majorana", (edited by A. Zichichi) pp. 192–234. Academic Press, NY, (1965).
5. J. D. Bjorken, Asymptotic sum rules at infinite momentum. *Phys. Rev.* **179**, 1547–1553 (1969).
6. C. N. Yang and R. L. Mills, Conservation of isotopic spin and isotopic gauge invariance. *Phys. Rev.* **96**, 191–195 (1954).
7. D. J. Gross and F. Wilczek, Ultraviolet behavior of non-Abelian gauge theories. *Phys. Rev. Letts.* **30**, 1343–1346 (1973).
8. H. D. Politzer, Reliable perturbative results for strong interactions?. *Phys. Rev. Letts.* **30**, 1346–1349 (1973).
9. S. Weinberg, A model of leptons. *Phys. Rev. Letts.* **19**, 1264–1266 (1967).
10. A. Salam, Weak and electomagnetic interactions, in *Elementary Particle Theory: Relativistic Groups and Analyticity* (Nobel Symposium No. 8), (edited by N. Svartholm) pp. 367–377. Almqvist and Wiksell, Stockholm (1968).
11. S. L. Glashow, Partial-symmetries of weak interactions. *Nucl. Phys.* **22**, 579–588 (1961); S. L. Glashow, J. Iliopoulos and L. Maiani, Weak interactions with lepton-hadron symmetry. *Phys. Rev.* **D2**, 1285–1292 (1970).
12. J. Bardeen, L. N. Cooper and J. R. Schrieffer, Theory of superconductivity. *Phys. Rev.* **108**, 1175–1204 (1957).
13. P. W. Anderson, Coherent excited states in the theory of superconductivity: gauge invariance and the Meissner effect. *Phys. Rev.* **110**, 827–835 (1958).
14. Y. Nambu, Quasi-particles and gauge invariance in the theory of superconductivity. *Phys. Rev.* **117**, 648–663 (1960).

15. Y. Nambu, Axial vector current conservation in weak interactions. *Phys. Rev. Letts.* **4**, 380–382 (1960).
16. J. Goldstone, Field theories with 'superconductor' solutions. *Nuovo Cim.* **19**, 154–164 (1961).
17. P. W. Higgs, Broken symmetries, massless particle and gauge fields. *Phys. Letts.* **12**, 132—133 (1964).
18. N. Cabibbo, Unitary symmetry and leptonic decays, *Phys. Rev. Letts.* **10**, 531–533 (1963).
19. G. t'Hooft, Renormalization of massless Yang-Mills fields. *Nucl. Phys.* **B33**, 173–199 (1971); Renormalizable Lagrangians for massive Yang–Mills fields. *Nucl. Phys.* **B35**, 167–188 (1971).
20. F. J. Hasert *et al.*, Observation of neutrino-like interactions without muon or electron in the Gargamelle neutrino experiment. *Phys. Letts.* **46B**, 138–140 (1973).
21. G. Goldhaber *et. al.*, Observation in e^+e^- annihilation of a narrow state at 1865 MeV/c^2 decaying into $K\pi$ and $K\pi\pi\pi$. *Phys. Rev. Letts.* **37**, 255–259 (1976).
22. G. Arnison *et. al.*, Experimental observation of isolated large transverse energy electrons with associated missing energy at $s^{1/2} = 540$ GeV. *Phys. Letts.* **122B**, 103–116 (1983).
23. G. Arnison *et. al.*, Experimental observation of lepton pairs of invariant mass around 95 GeV/c^2 at the CERN SPS collider. *Phys. Letts.* **126B**, 398–410 (1983).
24. J. C. Pati and A. Salam, Is baryon number conserved?. *Phys. Rev. Lett.* **31**, 661–664 (1973).
25. H. Georgi and S. L. Glashow, Unity of all elementary particle forces. *Phys. Rev. Letts.* **32**, 438–441 (1974).
26. H. Georgi, H. R. Quinn and S. Weinberg, Hierarchy of interactions in unified gauge theories. *Phys. Rev. Lett.* **33**, 451–454 (1974).
27. R. M. Bionta *et. al.*, Search for proton decay into $e^+\pi^0$, *Phys. Rev. Letts.* **51**, 27–30 (1983).
28. P. Fayet and S. Ferrara, Supersymmetry. *Phys. Rep.* **32C**, 249–334 (1977).
29. P. van Nieuwenhuizen, Supergravity. *Phys. Rep.* **68C**, 189–398 (1981).

APPENDIX: LISTING OF THE MAIN ELEMENTARY PARTICLES

In this appendix we collect together and tabulate the main elementary particles to which we have referred in the text.

(a) *The strongly interacting hadrons*
(i) *The proton family octet*

Name	mass in eV	spin	electric charge	isospin	strangeness	quark content
proton, p	938.3×10^6	1/2	1	1/2	0	uud
neutron, n	939.6×10^6	1/2	0	1/2	0	udd
Λ	1115.6×10^6	1/2	0	0	-1	uds
Σ^+	1189.4×10^6	1/2	1	1	-1	uus
Σ^0	1192.5×10^6	1/2	0	1	-1	uds
Σ^-	1197.3×10^6	1/2	-1	1	-1	dds
Ξ^0	1314.9×10^6	1/2	0	1/2	-2	uss
Ξ^-	1321.3×10^6	1/2	-1	1/2	-2	dss

(ii) *The pion family octet*

Name	mass in eV	spin	electric charge	isospin	strangeness	quark content
π^+	139.6×10^6	0	1	1	0	$d\bar{u}$
π^0	134.9×10^6	0	0	1	0	$\bar{u}u - \bar{d}d$
π^-	139.6×10^6	0	-1	1	0	$\bar{u}d$
η	548.8×10^6	0	0	0	0	$\bar{u}u + \bar{d}d - 2\bar{s}s$
K^+	493.7×10^6	0	1	1/2	1	$\bar{s}u$
K^0	497.7×10^6	0	0	1/2	1	$\bar{s}d$
K^-	493.7×10^6	0	-1	1/2	-1	$\bar{u}s$
$\bar{K}^0$	497.7×10^6	0	0	1/2	-1	ds

(iii) *The rho meson family octet*

Name	mass in eV	spin	electric charge	isospin	strangeness	quark content
ρ^+	767×10^6	1	1	1	0	$d\bar{u}$
ρ^0	770×10^6	1	0	1	0	$\bar{u}u - \bar{d}d$
ρ^-	767×10^6	1	-1	1	0	$\bar{u}d$
ω	783×10^6	1	0	0	0	$\bar{u}u + \bar{d}d$
K^{*+}	892×10^6	1	1	1/2	1	$\bar{s}u$
K^{*0}	896×10^6	1	0	1/2	1	$\bar{s}d$
K^{*-}	892×10^6	1	-1	1/2	-1	$\bar{u}s$
$\bar{K}^{*0}$	896×10^6	1	0	1/2	-1	ds

(b) *The weakly and electromagnetically interacting leptons*

Name	mass in eV	spin	electric charge
electron, e	0.5×10^6	1/2	-1
ν_e	<46	1/2	0
muon, μ	105.7×10^6	1/2	-1
ν_μ	$<0.5 \times 10^6$	1/2	0
τ	1784.2×10^6	1/2	-1
ν_τ	$<164 \times 10^6$	1/2	0

(c) *The vector gauge bosons*

Name	mass in eV	spin	electric charge	interaction mediated
photon, γ	$<3 \times 10^{-27}$	1	0	electromagnetic
W^+	$(80.8 \pm 2.7) \times 10^9$	1	1	weak
W^-	$(80.8 \pm 2.7) \times 10^9$	1	-1	weak
Z	$(92.9 \pm 1.6) \times 10^9$	1	0	weak
gluon	massless (theoretically)	1	0	strong

Comp. & Maths. with Appls. Vol. 12B, Nos. 1/2, pp. 185–196, 1986
Printed in Great Britain.

0886–9561/86 $3.00 + .00

SYMMETRY OF POINT IMPERFECTIONS IN SOLIDS

Ralph H. Bartram

Department of Physics and Institute of Materials Science, University of Connecticut,
Storrs, CT 06268, U.S.A.

Abstract—Symmetry principles are applied to condensed matter physics, with emphasis on point imperfections in crystalline solids, including both lattice defects and impurities. The theory of finite groups is reviewed, and the relation between group theory and quantum mechanics is established. The classification of electronic and vibronic states of point imperfections by the irreducible representations of crystallographic point groups is discussed. An example of spontaneously broken symmetry is provided by the Jahn–Teller effect.

1. INTRODUCTION

Symmetry principles pervade every branch of physics, and lend a methodological unity to the characterization of seemingly diverse phenomena. Group theory provides a common mathematical description. For example, the concept of spontaneous symmetry breaking, which underlies contemporary approaches to grand unification and supergravity theories in high energy physics and cosmology, was borrowed from condensed matter physics, where it finds particular application in the theories of ferromagnetism, superconductivity and phase transitions[1].

In this article, we shall explore the application of group theory to the classification of quantum states in condensed-matter physics, with emphasis on a particular sub-field: that of point imperfections in crystalline solids. Some of the relevant symmetries of these point imperfections are described by finite crystallographic point groups, rather than the Lie groups appropriate to isotropic systems and elementary particles. This topic provides a particularly simple and transparent example of the relation between group theory and quantum mechanics, as well as an example of spontaneously broken symmetry in the Jahn–Teller effect[2].

We begin with a discussion of the properties of crystallographic point groups, followed by a description of some of the point imperfections which occur in crystalline solids and of the experimental techniques available for their characterization. We next consider the role of symmetry in the quantum-mechanical description of point imperfections, and conclude with a discussion of the Jahn-Teller effect.

2. CRYSTALLOGRAPHIC POINT GROUPS

It is evident that mathematics provides a rich and versatile language for expression of the quantitative aspects of physics. We shall endeavor to demonstrate that qualitative aspects, such as symmetry, can find similar expression. In an effort to make this presentation at least minimally self-contained, we offer a concise summary of the rudiments of the theory of finite groups before proceeding to applications[3].

A *finite group* of order g is a set of g distinct elements, together with a law of composition (multiplication), such that: the set is closed under multiplication; it contains an identity element; every element has an inverse in the set; and multiplication is associative. In general, multiplication does not commute. If it does for all elements, the group is abelian. A finite group is defined by specification of its multiplication table. All of the elements can be expressed as products of a subset of elements called *group generators*. A subset of elements which obeys group axioms with the same law of composition is a *subgroup*. Group elements can also be organized in conjugate classes; elements a and b are mutually conjugate if the group contains another element u such that $b = uau^{-1}$. An invariant subgroup contains elements in complete classes. If every element of a group can be expressed as the product of elements of two invariant subgroups, then the group is the direct product of these subgroups. A many-to-one correspondence of the elements of two groups which preserves algebraic structure is a *homomorphism*; similarly, a one-to-one correspondence is an *isomorphism*.

Table 1. Multiplication table for point group C_{3v}.

C_{3v}	E	C_3	C_3^2	σ_a	σ_b	σ_c
E	E	C_3	C_3^2	σ_a	σ_b	σ_c
C_3	C_3	C_3^2	E	σ_c	σ_a	σ_b
C_3^2	C_3^2	E	C_3	σ_b	σ_c	σ_a
σ_a	σ_a	σ_b	σ_c	E	C_3	C_3^2
σ_b	σ_b	σ_c	σ_a	C_3^2	E	C_3
σ_c	σ_c	σ_a	σ_b	C_3	C_3^2	E

Among the various entities which can serve as group elements are those symmetry operations (rotations, reflections and translations) which leave a symmetrical object invariant in appearance. In the case of symmetry groups, the conjugate classes are subject to geometrical interpretation; e.g. rotations through the same angle about two different axes are conjugate elements if the group contains a symmetry operation which takes one axis into the other.

An ideal crystal is an infinite regular repetition of identical structural units in three-dimensional space. Each structural unit is associated with a lattice point, and the collection of such points is a crystal lattice. The set of translation elements which leave a crystal lattice invariant is a translation group. Translational symmetry imposes severe constraints on those point symmetry elements (rotations and reflections) which can leave a crystal structure invariant. In fact, only thirty-two different finite groups of point symmetry elements are compatible with the translational symmetry of a crystal lattice. These crystallographic point groups in turn impose restrictions on the possible translational symmetry, and on that basis they can be classified in seven symmetry systems and can be associated in various ways with the translation groups of fourteen Bravais lattices to form 230 distinct space groups. Although space groups have an infinite number of discrete elements, in practice they are converted to finite groups by the imposition of periodic boundary conditions. Their associated point symmetry is often manifest in crystal morphology.

As an example of a crystallographic point group, consider the group of symmetry operations which leave an equilateral triangle invariant without turning it over. This group, called C_{3v}, has six elements: an identity element E, rotations C_3 through 120 degrees and C_3^2 through 240 degrees about an axis perpendicular to the surface, and reflections σ_a, σ_b and σ_c in three vertical mirror planes; i.e. planes containing the three-fold rotation axis. The multiplication table for C_{3v} is shown in Table 1. The elements C_3 and σ_a can serve as group generators.

3. GROUP THEORY AND QUANTUM MECHANICS

The advent of quantum mechanics in the mid 1920s revolutionized the description of physical phenomena on an atomic scale[4]. The relevance of group theory to quantum mechanics, especially the theory of representations in relation to atomic structure, was recognized almost immediately[5], and the application to solid-state physics soon followed[6,7].

A matrix group is a set of square, non-singular matrices which satisfy group axioms with matrix multiplication as the law of composition. A matrix group which is a homomorphic image of a given group is said to be a matrix representation of that group. Every matrix group is equivalent (under a similarity transformation) to a group of unitary matrices; we will henceforth assume unitary representations. A matrix representation is said to be reducible when all of the matrices can be partitioned into a direct sum in the same way by the same unitary transformation; otherwise it is irreducible. The useful properties of irreducible representations follow from Schur's lemma[8] which states that only a constant matrix can commute with all of the matrices of an irreducible representation. The number of inequivalent irreducible representations of a finite group is equal to the number of its classes, and the sum of squares of their dimensions is equal to its order. Each element of an abelian group is necessarily in a class by itself, with the consequence that all of its irreducible representations are one-dimensional. The traces of the matrices are the same for all elements in a class. The set of traces for a given representation is called the character of the representation, and the set of characters of the irreducible representations of a given group is called the character table of that group.

Table 2. Irreducible representations of point group C_{3v}. The symbol ϵ denotes a cube root of unity: $\epsilon = \exp(i2\pi/3)$.

C_{3v}	E	C_3	C_3^2	σ_a	σ_b	σ_c
A_1	1	1	1	1	1	1
A_2	1	1	1	-1	-1	-1
E	$\begin{bmatrix} 1 & 0 \\ 0 & 1 \end{bmatrix}$	$\begin{bmatrix} \epsilon & 0 \\ 0 & \epsilon^2 \end{bmatrix}$	$\begin{bmatrix} \epsilon^2 & 0 \\ 0 & \epsilon \end{bmatrix}$	$\begin{bmatrix} 0 & 1 \\ 1 & 0 \end{bmatrix}$	$\begin{bmatrix} 0 & \epsilon^2 \\ \epsilon & 0 \end{bmatrix}$	$\begin{bmatrix} 0 & \epsilon \\ \epsilon^2 & 0 \end{bmatrix}$

As an example, consider the point group C_{3v} of Table 1. The identity element E is always in a class by itself. The remaining classes are $\{C_3, C_3^2\}$ and $\{\sigma_a, \sigma_b, \sigma_c\}$. Thus there are three inequivalent irreducible representations, with dimensions 1, 1 and 2. A set of irreducible representations is shown in Table 2. Note that the two-dimensional representation is not unique. The character table shown in Table 3 is unique, however, as a consequence of trace invariance.

In order to establish the connection between symmetry and quantum mechanics, we must also consider the isomorphic mapping of a symmetry group on a group of transformations R of a set of variables r, and on a group of operators O_R which transform functions $\psi(r)$ according to the recipe

$$O_R\psi(r) = \psi(R^{-1}r). \tag{1}$$

The quantum mechanical description of a physical system in a stationary state is accomplished by solution of the Schroedinger equation[4],

$$H\psi(k, i; r) = E(k)\psi(k, i; r), \tag{2}$$

where H is the Hamiltonian operator incorporating the kinetic energy and potential energy of interaction of the various particles which constitute the system, as well as their interactions with external forces. In general, H contains differential operators and Eq. (2) is a partial differential equation. The wavefunction $\psi(k, i; r)$ is a solution under specific boundary conditions; the index i distinguishes degenerate eigenfunctions corresponding to the same eigenvalue $E(k)$. The function $|\psi(k, i; r)|^2$ is interpreted as the probability density that the particles of the system have coordinates r, and the eigenvalues $E(k)$ are the possible results of a measurement of the total energy of the system. Typically, the eigenvalue spectrum includes several discrete values corresponding to bound states of the system plus a continuous range of values above a dissociation (or ionization) threshold.

The group G of the Hamiltonian H is the set of symmetry operations O_R which leave the Hamiltonian invariant,

$$O_R H O_R^{-1} = H. \tag{3}$$

It follows from Eqs. (2) and (3) that $\psi(k, i; r)$ and $O_R\psi(k, i; r)$ are degenerate eigenfunctions of H; accordingly, since the set of eigenfunctions can be shown to be complete, they are related by

$$O_R\psi(k, i; r) = \sum_j \psi(k, j; r) D(k, R; j, i). \tag{4}$$

It is readily demonstrated, by successive application of symmetry operations O_R and O_S, that the coefficients $D(k; R; j, i)$ are elements of the matrices $D(k; R)$ of a matrix representation

Table 3. Character table for point group C_{3v}.

C_{3v}	E	$\{C_3, C_3^2\}$	$\{\sigma_a, \sigma_b, \sigma_c\}$
A_1	1	1	1
A_2	1	1	-1
E	2	-1	0

of G. Application of O_S to both sides of Eq. (4) yields

$$O_S O_R \psi(k, i; r) = \sum_j \psi(k, l; r) \sum_l D(k; S; l, j) D(k; R; j, i). \tag{5}$$

On the other hand, it follows from Eq. (1) that application of O_S to the function $O_R \psi(k, i; r)$ gives, with the help of Eq. (4),

$$O_S O_R \psi(k, i; r) = O_R \psi(k, i; S^{-1}r) = \psi(k, i; R^{-1}S^{-1}r)$$
$$= O_{SR} \psi(k, i; r) = \sum_l \psi(k, l; r) D(k; SR; l, i) \tag{6}$$

Comparison of Eqs. (5) and (6) reveals that

$$D(k; SR; l, i) = \sum_j D(k; S; l, j) D(k; R; j, i), \tag{7}$$

which is just the rule for matrix multiplication. Since the algebraic structure of the group is also preserved in Eq. (7), the matrices $D(k; R)$ are elements of a matrix representation.

Typically, the representation is irreducible, and the degeneracy is said to be *symmetry-induced*; otherwise, it is called *accidental*. Thus, barring accidental degeneracy, the eigenfunctions and eigenvalues of the Hamiltonian operator H can be labeled by the irreducible representations of the group G of the Hamiltonian. This result has profound consequences, since a wide variety of physical properties and processes, such as selection rules for transitions between energy eigenstates, can be elucidated largely on the basis of symmetry arguments.

It is often the case that the Hamiltonian H can be expressed as a sum of two terms H_0 and H' where H_0 has a higher symmetry than H, and where H' is a small perturbation. In particular, H_0 may describe sub-systems, e.g. H_1 and H_2, which are invariant under independent transformations of their respective coordinates, and H' may describe an interaction of the sub-systems which is invariant only under simultaneous transformations of their coordinates. The group G_0 of the unperturbed Hamiltonian H_0 is then the direct product of its invariant subgroups G_1 and G_2, $G_0 = G_1 \times G_2$, and its irreducible representations are related to those of the subgroups by $D_0(k; R_1 R_2) = D_1(i; R_1) \times D_2(j; R_2)$, where " $\times$ " denotes a matrix direct product. The group G of the perturbed Hamiltonian is then the subgroup of G_0 with elements $R = R_1 = R_2$. The corresponding subset of matrices $D_1(i; R) \times D_2(j; R) = D(i \times j; R)$ is a subduced representation of G, called the Kronecker product representation, which is generally reducible; i.e.

$$D(i \times j; R) = \sum_k a(k) D(k; R), \tag{8}$$

where a direct sum of matrices is intended on the right-hand side. The coefficients $a(k)$, which are readily inferred from the character table, are non-negative integers indicating the number of times each irreducible representation appears. Reduction of the Kronecker product representation reveals how the degenerate eigenvalues of H_0 are split by the perturbation H', including the symmetry designations of the split-off levels and their residual degeneracies.

The Kronecker product representation is also useful when H' is an external perturbation with symmetry lower than that of H_0. In particular, H' may transform as a basis for an irreducible representation γ of G_0, other than the identity representation. The perturbation H' then splits a degenerate energy level of H_0 labeled by irreducible representation Γ, provided that Γ is contained in the Kronecker product representation $\Gamma \times \gamma$. We shall return to this point in Sec. 6.

4. POINT IMPERFECTIONS

Point imperfections in insulating and semiconducting crystals include isolated impurities, structural defects of atomic dimensions and impurity-related defects. They may be introduced deliberately by doping the starting material prior to crystal growth, by additive coloration (e.g.

heating the crystal in a metal vapor), by diffusion or by radiation damage; in any event, incorporation of structural defects is thermodynamically unavoidable during crystal growth. Even in minute concentrations, these point imperfections dominate the optical, magnetic and electrical properties of insulators and semiconductors. They are of enormous technological importance to devices such as solid-state lasers and semiconductor electronic circuits, both in lending useful characteristics to the materials on which these devices depend and in limiting their performance.

The incorporation of a point imperfection in a crystal entails a substantial reduction in symmetry. In particular, the translational symmetry is lost and the residual point symmetry may be reduced as well; thus the symmetry of the point imperfection is characterized by a subgroup of the original point group.

Experimental techniques which have been developed for investigation of point imperfections in solids include optical absorption and fluorescence, electron paramagnetic resonance (EPR), electron-nuclear double resonance (ENDOR), optically detected magnetic resonance (ODMR), two-photon absorption (TPA) and deep-level transient spectroscopy (DLTS). Measurements are made as functions of temperature and hydrostatic pressure, which tend to preserve symmetry, barring phase transitions; and as functions of uniaxial stress and external electric and magnetic fields, which do not. Analysis of symmetry as inferred from experimental spectra plays a major role in the identification and characterization of point imperfections. Computer modeling of imperfections also relies heavily on symmetry considerations.

Only a few of the myriad known imperfections will be mentioned here as examples. Structural defects in insulators include vacant sites and interstitial ions. These defects acquire interesting optical and magnetic properties when they trap excess electrons or holes. (A hole is the absence of an electron where there should be one; it should not be confused with a vacancy, which is a missing ion.) The prototype of such defects, which are called color centers, is the F center in alkali halides. It consists of a single electron trapped at a negative halogen ion (anion) vacancy, and it introduces an optical absorption band in the otherwise transparent crystal, lending it a color which depends on its composition. It also contributes paramagnetic susceptibility to an otherwise diamagnetic crystal, a feature which has facilitated its very detailed characterization by EPR and ENDOR. Other color centers in alkali halides include the F_2 or M center and the F_3 or R center, which are aggregates of two and three F centers, respectively. The V_K center is a self-trapped hole which is shared between two adjacent anions, and is more properly described as a small polaron, since no ionic rearrangement is involved. A related center is the H center in which a hole is shared by two anions occupying a single anion site. It is evident that these other color centers have lower symmetry than the F center.

Transition metals such as chromium, rare-earth metals such as neodymium, and heavy metals such as thallium can be incorporated in ionic crystals as substitutional impurities on positive-ion (cation) sites. Because of their special places in the periodic table of elements, these impurities also introduce optical absorption bands and contribute paramagnetic susceptibility. Impurity-related defects are important, as well. For example, KCl containing $Tl^0(1)$ centers (substitutional thallium atoms adjacent to anion vacancies) is a useful laser material.

Other defects and impurities are prominent in semiconductors. Substitutional phosphorous and aluminum in silicon contribute excess conduction electrons and valence holes, respectively, essential to the operation of semiconductor devices. These are called shallow impurities, in contrast to deep-level impurities such as chromium. Examples of structural defects are the neutral vacancy in silicon and the antisite pair in gallium arsenide; the latter consists of a gallium atom on an arsenic site adjacent to an arsenic atom on a gallium site.

5. APPLICATION OF GROUP THEORY TO POINT IMPERFECTIONS

The electronic structure of a perfect crystal is described in terms of its space group representations, which are specified with reference to its reciprocal lattice[7,9]. The position vectors of direct-lattice points are expressible as linear combinations of fundamental translation vectors $\mathbf{b}(1)$, $\mathbf{b}(2)$ and $\mathbf{b}(3)$, with integral coefficients. The fundamental translation vectors of the reciprocal lattice, $\mathbf{d}(1)$, $\mathbf{d}(2)$ and $\mathbf{d}(3)$, are defined by the condition that $\mathbf{b}(i) \cdot \mathbf{d}(j)$ equals 1 for $i = j$ and vanishes otherwise. The subgroup of pure translations is abelian, and its one-dimensional representations are labeled by a set of wave vectors which are consistent with the periodic boundary conditions, and which occupy the smallest volume of reciprocal space bounded by plane perpendicular bisectors of reciprocal lattice vectors, called the first Brillouin zone. The subset of point group operations which leave the wave vector invariant, modulo a reciprocal lattice vector, form a subgroup called the group of the wave vector. The irreducible representations of this group, together with the wave vector itself, label the space group representations. It is convenient to display the energy eigenvalues of the Hamiltonian as quasi-continuous, multivalued functions of the wave vector, called energy bands. The corresponding electronic wavefunctions are called Bloch functions.

Some semblance of the perfect crystal electronic bandstructure is retained in the electronic structure of shallow impurities in semiconductors, in that their wavefunctions are well approximated by a superposition of Bloch functions belonging to a single band. The sum over wave vectors is a reflection of the loss of translational symmetry. Many bands would be required in order to approximate the wavefunctions of more compact states; instead, the description of the electronic structure of impurity ions and color centers in insulators proceeds from a different starting point, as elaborated below.

At a certain level of approximation, the Hamiltonian for an isolated atom or ion at rest can be expressed as a sum of operators as follows:

$$H = T + V + H(\text{el}) + H(\text{so}), \tag{9}$$

where T is the electronic kinetic energy operator; V is an effective central potential energy function including the interaction of each electron with the positive nucleus and the spherically averaged charge distribution of the remaining electrons; $H(\text{el})$ is the residual mutual electrostatic interaction of the electrons; and $H(\text{so})$ is a term of relativistic origin which couples the orbital angular momentum of each electron with its intrinsic (spin) angular momentum. The last three interactions are written in descending order of strength, but each successive interaction entails a reduction of symmetry.

In the central-field approximation, $H \simeq T + V$, the Hamiltonian is invariant under independent rotations of the space coordinates of each electron. These rotations are elements of a continuous (Lie) group, the rotation group $O^+(3)$. They are generated by orbital angular momentum operators $\mathbf{l}$, and the operator which effects a rotation through an angle θ about an axis whose direction is specified by unit vector $\mathbf{n}$ is

$$O_R = \exp\left(-i\theta\mathbf{n}\cdot\mathbf{l}\right). \tag{10}$$

Irreducible representations of the rotation group are labeled by non-negative integers l, called orbital angular momentum quantum numbers. The rows of these $2l + 1$-dimensional representations are labeled by integers m_l which satisfy the inequalities $-l \le m_l \le l$. The Hamiltonian is also invariant under independent rotations of the spin coordinates of each electron. The spin angular momentum quantum number s, which has the value $1/2$ for every electron, actually labels a faithful representation of the universal covering group $SU(2)$, the unitary, unimodular group of transformations of two complex variables, whose elements have a two-to-one correspondence with the elements of $O^+(3)$. The rows of this representation, which may be regarded as a double-valued representation of the rotation group, are labeled by $m_s = + 1/2, -1/2$. Electronic configurations of the free ion in the central-field approximation are classified by the principal quantum numbers $n(i)$ and the angular momentum quantum numbers $l(i)$, and the degenerate eigenstates belonging to these electronic configurations are distinguished by the quantum numbers $m_l(i)$ and $m_s(i)$, where the index i distinguishes occupied one-electron wavefunctions (spin-orbitals). The Hamiltonian is also invariant under inversion of the space coordinates, and thus the states must have even or odd parity. This property is not independent, but is determined by the sum of the $l(i)$, which must also be even or odd. Finally, the Hamiltonian is invariant under independent permutations of both space and spin coordinates of the several electrons. The Pauli principle expresses an additional constraint on the acceptable solutions of the Schroedinger equation, however; only those wavefunctions which are totally antisymmetrical under simultaneous transpositions of the space and spin coordinates of any two electrons are physically meaningful.

When the mutual electrostatic interaction of the electrons is taken into account, there is a sharp reduction in symmetry. The Hamiltonian at this level of approximation, $T + V + H(\text{el})$, is no longer invariant under independent rotations of the space coordinates of different electrons, but it remains invariant under simultaneous rotations, generated by the total orbital angular momentum operator $\mathbf{L} = \mathbf{l}(1) + \mathbf{l}(2) + \ldots$. Furthermore, the Pauli principle is only compatible with simultaneous rotations of all of the spin coordinates. Thus the symmetry-induced degeneracy of the electronic configurations in the central-field approximation is partially removed, and they split into terms labeled by the angular momentum quantum numbers L and S (the multiplet structure). The quantum numbers $n(i)$ and $l(i)$ remain valid to the extent that $H(\text{el})$ is a small perturbation on the centralfield Hamiltonian, but the quantum numbers $m_l(i)$ and $m_s(i)$ are superceded by M_L and M_S, which distinguish degenerate states belonging to a single term. The parity of each term is the same as that of its parent configuration. Finally, when the still weaker spin-orbit interaction is taken into account, the Hamiltonian, $H = T + V + H(\text{el}) + H(\text{so})$, is invariant only under simultaneous rotations of all space and all spin coordinates, generated by the total angular momentum operator, $\mathbf{J} = \mathbf{L} + \mathbf{S}$. The terms are further split into fine-structure levels labeled by quantum numbers $n(i)$, $l(i)$, L, S and J, and the degenerate states within each fine-structure level are distinguished by M_J. Note that the quantum numbers S, M_S, J and M_J are integral or half-integral according to whether the number of electrons is even or odd.

From another point of view, the unitary transformations of a set of degenerate one-electron wavefunctions in the central-field approximation, $H \simeq T + V$, which transform as a basis for a $2l + 1 -$ dimensional irreducible representation of $O^+(3)$, are elements of a unitary, unimodular group $SU(2l + 1)$, of which $O^+(3)$ is a subgroup. When the interaction $H(\text{el})$ is included in H, irreducible representations of $SU(2l + 1)$ provide an additional classification scheme for different terms with the same values of L and S.[10] The higher $SU(n)$ groups also play a central role in the physics of elementary particles[1].

When the atom or ion is incorporated as an impurity in a crystalline solid, then, in principle, the Hamiltonian must be enlarged to encompass all of the particles which compose the solid. A major simplification is achieved by exploiting the large mass ratio of ions and electrons. In the Born–Oppenheimer approximation[11], the electronic problem is first solved for fixed nuclear coordinates Q to yield energy eigenvalues, $U_n(Q)$, which depend implicitly on the nuclear coordinates. These so-called adiabatic potential energy functions, together with the nuclear kinetic energy operators T_N, are subsequently used to determine the nuclear motions. In practice, the problem is rendered tractable by a phenomenological representation of the solid environment. In the harmonic approximation, the Hamiltonian of Eq. (9) is augmented by $H(\text{int})$, representing the interaction of the impurity with the host crystal, and $H(\text{cr})$, including the kinetic energy of the host ions and their mutual adiabatic potential energy of interaction to terms quadratic in their displacements from equilibrium. (The model can be elaborated to accommodate ionic polarizabilities as well, but that refinement will be omitted here.)

The classification of impurity eigenstates depends critically on the relative strengths of $H(\text{int})$ and the hierarchy of free-ion interactions V, $H(\text{el})$ and $H(\text{so})$. In the strong-field approximation, $H(\text{int}) \gg H(\text{el})$, the eigenstates of $T + V + H(\text{int})$ are labeled by $n(i)$, $l(i)$, $\gamma(i)$, $m_\gamma(i)$ and $m_s(i)$, where m_γ denotes a row of irreducible representation γ of the crystallographic point group appropriate to the impurity site in its equilibrium lattice configuration; i.e., with the ions in their equilibrium positions. The degenerate strong-field electronic configurations labeled by $n(i)$, $l(i)$ and $\gamma(i)$ are split by $H(\text{el})$ into terms labeled by Γ and S, whose degenerate eigenstates are distinguished by M_Γ and M_S in place of $m_\gamma(i)$ and $m_s(i)$. These terms are further split by $H(\text{so})$ into fine-structure levels, labeled by Γ', whose degenerate eigenstates are distinguished by M_Γ' in place of M_Γ and M_S. For impurities with an odd number of electrons, Γ' denotes an irreducible representation of a double point group, which bears the same relation to the crystallographic point group as $SU(2)$ bears to $O^+(3)$; i.e. there is a two-to-one homomorphism.

Two additional approximations are of interest. The free-ion multiplet structure is retained in the medium-field approximation, $H(\text{el}) \gg H(\text{int}) \gg H(\text{so})$. The free-ion terms are split into crystal-field terms by $H(\text{int})$, and these are further split into fine-structure levels by $H(\text{so})$, with symmetry designations $n(i)$, $l(i)$, L, S, Γ and Γ', whose degenerate eigenstates are distinguished by M_Γ'. Finally, in the weak-field approximation, $H(\mathrm{so}) \gg H(\mathrm{int})$, the free-ion fine-structure levels are retained as well. These levels are split into crystal-field fine-structure levels by $H(\mathrm{int})$, with symmetry designations $n(i)$, $l(i)$, L, S, J and Γ', whose degenerate eigenstates are distinguished by M_Γ'.

The symmetry of a point imperfection in a solid can be reduced further by imposition of external electric or magnetic fields or by uniaxial stress. These interactions are ordinarily very much weaker than the spin-orbit interaction, $H(\mathrm{so})$, and produce further splitting of the crystal-field fine-structure levels. The most important of these interactions is that with an external magnetic field (the so-called Zeeman interaction), since transitions between Zeeman levels are exploited in the EPR, ODMR and ENDOR techniques mentioned in Section 4. The dependence of these transitions on the orientation of the magnetic field with respect to the crystal axes provides useful diagnostic information concerning the point symmetry of the imperfection. Interaction with the magnetic moments associated with nearby atomic nuclei produces a hyperfine splitting of Zeeman levels which yields even more, often decisive, information in EPR, ODMR and, especially, ENDOR experiments.

Transition-metal impurities provide examples of both the strong-field and medium-field cases, corresponding to covalent and ionic complexes, respectively. The interactions $H(\mathrm{el})$ and $H(\mathrm{int})$ are actually comparable in either case, and should be considered simultaneously within each free-ion electronic configuration in a more rigorous treatment[12]. Rare-earth impurities have greatly enhanced spin-orbit interactions $H(\mathrm{so})$ by virtue of their positions in the periodic table of elements, and their interaction with the crystal field $H(\mathrm{int})$ is diminished by the compactness of their relevant wavefunctions. Consequently, they provide relatively unambiguous examples of the weak-field case[13]. Heavy-metal impurities such as thallium in alkali halides have even stronger spin-orbit interactions and diffuse wavefunctions as well. For them, $H(\mathrm{el})$, $H(\mathrm{int})$ and $H(\mathrm{so})$ are all comparable, and must all be considered simultaneously within each free-ion electronic configuration. The only valid symmetry designations are then $n(i)$, $l(i)$, Γ' and M_Γ'[14]. Finally, electron-excess color centers provide an extreme example of the strong-field case; for them, there is no central potential V distinguishable from $H(\mathrm{int})$, and the quantum numbers $l(i)$ are not appropriate, although they are retained in some F-center models which obscure the anisotropy of $H(\mathrm{int})$[15].

6. THE JAHN–TELLER EFFECT

The physics of point imperfections in solids is greatly enriched by the additional degrees of freedom associated with ion displacements. Symmetry arguments can go a long way toward elucidating the effects of these displacements. The adiabatic potential energy functions $U_n(Q)$ may be expressed in terms of symmetry-adapted combinations of ion displacements $Q(\gamma, \mu; j)$ which transform as bases for rows μ of irreducible representations γ of the point group G_0 of the electronic Hamiltonian H_0 in the perfect-lattice configuration. There is an adiabatic potential energy surface in the space of the symmetry-adapted ion displacements corresponding to each non-degenerate electronic state. Intersections of adiabatic potential energy surfaces in symmetrical lattice configurations are a consequence of symmetry-induced degeneracy.

The symmetrical static distortion around a point imperfection is the set of equilibrium displacements of ions from their perfect lattice positions; i.e. the displacements for which the adiabatic potential energy is minimum. When the nuclear kinetic energy operator T_N is included, a set of vibronic states is obtained for each non-degenerate electronic state, involving small vibrations of ions about their equilibrium displacements. The dependence of the static distortion on electronic state has a number of interesting consequences for transitions between vibronic states such as the Stokes shift in emission, the broadening of absorption and fluorescence spectra, and the occurrence of non-radiative transitions.

More exotic effects can occur in degenerate electronic states. It was demonstrated by Jahn and Teller[2] that, for each multidimensional irreducible representation Γ of a crystallographic point group G there exists another irreducible representation, γ, such that the Kronecker product representation $\Gamma \times \gamma$ contains Γ. The implication of this result is that, for each degenerate electronic state Γ in a symmetrical configuration of the lattice, there may exist an asymmetrical symmetry-adapted displacement $Q(\gamma, \mu; j)$ which removes the degeneracy. (See the discussion of the Kronecker product representation in Section 3.) It can be demonstrated further that at least one of the split-off levels has lower energy for small displacements; the consequent asymmetric distortion is an example of spontaneously broken symmetry. Its magnitude is limited by the elastic potential energy which is quadratic in $Q(\gamma, \mu; j)$ with a positive coefficient. The number of distortions which are equivalent under point symmetry operations, distinguished by the index μ, is equal to the dimension of irreducible representation γ. This static Jahn–Teller effect is manifest in the reduced symmetry of optical and EPR spectra. The Jahn–Teller effect applies to molecules as well, and the orbitally degenerate ground states of linear molecules provide the only known exception.

The effect of including the nuclear kinetic energy T_N depends on the strength of coupling to asymmetric displacements $Q(\gamma, \mu; j)$. In the strong coupling limit, the vibronic states simply correspond to small vibrations of the ions about their distorted equilibrium positions. However, the probability of tunneling between equivalent distortions is greatly enhanced in the weak-coupling limit, leading to the dynamic Jahn–Teller effect. In this limit, vibronic states labeled by irreducible representations of the group of the Hamiltonian in the symmetrical lattice configuration provide a more appropriate description. The dynamic Jahn–Teller effect may be viewed as a failure of the Born–Oppenheimer approximation for degenerate electronic states, since the vibronic wavefunctions are no longer separable into electronic and vibrational factors. Manifestations of the dynamic Jahn–Teller effect are much more subtle than for the static effect, since the ground vibronic state generally has the same symmetry as the ground electronic state. Thus the full symmetry of optical and EPR spectra is restored. There is a quantitative manifestation, however, known as the Ham effect[16]. The efficacy of operators such as the spin–orbit interaction may be substantially diminished in the ground vibronic state, in comparison with a purely electronic state of the same symmetry, as a consequence of the small overlap of vibrational wavefunctions.

As an example, we consider the electronic structure of the F_3 or R center in KCl[17,18]. The R center consists of three electrons trapped in three adjacent anion vacancies which form an equilateral triangle, as shown in Fig. 1. The point symmetry is C_{3v}, the example considered

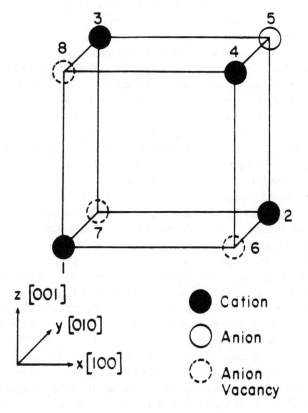

Fig. 1. Geometry of the R center in KCl.

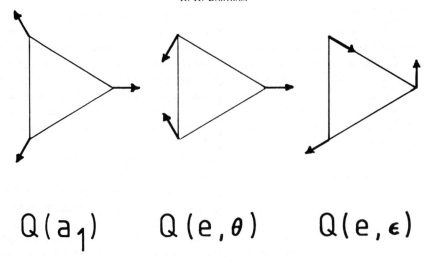

$$Q(a_1) \qquad Q(e,\theta) \qquad Q(e,\epsilon)$$

Fig. 2. Symmetry-adapted distortions of an equilateral triangle.

in Section 2, above. At the level of approximation $H \simeq T + H(\text{int})$, the one-electron state of lowest energy (the ground state) is an a_1 state, while the state of next lowest energy (the first excited state) is an e state. The Pauli principle permits at most two electrons to occupy a single, non-degenerate state, with spin projections $m_s = +1/2$ and $-1/2$. Consequently, the ground configuration of the R center is $a_1^2 e$. Since $A_1 \times A_1 \times E = E$, the degeneracy of the ground configuration is not removed by $H(\text{el})$, and there is a single term of E symmetry with $S = 1/2$.

The symmetry-adapted displacements which distort an equilateral triangle are illustrated in Fig. 2. The symmetrical dilatation, $Q(a_1)$, preserves the C_{3v} symmetry. The two distortions which transform as bases for the e representation, $Q(e, \theta)$ and $Q(e, \epsilon)$, are relevant to the Jahn–Teller effect. Of course, in the actual crystal, the surrounding ions are displaced rather than the vacancies, but the symmetry classification of the distortions remains valid.

The case of an electronic state of E symmetry coupled to lattice modes of e symmetry (an $E \times e$ system) has been treated exhaustively by Longuet–Higgins, Öpik, Pryce and Sack (L-HOPS)[19]. The adiabatic potential energy surfaces are illustrated in Fig. 3(a), in an approximation which retains only terms linear and quadratic in the displacements. In this approximation, there is an equilibrium value of the radial displacement $r = [Q(e, \theta)^2 + Q(e, \epsilon)^2]^{1/2}$, but all values of the polar angle $\phi = \tan^{-1}[Q(e, \epsilon)/Q(e, \theta)]$ are equally probable. As a consequence, the two-coordinate system has only one normal mode of vibration of finite frequency, associated with radial displacements; the unhindered azimuthal displacement may be viewed as a normal mode of zero frequency. This zero-frequency vibration is the analogue of the Goldstone massless boson in the theory of elementary particles[1,20,21]. However, the continuous symmetry in the present problem is an artifact of our approximation. When bilinear and cubic terms are retained in the expansion of the adiabatic potential energy in powers of symmetry-adapted displacements, the adiabatic potential energy surface is warped in such a fashion as to introduce three stable minima, 120 degrees apart in ϕ, as illustrated in Fig. 3b, and all vibration frequencies become finite.

The problem of the dynamic Jahn–Teller effect for an $E \times e$ system was also addressed by L-HOPS[19]. It is readily established from the character table for C_{3v}, Table 3, that $E \times E = A_1 + A_2 + E$. (The product of traces on the left-hand side should equal the sum of traces on the right-hand side for each conjugate class.) Thus the E-symmetry ground electronic state spawns vibronic states labeled by all three representations. L-HOPS established that the ground vibronic state has E symmetry as well. The linear Jahn–Teller coupling for the R center in KCl was found to be relatively strong, but the warping terms are sufficiently weak that there is no stabilization with respect to the polar angle ϕ and the dynamic description remains valid. In this system, the Ham effect reduces the effective spin-orbit interaction by an order of magnitude[18].

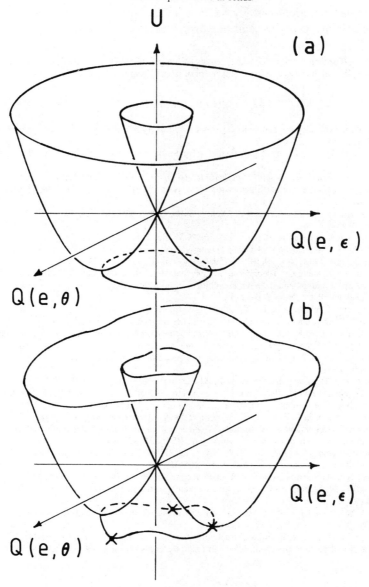

Fig. 3. (a) Adiabatic potential energy surfaces for an $E \times e$ Jahn–Teller system in the linear-coupling approximation; the "Mexican-hat" potential. (b) Adiabatic potential energy surfaces for the same system when bilinear and cubic warping terms are included.

A variant of the Jahn–Teller effect is the pseudo-Jahn–Teller effect, involving a near-accidental degeneracy of electronic states. This effect has been invoked to explain properties of the relaxed excited state of the F center[22], the positively-charged oxygen vacancy (E_1' center) in SiO_2[23] and small polarons[24].

7. EPILOGUE

Point imperfections in solids exemplify, in microcosm, the utility of symmetry principles in physics. They embody a hierarchy of interactions of successively diminishing symmetry, and they fully exploit the connection between group theory and quantum mechanics. The crystallographic point groups and double groups which characterize them are conceptually simpler than the Lie groups which describe continuous symmetries. They also provide examples of spontaneously broken symmetry in the static Jahn–Teller effect.

The dynamic Jahn–Teller effect, on the other hand, appears to be peculiar to point defects and molecules, as opposed to extended systems, since it depends on tunneling rather than thermally-activated transitions between rotationally equivalent distortions. Thus it may have no proper analogue in the theory of elementary particles.

Acknowledgements—The author gratefully acknowledges the support of the U. S. Army Research Office, Contract No. DAAG 29-82-K-0158, and the advice and encouragement of Istvan Hargittai, during the preparation of this manuscript.

REFERENCES

1. P. D. Mannheim, Symmetry and spontaneously broken symmetry in the physics of elementary particles. *CAMWA*, 000, 000 (1985) (this issue).
2. H. A. Jahn and E. Teller, Stability of polyatomic molecules in degenerate electronic states. *Proc. Roy. Soc.* **A161**, 220–235 (1937).
3. M. Lax, *Symmetry Principles in Solid State and Molecular Physics*. Wiley-Interscience, New York (1974).
4. I. Schroedinger, Quantisierung als Eigenwertproblem. *Ann. Physik* **79**, 361–376, 489–527; **80**, 437–490; **81**, 109–139 (1926).
5. E. P. Wigner, Einige Folgersungen aus der Schroedingerschen Theorie fur die Termstrukturen. *Zeits. f. Physik* **43**, 624–652 (1927).
6. H. Bethe, Termaufspaltung in Kristallen. *Ann. Physik* **3**, 133–208 (1929).
7. L. P. Bouckaert, R. Smoluchowski and E. P. Wigner, Theory of brillouin zones and symmetry properties of wave functions in crystals. *Phys. Rev.* **50**, 58–67 (1936).
8. I. Schur, Neue Begrundung der Theorie der Gruppencharactere. *Berl. Ber.* 406–432 (1905).
9. G. F. Koster, Space groups and their representations, in *Solid State Physics*, Vol. 5, (edited by F. Seitz and D. Turnbull). Academic Press, New York (1957).
10. G. Racah, Theory of complex spectra. IV. *Phys. Rev.* **76**, 1352–1365 (1949).
11. M. Born and J. R. Oppenheimer, Zur Quantentheorie der Molekeln. *Ann. Physik* (Leipzig) **84**, 457–484 (1927).
12. S. Sugano, Y. Tanabe and K. Kamimura, *Multiplets of Transition-Metal Ions in Crystals*. Academic Press (1970).
13. S. Hufner, *Optical Spectra of Transparent Rare Earth Compounds*. Academic Press, New York (1978).
14. L. F. Mollenauer, N. D. Viera and L. Szeto, Optical properties of the $Tl^0(1)$ center in KCl. *Phys. Rev.* **B27**, 5332–5346 (1983).
15. A. M. Stoneham, *Theory of Point Defects in Solids*. Oxford University Press (1975).
16. F. S. Ham, Dynamical Jahn–Teller effect in paramagnetic resonance spectra: orbital reduction factors and partial quenching of spin–orbit interaction. *Phys. Rev.* **138**, A1727–A1740 (1965).
17. R. H. Silsbee, R center in KCl; stress effects in optical absorption. *Phys. Rev.* **138**, A180–A197 (1963).
18. R. C. Kern, G. G. DeLeo and R. H. Bartram, R center in KCl. I. Point-ion electronic structure calculation with application to magneto-optical parameters. *Phys. Rev.* **B24**, 2211–2221 (1981).
19. H. C. Longuet-Higgins, U. Opik, M. H. L. Pryce and R. A. Sack, Studies of the Jahn–Teller effect II. The dynamical problem. *Proc. Roy. Soc.* **A244**, 1–16 (1958).
20. J. Goldstone, A. Salam and S. Weinberg, Broken symmetries. *Phys. Rev.* **127**, 965–970 (1962).
21. J. Sarfatt and A. M. Stoneham, The Goldstone theorem and the Jahn–Teller effect. *Proc. Phys. Soc.* **91**, 214–221 (1967).
22. F. S. Ham, Vibronic model for the relaxed excited state of the F center. I. General solution. *Phys. Rev.* **B8**, 2926–2944 (1973).
23. K. L. Yip and W. B. Fowler, Electronic structure of the E' center in SiO_2, *Phys. Rev.* **B11**, 2327–2338 (1975).
24. O. Schirmer, Small polaron aspects of defects in oxide materials. *J. Physique* **41**, C6-479-484 (1980).

Comp. & Maths. with Appls. Vol. 12B, Nos. 1/2, pp. 197–227, 1986
Printed in Great Britain.

0886-9561/86 $3.00 + .00
© 1986 Pergamon Press Ltd.

ON SYMMETRY AND ASYMMETRY IN LITERATURE†

B. PAVLOVIĆ

P.E.N.-Croatian Center, Trg Republike 7, 41000 Zagreb Croatia, Yugoslavia

and

N. TRINAJSTIĆ

The Rugjer Bošković Institute, P.O.B. 1016, 41001 Zagreb, Croatia, Yugoslavia

Abstract—An attempt is made to investigate whether or not the principles of symmetry and asymmetry operate in literature. The result of the analysis is that certain (near) symmetric and/or asymmetric patterns may be found in literary works. Commonly, the authors use some convenient symmetric (asymmetric) figure, either as a descriptive means or as an underlying core around which their stories unfold. A structure of a literary work may also be interpreted in terms of symmetric and/or asymmetric objects. Sometimes a graphic shape of a work, most often of a poem, possesses symmetry. The content of a literary work may likewise be associated with some symmetric or asymmetric configuration. The citations from several literary works are included.

> Symmetrisch oder nicht–
> das ist die Frage.
> Werner Gilde[1]

INTRODUCTION

Symmetry is an enormous subject[1–3]. The concept of symmetry is one of the most fundamental concepts in science[1–21], but also appears as a significant principle in nature, fine arts, and many human creative—and other—activities[1–3,22–27a]. Mathematics lies at its foundation. Symmetry, as an artistic conception and a medium of expression, has been with us since man's earliest attempts to communicate. Examples of symmetry in art and architecture are to be found everywhere[28]. In fact, the word symmetry[29] originated in art and is attributed to the Greek sculptor Polykleitos[2] (5th century B.C.).

Since symmetry is considered to be a universal principle[1–27,29], there is no reason why it should not be present, in some form, in literary works. Some examples can immediately be given. For instance, the following lines are taken from Chapter 1 of Daphne du Maurier's *Rebecca*[30a]:

> There was Manderley, our Manderley, secretive and silent as it had always been, the grey stone shining in the moonlight of my dream, the mullioned windows reflecting the green lawns and the terrace. Time could not wreck the perfect symmetry of those walls, not the site itself, a jewel in the hollow of a hand.

In Chapter 4 of the same book we find that the perfect symmetry of Manderley could not be removed even by the crude painting of it used for a picture postcard[30b]:

> It was the painting of a house crudely done of course and highly coloured, but even those faults could not destroy the symmetry of the building, the wide stone steps before the terrace, the green lawns stretching to the sea.

And again in Chapter Eleven we hear about the perfect symmetry of Manderley[30c]:

> We came around the sweep of the drive and Manderley was before us, serene and peaceful in the hollow of the lawns, surprising me as it always did, with its perfect symmetry and grace, its great simplicity.

---

† This essay is dedicated to Zagreb, the capital of Croatia and a beautiful city with a nice blend of historical core and modern parts, rich with evidence for all aspects of urban symmetry, asymmetry, and anti-symmetry.

As the next example we give two stanzas (the first and the last) from William Blake's beautiful poem on creation "The Tyger"[31]:

> Tyger! Tyger! burning bright
> In the forest of the night,
> What immortal hand or eye
> Could frame thy fearful symmetry?
>
> Tyger! Tyger! burning bright
> In the forest of the night,
> What immortal hand or eye,
> Dare frame thy fearful symmetry?

As the third example we give a few lines from Anna Wickham's "Envoi"[32] in which the poetess addresses God as the Divine Being of great symmetry:

> God, Thou great symmetry,
> Who put a biting lust in me
> From whence my arrows spring,
> For all the frittered days
> That I have spent in shapeless ways
> Give me one perfect thing.

What is symmetry? There are several possible answers to this question[1–3,14,27]. Intuitively we may say, in concordance with the meaning of the word, that a symmetry is a beauty of form, arising from harmony of its proportions. This is a rather vague definition, but often encountered (see the above citations). Another definition is as follows: Symmetry is the characteristic of an object that allows one to say that two or more parts of it, with reference to a point, a line, or a plane, are the same. In geometry symmetry is defined as the invariance of a configuration of elements under a group of automorphic transformations[2,27]. Shubnikov and Koptsik[27] offered a very general definition of symmetry. They defined symmetry as the law of composition of structural objects, or, more precisely, as the group of permissible one-to-one transformations preserving the structural integrity of the systems under consideration (p. 308 of ref. [27]).

Symmetry may possibly be considered as a very convenient basis for relating mathematics, physics, chemistry, biology, astronomy, art, sports, architecture, music, film, dance, linguistics, philosophy, archeology, geology, literature, etc., from the same standpoint[1–3]. Whilst several of the above fields (besides mathematics and natural sciences, e.g. art and architecture) have been studied, to some extent, from the symmetry point of view[1–3,22–27], literature, save poetry[1,2,22,33–36], is a rather neglected field in discussing symmetry[37]. Apparently, most humans find exact or near exact symmetry more appealing in the visual world than in music and poetry. However, we do find rhythm (translational symmetry in time) agreeable in music and poetry, at least as a background that generally has a much lower, but not absent, symmetry.

In the present essay we will center our discussion mainly on the symmetric and asymmetric objects and patterns that we hope to detect in fiction. Examples from poetry (metrics in poetry is closely related to symmetry) will also be used. Drama will be mentioned only cursorily. The number of examples will be rather limited because each of these fields (fiction, poetry, drama) would require separate lengthy studies. We will also attempt to interpret the literary works in terms of symmetric and/or asymmetric configurations. We will be concerned much more with the symmetry (and asymmetry) of the structure[38] and the content of a literary work than with that of its form. However, the symmetry or asymmetry of the form of literary works, especially poetry, is also an interesting problem to consider[39].

Let us point out at the end of this section that mathematics has enchanted artists since Egyptian and Greek times. Polykleitos, Plato (427–347 B.C.), Aristotle (384–322 B.C.), da Vinci (1432–1519), and Dürer (1471–1528) considered mathematical laws and symmetry (though not necessarily using the latter word) in art (and other fields of human interest). Poets and prose-writers have also been interested in mathematics. Let us just mention in this respect the essay by Musil entitled "The Mathematical Man"[40] (according to which mathematics is the luxury of a pure intellect) and his novel "The Man Without Qualities"[41] (for Ulrich, the main character in the novel, mathematics is the source of pleasure). A funny story by O. Henry entitled "The Chair of Philanthromathematics"[42] has little to do with mathematics, but without this word there would be no story.

Poe said, in his essay on the philosophy of the composition of a poem ("The Raven")[43]:

> It is my design to render it manifest that no one point in its composition is referable to accident or intuition—that the work proceeded, step by step, to its completion with the precision and rigid consequence of a mathematical problem.

He required that the literary work be balanced: There should be a mathematical relationship (ratio) between the length of all works of literary art and their merits (although it is difficult to see how this relationship can be quantified). Talking about the poem "The Raven", there is a near translational symmetry present in it through the repetition of the word "more" alone (six times) or as a part of the words "evermore" (once) and "nevermore" (eleven times). The rhythm of "The Raven" is trochaic, whilst the metre is octameter acalectic, alternating with heptameter catalectic repeated in the refrain of the fifth verse, and terminating with tetrameter catalectic.

Let us also mention here another interesting point concerning mathematics in poetry. There are many different stanza forms in use in poetry. Let us consider the sestina. The French sestina has a fixed poetic form: 6 stanzas, 6 lines each, envoy of 3 lines; usually unrhymed, but repeating as final words of the first stanza, in the following order (each letter represents the final word of a line):

$$
\begin{array}{c}
\quad\quad\quad 1 \quad 2 \quad 3 \quad 4 \quad 5 \quad 6 \\
\text{Stanzas}\quad
\begin{array}{c}
1 \\ 2 \\ 3 \\ 4 \\ 5 \\ 6
\end{array}
\left[
\begin{array}{cccccc}
A & B & C & D & E & F \\
F & A & E & B & D & C \\
C & F & D & A & B & E \\
E & C & B & F & A & D \\
D & E & A & C & F & B \\
B & D & F & E & C & A
\end{array}
\right]
\end{array}
$$

Envoy   B D F or A C E.

The above array has the structure of a quadratic (6×6) matrix where the even-labelled columns contain the first three elements, in increasing lexicographic order, from the row above, whilst the odd-labelled columns contain the last three elements, in reverse lexicographic order, from the above. The envoy consists of either the first three elements, in increasing lexicographic order, from the last row, or the second three elements, in reverse lexicographic order, from the last row of the sestina matrix. (Often the envoy uses all the final words, two a line: B E, D C, and F A).

Combinatorics is used repeatedly by poets. The use of combinatorics in fiction is much less frequent. A good example of combinatorics in the literature can be found in Joyce's *Ulysses*[44]:

> What anagrams had he made on his name in youth?
> Leopold Bloom
> Ellpodbomool
> Molldopeloob.
> Bollopedoom
> Old Ollebo, M.P.†

---

† In this line a letter "o" is missing in the edition we had at hand[44]. The same was detected in several other English editions and in the German translation published by Suhrkamp, Frankfurt/Main, 1975, Teil II, p. 857.

What acrostic upon the abbreviation of his first name had he (kinetic poet) sent to Miss Marion Tweedy on the February 14, 1888?

> Poets oft have sung in rhyme
> Of music sweet their praise divine.
> Let them hymn it nine times nine.
> Dearer far than song or wine,
> You are mine. The world is mine.

The curious interplay between the mathematics and poetry was described by Martin in his book on Tennyson[44a]:

The mathematician Charles Babbage† wrote to Tennyson after reading ''The Vision of the Sin''[44b] to say that he was bothered by the two lines:

> "Every minute dies a man,
> every minute one is born"

''I would therefore take liberty'', wrote Babbage, ''of suggesting that in the next edition of your excellent poem the erroneous calculation to which I refer should be corrected as follows:

> 'Every minute dies a man
> and one and a sixteenth is born.'

In the next edition 'minute' was changed to 'moment'.

This essay will consist of several sections. In the next section the basic concepts will be exposed in a rather elementary way. Then subsequent sections will report on symmetry and asymmetry of letters and words. The main part of the article is the section on symmetry and asymmetry in literature which will be partitioned into four subsections. The first subsection will be concerned with the symmetric and/or asymmetric objects in literary works and our attempt to assign the symmetric or asymmetric configurations to a given literary product.

The second subsection will report on symmetry characteristic of the content of literary works. In the third subsection a brief discussion about dualism in literature will be given. This section will end with some concluding remarks. The article will end with conclusions and with the list of references. The selection of references will be bounded by the availability of literature in foreign languages in Zagreb.

The quotations from a number of literary sources will be given. Note that all underlinings in quotations are the author's own.

BASIC CONCEPTS

The basic concepts requisite for this essay will be presented only qualitatively. The precise details can be found in several excellent expositions: symmetry by Weyl[2] and Shubnikov and Koptsik[27], diagrammatic approach and topology by Harary[45] and Alexandroff[46], respectively, and information theory by Brillouin[47]. Besides, all kinds of symmetries will also be reviewed in this special issue on symmetry by a variety of qualified authors[3].

A. *Bilateral symmetry*

A kind of symmetry we expect to detect in the literature is bilateral symmetry. This is so because bilateral symmetry[2], which refers to such operations as reflections, is the most often encountered concept of symmetry. In the world around us people, animals, plants, heavenly bodies, etc. usually exhibit (near) bilateral symmetry (and some of them other kinds of symmetries such as spherical symmetry and cyclindrical symmetry) in their external forms. Bilateral

---

†Charles Babbage (1792–1871) was analyst, statistician, and inventor. He was a prophet of the modern digital computing machines.

symmetry has also been present in the various products of human endeavours since the first stone axe made by cave-men. All around us are endless examples of man-made bilaterally symmetric shapes and patterns (chairs, tables, windows, doors, vases, lamps, bottles, earrings, coins, trademarks, plates, dishes, stamps, emblems, underwear, clothing, chandeliers, etc.). Many of these objects have such a symmetry for reasons of convenience, but numerous objects are designed bilaterally symmetric to make them attractive to the consumer. The aesthetic appeal of a given product is certainly of marketable value. The objects enumerated above posses either a vertical or horizontal plane of symmetry. Some possess both a vertical and horizontal plane synchronously.

As we have already pointed out, and as will be seen from the essays in this special issue on symmetry[3], symmetry is not restricted only to spatial objects: It has acoustical, colouristic, spiritual, etc., applications. Because of this we are allowed to employ terminology, otherwise reserved for use, say in geometry or physics, for identifying a particular (non-spatial) symmetry form in the literature. We will therefore do so just to keep the terminology uniform. However, in so doing we will run into an interesting situation. When we mention bilateral symmetry in literature it loses its precise meaning because we are not any more in the domain of physical reality. Now this bilateral symmetry is a strictly geometric and exact concept[2]: a spatial configuration is (bilaterally) symmetric with respect to a given plane σ if it is carried into itself by reflection in σ. Choose any line l perpendicular to the plane σ and any point p on l. There exists one, and only one, point p' on l which has the same distance from σ but lies on the other side of σ. Reflection in σ is that mapping of space upon itself, $S: p \rightarrow p'$, that carries the arbitrary point p into its mirror image p' with respect to σ. A mapping (symmetry operation) is defined whenever a rule is established by which every point p is associated with an image p'. The reflection in a plane is the basic operation of bilateral symmetry.

B. Some other kinds of symmetries

A circle and regular polygons, besides bilateral symmetry, also possess rotational symmetry. Note that a circle is carried into itself by the rotation around the center i for any angle. Similarly, regular n-polygons ($n = 3, 4, 5, \ldots$) are carried into themselves by the rotations around their centers for angles $\phi_k = (360/n)k(k = 1, 2, \ldots)$. Additionally, the reflections are carried out through symmetry planes passing through the centre and the n vertices of an n-polygon. Hence, an n-polygon possesses n-rotations and n-reflections. These symmetry operations form a group and this group determines the symmetry of the n-polygon. For example, the regular pentagon has ten symmetry operations (five rotations and five reflections) and, thus, it belongs to the group 5m of order 10.

In art, architecture, the organic and inorganic world appear many examples of cyclic symmetry, trigonal symmetry (the angle of rotation is 120°), tetragonal symmetry (90°), pentagonal symmetry (72°), hexagonal symmetry (60°), etc. Let us mention a few: the Roman amphitheatre in Pula, Croatia[28] (cyclic symmetry), the rosette of St. Pierre in Troyes, France (trigonal symmetry), city squares in the central Savannah, Georgia (tetragonal symmetry), the Pentagon building in Washington (pentagonal symmetry), and snow crystals (hexagonal symmetry).

Flowers are distinguished for their colour, pleasing scent, and rotational symmetry. The symmetry of 5 is very frequent amongst flowers. Families such as Caryophyllaceae or Rosaceae and many others possess pentagonal symmetry. Individual examples are Herniaria glabra L., Cydonia oblonga, etc. Some of the flower families (e.g. Papaveraceae, Cruciferae, or Rubiaceae) exhibit tetragonal symmetry, whilst others (e.g. Iridaceae or Liliaceae) possess hexagonal symmetry. Some families exhibit only bilateral symmetry such as, for example, Labiatae, whilst others like Compositae, n-gonal symmetry. Examples of this kind are Bellis perennis L., Helianthus annuus L., Matricaria chamomilla L., Calendula officinalis L., etc.

Similarly, as the regular n-polygons are connected with the finite groups of plane rotations, so the regular n-polyhedra are related to the finite groups of proper rotations around an axis in space. We should note here that there are unlimited regular n-polygons possible. But, there are only five regular n-polyhedra: the regular tetrahedron, the cube (or hexahedron), the octahedron, the dodecahedron, and the icosahedron. Note the following regularity in polyhedra (the Euler formula)[1]:

$$V - E + F = 2$$

where V is the number of vertices, E the number of edges, and F the number of faces.

The repetition of bilaterally symmetric figures leads to the construction of two-dimensional patterns such as the regular lattices (made up from either equilateral triangles, squares, or regular hexagons) and a variety of ornaments. Examples are wallpaper patterns, diaper patterns, floor tiles, honeycomb, Chinese, Egyptian and Arabic ornamentation, etc. The kind of symmetry that can be applied to ornaments is called *ornamental symmetry*[2]. The symmetry of ornaments is concerned with discontinuous groups of congruent mappings of the plane. Ornaments will be also discussed in this special issue on symmetry[3].

Crystals represent the geometric arrangements of atoms in three dimensions. In chemistry all kinds of compounds produce crystalline forms, e.g. boron-hydride polyhedra[48,49], co-ordination polyhedra of atoms in the structure of an alloy, RhBe[50], ice, diamond, etc., all of which possess crystallographic symmetry[2,14,19].

Another type of symmetry that is encountered in nature, art, architecture, music, poetry, etc., is translational symmetry. Let us consider an infinite set of points labelled by A, B, C, D, . . . positioned on a straight line l. The distance between the points is constant and the same for all adjacent pairs. This is illustrated in Fig. 1. The transformation relating two adjacent points A and B may be represented as

$$x_B = x_A + \Delta l$$

$$y_B = y_A.$$

A translation carries point A into the position of B, B into the position of C, C into the position of D, etc. Since there are an infinite number of points on the line, the translation changes nothing. This symmetry is called pure translational symmetry.

C. *Assymetry*

In order to introduce this concept let us consider for a moment the human hands. The left hand (or the right hand) is an asymmetric object: It does not possess any built-in symmetry elements. But asymmetry is not merely the absence of symmetry, because if we put the left hand in front of a mirror, the image of the left hand is produced by the mirror. These two objects, the left hand and its mirror-image, together form a symmetric figure. The left hand is an asymmetric unit of the figure. More generally, in the case of asymmetric objects, the object

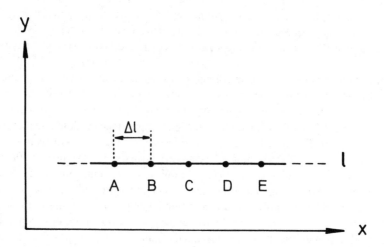

Fig. 1.

and its image together form a symmetric figure in that there is a one-to-one correspondence between any point p on the object and a twin point p' on the image (Fig. 2). Because of this objects of all kinds that bear a mirror relationship to one another are referred to as left- and right-handed objects. A mirror always converts a left-handed to a right-handed object. Asymmetry is the nonidentity of an object with its mirror image. A modern term for this property is *chirality*[51].

Objects exhibiting chirality are termed *chiral objects*. Chiral objects do not possess those symmetry elements which can superimpose the object on its mirror image. These are the reflection plane and the axis of improper rotation (or improper axis). A improper rotation (or a rotatory reflection) is a combined symmetry operation which consists of a rotation through a given angle about some axis and reflection in a plane perpendicular to this axis. (The order in which the rotation and reflection are performed is immaterial). Chiral objects may possess some elements of symmetry such as the axes of proper rotations. The objects identical to their mirror images are called *achiral objects*.

Two chiral objects like two hands or two feet, identical in shape, but not superimposable because one is right and one is left, are called *enantiomers*. Chirality is the necessary and sufficient condition for the existence of enantiomers. It is also a necessary, but not sufficient condition for optical activity.

There are many natural and hand-made objects which are chiral and which appear in two enantiomorphus forms. These are, for example, crystals of flint-stone, pine-cones, screws, snail shells, shoes, shells (though one of the forms may predominate depending on the habitat), etc. The daily rotation of our planet together with the direction of its axis from South to North pole is a right screw. Molecules of DNA, the basic material of the gene, consist of two right-handed chains and coil about each other in a helical form. Proteins present in plants and animals (man included) are made up (with rare exceptions) of only left-handed amino acids. (This fact so far lacks an explanation, though there are quite a few speculations on its origin). Modern biologists are even talking about asymmetric evolution[52].

Note that the human body is a symmetric (achiral) object, while the hands and feet are asymmetric (chiral) objects, respectively. Our left hand (left foot) and right hand (right foot) have opposite chirality. The heart of mammals is also chiral (asymmetric screw). Most people

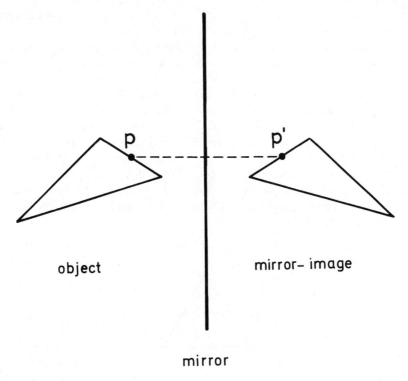

Fig. 2.

have their heart on the left side of their body. But, a few people have the heart on the right side and their intestines are inverted. This phenomenon is known in medical science as *situs viscernum inversus totalis*. When a position of only a single organ is inverted, a case which happens more often than the above, this is known as *situs viscernum inversus partialis*. When only the heart is displaced to the right, this is known as *dextrocardia* (''right'' heart).

Dorothy Leigh Sayers uses a character with a ''right'' heart in her story ''The Image in the Mirror''[53]:

> Wimsey obediently moved his hand across.
> 'I seem to detect a little flutter,' he said after a pause.
> 'You do? Well, you wouldn't expect to find it that side and not the other, would you? Well, that's where it is. I've got my heart on the right side, that's what I wanted you to feel for yourself.'
> 'Did it get displaced in an illness?' asked Wimsey sympathetically.

One way of getting the left and right sides reversed is described in the same story:

> Wimsey twisted his head round so as to get a view of the page. *'The Plattner Experiment,'* he said; 'that's the one about the schoolmaster who was blown into the fourth dimension and came back with his right and left sides reversed. Well, no, I don't suppose such a thing would really occur in real life, though of course it's very fascinating to play with the idea of a fourth dimension.'

The concept of chirality has found its greatest application in chemistry[50,54].

D. *Diagrams in literature*

The structure of literary works may be studied by means of diagrams[55–60], i.e. topological structures. The diagrammatic approach is well-established, for example, in natural sciences, linguistics, social sciences, and anthropology[61–67]. It is believed that diagrammatic analysis is a powerful method to use for the interpretation of work of literature. For example, the diagrams may be used to visualize the relationships (sometimes very complex) between the characters in a novel or a play. A diagrammatic analysis may be used to lay bare the basic structures around which the plot of a work is interwoven. In addition, an attempt to present a given literary work in diagrammatic form may bring forth some hidden structures and add a new dimension to the interpretation and understanding of the work.

Let us consider here by way of an example, a rather simple touching story ''Simon's Papa'' by Guy de Maupassant[68]. There are three main characters in the story: Simon, a seven-year old schoolboy (A), La Blanchotte, his un-wed mother (B), and Phillip Remy, the blacksmith and kind man (C). The graphical interpretation of the relationship between the characters may be given by Fig. 3 (a triangle). This is a variation of the love triangle: The intensity and kind of affection between the pairs AB (mother–son), BC (man–woman), and AC (son–father-figure) are obviously different. We will not quantify these relationships (this would be extremely difficult to do, if not impossible) and will only assume that the three distances A–B, B–C and A–C in the above triangle are not the same in order to point out that there are three kinds of relationships in the story. Thus, a convenient diagram to depict the relationships between the above three characters is the scalene triangle (Fig. 4). Therefore, the structure around which the story is built is an asymmetric (chiral) triangle.

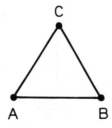

Fig. 3.

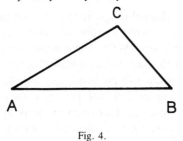

Fig. 4.

This technique of expressing diagrammatically the relationship between three people (for example, husband, wife, and lover) is used by Hercule Poirot in Christie's *Triangle at Rhodes*[69]:

> Susan looked up at Poirot.
> 'Well?' she said. "What do you make of this?'
> Hercule Poirot did not reply in words, but once again his forefinger traced a design in the sand. The same design—a triangle. 'The Eternal Triangle', mused Susan. 'Perhaps you're right. If so we're in for an exciting time in the next few weeks.'

This simple approach will be used later in the text in an attempt to present the structure of literary works in terms of symmetric and/or asymmetric objects.

A kind of topological structure (diagram) often employed in fiction, is the genealogical tree of family chronicles. An example of such a genealogical tree is taken from Marquez's *One Hundred Years of Solitude*[70] which is a chronicle of the Buendía family (Fig. 5). Family trees are, of course, asymmetric structures.

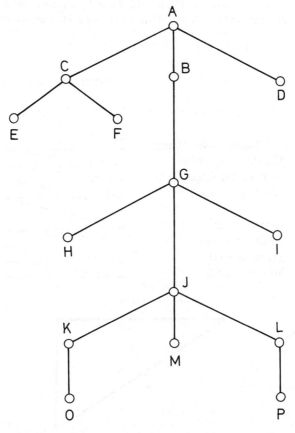

Fig. 5. The Buendía family tree. A = José Arcadio Buendía m. Úrsula Iguaran, B = José Arcadio m. Rebecca, C = Colonel Aureliano Buendía m. Remedios Moscote, D = Amaranta, E = Aureliano José (by Pilar Ternera), F = 17 Aurelianos, G = Arcadio (by Pilar Ternera) m. Santa Sofia de la Piedad, H = Remedios the Beauty, I = José Arcadio Segundo, J = Aureliano Segundo m. Fernanda del Carpio, K = Renata Remedios (Meme), L = Amaranta Úrsula m. Gaston, M = José Arcadio, O = Aureliano (by Mauricio Babilonia), P = Aureliano (by Aureliano).

Diagrams are also used by criminal story writers. At a certain point in a story they produce a diagram which should help the reader to speculate about the misdoer (see, for example, Christie's *The Murder of Roger Ackroyd*[71]). The construction of these diagrams is similar to that described earlier, though with many more additional details. The diagrams usually depict the scene of the crime and the position of the body, and sometimes the positions of characters involved in some way with the murdered person. However, these diagrams are not too helpful, because in this case the culprit would be discovered too early by the reader and the interest in the novel (and perhaps in the future works of the same author) may wane, unless the readers are true aficionados of this kind of fiction.

Poems may also be produced with graphic shapes of various kinds. A modern Croatian poet Slavko Jendričko conceived the poem "A Debatable Verse"[72] in the form of an asymmetric right-angled triangle (Fig. 6).

Croatian text	English text†
još samo idioti vjeruju u poeziju:	only idiots still believe in poetry:
samo idioti vjeruju u poeziju:	idiots still believe in poetry:
idioti vjeruju u poeziju:	still believe in poetry:
vjeruju u poeziju:	believe in poetry:
u poeziju:	in poetry:
poeziju:	poetry:

By the example of associating a graphic design with a poem, the following "zig-zag" diagram may serve to relate to the poem "Rondo" by Ante Stamać, another modern Croatian poet[73]:

Poem†   Corresponding diagram

RONDO

Ein Nichts in den Hauch
Ein Hauch ins Wort
Wort in den Wind
Wind in Gestalt

Gestalt in Fetzen
Fetzen in Schau
Schau in Wort
Wort in Hauch

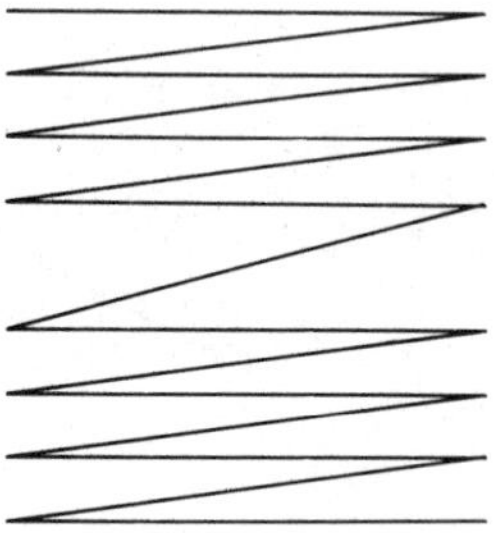

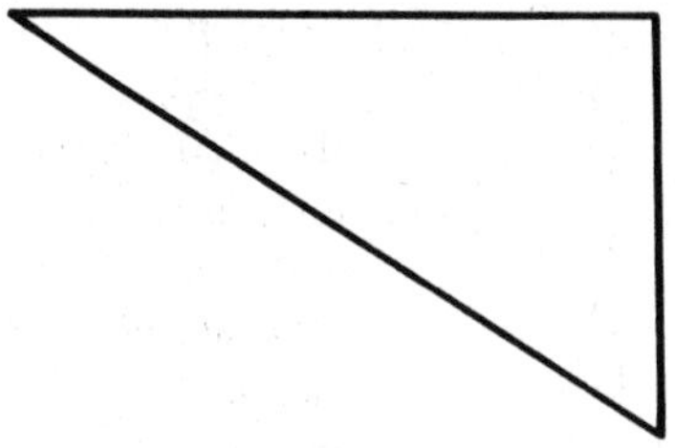

Fig. 6.

†Translation by the authors of this article.
†Translation by Ina Jun Broda.

E. *Information-theoretical concepts*

Information theory is a branch of probability theory[74] and was founded by Shannon[75]. The gain of information is defined by[47]

$$I = K \log_2 P$$

where I is the information (a dimensionless quantity), K is a constant (yet to be specified), and P is the total number of possibilities. As the most convenient unit, a system based on the binary digits or bits was selected. Thus, the logarithm is taken to base two for calculating the gain of information in bits. The above equation deals with equiprobable events.

The gain of information per element in the system $\bar{I}$ is given by the Shannon formula[75]

$$\bar{I} = - \sum_{i=1}^{n} p_i \log_2 p_i,$$

where $p_i = (P_i/P)$. P_i is the number of elements in the i-th set of elements. The equation is applicable to the cases with different probabilities.

The information connected to a specific physical problem, i.e. bounded information[47], is related to the entropy E. The bound information appears as a negative term in the total entropy of the physical system, and since the negative of the entropy is defined as the negentropy N, the following expression connects the bound information, the entropy, and the negentropy[47]

bound information I = decrease in entropy E = increase in negentropy N.

The above represents the negentropy principle of information.

Let us now consider the limiting cases:

(a) $P_i = 1, P = n, p_i = 1/n$

$$\bar{I} = \log_2 n = E_{max} = N_{min}$$

(b) $P_i = P = n, p_i = 1$

$$\bar{I} = - \log_2 1 = 0 = E_{min} = N_{max}.$$

The entropy is a measure of disorder within the system, whilst the information is the measure of a degrees of order within the system. When the information is at a maximum, the entropy is at a minimum, and vice versa[76].

SYMMETRY AND ASYMMETRY OF LETTERS AND WORDS

Before we apply the concept of symmetry to literature we will briefly discuss the symmetry and asymmetry of the capital letters in the English alphabet and in English words, since we prepared this article in the English language. (There would be only minor differences if our native Croatian tongue were employed, because our alphabet is also based on Latin. Some of the letters of the old Croatian alphabet, called the *Glagolithic* alphabet, also possess symmetry[77]. Letters and words of other languages may also be subjected to the following analysis.) We will perform this analysis because letters make words, and many define literature as the art of words, when they want to point out that words are building-blocks of a literary work. This fact also reveals the special place of literature among the arts. Words are not physical building materials—as, for example, stone is for sculpture—but very complex mental creations with many meanings. Therefore, the understanding of the structure of words is very important for the discernment of the nature of literature. Besides, letters and words are very suitable objects to illustrate the concepts of achirality and chirality.

Let us consider the vowels "A" and "E". The first letter has a vertical plane of symmetry, whilst the second has a horizontal plane of symmetry. The mirror-images of these letters are identical to the letters themselves. For example, the mirror-image "A" of the letter "A", can cover the original letter exactly in the plane. The same is true for the letter "E". Letters "A" and "E" are symmetric capital letters and they belong to the class of achiral objects. However, there are letters which are not symmetric and consequently they cannot be covered in the plane by their mirror-images. For example, the letter "L" belongs to this class. Its mirror-image "⅃" cannot cover "L". Letters like "L" are asymmetric capital letters and they belong to the class of chiral objects.

The classification of capital letters according to their symmetry or asymmetry characteristics is as follows:

(I) Achiral letters:
    (I.1) Achiral letters with a vertical symmetry plane:
        A, M, T, U, V, W, Y (26.9%).
    (I.2) Achiral letters with a horizontal symmetry plane:
        B, C, D, E, K (19.2%).
    (I.3) Achiral letters with both vertical and horizontal symmetry planes:
        H, I, O, X (15.4%).
(II) Chiral letters:
        F, G, J, L, N, P, Q, R, S, Z (38.5%).

Note that all achiral letters possess one or more axes of proper rotation, and the axes of improper rotation. Chiral letters, of course, do not possess improper axes. Three chiral letters: "N", "S" and "Z" have a two-fold axis.

Let us mention here that the letters alone may be used in poetry. A good example to illustrate this point is the following poem by Alexei J. Kruchonikh, written in 1912[77a]:

```
        o         e         u
   i         e         e         i
   o         e         e         e
```

Poem may be composed even without using letters (or words). An example of this kind is graphic poetry. One form of graphic poetry uses only punctuation marks. Below we give a graphic sonnet by the Serbian poet Dobrivoje Jevtić, prepared in 1978, in which the author made use of only slants (virgules, slashes), and which also reveals translation symmetry:

SONNET ABOUT THE RAIN

Nevertheless, most of the information that people use is communicated by language. In the spoken language the elementary symbols are the fundamental sounds. The written language consists of words spelled out in letters. Consider a sentence. The letters are symbols used to build the sentence. If these symbols were equally probable *a priori*, the information contained in the sentence of *N* letters would be

$$I = N \log_2 27 \text{ bits.}$$

The number 27 is made up of 26 letters and the spacing between the words as an additional symbol. The solution of the above equation is not satisfactory because we know empirically that different letters occur with different *a priori* probabilities in the language. The probability for the occurrence of the letters in the English language is given in Table 1[47]. In Croatian, the distribution is different[78]: The letters "A", "E", "I", "O", and "N" occur—in this order—most often. Note that the first four letters are vowels. This is one of the reasons why the Croatian language belongs to the group of "soft" languages.

The above analysis shows that the symmetry or asymmetry characteristics of letters apparently have little effect on the probability of their occurrence. Now that we are acquainted with achirality and chirality of letters, let us consider words. There are words in the English language that are bilaterally symmetric. These are the words that are the same when read forwards or backwards which possess a built-in vertical plane of symmetry. There are only a very few such words, usually names or acronyms: "AMMA" (Elmer L. Amma, Professor at Department of Chemistry, the University of South Carolina), "AMA" (American Medical Association), "AAA" (American Automobile Association), "AVA" (Ava Gardner, American Movie star). etc. These words belong to the class of palindromic words, words that spell the same in both directions: "ANNA", "HANNAH", "RADAR", "MALAYALAM" (the Dravidian language of the Malabar Coast of India, a branch of Tamil), "ROTATOR", "WASSA-MASSAW" (a swamp in Berkley County, South Carolina), etc. Logos that we find on the covers of the books, which are used as parts of registered trademarks of publishing companies, are either symmetric or asymmetric objects. An example of the (bilateral) symmetric logo is "ᗺB" standing for Ballantine Books, Inc. of New York. The portrayal of a rooster is an asymmetric logo which is used by Bantam Books, Inc. of New York. Note that the artistic name of the Swedish singing group "AᗺBA" possesses bilateral symmetry, which would be

Table 1.

Symbol	Probability of occurrence
Spacing	0.2
E	0.105
T	0.075
O	0.0654
A	0.063
N	0.059
I	0.055
R	0.054
S	0.052
H	0.047
D	0.037
L	0.029
C	0.023
F U	0.0225
M	0.021
P	0.0175
Y W	0.012
G	0.011
B	0.0105
V	0.008
K	0.003
X	0.002
J Q Z	0.001

lost if it were to be properly spelled out as ABBA. This is, of course, another palindromic word. Entire sentences may be composed that are palindromic by words. Even a short story exists that is palindromic by words[22].

In enigmatic poetry, only one case of a Latin palindromic matrix, symmetric about both diagonals is known:†

$$\begin{bmatrix} S & A & T & O & R \\ A & R & E & P & O \\ T & E & N & E & T \\ O & P & E & R & A \\ R & O & T & A & S \end{bmatrix}$$

There are bilaterally symmetric words that possess a horizontal plane of symmetry. Some examples are as follows: "BED", "BEDE" (Venerable Bede, 673–735, English scholar, historian, and theologian), "DICK" (familiar form of the name Richard), "BEE", "BECK", "BEDECK", "DIOXIDE", "BOX", etc. Both the palindromic words with a vertical plane of symmetry and bilaterally symmetric words with a horizontal plane of symmetry are achiral. All other words are chiral. "Symmetry" itself is also a chiral word.

It is hard to imagine a literary work that would be composed in such a manner that is palindromic by words, although the attempts along these lines are recorded[22,43]. The following verse which is palindromic by words is rather interesting[79]:

> Is it odd how asymmetrical is 'symmetry'?
> 'Symmetry' is asymmetrical how odd it is.

Both lines may be read word by word from the end to the beginning. They are invariant to the inversion: A change in the direction of reading with simultaneous change in the order of letters in the words produces no change in the meaning. Perhaps a computer will be able to perform such a task in the future. However, the "synthetic" (computerized) literary work is something that may appear sooner than we expect[80–82]. What the artistic values of such a prose (or poetry or plays) would be is highly uncertain, unless we find a machine-oriented criteria for human value of information.

Information is defined as the result of choice. It is not considered as a basis for a prediction, as a result that could be used for making another choice. The human value of information is completely ignored. Thus, a sentence of 100 letters from either "World Tennis" (leading international tennis monthly), "Hamlet", Einstein's book on relativity[83], or the text generated by a computer has exactly the same information value. In other words, "information" is defined as distinct from "knowledge" which so far has not been quantified in terms of numbers. The value of the information is completely ignored by the framework of information theory. Information is always measured by a positive quantity, whilst the value of information can, in certain cases, be regarded as negative. The idea of value refers to the possible use by an observer. Whilst information is an absolute quantity which has the same value for any observer, the human value of information is necessarily a relative quantity. It would have different values for different observers.

SYMMETRY AND ASYMMETRY IN LITERARY WORKS

Since symmetry in general and bilateral symmetry in particular have been permanent, and very visible, ingredients of man's culture from the dawn of human history, they are expected to permeate to some extent all kinds of literary works. In order to establish this, we will look for examples in a variety of works of literary art. Attention will be directed especially towards fiction.

---

†In English: The sower Arepo barely holds the wheel.

A. *The symmetric and asymmetric objects in literature*

It appears that in works of literature either the writer uses a symmetric and/or asymmetric objects as a descriptive means, or the basic structure of a literary product may be associated with a geometric object with some characteristic symmetry. In the first case the object possesses symmetry (or asymmetry) of some kind (bilateral, spherical, ornamental, crystallographic, etc.) and is employed as such by the author. A few examples will be mentioned. The story entitled "The Hands" by the Croatian writer Ranko Marinković[84], and the novel *Pentagram* by the Serbian writer Radomir Konstantinović[85] belong to the first category. Even poems have been written whose graphical shapes have bilateral symmetry. An example of such a poem is that produced by the German neoromantic poet Christian Morgenstern (1871–1914) entitled "Die Trichter" ("The Funnel")[86]:

> Zwei Trichter wandeln durch die Nacht.
> Durch ihres Rumpfs verengten Schacht
> fliesst weisses Mondlicht
> still und heiter
> auf ihren
> Waldweg
> u.s.
> w.

The shape of the poem in print is the funnel, a bilaterally symmetric object (Fig. 7).

The second case of the use of symmetry in literature is not as simple and clear-out as the above, because here we deal with the non-spatial objects to which we ascribe spatial features. Such an analysis (and interpretation) is ordinarily highly subjective and a given non-spatial configuration may be associated with the different geometric figures depending on the aim of the person carrying out such a study.

A variety of symmetric and asymmetric objects may be encountered in literature. Some of these objects that appear with a certain frequency in literary works will be listed below. In many instances the following symmetric (and asymmetric) two- and three-dimensional objects appear or may be identified which serve as the works' foundation: cycle, cross, polygons, polyhedra, etc. We add to them a centre as one-dimensional object, because the structure of

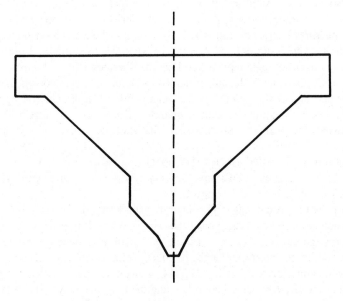

Fig. 7.

many a literary work is strongly centered. It may be considered as a centre of symmetry of such a literary work.

The centre symbolizes the starting point of all processes of efflux, emanation, and divergency, and the meeting point of all processes of return and convergency. Many writers use as a central point for their novels a (easily or not so easily recognizable, dynamic or static) place, date or idea. Let us mention here a few: Remarque's *The Arch of Triumph*[86a], Döblin's *Berlin Alexanderplatz*[87], Singer's *The Cafeteria*[88] (one cafeteria on Broadway in New York proudly displays the sign ''We are the cafeteria in Singer's *The Cafeteria*), Solzhenitsyn's *August 1914*[89], Metalious's *Peyton Place*[90], Chevalier's *Clochemerle*[91], Orwell's *1984*[92], etc. Very often authors choose a character to represent the (gravity) centre of the work. Selected examples are Puškin's *Evgeny Onegin*[93], Lagerlöf's *Gösta Berling*[94], London's *Martin Eden*[95], Goldman's *Marathon Man*[96], Voinovich's *The Life and Extraordinary Adventures of Private Ivan Chonkin*[97], Dickens's *Oliver Twist*[98], Farrell's *Studs Lonigan*[99] Tolstoy's *Anna Karenina*[100], Hašek's *The Good Soldier Schweik*[101], Pasternak's *Doctor Zhivago*[102], Solzhenitsyn's *One Day in the Life of Ivan Denisovich*[103], May's *Winnetou*[104], Voltaire's *Candide*[105], Balzac's *Eugènie Grandet*[106], Shakespeare's *Hamlet*[107], Hemingway's *The Old Man and the Sea*[108], etc. There are a great many novels (and other literary works) in this category.

Here belong also the autobiographies and biographies of ''famous'' people from all walks of life (statesmen, political and religious leaders, war heroes, movie and theatre actors and actresses, dancers, showbusiness and entertainment stars, scientists, opera singers, painters, poets, writers, artists, gangsters, band-leaders, indiscreet secretaries, composers, conductors, philosophers, murderers, executioners, kings and queens, ex-kings and ex-queens, sport stars, etc.), where by definition the central role is taken by the person who writes his/her autobiography or about whom the biography is being written. We list below some examples: Maurois's *The Life of Sir Alexander Fleming*[109], Birkenhead's *Rudyard Kipling*[110], Buckle's *Nijinsky*[111], Bonanno's *A Man of Honor*[112], La Mure's *Moulin Rouge*[113] (the life of Henry de Toulouse Lautrec, 1864–1901, French painter), Padover's *Karl Marx*[114], Louis's *Joe Louis: My Life*[115], Farago's *Patton*[116], Stone's *The Agony and Ecstasy*[117] (The life of Michelangelo), Zolotow's *Shooting Star*[118] (The biography of the movie star John Wayne), Clark's *The Life of Albert Einstein*[119], Solzhenitsyn's *The Oak and the Calf*[120] (an autobiographical account), Canneti's *Saved Tongue*[121] (the story of Elias Canneti's youth), Anne Frank's *The Diary of a Young Girl*[122], etc.

In this type of literary products one should include series of stories and novels, mostly mystery and adventure, about a single hero in a variety of situations such as the James Bond series by Ian Fleming, the Rabbi David Small series by Harry Kemelman, the Old Shatterhand series by Karl May, the C. Auguste Dupin series by Edgar Allan Poe, the Maigret series by George Simeon, the Hercule Poirot series by Agatha Christie, the Father Brown series by G. K. Chesterton, the Tarzan series by Edgar Rice Burroughs, the Sherlock Holmes series by Sir Conan Doyle, the Jeff Peters stories by O. Henry, the Nick Adams stories by Ernest Hemingway, etc. Some of these characters may have a side-kick, as Sherlock Holmes' most famous of them all: Dr. Watson. Here also qualify the numerous works about a given historical figure by various authors such as works about Alexander the Great, Gaius Julius Caesar, Charlemagne, Napoleon, etc. Similarly, in this class should be included countless narratives of Jesus Christ's life starting with the four gospels (according to St. Matthew, St. Mark, St. Luke, and St. John) in ''The New Testament''.

The type of novels classified as ''ich (I)'' novels may also be included in this group. The ultimate ''ich'' form of a novel is perhaps the very first novel in this genre: Rousseau's *Les Confessions*[123].

The circle possesses rotational (and bilateral) symmetry and is a symbol of unity, non-differentiable wholeness, endless time, and circular movements. It is also a symbol of protection. Example of this is the magician's circle. Before he began his conjuring, the magician would draw a circle round himself, inscribing on the periphery certain signs (of the Zodiac, for instance) and the Tetragrammaton (The four letters YHWH forming a Hebrew tribal name for the Supreme Being). So long as the circle remained unbroken and the magician stayed inside it, no evil spirit could harm him. See, for example, the description of the magician's circle by Marlowe (1564–1593) in "Doctor Faustus"[124], Act I, Scene iii:

FAUSTUS:

⋮

Within this circle is Jehovah's name,
Forward and backward anagrammatised:
Th' abbreviated names of holy saints,
Figures of every adjunct to the heavens,
And characters of signs and erring stars,
By which the spirits are enforced to rise:

⋮

Medieval castles, for example, often have round fortifications. The "protection" against evil forces of an individual used to be contained in rings, bracelets, necklaces, etc. These objects today serve merely as adornments. The marriage is "protected" by the wedding-ring. The Japanese flag contains a red circle on a white base. Some other nation's flags also contain a circle (e.g. South and North Korea, Tunis, Zaire, Brasil, etc.), but within the circles elaborate symbols appear, and the simplicity (and beauty) such as that of Japanese banner is diminished. Five circles are symbols of the Olympic movement representing the union of all the sportsmen on our planet.

The basic notion in philosophy of Heraclitus is circular motion. The whole philosophy of the mature Nietzsche is pervaded by the idea of the eternal (circular) return. The concept of circular motion may best be visualized by the self-returning walk along the cyclic path on a given structure where the start and the end are the same points.

Many literary works either use circles as symbols in some way or their structure is circular, e.g., Homer's *The Odyssey*[125], Dante's *Divine Comedy*[126], Verne's *Around the World in Eighty Days*[127], Brecht's *Caucasian Chalk Circle*[128], Solzhenitsyn's *The First Circle*[129], Tolkien's *The Lord of the Rings*[130], Bunyon's *The Pilgrim's Progress*[131], Maugham's *The Circle*[132], Matković's *Game Around Death*[132a], etc.

In many of the above cases such as *The Pilgrim's Progress,* and in others, the structure of the novel is truly spiral, i.e. it represents a combination of circular motion and translation. The Maelstrom, a spiral movement of (sea) water (whirlpool), is used by Poe in his stories "A Descent into the Maelstrom"[133] and "Narrative of A. Gordon Pym"[134].

The life of a human being also resembles, from a distance, a journey on a circular path. A closer look at this path shows that it is circular, but irregular, because in every life there are ups and downs. The autobiographies and biographies mentioned earlier may also be included in this group.

In Plato's *Symposium*[135] Aristophanes talks about round people:

You must begin your lesson with the nature of man and its development. For our original nature was by no means the same as it is now. In the first place, there were three kinds of human beings, not merely the two sexes, male and female, as at present: there was a third kind as well, which had equal shares of the other two, and whose name survives though the thing itself has vanished. For "man–woman"† was then a unity in form no less than name, composed of both sexes and sharing equally in male and female; whereas now it has come to be merely a name of reproach. Secondly, the form of each person was round all over, with back and sides encompassing it every way; each had four arms, and legs to match these, and two faces perfectly alike on a cylindrical neck. There was one head to the two faces, which looked opposite ways; there were four ears, two privy members, and all the other parts, as may be imagined, in proportion. The creature walked upright as now, in either direction as it pleased; and whenever it started running fast, it went like our acrobats, whirling over and over with legs stuck out straight; only then they had eight limbs to support and

†i.e. hermaphrodite.

speed them swiftly round and round. The number and features of these three sexes were
owing to the fact that the male was originally the offspring of the sun, and the female of the
earth; while that which partook of both sexes was born of the moon, for the moon also
partakes of both. They were globular in their shapes as in their progress, since they took
after their parents. Now, they were of surprising strength and vigour, and so lofty in their
notions that they even conspired against the gods; and the same story is told of them as
Homer relates of Ephialtes and Otus, that scheming to assault the gods in fight they essayed
to mount high heaven.

   Threat Zeus and their other gods debated what they should do, and were perplexed:
for they felt they could not slay them like the Giants, whom they had abolished root and
branch with strokes of thunder—it would be only abolishing the honours and observances they
had from men; nor yet could they endure such sinful rioting. Then Zeus, putting all his wits
together, spake at length and said: 'Methinks I can contrive that men, without ceasing to
exist, shall give over their iniquity through a lessening of their strength. I propose now to
slice every one of them in two, so that while making them weaker we shall find them more
useful by reason of their multiplication; and they shall walk erect upon two legs. If they
continue turbulent and do not choose to keep quiet, I will do it again,' said he; 'I will slice
every person in two, and then they must go their ways on one leg, hopping.' So saying, he
sliced each human being in two, just as they slice sorb-apples to make a dry preserve, or
eggs with hairs; and at the cleaving on each he bade Apollo turn its face and half-neck to
the section side, in order that every one might be made more orderly by the sight of the
knife's work upon him; this done, the god was to heal them up. Then Apollo turned their
faces about, and pulled their skin together from the edges over what is now called the belly,
just like purses which you draw close with a string; the little opening he tied up in the middle
of the belly, so making what we know as the navel.

Formally viewed, this story represents a nice example of the transition from spherical
(round people) to bilateral symmetry (two-legged people), and possibly to asymmetry (one-
legged people).

   The cross is the symbol of christianity, albeit it was known in the cultures of ancient
Knossos (Crete), China, India, Persia, and Egypt. Romans used to crucify the enemies of the
state on them. The cross was a symbol of the crusaders. Flags of several countries contain the
cross: Finland (blue cross), Greece (white cross), Iceland (red cross), Norway (blue cross),
Sweden (yellow cross), Tonga (red cross), Switzerland (white cross), etc. The international
charitable organization "Red Cross" has a red cross as the symbol. The cross appears in all
sorts of literary works. Let us mention here Sienkiewicz's *Quo Vadis*[136], Wallace's *Ben-
Hur*[137], Lagerkvist's *Barabbas*[138], Fast's *Spartacus*[139], Caldwell's and Stearn's *I, Judas*[140], Yerby's *Judas, My Brother*[141], Burgess's *Man of Nazareth*[141a], etc.

   Polygons (triangle, square, pentagon, etc) are used as symmetric (achiral) and asymmetric
(chiral) objects in literary works.

   The triangle is used in literature in many ways, and is connected with the number three.
The best example for this is Dante's *Divine Comedy*[126] with its three parts: Inferno, Purgatory
and Paradise, and the repeated use of the number three. This poem shows great structural and
formal symmetry. The composition is ideal: it is written in *terza rimae* (strophas consisting of
three hendecasyllables), each of the mentioned three parts contain 33 cantos of comparable
length which together with the introductory canto make a perfect number of 100 cantos.

   The most important triangle for Western civilization, the Holy Trinity, appears—depending
on interpretations—as a symmetric or an asymmetric triangle. It is an asymmetric triangle if
each apex of the figure is labelled by a different symbol representing the Holy Father, the Holy
Son, and the Holy Spirit. But, if the Holy Trinity is interpreted as the Divine Being in three
appearances, then the triangle is symmetric (i.e. equilateral triangle), because each of its apexes
should be labelled by the same symbol reflecting this unity.

   The triangle was important structure in Greek philosophy (see, for instance, works of Plato,
e.g. *Timaeus,* or Pitagora, 482–496 B.C.). Nietzsche in *Thus Spoke Zarathustra*[142] talks
about three evils, whilst Popper introduced three worlds (physical world, conscience, and world
of ideas, theories, and thoughts) in his works[143].

   Modern examples of the use of (asymmetric) triangles are Remarque's *Three Com-
rades*[144], Dos Passos's *Three Soldiers*[145], van Wyck Mason's *Three Harbours*[146], Christie's *Third Girl*[147], Greene's *The Third Man*[148], Jerome's *Three Men in a Boat*[149],
Wallace's *The Three Sirens*[150], Kästner's *Three Men in Snow*[151], Uris's *Trinity*[152], etc.

A poet may see in the triangle a form of the Universe. The Croatin poet Jendričko in his poem "Triangle" mentioned the ". . . . Triangular shape of Universe. . . ."[153].

In some works several triangles may be found. Various asymmetric triangles may be found, for example, in *The Illiad*[154]. The basic one is made up from the three goddesses Athene, Aphrodite and Here. Another important triangle consists of Menelaus, the King of Sparta; his wife, the beautiful Helen; and Paris, the abductor of Helen, one of many sons of Priam, the King of Troy (or Illium). In *The Odyssey*[125] besides the basic triangle of Odysseus, Penelope and Telemachus, there are also several others. A particularly moving story is related to the fate of Agamemnon (the leader of Greek expeditionary force in the *The Illiad*), who is murdered on his return from Troy by Clytaemnestra, his wife, and her lover, Aegisthus. Another triangle in *The Odyssey* is concerned with Odysseus, his old father Laertes, and his son Telemachus. One of many Shakespearen love/hate triangles is given in "Othello" and consists of Othello, Desdemona and Iago[155].

Love triangles may be detected in Dreiser's *An American Tragedy*[156], Cain's *The Postman Always Rings Twice*[157], or in London's *The Sea-Wolf*[158]. A classical love triangle is a basis of Tolstoy's *Anna Karenina*[100] which is made up of Anna, her husband Alexei A. Karenin, and her lover Alexei Vronski. A love triangle may also be found in Goethe's *The Sorrows of Young Werther*[159] consisting of Werther, Lotta, and Albert, respectively. Another love triangle is clearly described by Flaubert in *Madame Bovary*[160] with Madame Bovary (*née* Rouault), Charles Bovary, the husband, and baron Rodolphe, the lover, respectively.

Many writers have been and still are interested in love/hate triangles. The reason for the interest in this kind of human entanglement is perhaps related to the fact that the love/hate triangles (made up from all possibilities) are of frequent occurence in real life. Cain seems to be inordinately occupied with the (tragic) situations (pregnant with physical and emotional violence) induced by the love triangles (see any of his works, like *Double Indemity*[161] or *Mildred Pierce*[162]).

In some works may also be found several love/hate triangles. A good example to illustrate this point in fiction may serve du Maurier's *Rebecca*[30] where several love/hate triangles may be uncovered. The main triangle underlying the novel is made up of Maximilian de Winter, his second wife, and Rebecca (his first wife). This triangle is interesting per se because Rebecca is dead, and the reader slowly learns what kind of "real" relationship exists between these characters. Then appears the triangle from the past made up by de Winter, Rebecca, and Favell, Rebecca's lover. Finally, at the end of the novel another triangle appears, consisting of the Winter, his second wife, and Favell, offering the solution to the novel's plot. Since all these triangles are closely related, combined they lead to the three-dimensional structure: an asymmetric trigonal pyramid, which reveals another triangle depicting a relationship of de Winter's second wife, Rebecca (dead), and her former lover (alive) because both these characters caused her a lot of emotional trouble. However, there is another important character in the novel. This is Mrs. Danvers, the housekeeper. She is a part of triangles consisting of Rebecca, the second Mrs. de Winter, and her, then Rebecca, Mr. de Winter, and her, and finally the second Mrs. de Winter, Mr. de Winter, and her. Since, she was practically interacting very little with Favell, they may be assigned to the opposite vertices of the asymmetric trigonal bipyramid by which we may depict the relationship of the characters in *Rebecca* (Fig. 8).

The case of Dumas's *The Three Musketeers*[163] is interesting because here the basic figure is not the triangle, as one might assume from the title of the novel, but an asymmetric quadrangle consisting of three musketeers (making up a friendship triangle before the time of the novel): Aramis, Athos, and Porthos, and young Gascon (later to become a Musketeer) D'Artagnan.

The quadrangle (square) symbolizes fourfold division and is connected with the number four. According to the Aristotelian view the matter is differentiated into four primary elements: Earth, Air, Fire, and Water. With the four elements go four primary qualities: Hot, Cold, Wet, and Dry. The Corpuscularian philosophers (e.g. Newton, Boyle, Locke, Galileo, Decartes) considered that the corpuscles are wholly defined by their four features: size, shape, texture, and motion[164]. William Blake was preoccupied in his work with the four-fold man[34,165a,b]. A year is divided into four seasons. The Moon exhibits four phases. Many sacral, military, industrial, residential, etc., structures are shaped as quadrangles. Good examples of this struc-

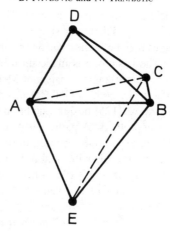

Fig. 8. A = Mr. de Winter, B = Mrs. de Winter, C = Rebecca, D = Favell, E = Mrs. Danvers.

ture, with different meanings, in literature are de Vinci's "Quadrifolium"[165c], Dumas's *The Three Musketeers*[163], Ibañez's *The Four Horseman of the Apocalipse*[166], Slavica's *The Fourth Horse*[167], Christie's *The Big Four*[168], Reymont's *Peasants*[169] (the novel is divided in four parts: each part is connected with one season of the year).

The pentagram appears in two forms: the pentagon and the five-pointed star and is an object of mani-fold symbolism related to the number five. In Greek times it was the Pithagorean sign of identification. The five-pointed star is the symbol of a human being. The five-pointed star was the symbol of the Third International, because it symbolizes the union of workers from all five continents. The (red, yellow, blue, or white) five-armed star is a symbol of many (socialistic) countries. Fifty white stars representing fifty states of the Union appear on the American flag. A number of five-armed stars is also a sign of good quality and excellence (see, for example, the ranking of walks and sights by Michelin, a tourist guide) and, by coincidence, of the ranking of generals in the American army.

An example of the direct use of a symmetric object in literature, i.e. pentagon, may be found Michener's *Poland*[170a]:

> . . . . , instead they came up with a chain made of woven hair from a cow's tail on which was suspended a curious pentagon-shaped medal dating back to some pre-Christian time.
> 'What's this?' a soldier asked, and Biruta said truthfully:
> 'We've always had it.'
> 'Why was it hidden?'
> 'It's our good-luck charm.'
> This was too complex, for the men, so they summoned Krumpf, and as soon as he saw it he surmised that it must be some early Germanic medallion, a souvenir of the time when Teutonic greatness began, and he snatched it from the soldier. As he stomped off with his prize, Biruta thought: 'How strange. A man from this village, centuries ago, took that medal from a pagan. Now the pagans have reclaimed it.'

The use of an asymmetric pentagon represents, for instance, the structure of Christie's *Five Little Pigs*[171].

The hexagon is also sometimes employed by writers, but much more by designers judging from the hexagonal patterns we see everywhere in our daily lives. Many naturally-occuring structures are built, entirely or partially, from hexagonal units (e.g., honey-comb, steroids, etc). It is interesting to note that one of the first scientific studies about hexagonal structures was performed by the Croatian scientist Rugjer Bošković (1711–1787). He studied the structure of the honey-comb and proved that the choice of a hexagonal structure was not by chance: Using such a structure, the bees economize on building material. Bošković published this in the study entitled *"De apium cellulis"* in 1760 in his commentaries on Volume II of Benedict Stay's (Stojković, 1714–1801, Croatian philosopher, poet, and orator) *Ten Books of Recent*

Philosophy in Verse. In this unique work Stay presented Newton's natural philosophy in verse[172,173].

An example of the use of hexagonal symmetry in the literature is provided by Mann's *Magic Mountain*[174], when he describes the disorder of the snow storm in which Hans Castorp, the hero of the novel, well-nigh dies. An hour before Hans Castorp goes skiing, he enjoys the play of the snow flakes "And among these myriads of enchanting little stars," so he philosophizes,

> in their hidden splendor, that was too small for man's naked eye to see, there was not one like unto another; an endless inventiveness governed the development and unthinkable differentiation of one and the same basic scheme, the equilateral, equiangled hexagon. Yet each in itself—this was the uncanny, the antiorganic, the life-denying character of them all— each of them was absolutely symmetrical, icily regular in form. They were too regular, as substance adapted to life never was to this degree—the living principle shuddered at this perfect precision, found it deathly, the very marrow of death—Hans Castorp felt he understood now the reason why the builders of antiquity purposely and secretly introduced minute variation from absolute symmetry in their columnar structures.

Polyhedra and crystal structures also appear in literary works. The regular polyhedra, often referred to as the Platonic solids, are prominent objects in Plato's philosophy. Plato, in *Timaeus*[175] associated the regular tetrahedron, octahedron, cube, and icosahedron with four elements: fire, air, earth, and water. In the dodecahedron he saw, in some sense, the image of the universe.

Ferdinand Baldensperger, in his capital work *La Littératura*[176], said that the contemporary philosophers like the scientific term "crystallization", because it well illustrates the transition from the saturated solution (i.e., the mind ready for creation) to the geometrical assembly (i.e., the basic idea for the future work).

Guy de Maupassant gives in his story *Love*[177] the following description of the ice-hut in which two hunters found shelter in the early morning cold in marshes just before starting the hunt:

> We made a pile in the middle of our hut which had a hole in the middle of the roof to let out the smoke, and when the flames rose up to the clear, crystal blocks they began to melt, gently, imperceptibly, as if they were sweating. Karl, who had remained outside, called out to me: 'Come and look here!' I went out of the hut and remained struck with astonishment. Our hut, in the shape of a cone, looked like an enormous diamond with a heart of fire, which had been suddenly planted there in the midst of the frozen water of the marsh. And inside, we saw two fantastic forms, those of our dogs, who were warming themselves at the fire.

The following lines are taken from Hemingway's "*The Old Man and the Sea*"[178] in which the prisms are used a descriptive means:

> He could not see the green of the shore now but only the tops of the blue hills that showed white as though were snow-capped and the clouds that looked, like high snow mountains above them. The sea was very dark and the light made prisms in the water. The myriad flecks of the plankton were annulled now by the high sun and it was only the great deep prisms in the blue water that the old man saw now with his lines going straight down into the water that was a mile deep.

A pleasing poetical description of the mystery and glory of the amber necklace (owned by Barbara Ossolinski) is given in Michener's *Poland*[170b]:

> Not harsh or brilliant like a challenging diamond,
> Nor stained with miner's blood like a throbbing ruby,
> Nor brazenly proclaiming its work like a cube of gold. . .
> You are an autumn moon rising over a field of ripened grain."

It is interesting to note that Erwin Schrödinger argued in his essay "What is Life?"[179] that aperiodic crystals are far more interesting structures than periodic crystals. This is so because the beautifully ordered structure of a periodic crystal—in its rigidity which does not allow much change—is dull in comparison with a structure which permits variations. If we leave inorganic world and consider, for example, the societal structures, then the structure of a totalitarian society—in its inflexibility—resembles a periodic crystal structure, whilst the structure of a democratic society, in its vitality and dynamism, bear likeness to an aperiodic structure. Aperiodic structures may evolve to different levels of organization whilst periodic structures are static and dead.

B. Symmetry characteristics of literary works

The definition of bilateral symmetry (reflection symmetry) that concerns such symmetry operations as reflections, loses its preciseness in the literature. This is so because in the novel (novelette, story) a spatial configuration (an object) is substituted, for example, by a given character, and its mirror-image by an identical character (a twin-brother or a double). The symmetry plane is the author's phantasy by which the two germane characters are related. A lot of fiction appears with bilaterally symmetric (in the above sense) characters. Identical twins appeared in novels such as *The Twin Lottas* by Kästner[180]. A good example of a double is the novel *A Double* by Dostoyevsky[181]. The explicit case of the convergency of a character and his image (his conscience) to a common unfortunate end may be found in Poe's story *William Wilson*[182]:

> At that instant some person tried the latch of the door. I hastened to prevent an intrusion, and then immediately returned to my dying antagonist. But what human language can adequately portray *that* astonishment, *that* horror which possessed me at the spectacle then presented to view? The brief moment in which I averted my eyes had been sufficient to produce, apparently, a material change in the arrangements at the upper or farther end of the room. A large mirror—so at first it seemed to me in my confusion—now stood where none had been perceptible before; and, as I stepped up to it in extremity of terror, mine own image, but with features all pale and dabbled in blood, advanced to meet me with a feeble and tottering gait.
>
> Thus it appeared, I say, but was not. It was my antagonist—it was Wilson, who then stood before me in the agonies of his dissolution. His mask and cloak lay, where he had thrown them, upon the floor. Not a thread in all his raiment—not a line all the marked and singular lineaments of his face which was not, even in the most absolute identity, *mine own!*
>
> It was Wilson; but he spoke no longer in a whisper, and I could have fancied that I myself was speaking while he said:
>
> *'You have conquered, and I yield. Yet, henceforward art thou also dead—dead to the World, to Heaven and to Hope! In me didst thou exist—and, in my death, see by this image, which is thine own, how utterly thou last murdered thyself.*

A variant of this theme is the story by Papini "Two Images in the Pool"[183] in which the character of the story appears as two identical people of differing age (the younger by seven years really represents the conscience). At the end of the story the younger counterpart is killed by the older who remains alive and thus represents the singular case of somebody who killed himself but is still alive. In these works by juxtaposing two parts the authors have transformed the inherent symmetry of left and right into something more complex and unique.

Sometimes the mirror-image is not like the character at all. This is so because the artist's creative phantasy acts as a distorting mirror that we all know from amusement parks. Examples are Stevenson's *The Strange Case of Dr. Jekyll and Mr. Hyde*[184] (this is the case of a character and his evil transformation), de Maupassant's *La Horla*[185] (the case of a character with a split-personality), Wilde's *The Picture of Dorian Gray*[186] (this is the case of a character staying young while his portrait is getting older), Poe's *The Oval Portrait*[187] (the reverse case from the above: a character is getting older fast whilst the portrait is being painted, and dies with the finishing stroke of the painter's brush), etc. The ultimate transformation of a character is described in Kafka's "The Metamorphosis"[188] though the mishap that happened to the protagonist of the story may have really occured only in his imagination.

A (political) satire also falls into the category of a literary work with a distorted mirror-image. A good example to illustrate this is Swift's *Gulliver's Travels*[189].

The first and the best-known part of the book "A Voyage to Lilliput" describes Lemuel Gulliver's involvement with the Lilliputians. The Lilliputians are a caricature of the English court and the government. English society is reduced below the six-inch level and its faults and flaws are made more ridiculous, more mean and foolish.

The variant of geometric symmetry which is concerned with such operations as rotations appears in fiction as a merry-go-round (a carrousel) of characters and events, real or fantastic. Many modern American novels reveal this kind of inner symmetry. Let us by way of example mention a few: Hemingway's *The Sun Also Rises*[190], Shaw's *The Young Lions*[191], Jones's *From Here to Eternity*[192], Mailer's *The Naked and the Dead*[193], Hailey's *Overload*[194], etc. In the literature of other nations this kind of novel may also be found. Examples are Kosinski's *The Painted Bird*[195], Kundera's *Joke*[196], or Škvorecký's *Armoured Battalion*[197]. A good example of this kind of novel is Hugo's *Les Misérables*[198]. All these novels treat human destiny in circumstances beyond the control of an individual.

As we have already pointed out, the concept of asymmetry may be best introduced with the help of the mirror. The mirror as a contraption which reflects reality has always fascinated writers. The mirror is an important device in fairy tales such as *Snow-white and the Seven Dwarfs*, fantasies such as Carroll's *Alice Through the Looking Glass*[199], straight fiction such as Papini's *Mirror Which Flees*[200], or in criminal stories such as Christie's *Dead Man's Mirror*[201]. Shakespeare[202] nicely said in "Hamlet":

> for anything so overdone is from the purpose of playing, whose end, both of the first and now, was and is, to hold, as 'twere, the mirror up to nature; to show virtue her own feature, scorn her own image, and the very age and body of the time his form and pressure.

An interesting use of the mirror in fiction is to be found in Sayers's *The Image in the Mirror*[203]. Let us use the author's own words:

> 'I was going along Holborn at lunch-time. I was still at Crichton's. Head of the packing department I was then, and doing pretty well. It was a wet beast of a day, I remember—dark and drizzling. I wanted a hair-cut. There's a barber's shop on the south side, about half-way along—one of those places where you go down a passage and there's a door at the end with a mirror and the name written across it in gold letters. You know what I mean.
>
> 'I went in there. There was a light in the passage, so I could see quite plainly. As I got up to the mirror I could see my reflection coming to meet me, and all of a sudden the awful dream-feeling came over me. I told myself it was all nonsense and put my hand out to the door-handle—my left hand, because the handle was that side and I was still apt to be left-handed when I didn't think about it.
>
> 'The reflection, of course, put out its right hand—that was all right, of course—and I saw my own figure in my old squash hat and burberry—but the face—oh my God! It was grinning at me—and then just like in the dream, it suddenly turned its back and walked away from me, looking over its shoulder. . .

In nature, mirror symmetry is observed when trees and other objects are reflected in a smooth lake or river. The myth of Narcissus is related to this kind of reflection symmetry: Nemesis caused Narcissus to fall in love with his own reflection in the water. His love was not returned and Narcissus died on the bank of a brook. From his body a flower, named after him *narcissus*, has grown. On account of the Narcissus myth there is psychoanalysis the term *Narcissus's complex* for a person who always watches himself (herself) in the mirror or, when on the street, in a store window.

The exact image of reality is the photograph. There are authors who record reality with photographic accuracy, such as the great masters of the written word Flaubert[160] and de Maupassant[68,177,185]. To this category also belongs Emile Zola[204,205], though not such an accomplished stylist as the former two. They perfectly recorded their times, but their works carry a recognizable personal signet of each of them. In making the literary "photos" they employed their own, highly sensitive, lenses.

In the history of literature there are many works (poems, stories, novelettes, novels, plays) which mirror-imaged the reality of a particular time. These are the cases when the literature plays the role of a (true or distorting) mirror. Folk tales and oral epics also played the role of a mirror. The reality and its mirror-image given by an author is at best near symmetric. However, much more often reality is distorted in literature. This is so because a creative author will always put into his work his experience or his impression of a particular historical incident in a very subjective way. Many rather critical stages in the history of mankind (war, famine, plague, revolution, the explosion of the first atomic bomb, the launching of the first man in outer space, the landing of the first man on the Moon, etc.) have been experienced by talented people, directly or indirectly, who recorded them by various means and have been concerned almost always with their impact on the fate of an individual and the inner life of his mind.

Great fiction which reflects the destiny of a man in cataclysmic incidents is rare, but it records or reconstructs a particular historical situation with the clarity of a magnifying glass. Let us here mention only several examples: *War and Peace*[206] (Russian people in the 1812 war against Napoleon), *Doctor Zhivago*[102] (Russian people in the October Revolution of 1917 and in early postrevolution days), *Chesapeake*[207] (American East, people and land, from 16th century to the present day), *The Bridge on the Drina*[208] (People of Bosnia under Turkish rule), *The Last Days of Pompeii*[209] (Pompeii in the time of the eruption of Vesuvius, 79 A.D.), *Masada*[210] (The defense of Masada by Jews against the Romans in 73 A.D.), *Goldsmith's Gold*[211] (People of Zagreb in the last quarter of 16th century), *Gone with the Wind*[212] (Georgia during the Civil War 1861–1865), etc. Why are these, broadly speaking, historical novels (distorted mirror-images, i.e. literature) and not history books (''true'' mirror-images, i.e. science)? Perhaps the reason is that the author of a ''historical'' novel can be as careless, or as meticulous, as the historian and yet reserve the right not only to rearrange events, but, more significantly, to ascribe motive, something the scrupulous historian should never do. Hence, the historical novels represent a blend of fact and fiction and reveal the spirit and drama of a particular point in the history through the author's mind.

In between historical fiction and historical works there are memoirs of historical figures such as Casanova's *Memoirs*[213], Eisenhower's *Crusade in Europe*[214] or Montgomery's *The Memoirs of Field Marshall the Viscount Montgomery of Alamein, K.G.*[215], and chronicles such as Ryan's *A Bridge Too Far*[216], Jones's *W W II*[217], and Tuchman's chronicle about XIV century very appropriately entitled *A Distant Mirror*[218]. In these works the general history is well-reflected. However, the interpretation of certain incidents and of roles played by various historical figures varies widely. In memoirs the central figure in everything is always the author, whilst in chronicles the sympathies of the author are usually centered only on some persons and incidents, and they are given more attention than the others.

Let us now consider the case of a large family chronicle, the plot of which is built around the genealogical tree. Family trees reflect the symmetry of growth, proposed by Jung[219], besides giving a mirror-image of the rise (and decline) of a given family in a certain period of time, and in characteristic social, economic, and political circumstances. Examples of this kind of novel are Mann's *Buddenbrooks*[220], Galsworthy's *The Forsyte Saga*[221], Krleža's *Glembays*[222], Singer's *The Family Moskat*[223], Marquez's *One Hundred Years of Solitude*[70], Haley's *Roots*[224], Michener's *The Covenant*[225], etc.

We already mentioned that the family trees are asymmetric structures. We can see the author making them symmetric if he wishes to do so, but on the expense of credibility. However, the science fiction authors can certainly devise a family which growth follows the ramifications of the symmetric tree by means of, for example, the genetic engineering. On the other hand, there are structures around us which may be depicted by symmetric trees. For instance, a year possesses a tree-like structure which is symmetric. This structure was also used in the literature.

Michener in his historical novel *Poland*[170c] gives the following description of the future Krzyztopor palace of the Polish magnates Ossolinskis, reflecting the symmetry of a tree-like structure of a year (this palace was later destroyed by the Swedish Army (1655), after only seven years of existence):

> And with that unfurled two rolls of paper on which architects had done much planning, and to his startled audience he disclosed the wild plans which preoccupied him:
> 'I am going to build nothing less than the grandest castle in Europe. See! It will have of one glorious tower representing the unity of God. It will have these four huge towers, each one—you will forgive me for saying, Cyprjan—larger than your castle here. They represent the four seasons of the year.
> 'We have inside seven major edifices—living area, guests, warehouses—representing the days of the week. We have twelve corridors for the months of the year and fifty-two separate rooms for the weeks. If you cared to count, you'd find three hundred and sixty-five windows plus this little one here for Leap Year.

The development of European literature (as well as art) itself generated a tree with roots in Greece. The development of human culture also resembles a tree-like structure (a cultural tree) with roots in the ancient civilizations of China, India, Persia, Egypt, Babylonia, Greece, Rome, South America, etc. Similarly, the growth of the Croatian culture represents a tree-like structure with roots in Dubrovnik, Split, Zadar, Šibenik, Zagreb, Varaždin, Požega, Vinkovci, Vukovar, Pula, Rijeka, and Kosinj[225a,226].

C. Dualism in literature

We now wish to discuss briefly dualism in literature. Dualism is related to opposing values such as good and evil (Christ and antichrist). The concept of dualism has been present in theology and philosophy since ancient times. The transition from unity to dualism (and later to pluralism) is described in the beginning of Lao Tsu's "Tao Te Ching"[227]:

> ONE
> The Tao that can be told is not eternal Tao.
> The name that can be named is not the eternal name.
> The nameless is the beginning of heaven and earth.
> The named is the mother of ten thousand things.
> Ever desireless, one can see the mystery.
> Ever desiring, one can see the manifestations.
> These two spring from the same source but differ in name;
>   this appears as darkness.
> Darkness within darkness.
> The gate to all mystery.

Dualism may be related to the principle of anti-symmetry. The anti-symmetry is a fundamental principle of quantum mechanics[228]: It reflects the very basic characteristic of fermions. Fermions are particles (electrons, protons, neutrons, positrons) for which only anti-symmetric states occur in nature. The wave function is anti-symmetric if the interchange of the positions of any two particles cause the function to change sign. The state corresponding to an anti-symmetric wave function is called an anti-symmetric state. Particles for which only symmetric states occur in nature are called bosons (for example, photons are bosons). The wave function is symmetric if the interchange of the positions of any two particles do not cause the function to change sign. The state corresponding to a symmetric wave function is called a symmetric state. It appears that all particles occuring in nature are either fermions or bosons. Note also that for every particle there is an anti-particle.

A pictorial way of describing anti-symmetry is by comparing positive and negative in photography: to each white point corresponds a black point. Let us denote a black point by p^- and a white point by p^+, respectively. (The signs, $+$ and $-$, are related to colours). Then the transformation $p^+ \rightarrow p^-$ (i.e., the colour change) corresponds to the reflection through the anti-symmetry plane. The anti-symmetry plane is a symmetry element as the symmetry plane except that the reflection through it is followed by the colour change. The notion of black-white or anti-symmetry is the simplest aspect of the principle of colour symmetry[27].

The concept of dualism in literature consists of building a story around two opposing characters (Cain and Abel), values (good and evil), powers (East and West), political systems (democracy and totalitarianism), doctrines (evolution and revolution), etc. A standard strategy of a writer is to build (in as many variations as there are writers) the conflict of antagonists up to a climax and then to resolve the clash in some way, and, in so doing, to deliver a message containing a moral lesson and in many cases (e.g., especially in nationalistic historical novels such as Sienkiewicz's *With Fire and Sword*[229]) a lot of optimism. In many a criminal story the conflict usually ends in the murder which oridinarily happens early in the story, and then the writer works through some 50 to 100 pages of text until the murderer is delivered.

Dualism may be, of course, detected in almost all literary works, but some are truly built on the opposite values. A few examples of this kind are given as follows: Tolstoy's *War and Peace*[206] Dante's *Divine Comedy*[126], Guareschi's *Don Camillo*[230], Shaw's *Rich Man, Poor Man*[231], Cervantes's *Don Quijote*[232], Dostoyevsky's *Crime and Punishment*[233], Beecher Stowe's *Uncle Tom's Cabin*[234], Stendhal's *Red and Black*[235], Turgenev's *Fathers and Children*[236], Tolkien's *The Lord of the Rings*[130], etc. The socialistic realism, the state sponsored art and literature movement, is a truly good representative of the principle of black and white (anti-symmetry).

D. *Concluding remarks*

Finally, we wish to say a few words about entropy and symmetry, as the opposing principles, in literary works. We have already pointed out that as symmetry increases, the information content increases, whilst the entropy decreases. Many of the writers we have cited in this essay composed their works with the aim to maximize the information they wished to transmit to the reader. In other words, the literary works *in toto* possess an inner (near) symmetry which takes care that the work is well-balanced and ultimately readable. Symmetry as an aesthetic factor will be discussed in this special issue on symmetry[237]. However, there is another stream of authors whose ideas differ from the above efforts. They bring into their works a certain amount of disorder, especially in connection with communication between the characters (see, for example, Ionesco's plays, e.g., ''The Lesson''[238]). The failure to communicate is an unfortunate feature of our times. This situation is mirror-imaged in literature. Thus, we have a case of an increase in entropy of a literary work, and a decrease in the related symmetry (symmetry within the language). However, the symmetry of the work as a (distorted) mirror of reality remains. Therefore, in literary work one may detect several levels of symmetry and some may be annihilated by the increasing entropy of their particular level, but some remain. Under the levels of symmetry in the literary product we consider, for example, the symmetry of a work as a whole (mirror-image of reality), order within the language, dualism, the use of symmetry objects, etc. So it may happen that a literary work with aesthetic appeal may also possess a high value of the entropy, especially concerning the communication between the characters. A very characteristic work in this respect is a story by Thomas Pynchon entitled appropriately ''Entropy''[239]. In this beautiful story (sic!) entropy refers to the failure of Saul (an information scientist) and his wife to communicate with each other and indirectly, the similar failure of everyone at the lease-breaking party (in Washington). Incidentally, in the story a number of scientific terms appear, e.g. Gibbs and Boltzmann statistics, cosmic heat-death, the theorem of Casius, and references to scientific literature sources such as ''Handbook of Chemistry and Physics'' (CRC Press Edition) and ''Scientific American''. Thus, the story ''Entropy'' requires that the readers have a certain knowledge of modern science.

CONCLUSIONS

This study shows that symmetry and asymmetry appear in works of literature. Their appearance is in several modes: the authors either use some symmetric (and/or asymmetric) objects directly in their works or the structure of their works may be interpreted in terms of symmetric (and/or asymmetric) figures, or a literary work itself may be considered as a symmetric or an asymmetric object. An important point is also that the literary products (near) mirror-image in some way the reality (of the past and present). The authors may also predict the future and then the work plays the role of the magic mirror. This, of course, is also true for looking back into the past.

It is necessary to emphasize that the preciseness of the definition of symmetry and asymmetry is lost in literary works. When we talk about symmetry in the literature, we should do this with great care, because in this area we are dealing with the symmetry of non-physical products, i.e. psychic produce. The somewhat loose and intuitive understanding of symmetry as harmony of proportions (or beauty of form) so perceptible in the visual world, translates in the literature to spiritual harmony which is susceptible to individual sensitivity[240]. If the process of writing consists of interpreting the images that appear in the mind of the author with words, then reading is a kind of reverse process. However, the image that the reader creates in his mind may not correspond to that one originated by the author; the difference may be related to the possibility of a great number of interpretations that the work of an author offers to the creative reader. This multitude of possible interpretations makes work universal with its meaning unbounded by space and time. This may best be seen when the same play is staged by different directors as is the case with the numerous productions of classical Greek dramae or Shakespearean plays. There are as many interpretations as directors: some more, some less successful with the public. Thus, the symmetry inherent in literary works may not always be easily discerned.

At this point we may ask it how it happens that the authors thought and wrote about the concept of symmetry in their minds. Was this done consciously or subconsciously? We are free to speculate how the artists learned about symmetry. Was it by observing symmetry in nature or did the creative artist's mind, in order to bring into being beauty, order, and perfection, intuitively follow the principle of symmetry? In other words, was the use of symmetry initiated from an empirical source or from an independent source? Perhaps the origin of symmetry in artists' works could be traced to both sources. It seems that evolution follows the symmetry rules. For example, the representatives of a given animal species in order to survive must have an optimum size and shape which is ordinarily very symmetric. The symmetry requirements for the spatial organization apparently appear indirectly in the genetic code of all living matter (see, for example, the studies in the spatial structures of viruses[241]). The artist is usually a very keen observer of nature. Thus, he surely noticed, and learnt about, many kinds of symmetries that appear in the Universe. This may be supported by simply observing the art of primitive cultures or medieval paintings. Symmetry in these works is too obvious. Order, perfection, and especially beauty are important ingredients in every work of art and literature[240]. This is particularly true for poetry according to Poe[43]. Poe's theory of poetry is based on a simple proposition. He sees the human self divided neatly into intellect, conscience, and soul. The first is concerned with truth, the second with duty, and the third with beauty. In poetry the third alone is in question. Oscar Wilde started Preface to *The Picture of Dorian Gray*[186] by stating: "The artist is the creator of beautiful things". The aesthetic effects resulting from symmetry of an object lie in the psychic process of perception, and this process is highly individualistic. However, the relationship between beauty and symmetry will be discussed by competent people elsewhere in this special issue on symmetry[3].

Acknowledgements—The authors would like to thank for their discussions and correspondence on symmetry, asymmetry, and antisymmetry in mathematics, natural sciences, arts, and literature B. M. Gimarc (Columbia, SC), Ž. Jeričević (Zagreb, Houston, TX), K. Horvatić (Zagreb), I. Hargittai (Storrs, CT), I. Bauer (Zagreb), N. Allegretti (Zagreb), J. Gašparac (Zagreb), Z. Slanina (Prague), L. Klasinc (Zagreb, Baton Rouge, LA), J. Herak (Zagreb), D. Bosanac (Zagreb), I. Mills (Reading), D. Petranović (Zagreb), V. Šunjić (Zagreb), D. Bonchev (Burgas), D. Mayer (Zagreb), V. Imper (Zagreb), M. Randić (Ames, IA), Z. Pavišić (Zagreb), Z. Trgovčević (Zagreb), and T. Cvitaš (Zagreb).

Early versions of the essay were examined by D. Bonchev, M. Randić, G. M. Gimarc, I. Hargittai, D. Jeričević, and L. Klasinc. This is a good opportunity to thank them for critical comments and many helpful suggestions.

REFERENCES

1. W. Gilde, *Gespiegelte Welt*. VEB Fachbuchverlag, Leipzig (1979).
2. H. Weyl, *Symmetry*. University Press, Princeton (1952).
3. See other contributions on symmetry in this special issue of *Computers & Mathematics with Applications*, edited by István Hargittai.
4. E. P. Wigner, *Group Theory and Its Applications to Quantum Mechanics of Atomic Spectra*. Academic, New York (1959).
5. M. Hamermesh, *Group Theory and Its Application to Physical Problems*. Addison-Wesley, Reading, Mass. (1967).
6. F. A. Cotton, *Chemical Applications of Group Theory*. Wiley, New York (1963).
7. L. Pauling and R. Hayward, *The Architecture of Molecules*. Freeman, San Francisco (1964).
8. R. McWeeny, *Symmetry: an Introduction of Group Theory and Its Applications*. Pergamon, Oxford (1964).
9. M. Tinkham, *Group Theory and Quantum Mechanics*. McGraw-Hill, New York (1964).
10. H. H. Jaffe and M. Orchin, *Symmetry in Chemistry*. Wiley, New York (1965).
11. R. M. Hochstrasser, *Molecular Aspects of Symmetry*. Benjamin, New York (1966).
12. R. B. Woodward and R. Hoffmann, *The Conservation of Orbital Symmetry*. Verlag-Chemie, Weinheim (1970).
13. D. S. Schonland, *Molecular Symmetry*. Van Nostrand & Rheinhold, London (1971), reprinted.
14. I. Bernal, W. C. Hamilton and J. S. Ricci, *Symmetry*. Freeman, San Francisco (1972).
15. A. D. Boardman, D. E. O'Connor and P. A. Young, *Symmetry and Its Applications in Science*. McGraw-Hill, London (1973).
16. D. B. Chestnut, *Finite Groups and Quantum Theory*. Wiley, New York (1974).
17. J. Rosen, *Symmetry Discovered*. Cambridge University Press, London (1975).
18. L. Klasinc, Z. B. Maksić and N. Trinajstić, *Symmetry of Molecules* (in Croatian). Školska knjiga, Zagreb (1979).
19. I. S. Dmitriev, *Symmetry in the World of Molecules*. Mir Publishers, Moscow (1979).
20. H. Primas, *Chemistry, Quantum Chemistry, and Reductionism*, Lecture Notes in Chemistry, No. 24. Springer, Berlin (1981).
21. J. Maruani and J. Serre (eds.), *Symmetries and Properties of Non-Rigid Molecules*. Elsevier, Amsterdam (1983).
22. M. Gardner, *The Ambidextrous Universe*. Basic Books, New York (1964); see also H. E. Huntley, *The Divine Proportion*. Dover, New York (1970) and *Patterns of Symmetry*, edited by M. Senechal and G. Fleck, University of Massachusetts Press, Amherst (1977).

23. C. H. MacGillavry, *Symmetry Aspects of M. C. Escher's Periodic Drawings*, 2nd Edn. Bohn, Scheltema & Holkema, Utrecht (1976).
24. V. A. Koptsik, *Shubnikov Groups*. University Press, Moscow (1966).
25. G. Kepes (ed.), *Module, Proportion, Symmetry, Rhythm*. Braziller, New York (1966).
26. M. C. Escher, *The Graphic Work of M. C. Escher*. Meredith, New York (1967).
27. A. V. Shubnikov and V. A. Koptsik, *Symmetry in Science and Art*. Plenum, New York (1974); (a) p. 308.
27a. I. Hargittai and M. Hargittai, *Szimmetria egy kémikus szemével*, Akadémiai Kiadó, Budapest (1983). Its revised English version will be published by Verlag Chemie, Weinheim, under the title *Symmetry through the Eyes of a Chemist*.
28. S. Batušić, *Pictorial History of Art* (in Croatian). Matica Hrvatska, Zagreb (1967).
29. The word "symmetry" is of Greek origin: It is constructed from the root μετρων (to measure) and the prefix συν (along or together) becoming συμ before the letter μ in the root. The meaning of the word συμμετρια is commensurable, well-proportioned, well-balanced.
30. D. du Maurier, *Rebecca*. Avon, New York (1979); (a) p. 2; (b) p. 23; (c) p. 133.
31. W. Blake, "The Tyger" in *The Portable Blake*, p. 109. Penguin, Harmondsworth, Middlesex (1974).
32. A. Wickham, "Envoi", in *Selected Poems*. Chatto and Windus, London (1971).
33. S. Lanier, *The Science of English Verse*. Scribner's, New York (1894).
34. N. Frye, *Fearful Symmetry*. University Press, Princeton (1972) third printing. The analytical study of the poet and the painter William Blake (1757–1827).
35. M. Franičević, *Some Problems of Our Rhythm* (in Croatian). Rad JAZU, No. 313, Zagreb (1958).
36. I. Slamnig, *Croatian Versification* (in Croatian). Liber, Zagreb (1981).
37. B. Pavlović and N. Trinajstić, *Mathematical Structures in Literature*, in preparation. Dr. Danail Bonchev (Burgas, Bulgaria) in his letter of November 8th, 1984, in which he commented on our essay, pointed out that 12 Cand. Sci. Theses have been produced on this topic in the U.S.S.R. Unfortunately, we were unable to obtain copies of any of these works.
38. See, for example, a discussion about meanings and uses of the term structure in M. Glucksman, *Structural Analysis in Contemporary Social Thought*. Routledge & Kegan Paul, London (1974), Chapter 2.
39. B. Pavlović and N. Trinajstić, "Symmetry in Croatian Art and Poetry" (in Croatian), preprint.
40. R. Musil, "Der matematisch Mensch", in *Prosa und Stücke*, p. 1004. Rowohlt, Reinbek bei Hamburg (1978).
41. R. Musil, *Der Man ohne Eigenschaften*. Rowohlt, Reinbek bei Hamburg (1978).
42. O. Henry, "The Chair of Philanthromathematics", in *Gentle Grafter*, p. 443. Octopus, London (1983).
43. E. A. Poe, "The Philosophy of Composition", in *Prose and Poetry*, p. 311. Raduga, Moscow (1983).
44. J. Joyce, *Ulysses*, p. 598. Penguin, Harmondsworth, Middlesex (1975).
44a. R. B. Martin, *Tennyson—The Unquiet Heart*, p. 462. Clarendon, Oxford (1980).
44b. A. Lord Tennyson, "The Vision of the Sin", in *Poems*, Vol. II, p. 124. Macmillan, London (1908).
45. F. Harary, *Graph Theory*. Addison-Wesley, Reading, Mass. (1971) second printing.
46. P. Alexandroff, *Elementary Concepts of Topology*. Dover, New York (1961).
47. L. Brillouin, *Science and Information Theory*. Academic, New York (1963) second printing.
48. W. N. Lipscomb, *Boron Hydrides*. Benjamin, New York (1963).
49. E. L. Muetterties and W. H. Knoth, *Polyhedral Boranes*. Dekker, New York (1969).
50. Q. Johnson, G. S. Smith, O. H. Krikorian and D. E. Sands, The crystal structure of $RhBe_{6.6}$. *Acta Cryst*. **B26**, 109 (1970).
51. V. Prelog, "Chirality in Chemistry", Nobel Lecture, December 12 (1975), Stockholm. Reprinted in *Croat. Chem. Acta* **48**, 195 (1976) and *Science* **193**, 17 (1976). The term "chirality" is derived from the Greek word χειρ for hand. Therefore, "chirality" means handedness in reference to a pair of non-superimposable objects we constantly have before us: our two hands. This term was proposed by Lord Kelvin (W. H. Thompson) in 1884. In the contemporary chemical literature the term chirality has largely replaced the older terms "dissymmetry" and "asymmetry".
52. e.g. D. M. Lambert, Specific-mate recognition systems, phylogenies, and asymmetrical evolution. *J. Theoret. Biol*. **109**, 147 (1984).
53. D. L. Sayers, "The Image in the Mirror", in *Great Tales of Detection*, p. 266. Everyman, London (1976).
54. K. Wiberg and J. Siegel, Stereoisomerism and local chirality. *J. Am. Chem. Soc*. **106**, 3319 (1984).
55. M. Rose, *Shakespearean Design*. Belknap, Cambridge, Mass. (1972).
56. T. J. Reed, *Thomas Mann—The Uses of Tradition*. Clarendon, Oxford (1974).
57. S. E. Grace, *The Voyage That Never Ends*. University of British Columbia Press, Vancouver (1982). The study of Malcolm Lowry's fiction.
58. S. Lasić, *The Structure of Krleža's 'Banners'* (in Croatian). Liber, Zagreb (1974).
59. V. Pavletić, *Ujević* (in Croatian), Liber, Zagreb (1978).
60. D. Suvin, Dramatic works of Ivo Vojnović in the European context (in Croatian), in *Croatian Literature in the European Context* (edited by A. Flaker and K. Pranjić), p. 413. Liber, Zagreb (1978).
61. C. Flament, *Applications of Graph Theory to Group Structure*. Prentice Hall, Englewood Cliffs (1963).
62. C. Lévi-Strauss, *Structural Anthropology*. Basic, New York (1963). Structural analysis in linguistics and anthropology.
63. F. Harary, Editor, *Graph Theory and Theoretical Physics*. Academic, London (1967).
64. P. Hage, A graph theoretic approach to the analysis of alliance structure and local grouping in highland New Guinea. *Anthropological Forum* **3**, 280 (1973).
65. P. Doreian, On the connectivity of social networks. *J. Math. Sociology* **3**, 245 (1974).
66. L. W. Beineke and R. J. Wilson, Editors, *Selected Topics in Graph Theory*, Academic, London (1978).
67. N. Trinajstić, *Chemical Graph Theory*. CRC Press, Boca Raton (1983).
68. G. de Maupassant, "Simon's Papa", in *The Great Short Stories of Guy de Maupassant*, p. 8. Pocket, New York (1959) 6th printing.
69. A. Christie, *Triangle at Rhodes*. Dell, New York (1971).
70. G. G. Marquez, *One Hundred Years of Solitude*. Avon Bard, New York (1971).
71. A. Christie, *The Murder of Roger Ackroyd*, P. 83. Pocket, New York (1974).

72. S. Jendričko, "A Debatable Verse", in *Title* (in Croatian), p. 17. Izdavački centar, Rijeka (1983).
73. A. Stamać, Rondo. *The Bridge* (Yugoslav Review of Croatian Literature) **1984**, 142.
74. A. Rényi, *Probability Theory*. North-Holland, Amsterdam (1970).
75. C. E. Shannon and W. Weaver, *The Mathematical Theory of Communication*. University of Illinois Press, Urbana, Illinois (1949).
76. I. Prigogine and G. Nicolis, *Symmetry and Thermodynamics*, Ref. 3.
77. B. Fučić, *Glagolitic Epigraphics* (in Croatian). Kršćanska sadašnjost, Zagreb (1982), see also *The First Croato-Glagolitic Primer*—1527. Grafički zavod Hrvatske, Zagreb (1983), reprint.
77a. Kruchonikh (b. 1883) told this poem, in 1964, to the Croatian poet Josip Sever when he visited him in Moscow. Sever later edited a collection of poems of Vladimir Mayakovsky (1893–1930) and there he told about this visit to, and his discussions with, Kruchonikh. See this recorded in V. Mayakovsky, *The Thirteenth Apostle* (translation into Croatian and comments by J. Sever), p. 239. Mladost, Zagreb (1982).
78. I. Škarić, Cybernetics and Language (in Croatian with the extended abstract in English). *Suvremena lingvistika* **5–6**, 17 (1973).
79. Ref. 27, p. 360.
80. e.g., L. T. Milic, *A Quantitative Approach to Style of Jonathan Swift*. Mouton, The Hague (1967).
81. See, for example, the issue of *bit international* **7** (1971) entitled "Dialogue with the Machine" containing papers reported at the International Symposium on Computers and Visual Research (Zagreb, May 5–6, 1969).
82. L. M. Branscomb and J. C. Thomas, Ease of use: a system design challenge. *IBM Systems J.* **23**, 224 (1984).
83. A. Einstein, *Relativity*. Bonanza, New York (1961).
84. R. Marinković, "Hände", in: *Erzählungen*, Steingrüben, Sttutgart (1961).
85. R. Konstantinović, *Pentagram* (in Serbian). Forum, Novi Sad (1966).
86. C. Morgenstern, "Die Trichter", in *Galgenlieder Gingganz und Horatius Travestitus*, Zbinden, Basel, 1972, p. 34.
86a. E. M. Remarque, *The Arch of Triumph*. Signet, New York (1984).
87. A. Döblin, *Berlin–Alexanderplatz*. Schleber, Kassel (1947).
88. I. B. Singer, "The Cafeteria" in *A Friend of Kafka and Other Stories*, p. 83. Fawcett Crest, New York (1980).
89. A. I. Solzhenitsyn, *August 1914*. Bantam, New York (1974).
90. G. Metalious, *Peyton Place*. Pan, London (1965) 15th printing.
91. G. Chevalier, *Clochemerle*. Penguin, Harmondsworth, Middlesex (1976).
92. G. Orwell, *1984*. Signet, New York (1961).
93. A. S. Puškin, *Eugene Onegin*. University Press, Princeton (1981).
94. S. Lagerlöf, *Gösta Berling*. Langen, Munchen (1912).
95. J. London, *Martin Eden*. Penguin, Harmondsworth, Middlesex (1980).
96. W. Goldman, *Marathon Man*. Dell, New York (1974).
97. V. Voinovich, *The Life and Extraordinary Adventures of Private Ivan Chonkin*. Penguin, Harmondsworth, Middlesex (1978).
98. C. Dickens, *Oliver Twist*. Penguin, Harmondsworth, Middlesex (1971).
99. J. T. Farrell, *Studs Lonigan*. Signet, New York (1965).
100. L. N. Tolstoy, *Anna Karenina*. Progress, Moscow (1982).
101. J. Hašek, *The Good Soldier Schweik*. Penguin, New York (1946) fourth printing.
102. B. L. Pasternak, *Doctor Zhivago*. Collins, London (1961).
103. A. I. Solzhenitsyn, *One Day in the Life of Ivan Denisovich*. Penguin, Harmondsworth, Middlesex (1963).
104. K. May, *Winnetou*. Continuum, New York (1977).
105. Voltaire, *Candide*, in *Romans et contes*. p. 145. Gallimard, Paris (1979).
106. H. de Balzac, *Eugènie Grandet*. Garnier, Paris (1965).
107. W. Shakespeare, "Hamlet", in *The Works of William Shakespeare* (edited by W. A. Wright), Vol. VII, p. 379. Macmillan, London (1904).
108. E. Hemingway, *The Old Man and the Sea*. Granada, London (1984).
109. A. Maurois, *The Life of Sir Alexander Fleming*. Cape, London (1959).
110. Lord Birkenhead, *Rudyard Kipling*. Star, London 1980.
111. R. Buckle, *Nijinsky*. Penguin, Harmondsworth, Middlesex (1980).
112. J. Bonanno with S. Lalli, *A Man of Honor*. Simon & Schuster, New York (1983).
113. P. La Mure, *Moulin Rouge*. Signet, New York (1954).
114. S. K. Padover, *Karl Marx*. Mentor, New York (1980).
115. J. Louis with E. Rust and A. Rust, Jr., *Joe Louis: My Life*. Berkeley, New York (1981).
116. L. Farago, *Patton*. Dell, New York (1976).
117. I. Stone, *The Agony and the Ecstasy*. Signet, New York (1961).
118. M. Zolotow, *Shooting Star*. Pocket, New York (1979).
119. R. W. Clark, *The Life of Albert Einstein*. Hodder and Stoughton, London (1973).
120. A. Solzhenitsyn, *The Oak and the Calf*. Harper and Row, New York (1980).
121. E. Canneti, *Die Gerettete Zunge*. Hanser, München (1977); *Die Iackel in Ohr*, Hanser, München (1980).
122. A. Frank, *A Diary of a Young Girl*. Cardinal, New York (1960).
123. J.-J. Rousseau, *Les Confessions*. Gallimard, Paris (1951).
124. C. Marlowe, *The Tragical History of the Life and Death of Doctor Faustus*, p. 15. Benn, London (1983).
125. Homer, *The Odyssey*. Greenwich, New York (1982).
126. Dante Alighieri, *Divine Comedy*. Pantheon, New York (1948).
127. J. Verne, *Around the World in Eighty Days*. French, London (1875).
128. B. Brecht, "Der kavkasische Kreidekreis", in *Gesammelte Werke*, Vol. 5, p. 1999. Suhrkamp, Frankfurt/Main (1967).
129. A. I. Solzhenitsyn, "The First Circle" (in Russian), *Collected Works*, Vol. 3. Posev, Frankfurt/Main (1969).
130. J. R. R. Tolkien, *The Lord of the Rings*. Unkin, London (1974).
131. J. Bunyon, *The Pilgrim's Progress*. Penguin, Harmondsworth, Middlesex (1979).
132. W. S. Maugham, *The Circle*. Heinemann, London (1932).

132a. M. Matković, *Game Around Death* (in Croatian). Zora, Zagreb (1955).
 133. E. A. Poe, "A Descent in the Maelstrom", in *The Fall of the House of Usher and Other Tales*, p. 32. Signet, New York (1960).
 134. E. A. Poe, "Narrative of A. Gordon Pym", ref. 133, p. 200.
 135. Plato, "Symposium", in *Plato in Twelve Volumes*, Vol. III, p. 73, Heinemann, London (1967); cited lines are taken from pp. 134–139.
 136. H. Sienkiewicz, *Quo vadis*. Airmont, New York (1968).
 137. L. Wallace, *Ben-Hur*. Pocket, New York (1959).
 138. P. Lagerkvist, *Barabbas*. Four Square, London (1960).
 139. H. Fast, *Spartacus*. Panther, London (1959).
 140. T. Caldwell and J. Stearn, *I, Judas*. Signet, New York (1978).
 141. F. Yerby, *Judas, My Brother*. Mayflower, St. Albans, Herts (1976).
141a. A. Burgess, *Man of Nazareth*. Bantam, New York (1982).
 142. F. Nietzsche, *Thus Spoke Zarathustra*. Schlechta, Munschen (1955).
 143. K. R. Popper, *La connaissance objective*. Editions Complexe, Bruxelles (1982).
 144. E. M. Remarque, *Drei Kameraden*. Desch, Munchen (1951).
 145. J. Dos Passos, *Three Soldiers*. Doran, New York (1921).
 146. F. van Wyck Mason, *Three Harbours*. Cardinal, New York (1952).
 147. A. Christie, *Third Girl*. Fontana, Glasgow (1977) seventeenth impression.
 148. G. Greene, *The Third Man*. Heinemann, London (1976).
 149. J. K. Jerome, *Three Men in a Boat, To Say Nothing of the Dog*. Collins, London (1957).
 150. I. Wallace, *The Three Sirens*. Signet, New York (1964).
 151. E. Kästner, *Drei Männer im Schnee*. Rascher, Zurich (1934).
 152. L. Uris, *Trinity*. Doubleday, Garden City (1976).
 153. J. Jendričko, *Triangle* (in Croatian), see Ref. 72, p. 20.
 154. Homer, *The Illiad*. Greenwich, New York (1982).
 155. W. Shakespeare, "Othello". Methuen, London (1969).
 156. T. Dreiser, *An American Tragedy*. Constable, London (1928).
 157. J. M. Cain, *The Postman Always Rings Twice*. Avenel, New York (1982).
 158. J. London, *The Sea-Wolf*. Avenel, New York (1980).
 159. J. W. Goethe, *Leiden des jungen Werthers*. Bibliographisches Institut, Leipzig (1902).
 160. G. Flaubert, *Madame Bovary*. Greenwich, New York (1982).
 161. J. M. Cain, *Double Indemnity*. Avenel, New York (1982).
 162. J. M. Cain, *Mildred Pierce*. Avenel, New York (1982).
 163. A. Dumas, père, *The Three Musketeers*. Penguin, Harmondsworth, Middlesex (1983).
 164. See the criticism of the Corpuscularian philosophy in R. Bošković, *Theoria Philosophiae Naturalis*. Remondini, Venice (1763); reprint, Liber, Zagreb (1974).
165a. W. Blake, *The Portable Blake*, (Edited by A. Kazin). Penguin, Harmondsworth, Middlesex (1974).
165b. C. A. Abrahams, *William Blake's Fourfold Man*. Bouvier, Bonn (1978).
165c. L. da Vinci, *Quadrifolium*. (translation into Croatian). Grafički zavod Hrvatske, Zagreb (1981).
 166. V. B. Ibáñez, *The Four Horsemen of the Apocalipse*. Four Square, London (1962).
 167. T. Slavica, *The Fourth Horse* (in Croatian). Grafički zavod Hrvatske, Zagreb (1982).
 168. A. Christie, *The Big Four*. Dell, New York (1968).
 169. W. S. Reymont, *Die Bauern*. Diederichs, Düsseldorf (1975).
 170. J. A. Michener, *Poland*. Corgi, London (1984) (a) p. 685, (b) p. 186, (c) p. 171.
 171. A. Christie, *Five Little Pigs*. Fontana, Glasgow (1981).
 172. Ž. Marković, *Ruđe Bošković* (in Croatian), part I, pp. 123–128. Yugoslav Academy of Sciences and Arts, Zagreb (1968).
 173. Ž. Dadić, *History of Exact Sciences by Croats* (in Croatian), part I, pp. 296–300. Liber, Zagreb (1982).
 174. T. Mann, *Magic Mountain*, p. 480. Vintage, New York (1969).
 175. Plato, "Timaeus", in *"Plato in Twelve Volumes*, Vol. IX, Heinemann, London (1966).
 176. F. Baldensperger, "La Littérature" (translation into Croatian), Chapter 1. HIBZ, Zagreb (1944).
 177. G. de Maupassant, *Love*, in Ref 68, p. 1.
 178. Ref. 108, p. 32.
 179. E. Schrödinger, *What is Life?*. Cambridge University Press, London (1967).
 180. E. Kästner, *Das doppelte Lottchen*. Atrium, Zürich (1949).
 181. F. M. Dostoyevsky, *A Double* (in Russian), p. 109. Collected Works, Vol. 7. Nauka, Leningrad (1972).
 182. E. A. Poe, "William Wilson—A Tale", in *The Unabridged Edgar Allan Poe*. p. 549. Running Press, Philadelphia (1983).
 183. G. Papini, "Due immagini in una vasca", in *Poesia e Fantasia*, p. 571. Mondadori, Milano (1958).
 184. R. L. Stevenson, "The Strange Case of Dr. Jekyll and Mr. Hyde", in *The Great Short Stories of Robert Louis Stevenson*. 6th printing, p. 1. Pocket, New York, (1959).
 185. G. de Maupassant, *La Horla*. D'Art H. Piazza, Paris (1972).
 186. O. Wilde, *The Picture of Dorian Gray*. Keller-Farmer, London (1907).
 187. E. A. Poe, *Life and Death (The Oval Portrait)*. in Ref. 182, p. 734.
 188. F. Kafka, "Metamorphosis", in *German Stories and Tales* (edited by R. Pick) p. 266. Pocket, New York (1955).
 189. J. Swift, *Gulliver's Travels*. Blackwell, Oxford (1965).
 190. E. Hemingway, *The Sun Also Rises*. Scribner's, New York (1970).
 191. I. Shaw, *The Young Lions*. Pan, London (1958) fourth printing.
 192. J. Jones, *From Here to Eternity*. Avon, New York (1975).
 193. N. Mailer, *The Naked and the Dead*. Signet, New York (1976).
 194. A. Hailey, *Overload*. Bantam, New York (1980).
 195. J. Kosinski, *The Painted Bird*. Bantam, New York (1979) 11th printing.

196. M. Kundera, *Joke* (in Czech). Československý spisovatel, Praha (1967).
197. J. Škvorecký, *Armoured Battalion* (in Czech). Československý spisovatel, Praha (1969).
198. V. Hugo, *Les Misérables*. Hetzel & Quantin, Paris (1881).
199. L. Carroll, "Through the Looking-Glass and what Alice found There", in *The Penguin Complete Lewis Carroll*, p. 121. Penguin, Harmondsworth, Middlesex (1983).
200. G. Papini, *Lo specchio che fugge*. Ricci, Milano (1975).
201. A. Christie, *Dead Man's Mirror*. Dell, New York (1981).
202. Ref. 107: The cited lines were said by Hamlet in Act III, Scene II, lines 19–24.
203. Ref. 53, p. 275.
204. E. Zola, *Germinal*. Bernouard, Paris (1928).
205. E. Zola, *Nana*. Bernouard, Paris (1928).
206. L. N. Tolstoy, *War and Peace*. Penguin, Harmondsworth, Middlesex (1984).
207. J. A. Michener, *Chesapeake*. Random, New York (1978).
208. I. Andrić, *The Bridge on the Drina*. University of Chicago Press, Chicago (1977).
209. E. G. Bulwer-Lytton, *The Last Days of Pompeii*. Rutledge, London (1850).
210. E. K. Gann, *Masada*. Coronet, London (1980).
211. A. Šenoa, *Goldsmith's Gold* (in Croatian) Globus, Zagreb (1978).
212. M. Mitchell, *Gone with the Wind*. Macmillan, New York (1948).
213. Casanova, *Mémoires*. Gallimard, Paris (1958).
214. D. D. Eisenhower, *Crusade in Europe*. Heinemann, London (1948).
215. B. Montgomery, *The Memoirs of Field Marshall the Viscount Montgomery of Alamein, K. G.* Collins, London (1958).
216. C. Ryan, *A Bridge Too Far*. Coronet, London (1977), sixth impression.
217. J. Jones, *W W II*. Ballantine, New York (1977).
218. B. W. Tuchman, *A Distant Mirror*. Penguin, New York (1978).
219. C. G. Jung, *Alchemical Studies*, Chapter 5. University Press, Princeton, 1976, third edition.
220. T. Mann, *Buddenbrooks*. Fischer, Berlin (1960).
221. J. Galsworthy, *The Forsyte Saga*. Heinemann, London (1935).
222. M. Krleža, *Glembays* (in Croatian). Zora, Zagreb (1950).
223. I. B. Singer, *The Family Moskat*. Farrar, Straus, & Giroux, New York (1950).
224. A. Haley, *Roots*. Doubleday, Garden City, N.Y. (1976).
225. J. Michener, *The Covenant*. Corgi, London (1982).
225a. Z. Črnja, *Cultural History of Croatia* (in Croatian, but there is also an English translation available). Epoha, Zagreb (1965).
226. J. Horvat, *Croatian Culture through 1000 Years"* (in Croatian), Vol. 1, Vol. II. Globus, Zagreb (1980).
227. Lao Tsu, *Tao Te Ching*. Vintage, New York (1972).
228. P. A. M. Dirac, *The Principles of Quantum Mechanics*, reprinted fourth edition, Chapter 9. Clarendon, Oxford (1962).
229. H. Sienkiewicz, *With Fire and Sword* (translation in Slovenian). Državna založba Slovenije, Ljubljana (1975).
230. G. Guareschi, *Don Camillo*. Rizzoli, Milano (1965).
231. I. Shaw, *Rich Man, Poor Man*. Delacorte, New York (1970).
232. M. de Cervantes Saavedra, *El ingenioso hidalgo don Quijote de la Mancha*. Espasa—Calpe, Madrid (1967).
233. F. M. Dostoyevsky, *Crime and Punishment* (in Russian), *Collected Works*. Nauka, Leningrad (1973).
234. H. Beecher Stowe, *Uncle Tom's Cabin*. Penguin, New York (1983).
235. Stendhal, *Le Rouge et le Noir*. Kraus, Nendeln, Liechtenstein (1968).
236. I. S. Turgenev, *Fathers and Children* (in Russian). Nauka, Moscow (1981).
237. H. Osborne, *Symmetry as an Aesthetic Factor*, Ref. 3.
238. E. Ionesco, *La Leçon*. Gallimard, Paris (1975).
239. T. Pynchon, "Entropy", in *12 from the Sixties* (edited by R. Kostelanetz), fifth printing, p. 22. Dell, New York (1974).
240. A. Haler, *The Experience of Beauty* (in Croatian). Matica Hrvatska, Zagreb (1943).
241. R. W. Hendrix, J. W. Roberts, F. W. Stahl and R. A. Weisberger (eds.), *Lambda II*. Cold Spring Harbor Laboratory, Cold Spring Harbor (1983).

Comp. & Maths. with Appls. Vol. 12B, Nos. 1/2, pp. 229–236, 1986
Printed in Great Britain.

0886–9561/86 $3.00 + .00

SYMMETRY RULES FOR CHEMICAL REACTIONS

RALPH G. PEARSON

Chemistry Department, University of California, Santa Barbara, CA 93106, U.S.A.

Abstract—The symmetry properties of molecular orbitals and of reaction coordinates can be used to decide on the feasibility of selected chemical reaction mechanisms. Some reaction paths are shown to have a large energy barrier and are said to be "forbidden by orbital symmetry." The reactions of molecules with no symmetry can also be analyzed by being compared to related symmetric molecules, where the molecular orbitals are topologically identical.

INTRODUCTION

Physical science is deeply dependent on symmetry. In the sense of order, pattern and regularity, it is clear that it would be hopeless to try to understand nature, unless such order existed. The general experience in science has been that early recognition of as much symmetry as possible in any problem will lead to the quickest solution to the problem. In many cases an immediate, but incomplete, answer can be obtained. Crystal field theory and Huckel molecular orbital theory are chemical examples where the greater part of the useful results depend on symmetry arguments alone.

A very powerful mathematical tool has been developed that enables us to exploit symmetry in an exact manner. This tool is group theory[1]. For most chemical problems dealing with molecules, it is point group theory that is used. The nuclei of the molecule are represented by a set of points, which are interchanged by the various symmetry operations. Each molecule belongs to one of the point groups, for which symbols such as T_d, C_v, O_h and so on, are used. Each point group is defined by the number and kind of symmetry elements that it possesses. These elements are planes, axes, centers and improper axes of symmetry.

Each point group is also characterized by a certain number of symmetry species associated with it. The species are represented by symbols such as A_1, E, T_{2g} and so on. Things or functions associated with a molecule must be resolvable into these various symmetry species.

There are two important conclusions that can be drawn by the application of group theory to a molecule. One has to do with the wave function ψ which is a solution to the Schrodinger equation

$$H\psi_i = E_i\psi_i, \qquad (1)$$

where H is the molecular Hamiltonian. A series of wave functions, or eigenfunctions ψ_i exists as solutions to (1). Correspondingly, there are a series of energy values, or eigenvalues E_i. ψ_0 and E_0 are the ground state quantities.

Now it can be shown that each wave function ψ_i must belong to one of the symmetry species of the group. This applies to the exact solutions of (1), which are usually unknown. Fortunately, it also applies to molecular orbital theory, where a molecular orbital (MO) is written as a linear combination of atomic orbitals (LCAO). Each MO ϕ_i satisfies the equation

$$h\phi_i = \epsilon_i \phi_i \qquad (2)$$

where h is the averaged one electron Hamiltonian. Each ϕ_i must be a proper symmetry orbital. This puts strong restraints on the LCAO's that may be used.

The second important conclusion that may be drawn from group theory has to do with nuclear motions. These motions are usually classified as translation and rotations of the molecule as a whole, and vibrations within the molecule. Translation along the three Cartesian coordinates and rotations about the two or three axes of rotation each have a symmetry label. The vibrational motions are resolvable into a set of normal modes of vibration, each with a definite symmetry. The above conclusions are also valid if two or more molecules interacting with each other are considered. A super molecule is formed which still belongs to one of the point groups. Now chemical reaction between two molecules requires changes in the initial nuclear positions, characteristic of the reactants, to new nuclear positions, characteristic of the products. These changes can always be resolved into normal modes of vibration of the super molecule, each belonging to a definite symmetry species.

At the same time the wave function must change. Since ψ^2 is the electron density function, it must change from that of the reactants to that of the products. In MO theory, the valence shell orbitals of all of the atoms involved will change from one set of LCAO's to another. The same set of AO's will be used as building blocks for both reactants and products.

Since chemical bonds are broken and new bonds are formed in a reaction, and since MO's are bonding, anti-bonding or non-bonding between various atoms, it is convenient to think of reaction as occurring by a partial transfer of electrons from certain filled MO's in the reactants to certain empty MO's, also in the reactants. This is a mechanism for generating the new LCAO's of the products. Fukui pioneered this method of looking at chemical reactions.

Woodward and Hoffman are chiefly responsible for the realization of the important role played by orbital symmetry in chemical changes. They introduced much of the language that is used today. However they did not use point group theory as such in their papers. Fukui and Hoffman shared the Nobel Prize in Chemistry for 1981 as a result of their separate contributions.

APPLICATION OF PERTURBATION THEORY

We will describe two apparently quite different, but still equivalent, methods for developing symmetry based rules for chemical reactions (2). First we will apply quantum mechanical perturbation theory to a collection of nuclei and electrons in the act of chemical change. Figure 1 is a plot of the potential energy of this system (electronic energy plus nuclear repulsion energy) as a function of the so called "reaction coordinate." This is a collective name given to the set of changing nuclear positions which correspond to going from reactants on the left to products on the right. *En route* one or more potential energy maxima, such as B, may occur. At these points the system is said to be an activated complex, existing in a transition state. Local minima may also exist such as C, defining reactive intermediates if the minimum is shallow, or stable products, if the minimum is deep.

We now use quantum mechanics in the form of perturbation theory to relate the potential

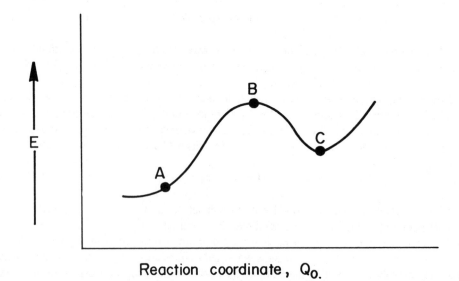

Reaction coordinate, Q_o.

Fig. 1. Plot of potential energy as a function of the reaction coordinate. Points A, B and C are discussed in the text.

energy E to the reaction coordinate. First the Hamiltonian is expanded in a Taylor–Maclaurin series about the point Q_0, corresponding to the original configuration with Hamiltonian H_0:

$$H = H_0 + \left(\frac{\delta U}{\delta Q}\right) Q + \frac{1}{2} \left(\frac{\delta^2 U}{\delta Q^2}\right) Q^2, \tag{3}$$

Here Q represents the reaction coordinate and also the magnitude of the small displacement from Q_0. For convenience, we consider only one normal mode at a time.

Since the Hamiltonian must be invariant to all the symmetry operations of the pseudo-molecule, it follows that Q and $(\delta U/\delta Q)$ have the same symmetry. Their direct product is totally symmetric. Since Q^2 is symmetric, it follows next that $(\delta^2 U/\delta Q^2)$ is also symmetric. U is the nuclear–electronic and nuclear–nuclear potential energy. The kinetic energy of the electrons and electron–electron potential energy are not functions of the nuclear coordinates, to the first order.

The last two terms in (3) represent the perturbation. Using standard second-order perturbation theory, we now solve for the new wave functions and energies. For the ground electronic state, the energy becomes

$$E = E_0 + Q \left\langle \psi_0 \left| \frac{\delta U}{\delta Q} \right| \psi_0 \right\rangle + \frac{Q^2}{2} \left\langle \psi_0 \left| \frac{\delta^2 U}{\delta Q^2} \right| \psi_0 \right\rangle + Q^2 \sum_k \frac{\left[\left\langle \psi_0 \left| \frac{\delta U}{\delta Q} \right| \psi_k \right\rangle \right]^2}{E_0 - E_k} \tag{4}$$

where E_0 is the energy at Q_0, the next two terms are the first-order perturbation energy, and the last term is the second-order perturbation energy. While (4) is valid only for Q very small, we can select Q_0 anywhere on Fig. 1. Hence (4) is general for the purpose of displaying symmetry properties.

The symbol $\langle \ldots \rangle$ represents integration over the electron coordinates, covering all space. We can now use a group theory rule to decide whether the integrals in (4) are exactly zero or not. The rule is that the direct product of three functions must contain the totally symmetric species, or the integral over all space is zero.

Let us consider the term in (4) which is linear in Q. At any maximum or minimum in the potential energy curve, $\delta E/\delta Q = 0$ and therefore the integral must be identically zero, independent of symmetry. At all other points this term must be the dominant one, since Q is small. If ψ_0 belongs to a degenerate symmetry species (E or T), the term usually leads to the first-order Jahn–Teller effect[3] which removes the degeneracy. Since this is not important in the present context, we will assume that ψ_0 is nondegenerate.

Since the direct product of a nondegenerate species with itself is always totally symmetric, we derive our first symmetry rule: all reaction coordinates belong to the totally symmetric representation. That is $(\delta U/\delta Q)$ and also Q, must be totally symmetric, otherwise its product with ψ_0^2 will not be symmetric and the integral will be zero. However, it must be nonzero for all of the rising and falling parts of Fig. 1. This means that once a reaction embarks on a particular reaction path it must stay within the same point group until it reaches an energy maximum or minimum. A totally symmetric set of nuclear motions can change bond angles and distances, but it cannot change the point group.

We now consider point A in Fig. 1. The integral $\langle \psi_0 | \delta Q | \psi_0 \rangle$ has a positive value since the reaction has a positive activation energy. Instead of trying to evaluate the integral we accept that its value is the slope of Fig. 1 at the point A. The terms in Q^2 in (4) now become important. Their sum determines the curvature of the potential energy plot. For a reaction with a small activation energy, the curvature should be as small as possible (or negative).

The integral $\langle \psi_0 | \delta^2 U/\delta Q^2 | \psi_0 \rangle$ has a nonzero value by symmetry since $(\delta^2 U/\delta Q^2)$ is totally symmetric. Furthermore, it will be positive for all molecules. It represents the force constant which resists the move of any set of nuclei away from an original configuration for which ψ_0^2 is the electron density distribution. The last term in (4) represents the change in energy that results from changing the electron distribution to one more suited to the new nuclear positions determined by Q. Its value is always negative since $E_0 - E_k$ is a negative number.

This can be seen more easily if the equation for the wave function is written down from perturbation theory:

$$\psi = \psi_0 + \sum_k \frac{\left\langle \psi_0 \left| \dfrac{\delta U}{\delta Q} \right| \psi_k \right\rangle}{E_0 - E_k} \psi_0. \tag{5}$$

The summations in (4) and (5) are over all excited states. Each excited-state wave function is mixed into the ground-state wave function by an amount shown in (5). The wave function is changed only because the resulting electron distribution ψ^2 is better suited to the new nuclear positions.

Now we can use group theory to show that only the excited-state wave functions ψ_k that have the same symmetry as ψ_0 can mix in and lower the potential energy barrier. This follows because we already have shown that $(\delta U/\delta Q)$ must be totally symmetric. Hence the direct product of ψ_0 and ψ_k must be totally symmetric, but this requires that they have the same symmetry. We can conclude that, for a chemical reaction to occur with a reasonable activation energy, there must be low-lying excited states for the reacting system of the same symmetry as the ground state. Such a reaction is said to be symmetry allowed. A symmetry-forbidden reaction is simply one which has a very high activation energy because of the absence of suitable excited states.

Equations (4) and (5) are exact, as are the symmetry rules derived from them. For practical applications, some rather drastic assumptions must now be made. One is that LCAO-MO theory will be used in place of the exact wave functions ψ_0 and ψ_k. Since we are interested only in the symmetry properties, this creates no serious error, since MO theory has the great virtue of accurately showing the symmetries of the various electronic states.

The second assumption is more serious, since we will replace the infinite sum of excited states in (1) and (2) by only a few lowest lying states. This procedure will work because we are not trying to evaluate the sum but only to decide if it has a substantial value. It can be shown that the various states contributing to (6) and (4) fall off very rapidly as the difference, $|E_0 - E_k|$, becomes large. This is because the integral $\langle \phi_0 | \delta U/\delta Q | \phi_k \rangle$ decreases very rapidly for two wave functions of quite different energy.

We use MO theory to represent the ground and excited states that are needed. The symmetry of $\psi_0 \psi_k$ is replaced by $\phi_i \phi_f$, where ϕ_i is the occupied MO in the ground state and ϕ_f is the MO occupied in its place in the excited state. Positions of special importance are occupied by the highest occupied and lowest unoccupied molecular orbitals, since excitation of an electron from HOMO to LUMO defines the lowest excited state.

It is helpful to point out that the requirement that two orbitals, ϕ_i and ϕ_f, have the same symmetry is the same as saying that they must have a net positive overlap. Two molecular, or atomic, orbitals of different symmetry species have exactly zero overlap.

We now go to a consideration of points B and C in Fig. 1. B refers to an activated complex and C to a single molecular species, which is unstable with respect to isomerization, or breakdown to other products. In either case, the theory is changed somewhat from that of the bimolecular reactions discussed earlier.

The term linear in Q in (3) now vanishes, since we are at an extremum in the potential energy plot. As before, the first quadratic term is positive, and the second one is negative. Clearly at a maximum, point B, the second term is larger than the first. At a minimum, point C, the first term dominates, but the magnitude of the second term determines whether we lie in a deep potential well or a shallow one.

Again the existence of low-lying states ψ_k of the correct symmetry to match with ψ_0 is critical. Now there is no restriction on the reaction coordinate which forces it to be totally symmetric. However, ψ_0, $(\delta U/\delta Q)$ and ψ_k are still bound by the symmetry requirement that their direct product must contain the totally symmetric representation.

If we consider rather symmetrical molecules to begin with, it will usually be found that the reaction coordinate, and $(\delta U/\delta Q)$, are nonsymmetric. The reason for this is that maximum and minimum potential energies are usually found for nuclear arrangements with a high degree of symmetry. Any disturbance of the nuclear positions will now reduce the symmetry. However, this corresponds to a change in the point group, which can only come about by a nonsymmetric vibrational mode.

Conversely, it may be pointed out that a number of point groups depend upon a unique value of Q_0 in Fig. 1. For example, a tetrahedral molecule has uniquely determined bond angles. All such cases must correspond to either maxima or minima in Fig. 1 if the reaction coordinate is taken either as the bond angles or relative bond distances.

In molecular orbital theory the product $\psi_0\psi_k$ is again replaced by $\phi_i\phi_f$, where both the occupied and empty MO's must be in the same molecule. Electron transfer from ϕ_i to ϕ_f results in a shift in charge density in the molecule. Electron density increases in the regions where ϕ_i and ϕ_f have the same sign (positive overlap) and decreases where they have opposite signs (negative overlap). The positively charged nuclei then move in the direction of increased electron density. The motion of the nuclei defines a reaction coordinate. The symmetry of Q is the same as that of the product $\phi_i \times \phi_f$. We may write this as

$$\Gamma\phi_i \times \Gamma\phi_f \subset \Gamma Q \qquad (6)$$

where Γ means symmetry species.

The size of the energy gap between ϕ_i and ϕ_f is critical. A small gap means an unstable structure, unless no vibrational mode of the right symmetry exists for the molecule capable of changing its structure. A large energy gap between the HOMO and the LUMO means a stable molecular structure. Reactions can occur, but only with a high activation energy.

For an activated complex (point B) there must necessarily be at least one excited state of low energy. The symmetry of this state and the ground state then determines the mode of decomposition of the activated complex. This was the subject of the first application of (4) to chemical reactions by Bader[4].

When a molecule lies in a shallow potential well (point C), the activation energy for unimolecular change is small. In this case we can again expect a low-lying excited state. The symmetry of this state and the ground state will determine the preferred reaction of the unstable molecule. For a series of similar molecules, we expect a correlation between the position of the absorption bands in the visible–uv spectrum and the stability.

As an example, consider the similar molecules ozone and sulfur dioxide. The former is blue in color and is highly unstable. Sulfur dioxide is colorless and is also much more stable toward dissociation into SO and O. We correlate the instability of O_3 with the fact that it absorbs light of lower energy than SO_2 does because of lower lying excited states.

THE ORBITAL CORRELATION METHOD

Before giving forth examples of the perturbation method, it is convenient to give the details of our second method, that of orbital correlation. The phrase "conservation of orbital symmetry," is often used, following Woodward and Hoffmann[5]. The method grew out of earlier work on state correlation. Wigner and Witmer developed rules for predicting the possible states of a diatomic molecule formed by combining two atoms.

Attempts to extend these rules to the states of reacting molecules and their product states did not prove very useful. State correlation did not provide enough information. A more useful method was initiated as early as 1927 by Hund, and later by Mulliken. This was a correlation of the MO's of a diatomic molecule with the AO's of the separated atoms on the one hand, and with the AO's of the so-called united atom on the other.

First the AO's of the separated atoms were combined to form MO's of the diatomic molecule. This is simply done by taking sums and differences of equivalent, or identical, orbitals; $1s_A \pm 1s_B$, etc. These MO's each have a symmetry label in the $D_{\infty h}$ (like nuclei) or $C_{\infty h}$ (unlike nuclei) point groups. The MO's were then correlated with the AO's of the united atoms (coalesced nuclei) to determine if their energies increased or decreased as the nuclei approached each other.

The correlation depended on reduction in symmetry tables, so that a p_x atomic orbital of the united atom could be clarified in the $D_{\infty h}$ point group of the molecule.

There are three correlation rules:

1. The symmetry of each MO remains unchanged as it becomes an AO of the united atom.
2. MO's of the same symmetry cannot cross.
3. No orbitals are gained or lost.

Surprisingly, there was no attempt to use the orbital correlation method for the reactions of polyatomic molecules until 1965[5]. The procedure is straight-forward. A reaction coordinate is selected which preserves one or more elements of symmetry as the nuclei rearrange from reactants to products. This generates a point group in which these MO's of the reactants, products and all intermediate steps may be classified (if there are identical MO's in two molecules, their sum and difference must be taken).

The same three correlation rules are used to connect the MO's of the reactants with those of the products. Energies are estimated by knowing whether a given MO is bonding, non-bonding or anti-bonding. The new information that is gained is an estimate of the energy of the transition state. If this energy is higher than that of the reactants, as in Fig. 1, there will be an appreciable activation energy. The reaction will be slow because of this.

The reaction is said to be "forbidden by orbital symmetry." This is too strong a statement since all that is predicted is an appreciable barrier. The reaction may still occur at a reasonable rate for many purposes. A reaction "allowed by orbital symmetry" is one where no barrier is predicted. It may be slow for other reasons. Nevertheless, analysis of possible reaction mechanisms by orbital correlation has proved to be very useful[2].

We will give an example by considering the exchange reaction between two hydrogen molecules occurring by a broad-side collision:

$$H_2 + D_2 = 2HD. \tag{7}$$

Deuterium would be used to show that a reaction actually took place. For simplicity we will consider two H_2 molecules.

$$\begin{array}{ccc}
\text{H} \quad \text{H} & \text{H}\!-\!-\!\text{H} & \text{H}\!-\!\text{H} \\
| \quad | & | \quad\quad | & \\
\text{H} \quad \text{H} & \text{H}\!-\!-\!\text{H} & \text{H}\!-\!\text{H.}
\end{array} \tag{8}$$

In such a mechanism three planes of symmetry are preserved, three twofold axes, and an inversion center. These elements define the D_{2h} point group. If the transition state is assumed to be a symmetric square, the point group is D_{4h}, but only at one point on the potential energy surface.

The starting MO's are the filled σ_g orbitals of the hydrogen molecules, and the empty σ_u orbitals. In D_{2h}, we must take symmetry adapted linear combinations of these. These SALC's are simply the sums and differences of the MO's. Figure 2 at the bottom, shows how the sums, $\sigma_g(1) \pm \sigma_{(g)}(2)$, generate species of a_g and b_{2u} symmetry in the D_{2h} point group. The combinations, $\sigma_u(1) \pm \sigma_u(2)$, generates species of b_{3u} and b_{1g} symmetry.

Figure 2 shows how the energies of these four MO's change as we go from reactants to products. No calculations need be made. All we need to know is that at the start of the reaction, the electrons must be in the orbitals a_g and b_{3u}, since we now bond together the two top H atoms and the two bottom H atoms of (8). For the products, then, a_g and b_{3u} are the low energy orbitals. This defines the orbital diagram. There is a crossing of the b_{2u} and b_{3u} orbitals. This crossing of an originally filled and an originally empty MO makes the reaction forbidden by orbital symmetry.

The meaning of this phrase, of course, is that the transition state is of high energy. The two electrons in the b_{2u} orbital must rise in energy as the nuclei move along the reaction coordinate. Notice that in the TS, the b_{2u} and b_{3u} orbitals become degenerate. This is the result

(a)

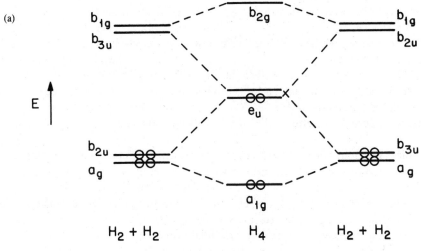

E

$H_2 + H_2$ H_4 $H_2 + H_2$

(b)

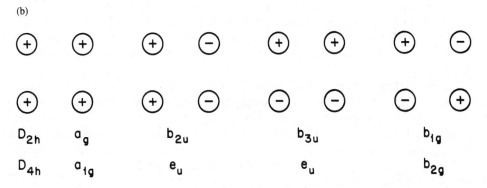

D_{2h} a_g b_{2u} b_{3u} b_{1g}

D_{4h} a_{1g} e_u e_u b_{2g}

Fig. 2. (a) Molecular orbital correlation diagram for reaction $H_2 + H_2 \rightarrow H_2 + H_2$ with square H_4 as the activated complex. (b) Linear combinations of four $1s$ orbitals and their species in D_{2h} and D_{4h} point groups. The plane of the molecule is the x–y plane.

of the higher D_{4h} symmetry assumed at this point. Symmetry produces degeneracies. After the TS, the electrons appear in the more stable b_{3u} orbital. This occurs because of the process known as configuration interaction. This mixes together the $(b_{2u})^2$ and $(b_{3u})^2$ configurations even before the TS is reached.

We can use the same example to illustrate the perturbation theory method. The filled MO's are of ag and b_{2u} species in the D_{2h} point, as mentioned. The only low energy empty MO's are of b_{1g} and b_{3u} species. Therefore there are no low-lying excited states of the same symmetry as the ground state, and the energy barrier must be large.

ORBITAL TOPOLOGY

Experimentally, the isotope exchange reaction (8) does not occur by the mechanism discussed. Instead a free atom chain mechanism is followed. Instead of the symmetric example that we chose, any four-center reaction mechanism could be selected. For example.

$$
\begin{array}{ccccc}
\text{H} & \text{F} & & \text{H--\!-F} & \text{H--F} \\
| & | & & | \quad\ | & | \\
\text{I} & \text{Cl} & \longrightarrow & | \quad\ | & \longrightarrow \\
& & & \text{I--\!-Cl} & \text{I--Cl .}
\end{array}
\tag{9}
$$

The same results of a high energy transition state may be anticipated. However, we do not have the convenient symmetry labels to use as a guide.

The symmetry of a wavefunction essentially is the way in which the mathematical sign of the wavefunction changes from plus to minus in different parts of the molecule. Thus, the nodal properties of the orbital are defined. A nodal surface raises the electronic energy by increasing the kinetic energy. If the node exists in a region where the orbital has a small value in any case, the effect is minimal. An example would be an anti-bonding MO for two atoms very far apart. However as the two atoms approach each other, the effect of the node becomes very large. Any two polyelectronic molecules, if formed close together, will have a large positive electronic energy because of this effect. This may be called repulsion due to the Pauli exclusion principle.

In a chemical reaction, where atoms change their relative positions, some orbitals go up in energy and some go down. This depends on whether the orbitals have, or do not have, a node between the atoms which approach each other. As the nuclei move about, the nodal surfaces will be deformed and distorted. However, their number and kind will not change. Also for symmetry elements that are conserved, the symmetry species defined by these nodes will be maintained. Hence we have "conservation of orbital symmetry."

In a molecule which has no symmetry, we can still see that the molecular orbitals will have nodes between certain atoms. In fact for a molecule like CO, which is less symmetric than N_2, the MO's can be obtained from those for N_2 by a process of deformation. The orbitals of CO are topologically identical to those of N_2. Two things are topologically identical if they can be interconverted by continuous twisting, stretching, and other deformation. Cutting, tearing, and piercing are not permitted. In a chemical reaction, the wavefunctions will maintain their topological identity (6). This is the more generalized statement of the conservation of orbital symmetry.

An energy barrier due to orbital symmetry results from a situation where electrons are trapped in an orbital which is going up in energy very rapidly because of the close approach of atoms separated by a nodal surface. They are trapped because no orbital of lower energy exists into which they can escape. Empty orbitals of higher energy exist and these may be useful because they can be mixed with the original orbital to produce a new orbital where the effect of the nodal surface is alleviated. However, only empty orbitals of the same symmetry can mix with the original orbital.

All of these effects would exist in reactions where there is no symmetry. We would not have symmetry labels to use as handy tags. Fortunately we can use other criteria. For example, a net positive overlap of two orbitals is all that is needed to allow them to mix. A convenient procedure is to use some related, but more symmetric, set of reactants as a model. This is what is done in predicting that (9) will be a slow reaction.

It is sometimes thought that the energy barriers deduced from symmetry can be circumvented by a reaction mechanism devoid of symmetry elements. For example, in reaction (8), if one H_2 molecule is canted a little to left or right and then twisted a little out of the plane, all elements of symmetry are lost. The point group becomes C_1, and all MO's have the same A symmetry, and can mix. However all we have succeeded in doing is destroying the predictive powers of symmetry. As far as energy is concerned, there will be little change caused by small displacements of the hydrogen atoms. In fact the energy may be increased, since the symmetric reaction path is the one that leads most smoothly from reactants to products.

REFERENCES

1. F. A. Cotton, *Chemical Applications of Group Theory*, 2nd Edn. Wiley-Interscience, New York (1971), is a good introduction.
2. R. G. Pearson, *Symmetry Rules for Chemical Reactions*. Wiley-Interscience, New York (1976), is a comprehensive review of these methods.
3. H. A. Jahn and E. Teller, Stability of polyatomic molecules in degenerate electronic states. *Proc. Roy. Soc., Ser. A.* **161**, 220–235 (1937).
4. R. F. W. Bader, Vibrationally induced perturbations in molecular electron distributions. *Can. J. Chem.* **40**, 1164–1175 (1962).
5. R. B. Woodward and R. Hoffmann, *The Conservation of Orbital Symmetry*. Verlag Chemie, Gmbh. Weinheim/Bergstrasse (1970).
6. C. R. Trindle, Mapping analysis of concerted reactions. *J. Am. Chem. Soc.* **92**, 3251–3260 (1970).

Comp. & Maths. with Appls. Vol. 12B, Nos. 1/2, pp. 237–269, 1986
Printed in Great Britain.

0886–9561/86 $3.00 + .00

SYMMETRY OF BIOLOGICAL MACROMOLECULES AND THEIR ASSOCIATIONS

B. K. VAINSHTEIN
Institute of Crystallography, the U.S.S.R. Academy of Sciences, Moscow, U.S.S.R.

Abstract—The structure of biological macromolecules and their associations obeys the symmetry theory principles. The biological macromolecules are composed of asymmetric small molecules and are found to be chiral. This may be explained by specificity of biochemical reactions. The symmetry groups of biomolecules and their associations have been considered at different hierarchical levels of organization of biostructures—polypeptides, globular proteins, quaternary structure of proteins, viruses, nucleic acids, tubular crystals and ordinary three-dimensional crystals. The data are given on noncrystallographic point symmetry of packing of globular proteins in the crystal lattice, and on the statistics of protein crystal distribution over space groups.

1. GENERAL PRINCIPLES. CHIRALITY OF BIOMOLECULES.

Introduction

The symmetric approach to the analysis of structure of most diverse objects of nature originated in the geometry and aesthetics of the ancients. The theory of symmetry grew out of the study of crystals—their external shape and internal structure. The 20th century brought about the intensive development of this theory, a deep penetration of the ideas of symmetry into many fields of physics, the application of its methods for describing not only geometrical, but also nongeometrical properties of objects[1–3].

Symmetry is also inherent in animate nature—both at a macrolevel of animal and plant organisms and at a microlevel of the structure of biomolecules and their various associations. The development of structural molecular biology involves the use of symmetry principles, elucidation of the causes of symmetry manifestation in the hierarchy of living systems, and gives an impetus to development of the symmetry theory itself.

Symmetry is the invariance, self-equality of objects

A symmetry transformation of a symmetric object changes nothing in it—the object remains equal to itself, invariant to this transformation (Fig. 1). The object—its structure and properties may be depicted by a function $F(x_1, x_2, x_3 \ldots) = F(\mathbf{x})$ in space of the variables used for its description $x_1, x_2, x_3, \ldots = \mathbf{x}$. For geometric symmetry in three-dimensional space $x_1 x_2 x_3 = \mathbf{x}$ are the Cartesian coordinates. The symmetry operation g^i transforms the coordinates: $g^i(x_1 x_2 x_3) = x_1^i, x_2^i, x_3^i; g^i(\mathbf{x}) = \mathbf{x}^i$. Thus, an object is symmetric, if

$$F(\mathbf{x}) = F(\mathbf{x}_i) \equiv F[g_i(\mathbf{x}_1)]. \tag{1}$$

Mathematically, a set of g_i for a symmetric object makes a *group* $G = G[g_1, \ldots, g_n]$. The operation of identity "immobility" of an object is the unit $g_1 = 1$ of the group.

From (1) it follows that important for symmetry are not the specific function values F at different $\mathbf{x}$, but the regular relationships, i.e. the transformation operations acting on variables $\mathbf{x}$ to which F is invariant—this is the invariability which makes the object symmetric.

Asymmetric unit

The transformation g of an object into itself implies that its parts disposed in one place will be brought into coincidence, after transformation, with the parts located in another place. It means that the object may be divided into equal parts A_i (Fig. 1(b)).

The smallest part of a symmetric object F which, when being multiplied by all the operations g_i of the group G forms the entire object, is called an independent or asymmetric unit—a "stereon" in three-dimensional space. The number of stereons in an object is equal to the group

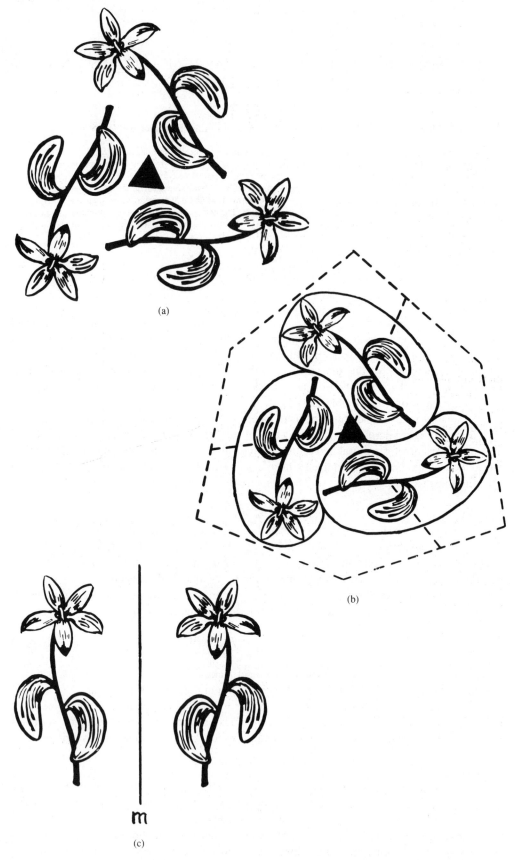

Fig. 1. (a) Symmetric object with the three-fold axis of symmetry; (b) the object may be divided into equal parts—stereons, by the infinite number of ways. Dashed line shows an example of an arbitrary division, solid line—division into "physically isolated" stereons; (c) mirror-equal asymmetric figures.

order n. Each stereon A_i may be obtained from stereon A_1 chosen as the initial one, by "its own" operation g_i:

$$g_i[\mathbf{x}_{(A_1)}] = \mathbf{x}_{(A_i)}^i \quad i = 1, 2, \ldots, n. \tag{2}$$

In nearly all the groups there exists some arbitrariness in the choice of stereon boundaries, i.e. in its shape (Fig. 1(b)); at such a choice only the point symmetry elements of an object—axes, planes and the inversion center—are fixed.

Origin of symmetry in nature

Why does symmetry permeate almost all constructions and laws of animate and inanimate nature? This fundamental question has received no general answer as yet. Mathematicians postulate, axiomatize and analyze various symmetries, physicists find it in ever-increasing number of natural phenomena and make use of predictive power of the group theory, biologists certify primordial asymmetry of small molecules of life holding together in highly symmetric associations. But what is in common in all of this?

Apparently, initial is the concept on a *finite set* of sorts and *equality* of elementary entities of which larger units at a definite matter organization level are composed. At the lowest level these are elementary particles and quarks they are made of, and the fields acting between them which can be represented again as particles of which everything existing is formed. Further, their association into nuclei and atoms leads to a new level of elementary entities whose number is also finite, and whose special aggregations provide a practically infinite set of various objects. Among these objects are molecules and crystals, and the molecules may be used anew to build more complex molecules and their associations.

However, the geometrical equality as such is a necessary, but insufficient condition for the symmetry origin. There must also be the equality in interaction, geometrically, it should be manifested in mutual arrangement of equal parts. Being governed by principles of conservation and thermodynamics, the nature imposes restrictions on the infinite number of interaction modes. It may well be that the principle of energy minimum is the most important one as regards the appearance of symmetry in three-dimensional objects[3].

If the energy minimum has been achieved at a certain configuration of an isolated subsystem, the configuration should be the same for all similar subsystems. One may believe that the energy minimum in a system composed of finite or infinite number of such subsystems is also attained at their symmetric combination with the account for their interaction energy. Otherwise, the inequality of interaction, the inequality of arrangement (when only geometry is taken into account) will cause a difference between the subsystem energies, which would be in conflict with the subsystem energy equality condition. Just in such a way we may explain, in particular, the symmetry of molecules composed of several sorts of identical atoms. If the number of atoms is very large or infinite, this approach allows one to understand the origin of translational symmetry in crystals or in polymer molecules.

As already mentioned (see Fig. 1(b)), the choice of an asymmetric unit—the stereon A of a symmetric object F, is, geometrically, ambiguous. But from the viewpoint of energetic stability or structural isolation, one can often choose a "physical stereon" (Fig. 1(b) solid line). Thus in the case of molecular biostructures it seems physically justified to take, as a stereon, an individual molecule together with some space around it, since the molecule is a firm structural unit: the atoms in it are covalently bound whereas the van der Waals forces between the molecules are weak. The case of asymmetric molecules is especially favourable for "physical" choice of a stereon, although, geometrically, the arbitrariness still remains. It should be noted that such a "physical" choice is not always possible, e.g. when the interaction between "parts" of a system is of the same order as inside these "parts." As an example one may consider the complex inorganic crystals in which a continuous spatial network of approximately equivalent bonds between atoms exists[3].

Pseudosymmetry and morphological symmetry

For some objects the symmetry conditions are not fulfilled precisely. Then, although, approximate, the equality in some parameters describing the system or its parts, enables the manifestation of symmetry laws. For instance, the molecules, similar in structure, can form pseudosymmetrically packed associations (see Fig. 18 below).

An interesting and important case is the morphological symmetry—the symmetry of an external shape. Human body is symmetric externally, but asymmetric in its internal organs. A symmetry of the form in this, as well as in other cases, is determined by the interaction with the environment. On the contrary, the ideal symmetry of the crystal habit is fully defined by the inner atomic structure of crystals, but the real habit and its symmetry turn out to be dependent on the crystal growth conditions.

The formation of biological structures

A physical approach to elucidate the structural laws of atomic systems and their symmetry is universal and may also be applied to biological macromolecules. However, the biological systems possess their own, specific laws of structural organization which, acting within the framework of general physical laws, modify the manifestation mode of the latter. Moreover, the combination of atoms and small molecules in a macromolecule and the subsequent self-organization into the higher-order structures proceed not as the result of random events of physico-chemical interactions between them, but according to a *compulsory program* stored in the DNA genetic information. As already mentioned, the energy needed to accomplish this program is available.

Biological systems possess many very important specific physical features. These systems are *open* thermodynamically and exist in a free energy flow through themselves, this flow being provided by chemical metabolic reactions. The biological systems, according to E. Schrödinger[4], "consume" negative entropy (negentropy) which, contrary to the entropy, a measure of disorder, is a measure of order. This means that a mechanism of ordering is operating within each biological system.

Thus, the biosystems exist to continue their existing, self-reproducing and developing.† We can say that the category of causality permeating the physical laws that realize it in inanimate nature, may be interpreted in biological systems as a category of purposefulness of functioning of all its micro- and macroscopic parts.

What conclusion can be drawn from the above to understand the symmetry of biomolecules and their associations? Symmetry bears in itself the principle of economy of the number of certain main building blocks in the process of biosystem formation and arises where it may be of help. In the conditions of cellular medium the symmetry of macromolecules depends on local physical conditions. Genetic information governing the appearance of macromolecules with a definite chemical structure "takes into account" the conditions in which the macromolecule will exist and through them exerts influence on the arising spatial structure. On the other hand, the structure and symmetry of macromolecules during their interaction with each other in a given medium determine the construction of higher-level associations built of them.

Asymmetry of small living protomolecules

One has to begin the specific consideration of the symmetry of biomolecules with the fact that the small, primary biomolecules—so-called protomolecules of which large molecules and their associations are built, are asymmetric, i.e. devoid of any symmetry: the protomolecules are depicted by the only asymmetric group $G = 1$. Protomolecules are the amino acids of which proteins consist and the nucleotides of which nucleic acids are built. Amino acids

$$H_2N-\overset{\overset{\displaystyle H}{\displaystyle |}}{\underset{\underset{\displaystyle R}{\displaystyle |}}{C}}-COOH \tag{3}$$

†We cannot dwell here into mechanisms of the accumulation and transfer of information in biosystems, into the origin and evolution of living systems on the basis of variability and natural selection; this would lead us far away from the topic of this article.

(there are twenty sorts of main amino acids) vary in their side radical R. Nucleotides contain pyrimidine or purine bases B of four sorts: adenine A, thymine T, guamine G, cytosine C (in RNA thymine is replaced by uracil U) attached to a sugar-phosphate backbone

$$
\begin{array}{c}
\quad\quad\quad\quad\quad\quad\quad\quad\quad \text{B} \\
\quad\quad\quad\quad\quad\quad\quad\quad\quad | \\
\quad\quad\quad\quad\quad\quad\quad\quad \text{CH} \\
\quad\quad\quad\quad\quad\quad\quad\swarrow\quad\searrow \\
\quad\quad \text{O}\quad\quad\quad\quad \text{O}\quad\quad\quad \text{CH}_2 \\
\quad\quad \| \quad\quad\quad\quad\searrow\quad\swarrow \\
-\text{O}-\text{P}-\text{O}-\text{CH}_2-\text{CH}-\text{CH}- \\
\quad\quad | \\
\quad\quad \text{OH}
\end{array}
\qquad (4)
$$

The carbon atom in an amino acid is surrounded tetrahedrically by four different atoms (atomic groups) and is, with respect to the surroundings, asymmetric (Fig. 2(a)). Ribose of nucleotide also contains asymmetric C atoms.

Symmetry groups of biomolecules and their associations—groups of the first kind

Symmetry operations congruently bringing a figure and its asymmetric parts to coincidence as in Fig. 1(a), are called operations of motion or the first-kind operations g^I. These are parallel displacements, rotations and their combinations: screw motions. The corresponding groups are the first-kind groups G^I. At the second kind operations g^{II} asymmetric parts of a figure are equal in the general meaning of symmetric equality[1,2], they are mirror-equal (Fig. 1(c)), but no motion can bring these parts to coincidence, i.e. they are not equal congruently, i.e. physically. The groups containing the mirror-equal operations g^{II} and other operations of the type g^{II} arising from them, e.g. the center of inversion, are the groups G^{II}.†

Any asymmetric object or the object depicted by the first-kind group G^I may have a mirror-equal figure—enantiomorphic one (Fig. 1(c)). The property of such objects, say, molecules, to exist in two mirror-equal, enantiomorphic forms is called chirality. As for the objects depicted by the groups G^{II}, they are chiral to themselves. Usually, one of the pairs of the chiral objects is called "left-handed," the other "right-handed" by analogy with the right and left hand. The "left," "l," tetrahedral C atom (exactly speaking, surrounded by the left mode), specific for living matter, is shown in Fig. 2(a). Its chiral analog, the "d" C atom that can also be found in biomolecules is shown too.

Thus, all biostructures, i.e. the living matter as a whole, is chiral, because chiral are the protomolecules it is formed of. Chirality of protomolecules does not preclude their joining into the symmetric higher-order structure, but the symmetry of thus obtained associations is only the symmetry of the first kind G^I.

The cause of chirality of living molecules is usually assumed to lie in the origin of life, in general, e.g. in chirality of some nonbiological structures, minerals, on which the first

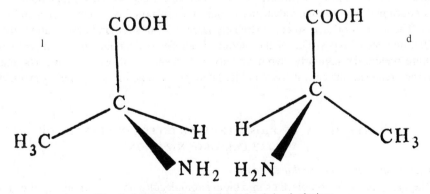

Fig. 2. Left-handed—1, and right-handed—d, alanines.

†The groups G^{II} may contain g^I operations, too. No combination of motions, i.e. of operations g^I can produce the operation g^{II}, while combinations of even number of g^{II}, give an operation of the type g^I.

biological reactions might take place, in polarization of solar radiation[5]. Recently, the ever-increasing development have received the hypotheses connecting the asymmetry of protomolecules with the fundamental physical asymmetry of elementary particles and their interactions, although the manifestation of such interactions at the atomic level in chemical bond between atoms during the molecule formation, is quite negligible.

Apparently, chirality is associated substantially with exceptional specificity of all bioorganic reactions. The right and left molecules (the objects with symmetry G^I) are chemically and physically equivalent when interacting with non-chiral objects having the second-kind symmetry G^{II}. But they appear to be different, nonequivalent in interaction with the objects which are chiral themselves, possessing symmetry G^I.† If we denote a level of interaction, its specificity, by E (it may be, e.g. the interaction energy, constants of reactions etc.) then

$$E(G^{I,L}_{(1)}, G^{II}_{(2)}) = E(G^{I,R}_{(1)}, G^{II}_{(2)}) \qquad (5a)$$

$$E(G^{I,L}_{(1)}, G^{I}_{(2)}) \neq E(G^{I,R}_{(1)}, G^{I}_{(2)}) \qquad (5b)$$

$$E(G^{I,L}_{(1)}, G^{I,L}_{(2)}) = E(G^{I,R}_{(1)}, G^{I,R}_{(2)}) \neq \qquad (5c)$$

$$E(G^{I,L}_{(1)}, G^{I,R}_{(2)}) = E(G^{I,R}_{(1)}, G^{I,L}_{(2)}) \qquad (5d)$$

$$E(G^{II}_{(1)}, G^{II}_{(2)}) = E(G^{II}_{(1)}, G^{II}_{(2)}). \qquad (5e)$$

Thus, only in case (5b) which is examplified by variants (5c) and (5d), the interaction proves to be specific. The specificity in the interaction of biosystem molecular components with each other or with the environment contributes to its autonomy, uniqueness of structure and peculiarity of the ways in which some or other reactions proceed in it, and serves as a tool providing protection and selectivity at a contact with chemical substances of the environment.

Condition (5c) in which the terms written on the left and right are "mirror image" of one another shows that the mirror-equal interactions between pairs of the L and R-objects are possible (the same holds true for LR- and RL pairs). Condition (5c) shows that the R-world analogous to our L-world of animate nature is possible. However, according to the main condition (5b) such a world R would be incompatible with the existing world L and vice versa. Also, they could not interact correctly between themselves. Therefore, if at the early stages of the origin of life the L and R (or LR) systems might come into existence, in the competition only one type out of them survived. More unstable, indifferent to reactions not only with the chiral (a), but also with the non-chiral objects would be the world of the type (5e).

Principle of hierarchy and of small number of elements

Living systems are built up according to the hierarchical principle. The first (lowest) steps in this hierarchy are small protomolecules. Important is the fact that there is only a small number of their sorts, but it allows an infinite number of chemical and spatial combinations. The next step is concerned with the individual macromolecules of proteins, nucleic acids and polysuccarides. The number of their sorts is extremely large, but, in structural organization, they form a limited number of types. Out of the unlimited number of combinations of macromolecules the nature prefers the regularly, symmetrically, expediently built associations. The main unit of a living organism, the cell, is asymmetric, although in some of its organella a symmetry can be observed.

2. SYMMETRY OF BIOSYSTEMS AT DIFFERENT LEVELS OF MOLECULAR ORGANIZATION

Symmetry groups of biomolecular systems

The grouping of protomolecules into biomacromolecules and the arrangement of macromolecules into associations can be described by groups of symmetry of three-dimensional space G^3 and only by the first-kind groups $G^{3,I}$[1,3].

---

†One cannot get the right boot on the left foot, it is merely awkward. Enantiomorphic analogs of useful medicine substances may prove to be indifferent or harmful.

These groups may contain only the following operations and symmetry elements corresponding to them (Fig. 3(a–c)):

Rotations N around the N-order axes through $2\pi/N$

Parallel translations t through the period c

Screw (helical) rotations S_M: combination of a translation along t on $c' = c/M$ with the rotation on $\alpha = 2\pi/M$ around the axis of translation. M: the parameter of a screw rotation may be the integer $M = N$ or fractional: $M = p/q$, p: the number of rotations per q turns,

$$\alpha = 2\pi\frac{p}{q}, \quad c' = \frac{qc}{p}.$$

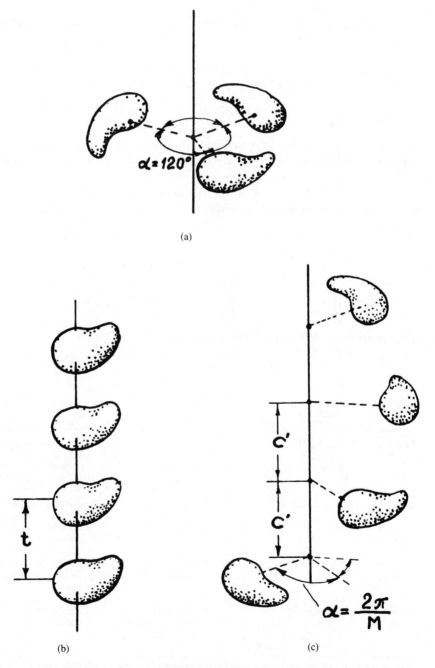

Fig. 3. Symmetry operations of the first kind: (a) rotation ($N = 3$); (b) translation t; (c) screw rotation S_M.

If the number of particles being joined together is finite, such association is described by the point groups $G_0^{3,l}$ containing only rotations. At the infinite number of particles in an association there always takes place periodicity (the lower index in the group notation). The groups G_1^3 with periodicity in one direction depict the chain and helical structures. The twice periodic groups G_2^3 depict layers[3,6]. Triple periodic groups G_3^3 are Fedorov space groups of symmetry of crystals. The characteristics of groups $G_m^{3,l}$ are given in Table 1.

It should be emphasized that the order of rotations N or screw rotations M in biostructures may be any one in all cases, except biocrystals. In crystals, due to the presence of the lattice only simple or screw rotations of order 2, 3, 4 and 6 are possible. In biostructures also the axes of the 5th, 7th and highest orders as well as screw rotations with fractional M are not forbidden.

Let us consider various molecular biostructures. The information on their structure has been, mainly, obtained with the aid of X-ray structure analysis[7].

Peptides, polypeptides

The protomolecules (3) are asymmetric, point group 1. Linking of amino acid residues by peptide bonds gives a chain

$$
\xrightarrow[2_1]{}
\begin{array}{c}
\text{H} \quad \text{O} \quad \text{R}_2\text{H} \quad \text{H} \quad \text{O} \\
\text{N} \quad \text{C} \quad \text{C}_\alpha \quad \text{N} \quad \text{C} \\
\quad \text{C}_\alpha \quad \text{N} \quad \text{C} \quad \text{C}_\alpha \\
\quad \text{R}_1\text{H} \quad \text{H} \quad \text{O} \quad \text{R}_3\text{H}
\end{array}
\Rightarrow \tag{6}
$$

The polypeptide chain is polar, the running $\text{NH—C}_\alpha\text{RH—CO}\!\Rightarrow$ in one direction is different as compared with the running in the opposite direction. The chain has the symmetry $S_2 = 2_1$ (two-fold screw rotation), denoted by $\rightarrow$.

Therefore, small peptides of an open chain are always asymmetric (group 1). On the contrary, the cyclic peptides—closed chains and related molecules may have the N-order axis, e.g. the cycle

$$
\begin{array}{ccc}
 & \text{R}_1 & \\
\text{R}_2 & & \text{R}_2 \\
\text{R}_1 & & \text{R}_1 \\
 & \text{R}_2 &
\end{array}
$$

(we designate the amino acid residue simply by its radical) has the axis of symmetry 3. An example of such a molecule is given in Fig. 4[8]. The combination of two similar rings, lying

Table 1. Symmetry groups $G_m^{3,l}$

Groups	Types of groups	Number of groups	Representatives in biosystems
Point $G_1^{3,l}$	$N, N2, 3/2, 3/4, 3/5$ All symmetry elements intersect at a special point	$N + N2 + 3$ Infinite, depends on order N	Small protomolecules, oligonucleotides, globular proteins, RNA, viruses, nucleoproteids
Chain $G_1^{3,l}$	$Nt, Nt2, S_M N, S_M N2$	Infinite, depends on values N and M	Secondary structure of proteins, fibrous proteins, DNA, tubular crystals, rod-shaped viruses
Layer $G_2^{3,l}$	Combination of $t_1 t_2 2 2_1, 2$	9 (out of the total number G_2^3 80)	Secondary structure of proteins, layered biocrystals, membranes
Crystal $G_3^{3,l}$	Combinations of $t_1 t_2 t_3$ $N = 1, 2, 3, 4, 6$ S_M at $M = 2, 3, 4, 6$	65 (out of the total number of Fedorov space groups 230)	Crystals of proteins, oligonucleotides, viruses

2, 3, 4, 6 are the axes of rotation of the corresponding order, Nt—parallelism N to translation t, $N2$—axes 2 are perpendicular to N.

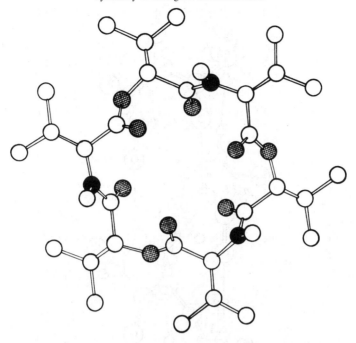

Fig. 4. Cyclic hexadepsipeptide with the three-fold axis of symmetry.

one above the other, but with a different direction of polarity gives symmetry $N2$. In such a way, the symmetry groups of peptides are: 1, N, $N2$. If radicals R are different, the symmetry (may be slightly distorted) is held only for the chain backbone—NH—CH—CO—while the arrangement of radicals is pseudosymmetric.

Secondary structure, polypeptides, fibrous proteins

When a large number of amino acid residues are joined together, some stable conformations of polypeptide chains arise which are called the secondary structure. One is the famous α-helix of Pauling and Corey (Fig. 5), the other is the β-structure of pleated sheets[9]. The α-helix is shown in Fig. 5. Its symmetry group is $G_1^3 = S_M$, $M = 18/5$, (i.e. 18 residues per 5 turns), the pitch of the helix 5,4 Å, projection of a residue on the axis $c' = 1,5$ Å, period of translation is $5,4 \times 5 = 1,5 \times 18 = 27$ Å. The α-helix is stabilized by hydrogen bonds NH—O between 1–4 residues along the chain. If all the residues are identical, the symmetry $S_{18/5}$ is a true one (this is the case of synthetic polypeptides). If the radicals are different, as in real proteins, this symmetry is observed only for the backbone, while different R_i are not equal to one another symmetrically, but their locations remain symmetrically related.

The α-helix shown in Fig. 5, is right-handed, it is thus called because it follows the right screw (amino acid residues are left-handed). One can build up the left-hand (i.e. left-screwed) helix, but it is less favourable energetically due to the packing in it of left residues, and is not observed in nature.

Many fibrous proteins—keratin, myosin and others—are built on the basis of the α-helix. But when the α-helices are packed into fibres, they are found to be slightly distorted, and helical superstructures of higher order are formed (Fig. 6)[10]. The triple chain collagen molecule represents one more type of helical protein structure[11].

The β-structure of polypeptide chains is shown in Fig. 7(a, b). The chains are arranged parallel to each other and are linked by hydrogen bonds. These structures are described by groups of layers $G_2^{3,1}$. The symmetry of an extended polypeptide chain with identical radicals is $S_2 = 2_1$; in crystallographic notation 2_1 is a two-fold screw axis, symmetry of the parallel β-pleated sheet is $2_1 t_1 t_2$ where $t_1 t_2$ are two translations in a sheet (Fig. 7(a)). The symmetry of an antiparallel β-pleated sheet (Fig. 7(b)) is $2_1 t_1 2_1 t_2 2$, screw axes 2_1 also arise along the second translation, whereas the rotation axes 2 are arranged perpendicularly to the sheet "piercing" it through. For example, silk fibroin is built according to the β-structure type.

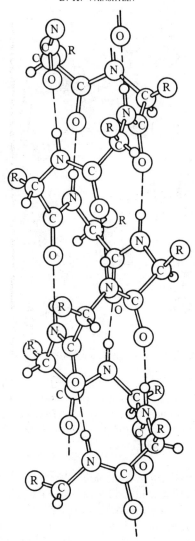

Fig. 5. The Pauling–Corey α-helix. R is a side radical of an amino acid residue.

Globular proteins, their secondary structure, supersecondary and tertiary structure

The molecules of globular proteins are the most complex atomic constructions of animate nature. A globular protein represents a single polypeptide chain (Fig. 8)[12] folded in a specific way into a globule, or an aggregation of several globules. The number of residues in a chain of various proteins ranges from dozens to several hundreds, the number of atoms in them, from hundreds to tens of thousands. Since a polypeptide chain consists of asymmetric amino acid residues and is folded in a certain way, a protein globule is always asymmetric, its symmetry being 1. However, such a globule is rich in pseudosymmetry and local symmetry. The structure of a globule is defined by the chemical sequence of radicals in a chain (8); this sequence is called the *primary* structure. After synthesis of the chain in a ribosome according to the genetic code, the chain, under cell conditions, folds spontaneously into a unique conformation which is inherent only in the given protein. All millions of molecules of the given protein are identical. It should be noted that some proteins contain molecules of non-protein nature—cofactors.

For the structure of a protein globule characteristic are:

(1) a strict periodicity repetition of the sequence of the backbone atoms $NC_\alpha C$, and the same periodicity of the attachment of side radicals. But this cannot be called the exact translational symmetry, because the chain is not straight, winding in a different way in space; it may be treated as topological translational symmetry;

(2) the stable symmetric α- or β-secondary structure arises locally on separate segments.

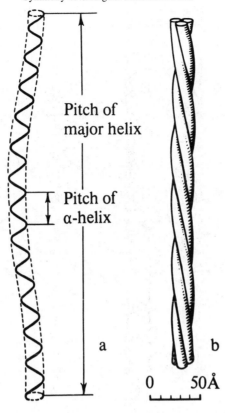

Fig. 6. Scheme of the structure of α-keratin: (a) the supercoiled α-helix; (b) three-stranded subfibril.

Between the segments of the secondary structure the chain has an irregular conformation;

(3) the segments of the β or α secondary structure often aggregate into the secondary superstructure (Fig. 9);

(4) sometimes, separate parts of a folded chain, domains, can be observed in a globule.

The spatial organization of a protein globule, as a whole, is called tertiary structure. The convenient way of describing the structure of protein molecules is the representation of β-chains by arrows, α-helices by helical ribbon or a cylinder, irregular chain by a lace (Fig. 9(a-c))[13,14]. Figs. 10–12 demonstrate some representatives of protein globules—hemoglobin[15], hem-erythrin[16] (α-proteins), γ-crystallin[17] (β-protein), carboxypeptidase[18], catalase[19] (α/β proteins). An interesting variant of the secondary β-structure—the pleated β-sheet twisted into a propeller—is observed in carboxypeptidase (Fig. 12) and some other proteins. Catalase (Fig. 13) is a protein with the clearly expressed domain structure. Figure 14[14] shows some other variants of the superstructure of domains possessing (if the details of the arrangement of radicals are neglected and only the idealized course of the chain is considered) the fine symmetry whose analogs can be found on the ornaments of paintings of the ancients.

So, we can see, that in the structure of globular proteins there is a certain hierarchy. The primary elements of the structure are small protomolecules of amino acids. On the basis of chemical sequence of amino acids the spatial superstructure and the secondary structure are formed. As the result, the domains and, finally, tertiary structure of the entire globule come into being. The packing of atoms along the chains in the entite globule is rather dense (Fig. 14).

It should also be noted that the protein molecule is in a thermal motion, its vibrations are made up in such a way as to enhance the fulfillment of the biological function. When the molecule is functioning, its parts, especially at the active center, experience slight (0.5–1 Å), or more considerable (up to 5–10 Å) conformational shifts. The expedient, unique structure of protein molecules has been elaborated in the course of many hundreds of billions years of biological evolution on the Earth.

B. K. Vᴀɪɴꜱʜᴛᴇɪɴ

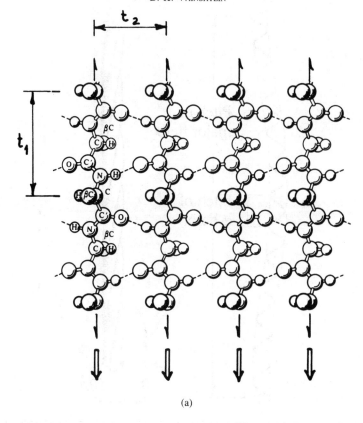

(a)

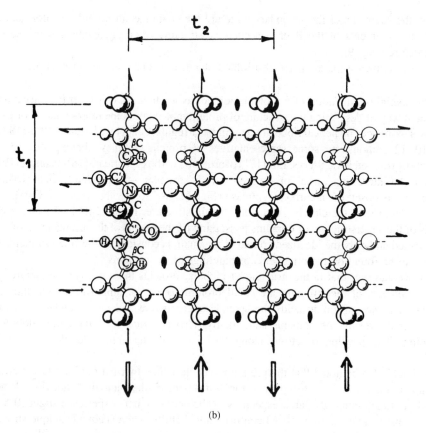

(b)

Fig. 7. Parallel (a) and antiparallel (b) β-structures. ↑, ● —two-fold axes, ↑ —screw two-fold axes, ⇑ —direction of chain.

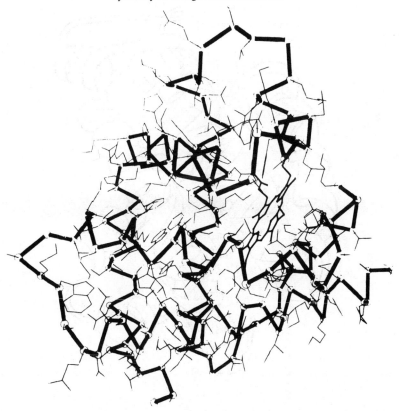

Fig. 8. Skeletal structure of the leghemoglobin molecule.

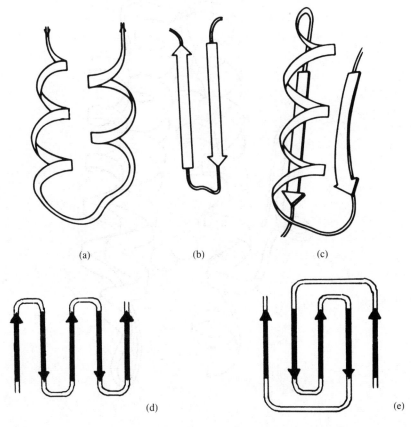

(a) (b) (c)

(d) (e)

Fig. 9. Some elements of the supersecondary structure: αα (a); ββ (b); βαβ (c); typical variants of the β-structures: méandre (d); Greek key (e).

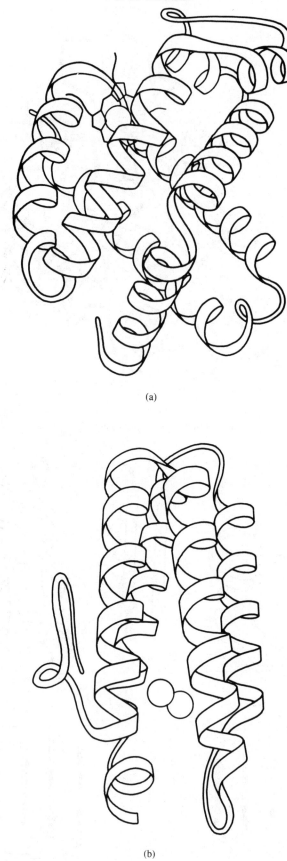

(a)

(b)

Fig. 10. Polypeptide chain folding in the structures of α-proteins: hemoglobin β-subunit (a) and hemerythrin (b).

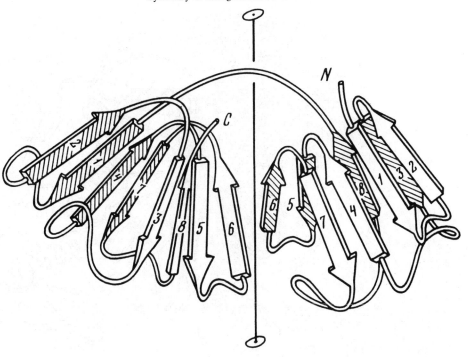

Fig. 11. γ-crystallin. An example of β-protein.

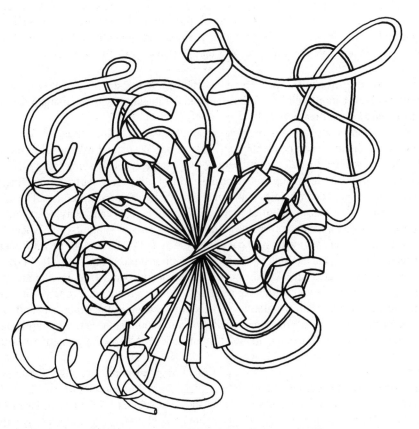

Fig. 12. Carboxypeptidase. An example of α/β-proteins.

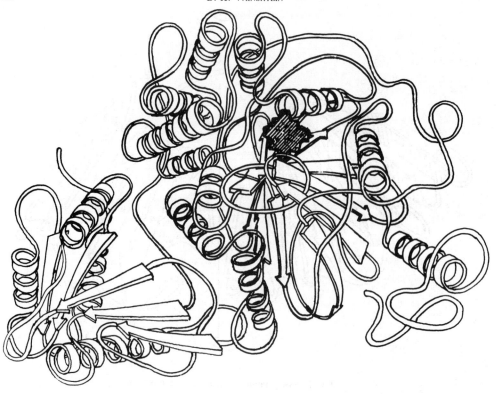

Fig. 13. A subunit of α/β-protein catalase Penicillium Vitale consisting of three domains.

The quarternary structure

The quaternary structure is the aggregation of several globules into one symmetric association. This level of organization, next to the tertiary structure, is observed in many proteins. In this case, the globules being joined together are called subunits, the form of their contacting surface is complementary. The proteins with the quaternary structure are depicted by the point groups $G_0^{3,I}$.

The quaternary structure of proteins is revealed by X-ray structure analysis; the effective method of its investigation is electron microscopy in combination with the mathematical method of three-dimensional reconstruction[3].

The formation of associations of protein globules is due to the attractive forces between them. These interactions may be electrostatic or van der Waals and hydrophobic ones. Figure 15(a, b) shows a scheme for interaction leading to the appearance of symmetry. It is facilitated by complementarity: the intersupplement of the form of contacting parts. Complementarity is one of the clearly expressed principles of organization of biomolecular structures. The association of globules enables the improvement of protein functioning. This can easily be explained when the active center of a molecule is formed on the adjacent parts of subunits, i.e. when the amino acid residues of different segments of the chain in the subunit take part in the fulfillment of the active center function. One example is aspartate aminotransferase (Fig. 16)[20]. In other cases, the active center is located far away from the boundaries of contact, as in catalase (Fig. 17), but the aggregation of subunits still exerts influence on the function, probably on account of electron structure, electrostatical potential of subunits, change in their thermal motion. This is clearly demonstrated by the fact that, on dissociation into subunits, the activity sharply decreases.

The most frequently encountered point symmetry groups of the quaternary structure of proteins are: 2,222 (tetrahedral), 32, 42.

Very interesting is the case when protein functioning prompts a change in the quaternary structure. The classical example is hemoglobin consisting of four pairwise identical subunits[21]. The exact symmetry of the molecules is 2, but, since the subunits are very similar in structure, the pseudosymmetry is tetrahedral 222 (Fig. 18). In hemoglobin, owing to the cooperative steric interaction, the function of binding and transfer of O_2 is much improved as compared with monomeric (consisting of one subunit) proteins of this type. An oxygen attachment involves

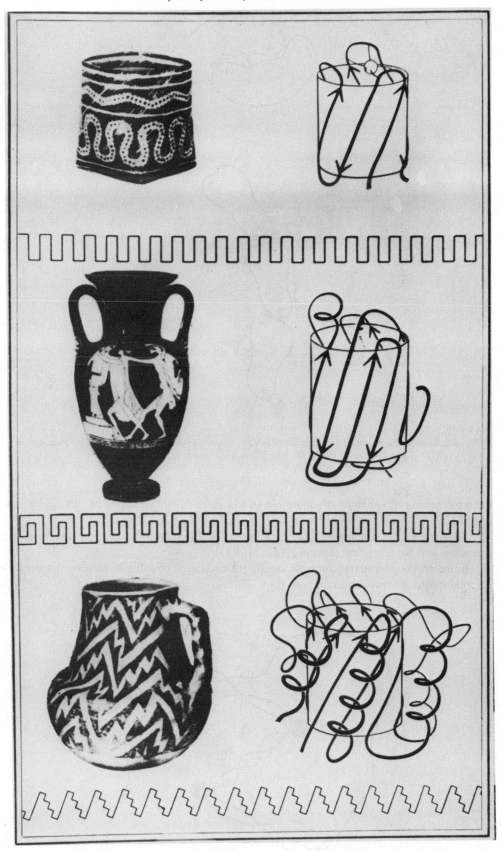

Fig. 14. Geometrical motifs in cylindrical β-sheets and comparison with the ornaments on paintings of the ancients. Top: β-sheet in rubredoxin, middle: in preabumin, bottom: β-sheet covered by α-helices in triose-phosphate isomerase.

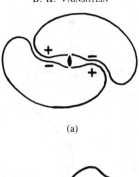

(a)

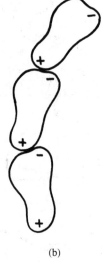

(b)

Fig. 15. Electrostatic and other forces of interaction of the surfaces of molecules may promote their joining into finite (a) or infinite (b) associations ($+$, $-$ charged parts of a surface).

the displacement of subunits, the molecule stands out as if "breathing," preserving its symmetry 2.

Sometimes, the subunits of a molecule serve for different tasks, e.g. some subunits are regulatory and the others are functional (Fig. 19)[22].

Some proteins form very intricate complexes composed of subunits of several sorts with the total molecular mass up to millions (Fig. 20)[23].

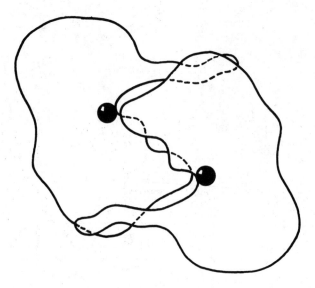

Fig. 16. Aspartate aminotransferase dimer, circles active center (model at 5 Å resolution).

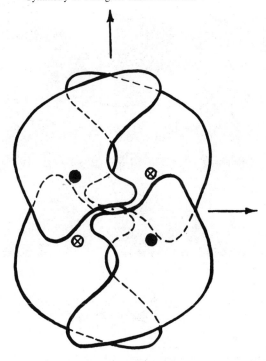

Fig. 17. Scheme of a catalase tetramer, circles active centers (cf. the monomer structure, Fig. 13).

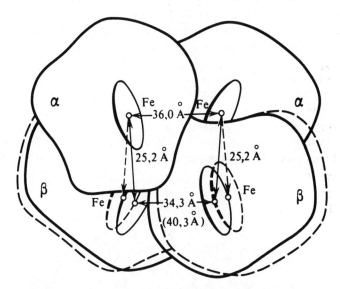

Fig. 18. A change in the quaternary structure of a hemoglobin tetramer in the course of oxygenation (solid line—deoxygenated form).

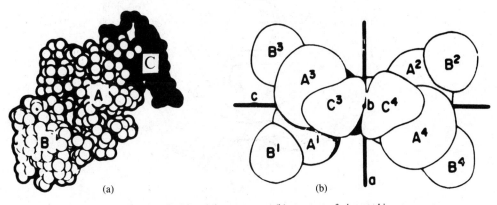

(a)   (b)

Fig. 19. The domain (a) and the quaternary (b) structure of piruvate-kinase.

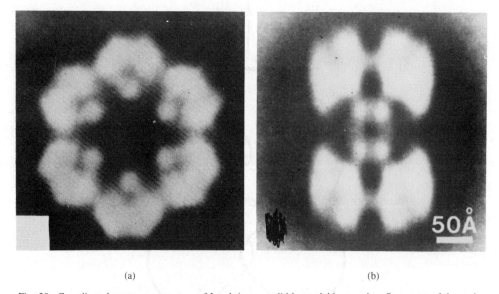

(a)   (b)

Fig. 20. Complicated quaternary structure of Lumbricus annelid hemoglobin complex. Symmetry of the molecules—62, pseudosymmetry 6mm; (a) view along axis 6, (b) view along axis 2.

Some evidences indicate that the joining of the substrate molecules to the subunits does not occur simultaneously. In these cases the ideal symmetry of a quaternary structure is disturbed, but, pseudosymmetry is, certainly, retained. It should also be remembered that the "ideal" symmetry is really the averaged symmetry on account of the thermal motion of the atoms and their groups around their equilibrium positions in a protein molecule.

Spherical viruses

The "task" of a virus particle is to introduce the nucleic acid into the host cell, which, making use of a protein-synthesizing machinery of the cell, makes it to produce not its own proteins, but proteins of the virus particle.

The simplest viruses—"spherical" or, to be more precise, icosahedral—consist of a protein capsid (the container), and RNA stored in it. The protein shell should be constructed most

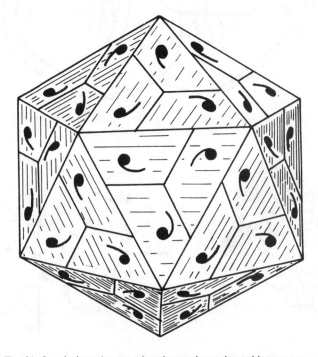

Fig. 21. Icosahedron. Asymmetric units are shown denoted by commas.

economically, i.e. should embrace the largest volume at the smallest surface. This is achieved when its shape approximates most closely the spherical one[24]; this requirement is best satisfied by shells with icosahedral point symmetry 532 (Fig. 21).

The order of this group is $n = 60$, therefore the number of asymmetric protein subunits S is a multiple of 60. The subunits join into pentamers $P = 5S$ which lie on the exit points of axes 5, the number of pentamers being 12. In simplest case, the virus is made only of 12 pentamers. In other cases, the virus also possesses hexamers $H = 6S$, the number of hexamers is $10(T - 1)$, T is the so-called triangulation number which may be equal to 1, 3, 4, 7 . . .[24]. In the tomato bushy stant virus there are 180 subunits. The hexamers lies on the exit points of axes 3 and between them there arise additional axes of quasisymmetry 3^q and 2^q (Fig. 22)[25].

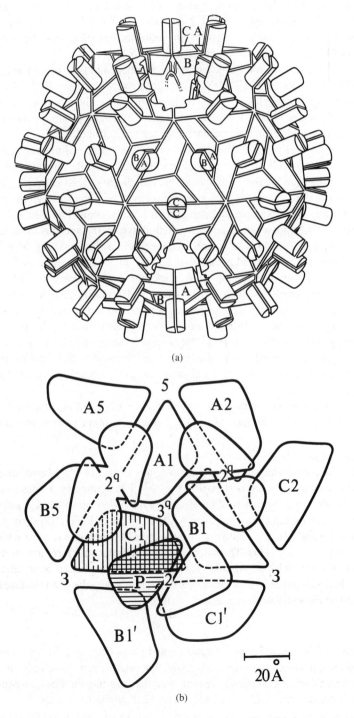

(a)

(b)

Fig. 22. Tomato bushy stunt virus: (a) general scheme of subunit packing, (b) the arrangement of subunits between axes 5, 3, 2; quasiaxes 2^q and 3^q are shown.

Recently, it has been found that some icosahedral viruses may be made completely of pentamers P which are arranged not only on axes 5, e.g. the polyoma virus consists of 360 pentamers[26]. The RNA chain inside the virus is asymmetric, but, apparently, its loops fit, to some extent, the regular packing of subunits in a capsid.

The shape of protein subunits in icosahedral viruses is specially designed for their complementary packing just into an icosahedral shell. (The attempts to crystallize the virus subunits into true three-dimensional crystals have so far failed.)

Helical packing of protein molecules

Globular proteins often aggregate into associations with helical symmetry S_M. On the basis of the relation of the particle size to the radius of the structure R they can be subdivided into thread-like $d/R \sim 1$, rod-shaped $d/R \sim 1.2–1.5$ and tubular $d/R \sim 2–3$ ones (Fig. 23).

The examples of thread-like structures are protofibrillae of actin—the protein of muscle forming a two-pitch helix with symmetry S_{13} (Fig. 24)[28].

The classical example of rod-shaped structure is tobacco mosaic virus TMV[29]. The 2140 elongated protein subunits of TMV are packed into a rod with $M = 49/3$. The RNA chain is found to be helically folded closer to the rod axis (Fig. 25). The conditions may be provided in which the subunits are packed into discs with the 17th-fold axis of symmetry ($49/3 \approx 17$), such discs are considered as an intermediate stage of rod formation[30].

The function of viral protein subunits is not only the storage of nucleic acids, but also the interaction with the host cell in order to adhere to it and penetrate into the cell. This function finds its vivid expression in the structure of bacteriophages representing a complex mobile molecular apparatus constructed on the clearly expressed principles of symmetry.

Let us consider, as an example, the structure of bacteriophage Phy 1 E. coly (Fig. 26)[31]. Its head contains DNA, possesses pseudoicosahedral symmetry and has (attached to it by the neck) a tail consisting of a rod with an inner channel and a sheath whose protein subunits are dimers. The symmetry of particle packing in the tail in its intact state is $S_{7/2}.6$, the sheath appears to consist of flat discs about 40 Å thick with six subunits in each. The discs are superimposed one on another with a 103° rotation and a period 252 Å (Fig. 26(b, c)). The phage is attached to the cell by means of the basal plate and tail fibers; this device also has the 6th-fold symmetry (Fig. 26(d)). When coming into contact with the cell, the basal plate rearranges, still preserving symmetry 6 (Fig. 26(c)) and initiates the sheath contraction: the disc subunits rotate and the discs enter one into another more closely, the sheath symmetry is $S_{11}6$. As for the rod, it preserves its structure and enters the cell; through its channel DNA stored in the head is ''injected'' into the cell.

In larger viruses the symmetry is not revealed so clearly as in the viruses described above, but pseudosymmetry in the structure of the shell and some other parts is preserved.

Tubular crystals of proteins

It has been found[32] that some globular proteins can be associated into tubular structures, the geometrical scheme of which is shown in Fig. 27. The symmetry of these structures is S_M, they can also be described as a two-dimensional layer rolled up into a cylinder. Owing to a high regularity such structures may be called ''tubular crystals.'' Fig. 28(a) shows an image of a phosphorylase b tube[33], Fig. 28(b) represents three-dimensional reconstruction of catalase tubes, symmetry $G_1^{3,1} \approx S_{92/11}$[34]. The natural, *in vivo*, tubular structures are known for tubulin. The formation of a tube with monomolecular walls may be explained by a selective character of interaction between protein molecules, ''sidelong'' (along the wall surface) attraction of molecules and the respective complementary shape.

Layers

The number of proteins, including catalase and phosphorylase[33] form two-dimensional plane monomolecular layers with symmetry $G_2^{3,1}$; but such layers arise only on flat supports. The most important example of native layered structures are membranes consisting of a double layer of lipid molecules (Fig. 29). The twice-periodical symmetry is expressed in membranes, but only roughly, it approximates the statistical symmetry of smectic liquid crystals[36].

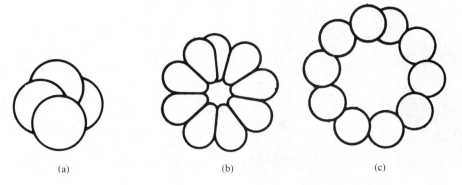

Fig. 23. Scheme of the subunit packing projection along the helical axis in thread-like (a), rod-shaped (b) and tubular (c) helical structures.

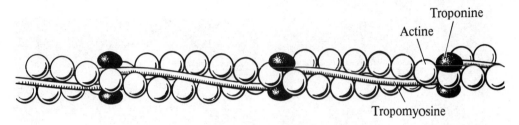

Fig. 24. Actin protofibrillae with proteins laid out on them—troponin and tropomyosin.

Fig. 25. Packing of protein subunits in tobacco mosaic virus.

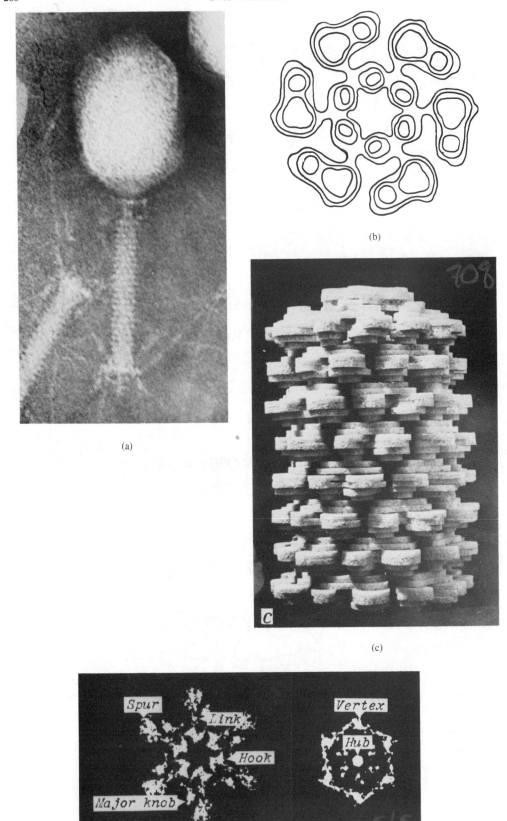

Fig. 26. (a) Electron micrograph of phage Phi-1 in intact state; (b) scheme of one disc of the rod and sheath;
(c) three-dimensional reconstruction of the rod and sheath; basal plate in intact (d) and contracted (e) states.

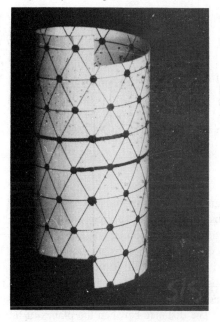

Fig. 27. Geometrical scheme of a tubular structure.

(a)   (b)

Fig. 28. Tubular crystals of proteins: (a) phosphorylase b, an electron microscopic image, optically filtered;
(b) ox-liver catalase, three-dimensional reconstruction.

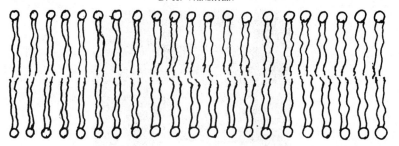

Fig. 29. Double layer of molecules in a membrane.

The structure of DNA

The secondary structure of DNA is the famous Watson–Crick double helix (Fig. 30). Its symmetry clearly emphasizes and reveals the main DNA function: the storage and possibility of reading out and duplicating genetic information. Two phosphate-sugar (4) chains are strictly periodic and are linked by complementary hydrogen-bonded pairs of the A—T and G—C bases. Symmetry of the structure is S_M2, axes 2 are perpendicular to the axis of the helix, so that both polar chains run in the opposite direction. The double DNA helix is right handed. For the bases, symmetry S_M2 is the pseudosymmetry, it only indicates the position and orientation of an "averaged" base in space. The distance between the planes of the base pairs is equal to 3, 4, Å. In the main B-form $M = 10$, i.e. the double helix makes the complete revolution with the period 34 Å. In the other A-form the base planes are not perpendicular to the main axis of the helix, $M = 11$.

Recently, it became possible to synthesize short oligonucleotides with the number of bases from 4 to 12[37,38]. Such oligonucleotides are of special interest when they are self-complementary, as, e.g., in a dodecamer

CGCGAATTCGCG

GCGCTTAAGCGC.

Such a molecule represents a palindrome—it is identical when read out in both directions. Oligonucleotides may be crystallized, they form a three-dimensional crystal structure which renders to an accurate X-ray diffraction analysis (Fig. 31). Symmetry S_{10} is preserved for the right-handed double chain, and in the case of self-complementarity there exists one true axis 2 which passes through the center of the oligonucleotide[37]. There has also been found an unusual, left-handed, with respect to helicity DNA form—the so-called Z-DNA[39].

In the cases where DNA or RNA is arranged into a tertiary structure inside viruses, chromosomes or ribosomes, they have, in some parts, the double-helical secondary structure described above, but at other parts they also exhibit an irregular conformation with an unbraided single chain. The asymmetric globular molecule of *t*-RNA is built up in the same way (Fig. 32)[40].

In the quaternary structure of elements of chromosomes—nucleosome—the special globular proteins, histones, are attached to the double RNA helix consisting of 140 pairs of nucleotides, DNA is bent over its main axis. This bent molecule forms a sloping superhelix with $1\frac{3}{4}$ turns and pseudosymmetry 2, axis 2 being perpendicular to the superhelix axis (Fig. 33)[41].

Biomolecular crystals

One can manage to crystallize many proteins, oligonucleotides, spherical viruses into true crystals with three-dimensional periodic packing $G_3^{3,1}$. The biocrystal formation can hardly assist the molecule in the fulfillment of some or another biological function. Therefore, such crystals in vivo are rarely observed—only as "stores" of the cell products which, when necessary, are utilized. For instance, it is known that the pancreas cells contain insulin crystals[42], some viruses are also crystallized in cells (Fig. 34).

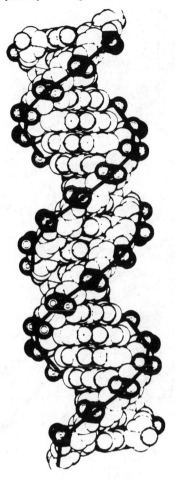

Fig. 30. DNA structure, B form.

At the same time, the purified preparations of biomolecules may be crystallized in vitro (Figs. 35, 36). The crystals contain, in the space between the molecules, the mother liquor and are stable only in the presence of this liquor, most frequently, this is water with some ions (Fig. 37). The water content in protein crystals constitutes from 30 to 70%, in crystals of t-RNA or oligonucleotides the solvent content reaches 90%. Thus the crystals of biomacromolecules are very peculiar systems which combine strict spatial periodicity and orientation of these molecules with liquid disordering of solvent molecules.

Such inwardly-two-component crystalline systems are not known to exist in inorganic and simple organic compounds. As some kind of analogy, we may take metals in which ionic skeltons of atoms are in the "gas" of free electrons surrounding them. The symmetry groups of crystals are Fedorov space groups $G_3^3 \equiv \Phi$, 230 in number. Of them, only 65 groups of the first kind $G_3^{3,1}$, containing only the simple or screw axes of symmetry (groups C and D, according to Schönflies nomenclature) are possible for biocrystals. Statistics of X-ray structural works on proteins (at present, such investigations run to hundreds) shows that there are only several Φ^I-groups which are favourable for proteins. These groups are the following: $P2_12_12_1$, (D_2^4)—23%, $C2$, (C_2^3)—13%, $P2_1$, (C_2^2)—11%, $P3_121$, (D_3^5) and $P3_221$, (D_3^6)—10%, $P2_12_12$, (D_2^3)—5%, $P4_122$, (D_4^4) and $P4_322$, (D_4^8)—4%, PI, (C_1^1)—4%. These groups depict 70% of protein crystals. Obligatory for all the groups (except asymmetric group $P1$ containing only translations) is the presence of screw axes, mainly, 2_1, which may be combined with other screw or rotational axes. The groups containing only the rotational axes, e.g. $P222$, are not, in general, observed.

A definite preferable population of some space groups is also known for inorganic and simple organic compounds. For them, such a population can be explained, mainly, by the principle of maximum filling (close packing) of the crystal volume by atoms and molecules[36,43]. In the case of proteins, these considerations are not decisive due to the above

Fig. 31. Oligonucleotide duplex of 12 pairs of bases.

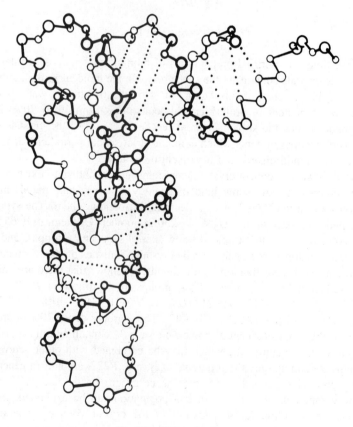

Fig. 32. tRNA molecule.

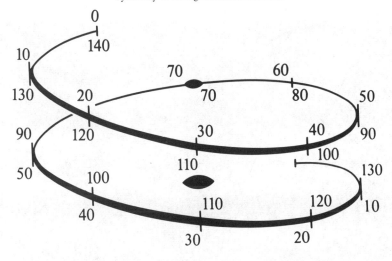

Fig. 33. Scheme of the arrangement of two-helical DNA strand in nucleosome.

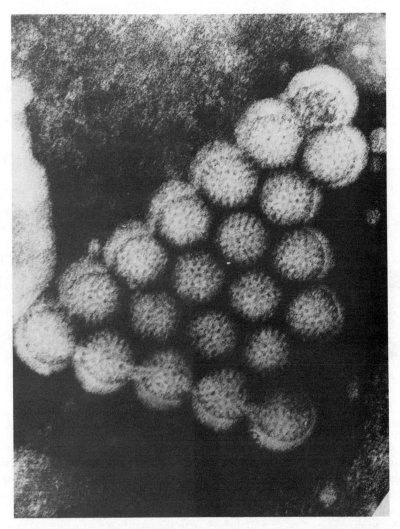

Fig. 34. Crystalline packing of particles of human rota virus. Electron micrograph, × 200000. (Courtesy of M. B. Korolev.)

(a)

(b)

Fig. 35. Protein crystals: (a) histidinedecarboxylase, (b) pyrophosphatase ($\times$ 30).

mentioned peculiarity in their structure—the possibility of filling up the space between the molecules by solveint. The forces which are acting when the molecules pack themselves into a crystal are, for the most part, electrostatic, and, in this case, the arising of screw arrangements is most probable (Fig. 15b).

In the arrangement of protein molecules with the quaternary structure into a crystal, one can often observe the discrepancy between the proper symmetry of the molecule and the

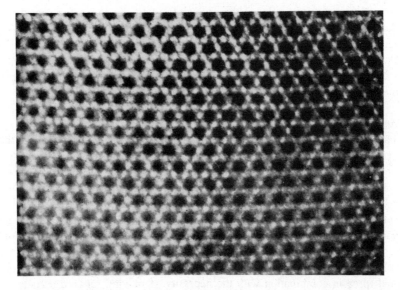

Fig. 36. Packing of the molecules in ox-liver catalase crystal. Electron micrograph, × 400000.

symmetry of its position defined by a space group. For instance, the point symmetry of catalase is 222 (Fig. 17), space group $P3_121$; when packing into a crystal, only one of the three axes 2 of the molecule coincides with axis 2 of the crystal, while the other two only depict the molecule. It turns out that in the asymmetric unit of the crystal A_{cr} there are arranged two symmetrically-equal (on the point symmetry of the molecule) parts A_M of the molecule, the stereon of the crystal is twice as large, by volume, as the stereon of the molecule: $V_{A_{cr}} = 2V_{A_M}$. A crystalline modification of catalase is known in which the entire molecule is placed in A_{cr}, i.e. none of the elements of point symmetry 222 of the molecule coincides with the elements of symmetry of the crystal, here $V_{A_{cr}} = 4V_{A_M}$[19]. Such cases are called noncrystallographic symmetry of the molecule in a crystal. Another example is aspartate aminotransferase (Fig. 16)[20] with space group $P2_12_12_1$, a symmetric dimer $2A_m$ with axis 2 is arranged in A_{cr}. Sometimes A_{cr} contains two (or more) asymmetric molecules of protein having no quaternary structure. And here $V_{A_{cr}} = 2V_{A_M}$, too, but the molecules do not transform one into another by some or other operation of point symmetry.

On the other hand, the symmetry of position in a crystal may coincide exactly with the point symmetry of the molecules. The cause of the arising of noncrystallographic symmetry or

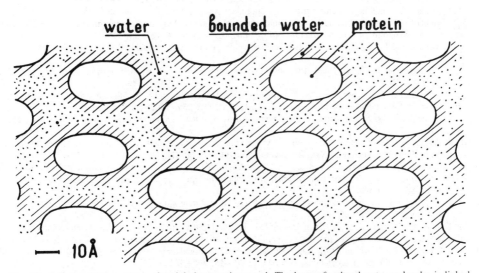

Fig. 37. Scheme of the structure of a globular protein crystal. The layer of ordered water molecules is linked with a protein molecule (hatched), the solvent molecules are shown by dots.

the presence of a pair of molecules in the asymmetric unit lies in the fact that dominating for a crystal is the attainment of the molecular packing giving the energy minimum. And this can be achieved both with use of proper point symmetry of molecules, when it partly or completely coincides with the symmetry of the crystal, and without such a coincidence. Noncrystallographic symmetry is one of the expressions of the symmetrization-dissymmetrization principles of Curie and Neumann for component physical systems[1,44].

Conclusion

We have seen that the symmetry principles manifest themselves distinctly in the structure of biomolecules and their associations. Despite the initial asymmetry of small protomolecules, either the exact symmetry or pseudosymmetry stands out at further levels of organization in the primary, secondary and quaternary structure. Symmetry in biostructures may be explained in terms of expediency of their mechanism for fulfilling some or other functions, this expediency has been elaborated by the evolutionary process. The general physical principles enabling the symmetry to manifest itself are performed by the genetically determined frameworks of biological processes.

Symmetry of macroorganisms obeys their interaction with the environment. The Earth's gravitational force in combination with the necessity of moving forward defines not only the morphological symmetry *m* of the majority of the organisms living on the Earth, but also the symmetry of propulsion devices created by a human—bicycle, airplane etc. Plants and animals living in the Ocean, that do not move on their own, are often found to possess the axial symmetry.

REFERENCES

1. A. V. Shubnikov and V. A. Koptsik, *Symmetry in Science and Art*. Plenum Press, New York (1974).
2. H. Weyl, *Symmetry*. Princeton University Press, Princeton, New Jersey (1952).
3. B. K. Vainshtein, *Modern Crystallography. I. Symmetry of Crystals, Methods of Structural Crystallography*, Springer Series in Solid-State Sciences, Vol. 15. Springer-Verlag, Berlin, Heidelberg, New York (1981).
4. E. Schrödinger, *What is Life? The Physical Aspects of the Living Cell*. The University Press, Cambridge (1945).
5. J. D. Bernal, *The Origin of Life*. Weidenfeld and Nicolson, London (1967).
6. B. K. Vainshtein, *Diffraction of X-rays by Chain Molecule*. Amsterdam, London, New York (1966).
7. T. L. Blundell and L. N. Johnson, *Protein Crystallography*. Academic Press, New York (1976).
8. N. E. Zhukhlistova and G. N. Tishchenko, Kristallicheskaya struktura Na, Ni-kompleksa enniatina B. *Kristallografiya* **26**, 1232–1239 (1981).
9. L. Pauling and R. B. Corey, H. R. Branson, The structure of proteins: two hydrogen-bonded helical configurations of the polypeptide chain, *Proc. Nat. Acad. Sci.* **37**, pp. 205–211 (1951).
10. F. H. C. Crick and J. D. Watson, The complementary structure of deoxyribonucleic acid. *Proc. Roy. Soc., London* **A223**, 80–96 (1954).
11. G. H. Ramachandran and G. Kartha, Structure of collagen. *Nature, London.* **176**, 593–595 (1955).
12. E. G. Harutyunyan, Struktura Leggemoglobina, *Molekulyarnaya Biologiya* **15**, 27–44 (1981).
13. M. Levitt and C. Chothia, Structural patterns in globular proteins. *Nature, London* **261**, 552–558 (1976).
14. J. S. Richardson, β-sheet topology and the relatedness of proteins. *Nature, London* **268**, 495–500 (1977); The anatomy and taxonomy of protein structure. *Adv. Protein Chemistry* **34**, 167–339 (1981).
15. H. C. Watson and J. C. Kendrew, The stereochemistry of the protein myoglobin. *Prog. Stereochem.* **4**, 299–321 (1969).
16. W. A. Hendrickson and K. B. Ward, Pseudosymmetry in the structure of myohemerythrin. *J. Biol. Chem.* **252**, 3012–3018 (1977).
17. Yu. N. Chirgadze, Yu. V. Sergeyev, N. P. Fomenkova, V. D. Oreshin and S. V. Nikonov, Struktura γ-kristallina IIIb iz glaznoy linzy telenka pri razreshenii 3 Å. *Dokl. AN SSSR* **259**, 1502–1505 (1981).
18. F. A. Quiocho and W. N. Lipscomb, Carboxypeptidase A: A protein and an enzyme. *Adv. Protein Chem.* **25**, 1–78 (1971).
19. B. K. Vainshtein, W. R. Melik-Adamyan, V. V. Barynin, A. A. Vagin and A. I. Grebenko, Three-dimensional structure of the enzyme catalase. *Nature, London* **293**, 411–412 (1981).
20. V. V. Borisov, S. N. Borisova, N. I. Sosfenov, A. A. Vagin, Yu. V. Nekrasov, B. K. Vainshtein, V. M. Kochkina and A. E. Braunshtein, Prostranstvennyi khod polipeptidnoy tsepi v molekule aspartat-transaminazy. *Dokl. AN SSSR* **250**, 988–991 (1980).
21. M. F. Perutz, Structure and function of haemoglobin. I. A tentative atomic model of horse oxyhaemoglobin. *J. Mol. Biol.* **13**, 646–668 (1965).
22. M. Levine, H. Muirhead, D. K. Stammers and D. I. Stuart, Structure of pyruvate kinase and similarities with other enzymes: possible implications for protein taxonomy and evolution. *Nature, London* **271**, 626–630 (1978).
23. A. V. Crewe, D. A. Crewe and O. H. Kapp, Inexact three-dimensional reconstruction of a biological macromolecule from a restricted number of projections. *Ultramicroscopy* **13**, 365–372 (1984).
24. D. L. D. Caspar and A. Klug, Physical principles in the construction of regular viruses. *Cold Spring Harbor Symp. Quant. Biol.* **27**, 1–24 (1962).

25. S. C. Harrison, A. J. Olson, C. E. Schutt, F. K. Winkler and G. Bricogne, Tomato bushy stunt virus at 2.9 Å resolution. *Nature, London* **276**, 368–373 (1978).

26. I. Rayment, T. S. Baker, D. L. D. Caspar and W. T. Murakami, Polyoma virus capsid structure at 22.5 Å resolution. *Nature, London* **295**, 110–115 (1982).

27. B. K. Vainshtein, Tubular crystals of globular proteins, *Mat. Res. Bull.* **7**, 1347–1356 (1972).

28. I. Ohtsuki. Localization of troponin in thin filament and tropomyosin paracrystal. *J. Biochem.* **75**, 753–763 (1974).

29. A. Klug and D. L. D. Caspar, The structure of small viruses. *Adv. Virus Res.* **7**, 225–325 (1960).

30. A. C. H. Durham, J. T. Finch and A. Klug, States of aggregation of tobacco mosaic virus protein. *Nature, London* **229**, 37–50 (1971).

31. B. K. Vainshtein, A. M. Mikhailov, A. S. Kaftanova, M. Platzer and I. A. Andriashvili, Electron microscopic studies of spatial structure of phage Fi-I, Electron Microscopy 1980, v.2. Biology, Proc. 7-th European Congress on Electron Microscopy, Netherlands, 600–601 (1980).

32. B. K. Vainshtein, N. A. Kiselev and V. L. Spitzberg, Kristallizatsiya katalazy v truby s monomolekulyarnymi stenkami. *Dokl. AN ASSR* **167**, 211–214 (1966).

33. N. A. Kiselev, F. Ya. Lerner, and N. B. Livanova, Electron microscopy of muscle phosphorylase a. *J. Mol. Biol.* **86**, 587–599, (1974).

34. V. V. Barynin, B. K. Vainshtein, O. N. Zograf and S. Ya. Karpukhina, Trekhmernaya rekonstruktsiya trubchatykh kristallov katalazy. *Molekulyarnaya Biologiya* **13**, 1189–1197 (1979).

35. L. K. Tamm, R. H. Crepau and S. J. Edelstein, Three-dimensional reconstruction of tubulin in zinc-induced sheets, *J. Mol. Biol.* **130**, 473–492 (1979).

36. B. K. Vainshtein, V. M. Fridkin and V. L. Indenbom, *Modern Crystallography. II. Structure of Crystals,* Springer Series in Solid-State Sciences, vol. 21. Springer-Verlag, Berlin, Heidelberg, New York (1982).

37. R. E. Dickerson, H. R. Drew, B. N. Conner, R. M. Wing, A. V. Fratini and M. L. Kopka, The anatomy of A-, B-, and Z-DNA. *Science* **216**, 475–485 (1982).

38. Z. Shakked, D. Rabinovich, W. B. T. Cruse, E. Egert, O. Kennard, G. Sala, S. A. Salisbury and M. A. Viswamitra, Crystalline A-DNA: the X-ray analysis of the fragment d(G-G-T-A-T-A-C-C). *Proc. Roy. Soc. London* **B213**, 479–489 (1981).

39. A. H.-J. Wang, G. J. Quigley, F. J. Kolpak, J. L. Crawford, J. H. von Boom, C. van der Mazel and A. Rich, Molecular structure of a left-handed double helical DNA fragment at atomic resolution. *Nature, London* **282**, 680–686 (1979).

40. S. H. Kim, F. L. Suddath, G. J. Quigley, A. McPherson, J. L. Sussman, A. H.-J. Wang, N. C. Seeman and A. Rich, Three-dimensional tertiary structure of yeast phenylalanine transfer RNA. *Science* **185**, 435–440 (1974).

41. J. T. Finch, L. C. Lutter, D. Rhodes, R. S. Brown, B. Rushton, M. Levitt and A. Klug, Structure of nucleosome core particles of chromatin. *Nature, London* **269**, 29–36 (1977).

42. D. Hodgkin, Mgnoveniya otkrytii. *Kristallografiya* **26**, 1029–1045 (1981).

43. A. I. Kitaigorodsky, *Organic Chemistry and Crystallography.* Consultants Bureau, New York (1961).

44. V. A. Koptsik, Printsipy simmetrizatsii-dissimetrizatsii Shubnikova-Curie dlya sostavnykh fizicheskikh sistem, in: *Problemy sovremennoy kristallografii* p. 42–61. Nauka, Moskva (1975).

Comp. & Maths. with Appls. Vol. 12B, Nos. 1/2, pp. 271–290, 1986
Printed in Great Britain.

0886–9561/86 $3.00 + .00
© 1986 Pergamon Press Ltd.

MATCHING AND SYMMETRY OF GRAPHS

Haruo Hosoya

Department of Chemistry, Ochanomizu University, Bunkyo-ku, Tokyo 112, Japan

Abstract—Matching is a mathematical concept that deals with the way of spanning a given graph network with a set of pairs of adjacent points. It is pointed out that in many different areas of science and culture (e.g., physics, chemistry, games, etc.), computing the perfect and imperfect matching numbers is commonly performed but under different names, such as partition function for dimer statistics, Kekulé structures of molecules, paving domino problem. This paper demonstrates the mathematically beautiful but somewhat mystic relation between the symmetry of a graph and the factorable nature of its perfect matching number. There is introduced another interesting relation between the certain series of graphs and a family of orthogonal polynomials through the matching polynomial and topological index that are defined for counting the matching numbers.

1. INTRODUCTION

Matching is an important concept both in the graph theory and combinatorial theory[1–4]. Although different terminologies are given, this is also the case in chemistry and physics—in realization and quantification of indistinguishable objects of one or two species[5,6]. Chemists have long used the concept of the electron pair for describing and predicting the structure of a molecule[7]. For example, non-existence or very low stability of molecular species Fig. 1 and Fig. 2 can be predicted easily just by noticing that these two graphs have no Kekulé structure or perfect matching, or are not 1-factorable.† The diagrams are drawn so as to span as many points (atoms) as possible with double bonds but are not necessarily determined uniquely (See Fig. 1(a) and Fig. 1(b)).

The dot in these diagrams represents an unpaired electron, which cannot contribute to the stability of the molecular species but also can easily react with other chemical species. Further, one may safely conclude that the molecule in Fig. 3 is more stable than the isomer Fig. 4 just by comparing the number $K(G)$ of the Kekulé structures.

A number of chemistry papers have been published on the systematic enumeration of $K(G)$, especially for polyhex graphs (polycyclic aromatic hydrocarbons), which are composed only of hexagons[8–15]. The followings are interesting examples:

$$K(P_{m,n}) = \binom{m + n}{n} = \frac{(m + n)!}{m!\, n!} \quad \text{(Fig. 5)} \tag{1}$$

$$K(H_{m,n}) = \prod_{k=m}^{2m-1} \binom{n + k}{n} \bigg/ \prod_{k=1}^{m-1} \binom{n + k}{n} \quad \text{(Fig. 6)} \tag{2}$$

On the other hand, solid state physicists are very eager to count the number of perfect (and also imperfect) matchings, such as Fig. 7(a), of the network of rectangular lattices (or polyominoes)[16–19]. Each matching pattern represents a distinctive mode of nearest-neighbor interaction among the α and β spins alternately located in the anti-ferromagnetic metal lattice (Fig. 7(b)). The set of perfect and imperfect matchings for these types of graphs, when extrapolated to an infinitely large network, gives the partition function for the magnetic and thermodynamic properties of the crystal and also the function for the kinetics of adsorption of diatomic, say, oxygen molecules onto a clean surface of metal, while each matching pattern can be deemed to represent a pattern of domino tiling or that of *tatamis* (flooring mats) for a room of a typical Japanese house (Fig. 7(c)).

To avoid confusion in this paper, we will consider mainly the "matching pattern" rather than the "tiling pattern". Temperley and Fisher[20] and Kasteleyn[21] independently discovered

---

†The diagram Fig. 1(a) is a short-hand notation of the molecular structure Fig. 1(c), but is more convenient.

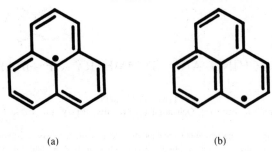

(a)   (b)

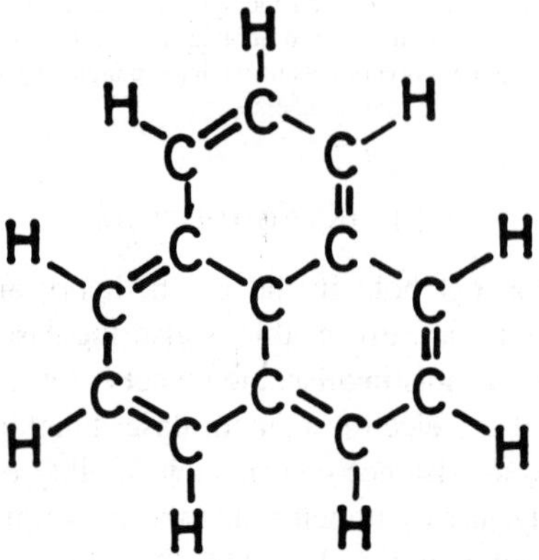

(c)

Fig. 1.

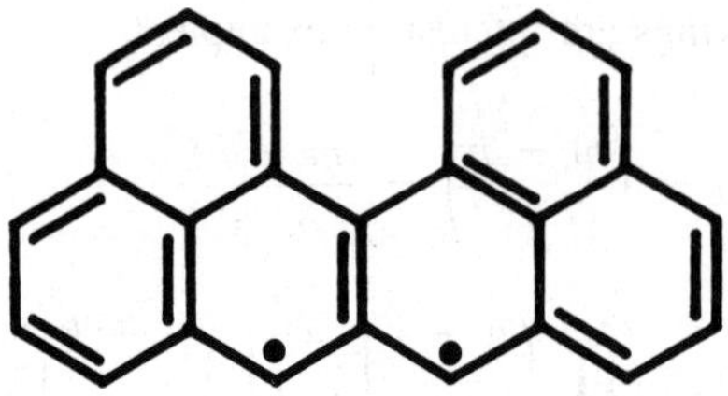

Fig. 2.

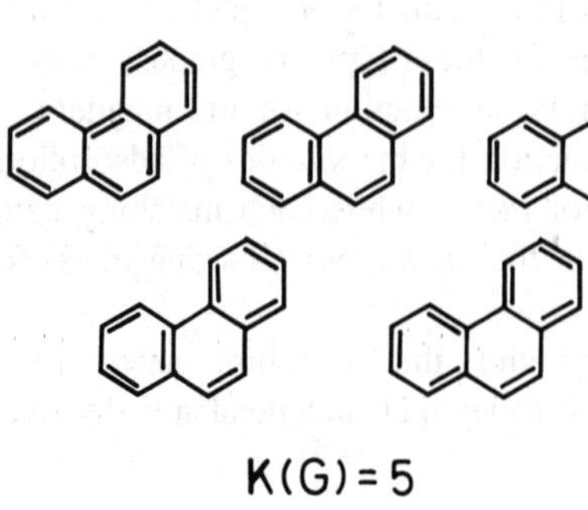

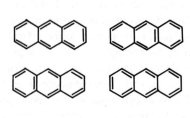

K(G) = 5   K(G) = 4

Fig. 3.   Fig. 4.

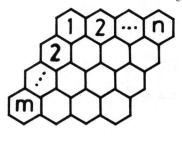

Fig. 5.

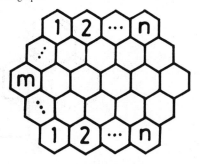

Fig. 6.

fascinating formulas of $K(G)$ number for $n \times m$ rectangular lattice by using the technique of Pfaffian as follows:†

$$K(2m \times 2n) = 2^{2mn} \prod_{k=1}^{m} \prod_{l=1}^{n} \left(\cos^2 \frac{k\pi}{2m+1} + \cos^2 \frac{l\pi}{2n+1} \right) \qquad (3)$$

$$K(2m-1 \times 2n) = 2^{2mn} \prod_{k=1}^{m} \prod_{l=1}^{n} \left(\cos^2 \frac{k\pi}{2m} + \cos^2 \frac{l\pi}{2n+1} \right), \qquad (4)$$

which include the famous Fibonacci sequence for the series of graphs $\{2 \times n\}$ as

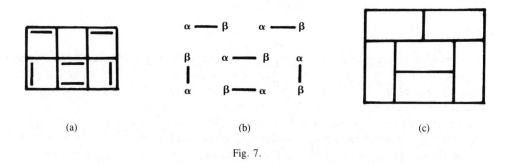

$\{2 \times n\} \quad \phi$

n	0	1	2	3	4	5
$K(2 \times n)$	1	1	2	3	5	8

with the recursion relation of

$$K(2 \times n) = K_n = K_{n-1} + K_{n-2} \qquad (5)$$
$$K_0 = K_1 = 1,$$

where ϕ denotes the vacant graph.

It has been observed that $K(G)$ of square $2n \times 2n$ graph is either a perfect square or twice a perfect square[22] as follows:

n	1	2	3	4	5
$K(2n \times 2n)$	2	6^2	2×58^2	$(4 \times 17 \times 53)^2$	$2 \times (4 \times 241 \times 373)^2$.

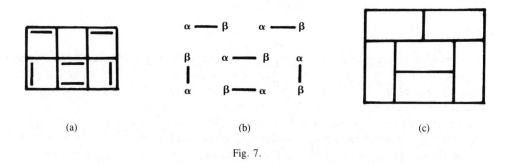

(a)   (b)   (c)

Fig. 7.

†In [22] it is stated that Knuth has derived an equation quite similar to (3) and (4) for the first time, but they seem to have been unaware of the physics papers[20,21].

H. Hosoya

Fig. 8.

This can be explained by the form of (3) and (4). The number of ways for the domino tiling on a chessboard is thus turned out to be 12988816, while the number of perfect matchings for a "go" board (18 × 18) is estimated to be as large as a 40-digit number.

From the experimental point of view, no dramatic difference can be expected between the properties of two compounds with prime and highly composite $K(G)$ numbers. However, as will be shown later in this paper, the distribution of the numbers of perfect and imperfect matchings for certain series of graphs has very interesting mathematical consequences. It is one of the main motivations of this work to look into the details of the relation between the matching and symmetry of graphs. Thus, all the discussions in this paper have been performed from purely mathematical and, if possible, aesthetic points of view.

2. PERFECT MATCHING

Various methods for the enumeration of the perfect matching or $K(G)$ number are introduced elsewhere[14]. Let us here expose the $K(G)$ numbers of several series of graphs of relatively high symmetry.

Polyomino graphs

The $K(G)$ numbers for smaller $n \times m$ rectangular lattices ($n \times m \leq \sim 150$) have been tabulated[22–24], each line of which ($K(n \times m)$'s with a fixed n or m) is known to obey an individual recursion formula such as (5) for $2 \times n$ graphs. Inspection of these tables reveals that although these $K(G)$ numbers are highly factorable in general no specially interesting series of graphs come out except for the square lattices introduced above. If a pair of corner points facing diagonally are deleted from a $2n \times 2n$ lattice, no perfect matching can be attained. This is easily explained by the different numbers of α and β, or "starred" and "unstarred" points as in Fig. 8:

8 starred points
− 6 unstarred points

non-zero → no perfect matching

Then what if an equal number of α and β points are deleted? Let us consider those subgraphs of square lattices which have the same number of α and β points and still retain the D_{4h} (square) symmetry. Remarkably interesting results come out as in Fig. 9(a–c). All the studied graphs of this category have $K(G)$ of either a perfect square or its double.

Especially, the $K(G)$ of the graphs denoted by $\{X_n\}$ in Fig. 9(b) was found to be expressed simply as $2^{n(n+1)/2}$, which, however, could not yet be proved. For the series of graphs $\{O_n\}$ the $K_n = K(G)$ was found to be expressed as $(2a_n)^2$, where a_n obeys the following recursion relation

$$a_n = 2 a_{n-1} + 2 a_{n-2} - a_{n-3}. \tag{6}$$

By using the three roots of the equation $x^3 - 2x^2 - 2x + 1 = 0$ one can obtain the general

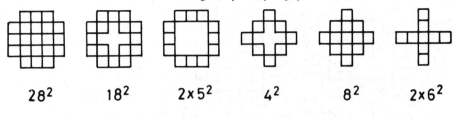

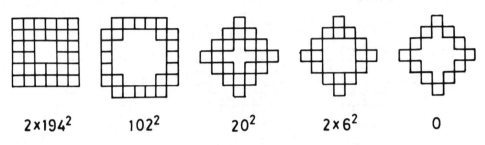

Fig. 9(a).

expression for a_n as

$$a_n = \frac{1}{5}\left[2\left(\frac{3 + \sqrt{5}}{2}\right)^{n-1} + \left(\frac{3 - \sqrt{5}}{2}\right)^{n-1} - (-1)^n\right]. \tag{7}$$

All the $K(G)$'s of the subgraphs a–f of X_5 shown in Fig. 9(c) also have either a perfect square or its double. Among them the values for the networks a and b are smaller than that of smaller graph c. It is possibly due to the fact that both the former two graphs contain as many as four odd-square units of 3×3.

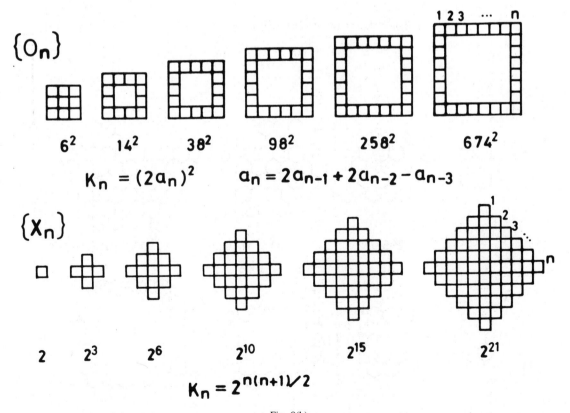

Fig. 9(b).

Fig. 9. Perfect matching numbers of polyomino graphs with square symmetry.

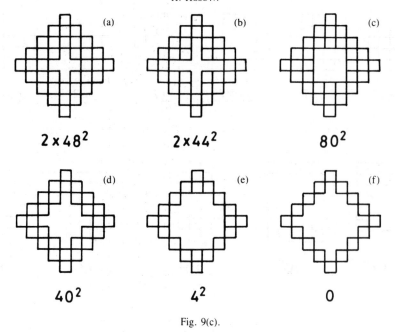

$$2 \times 48^2 \qquad 2 \times 44^2 \qquad 80^2$$

$$40^2 \qquad 4^2 \qquad 0$$

Fig. 9(c).

Polyhex graphs

As has been demonstrated in (1) and (2), several series of polyhex graphs have highly composite $K(G)$ numbers. Two examples are shown in Fig. 10, where $K(G)$'s of hexagonal honeycombs $\{H_{n,n}\}$ are directly obtained from (2), and $K(G)$'s of hexagonal rings $\{R_n\}$ are found to be expressed by

$$K(R_n) = K_n = (n^2 - 2n + 5)(n^2 - 2n + 2)^2 \qquad (8)$$

$$= (n - 1)^6 + 6(n - 1)^4 + 9(n - 1)^2 + 4. \qquad (8')$$

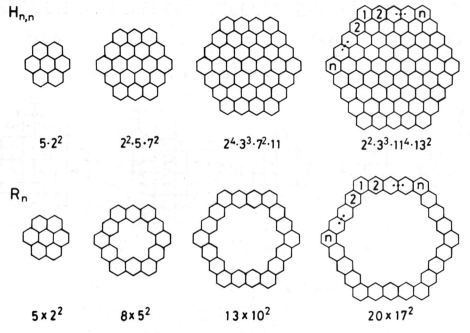

$H_{n,n}$

$$5 \cdot 2^2 \qquad 2^2 \cdot 5 \cdot 7^2 \qquad 2^4 \cdot 3^3 \cdot 7^2 \cdot 11 \qquad 2^2 \cdot 3^3 \cdot 11^4 \cdot 13^2$$

R_n

$$5 \times 2^2 \qquad 8 \times 5^2 \qquad 13 \times 10^2 \qquad 20 \times 17^2$$

Fig. 10. Perfect matching numbers of polyhex graphs with hexagonal symmetry.

Expression (8') reminds us of the characteristic polynomial[†],

$$x^6 - 6x^4 + 9x^2 - 4,$$

of benzene, or a hexagonal graph. The reason for this mystic coincidence, however, is not known. It is straightforward to get the recursion formula of K_n from (8'):

$$K_n = 6K_{n-1} - 15K_{n-2} + 20K_{n-3} - 15K_{n-4} + 6K_{n-5} - K_{n-6} + 720. \qquad (9)$$

The hydrocarbon molecules corresponding to $R_2 \equiv H_{2,2}$ and R_3 have been synthesized and found to be stable. They are respectively named as coronene (after corona) and kekulene (after famous organic chemist Kekulé). One can extend the polyhex network of D_{6h} symmetry by a symmetrical addition to hexagons to the coronene skeleton. Among them hexabenzocoronene (Fig. 11(a)) is a highly stable compound. Its melting point is reported to be higher than 700°C but not accurately determined yet[25]. The glass capillary for the melting point measurement had melted away before the crystal of Fig. 11 in it did not undergo any change. According to Clar this extraordinarily high stability is due not only to its large $K(G)$ value of 250 but also to its large number of conjugated "aromatic sextets" as depicted in Fig. 11(b), where a circle represents a cooperative movement of six free electrons stabilizing the hexagonal structure[25]. This classical theory has been proved and reinterpreted both by quantum chemical and graph theoretical analyses[26–30].

Further extension of the network Fig. 11 gives a variety of snowflake graphs such as Figs. 12–14. Note their highly composite $K(G)$ numbers. On the other hand, the network Fig. 14 does not have any perfect matching. Thus the corresponding hydrocarbon molecule, if ever, is predicted to be unstable. To the author's knowledge this is the smallest polyhex graph of D_{6h} symmetry with the same number of α and β points. An algorithm for designing this type of graph will be published elsewhere[31].

No polyhex network of D_{6h} symmetry composed of an even number of points but with different numbers of α and β points is possible, whereas the molecule of D_{3h} symmetry in Fig. 15 is predicted to be unstable due to the different numbers of α and β points. This hypothetical molecule is called as triangulene or Clar's hydrocarbon.

Polycube graphs

In Fig. 16 are given some examples of the $K(G)$ numbers for the polycube graphs of $2 \times m \times n$ type. Although a limited number of $K(G)$'s for 3-dimensional lattices have been obtained, it seems that the higher the symmetry of a graph the higher the $K(G)$ is factored. Up to now recursion formulas of $K(G)$'s have been obtained only for the following two series of graphs:

$$K(2 \times 2 \times n) = K_n = 3K_{n-1} + 3K_{n-2} - K_{n-3} \qquad \text{(Ref. 32)} \qquad (10)$$

$$K(2 \times 3 \times n) = K_n = 6K_{n-1} + 21K_{n-2} - 42K_{n-3}$$
$$- 89K_{n-4} + 68K_{n-5} + 89K_{n-6} - 42K_{n-7}$$
$$- 21K_{n-8} + 6K_{n-9} + K_{n-10}. \qquad \text{(Ref. 24)} \qquad (11)$$

(a)

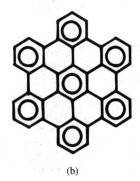
(b)

Fig. 11.

[†]Characteristic polynomial $P_G(x)$ is defined as follows for graph G composed of N points being connected as in the adjacency matrix E: $P_G(x) = (-1)^N \det(A - xE)$, where E is the $N \times N$ unit matrix. The matrix element of A is defined as $A_{ij} = 1$ for neighbors ij and 0 otherwise.

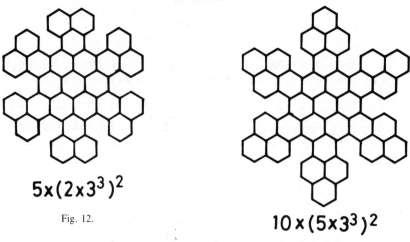

$5\times(2\times3^3)^2$

Fig. 12.

$10\times(5\times3^3)^2$

Fig. 13.

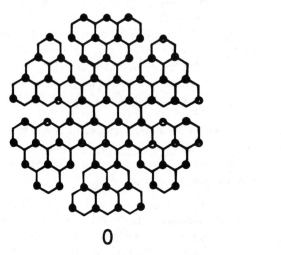

0

Fig. 14.

Fig. 15.

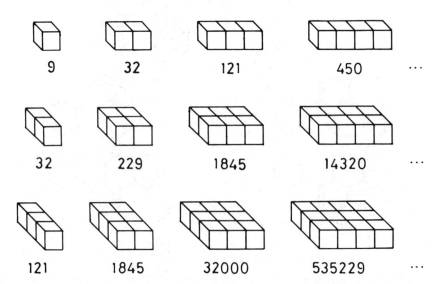

9	32	121	450 ⋯
32	229	1845	14320 ⋯
121	1845	32000	535229 ⋯

Fig. 16. Perfect matching numbers of $2\times m\times n$ polycube graphs.

Further, we found that $K(2 \times 2 \times n)$ can be expressed as

$$K(2 \times 2 \times 2m) = \left\{ 2^{2m} \prod_{k=1}^{m} \left(\cos^2 \frac{k\pi}{2m + 1} + \frac{1}{2} \right) \right\}^2 \tag{12}$$

$$= 2^{4m} \prod_{k=1}^{m} \left(\cos^2 \frac{k\pi}{2m + 1} + \cos^2 \frac{\pi}{2 + 1} + \cos^2 \frac{\pi}{2 + 1} \right)^2$$

$$K(2 \times 2 \times 2m - 1) = 2^{4m-1} \prod_{k=1}^{m} \left(\cos^2 \frac{k}{2m} + \frac{1}{2} \right)^2. \tag{13}$$

However, attempts to get these types of formulas for larger series of polycube lattices have not yet succeeded. It is pointed out that for non-planar graphs, straightforward application of the Pfaffian method fails and no simple formula for $K(G)$ can be expected[16]. The lattice $\{2 \times 2 \times n\}$ is planar, while $\{2 \times 3 \times n\}$ is not planar for $n \geqq 3$.

In Fig. 17 are given the $K(G)$ numbers of three graphs with O_h (cubic or octahedral) symmetry. The $K(G)$ values for them are all highly composite. Among them the value 3^{10} for Fig. 17(c) is remarkable. Graph Fig. 17(a) is a non-planar graph having a cube inside the skeleton of (c), but still has a rather composite $K(G)$ number of $3^2 \times 13^4$. The $K(G)$ numbers for such larger graphs as Fig. 17(b) and $4 \times 4 \times 4$ (Rubik's cube) are very likely to be highly composite, but not yet obtained.

All the regular (Platonic) polyhedra also have highly composite $K(G)$'s except for the tetrahedron, and this is also the case with the semiregular (Archimedian) polyhedra. However, the results will be introduced later together with their imperfect matching numbers.

3. IMPERFECT MATCHING AND TOPOLOGICAL INDEX

The numbers of imperfect matching of a given graph also provide us of important information on the mathematical and physicochemical features of the graph. The present author defined the non-adjacent number $p(G, k)$ for graph G as the number of ways for choosing k disjoint lines for characterizing the topological nature of the structural isomers of saturated

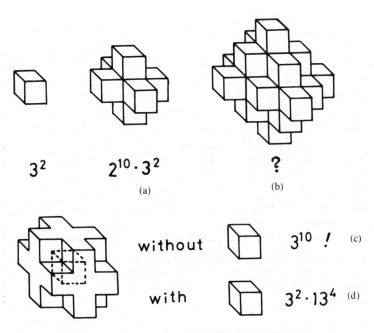

$$3^2 \qquad 2^{10} \cdot 3^2 \qquad ?$$

(a) (b)

without $\quad 3^{10}$! (c)

with $\quad 3^2 \cdot 13^4$ (d)

Fig. 17. Perfect matching numbers of polycube graphs with cube symmetry.

hydrocarbon molecules[33–35]. With the set of $p(G, k)$'s the Z-counting polynomial $Q_G(x)$ and topological index Z_G are defined as

$$Q_G(x) = \sum_{k=0}^{m} p(G, k)\, x^k \tag{14}$$

$$Z_G = \sum_{k=0}^{m} p(G, k) = Q_G(1) \tag{15}$$

with $m = [N/2]$. Later it was found that several researchers in statistical physics have also independently proposed quite the same ideas for obtaining the partition function for the lattice dynamics problems[36–39].

The matching polynomial $\alpha_G(x)$,

$$\alpha_G(x) = \sum_{k=0}^{m} (-1)^k\, p(G, k)\, x^{N-2k} \tag{16}$$

has also a similar history of competition. Aihara[40] and Gutman et al.[41], respectively, with the name of reference polynomial and acyclic polynomial, proposed (16) for characterizing the stability of aromatic hydrocarbon molecules. From the graph-theoretical point of view, Farrell[42] extended the theory of the matching polynomial[43].

These quantities have been shown to be effective not only for characterizing the topological features of graphs, networks, and molecules, but also for correlating them with various physicochemical properties. The matching polynomial for a tree graph is identical to its characteristic polynomial[33]. For non-tree graphs the latter can be shown to be constructed by adding the ring correction terms to the former: all the correction terms are also the matching polynomials of a certain set of subgraphs[34, 44]. By the aid of several recursion formulas[24, 33–35] and with an effective algorithm[15] one can obtain these quantities for fairly large networks. The results reveal that although the topological index Z_G does not seem to reflect the symmetry of a graph the individual $p(G, k)$ numbers for a highly symmetrical graph are highly composite as well as the $p(G, m)$, or the perfect matching number $K(G)$.

Regular and semiregular polyhedra

In Fig. 18 and Table 1 and Fig. 19 are given the coefficients $\{p(G, k)\}$ of the Z-counting polynomial $Q_G(x)$ or the matching polynomial $\alpha_G(x)$ for the regular (Platonic) and semiregular (Archimedian) polyhedra. Except for the tetrahedron (Fig. 18(a)) the $K(G)$ number is either a perfect square or its multiple. Further all the $p(G, k)$ numbers studied except for $k = 0$ are composite. Among them the $K(G)$ values for the octahedron (Fig. 18(c)), truncated tetrahedron (Fig. 19(a)), cuboctahedron (Fig. 19(b)), truncated cube (Fig. 19(c)), icosidodecahedron (Fig. 19(f)), truncated cuboctahedron (Fig. 19(g)), and truncated dodecahedron (Fig. 19(h)) are all 2^n. Truncation of the apices of a polyhedron naturally yields a fair amount of increase in the numbers of the apices and edges, but it does not necessarily mean an acceleration of factorization as observed in the above results. For example, the effect of truncation of the dodecahedron (Fig. 18(d)) and icosahedron (Fig. 18(e)) yields remarkable changes in $K(G)$ value as from 6^2 to 2^{11} and from 5^3 to 2^{11}, respectively. By the way, it is interesting to note that the perfect matching number for the soccer ball network (Fig. 19(i)) is as large as $12500 = 2^2 \cdot 5^5$.

Besides these regular and semiregular polyhedra, the rhombo-dodecahedron in Fig. 20 and rhombotriacontahedron in Fig. 21 are known for their special feature of symmetry. Namely, they are respectively composed of the equivalent rhombi but with two kinds of apices of different degrees. As evident from Figs. 20 and 21 these two kinds of apices are alternately joined but differ in number. This is the reason why these two rhombohedra do not have any perfect matching.

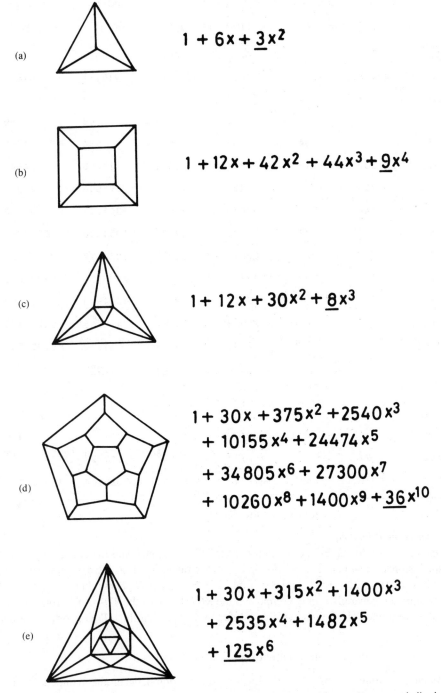

(a) $1 + 6x + \underline{3}x^2$

(b) $1 + 12x + 42x^2 + 44x^3 + \underline{9}x^4$

(c) $1 + 12x + 30x^2 + \underline{8}x^3$

(d)
$$1 + 30x + 375x^2 + 2540x^3$$
$$+ 10155x^4 + 24474x^5$$
$$+ 34805x^6 + 27300x^7$$
$$+ 10260x^8 + 1400x^9 + \underline{36}x^{10}$$

(e)
$$1 + 30x + 315x^2 + 1400x^3$$
$$+ 2535x^4 + 1482x^5$$
$$+ \underline{125}x^6$$

Fig. 18. Z-counting polynomials of regular polyhedra. The perfect matching numbers are underlined.

Table 1. Non-adjacent numbers of semiregular polyhedra (the perfect matching number is underlined and factored out below). See Fig. 19

Name	Truncated Tetra- hedron	Cubocta- hedron	Truncated Cube	Truncated Octa- hedron	Rhombi- cubocta- hedron	Icosi- dodeca- hedron
	Fig. 19(a)	Fig. 19(b)	Fig. 19(c)	Fig. 19(d)	Fig. 19(e)	Fig. 19(f)
Sym- metry	T_d	O_h	O_h	O_h	O_h	I_h
k	$p(G,k)$	$p(G,k)$	$p(G,k)$	$p(G,k)$	$p(G,k)$	$p(G,k)$
0	1	1	1	1	1	1
1	18	24	36	36	48	60
2	117	204	558	558	984	1590
3	332	744	4876	4884	11288	24540
4	390	1116	26421	26619	79806	244965
5	144	528	92016	94008	361248	1661220
6	$\underline{8}$	$\underline{32}$	206676	217172	1054328	7825660
7			292872	323976	1951272	25749300
8			249420	301203	2196753	58723620
9			117600	163444	1394608	90939600
10			26928	46182	436608	92286432
11			2304	5508	51552	58006560
12			$\underline{32}$	$\underline{169}$	$\underline{1088}$	20645280
13						3573120
14						222720
15						$\underline{2048}$
Factor	2^3	2^5	2^5	13^2	$17 \cdot 2^6$	2^{11}
Z_G	1010	2649	1019740	1183760	7539584	359906716

Four-dimensional polyhedra

One can go on to survey the relation between the symmetry and matching polynomial for four-dimensional networks. The results on the 4D-tetrahedron (Fig. 22(b)), cube (Fig. 22(a)) and octahedron (Fig. 22(c)) are shown in Table 2. While the $p(G, k)$ numbers ($k \geq 1$) for these 4D-polyhedra are generally highly composite, the $K(G)$ value of the 4D-octahedron (Fig. 22(c)) is found to be prime. This might possibly be due to the fact that this 4D-network is not planar.

4. ORTHOGONAL POLYNOMIALS

As the final topics of this essay the most mystic relation between the symmetry of a graph and the matching polynomial or the Z-counting polynomial will be presented[37,43,45–47].

The Chebyshev polynomials of the first and second kinds are defined for non-negative integer n as

$$T_n(\cos \theta) = \cos n\theta \qquad \text{(1st kind)} \qquad (17)$$

$$U_n(\cos \theta) = \sin(n + 1)\theta / \sin \theta \qquad \text{(2nd kind)}. \qquad (18)$$

Table 1 (*Continued*)

Name	Truncated Cuboctahedron	Truncated Dodecahedron	Truncated Icosahedron
	Fig. 19(g)	Fig. 19(h)	Fig. 19(i)
Symmetry	Oh	Ih	Ih
k	p(G,k)	p(G,k)	p(G,k)
0	1	1	1
1	72	90	90
2	2412	3825	3825
3	49944	102100	102120
4	716250	1920480	1922040
5	7555128	27073548	27130596
6	60763172	297017670	298317860
7	381211272	2599271940	2619980460
8	1893125565	18452804370	18697786680
9	7509594912	107509368860	109742831260
10	23910938376	518092164744	534162544380
11	61201444128	2075424449400	2168137517940
12	125732080364	6929555927025	7362904561730
13	206409077952	19297656051090	20949286202160
14	268773016464	44774805188205	49924889888850
15	274626007968	86315702921360	99463457244844
16	216971087556	137639652148260	165074851632300
17	129962010528	180432784692900	227043126274260
18	57489983904	192895567767700	256967614454320
19	18129337440	166490504865960	237135867688980
20	3881510208	114582353107800	176345540119296
21	525971200	61926709855920	104113567937140
22	40481856	25792171457280	47883826976580
23	1465536	8085072744000	16742486291340
24	<u>16384</u>	1850294700320	4310718227685
25		296798234112	783047312406
26		31509790080	94541532165
27		2030914560	6946574300
28		68820480	269272620
29		921600	4202760
30		<u>2048</u>	<u>12500</u>
Factor	2^{14}	2^{11}	$2^2 \cdot 5^5$
z_G	1397507448592	1050065644287728	1417036634543488

The modified polynomials $C_n(x)$ and $S_n(x)$ are defined as[48,49]

$$C_n(x) = 2 T_n(x/2) \qquad (19)$$

and

$$S_n(x) = U_n(x/2). \qquad (20)$$

Interestingly enough, these two polynomials not only coincide with the matching polynomials

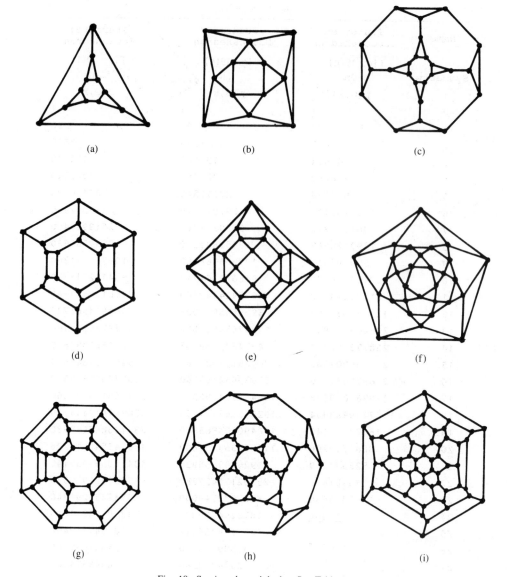

(a)

(b)

(c)

(d)

(e)

(f)

(g)

(h)

(i)

Fig. 19. Semiregular polyhedra. See Table 1.

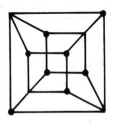

8 starred and
6 unstarred

Fig. 20.

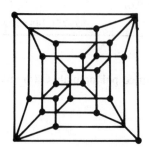

20 starred and
12 unstarred

Fig. 21.

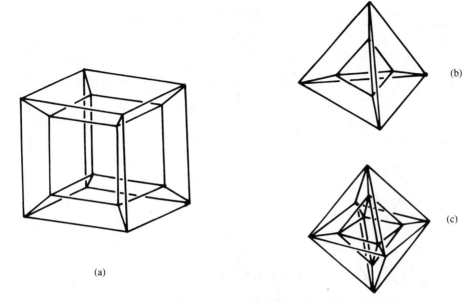

Fig. 22. Four-dimensional polyhedra. See Table 2.

$\alpha_G(x)$ of the cycle graph C_n and path graph S_n as shown in Tables 3 and 4, respectively, but also have exactly the same symbols as those currently used for these series of graphs[3],

$$\alpha_{C_n}(x) = C_n(x) \tag{21}$$

$$\alpha_{S_n}(x) = S_n(x). \tag{22}$$

It is known that the sums of the absolute values of the coefficients of these two polynomials form the well-known Fibonacci and Lucas series[33,35].

Table 2. Non-adjacent numbers of four-dimensional polyhedra. See Fig. 22

Name	4D-Tetrahedron Fig. 22(b)	4D-Cube Fig. 22(a)	4D-Octahedron Fig. 22(c)
k	p(G,k)	p(G,k)	p(G,k)
0	1	1	1
1	16	32	30
2	72	400	315
3	88	2496	1404
4	16	8256	2571
5		14208	1518
6		11648	137
7		3712	
8		272	
Factor	2^4	$2^4 \cdot 17$	137
z_G	193	41025	5976

Table 3. Matching polynomials of path graphs and Chebyshev polynomials of the second kind

n	S_n	$\alpha_{S_n}(x) = U_n(x/2)$	$U_n(x)$	Z_G
0	ϕ	1	1	1
1	•	x	$2x$	1
2	•—•	$x^2 - 1$	$4x^2 - 1$	2
3	∧	$x^3 - 2x$	$8x^3 - 4x$	3
4	∿	$x^4 - 3x^2 + 1$	$16x^4 - 12x^2 + 1$	5
5	∿	$x^5 - 4x^3 + 3x$	$32x^5 - 32x^3 + 6x$	8
6	∿	$x^6 - 5x^4 + 6x^2 - 1$	$64x^6 - 80x^4 + 24x^2 - 1$	13

The next example is the Hermite polynomial, which is defined as,

$$H_n(x) = (-1)^n \exp(x^2) \frac{d^n}{dx^n} \exp(-x^2) \tag{23}$$

or as[48]

$$h_n(x) = 2^{-n/2} H_n(x/\sqrt{2}). \tag{24}$$

As is evident from Table 5, this is nothing else but the matching polynomial of the complete graph K_n, which is composed of n points and all the possible $n(n - 1)/2$ lines joining them,

$$\alpha_{K_n}(x) = h_n(x). \tag{25}$$

Table 4. Matching polynomials of cycle graphs and Chebyshev polynomials of the first kind

n	C_n	$\alpha_{C_n}(x) = 2T_n(x/2)$	$T_n(x)$	Z_G
0	ϕ	2	1	2
1	•	x	x	1
2	◯	$x^2 - 2$	$2x^2 - 1$	3
3	△	$x^3 - 3x$	$4x^3 - 3x$	4
4	▢	$x^4 - 4x^2 + 2$	$8x^4 - 8x^2 + 1$	7
5	⬠	$x^5 - 5x^3 + 5x$	$16x^5 - 20x^3 + 5x$	11
6	⬡	$x^6 - 6x^4 + 9x^2 - 2$	$32x^6 - 48x^4 + 18x^2 - 1$	18

Table 5. Matching polynomials of complete graphs and hermite polynomials

n	K_n	$\alpha_{K_n}(x) = h_n(x)$	$H_n(x)$
0	ϕ	1	1
1		x	$2x$
2		$x^2 - 1$	$4x^2 - 2$
3		$x^3 - 3x$	$8x^3 - 12x$
4		$x^4 - 6x^2 + 3$	$16x^4 - 48x^2 + 12$
5		$x^5 - 10x^3 + 15x$	$32x^5 - 160x^3 + 120x$
6		$x^6 - 15x^4 + 45x^2 - 15$	$64x^6 - 480x^4 + 720x^2 - 120$

Table 6. Matching polynomials of complete bipartite graphs and Laguerre polynomials

n	$K_{n,n}$	$\alpha_{K_{n,n}}(x) = (-1)^n n! \, L_n(x^2)$
0	ϕ	1
1		$x^2 - 1$
2		$x^4 - 4x^2 + 2$
3		$x^6 - 9x^4 + 18x^2 - 6$

m	n	$K_{m,n}$	$\alpha_{K_{m,n}}(x) = (-1)^n n! \, x^{m-n} L_n^{m-n}(x^2)$
2	1		$x^3 - 2x$
3	1		$x^4 - 3x^2$
3	2		$x^5 - 6x^3 + 6x$
4	2		$x^6 - 8x^4 + 12x^2$

Table 7. Correspondence between the orthogonal polynomials and special series of graphs through matching polynomials

Orthogonal Polynomial	Series of Graphs	Quantum Chemical Aspects
Chebyshev (1st) T_n, C_n	Cycle graph	HMO[a] of cyclo-polyene
Chebyshev (2nd) U_n, S_n	Path graph	HMO of linear polyene
Hermite H_n, h_n	Complete graph K_n	Harmonic oscillator
Laguerre L_n	Complete bipartite graph $K_{n,n}$	Radial part of the H-atom wavefunction
Associated Laguerre L_n^{m-n}	Complete bipartite graph $K_{m,n}$	
Legendre P_n	Not known	Angular part of the H-atom wavefunction

[a] Hückel molecular orbital.

The complete bipartite graph $K_{m,n}$ is defined as the graph composed of two distinct sets of m and n points and all the possible mn lines joining the two sets. The matching polynomial of $K_{m,n}$ is found to be identical to the associated Laguerre polynomial $L_n^{m-n}(x^2)$ as shown in Table 6,

$$\alpha_{K_{m,n}}(x) = (-1)^n \, n! \, x^{m-n} \, L_n^{m-n}(x^2) \quad (m \geqq n), \tag{26}$$

where

$$L_n^\alpha(x) = \sum_{k=0}^{n} (-1)^k \binom{n+\alpha}{n-k} x^k / k!. \tag{27}$$

For the special case of $m = n$ the matching polynomial $\alpha_{K_{n,n}}(x)$ represents the graphical interpretation of the Laguerre polynomial $L_n(x)$.

The perfect matching numbers for the complete graph K_n and complete bipartite graph $K_{n,n}$ are obtained as follows:

$$K(K_{2n}) = (2n - 1)(2n - 3)(2n - 5) \ldots 3 \cdot 1 = (2n - 1)!! \tag{28}$$

$$K(K_{2n+1}) = 0 \tag{29}$$

$$K(K_{n,n}) = n!. \tag{30}$$

These novel coincidence between the matching polynomials of special series of graphs and the typical orthogonal polynomials are summarized in Table 7, where the quantum chemical aspects of these orthogonal polynomials are also given. However, no special series of graphs has even been known whose matching polynomial is identical to the Legendre polynomial, the most typical of the family of the orthogonal polynomials. This is an open question.

Acknowledgments—The author wishes to thank Mrs. Keiko Takano for assisting him in calculating the matching numbers. All his students who have been engaged in the relevant work are also greatly acknowledged. A part of this work was supported by the Joint Studies Program (1984) of the Institute for Molecular Science.

REFERENCES

1. W. T. Tutte, The factorization of linear graphs, *J. London Math. Soc.* **22**, 107–111 (1947).
2. C. Berge, *Graphs et Hypergraphes*. Dunod, Paris (1970).
3. F. Harary, *Graph Theory*. Addison-Wesley, Reading, MA (1969).
4. D. Cvetković, M. Doob and H. Sachs, *Spectra of Graphs, Theory and Applications*. Academic Press, Berlin (1979).
5. E. Ising, *Z. Physik* **31**, 253 (1925).
6. H. A. Bethe, Statistical theory of superlattices. *Proc. Roy. Soc.* (London) **A150**, 552–575 (1935).
7. L. Pauling, *The Nature of the Chemical Bond*. Cornell Univ. Press, Ithaca (1960).
8. G. Wheland, The number of canonical structures of each degree of excitation for an unsaturated or aromatic hydrocarbon. *J. Chem. Phys.* **3**, 356–361 (1935).
9. M. Gordon and W. H. T. Davison, Theory of resonance topology of fully aromatic hydrocarbons. *J. Chem. Phys.* **20**, 428–435 (1952).
10. T. F. Yen, Resonance topology of polynuclear aromatic hydrocarbons. *Theor. Chim. Acta* (Berl.) **20**, 399–404 (1971).
11. W. C. Herndon, Enumeration of resonance structures, *Tetrahedron* **29**, 3–12 (1973).
12. M. Randić, Enumeration of the Kekulé structures in conjugated hydrocarbons. *J. Chem. Soc. Faraday Trans. II* **72**, 232–243 (1976).
13. J. Yuansheng (Kiang Yuan-Sun), Graph theory of molecular orbitals. *Scientia Sinica* **23**, 847–861 (1980).
14. N. Trinajstić, *Chemical Graph Theory*. CRC Press, Boca Raton, Florida (1983).
15. R. Ramaray and K. Balasubramanian, *Computer Generation of Matching Polynomials of Chemical Graphs and Lattices*, to be published.
16. P. W. Kasteleyn, *Graph Theory and Theoretical Physics* (F. Harary ed.), p. 43. Academic Press, London (1967).
17. J. K. Percus, Combinatorial Methods. *Courant Inst. Math. Sci.* New York Univ., New York (1969).
18. H. N. V. Temperley, Phase Transition and Critical Phenomena (C. Domb and M. S. Green ed.), Vol. 1, p. 227. Academic Press, London (1972).
19. H. N. V. Temperley, *Graph Theory and Applications*. Ellis Horwood, Chichester, Sussex (1981).
20. H. N. V. Temperley and M. E. Fisher, Dimer problems in statistical mechanics. An exact result, *Phil. Mag.* **6**, 1061–1063 (1961).
21. P. W. Kasteleyn, The statistics of dimers on a lattice. *Physica* **27**, 1209–1225 (1961).
22. D. Klarner and J. Pollack, Domino tilings of rectangles with fixed width. *Discrete Math.* **32**, 45–52 (1980).
23. R. C. Read, A note on tiling rectangles with dominoes. *Fibonacci Quart.* **18**, 24–27 (1980).
24. H. Hosoya and A. Motoyama, An effective algorithm for obtaining polynomials for dimer statistics. *J. Math. Phys.* **26**, 157–167 (1985).
25. E. Clar, *The Aromatic Sextet*. John Wiley, London (1972).
26. H. Hosoya and T. Yamaguchi, Sextet polynomial. A new enumeration and proof technique for the resonance theory applied to the aromatic hydrocarbons. *Tetrahedron Lett.* (1975), 4659–4662; N. Ohkami, A. Motoyama, T. Yamaguchi, H. Hosoya, and I. Gutman, Graph-theoretical analysis of the Clar's aromatic sextet. *Tetrahedron* **37**, 1113–1326 (1980).
27. M. Aida and H. Hosoya, MO- and VB-Benzene characters. Analysis of the ''Clar's aromatic sextet'' in polycyclic aromatic hydrocarbons. *Tetrahedron* **36**, 1317–1326 (1980).
28. N. Ohkami and H. Hosoya, Topological dependency of the aromatic sextets in polycyclic benzenoid hydrocarbons. *Theor. Chim. Acta* (Berl.) **64**, 153–170 (1983).
29. I. Gutman, Topological properties of benzenoid molecules. *Bull. Soc. Chim. Beograd* **47**, 453–471 (1982).
30. S. El-Basil, Graphical enumeration of the sextet polynomials. *Bull Chem. Soc. Jpn.* **56**, 3152–3158 (1983); I. Gutman and S. El-Basil, Topological properties of benzenoid systems. XXIV. Computing the sextet polynomial. *Z. Naturforsch.* **39a**, 276–281 (1983); S. El-Basil and N. Trinajstić, Application of the reduced graph model to the sextet polynomial, to be published.
31. H. Hosoya, ''How to design non-Kekulé even alternant hydrocarbons,'' presented at the 1st Japan-Korean Symposium in Theor. Chem., Okazaki, Japan (1984); to be published in *Croat. Chem. Acta.*
32. J. L. Hock and R. B. McQuistan, The occupation statistics for indistinguishable dumbbells on a $2 \times 2 \times N$ lattice space. *J. Math. Phys.*, **24**, 1859–1865 (1983).
33. H. Hosoya, Topological index. A newly proposed quantity characterizing the topological nature of structural isomers of saturated hydrocarbons, *Bull. Chem. Soc. Jpn.* **44**, 2332–2339 (1971).
34. H. Hosoya, Graphical enumeration of the coefficients of the secular polynomials of the Hückel molecular orbitals. *Theor. Chim. Acta* (Berl.) **25**, 215–222 (1972).
35. H. Hosoya, Topological index and Fibonacci numbers with relation to chemistry, *Fibonacci Quart.*, **11**, 255–266 (1973).
36. O. J. Heilman and H. Lieb, Monomers and dimers. *Phyd. Rev. Lett.* **24**, 1412–1414 (1970).
37. O. J. Heilman and H. Lieb, Theory of monomer-dimer systems. *Comm. Math. Phys.* **25**, 190–232 (1972).
38. H. Kunz, Location of the zeros of the partition function for some classical lattice systems. *Phys. Lett.* **32A**, 311–312 (1970).
39. C. Gruber and H. Kunz, General properties of polymer systems. *Comm. Math. Phys.* **22**, 133–161 (1971).
40. J. Aihara, A new definition of Dewar-type resonance energy. *J. Am. Chem. Soc.* **98**, 2750–2758 (1976).
41. I. Gutman, M. Milun and N. Trinajstić, Non-parametric resonance energy of arbitrary conjugated systems. *J. Am. Chem. Soc.* **99**, 1692–1704 (1977).
42. E. J. Farrell, An introduction to matching polynomials. *J. Comb. Theory* **B27**, 75–86 (1979).

43. C. D. Godsil and I. Gutman, On the theory of the matching polynomial. *J. Graph Theory* **5**, 137–144 (1981).
44. H. Hosoya and K. Hosoi, Topological index as applied to electronic systems. III. Mathematical relations among various bond orders. *J. Chem. Phys.* **64**, 1065–1073 (1976).
45. I. Gutman and H. Hosoya, On the calculation of the acyclic polynomials. *Theor. Chim. Acta* (Berl.), **48**, 279–286 (1978).
46. I. Gutman, The matching polynomial. *MATCH* **6**, 75–91 (1979).
47. H. Hosoya, Graphical and combinatorial aspects of some orthogonal polynomials. *Natural Sci. Rept. Ochanomizu Univ.*, **32**, 127–138 (1981).
48. L. A. Lyusternik, O. A. Chervonenkis and A. R. Yanpol'skii, *Handbook for Computing Elementary Functions*, pp. 163. Pergamon Press, Oxford (1965).
49. M. Abramowitz and I. A. Stegun, *Handbook of Mathematical Functions*. Dover, New York (1972).

Comp. & Maths. with Appls. Vol. 12B, Nos. 1/2, pp. 291–301, 1986
Printed in Great Britain.

0886–9561/86 $3.00 + .00

SYMMETRY-MAKING AND -BREAKING IN VISUAL ART

V. Molnar and F. Molnar
Centre de Recherche Expérimentale et Informatique des Arts Visuels,
Université de Paris I Paris, France

Abstract—Symmetry has always been considered an important factor in Visual Aesthetics. But perceptual symmetry is not always identical to the symmetry defined by the mathematicians. A symmetrical picture is not necessarily symmetrical in the mathematical sense. Using a well defined abstract picture we try to determine the principles governing the aesthetical effects of symmetrical images.

Regarding symmetry, we can speak of at least four different points of view which are closely correlated: the physical, mathematical, psychological and aesthetical points of view. We intend to examine the question of symmetry in aesthetics and by a more restricted approach, the question of symmetry in aesthetics of the visual arts.

Since the aesthetical point of view is strictly linked to the perceptive system, in examining the problems of aesthetics we find ourselves dealing with two distinct groups of problems:

(1) the problem of the perception of symmetry;
(2) the aesthetical effect of the perception of a symmetrical pattern.

THE PROBLEM OF THE PERCEPTION OF SYMMETRY

Perceived symmetry only rarely coincides with the symmetry defined in mathematics. Often what is considered by an analytical science as symmetrical is, in fact, far from being symmetrical in our perception. Likewise, many objects judged as symmetrical are not, according to the mathematical definition of symmetry.

The pattern in Fig. 1 is undoubtedly symmetrical according to the mathematical definition, since it presents a series of translations. All these equivalent points are related by translation: by the simplest direct isometry operation which leaves a figure invariant. This operation is called a *symmetry-operation*. However, this pattern would be perceived as asymmetrical by all normal human beings. Raphael's famous painting the "Sposalitio" (Fig. 2), often referred to as a classical example of symmetrical composition, is obviously not symmetrical. The relation between the two sides of the picture is not isomorphic. We cannot detect any translation, any rotation or any guided reflection in respect to an axis of symmetry. Nor is Manet's "Olympia" symmetrical (Fig. 3). It was however described as a "vulgar symmetrical playing-card" after it's first scandalous exhibition.

Certainly, the shapes on the two sides of the picture, the two human figures, Olympia and her servant, can suggest a certain symmetry. In fact, they are slightly similar but are far from being isomorphous. The servant's bust is not a translation eventually obtained by the rotation of Olympia's body. Likewise patterns made up of simple geometrical forms can seem more or less symmetrical without being symmetrical in the strict mathematical sense. Besides, from the strict mathematical point of view, the expression "more or less symmetrical" does not have much significance. A figure is either symmetric or asymmetric regardless of how many symmetric operations may be performed on it. On the other hand, as far as human perception and perceptive judgement are concerned, to evaluate hierarchy of symmetry is a legal procedure. A symmetrical pattern can be perceived as more symmetrical than another. In a first analysis, one can base the perceptive evaluation of symmetry on the number of symmetrical axes without taking into account that one is or is not aware of the number of these axes. In any case, symmetry is psychologically a continuous variable. A picture may be rich or poor in symmetry depending on how many types of symmetrical operations may be performed on it without affecting it. The letter "A", for instance, will remain unchanged only if reflected in one axis. As far as the rotation is concerned the letters "S", "I", "H" or "O" possess two-fold symmetry: these appear the same after 180° rotation. The number of symmetrical axes is undoubtedly only one

Fig. 1. Translational symmetry. (after A. L. Loeb (3).)

of the intervening variables in the perception of symmetry. The reflection around the vertical axis seems to provoke a stronger perception of symmetry than the reflection around the horizontal · one. So, the spatial position of the pattern influences the perception of symmetry. Another important variable concerning the perceptual symmetry is the *context*.

Under normal circumstances, Raphael's painting represented in Fig. 2 is perceived as symmetrical. It would not be symmetrical if the spectator were requested to examine the picture not as a work of art or as a scenery, but as an arrangement of colors and forms. In this respect one can also distinguish spontaneous perception with or without scrutinization. The scrutinising perception can go as far as the precise examination of the picture which corresponds roughly to a scientific analysis of a visual phenomenon.

In spite of at least a century of intensive research, psychologists are still unable to link the level of perceived symmetry to the physical elements of the picture. In a statistical processing of a picture the skewness of a one-dimensional distribution can be measured, but skewness means degrees of deviation from symmetry. So, skewness measures asymmetry. The same concept may be applied to a two-dimensional distribution and the asymmetry of a shape may also be measured on a continuum. In this approach the parameter for symmetry is considered to be an analogy of the "third moment" of the statistical distribution of the elements[1].

THE PROBLEM OF THE AESTHETIC ASPECT OF SYMMETRY

Symmetry, besides harmony and rhythm, is one of the most indistinct concepts on aesthetic vocabulary. Moreover, the semantic field of these three words have a very large intersection. Symmetry, in the olden days from Plato to Vitruvius and Alberti was a concept which has very little relation to it's actual meaning. In Plato's system symmetry of a phenomenon is the result of "commoduatio" joining all the elements with the whole by means of a standard measure called "modulus". Vitruvius, the theoretician of Roman architecture used the word almost in the same sense. Symmetry, he said, consists in the agreement of the measure between the distinct elements and the whole.

Alberti shared this opinion at the dawn of the Renaissance. Associated with order and unity, symmetry becomes an important aesthetic category. Diderot treats it as such. Still today, it remains an aesthetic category, at least in the mind of certain theoreticians of visual art. Some decades before Diderot, Alciphon, the "minute philosopher" of Berkeley's declared that "beauty is a fugacious charm which is felt." Indeed, what quality is felt? one may ask. It is "Symmetry

Fig. 2. Raphael. *Sposalitio*. (Photo Bulloz, Paris.)

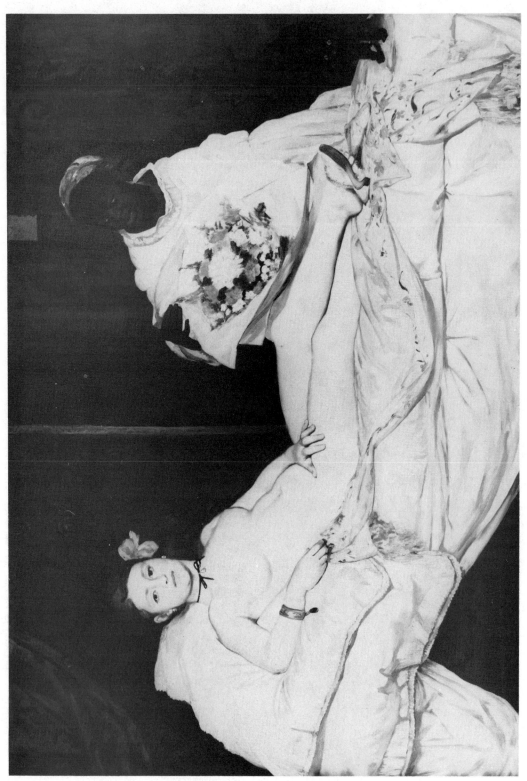

Fig. 3.   Manet. *Olympia.* (Photo Bulloz, Paris.)

and proportion as they please the eye''[2]. But what does this symmetry, which is so pleasing to the eye, signify?

This is one of the main problems of scientific aesthetics and of many artists conscientious of the inherent difficulties of modern art. There are several well-defined ways to carry out research in this field. We do not want to follow the so promising way chosen by MacGillavry in her ''Symmetry aspects of M. C. Escher's periodic drawings''.[8] Nor can we choose the royal route of topology laid down by the painter and topologist A. Hill[4,5]. This researcher in two excellent studies showed that the painter Mondrian, the sworn enemy of all kinds of symmetry, created symmetrical works all his life. In fact, Mondrian's asymmetrical works transform themselves into symmetrical ones in a certain topology. But as soon as one adds a distance to the topology, Mondrian's paintings become asymmetrical again. We stated above that two great-world-wide works of art spontaneously judged as being symmetrical were not symmetrical in the mathematical sense. But in these pictures the distributions of the great shapes on the two sides are quite symmetrical. There are the more precise forms, the details, which break the symmetry of the whole. Of course, there are many experimental evidences showing that, for perceptual symmetry, the lower spatial frequencies are more important than the higher spatial frequences. However, in realistic and representative works, the exact geometrical symmetry cannot exist. The human body is undoubtedly symmetric at a first glance. But the left side of our body is not identical to the right side. The apparent symmetry of the human face is lightened by an asymmetry of the details. There are differences in these details which make up an interesting face. Nothing is stranger, more bizarre and even uglier than those symmetrical faces fabricated artificially by copying one side of the face opposite the other side in a mirror-symmetrical fashion.

It is relatively easy to study the aesthetical effect of symmetry linked to perception by patterns built up of simple geometrical forms organised or randomly arranged (random patterns). Such are the patterns which have been constructed by Green[6] and by Julesz[7] to study perception and by Morellet or by the authors to study the aesthetical effect.

Figure 4 shows some more or less symmetrical random patterns constructed by Julesz in 1965. Certainly ''de gustibus et coloribus non disputandum''. However the experts are in agreement with the majority of people in stating that the strictly symmetrical figures are less satisfying aesthetically. In another context (for example in ornamental art) symmetry would be, on the contrary, judged as attractive. Concerning the purely aleatory random pattern, there is hardly any interest either from an aesthetical or from a perceptive point of view. Moreover, perception demands a certain level of physical organization. Faced with white noise the perceptive mechanism itself introduces a minimum of order in grouping several elements together according to certain rules which psychologists are beginning to know and at the heart of which symmetry seems to play a role.

The two pictures presented in Figs. 5 and 6 are works of art which can please or displease

(a)   (b)

Fig. 4. Symmetrical patterns; (a) two-fold-symmetry with clusters broken, (b) twofold symmetry perturbed. (after B. Julesz).

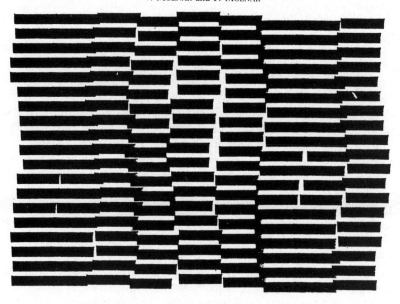

Fig. 5. V. Molnar, *Hommage à Gabo* I.

a spectator according to his personal taste. However, they are undoubtedly works of art since they have been exhibited several times in museums or avant'garde art galleries and reproduced in art reviews.

These pictures are not symmetrical either perceptually or mathematically even after a close examination. Nevertheless, after a short contemplation they cannot be judged as being totally asymmetrical. Each of these paintings is made up of 171 rectangles at regular intervals in 9 columns in 19 lines. This arrangement of rectangles has a certain symmetry, even a twofold symmetry (see Fig. 7). The pattern represented in Fig. 7(a) may be considered, in a certain context, as a work of art by its simplicity and not by virtue of its physical properties or its physical features. For reasons which we cannot develop here, this regular picture extracted from its context does not have a great aesthetical value. To render the image more aesthetical the constituent elements (the rectangles) are displaced according to a sine line whose parameters are arbitrarily chosen and arranged symmetrically around a vertical axis (Fig. 7(b)).

The pattern thus organised obviously loses its simplicity and becomes displeasing—aesthetically negative even in a decorative context. In order to lighten the sudden change around

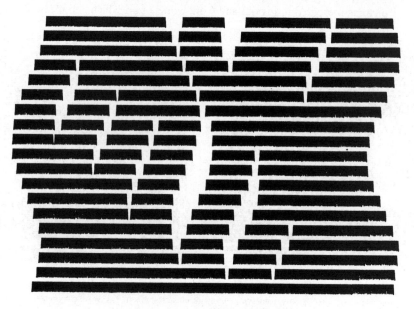

Fig. 6. V. Molnar, *Hommage à Gabo* II.

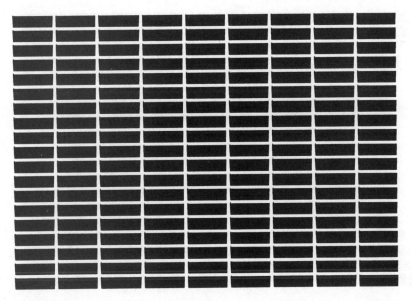

Fig. 7. (a) Regular distribution.

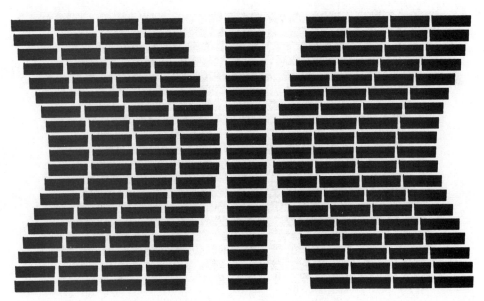

Fig. 7. (b) Symmetric sinusoidal distribution.

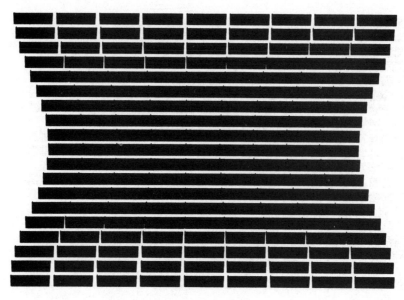

Fig. 7. (c) Symmetric sinusoidale distribution with variable amplitude (between $a = -8$ and $a = 8$).

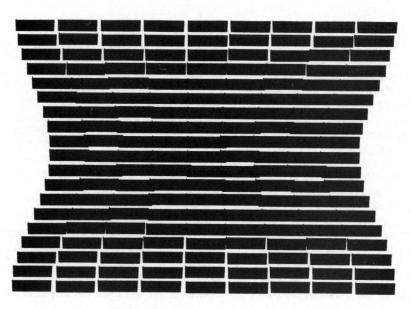

Fig. 7. (d) Similar to $c + 2\%$ of noise.

Fig. 7. (e) Similar to c + 8% of noise.

Fig. 7. (f) Distribution sinusoidale with displacement of phase.

the vertical axis, we decrease the amplitude of the sine curve progressively to the amplitude 0, then we make that amplitude increase in a symmetrical way (see Fig. 7(c)). If we note a_0 the first amplitude, the following ones may be written

$$a = a_0 \left| 1 - \frac{2k}{n-1} \right|$$

where n is the number of columns and k the rank of the column we consider (n is odd).

The picture thus obtained is perfectly symmetrical around horizontal and vertical axes. It acquires a certain aesthetical and decorative value which is nevertheless judged insufficient to construct a work aesthetically autonomous. It should be pointed out that the successive changes introduced in the organization of Fig. 7(a–c) have modified their aesthetical content without affecting the number of axes of symmetry. To break the aesthetically prejudicial monotony of the picture, we introduced a certain quantity of well-controlled disorder displacing the rectangles in an aleatory way.

The main aim of this operation is to destroy the symmetry of the pattern. The picture hence obtained (Fig. 7(c)) is not symmetric. The right side of the image is the exact reproduction of the left side. According to our arbitrary rule of construction, each rectangle of the picture could be deviated at random horizontally and vertically by between 0 and 6 elementary units. (The size of a unit, the point, was variable). Any position of a rectangle had then a 1/36 chance of materializing.

With a probability $p = 0.027$, the chance that 76 pairs of rectangles selected at random take identical positions in relation to the central axis was practically non-existent. Although the pattern seems symmetrical at least with the perception "without scrutinizing". A closer examination will obviously reveal that each rectangle is not an exact reflection, in relation to the

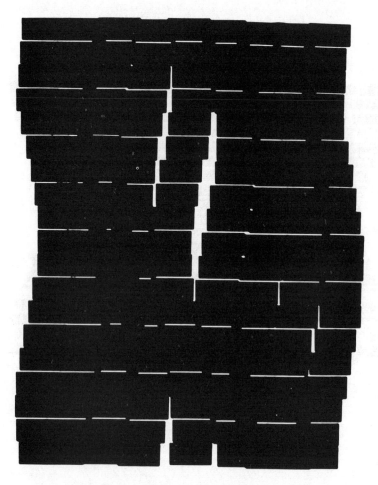

Fig. 8. V. Molnar, *Hommage à Gabo* III.

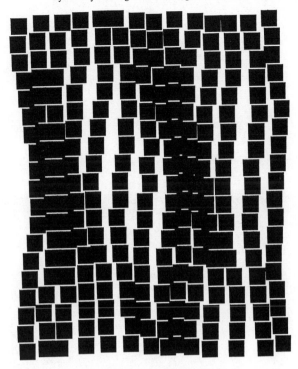

Fig. 9. V. Molnar, *Hommage à Gabo* IV.

axis of symmetry, of another rectangle. It is important to observe that in increasing the size of the elementary unit of the displacement we modified the level of symmetry without altering the probability structure of the pattern. The picture obtained in this way is still not aesthetically satisfactory.

In our efforts to break a carefully established symmetry in order to obtain higher aesthetical results, the next and provisionally the last step consists of the manipulation of the phase of the sine. Our original sinusoidal distribution was extending over a period of 2π. To obtain a sinusoidal distribution symmetrical with respect to a symmetry axis for a period of 2π, we have used the function

$$ y = a \sin\left(t - \frac{\pi}{2}\right), $$

where a is the amplitude. Then we have begun to modify the phase of the sine which governs the distribution of elements in displacing, in a systematic or random manner, the phase of each column. This operation modifies the perceptual organisation of the pattern which produces a kind of aesthetically satisfactory hidden symmetry.

By modifying the variables studied above and by systematically or randomly varying others, such as the proportion of constituting elements or the distance separating them, we obtain images aesthetically satisfactory like those, for example, shown in Figs. 5, 6, 8 and 9.

REFERENCES

1. L. Zusne, *Visual Perception of Form*. Academic Press, New York (1970).
2. K. E. Gilbert and H. Kuhn, *A History of Esthetics*. Indiana University Press, Bloomington (1954).
3. A. L. Loeb, *Color and Symmetry*. Wiley and Sons, New York (1971).
4. A. Hill, *Programme paragram structure*. in *Data*, (A. Hill (Ed)). Faber and Faber, London (1968).
5. A. Hill, A view of non-figurative art and Mathematics and an analysis of a structural relief. in *Visual Art Mathematics and Computers*, (F. J. Malma (Ed)). Pergamon Press, Oxford (1979).
6. B. F. Green, Jr., The use of high-speed digital computers in study of form perception. In *Form discrimination related to military problems*. (J. W. Wielfec and J. H. Taylor (Ed.)). Nat. Acad. Sci.- Nat. Res. Com., Washington (1957).
7. B. Julesz, *Foundation of Cyclopean Perception*. University of Chicago Press, Chicago (1971).
8. C. H. MacGillavry, *Symmetry Aspects of M. C. Escher's Periodic Drawings*. Bohn, Scheltema and Holkema, Utrecht (1976).

Comp. & Maths. with Appls. Vol. 12B, Nos. 1/2, pp. 303–314, 1986
Printed in Great Britain.

0886-9561/86 $3.00 + .00
© 1986 Pergamon Press Ltd.

SYMMETRY AND NOTATION: REGULARITY AND SYMMETRY IN NOTATED COMPUTER GRAPHICS

John G. Harries

Research Centre for Movement Notation, Faculty of Visual and Performing Arts, Tel Aviv
University, Tel Aviv, Israel

Abstract—A comprehensive cognitive method for dealing with shapes and their formal organisation in visual art is derived from the model of movement which finds its expression in Eshkol-Wachman Movement Notation. In this application, shapes are conceived as being swept out by movements which are analysed and recorded in the notation. This method is quantitative and ideally suited for computer input, while the computer is ideally suited for carrying out in detail and with precision the instructions compactly expressed in the symbol code of the notation. Using computer and notation together, regularities and symmetries can be observed in displayed shapes or sequences of shapes, and in their notation; some of these exemplify static symmetry, others dynamic symmetry, and others again, combinations of assymmetry in individual shapes together with symmetry among variants of the motif, related through the (symmetrical) structure of the process from which they arise. Families of transformations of a motif can be generated, which possess a unity that may be intuitively perceptible and will always be objectively verifiable.

FRAMES OF REFERENCE

Figurative representation in visual art, besides serving ritual, pictorial or anecdotal ends, also provides a unifying framework. By reference to, and comparison with, the "real" physical world, shapes are understood through their degree of conformity with the accepted laws of that world and to the expectations aroused by the evident adoption of a realistic style. The extent and kinds of transformation of a shape are limited to those which allow its identification as a representation of the same physical object wherever variants of the shape occur. The organisation of areas so as to "copy" nature entails rules governing the way they may obscure one another and their relative sizes, and the colour partitions of a given organisation of a picture space are confined to those which suggest natural scenes. This is a well-tried method of preserving unity, but its rules are derived from only a restricted area of visual form and formative processes.

When more "abstract" forms are employed and there is no figurative or specifically associative intention, unity and coherency are often provided through the use of patterns created by the more or less regular organisation of motifs. To ensure intelligibility these are often rather familiar and simple regular shapes of the kind frequently called "geometrical"—i.e. circles, rectangles etc. This imposes limits upon the repertoire of shapes which can serve as motifs. Given these circumstances, any greater subtlety can only be obtained by resorting to improvisation over which there is reduced conscious control.

Whether the approach is realistic or abstract, the formal aspect reveals a greater or less degree of planning, varying from simple decorative repetition to sophisticated conscious design rich in suggestion. This planning is *the way in which the constituent formal elements are related to the whole*, and this will serve, indeed has served, as a first definition of what is meant by symmetry. Without symmetry in this general sense, there can be no design[1].

This rather schematic survey of alternative ways of regarding or ignoring design is perhaps enough to suggest that in visual art there is no generally accessible, comprehensive cognitive method for dealing with shapes and their formal organisation—one in which all intentional forms and transformations are admissable.

The concepts of shape which form the basis for what follows, are derived from the model of movement in three-dimensional space which finds its expression in Eshkol–Wachman Movement Notation. This method, devised for human movement, transcends limitations which had dogged previous attempts at formulating a useful system in that field. An appropriately modified version of this system is perfectly suited to shape description, where too its effectiveness is unhampered by gratuitous extraneous restrictions.

ESHKOL–WACHMAN MOVEMENT NOTATION AND SHAPE DESCRIPTION

In EW movement notation, the movements of a single axis are equated with the easily recognised simple shapes they produce in space, or with rotation about its own length—a movement which produces no shape[2,3]. The movements of chains of two or more articulated axes might appear much harder, or even impossible, to define. But in fact, by treating each axis as if it moved in isolation, the notation enables us to express unambiguously the compound paths of a chain of simultaneously moving axes. Entire three-dimensional *space chords* are defined, using only the three types of movement: plane, conical and rotatory. Symbols are assigned to these types, which together with numerals indicating quantities, constitute the symbolic expression of movements. It is not difficult to conceive intuitively the basic forms, and the corresponding movements of a line which sweeps them out; and it is as easy to pass from shape to movement as it is to pass from movement to shape. This easily grasped *two-way* correspondence is crucial in the present application. Not only is the equivalence effective in the analysis of movement by identifying paths of movements with the shapes of the paths, but the elementary shapes can be identified with the movements which produce them. These *shapes* can therefore be expressed using the associated *movement* symbols (see Fig. 1).

Still using this equivalence, we can, by indicating simultaneous movements of a chain of articulated axes, give unambiguous expression to the compound shape traced by them. Indeed it is possible to define the shape swept out by any line moving in two dimensions or in three. Such a shape will often bear little resemblance to the basic shapes from which it is synthesized and to which it can be reduced. The examples in what follows are all two-dimensional, but the ideas are equally applicable in three dimensions.

Armed with the concepts expressed in the language of movement notation, it becomes possible to discourse intelligibly about components of visual experience encountered in visual art, such as the shapes of lines and areas, including unfamiliar figures not normally dealt with in geometry. We can define the way in which they vary from place to place and from instant to instant, and the ways in which they interact under varying conditions. The introduction of real movement in explicitly "kinetic" work is indicated in the notation by means of only quite minor differences of writing, and the phenomena which can then be dealt with include observable *processes* of growth and decay, displacement, and arrays of shapes interacting with one another[4–6].

With a proper notation all shapes, including unfamiliar or complex ones, are clear, and it is not necessary to be restricted to obviously regular shapes in order to maintain a firm frame of reference. When shapes and their transformations are defined with precision, our sensitivity to them is greatly increased.

The use of the notation gives practicability to compositional approaches which have not in the past been associated with visual art, but which seem to be aspired to in the more recent systemic and serialist work. The use of movement notation favours an approach wherein processes of growth and transformation are chosen which entail the unfolding of what is latent in the form of a given motif. "Variety within unity" is preserved by a means which constitutes an exploration of what is involved, visually and structurally, in having made an initial choice.

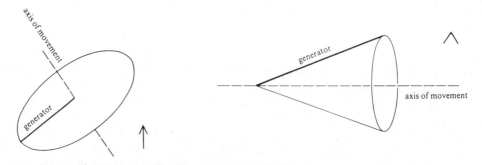

Fig. 1. (a) Movement of the generating axis at an angle of 90° to the axis of movement, producing a plane. The symbol is an arrow. (b) Movement of the generator at an angle of less than 90° to the axis of movement, producing a conical shape. The symbol is a circumflex.

This choice may have been made according to personal preference, a figurative analogy, narrative-pictorial idea, or any other urge or fancy. A particular preference is conscious inasmuch as it is based upon knowledge of the alternatives. The advantage of this way of preserving continuity is that being neither arbitrarily superimposed nor a result of improvisation based upon habit, a formally appropriate procedure can reveal whatever is formally valid in the original idea, untrammelled by preconceived notions of what it *should* be. With even quite a little practical experience, one passes easily from one formal interpretation of the basic idea to another, and at the same time the notation provides a thread which can be confidently retraced through the labyrinth of possible exploratory developments of a motif.

COMPUTER LANGUAGE AND A LANGUAGE FOR SHAPE

EW movement notation is then a tool exact enough to be practical and at the same time flexible enough to be used in all visual media. Complex sequences of shape and movement expressed in it can be carried out intuitively with some accuracy by experienced practioners; this is in fact the way EW notation is used in the performance of dance by human dancers. The precise execution of shapes, using exact measurements and instruments such as compass and straightedge, however, requires calculations which are not necessarily very difficult, but usually very repetitive. This is the lowest (but not insignificant) level at which the microprocessor now provides ideal new instruments for the visual artist.

Instructions formulated in movement notation symbols are primarily humanly oriented, making possible the holistic grasp of basically complex instructions; at the same time, the conciseness and absence of ambiguity of this notation make it an ideal medium for inputting data to a computer. The unfolding in all its detail, of the explicit display of what is implicit in the data, is carried out when a suitably programmed computer processes the compactly expressed symbol code of the notation. The automated realisation of composed shape in a visual display will often be a decisive factor in determining the feasibility of projects which might not be practicable working by hand alone. A computer with high resolution graphics facility can almost instantly display shapes, so that variants of a shape or a sequence can be tried out—at speeds which could never be attained by hand—before embarking on their manual execution in a chosen medium. By using the computer to review systematically the consequences of choosing certain defined conditions, we can explore in detail and in an acceptable length of time, many more strands of development than it would otherwise be reasonable to undertake; and it will always be possible to refer these back to the symbols of the notation through which they were generated[4,7].

Since one of the effects of using a notation is to encompass more extensive structure than the unaided immediate memory could cope with, the combination of movement notation and computer provides an instrument which can serve both as an extension of the hand and of the scope of the mind. As an instrument in its own right, the computer can be treated as a means of performing compositions, using the monitor screen as an active surface upon which real events take place, not a pictorial representation of the world but part of it. The conditions are thus provided for a kinetic visual art composed in a symbolic language adapted specifically for use with visual images[8].

Assuming that a microcomputer is the sort of machine most likely to be accessible to the potential user, the remaining discussion and examples will be directly related to the use of a BBC(B) microcomputer programmed in BASIC. The software can of course be adapted and expanded for use with a large computer[9].

As already described, we regard shape as the path swept out by a line segment or a series of articulated segments—a chain, free at one end. The components of a movement generating a shape are then:

(1) *Length*. The links have specified lengths, and these may change during the movement, or while in a position.

(2) *Sense of movement*. A link may move positively, or negatively. Positive movement in a plane is notated ↑ and for present purposes will be assumed to be always counterclockwise. Negative (clockwise) movement is notated ↓.

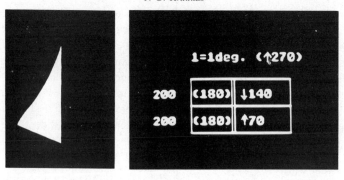

Fig. 2. A shape is generated on the screen, when the data has been keyed in and displayed on the screen as shown.

(3) *Amount of movement.* This is the angular change of position of a link *in relation to the link which carries it.* This magnitude is absolute only in the case of the first link—i.e. the one which "carries" all the others. (The amount can be written in units of a specified number of degrees, but here we shall use one degree as the unit.)

(4) *Time.* When the linkage consists of more than one link, the synchronisation of their movements is vital in determining the total shape swept out; this applies even in the case of a static result. The values are given in terms of time units, seen as a number of vertical columns.

Apart from the positive and negative signs ↑ and ↓ for sense and + and − for length, these components are expressed as numerical quantities.

The program used allows for several modes of interactive operation. The simplest of these, intended for the user familiar with the concepts but not with their symbolic representation in the notation, prompts the supply of data through a dialogue, beginning with the questions:

How many movements? [i.e. the number of movements comprising the sequence.]
Increment?
Number of links? [in the articulated chain of generators]
Length of first link?

. . . and so on, until all the necessary data has been provided, whereupon the screen is cleared and the shapes are produced upon it.

Alternatively, for the reader of movement notation, after the first three questions, the screen is treated as a ready-ruled page upon which movement notation is typed from the terminal keyboard. Again, once the "score" is complete, the display is cleared and the notated shapes are generated on the screen.

Finally, the user can, if he wishes, recall the data entered, facilitating changes or repetitions of the same shapes or sequence. The program can also be used to return points in *x* and *y* for plotting when the computer is being used as an adjunct to manually executed work.

Figure 2 shows a simple shape, together with its notation. The two links of length 200 form a linkage; both are initially in position (180)—i.e. vertically upward on the screen. The link which has one end at the origin occupies the lowest space of the score; it is seen to move positively (counterclockwise) through 70 degrees. At the same time, the link which it carries (since the two are articulated), moves in *negative* sense through 140 degrees in relation to the link by which it is carried. The shape shown is the trace swept out by the two generating links in their simultaneous movements[4–6].

REGULARITY, SYMMETRY

In Figure 3 the shape of Fig. 2 is repeated, together with its mirror image, shown this time as a sequence of discrete positions of the two links, demonstrating how the trace is formed. That the two traces are indeed reflectively symmetrical could be seen by physically cutting out the figure and folding it, or using the computer to check as many points *x*, *y* as we wish,

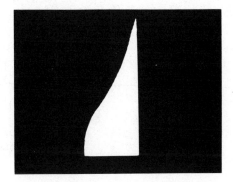

Fig. 3.   Fig. 4.

Fig. 3. The shape shown in Fig. 2, with its mirror image displayed as a sequence of positions. Fig. 4. The notation of this shape expresses its similarity to, and difference from, that of Fig. 2.

for distance from the starting position. Or, by observing that the notation for the right-hand trace is

200	(180)	↑140
200	(180)	↓70

(a)

That is to say that the movements, starting from the same positions, are different only in *sense* of movement; amount of movement and all else, remain the same. The notation thus not only provides the instructions for the generation of the shapes, and conversely an exact description of them, but also reflects the fact of their symmetry.

Figure 4 shows a figure whose similarity to, and difference from, the first, can be seen in the notation:

200	(180)	↓180
200	(180)	↑90

(b)

A trace which is reflectively symmetrical will be generated as before, by inverting the senses. Reflective symmetry about the *end* position will be produced by continuing the movements in the same sense and through the same amount of movement. The presence of reflective symmetry is revealed if we write this as two separate traces in opposite senses away from the "middle" position (axis of reflective symmetry):

200	(270)	↑180
200	(90)	↓90

(c)

200	(270)	↓180
200	(90)	↑90

(d)

If the movement is continued in the same sense through a complete cycle (See Figure 5.), i.e.:

200	(180)	↑720
200	(180)	↓360

(e)

then resulting trace is found to have reflective symmetry about two axes, and also two-fold rotational symmetry in the plane. This can be verified by moving the completed figure as a rigid shape ↑180 and ↑360 about its central origin. (A completed path of movement may be treated as a "frozen trace" which can be moved in ways which are notated in the score[4].) These symmetries could of course be classified in terms of type and degree of symmetry without using EW notation, but this would indicate nothing about the specific way in which they were generated.

The notation can be used to indicate, and the computer to execute, transfer of the relation between the linkage and the fixed origin, resulting in displacement of the "frozen trace". See Figs. 6–9, which exhibit three-, four-, six- and eight-fold rotational symmetry. By the addition of an "empty" ("pen up") carrying link, any displacement can be expressed. The "carrying" link in Fig. 6 has the movement sequence ↑120 ↑120 ↑120; in Fig. 7, ↓90 ↓90 ↓90 ↓90 . . . and similarly with the others.

Orthogonal translation is effected by change of length of the "empty" carrying link, as in Figs. 10–12. The motif shown in Fig. 10(a) allows for close packing, and an infinitely extendable tessellation is obtained by regularly adding carrying links and increasing the number of repetitions of the motif; see Fig. 10(b). This is presented in horizontally and vertically alternating black and white. Figures 11 and 12 show orthogonal translation of another motif; in Fig. 12 this is produced by the simultaneous movements of *two* empty carrying links:

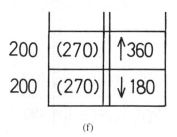

200	(270)	↑360
200	(270)	↓180

(f)

But here the displacements are not equally spaced. Neither must regular angular displacements in simple plane movement of a single link necessarily be of equal magnitude, or necessarily produce self-coincidence. In Fig. 13 the seven appearances of the motif are placed by movements of unequal (but related) size, of the empty carrying link. In Fig. 14 the successive repetitions of the motif are at unequally increasing distances from the origin. In Fig. 15 both distances and movements are determined by this unequal series, which is the well-known Fibonacci series 1, 1, 2, 3, 5, 8, 13 . . . Here we have passed from instances of "static" symmetry, which is dependent upon division of the whole into even multiple parts, to dynamic symmetry; we describe both in terms of shape and movement, without need of changing the adopted frame of reference. Static symmetry may be seen as a special case of dynamic symmetry; but while the first can be used either consciously or intuitively, dynamic symmetry can hardly be used intuitively. "The world," said Hambidge, "cannot always regard the artist as a mere medium who reacts blindly, unintelligently, to a productive yearning. There must come a time when instinct will work with, but be subservient to, intelligence."[1].

We saw earlier how a shape is the result of a given amount of positive or negative movement

Fig. 5. The preceding trace is continued through a complete cycle.

of a generating link (or links) of constant or changing length about the origin. The specific shape is determined by the attribution of values to these components. A patterned sequence of values may be uniform, or it may have the regularity of a non-monomial progression such as we saw in Figs. 13 to 15; or a more complex regularity may be generated, as Joseph Schillinger suggested[10], by the combination of simple types of periodicity. For example, a sequence such as 2 1 1 2 may be formed by the synchronisation of the two monomial periodicities 2 and 3— the simplest relation which yields a non-monomial result. By distributive involution, a more extended series is developed: 4 2 2 4 2 1 1 2 2 1 1 2 4 2 2 4, preserving the same ratios between the groups as obtained between the single values of the original motif[10].

We now interpret the series as the (total) *length* of the generator at the start of each movement, and as *amount of movement* of a single generating axis, taking 10° as the unit; thus:

Length of generator: 4 2 2 4 . . . etc.
Movement of generator: ↑ 4 2 2 4 . . . etc.

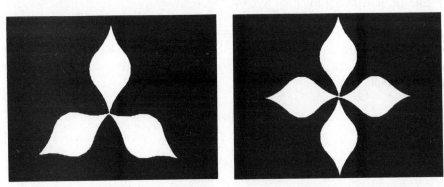

Fig. 6.   Fig. 7.

Fig. 8.   Fig. 9.

Fig. 6–9. These four figures exhibit displacements of the motif of Fig. 5, resulting in three-, four-, six- and eight-fold symmetry.

J. G. HARRIES

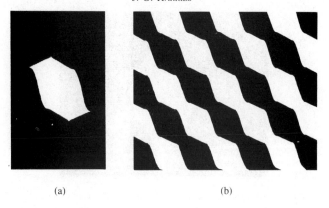

(a)   (b)

Fig. 10. The motif (a) is shown in (b) translated by the changing lengths of (unseen) additional carrying links, and displayed alternately as black and as white.

If we next apply circular permutation to the series of lengths, a family of shapes is obtained—schematically:

Lengths (constant):   4 2 2 4 2 1 1 2 2 1 1 2 4 2 2 4
Positive movement: 4 2 2 4 2 1 1 2 2 1 1 2 4 2 2 4
                      2 2 4 2 1 1 2 2 1 1 2 4 2 2 4 4
                      2 4 2 1 1 2 2 1 1 2 4 2 2 4 4 2
                             ⋮
                  4 4 2 2 4 2 1 1 2 2 1 1 2 4 2 2
                  4 2 2 4 2 1 1 2 2 1 1 2 4 2 2 4

(See Figure 16, progressing from left to right, top to bottom.)

We may now make the following observations about this sequence:

(1) The parts of the series of values, as already remarked, are interrelated according to a defined ratio—i.e. the 2 1 1 2 motif is preserved, and the extended series too has symmetry about its mid-point:

$$4\ 2\ 2\ 4\ 2\ 1\ 1\ 2 \quad\Big|\quad 2\ 1\ 1\ 2\ 4\ 2\ 2\ 4$$
axis

Fig. 11. Orthogonal translation of a motif.

Fig. 12. Orthogonal translation of the motif of Fig. 11, produced by the simultaneous movement of two (unseen) carrying links.

Fig. 13. The motif is displaced by movements of unequal angular size.

(2) The series is also interpreted as initial lengths of the moving generator—therefore also as the final lengths arrived at with the close of each movement. Thus the *movement* is concurrent with a *change of length* from one term of the series to the next. Synchronising the two series in this specific manner, we obtain shapes none of which is itself symmetrical in the sense of being in any way capable of being brought into self-coincidence by movement as a rigid form.

(3) Considering the whole sequence of shapes, we find the following: Two adjacent pairs (only) in the sequence are symmetrical in relation to each other, in that reflective symmetry is exhibited by the two middle shapes, and the first and last—which we may call adjacent in view of the cyclic nature of the sequence. Other pairs of reflectively symmetrical shapes are distributed (symmetrically) on either side of the two middle terms. Thus, the successive shapes are phases in a process which produces reflective symmetry over 16 terms.

Fig. 14. The motif is displaced by movements of equal size but unequally increased distances from the origin.

Fig. 15. Distances and movements are unequal.

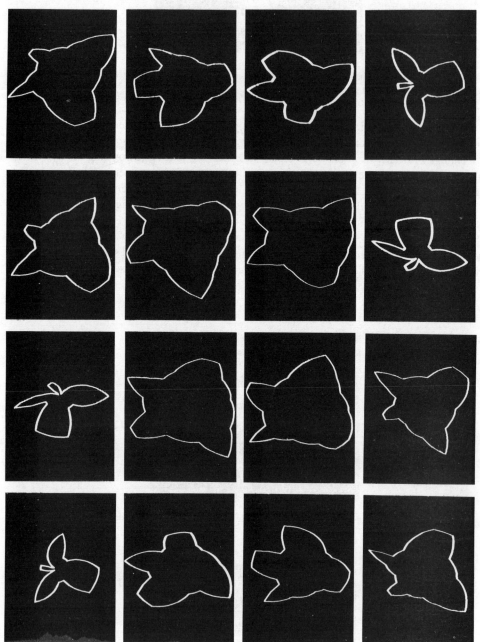

Fig. 16. Applying synchronisation of periodicities, distributive involution and circular permutation, a sequence of sixteen cycles is developed (above). The order is from left to right, top to bottom.

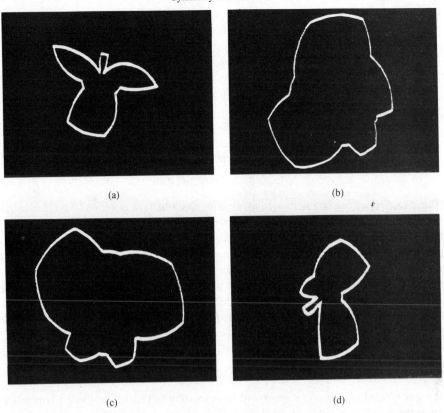

Fig. 17. Four variations resulting from raising to the second power circular permutations of the basic motif of Fig. 16.

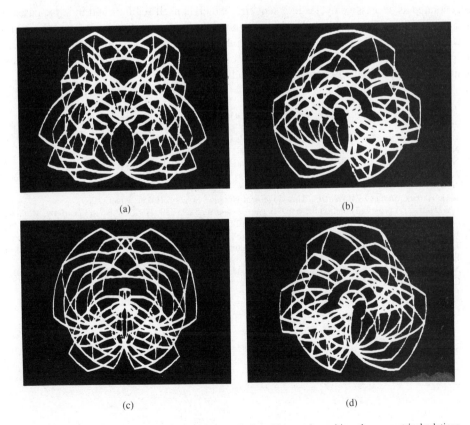

Fig. 18. When the "family" of each variant of Fig. 17 is displayed in superimposition, the symmetrical relations between and within them are clearly perceptible.

Circular permutation of the basic motif 2 1 1 2, when raised to the second power, yields four related series including the one we have just discussed:

$$2\ 1\ 1\ 2, \text{ raised to the second power } =$$
$$4\ 2\ 2\ 4 \quad 2\ 1\ 1\ 2 \quad 2\ 1\ 1\ 2 \quad 4\ 2\ 2\ 4 \ldots \text{ a (Fig. 17(a))},$$
$$1\ 1\ 2\ 2, \text{ raised to the second power } =$$
$$1\ 1\ 2\ 2 \quad 1\ 1\ 2\ 2 \quad 2\ 2\ 4\ 4 \quad 2\ 2\ 4\ 4 \ldots \text{ b (Fig. 17(b))},$$
$$1\ 2\ 2\ 1, \text{ raised to the second power } =$$
$$1\ 2\ 2\ 1 \quad 2\ 4\ 4\ 2 \quad 2\ 4\ 4\ 2 \quad 1\ 2\ 2\ 1 \ldots \text{ c (Fig. 17(c))},$$
$$2\ 2\ 1\ 1, \text{ raised to the second power } =$$
$$4\ 4\ 2\ 2 \quad 4\ 4\ 2\ 2 \quad 2\ 2\ 1\ 1 \quad 2\ 2\ 1\ 1 \ldots \text{ d (Fig. 17(d))}.$$

These four series can be interpreted as we have described, to generate four clearly different shapes including the original: Fig. 17; and each can be developed into a distinct "family" like that of Fig. 16. It will be noticed that the series of values (c) has symmetry similar to that of (a) (the original form), whereas (b) and (d) have not; (b) and (d), however, are seen to be symmetries of each other. These characteristics of the four series are very clearly seen if all the cycles of each are retained on the screen, as in Fig. 18. They can be further verified by obtaining points in x and y as mentioned previously.

Thus not only is each cycle in such a sequence of permutations a stage arrived at and departed from; each sequence is in turn also the starting point for new departures.

SUMMARY

We have found that regularities of and between shapes can be explored by means of the computer and movement notation, by virtue of the detailed quantitative approach which has thereby been made possible. Symmetries are uncovered not all of which are conveniently classifiable in terms of static (intuitive) symmetry. Families of transformations of a motif can be generated as successive cycles in a sequence which is itself regular, and in some cases [like (a) and (c) above] symmetrical. Neither the individual variants of the motif nor the relations between them necessarily possess symmetry in themselves, but the regularity of the whole, and the proportions preserved in the passage from shape to shape bestow a unity which may be intuitively perceptible, and is always objectively verifiable.

REFERENCES

1. J. Hambidge, *The Elements of Dynamic Symmetry*. Dover Publications Inc., New York (1967).
2. N. Eshkol and A. Wachman, *Movement Notation*. Weidenfeld and Nicolson, London (1958).
3. N. Eshkol, *Diminishing Series*. The Movement Notation Society/Tel Aviv University, Tel Aviv (1978).
4. J. G. Harries, *Language of Shape and Movement*. The Movement Notation Society/Tel Aviv University, Tel Aviv (1983).
5. J. G. Harries, *Shapes of Movement*. The Movement Notation Society, Holon, Israel (1969).
6. J. G. Harries, A proposed notation for visual fine art. *Leonardo* **8** 295–300 (1975). Also in *Visual Art, Mathematics and Computers* (ed. F. J. Malina). Pergamon, Oxford (1978).
7. J. G. Harries, Personal computers and notated visual art. *Leonardo* **14**, 299–301 (1981).
8. J. G. Harries, "Eshkol-Wachman Movement Notation and Kinetic Visual Art," unpublished paper read at the First International Congress on Movement Notation, Israel (1984).
9. N. Eshkol, P. Melvin, J. Michl, H. von Foerster and A. Wachman, *Notation of Movement* (report) Biological Computer Laboratory, University of Illinois, Urbana (1970).
10. J. Schillinger, *The Mathematical Basis of the Arts*. Philosophical Library, New York (1966).

Comp. & Maths. with Appls. Vol. 12B, Nos. 1/2, pp. 315–328, 1986
Printed in Great Britain.

0886–9561/86 $3.00 + .00
© 1986 Pergamon Press Ltd.

SYMMETRY IN MATHEMATICS

PETER HILTON

Department of Mathematical Sciences, SUNY-Binghamton, NY, U.S.A.

and

JEAN PEDERSEN

Department of Mathematics, University of Santa Clara, CA, U.S.A.

Abstract—The role of symmetry in geometry is universally recognized. The principal purpose of this article, on the other hand, is to show how it plays many significant, but varied, roles throughout the whole of mathematics. We illustrate this fact through characteristic examples; in most of these examples the mathematics is well known, but the symmetry aspects of the arguments have not been rendered explicit as a guiding principle. In one example, however, we do look at an unfamiliar geometrical construction of regular star polygons, which we relate to number-theoretical properties. In this example we draw attention to the presence of symmetry of an unexpected and untraditional nature, not obviously related to the regularity of the polygons.

We have attempted to identify specific principles illustrated by our examples of symmetry in mathematics. We draw attention to the *Halfway Principle* (Sec. 3) and the principle of the *symmetric definition of symmetric concepts* (Sec. 4).

INTRODUCTION

It is a commonplace—though a vitally important commonplace—that symmetry plays a crucial role in our apprehension of pattern, and that it is, in its geometrical aspect, central to our comprehension of the real world. It would be presumptuous for us to attempt to improve on the available texts which deal with the philosophical[15] and technical[1] aspects of geometric symmetry, and this is not our purpose in this article. Rather, we wish to explore the extent to which symmetry considerations permeate the whole of mathematics. We make no global analysis of this universal penetration of mathematics, but use a number of examples to probe the nature and depth of that penetration.

Our first example, described in Sec. 1, is indeed drawn from geometry, but from a very unconventional, unclassical part of geometry. We refer to the procedure, described in [5–8], for folding a straight strip of paper to construct, in a systematic way, an arbitrarily good approximation to a regular star {b/a}-gon, in the nomenclature of Coxeter[1]. The principal interest, both practically and mathematically, centers on the case where a, b are both odd, and, of course, a is prime to b with $a < b/2$.

The construction leads to number-theoretical questions and results of independent interest; but here we emphasize how one exploits, throughout the investigation, many different facets of the symmetry notion. These facets rise to the surface in response, it seems, to the natural dynamic of the mathematical argument; they appear spontaneously and inevitably, and are not simply derived from a *post hoc* summation of that argument.

This 'geometrical' example is really a very general type of mathematical example. The argument illustrates the unity of mathematics and we view symmetry as a fundamental unifying concept. We entirely agree with Grünbaum[2]—perhaps taking his argument further than he would wish—that symmetry is not to be subsumed under group theory.

However, in Sec. 2, we discuss an important symmetry concept in algebra which is certainly related to the notion of the symmetric group; that is, we present the elementary symmetric polynomials as a free generating set for the set of all symmetric polynomials. This notion is, of course, classical, but we depart from the classical line in presenting an application to modern algebraic topology.

Section 3 is concerned with the exploitation of symmetry in mathematical proof and describes a particular proof-strategy which is appropriate to such exploitation. We instance 4 examples, drawn from the geometry of vectors, from linear algebra, from homological algebra, and from differential topology. We claim that these very diverse examples are really only united by the symmetry thread running through the proofs of the assertions, a thread which, of course, inheres in the assertions themselves.

Finally, in Sec. 4, we take up a very different aspect of symmetry in mathematics. It is our contention that where the significance of a concept depends on its symmetry, it is perverse and obscurantist to offer an unsymmetrical definition of that concept. This might seem to be so obvious as not to require stating, especially if one stays with geometrical examples like the circle or the square. However, we contend that the rule we propose is systematically flouted in our textbooks. Consider the example of a one-one correspondence f from the set A to the set B. This is traditionally defined as a function f which is one-one and maps A onto B. However the *significance* of one-one correspondences is that they set up an equivalence between A and B, that is, that A and B appear in symmetric roles; were this not so, one-one correspondences, so defined, would be extremely uninteresting. Thus we contend that we should *define* $f: A \to B$ to be an equivalence if there exists $g: B \to A$ such that gf and fg are both identity maps; it is then an interesting and important fact that equivalences between sets are characterized as one-one functions mapping onto their range. We develop arguments along these lines in Sec. 4, showing how the *symmetric* notion of an equivalence recurs in every category of mathematical discourse, while the nature of a set of unsymmetric characteristics of such an equivalence varies with the mathematical objects under discussion.

Our examples, in all sections of this paper with the arguable exception of Sec. 1, are drawn exclusively from 'pure' mathematics. This again is because the wonderful symmetries in nature are discussed in other contributions to this volume. Of course, any mathematical model reflecting a symmetry in nature must itself feature the model of that symmetry; and the mathematical model is then indistinguishable from a domain of pure mathematics and therefore, we make bold to hope, possessed of features to which our remarks below may apply.

We go into considerable detail in Sec. 1 because no part of the material described there is to be found in any text.† We allow ourselves to be considerably more succinct in the later sections since the concepts we discuss (though not the discussion itself!) are standard within the mathematical disciplines to which they belong.

1. SYMMETRY IN GEOMETRY

There is no need to stress the importance of symmetry in geometry. The literature is full of sources in which discussions of the various aspects of symmetry related to specific geometric configurations are treated at length and in depth.

In this section we discuss a less conventional example of symmetry in geometry involving a straight strip of paper, for which we describe a precise folding procedure that may be used to make the top edge of the strip approximate (to any desired degree of accuracy) a regular star $\{b/a\}$-gon‡, where a and b are mutually prime integers and $a < b/2$. The procedure is systematic, easy to execute, and leads not only to the construction of beautifully symmetric star polygons, but also, as we will show, to surprising *symmetrical* relationships in the number theory that naturally arises from the construction. These number-theoretical symmetries are, however, quite unrelated to the symmetries of the regular polygons we construct.

For the moment assume that we have a straight strip of paper that has *creases* or *folds* along straight lines emanating from marked vertices A_i, $i = 0, 1, \ldots$, at the top and bottom edges, and that, for a fixed k, those at the vertices A_{nk}, $n = 0, 1, 2, \ldots, b$, which are on the top edge, form identical angles $a/b\ \pi$ (as shown in Fig. 1(a)). Suppose further that these vertices are equally spaced (we describe below how you might obtain such a strip). If we fold this strip on $A_{nk}A_{nk+2}$ (as shown in Fig. 1(b)) and then on $A_{nk}A_{nk+1}$ (as shown in Fig. 1(c)), the direction of the *top edge* of the tape will be rotated through an angle of $2(a/b)\pi$ and the tape will be oriented the same way, with respect to the center of the polygon being delineated by

†The definitions and details may be found in [8], but we regard it as unfair and unrealistic to expect the reader to consult a technical research publication in order to understand what we're talking about—especially when the article in question has not even been published at the time of writing!

‡A closed sequence of b edges that visit, in order, every a^{th} vertex (mod b) of a bounding regular convex b-gon. We include, among the regular star polygons, the special case of the regular convex b-gon obtained when $a = 1$.

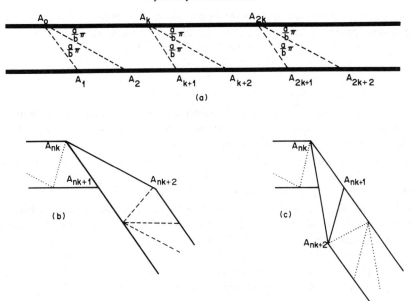

Fig. 1. Folding a star $\{b/a\}$-gon.

its top edge. We call these two folds through A_{nk}, in that order, a $2(a/b)\pi$-*twist* at A_{nk}, and observe that, if a $2(a/b)\pi$-twist is performed at A_{nk} for $n = 0, 1, 2, \ldots, b\text{-}1$, the top edge of the tape will have turned through an angle of $2a\pi$ and the point A_{bk} will then be coincident with A_0. Thus the top edge of the tape will have visited every a^{th} vertex of a bounding regular convex b-gon, and hence determines a regular star $\{b/a\}$-gon.

We now describe how we can obtain the desired crease lines on the strip of tape in the first place. We assume only that a, b are odd with $a < b/2$ and that we wish to have a strip of paper on which the angle $(a/b)\pi$ appears at regular intervals along the top edge. We designate the direction from left to right as the *forward* direction on the tape. We begin by marking a point A_0 on the top of the tape and making an *initial* crease line going in the downward forward direction from A_0 to A_1 at the bottom of the tape, and *assume* that the angle it makes with the top edge is $(a/b)\pi$; we call this the *putative* angle. Then we continue to form new crease lines according to the following four rules:

(1) The first new crease line emanates from the vertex A_1.
(2) Each new crease line goes in the forward direction along the strip of paper.
(3) Each new crease line always *bisects* the angle between the last crease line and the edge of the tape from which it emanates.
(4) The bisection of angles at any vertex continues until a crease line produces a putative angle of the form $(a'/b)\pi$ where a' is an *odd* number; then the folding stops at that vertex and commences at the intersection point of that last crease line with the other side of the tape.

Let us consider an example, say, $b = 11, a = 3$. It is clear that if we begin with an angle of $(3/11)\pi$ at P_0 and adhere to the above rules we will obtain a strip of tape with the angles and creases (indicated by dotted lines) shown in Fig. 2. We could denote this folding process as

$$\{1, 3, 1, 1, 3, 1\}.$$

In this symbol the first "1" refers to the one bisection (producing a line in a downward direction) at P_{6n} (for $n = 0, 1, 2, \ldots$) on the top of the tape; the "3" refers to the 3 bisections (producing creases in an upward direction) made at the bottom of the tape through P_{6n+1}; etc.† However, since the folding process is *duplicated* halfway through, we can abbreviate the notation and write simply

$$\{1, 3, 1\},$$

with the understanding that we alternate folding from the top and bottom of the tape as described, with the *number* of bisections at each vertex running, in order, through the values 1, 3, 1, . . . We call this procedure a primary folding procedure *of period* 3.

It is a surprising, but easily verifiable, fact that starting with any putative angle $(a/b)\pi$ (a, b odd, mutually prime, $a < b/2$), we will always obtain by our rules a primary folding procedure $\{k_1, k_2, \ldots, k_r\}$ which 'produces' (that is, converges to) the desired angle (see [8] for the detailed proof). We further note that the tape described above and in Fig. 2, whose crease lines were chosen to produce a regular star $\{11/3\}$-gon, could also be used to fold a regular convex 11-gon and a regular star $\{11/5\}$-gon. More still is true; for, as we see, if there are crease lines enabling us to fold a star $\{11/a\}$-gon, there will be crease lines enabling us to fold star $\{11/2^k a\}$-gons, where $k \geq 0$ takes all values such that $2^{k+1} a < 11$. These features, as described for $b = 11$, $a = 3$, apply for any odd number b and any a relatively prime to b with $a < b/2$. However, this tape has a special symmetry as a consequence of its *odd* period; namely, if it is "flipped" about the horizontal line half way between its parallel edges, the result is a *translate* of the original tape. As a practical matter this special symmetry of the tape means that we can use either the top edge or the bottom edge of the tape to construct our star polygons. On tapes with an *even* period the top edge and the bottom edge of the tape are not translates of each other (after the horizontal flip), which simply means care must be taken in choosing the edge of the tape used to construct a specific polygon.

It is natural to ask. "What is the relation of the putative angle to the true angle?" It turns out—the easy proof was given in [6]—that if we repeat the folding rules, starting at the successive iterates of P_0 (thus at P_0, P_3, P_6, . . . in Fig. 2), then *the actual angle rapidly converges to the putative angle*. Thus we obtain arbitrarily good approximations to the regular star-polygons produced by this tape by starting sufficiently far along the tape. Figure 3 shows the completed star $\{11/3\}$-gon formed by making a 2 $(3/11)\pi$-twist at P_{6n} ($n = 0, 1, \ldots , 10$). Similarly, Fig. 4 illustrates a completed $\{11/4\}$-gon formed by making a $2(4/11)\pi$-twist at P_{6n+1} ($n = 0, 1, \ldots , 10$). In this case there is excess tape that would 'stick out' at each vertex—and this has been folded under at each vertex to make the finished model more symmetrical.

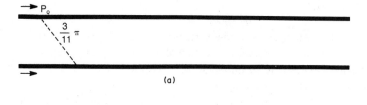

(a)

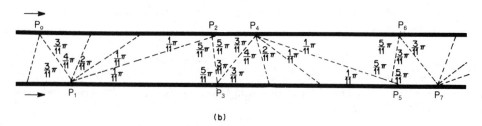

(b)

Fig. 2. The fold lines on a tape which can produce various regular star 11-gons.

†We can, if we like, imagine an arbitrary fold line terminating in P_0 which initiates the process (see Fig. 2). We have renamed the vertices, compared with Figure 1, to simplify our description.

Step 1   Step 2

Fig. 3. (*Top*) The two steps in making the $2(3\pi/11)$-twist. (*Bottom*) The completed $\{11/3\}$-gon.

Now let us look at the patterns in the *arithmetic* of the computations when $a = 3$ and $b = 11$. Referring to Fig. 2 we observe that

the angle to the right of P_n where n =	is of the form $(a_n/11)\pi$ where a_n =	and the number of bi-sections at P_{n+1} =
0	3	3
1	1	1
2	5	1
3	3	3
4	1	1
5	5	1

We could write this in shorthand form as follows:

$$(b=)11 \begin{vmatrix} (a=)3 & 1 & 5 \\ 3 & 1 & 1 \end{vmatrix}. \tag{1.1}$$

Observe that had we started with the putative angle of $(1/11)\pi$ then the symbol (1.1) would

Fig. 4. An {11/4}-gon obtained from the tape of Fig. 2(b).

have taken the form

$$(b=)11 \begin{vmatrix} (a=)1 & 5 & 3 \\ & 1 & 1 & 3 \end{vmatrix}. \tag{1.2}$$

In fact, it should be clear, that we can *start anywhere* (with $a = 1$, 3 or 5) and the resulting symbol, analogous to (1.1), will be the cyclic permutation of the interior of the symbol that places our choice of a in the first position along the top row.

Of course the process generalizes so that, by the symbol

$$b \begin{vmatrix} a_1 & a_2 & \ldots & a_r \\ k_1 & k_2 & \ldots & k_r \end{vmatrix} \tag{1.3}$$

we understand that b is an odd positive integer, that a_i is an odd positive integer $< b/2$, $i = 1, 2, \ldots, r$, and that $k_1, k_2, \ldots, k_r$ are positive integers such that

$$b = a_i + 2^{k_i}a_{i+1}, \quad i = 1, 2, \ldots, r, \quad a_{r+1} = a_1. \tag{1.4}$$

Then the symbol (1.3) is *cyclically* (or *rotationally*) *symmetric* in the sense that we may rotate the symbol to produce a valid symbol encoding exactly the same information. Notice that this symmetry is not in any way related to the symmetry of the star-polygons which may be obtained from paper folded according to the instructions coded in (1.3).

Let us agree, where convenient, to define a_i for all integers i by making a_i periodic in i, with period r, and similarly for k_i. We note that, given odd positive integers a, b with $a < b/2$, there is always a symbol (1.3) with $a_1 = a$, and that the symbol is unique up to *iteration*; here we say that (1.3) arises by iteration if there exists $s|r$, such that $a_{i+s} = a_i$, $k_{i+s} = k_i$, for all i. A *proper* iteration, that is, one in which $s \neq r$, is called a *repetition*. If there is no repetition, we say that the symbol (1.3) is *reduced*.

Now the equations (1.4) have the unique solutions, in the 'unknowns' a_i, given by

$$Ba_i = bA_i, \quad i = 1, 2, \ldots, r, \tag{1.5}$$

$$\text{where } B = 2^k - (-1)^r, \quad k = \sum_{i=1}^{r} k_i, \tag{1.6}$$

and $A_i = 2^{k-k_{i-1}} - 2^{k-k_{i-1}-k_{i-2}} + \ldots + (-1)^r 2^{k_i} - (-1)^r, \quad i = 1, 2, \ldots, r.$ (1.7)

Observe that this formula is *true for any i* and that A_i is independent of k_{i-1}. We also remark that the solutions (1.5) of the equations (1.4) always exist, but that (for a given odd positive integer b) the numbers a_i given by (1.5) may fail to be integers. Indeed, we have

PROPOSITION 1.1

    (i) *The solutions of* (1.4) *are rational numbers* a_i *satisfying* $0 < a_i < b/2$;
    (ii) *if any* a_i *is an integer, then all* a_i *are odd integers.*

Once again, Proposition 1.1 (especially part (ii)) dispays a crucial symmetry in our symbol (1.3).

Further details of the number-theoretical properties are developed in [8]. Since we are here concerned with the aspects of that development that are related to symmetry, we will list just two features of our symbols (1.3). The first is easily stated—and easily proved.

PROPOSITION 1.2

    In the symbol (1.3), $\gcd(b, a_i)$ *is independent of* i.

For the second, we revert to the notion of a symbol (1.3) obtained by repetition. Such a repetition may be viewed—indeed, should be viewed—as an instance of *translational symmetry* of the given symbol; that is, translating by a step of length s preserves the symbol. Let us call this translational symmetry *of length s*. We then have

PROPOSITION 1.3

    The following statements are equivalent:

    (i) *The symbol* (1.3) *has translational symmetry of length* s;
    (ii) *The sequence* $(a_1, a_2, \ldots, a_r)$ *has translational symmetry of length* s.
    (iii) *The sequence* $(k_1, k_2, \ldots, k_r)$ *has translational symmetry of length* s.

The equivalence of (i) and (iii) above is crucial to the arguments adduced in [8]. Those arguments lead to the following striking result which may be regarded as having an algorithmic nature.

QUASI-ORDER THEOREM

    Let b *be an odd positive integer, and let* a_i *be an odd positive integer with* $a_i < b/2$ *and* a_i *prime to* b. *Then, given the reduced symbol*

$$b \begin{vmatrix} a_1 & a_2 & \ldots & a_r \\ k_1 & k_2 & \ldots & k_r \end{vmatrix}$$

with $\Sigma_{i=1}^{r} k_i = k$, *we have*

    (i) k *is the minimal* l *such that* $b | 2^l \pm 1$,
    (ii) $b | 2^k - 1$ *if* r *is even,* $b | 2^k + 1$ *if* r *is odd.*

Let us take as an example the case $b = 641$, $a_i = 1$. In that case we obtain, by our algorithm, the reduced symbol

$$641 \begin{vmatrix} 1 & 5 & 159 & 241 & 25 & 77 & 141 & 125 & 129 \\ 7 & 2 & 1 & 4 & 3 & 2 & 2 & 2 & 9 \end{vmatrix}.$$

Thus we can infer, since $k = 32$, $r = 9$, that†

$$641 \mid 2^{32} + 1.$$

Moreover, we know from (1.5,6,7) that

$$2^{32} + 1 = 641A_1,$$

where

$$A_1 = 2^{23} - 2^{21} + 2^{19} - 2^{17} + 2^{14} - 2^{10} + 2^9 - 2^7 + 1$$
$$= 6700417.$$

By this method we produce (effortlessly!) Euler's famous factorization showing that $2^{2^5} + 1$ is not a (Fermat) prime (see [13]).

2. SYMMETRY IN ALGEBRA

Of course, symmetry appears in algebra in numerous *ad hoc* and informal ways. We discuss in this section a way in which it enters more formally. The reader is referred to, say. [3] and [12] for further details.

Naturally, we can say that it enters formally through the symmetric groups S_n. Since every finite group may be embedded in some symmetric group—for example, the regular representation embeds G in S_n, where $n = |G|$, the number of elements in G—it is reasonable to say that finite group theory is the study of subgroups of the symmetric groups. However, we wish here to emphasize a different, though related, aspect of symmetry in algebra.

Let us consider polynomials over the field $\mathbf{C}$ of complex numbers. If $x_1, x_2, \ldots, x_n$ are n indeterminates, we may form the polynomial ring $\mathbf{C}[x_1, x_2, \ldots, x_n]$. In this ring we distinguish the *elementary symmetric polynomials* $\sigma_0, \sigma_1, \sigma_2, \ldots, \sigma_n$, defined by

$$\sigma_0 = 1,$$
$$\sigma_1 = x_1 + x_2 + \cdots + x_n,$$
$$\sigma_2 = x_1 x_2 + x_1 x_3 + \cdots + x_{n-1} x_n,$$
$$\cdots$$
$$\sigma_n = x_1 x_2 \cdots x_n.$$

Alternatively we may define the elementary symmetric polynomials by forming the polynomial ring $\mathbf{C}[t, x_1, x_2, \ldots, x_n]$ and then expanding the polynomial $\Pi_{i=1}^n (t - x_i)$ in powers of t as $\Sigma_{i=0}^n (-1)^i \sigma_i t^{n-i}$.

Of course, there are many other symmetric polynomials in the ring $\mathbf{C}[x_1, x_2, \ldots, x_n]$; for example, the polynomial $x_1^2 + x_2^2 + \cdots + x_n^2$. However, one easily sees that

$$x_1^2 + x_2^2 + \cdots + x_n^2 = (x_1 + x_2 + \cdots + x_n)^2$$
$$- 2(x_1 x_2 + x_1 x_3 + \cdots + x_1 x_n) = \sigma_1^2 - 2\sigma_2. \quad (2.1)$$

The relation (2.1) suggests a general result which we may enunciate as follows. We first give a precise definition of a symmetric polynomial.

†The Quasi-Order Theorem tells us more, namely, that if $|l < 32$, then $641 | 2^l \pm 1$.

DEFINITION 2.1

Let the symmetric group S_n act on the polynomial ring $\mathbf{C}[x_1, x_2, \ldots, x_n]$ by the rule

$$\tau(x_i) = x_{\tau(i)}, \quad 1 \leq i \leq n, \quad \tau \in S_n.$$

Then a polynomial f in $\mathbf{C}[x_1, x_2, \ldots, x_n]$ is *symmetric* if $\tau(f) = f$ for all $\tau \in S_n$.

THEOREM 2.2

The symmetric polynomials in $\mathbf{C}[x_1, x_2, \ldots, x_n]$ form a subalgebra freely generated by the elementary symmetric polynomials.

We will not give a complete proof, but we indicate a line of proof. Let $x_1^{m_1} x_2^{m_2} \ldots x_n^{m_n}$ be a monomial appearing in the symmetric polynomial f. Then every monomial obtained from this one by permuting the x_i, or, *equivalently, the m_i*, must occur in f with the same coefficient. Two consequences now follow:

(i) we need only consider monomials $x^{m_1} x^{m_2} \ldots x^{m_n}$ with $m_1 \geq m_2 \geq \ldots \geq m_n$;

(ii) if we write $s(m_i, m_2, \ldots, m_n)$ for the symmetric polynomial

$$\sum_{\tau \in S_n} \tau(x_1^{m_1} x_2^{m_2} \ldots x_n^{m_n}),$$

where τ acts on such monomials in the agreed way, then every symmetric polynomial is a linear combination of such polynomials[†] $s(m_1, m_2, \ldots, m_n), m_1 \geq m_2 \geq \ldots \geq m_n$.

Thus, to show that every symmetric polynomial f is in the algebra generated by the elementary symmetric polynomials, it suffices to show this for the polynomials $s(m_1, m_2, \ldots, m_n)$. This enables us to set up a proof by induction, adopting the lexicographical ordering of the n-tuples $(m_1, m_2, \ldots, m_n), m_1 \geq m_2 \geq \ldots \geq m_n$, namely,

$$(m_1, m_2, \ldots, m_n) > (m_1', m_2', \ldots, m_n')$$

if there exists a k such that $m_i = m_i'$, $1 \leq i \leq k$, but $m_{k+1} > m_{k+1}'$.

Now if $m_1 = m_2 = \ldots = m_n = m$, then $s(m_1, m_2, \ldots, m_n) = \sigma_n^m$. Thus assume $m_1 = m_2 = \ldots = m_j > m_{j+1}$ for some $j < n$, and write $k = m - m_{j+1}$, where $m = m_1 = \ldots = m_j$. Then

$$s(m_1, m_2, \ldots, m_m) = \sigma_j^k s_1 + s_2,$$

where it may be shown that s_1 is earlier in the lexicographic ordering than s and s_2 is a sum of symmetric polynomials of type (ii) above, earlier in the lexicographic ordering than s.

Two important points emerge from the argument. Obviously the field $\mathbf{C}$ may be replaced by any other field; and, second, if our original symmetric polynomial only involved *integer* coefficients, then so does its expression in terms of the elementary symmetric polynomials. We may thus deduce

THEOREM 2.3

Let $\alpha_1, \alpha_2, \ldots, \alpha_n$ be the algebraic integers which are the zeros of a polynomial of degree n in $\mathbf{Z}[x]$. Then any symmetric polynomial in $\alpha_1, \alpha_2, \ldots, \alpha_n$ with rational integer coefficients is a rational integer.

This is a key result in algebraic number theory. A second application shows that if $\mathbf{C}(x_1, x_2, \ldots, x_n)$ is the field of rational functions and if L is the subfield of symmetric rational

---

[†]There will, of course, be repetitions if, for some i, $m_i = m_{i+1}$. We would suppress such repetitions. Thus $s(2, 2, 1) = x_1^2 x_2^2 x_3 + x_1^2 x_2 x_3^2 + x_1 x_2^2 x_3^2$. We can think of this as fixing the x_i's and attaching the exponents in all possible *distinct* ways.

functions, then $L = \mathbf{C}(\sigma_1, \sigma_2, \ldots, \sigma_n)$ and the Galois group of $\mathbf{C}(x_1, x_2, \ldots, x_n)$ over L is precisely the symmetric group S_n. Here again we could replace $\mathbf{C}$ by any field.

Let us mention now a remarkable application to topology. Theorem 2.2 tells us that, for each $k \geq 1$, there is a unique expression for $\Sigma_{i=1}^n x_i^k$ in terms of the elementary symmetric functions $\sigma_1, \sigma_2, \ldots, \sigma_n$. Thus

$$\sum_{i=1}^n x_i^k = N_k(\sigma_1, \sigma_2, \ldots, \sigma_n),$$

and N_k is called the kth *Newton polynomial*. Thus

$$N_1(\sigma_1, \sigma_2, \ldots, \sigma_n) = \sigma_1,$$
$$N_2(\sigma_1, \sigma_2, \ldots, \sigma_n) = \sigma_1^2 - 2\sigma_2,$$
$$N_3(\sigma_1, \sigma_2, \ldots, \sigma_n) = \sigma_1^3 - 3\sigma_1\sigma_2 + 3\sigma_3.$$

Now let $B(x)$ be the set of equivalence classes of vector bundles over the space X. The exterior power operation Λ^p is thus defined on $B(x)$. We now define an operation ψ, called the kth *Adams operation* by the formula

$$\psi_k(z) = N_k(\Lambda^1 z, \ldots, \Lambda^k z).$$

It is a beautiful theorem that the operations ψ_k are not only additive but also multiplicative (with respect to tensor product) and are therefore *universally defined operations in* K-*theory*. As such they have proved crucial in applications of topological K-theory, for example, in showing that, among the spheres, only S^1, S^3, S^7 carry continuous multiplications† with 2-sided identity; and in computing the number of independent vector fields on S^n (see [4,11]).

3. SYMMETRY IN MATHEMATICAL DEDUCTION

In this section we give four examples where the concept of symmetry may be used to simplify a piece of mathematical deduction. After describing the examples, we will endeavor to enunciate the symmetry principle invoked.

Example 3.1

If $\mathbf{A}$, $\mathbf{B}$ are vectors and if $a = |\mathbf{A}|$, $b = |\mathbf{B}|$, then we claim that the vector $\mathbf{C} = a\mathbf{B} + b\mathbf{A}$ bisects the angle between $\mathbf{A}$ and $\mathbf{B}$. We use the fact that

$$\mathbf{A} \cdot \mathbf{C} = |\mathbf{A}|\,|\mathbf{C}|\cos\theta,$$

where θ is the angle between $\mathbf{A}$ and $\mathbf{C}$. Notice that the expression for $\mathbf{C}$ is symmetrical in $\mathbf{A}$ and $\mathbf{B}$. By a straightforward computation we obtain

$$\cos\theta = \frac{\mathbf{A} \cdot \mathbf{C}}{|\mathbf{A}|\,|\mathbf{C}|} = \frac{(\mathbf{A} \cdot \mathbf{B} + ab)}{|\mathbf{C}|}.$$

Since the expression for $\cos\theta$ is symmetrical in $\mathbf{A}$ and $\mathbf{B}$ we know that we must get the same value for $\cos\phi$, where ϕ is the angle between $\mathbf{B}$ and $\mathbf{C}$. Thus the proof is complete.

Example 3.2

Let A be an $(m \times n)$-matrix and let B be an $(n \times m)$-matrix. The $AB = C$ is an $(m \times m)$-

---

†Actually, we get a stronger result—the homotopy group $\pi_{2n-1}(S^n)$ has an element of Hopf-invariant 1 if and only if $n = 2$, 4 or 8. This deep result, and the solution of the vector fields problem, are due to J. F. Adams.

matrix and $BA = D$ is an $(n \times n)$-matrix. We claim that trace $C = $ trace D. For (in an obvious notation)

$$\text{trace } C = \sum_{i=1}^{m} c_{ii} = \sum_{i=1}^{m} \sum_{j=1}^{n} a_{ij} b_{ji}.$$

This completes the proof! For the expression $\sum_{i=1}^{m} \sum_{j=1}^{n} a_{ij} b_{ji}$ is obviously symmetrical with respect to A and B and is thus, equally well, trace D.

Example 3.3

Let A, B be abelian groups. There is a construction to yield an abelian group Tor(A, B) which proceeds as follows. Let

$$O \to R \to F \to A \to O \tag{3.1}$$

be a free presentation of A, that is, F is a free abelian group, R is a subgroup of F, and $A = F/R$. We consider the homomorphism $\kappa\colon R \otimes B \to F \otimes B$ induced by the inclusion of R in F. Then

$$\text{Tor}(A, B) = \ker \kappa. \tag{3.2}$$

Now there are two important theorems of homological algebra relating to this construction[†]:

THEOREM 3.1

Tor (A, B) *is independent, up to isomorphism, of the choice of presentation* (3.1).

THEOREM 3.2

Tor (A, B) *is balanced; that is, if*

$$O \to S \to G \to B \to O \tag{3.3}$$

is a free presentation of B, *and if* $\bar{\kappa}\colon A \otimes S \to A \otimes G$ *is induced by the inclusion of* S *in* G, *then*

$$\ker \kappa \cong \ker \bar{\kappa}. \tag{3.4}$$

Of course, Theorem 3.1 is a consequence of Theorem 3.2, although it is not customary for texts on homological algebra to establish Theorem 3.1 in this way (see [10]).

Now a very attractive proof of Theorem 3.2, essentially due to J. Lambek, proceeds by showing that

$$\ker \kappa \cong (F \otimes S + R \otimes G)/(R \otimes S), \tag{3.5}$$

where all groups on the right are regarded as subgroups of $F \otimes G$. Then (3.5) completes the proof! For the expression on the right of (3.5) is obviously symmetrical with respect to A and B and is thus equally well isomorphic to $\ker \bar{\kappa}$.

Example 3.4

A celebrated—and surprising—theorem due to Smale asserts that the 2-sphere may be turned inside out smoothly through immersions in $\mathbf{R}^3$ (this is in marked contrast to the circle which cannot be turned inside out smoothly in $\mathbf{R}^2$). A French mathematician, Bernard Morin,

[†]Of course, these theorems, and their proofs, generalize immediately to modules over unitary rings. Recall that $R \otimes B$ is the *tensor product* of the abelian groups R, B; this generalizes to the tensor product of a right-module and a left-module.

was the first to give an explicit description of a suitable isotopy; and it was a great triumph that Nelson Max was able to make a film[14] showing Morin's process. At the initial stage the inside and the outside of the sphere are colored, say, red and blue; and at a subsequent stage of the process, the red and blue parts are seen to be playing symmetrical roles. This completes the proof! For we simply have to interchange the roles of red and blue and reverse the steps in the Morin list of instructions up to that stage to complete the process of turning the sphere inside out.

The principle adduced in these four examples may be described as follows.

Symmetry Principle

Let Σ be a set and let E be an equivalence relation on Σ. Suppose that P, Q are elements of Σ each depending on parameters A, B. We say that Q is E-symmetric if $Q(A, B) \overset{E}{\sim} Q(B, A)$. Then if $P \overset{E}{\sim} Q$, P is also E-symmetric.

Of course, the principle is *useful* only when Q is visibly E-symmetric while P is not. In practice—thus, in all our examples—we apply the principle to enable us to terminate a mathematical argument at what appears to be the halfway stage. Thus we might describe this proof-strategy as the *Halfway Principle*! An application of the Halfway Principle to the study of patterns in Pascal's Triangle is to be found in [9].

4. SYMMETRY IN THE FORMULATION OF MATHEMATICAL CONCEPTS

In this section we discuss the importance of *symmetrical definitions of symmetrical concepts* in mathematics. In fact, the concepts discussed here will also be reflexive and transitive, but this feature will not be pertinent to our main point (though it is referred to in our closing remark).

First let X, Y be sets and let $f: X \to Y$ be a function. It is traditional to pick out a particularly important class of functions f, the *one-one correspondences*, and to define the class by the property that f is *one-one* and maps X *onto* Y. It is then proved that, if f is a one-one correspondence, it is *invertible*, that is, there exists a function $g: Y \to X$ such that $gf = 1_X$, the identity on X and $fg = 1_Y$, the identity on Y. We claim that this approach is highly misleading. We would advocate proceeding as follows.

Definition 4.1

Let $f: X \to Y$ be a function from the set X to the set Y. Then f is *invertible* if there exists a function $g: Y \to X$ such that $gf = 1_X, fg = 1_Y$.

Notice that (a) g is uniquely determined by f, so we may write $g = f^{-1}$; (b) g is invertible and $g^{-1} = f$. Thus the concept that there exists an invertible function from X to Y is a *symmetric* relation in the category of sets.

At this point we may remark that there is nothing special here about the category of sets. We may replace this category by any other category. For example, we may consider the category of groups (and homomorphisms) or the category of topological spaces (and continuous maps). Thus we generalize Definition 4.1 as follows.

Definition 4.1G

Let $f: X \to Y$ be a morphism from the object X to the object Y in the category C. Then f is *invertible* if there exists a morphism $g: Y \to X$ in C such that $gf = 1_X, fg = 1_Y$.

We now 'notice' exactly the same two things: (a) g is uniquely determined by f, so we may write $g = f^{-1}$, and (b) g is invertible and $g^{-1} = f$. Again we have a *symmetric* relation between X and Y.

Symmetric relations are obviously important, and we would like to be able to characterize this particular one. Thus the following theorems are significant.

Theorem 4.2

In the category of sets, a function is invertible if and only if it is a one-one correspondence.

Proof. Let $f: X \to Y$ be invertible with inverse g. Then, for any $y \in Y$, $y = fgy$, so f maps X onto Y. Now let $fx = fx'$. Then $x = gfx = gfx' = x'$, so f is one-one.

Conversely, let $f: X \to Y$ be a one-one correspondence. Then, for each $y \in Y$, there exists a unique $x \in X$ such that $fx = y$. Define $g: Y \rightarrow X$ by $gy = x$. It is then evident that $gf = 1_X$, $fg = 1_Y$.

At this point, we may have the feeling that the first part of the proof is more general than the second, in the sense that it should apply, suitably understood, in any category. The converse argument depended on our *defining* a function g, and this could be more special. Let us pursue this line of thought.

In any category C, we call a morphism $g: A \rightarrow B$ a *monomorphism* if $g\alpha = g\beta \Rightarrow \alpha = \beta$; and an *epimorphism* if $\gamma g = \delta g \Rightarrow \gamma = \delta$. It is then not difficult to prove that, in the categories of sets, groups or topological spaces, the monomorphisms are precisely the *injective* (one-one) maps and the epimorphisms are precisely the *surjective* maps (mapping the domain *onto* the range). It is now trivial to show

THEOREM 4.3

In any category an invertible morphism is both a monomorphism and an epimorphism.

Thus the crux of Theorem 4.2 is the assertion that, in the category of sets, the converse of the assertion of Theorem 4.3 holds. Let us look at the categories of groups and topological spaces.

THEOREM 4.4

In the category of groups, every homomorphism which is both a monomorphism and an epimorphism is invertible.

Proof. Let $f: X \rightarrow Y$ be a homomorphism of groups. Construct $g: Y \rightarrow X$ just as in Theorem 4.2. Then g is a homomorphism. For if $y_1, y_2 \in Y$ and $fx_1 = y_1, fx_2 = y_2$, then $f(x_1 x_2) = y_1 y_2$, so $g(y_1 y_2) = x_1 x_2 = g(y_1)g(y_2)$. Thus f is invertible.

However, the situation is very different in the category of topological spaces.

THEOREM 4.5

In the category of topological spaces, there are continuous functions f: $X \rightarrow Y$ *which are both monomorphisms and epimorphisms without being invertible.*

Proof. Let X be a set and let T_1, T_2 be two topologies on X with $T_1 \geq T_2$ but $T_1 \neq T_2$. For example we may give the set of real numbers $\mathbf{R}$ its usual topology (T_1) or the topology (T_2) in which only $\mathbf{R}$ and the empty set are open. Let X_i, $i = 1, 2$, be the set X furnished with the topology T_i, and let $f: X_1 \rightarrow X_2$ be the identify function. Then f is continuous, so f is a morphism of our category which is both a monomorphism and an epimorphism. However, f is not invertible, since the identity function $g: X_2 \rightarrow X_1$ is *not continuous*. This, then, is the point. In general, if $f: X \rightarrow Y$ is a continuous function which is also a one-one correspondence, then the inverse function $g: Y \rightarrow X$, constructed as in Theorem 4.2, may fail to be continuous.

Remark. A particularly vivid example is formed by the continuous function f which takes the half-open interval $0 \leq t < 1$ and wraps it round the unit circle of radius 1:

$$\overset{\xrightarrow{f}}{\underset{0 \qquad 1}{\bullet\!\!-\!\!\!-\!\!\!-\!\!\!-}} \xrightarrow{} \quad \bigcirc \!\! \uparrow f(0)$$

Then the inverse function g fails to be continuous at $f(0)$, so f is not a homeomorphism. A natural formula for f is $f(t) = e^{2\pi i t}$.

The moral of this section, we repeat, is that symmetric notions should, whenever possible, be given symmetric definitions. We should simply define one-one correspondences (of sets), isomorphisms (of groups), and homeomorphisms (of spaces) as the invertible morphisms of their respective categories. It is then important that we may *recognize* the invertible set-functions and group-homomorphisms as being the injective and surjective mappings in those categories; while, in topology, there are injective-surjective mappings which are not homeomorphisms. For we contend that the importance of a one-one correspondence of sets or a bijective homomorphism of groups lies precisely in the fact that they are invertible and thus set up a symmetric relation between their domain and range; and that they would not be significant ideas if this were not so. Likewise, the notion of a bijective continuous function from X to Y is not important precisely because it does not set up a symmetric relationship between X and Y. It is thus very

328 P. HILTON and J. PEDERSEN

unfortunate to find many texts *defining* an isomorphism between the groups G and H to be a one-one homomorphism of G onto H. Further, it is ludicrous—as is done in many texts—to *define* a homeomorphism from the space X to the space Y as a one-one continuous function from X onto Y with continuous inverse. For any continuous function from X to Y with continuous inverse, that is, any invertible continuous function from X to Y must necessarily be one-one and onto Y.

Of course the relation between objects X and Y of a given category that there exists an invertible morphism $f: X \rightarrow Y$ is not only a symmetric relation. It is also reflexive and transitive, so that it separates the objects of the category into equivalence classes. However, these other two properties of the relation are not the significant ones, since the existence of *any* morphism from X to Y defines a relation which is reflexive and transitive. It is precisely the *symmetry* of the relation which is guaranteed by the invertibility of the morphism in question.

REFERENCES

1. H. S. M. Coxeter, *Regular Polytopes*. Methuen (1948).
2. Branko Grünbaum, The Emperor's New Clothes: Full regalia, G-string, or nothing? *The Mathematical Intelligencer* **6**, 47–53 (1984).
3. I. N. Herstein, *Topics in Algebra*. Blaisdell (1964).
4. P. Hilton, *General Cohomology Theory and K-theory*. Cambridge University Press (1971).
5. P. Hilton and J. Pedersen, Approximating any regular polygon by folding paper: An interplay of geometry, analysis and number theory. *Math. Magazine* **56**, 141–155 (1983).
6. P. Hilton and J. Pedersen, Regular polygons, star polygons and number theory. *Coxeter Festschrift*. Math. Sem. Giessen 164, 217–244 (1984).
7. P. Hilton and J. Pedersen, Folding regular star polygons and number theory. *The Mathematical Intelligencer* **7**, 15–26 (1985).
8. P. Hilton and J. Pedersen, On certain algorithms in the practice of geometry and the theory of numbers, *Publicacions*, Sec. Mat., U.A.B. (to appear).
9. P. Hilton and J. Pedersen, On a pattern in Pascal's Triangle (to appear).
10. P. Hilton and U. Stammbach, *A Course in Homological Algebra*. Graduate Texts in Mathematics, Springer (1971).
11. D. Husemoller, *Fibre Bundles*. McGraw-Hill (1966).
12. B. W. Jones, *An Introduction to Modern Algrbra*. Macmillan (1975).
13. *Mathematical Intelligencer*, **6**, No. 3 (1984), front cover.
14. Nelson Max, "Turning a sphere inside out" (23 min color), International Film Bureau Inc., 332 South Michigan Avenue, Chicago, Illinois 60604.
15. Hermann Weyl, *Symmetry*. Princeton University Press (1952).

Comp. & Maths. with Appls. Vol. 12B, Nos. 1/2, pp. 329–361, 1986
Printed in Great Britain.

0886–9561/86 $3.00 + .00

MOIRÉS

Hans Giger
Bueglenstrasse 67, 3006 Berne, Switzerland

Abstract—Moiré phenomenon with its applications is presented and its global geometric properties are analysed and established.

1. INTRODUCTION

Moiré originally was a fabric design with a variable play of lustre. The name is derived from the French word for "watered". The Moiré-technique, discovered and developed in China, was introduced into France in 1754 by the English manufacturer Badger (or Badjer).

Natural or true Moiré is produced by two fabrics which are pressed one onto the other with their ribbed sides or by one fabric which on top of its ribbed side is ribbed by a barrel. False Moiré is a design impressed on a smooth fabric.

One of the first to investigate the Moiré phenomenon was Lord Raleigh (1845–1919), the wellknown English physicist. The Moiré phenomenon is based on purely geometric principles and its scientific applications are therefore rather rare. Moirés are used as analog computers, as models of physical patterns and they are applied to optics and to the investigation of perception.

The sense of aesthetic delight derives from the fact, that the Moiré phenomenon produces order out of unobserved order or even order out of chaos. This property also seems to be its mathematical content because, to some extent, the whole object of mathematics is to create order where previously chaos seemed to reign, to extract structure and invariance out of the midst of disarray and turmoil, or in one word, to establish symmetry.

Furthermore this property of the Moiré phenomenon shows an amazing relationship to our senses: a Moiré is an undifferential, global Gestalt phenomenon which to some extent depends on the corresponding properties of the eye. In this article, the Moiré phenomenon with its applications is presented and its global geometric properties are analysed and established.

2. THE MATHEMATICAL MOIRÉ

Moiré patterns occur quite often in daily life. If we observe the folds of a slightly moving nylon curtain with small mesh we can see these patterns moving about.

Figure 1 is the photograph of a Moiré-silk. The special treatment gives the fabric a design which resembles the surface of water, hence the name.

Figure 2 shows two different ways of producing Moiré-patterns. In Fig. 2(a) the two black and white components of textures are geometrically superposed by forming the union of the two pointsets. In Fig. 2(b) the components of Fig. 2(a) are overlaid in the sense of geometric intersection, i.e. the white elements may be interpreted as apertures in the black areas of the two components. Obviously, the components in Fig. 2(a) and 2(b) are themselves Moirés in the sense of the described superposition. Such patterns are used as models of random point sets (colloids with gel structure, liquids, mineralogic deposits) and their mathematical properties are known to a great extent.

Often the texture components are systems of black lines or opaque screens with holes arranged along the u- and v-curves of a general mathematical net of coordinates. In each point (u, v) of the net, an u- and a v-curve intersect. Any algebraic relation between u and v, such as $u + v = k$ or $u - v = l$, where k and l are integers, defines a curve of the Moiré.

In Fig. 3 the points with a constant sum or difference of their coordinates are linked by a pointrow. If k or l runs over some subset of the integers, a set of Moiré curves is defined. They produce the Moiré-texture of the corresponding algebraic relation. The freedom in the choice of the algebraic relation brings discretion into the concept of Moiré-texture but makes it convenient

Fig. 1.

for mathematical investigations[1]. A more synthetic interpretation of Moiré-textures will be given in Sec. 4.4.

3. PSYCHOLOGICAL MOIRÉS

The arbitrariness in the mathematical concept of Moiré given in the last Section becomes obvious, if the laws of seeing are taken into account. The human eye, on its lowest levels of image processing, is ordering points according to their neighbourhood. In each quadrangle of Fig. 3 the smaller diagonals seem to be emphasized as shown in Fig. 4 and therefore the discussed algebraic relation $u - v = l$ stands informationally out.

Generally human perception seems to extract order out of static or dynamic disorder or impresses order on chaos by investigating or creating correlations between the percepted ele-

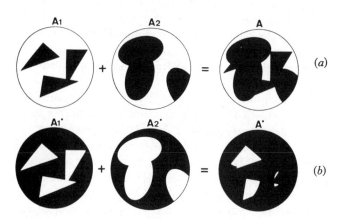

Fig. 2.

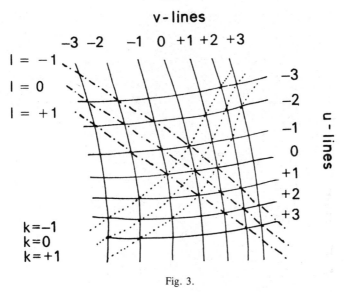

v-lines

Fig. 3.

ments. In a general sense, these faculties of the eye seem to correspond to the capacities of the ear, which is capable of extracting overtones from a tone.

Figure 5 shows the superposition of two identical random-dot fields after Klein[2]. The copy is first covered by the original field and the latter is then turned by a small angle around a fixed point relative to the copy. The eye sees a Moiré-texture of concentric circles. Similar effects of ordering after MacKay[2] occur in dynamic processes. If a white random noise produced on a television screen is looked at through a fine ruling, a radial order is observed.

Observing Moiré-textures we often see physiological colors, as they appear on a rotating Benham disk (Fig. 6). The ever moving Moiré-fringes, transformations caused by the movement of the object or the observer, can be answered by the eye with the complement of the transformation. If we fix the moving surface of a river for a certain time and then suddenly look at the bank we see it moving river upwards. This effect can be the cause of uncertainty or even dizziness.

4. SCIENTIFIC APPLICATIONS

Moiré textures have different scientific applications, the most important of which will be described in this section.

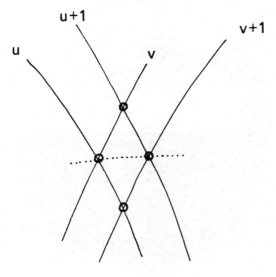

Fig. 4.

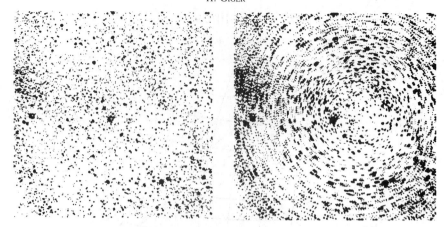

Fig. 5.

4.1 *Lensless enlargement*

Printers often have problems with Moiré-patterns formed by two superposed screens which are not correctly combined. Figure 7(a) shows such a superposition. In the resulting pattern the constituting identical figures (e.g. squares, hexagons) of the screens emerges in an enlargement depending on the relative position of the superposed components. In Fig. 7(b) the superposition of the two rulings results in the same effect. It can easily be observed if two identical small tooth combs are superposed. Lord Raleigh proposed this effect in 1874 for the control of optical grids. The time analogy of the enlargement effect can be produced by a stroboscope. In film the speed of wheels is slowed down or even turned back by this effect. The enlargement effect of Moirés was used by V. F. Holland in about 1960 to make the arrangement of the particles in a crystal visible. W. H. Bragg and W. L. Bragg used this effect in 1915 by superposing several X-ray diffraction pictures to reconstruct the molecule hexamethylbenzol.

4.2. *Correlations in random textures*

If two black and white textures are superposed on each other or on a ruling, relations in the texture can be made visible by the Moiré-effect. Figure 8 shows an anisotropic texture. If a ruling is superposed in different directions the restorder in the texture can be demonstrated.

The Moiré effect can be used to demonstrate the hidden periodicity in the following texture: If a handfull of pebbles is arranged in a row so that two successive pebbles are in contact, this

Fig. 6.

a

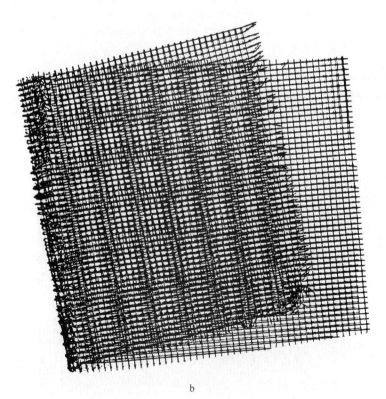

b

Fig. 7.

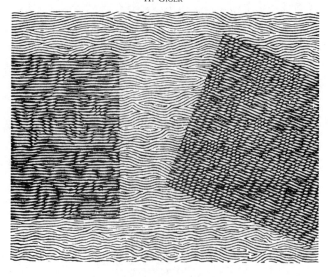

Fig. 8.

chain has an inner periodicity the period of which equals the mean length of the pebbles in the direction of the row. In Fig. 9 this pattern of pebbles is represented by vertical lines with varying spacings. The variation of the spacing is a certain percentage, say 20%, of the mean distance between the lines. The obtained ruling is now inclined at a certain angle so that the lines of the first and second ruling intersect on a horizontal line. In the resulting Moiré-pattern the hidden periodicity of the random texture appears in the horizontal Moiré fringes which get more and more disturbed if the distance from the intersection line grows[3].

4.3 *Pattern simulation*

Moirés can be used to simulate different patterns. This can be shown by two examples. Figure 10 shows the correct representation of the electric field in a plane through an electron and a proton. In a similar way the streamlines in a plane of a streaming liquid can be simulated. Figure 11 is the simulation of the interfering waterwaves occuring when the surface of a pond is excited simultaneously at two different points.

4.4 *Representation of planes curves and surfaces*

With a plotter governed by a computer, a plane curve can approximately be represented as a point-row, the points of which are often linked by straight segments, or as an envelope to

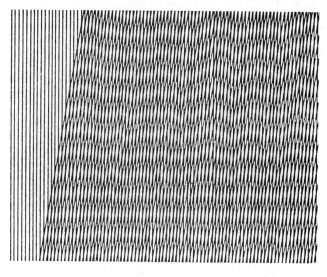

Fig. 9.

Fig. 10.

Fig. 11.

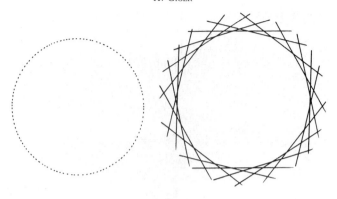

Fig. 12.

the curve (Fig. 12). Less known is the possibility of representing the graph of a function of one or two variables by a Moiré-pattern. A possibly new method will be described in this section. Because this construction may be of broader interest, technical details are omitted and will be given in Appendix 1.

We restrict the concept of Moiré to the superposition of two linefields L_1 and L_2. The resulting Moiré $L_1 \cup L_2$ can be understood on the basis of two theorems which are the consequence of interpreting the middle lines of the Moiré-fringes as a selection of the contours of the graph of a function of two variables $z = z(x, y)$ in exactly the same way as contours of a geographical map convey information about the height of the land. Figure 13 gives the contours of a landscape. The contours in the surrounding of a point are the straight evenly spaced contours of the tangent plane of the land. Figure 14 shows the contours of a selection of functions $z = z(x, y)$.

It is also possible that the contours of $z = z(x, y)$ are radial fields of lines, Fig. 15(a). Two examples are given in Fig. 15(b, c). One of the graphs can be described by $z = \varphi$, the other by $z = \ln [\mathrm{abs}(\tan (\varphi/2))]$, if the parameter φ is the angle of the polar coordinates in the x-y-plane.

The question arises: which fields of lines L_1, L_2 have to be chosen to obtain a certain contour Moiré $L_1 \cup L_2$, the fringes of which are the contours of the graph of a given function $z = z(x, y)$. Pairs of linefields L_1 and L_2 which give rise to the same Moiré will be termed *moiré-equivalent*. The answer to the question is given by the following construction that can be generalized (Appendix 1).

Figure 16 shows the graphs of two different functions $z_1 = z_1(x, y)$ and $z_2 = z_2(x, y)$. If these graphs are cut by parallel equidistant planes the projection of the lines of intersection onto the x-y-plane produces linefields L_1 and L_2. In the limiting case H the linefields are the projections

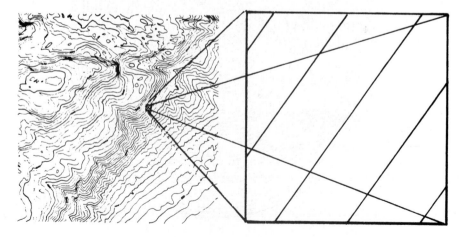

Fig. 13.

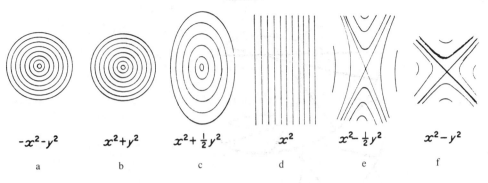

$-x^2-y^2$	x^2+y^2	$x^2+\frac{1}{2}y^2$	x^2	$x^2-\frac{1}{2}y^2$	x^2-y^2
a	b	c	d	e	f

Fig. 14.

of the contours of the graphs of z_1 and z_2. In the other limiting case G the intersection lines on the graphs project in the same ruling on the x-y-plane. The case G is to be excluded.

The first statement is given by

THEOREM 1

The Moiré $L_1 \cup L_2$ does not depend on the direction and inclination of the cutting planes. This fact raises the question in what way the Moiré $L_1 \cup L_2$ depends on z_1 and z_2. In the limiting case H the answer can be given: If z_1 is a horizontal plane, z_1 = constant, L_1 is void. Therefore $L_1 \cup L_2 = L_2$ is the contourmap of the graph of z_2 with equidistance determined by the horizontal cutting planes.

With Theorem 1 we obtain

COROLLARY

If z_1 is a horizontal plane the Moiré $L_1 \cup L_2$ is the contourmap of the graph of z_2 for each system of cutting planes.

In the general case the preceding question is answered by

THEOREM 2

The Moiré $L_1 \cup L_2$ is the contourmap of $z = z_2 - z_1$.

The well-known indefiniteness of a geographical map, where the contours of a hill can be interpreted as the contours of a pit and vice versa, must also be taken into consideration with Moirés; i.e. z can also be described by $z = z_1 - z_2$. The graph of the difference function z is represented by Fig. 17. The vertical rods between z_1 and z_2 must be lowered till the deeper ends meet the x-y-plane.

A consequence of Theorem 2 indicated by Fig. 18 is of special interest. The Moiré works as analog computer and differential analyzer. Beside the superposition of two different linefields L_1, L_2, it is also possible to overlay two identical fields $L = L_1 = L_2$ of $z = z_1 = z_2$ in translating or rotating one of the fields relative to the other or in superposing the stretched field λL with L. Translation is denoted by L^τ, rotation by L^φ.

The effect of selfsuperposition is described by

THEOREM 3

The Moiré $L_1 \cup L_2$ with $L = L_1 = L_2$ in the case of

- a small translation is the contourmap of the partial derivative of z in the given direction of the translation;
- a small rotation is the contourmap of the rotational derivative of z;
- small stretching is the contourmap of the radial derivative of z.

The word "small" in the theorem is used in the sense of calculus. A strict formulation has to take into account that in order to get a contourmap with fixed equidistance the distance between the cutting planes has to be taken infinitesimal if the transposition nears identity. Demonstrations of the three theorems are given by the following figures.

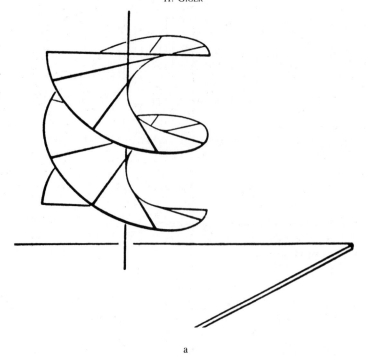

a

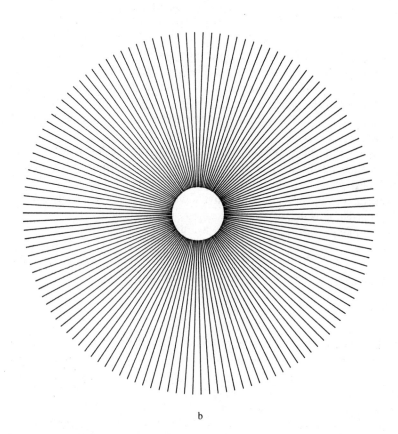

b

Fig. 15(a–c).

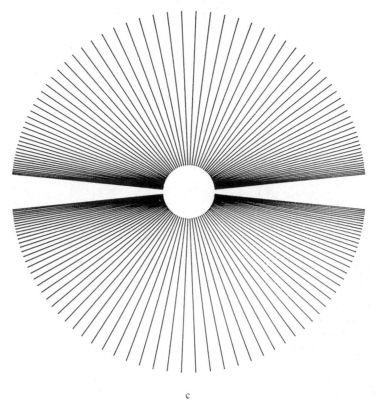

c

Fig. 15. (*Con't*)

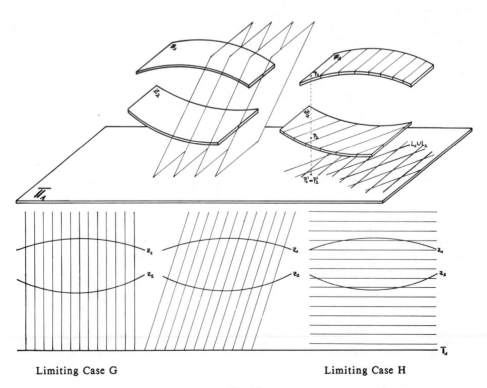

Limiting Case G   **Limiting Case H**

Fig. 16.

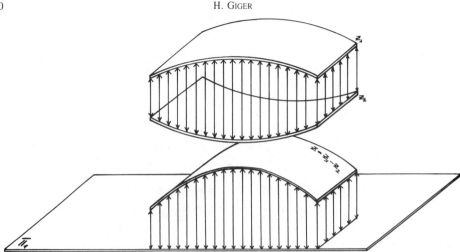

Fig. 17.

Figure 19(a) explains the formula used in the Appendix 1 that relates the spacings a and b of the two superposed rulings and the angle φ between a straight line of the first and one of the second ruling to the distance d between adjacent Moiré-fringes. Figures 19(a) and 19(c) demonstrate the two possibilities of superposing two rulings that, according to Fig. 16, can be interpreted as the intersection lines L_1 and L_2 when two planes given by $z_i = a_i x + b_i y + c_i$, ($i = 1, 2$), are cut by a system of the equidistant planes. The resulting Moiré is according to Theorem 2 the contourmap of the plane $z = z_1 - z_2$.

Figure 20(a) gives the contour Moiré $L_1 \cup L_2$ of a hill or a pit. L_1 is constructed according to Fig. 16 as the linefield of $z_1 = \hat{x}^2 + \hat{y}^2 + ax + by + c$; L_2 as that of $z_2 = ax + by + c$. The resulting Moiré is the contourmap of the paraboloid $z = z_1 - z_2 = \hat{x}^2 + \hat{y}^2$. Figure 20(b) shows the Moiré of the selftransposition $L \cup L^\tau$, where L is the contourmap of $z = \hat{x}^2 + \hat{y}^2$. The resulting contourmap is, according to Theorem 3, the map of the plane $Dz = 2\alpha x + 2\beta y$, i.e. the directional derivative of z in the direction $\tau = (\alpha, \beta)$, $\hat{\alpha}^2 + \hat{\beta}^2 = 1$. Figure 20(c) gives the superposition of the contourmap $z_1 = \hat{x}^2 + \hat{y}^2$ on a ruling, the contourmap of a plane $z_2 = ax + by + c$. The resulting Moiré seems to contradict Theorem 2. The periodic appearances of systems of concentric Moiré-circles can be explained as a consequence of indefiniteness of the ruling as a map of contours of a plane, because according to the construction of Fig. 16 a set of quantisized planes give the same ruling.

This effect raises the question which linefields L_1 and L_2 should be chosen to get the best

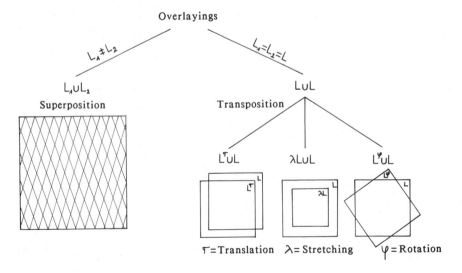

Fig. 18.

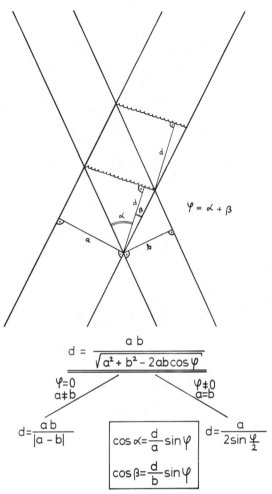

$$\varphi = \alpha + \beta$$

$$d = \frac{a\,b}{\sqrt{a^2 + b^2 - 2ab\cos\varphi}}$$

$\varphi = 0$
$a \neq b$

$\varphi \neq 0$
$a = b$

$$d = \frac{a\,b}{|a - b|}$$

$$\cos\alpha = \frac{d}{a}\sin\varphi$$

$$\cos\beta = \frac{d}{b}\sin\varphi$$

$$d = \frac{a}{2\sin\frac{\varphi}{2}}$$

a

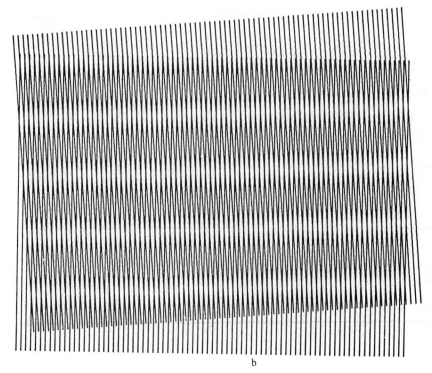

$a = b$
$\varphi \neq 0$

b

Fig. 19(a–c).

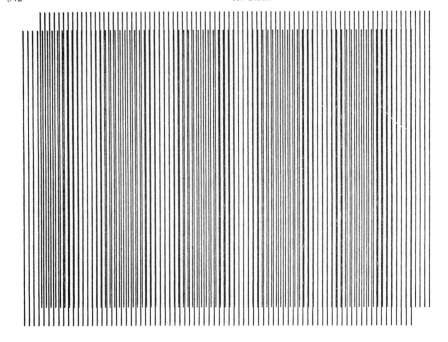

$$a \neq b$$
$$\varphi = 0$$

c

Fig. 19. (Con't)

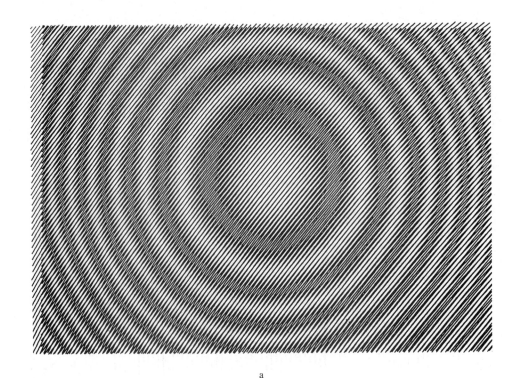

a

Fig. 20(a–c).

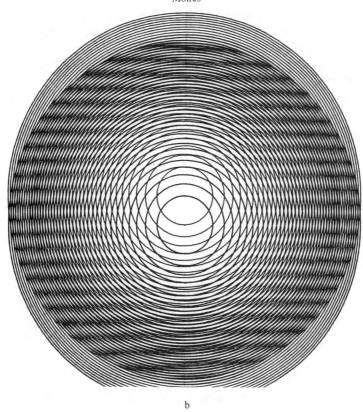

b

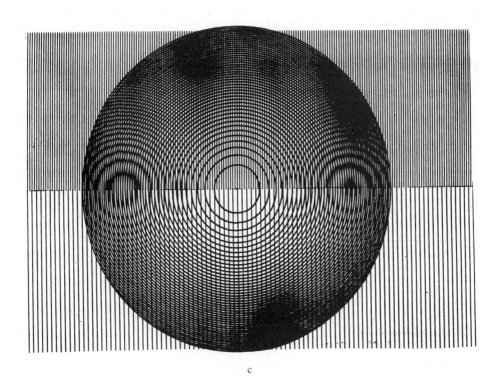

c

Fig. 20. (*Con't*)

representation of a planned contour Moiré $L_1 \cup L_2$. The following answer is substantiated in Appendix 1: The linefields L_1 and L_2 must be constructed such that locally the lines of L_1 and L_2 are nearly parallel to the contours of the desired Moiré.

Figure 21(a) is the contour Moiré of the saddle $z = \hat{x}^2 - \hat{y}^2$ constructed like that of Fig. 20. Figure 21(b) shows the Moiré of the transposition $L \cup L^\tau$, where L is the contourmap of $z = \hat{x}^2 - \hat{y}^2$. As in Fig. 20(b), we obtain the contourmap of $Dz = 2\alpha x - 2\beta y$, the directional derivative of z.

Figure 22(a) is the contour Moiré $L_1 \cup L_2$ of $z = \hat{x}^3 + \hat{y}^3$ constructed like Figure 20(a). Figures 22(b) and 22(c) show two different transportations $L_1 \cup L_1^\tau$, where L_1 is the linefield of $z = \hat{x}^3 + \hat{y}^3 + ax + by + c$. The contour Moiré is the map of the directional derivative $Dz_1 = 3\alpha\hat{x}^2 + 3\beta\hat{y}^2 + k$ and corresponds therefore for appropriate translations $\tau = (\alpha, \beta)$, $\alpha^2 + \beta^2 = 1$, to the maps of Fig. 20(a) and Fig. 21(a). Figures 22(d) and 22(e) give two translations $L \cup L^\tau$, where L is the contourmap of $z = \hat{x}^3 + \hat{y}^3$. The resulting contour Moirés are those of Fig. 22(b) and 22(c). Figure 22(f) is the rotational transposition $L \cup L^\varphi$, where L is the contourmap of $z = \hat{x}^3 + \hat{y}^3$. The origin of the system of coordinates is the fixpoint of the rotation. The contour Moiré is the map of the monkey-saddle $Dz = -3\hat{x}^2 y + 3x\hat{y}^2$.

Figure 23(a) gives the correct representation of a function $y = f(x)$, here $f(x) = \exp(-\hat{x}^2/2) + c$. L_1 is the linefield of $z_1 = f(x) - xy$, L_2 those of $z_2 = -xy + y$. (For the reason of this setting see Fig. 23(b). The resulting contour Moiré $L_1 \cup L_2$ is the map of $z = z_1 - z_2 = f(x) - y$ and gives the desired representation. Figure 23(b) gives the transposition $L_1 \cup L_1^\tau$, where L_1 is the linefield of Fig. 23(a) and $\tau = (1, 0)$, a translation in the direction of the x-axis. The resulting Moiré corresponds to $Dz_1 = f'(x) - y$, the derivative of the given function. In the following figures the radial linefields L_1 and L_2 of Fig. 15 are transposed.

Figure 24(a) gives the transposition $L_1 \cup L_1^\tau$, which for $\tau = (0, 1)$ is the Moiré $Dz_1 = \cos \varphi/\rho$ of $z_1 = \varphi$, a system of circles with a common tangent at the origin of coordinates. ρ and φ are the polar coordinates. Figure 24(b) shows the transposition $L_2 \cup L_2^\tau$. For the chosen translation the contour Moiré is a system of concentric circles with centres at the origin of the system of coordinates.

The next figures are concerned with the functions $z_1 = \rho \cdot \varphi$ and $z_2 = \rho \cdot \exp(\varphi)$ given by polar coordinates ρ, φ [Figure 25(a)]. The contour-maps L_1 and L_2 of these functions are shown in translative, rotative and stretched trans- and superposition according to Fig. 18. Figures 25(b) and 25(c) show $L_1 \cup L_1^\tau$ and $L_2 \cup L_2^\tau$, which as directional derivatives give radial Moiré-fringes.

Figures 26(a) and 26(b) give $L_1 \cup L_1^\varphi$ and $L_2 \cup L_2^\varphi$ the first of which as rotational derivative is a contour Moiré of concentric circles, whereas the second reproduces as contour Moiré the linefield L_2.

Figure 27 demonstrates the superposition $L_1 \cup \lambda L_1$, where λL_1 denotes the stretched linefield L_1, the centre of the stretching lies at the origin. The resulting contour Moiré reproduces the linefield L_1. The same effect is observed with L_2.

Figure 28 represents the two linefields L_1 and L_2 of Fig. 21(a), the superposition of which is the contour Moiré of the saddle $z = \hat{x}^2 - \hat{y}^2$. If this figure is observed with the red–green spectacles used to view stereoscopic pictures, the saddle can be seen as three-dimensional surface. If red and green are interchanged the represented function changes its sign. The stereoscopic effect can also be observed when the two linefields are not combined in the same picture and therefore are not expoundable. This nonrandom pictures like the well-known random dot pictures of Bela Jules can only be interpreted in stereological presentation.

5. SPATIAL MOIRÉS

A spatial Moiré results if two sets of points or lines each of which is located in the three-dimensional space on the graph of one of the functions or relations $z_1 = z_1(x, y)$, $z_2 = z_2(x, y)$ are looked at in diffuse illumination. The objects causing these patterns will also be termed *spatial Moirés*.

If understanding is interpreted in the constructivist sense, that is, a phenomenon is explained if it can be classified, planned and constructed, there is not much insight in the various patterns

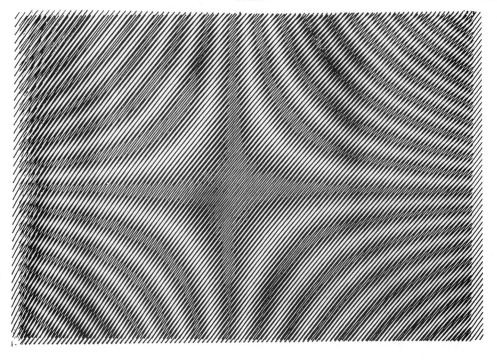

a

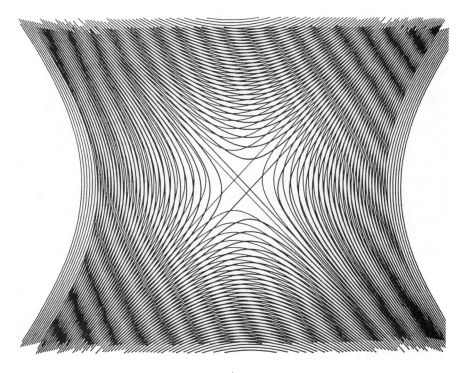

b

Fig. 21(a–b).

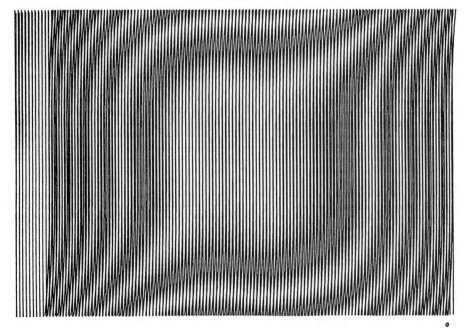

a

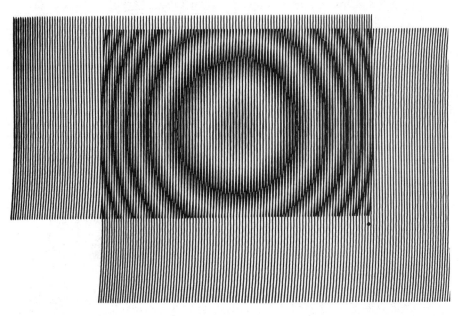

b

Fig. 22(a–f).

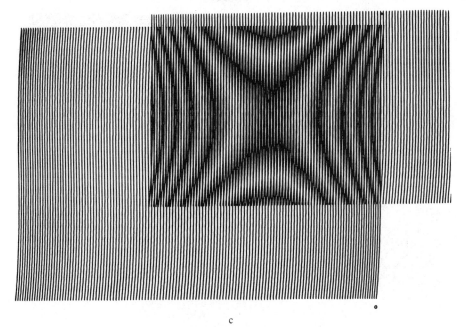

c

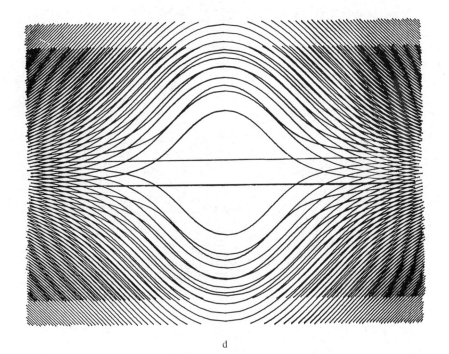

d

Fig. 22. (*Con't*)

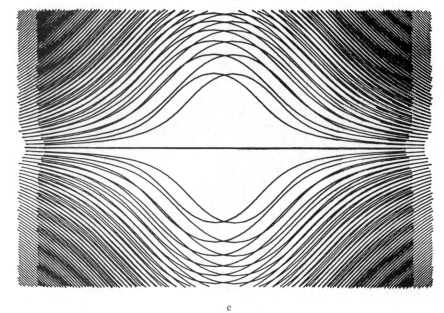

e

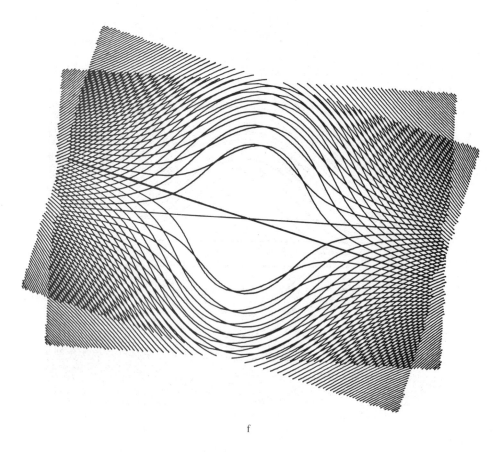

f

Fig. 22. (*Con't*)

a

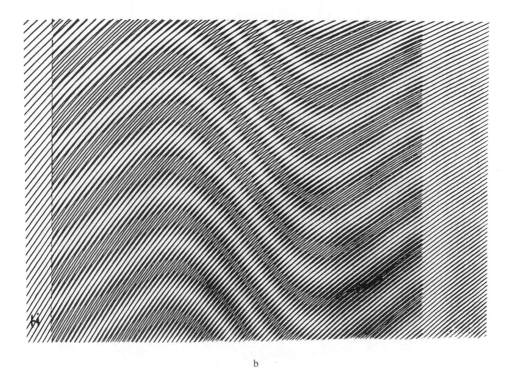

b

Fig. 23.

a

b

Fig. 24.

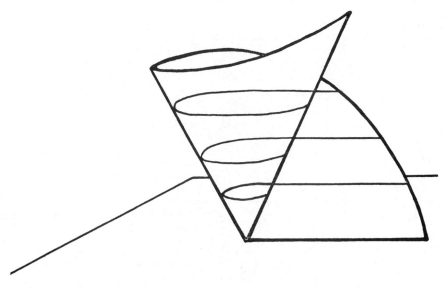

a

b

Fig. 25(a–c).

c

Fig. 25. (*Con't*)

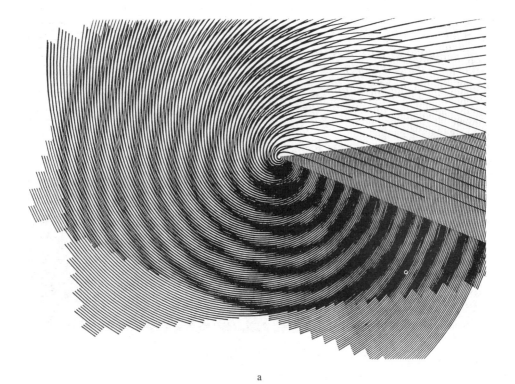

a

Fig. 26(a–b).

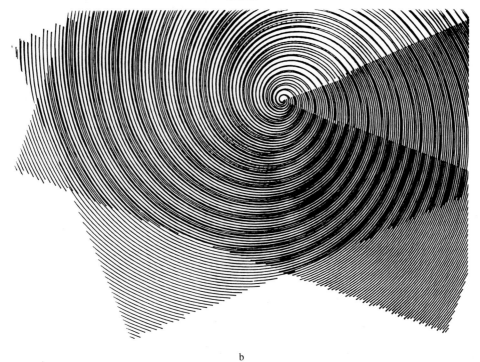

b

Fig. 26. (*Con't*)

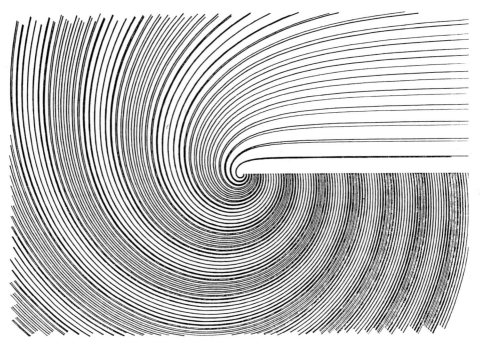

Fig. 27.

Fig. 28.

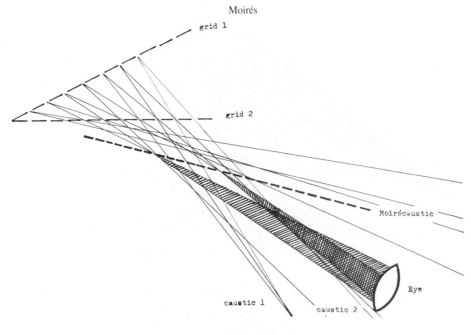

Fig. 29.

that can be observed on real Moiré objects. Two interpretations of spatial Moiré are proposed in this section.

The first interpretation explains spatial Moiré locally as a plane Moiré. If the two systems of discrete lines, the u- and v-lines, on the graphs of two functions are projected from a center, the eye of the observer, onto a reference plane, that may be the background of the eye, a spatial Moiré results. The Figs. 30(a), 31(a) show spatial Moirés constructed according to this interpretation. The designed axes of the system of coordinates indicate the direction of the line of

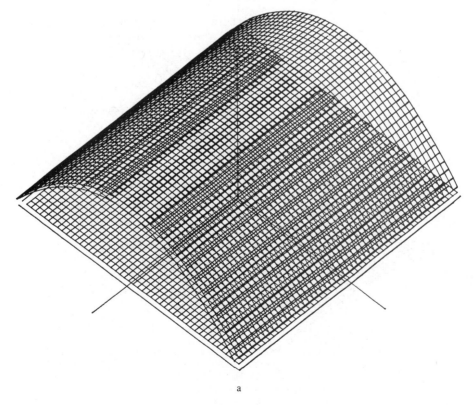

a

Fig. 30(a–c).

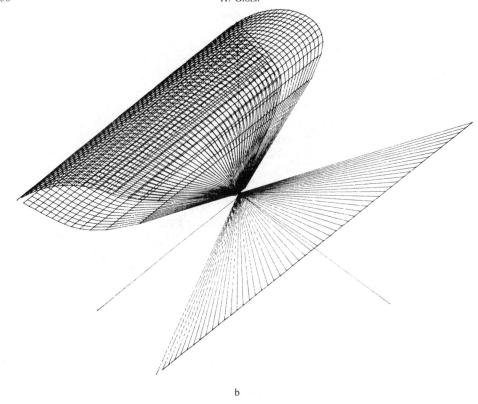

b

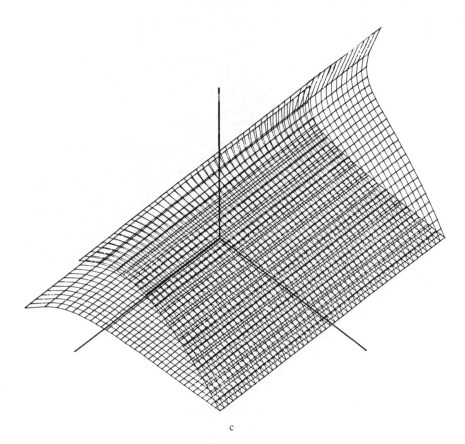

c

Fig. 30. *(Con't)*

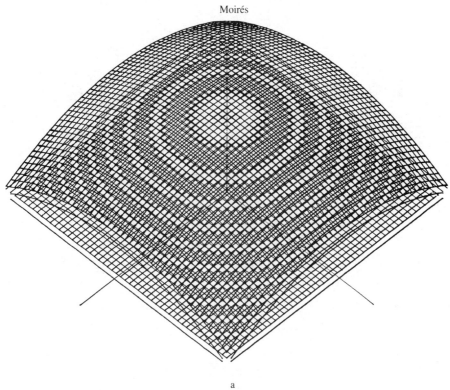

a

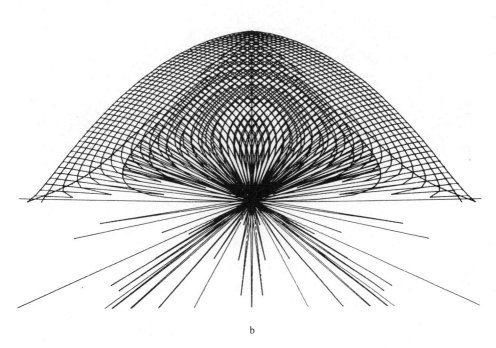

b

Fig. 31(a–d).

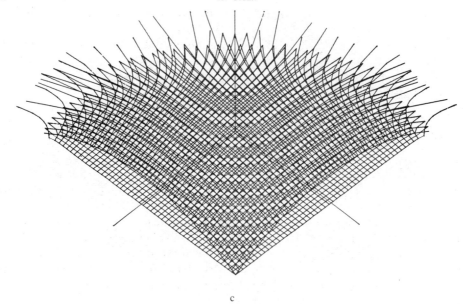

c

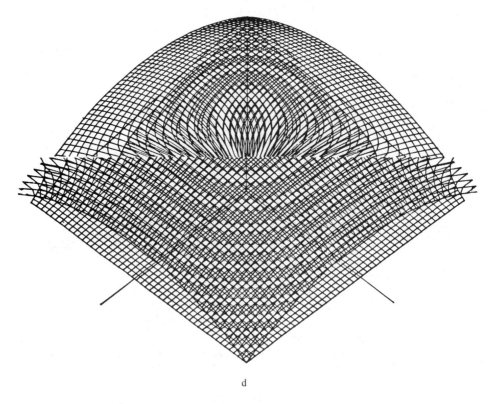

d

Fig. 31. (Con't)

sight. It must be mentioned that little changes in the positions of the causing screens have dramatic effects in the Moiré.

The second interpretation was given by Theocaris *et al.*[4] and will be described in Appendix 2 in a modified way which is elementary, computer-adapted and does not use line geometry[5]. This interpretation answers the question where the spatial Moiré is located.

Figure 29 explains the construction of the locus of spatial Moiré for the most simple situation. Corresponding lightrays through the holes of the plane screens or grids shown in profile define the different caustics. The eye of the observer sees a lightpoint in the real or virtual point of intersection of lightrays entering the eye. In the given situation the different Moiré-lines which are perpendicular to the plane of the Fig. 29 seen by the observer too lie on a plane, the Moiré-caustic.

We term the spatial Moiré *real* if the Moiré-caustic is in front of, *virtual* if it is behind the causing screens relative to the observer.

If the nets of lines on z_1 and z_2 of Figs. 30(a) and 31(a) are mapped onto the Moiré-caustic we obtain Fig. 30(b, c) and Fig. 31(b, c). Figure 31 (b) gives real, Fig. 31(c) virtual Moiré-caustic. It should be noted that the two Moirés of Figure 31(b) and 31(c) must be combined additively, i.e. white and black gives white. This combination is shown for Figs. 31(b) and 31(c) in 31(d). The resulting pattern differs greatly from Fig. 31(a) and demonstrates the difficulties in interpreting spatial Moirés.

REFERENCES

1. G. Oster and Y. Nishijima, Moiré patterns. *Scientific American* **208** (1963).
2. J. Walker, Visual illusions in random-dot patterns and television snow. *Scientific American* **242** (1980).
3. H. Giger, Moirétexturen, in *Catalog of the Kunstmuseum of Berne*, Werner Witschi, Moirés, 5.12.–7.2. 37–44 (1982).
4. P. S. Theocaris, A. P. Vafiadakis, and C. Ziakopoulos, Theory of spatial Moiré fringes, *J. Optical Soc. Amer.* **8**, 1092–1099 (1968).
5. W. Blasdike, Differential Geometrie I, Springer, 261–304, (1929).

APPENDIX 1

THEOREM   The Moiré $L_1 \cup L_2$ of two superposed systems L_1 and L_2 of line-fields, each given as the intersection lines of a system of parallel planes with equidistant spacings (the planes not normal to the x-y-plane of the system of coordinates) with the graph of the functions $z_1 = z_1(x, y)$ and $z_2 = z_2(x, y)$ is the contour Moiré of $z = |z_2 - z_1|$. The contours are defined as the middlelines of the Moiré-fringes.

We prove this theorem using the formula of Fig. 19(a) and refering to the symbols explained in this figure.

First suppose that z_1, z_2 are different functions, the graphs of which are planes with normals $N_i = (a_i, b_i, -1)$ and the equations $z_i = a_i x + b_i y + c_i$, $(i = 1, 2)$. Each of these planes are cut by the parallel and equidistant planes with the normal $N = (\alpha, \beta, \gamma)$, $\alpha^2 + \beta^2 + \gamma^2 = 1$, $\gamma \neq 0$, whose distance from the origin of the system of coordinates is p, where p denotes an arithmetic series with positive difference Δp. The equation of this system of planes is therefore $E(p)$: $\alpha x + \beta y + \gamma z - p = 0$.

The direction of the intersection lines of $E(p)$ and z_i is given by

$$s_i = (-\gamma b_i - \beta, \gamma a_i + \alpha, \alpha b_i - \beta a_i), \quad i = 1, 2. \tag{A.1.1}$$

If these vectors are projected onto the x-y-plane we get the vectors $s_i' = (-\tau_i, \sigma_i)$ with

$$\sigma_i = \gamma a_i + \alpha, \quad \tau_i = \gamma b_i + \beta, \quad i = 1, 2. \tag{A.1.2}$$

The plane $z_i = a_i x + b_i y + c_i$ is cut by $E(p)$ in the straight line $g_i(p)$ whose projection $g_i'(p)$ onto the x-y-plane is given by

$$g_i'(p): [\sigma_i x + \tau_i y - (p - \gamma c_i)]\sqrt{\sigma_i^2 + \tau_i^2} = 0, \quad i = 1, 2. \tag{A.1.3}$$

The distance q_i of $g_i'(p)$ from the origin is therefore given by $q_i = (p - \gamma c_i)/|s_i'|$. Therefore, the spacing Δq_i of the ruling $g_i'(p)$ is

$$\Delta q_i = \Delta p/|s_i'|, \quad i = 1, 2. \tag{A.1.4}$$

The angle φ between the two rulings $q_i'(p)$ is determined by the vectors s_i', $(i = 1, 2)$. Without any restriction of generality we may suppose s_i' oriented such that $0 < \varphi < = \pi/2$. Therefore, $\cos \varphi = (s_1' \cdot s_2')/|s_1'|/|s_2'|$. With the formula of Fig. 20 we get

$$d = \Delta p/|s_2' - s_1'|. \tag{A.1.5}$$

The vectors s_i' are related to the vectors t_i spanning the parallelogram of Fig. 19(a). Because of $|t_1| = \Delta q_2/\sin \varphi$ and $|t_2| = \Delta q_1/\sin \varphi$ and $t_i = \kappa_i \cdot s_i'$, $(i = 1, 2)$, from $|t_1 \times t_2| = \kappa^2|s_1' \times s_2'|$ and $|s_1' \times s_2'| = |\gamma||s_1 \times s_2|$ we obtain $\kappa = \kappa_1 = \kappa_2 = \Delta p/|\gamma|/|s_1 \times s_2|$.

Therefore, with (A.1.4)

$$\mathbf{t}_i = \Delta p \mathbf{s}_i' / |\mathbf{s}_1' \times \mathbf{s}_2'|, \quad i = 1, 2. \tag{A.1.6}$$

According to (A.1.1) and (A.1.2) the vector $\mathbf{s}' = \mathbf{s}_2' - \mathbf{s}_1'$ is given by

$$\mathbf{s}' = \gamma(-b_2 + b_1, a_2 - a_1). \tag{A.1.7}$$

This vector is normal to the vector $\mathbf{M}' = (a_2 - a_1, b_2 - b_1)$ which is the projection of $\mathbf{M} = (a_2 - a_1, b_2 - b_1, -1)$ onto the x-y-plane. But $\mathbf{M}$ is the normal of $z = z_2 - z_1$.

If we set $\mathbf{M} = (a, b, -1)$ with $a = a_2 - a_1$, $b = b_2 - b_1$ for the angle ϵ of the inclination of z we get with (A.1.5) $\tan \epsilon = \sqrt{a^2 + b^2} = |\mathbf{s}'|/|\gamma| = \Delta p/d/|\gamma|$. Therefore, the equidistance H of the contour Moiré of z is

$$H = d \tan \epsilon = \Delta p/|\gamma|. \tag{A.1.8}$$

In the general case where z_1 and z_2 are smooth functions, for each point $P(x, y, 0)$ the tangent planes of z_1 and z_2 have to be cut by planes $E(p)$. In the vicinity of $P(x, y, 0)$, the Moiré of the corresponding line systems L_1 and L_2 is therefore the contourmap of $z = z_2 - z_1$. Because this result is valid for all points $P(x, y, 0)$ the theorem is proven.

It should be mentioned that the planes $E(p)$ can locally be chosen at our discretion if according to (A.1.8) H is taken an constant. If $\mathbf{N}$ is chosen so that it is linearly dependent on $\mathbf{N}_1$ and $\mathbf{N}_2$ the parallelogram of Fig. 19(a) spanned up by the vectors (A.1.6) degenerates. This means that the two rulings are locally parallel to the contours of $z = z_2 - z_1$. This contour Moiré may be called ideal.

APPENDIX 2

A spatial Moiré results in the eye of the observer as union or intersection of two systems of lines or lattices of points, each of which is located on one of the graphs of two functions $z_1 = z_1(x, y)$ and $z_2 = z_2(x, y)$.

Here we describe the spatial Moirés where $z_2 = 0$ and the corresponding lattice of points is given by the points with integer coordinates in the x-y-plane. Each point $A(x, y, 0)$ of this grid shall be mapped on the point $B(x, y, z(x, y))$ of the graph of the function $z_1 = z(x, y)$. The set of these points defines the second lattice. As a realisation we can think of two screens with holes. It is the purpose of the following consideration to show that a spatial Moiré can be interpreted as catastrophe in the sense of Catastrophe Theory.

First we consider the straight lines running through the two points $A(x + a, y + b, 0)$, $B(x, y, z(x, y))$, where a and b are fixed integers. If the line AB is cut by a plane $Z = $ constant, the position of the point of intersection $S(X, Y, Z)$ depends on x and y:

$$X = x + a - aZ/z(x, y) \tag{A.2.1}$$
$$Y = y + b - b Z/z(x, y).$$

The principle of geometric optics—postulating that the eye sees a lightpoint in the real or virtual point of intersection of light-rays entering the eye—suggests the answer to the question of where spatial Moiré is located.

If $A(x(t), y(t), 0)$ is any interpolating curve through the points of the grid in $z_2 = 0$ the point $S(X, Y, Z)$ of (A.2.1) runs through a corresponding line in the plane $Z = $ constant. For those t for which $X(t)$ or $Y(t)$ changes direction an intersection of the corresponding straight lines is realized. Therefore $dX/dt = 0$ and $dY/dt = 0$ are necessary conditions a straight line must fulfill so that in its vicinity such a particular lightray passing the two pointsystems can be found. By differentiating (A.2.1) we get the conditions

$$0 = \dot{x} + aZ/z^2 \cdot (z_x \dot{x} + z_y \dot{y}) \tag{A.2.2}$$
$$0 = \dot{y} + bZ/z^2 \cdot (z_x \dot{x} + z_y \dot{y}),$$

where z_x and z_y are the partial derivatives of z.

Now we suppose Z to be chosen in such a way that both of the equations (A.2.2) are fulfilled simultaneously. Eliminating dx/dt and dy/dt in (A.2.2) we get

$$Z/z = -z/(az_x + bz_y) \tag{A.2.3}$$

and therefore, if we write $S(U, V, W)$ for S in this particular position, by substituting (A.2.3) into (A.2.1) we obtain

$$U = x + a + az/(az_x + bz_y)$$
$$V = y + b + bz/(az_x + bz_y) \tag{A.2.4}$$
$$W = -z^2/(az_x + bz_y).$$

For each pair (a, b) the set of points $S(U, V, W)$ defines a caustic in the sense of Catastrophe Theory.

If one of the eyes of an observer is at point $E(P, Q, R)$ we select the lightrays that enter the eye and have in their vicinity a tangent of caustic (A.2.4). If $B(x, y, z(x, y))$ is a point of the lattice on the graph of $z_1 = z(x, y)$ and if we join it with the eye by a straight line, the line EB cuts the plane $z_2 = 0$ at a point $A(x_0, y_0, 0)$. Therefore, if we calculate x_0 and y_0 and if in (A.2.4) we set $a = x_0 - x$, $b = y_0 - y$ the point $F(U_1, V_1, W_1)$ on the caustic which corresponds to B is given by

$$U_1 = x - (P - x)z/(R - z) + (P - x)z/((P - x)z_x + (Q - y)z_y)$$
$$V_1 = y - (Q - y)z/(R - z) + (Q - y)z/((P - x)z_x + (Q - y)z_y) \tag{A.2.5}$$
$$W_1 = (R - z)z/((P - x)z_x + (Q - y)z_y).$$

Furthermore, if $A(x, y, 0)$ is a point of the grid in $z_2 = 0$, we join it with E by a straight line. This line EA cuts the graph of $z_1 = z(x, y)$ at a point $B(x_0, y_0, x(x_0, y_0))$. If we set $a = x - x_0$, $b = y - y_0$ we obtain the following system of equations for a and b:

$$a/(P - x) = -z(x - a, y - b)/R = b/(Q - y). \qquad (A.2.6)$$

Therefore, if we replace x by $x_0 = x - a$ and y by $y_0 = y - b$ in z, z_x, z_y from (A.2.5) we get the point $G(U_2, V_2, W_2)$ on the caustic which corresponds to $A(x, y, 0)$ in $z_2 = 0$.

It's obvious that the points F and G lie on the same graph which we shall name Moiré-caustic. F and G are the pictures of the corresponding points B and A of the lattice in $z_1 = z(x, y)$ and of the grid in $z_2 = 0$. Therefore, the two systems (A.2.5) and (A.2.6) can be mapped onto the Moiré-caustic, where they produce the spatial Moiré that may be seen by the eye at E.

If we specify $z_1 = mx$, the two mappings can be explicitly calculated. If the observer is far away from the observable piece of Moiré-caustic the observed Moiré is a plane. This result is given in[4]. It can easily be obtained by specializing (A.2.5) and (A.2.6). If we set $\lambda = P/R$ and $\mu = Q/R = 0$, for R tending to infinity, we obtain the following mappings of the two point systems on the Moiré-caustic:

$$U_1 = (2 - \lambda m)x$$
$$V_1 = y \qquad (A.2.7)$$
$$W_1 = x/\lambda.$$

$$U_2 = (2 - \lambda m)x/(1 - \lambda m)$$
$$V_2 = y \qquad (A.2.8)$$
$$W_2 = x/\lambda.$$

By eliminating x from (A.2.7) and (A.2.8) we get the function whose graph is the Moiré-caustic

$$W = U/\lambda/(2 - \lambda m). \qquad (A.2.9)$$

The formula also can be obtained with the methods of line geometry [5]. Discussion of (A.2.9) substantiates the following rule: If the two plane grids form a wedge and if from the side of its edge the spatial Moiré is seen, the Moiré-caustic is virtual. If the Moiré is seen from the opposite side and the eye E is not enclosed by the two planes of the grids, the Moiré-caustic is real. The difference in the observation can easily be tested if the wedge is laid with one side on a white background. Figures 30(b, c) and 31(b, c) are constructed with (A.2.5) and (A.2.6).

Comp. & Maths. with Appls. Vol. 12B, Nos. 1/2, pp. 363–378, 1986
Printed in Great Britain.

0886–9561/86 $3.00 + .00
© 1986 Pergamon Press Ltd.

MOIRÉS

Werner Witschi
Brunnenhofstrasse 13, 3065 Bolligen, Switzerland

Abstract—Graphic arts of Moiré patterns are presented. Eleven plates show a great variety of applications, some of which are mentioned specifically in the text.

Before I turned to kinetic and Moiré objects, experiencing space was the essential, fascinating, driving element for me; space, or rather interspace (arranged in layers or "boxed-in") whose relationship to the element of air I felt to be something indefinite, transitory, suspended. Movement has always been inherent in these spatial sculptures. Then, however, rhythmical movement was added. When I became attracted by the moving Moirés, above all looking-through, transparency, became the driving element leading me on to new experiments. Thus I entered a new world, which took me by surprise and mysteriously touched me time and time again.

In Milan and Kiel the Moiré patterns were correctly called spatial graphic arts (graphiche spaziali), because they were demonstrated by three-dimensional, partly mobile objects. Besides, part of these objects show clearly space deceptions which go beyond the (back or frontside of the) object. The reaction of these Moiré patterns in the objects to the slightest movement of the spectator are so vivid that we tend to regard them as independently acting subjects. They directly animate the spectator to take part in the game: he will move, have a look from different sides, move the pendulums, so that action and reaction become interchangeable. You can feel—for once in a positive sense—that technique has become independent and removed from our control. Moiré objects belong so to speak to two branches of the arts: on one hand they are objects and are therefore part of the art of sculptures; on the other hand they contain an additional element which rather relates them to the arts of painting, graphics or music. Their Moiré patterns make them part of the graphic arts, their time element links them to music. This is above all true for the mobile objects. Due to the fact that they are part of two branches of the arts, these objects seem to be instable, uncertain; they are the children of a destabilized world. The transitory effect of the Moiré patterns stresses this very characteristic. If you look at Moiré objects with both eyes very closely, your senses will be puzzled, because each eye conceives an essentially different picture. You had better use one eye only in such a case.

This outward uncertainty of Moirés is a direct contrast to the fact that they are also objects of transparency; transparency, however, stands symbolically for transcendence: you can look through the objects and beyond them (trans), and this is felt to be a symbol of the mind. Thus, on one hand, the effect of Moiré objects on the spectator is vividly animating, even disconcerting, on the other hand it produces a feeling of happiness, comfort and calmness in him.

To look at Moiré patterns is quite a complicated process. Although the patterns develop according to mathematical laws, they often remain completely inexplicable to the common spectator and therefore take him by surprise. For example, if you look through two convex wire screens, you will see an unexpected number of Moiré circles [see Appendix (A8)]. Even more surprising is the Moiré effect of rotating wire screens.

If you move only one of the screens, you will perceive a unique and inexplicable phenomenon: namely fast rotating screens produce clear patterns, whereas slowly rotating ones produce patterns which are more blurred and suspended on the wire screen. These patterns are strikingly colourful, although I use only black screens as a rule, because their transparency against a light background is better.

I must add that, if you look through a wire-screen object for a certain time, you will be able to perceive colours, too. These are particularly clear to the eye, if a filmed object is projected on a big screen. These are the so called Benham-Fechner colours.

We tend to call the patterns we see on one rotating wire screen physiological. Yet we are puzzled to perceive the same patterns in a picture taken by a camera; and this is in spite of the fact that a camera produces objective pictures only.

The patterns of the pictures correspond to the structure of the wire screens: a screen with quadrangular perforation produces a cross (A55), one with rhombic perforation an X (A54) and one with round perforation six rays (A53), because the holes are arranged in a hexagon.

I divide my Moiré objects into positive and negative Moirés: *Negative* patterns are created by sheet metals with punched round or rectangular perforation. They are black in order to form a sharper contrast against a light background or seen against the light.

The *positive* Moirés are produced by forms printed or stuck on acrylic glass. Here I can use colours. However, I have restricted myself to the most marked prime colours red and blue to limit the endless possibilities. Strange variations occur: even the stable objects and the Moiré prints show an alteration in colour depending on the neighbouring colours. The movement of the rotating objects, however, produces even more striking effects. They are so varied and numerous that they would merit a special study in themselves.

I am going to give only the description of a particularly interesting object (A49–A52), the "Rotating Moiré Pendulum 1981 with red and blue rectangles 58 × 50 cm".

The transparent colours change immediately when the two oblong acrylic glasses rotate. If the rotation is fast, you will unexpectedly perceive all six spectral colours; in addition, of course, you will see red and blue and the mixed colour purple. On the pane with the blue rectangles you can see an additional green colour, on the pane with the red rectangles, orange ranging to yellow. Perhaps this effect can be explained by the fact that the eye has the ability to visualize automatically the complementary colour to any colour it perceives.

It is known that we are able to perceive shapes more clearly if we look through a small hole (e.g. by forming a peephole with our hands). In the same way colours become strikingly more distinct and brighter.

Above I have mentioned variations of forms with reference to single or various wire screens moving. A rectangular shape approaches a round shape if rotated (A48) as the edges become blurred. An oblong rectangle changes into an oval shape. Stripes become thinner and pointed towards the end (A51). Again the countermovement of two rotating screens is particularly interesting: round or rectangular shapes are—so to speak—cut by rays.

On one hand, broad- or narrow-meshed screens produce lively and varied Moiré patterns, on the other hand, however, quiet ones which change only slowly. Thus textile Moirés produced by superimposing two textiles, create very lively patterns, similar to those produced by two very finely perforated nickel sieves (60 or 80 mesh) (A5). In contrast to this, the pattern of screens with two or more cm meshes are calm, still, changing only slowly. They stimulate the spectator to meditation. (A2)

The Moiré movement of rotating wire screens or acrylic panes with stripes or round, respectively, rectangular dots is extremely striking indeed. Seen more or less from the side they show, among other things, strange, wandering, appearing and disappearing whirls, which however are not visible in a photo. Are these perhaps purely physiological pictures?

If two or three screens are rotated against each other, rays will appear from the centre. These vary according to the kind of screen used.

I have used different techniques to make by two-dimensional Moirés: printed with a roller over a stencil on paper; then Moiré screen prints (A59) and finally material Moiré prints made by printing the perforated steel sheets directly on to the paper by hand (A60).

Moiré patterns show the most vivid reaction to the movement of the spectator, if a Moiré object is placed in front of a convex mirror. Even the slightest movement of the spectator's head causes a vivid change in the pattern. Reflexions may appear reaching, so to speak, out of the mirror: e.g. "Moiré Object with Slanted Screens," 1978–1979 (A24, 25, 26, 27).

APPENDIX

I Stable Moiré-Objects (A1–A11)

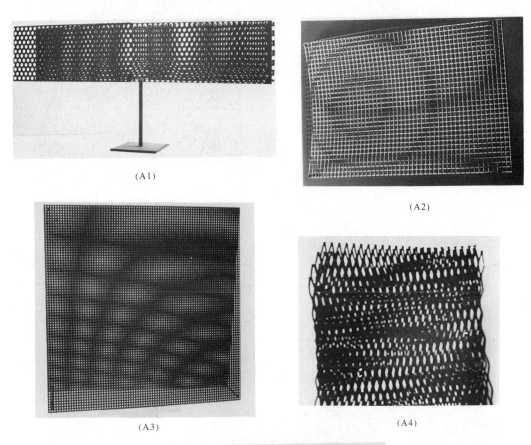

(A1)

(A2)

(A3)

(A4)

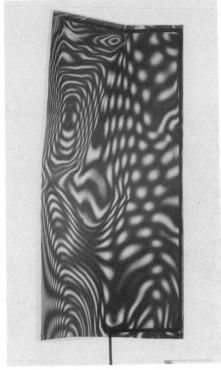

(A5)

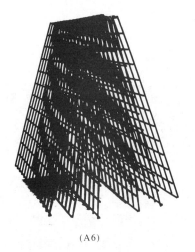

(A6)

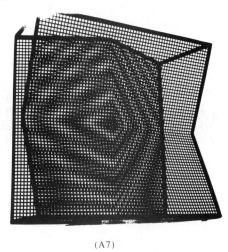

(A7)

(A8)

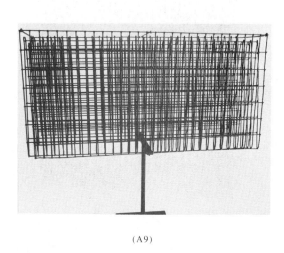

(A9)

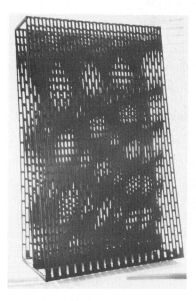

(A10)

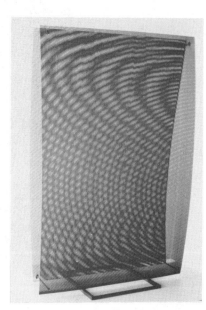

(A11)

II Great Moiré, University of Berne (A12–A18)

(A12)

(A13)

(A14)

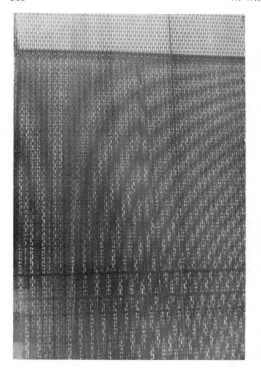

(A15)

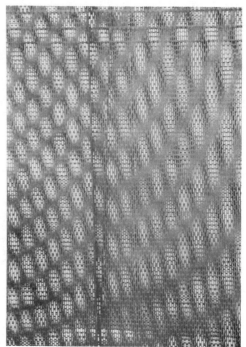

(A16)

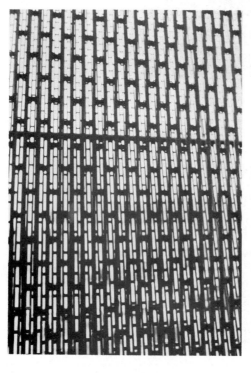

(A17)

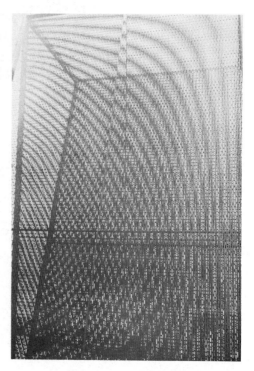

(A18)

III Solid Moiré-Objects (A19–A23)

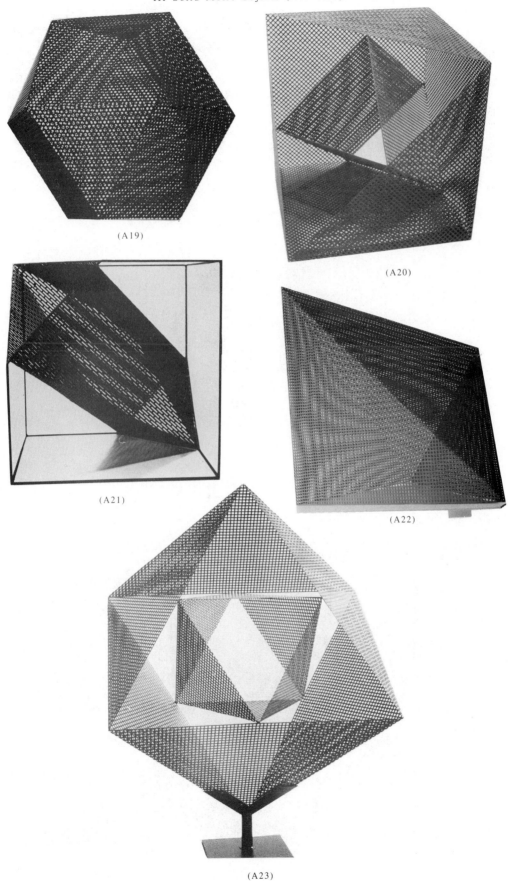

(A19)

(A20)

(A21)

(A22)

(A23)

IV Moiré-Objects in front of an inflected mirror (A24–A27)

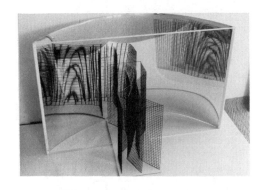

(A24)

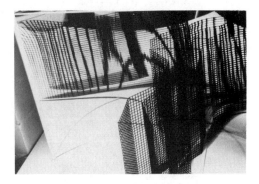

(A25)

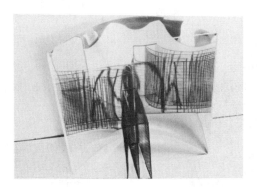

(A26)

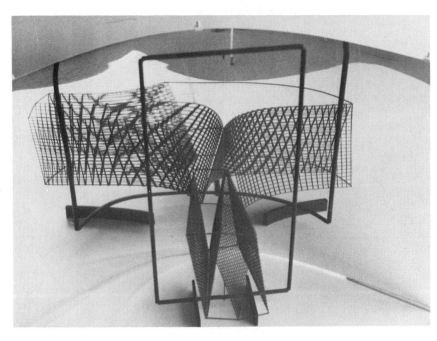

(A27)

V Moiré-Objects hanging and turning (A28–A32)

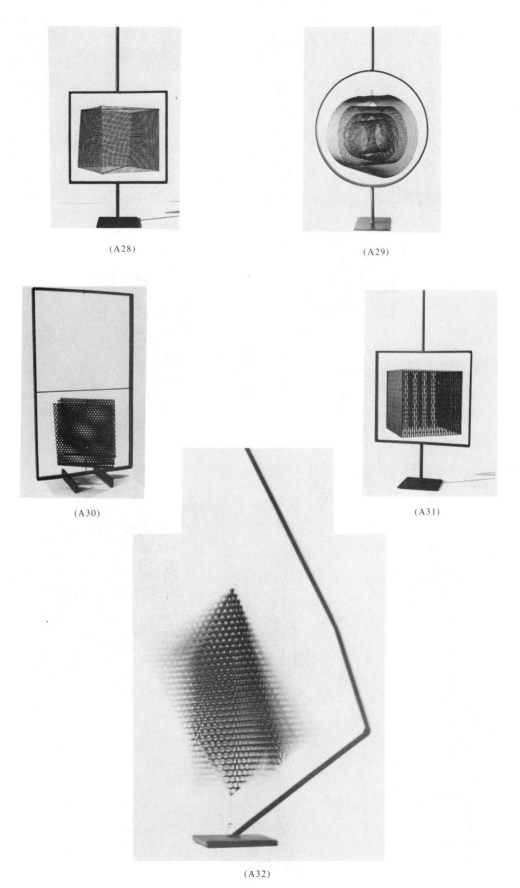

(A28)

(A29)

(A30)

(A31)

(A32)

VI Moiré-Pendulums (A33–A41)

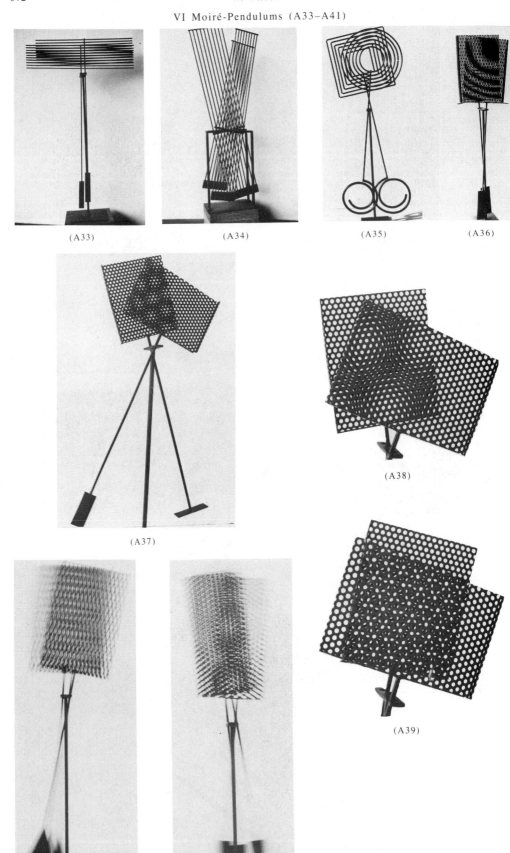

(A33)   (A34)   (A35)   (A36)

(A38)

(A37)

(A39)

(A40)   (A41)

VII Moiré-Objects rotating and oscillating (A42–A48)

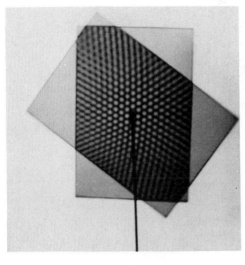

(A42)

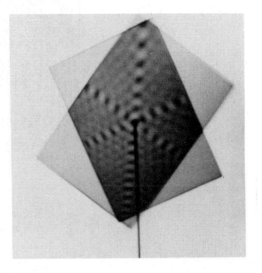

(A43)

(A44)

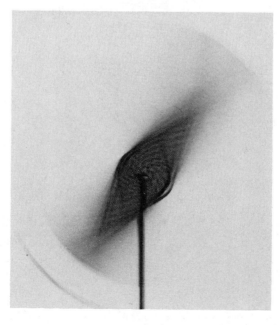

(A45)

(A46)

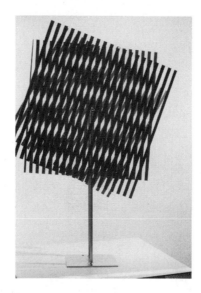

(A47)

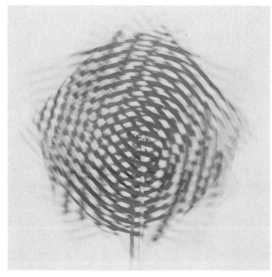

(A48)

VIII Moiré-Objects in red and blue, rotating and oscillating (A49–A52)

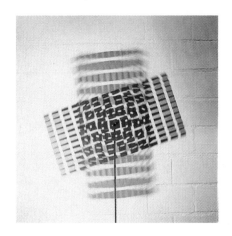

(A49)

(A50)

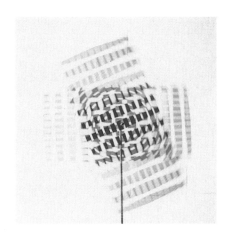

(A51)

(A52)

W. WITSCHI

IX Rotating different lattices (A53–A55)

(A53)

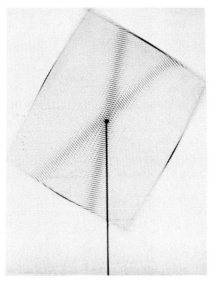

(A54)

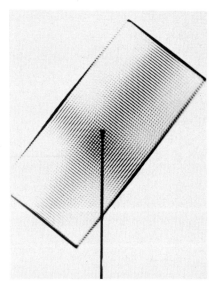

(A55)

X Stable Moirés (A56–A58)

(A56)

(A57)

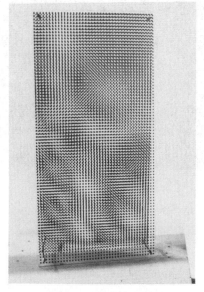

(A58)

XI Moiré-Serigraphy, material printing (A59, A60)

(A59)

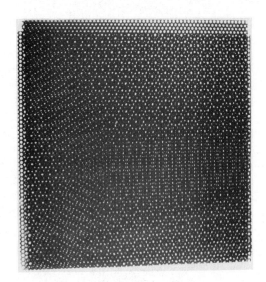

(A60)

Comp. & Maths. with Appls. Vol. 12B, Nos. 1/2, pp. 379–405, 1986
Printed in Great Britain.

0886–9561/86 $3.00 + .00

SYMMETRY OF SYSTEM AND SYSTEM OF SYMMETRY

Yu. A. Urmantsev
Institute of Plant Physiology, Moscow 127276, U.S.S.R.

Abstract—Relation "symmetry–system" is investigated from the point of view of the author's version of general systems theory (GST(U)). The symmetry of system is explicated in three ways: (i) in the form of the theory of groups of nonevolutionary and evolutionary system transformations and their invariants, (ii) in the form of a proof of symmetry of the system as itself, (iii) in the form of a proof of the group nature of systems of 2-, 1-, 0-sided actions and relationships. The system of symmetry is described both as a special object-system, and as a specific system of objects of the same kind (in particular, as a system of 64 fundamental and 54 structural symmetries). Premises, the basic concepts of GST(U), and the laws of system transformations, correspondence, symmetry, and the system similarity are considered in some or other connection. The general theory of isomerism is presented in brief; the relations "isomerism–symmetry", "system isomorphism–symmetry", "equality–symmetry" are explicated.

INTRODUCTION

The study of symmetries in various systems and the study of systematics of various symmetries lead us inevitably to a fundamental problem: to explicate the symmetry of a general system, on the one hand, and the system of symmetries in general, on the other hand. The immediate result is the General System Theory (GST), or systemonomy, i.e. a science concerning the origins, existence, evolution and development of systems in nature, society and thinking. Evidently, such general problems may find solutions only within the framework of GST. There are several versions of GST at present; first of all we should mention the approaches developed by M. Mesarovic[1], A. I. Uyomov[2] and Yu. A. Urmantsev[3,4]. It is only in the latter approach (referred to as GST(U)) that the "symmetry–system" relation is under investigation, and this is done in a manner described concisely by the title of the present article.

GST(U) is still rather far from being completed. In the present state of the theory, it explicates the symmetry of a system in three ways: first, as the theory of groups of non-evolutionary and evolutionary system transformations and the group invariants, second, as a proof of symmetry properties of the system itself, third, as a proof of group-theoretical nature of systems of two-, one-, and zero-sided actions and relationships which govern the origin, existence and evolution of individual systems.

As to the system of symmetry, in GST(U) it is expressed in two aspects: as a special type of object-system (the mathematical group as an object-system) and as a peculiar system of object-systems belonging to the same type (in particular, as a system of 64 fundamental symmetries and 54 structural symmetries).

This approach results in that the *symmetry* is represented in GST(U) as *a system category implying a coincidence of systems* S *in features* F *with an account of modifications* M; that is to say, the concept of symmetry acquires a relative meaning. In fact, the same systems may turn out to be dissymmetric (or even asymmetric, in a limiting case) in the same features but with respect to different modifications, as well as with respect to the same modifications, but in different features. Take for example a geometrical object composed of left-handed black tetrahedron and its white mirror image. Its configuration has a reflection symmetry. However, if one takes into account not only the geometry, but also the colour, the object must be considered as having no reflection symmetry with respect to the usual plane, while being symmetrical under reflections in the unusual (anti-symmetrical) plane. The latter one is known as an "antiplane" in the antisymmetry theory; it changes not only left to right and right to left, but also transforms black to white and white to black. Thus the whole composite body (in the case in view it is a combination of a left blank tetrahedron and right white tetrahedron) is transformed into itself.

In other words, the symmetry has a necessary complement and opposite that is a corresponding asymmetry.

The asymmetry is an opposite to the symmetry. Within the framework of GST(U) *the asymmetry is a system category indicating a non-coincidence of systems* S *in features* F *with an account of modifications* M. This implies that the asymmetry has also a relative meaning. Besides, the asymmetry has "its proper" symmetry as a necessary complement and opposite.

In view of the above discussion, in GST(U) systems appear as unities of opposites; any system is symmetric with respect to a set of features and transformations (modifications) and asymmetric with respect to another set of features and transformations (modifications).

The relativity of symmetry and asymmetry influences the ways of the scientific progress; having detected a symmetry in some phenomena observed in nature or society one should look for facts violating the symmetry. Conversely, when the asymmetry is found, one should search for such a new symmetry which would be able to interpret both the old symmetry violation and the original symmetry itself as some particular cases of the new symmetry, and so on. All this can be done by means of an investigation of new objects, features and modifications.

Thus, the title of the present paper is not just a harmonical play of words, it is a manifestation of a dialectical device which takes place in GST(U), namely, an inversion of viewpoints that enables one to obtain two essentially different and complementary results.

One more remark seems in order. We believe that the group nature of a certain set of elements with respect to a composition law introduced in it is just a mathematical expression of an intrinsic symmetry of the set in view. Actually, each of the four axioms of group theory permits us to claim that an arbitrary group $\mathcal{G}$ is symmetric since

(i) for any two group elements a, $b \in \mathcal{G}$ with a given composition law T, their composition aTb does also belong to the group $\mathcal{G}$, and all the possible pair products is a mapping of the group to itself (the "closure" axiom),

(ii) for any three group elements a, b, $c \in \mathcal{G}$ the equality $(aTb)Tc = aT(bTc)$ holds, i.e. there is an invariance of products of three elements with respect to different positions of the parentheses,

(iii) there exists such a (unique) element $e \in \mathcal{G}$ that $aTe = eTa = a$ for any group element a, i.e. any group element coincides with its product by e,

(iv) for any group element $a \in \mathcal{G}$ there exists a (unique) group element b which is symmetric (reciprocal) to it, so that $aTb = bTa = e$, i.e. the composition of mutually symmetrical elements is the so called "neutral" element e which is by itself the first-order group with respect to the composition law T.

The symmetry of the abstract group is the reason why we treat the group nature of the set of system transformations, the nature of the system of actions and relations realized in the transformations, and the system itself associated with some composition laws, as a manifestation of their symmetry.

Seemingly, we are in position to present a proof of the existence and significance of all mentioned symmetries. Nevertheless, first of all we must introduce premises and basic concepts (dealing with systems as primary objects) which are principal in GST(U) and without which no such proofs can be formulated.

Premises of GST(U)

The concrete subject of any theory depends on its primary categories, i.e. the premises. In our case, in order to provide the maximum generality (universality) of the theory we have selected the following philosophical categories: (1) existence, (2) multitude of objects, (3) universality, (4) unity and (5) sufficiency.

The requirement (1) is essential, as the existence is a fundamental characteristic of any system, its attribute. In accordance with the philosophical tradition, we reduce the category of the existence to its seven forms, three of them being fundamental and other four secondary: (a) spatial, (b) temporal, (c) dynamical, (d) space-time, (e) space-dynamical, (f) time-dynamical and (g) space-time-dynamical. Motion is a form of existence of matter which is of the most importance to us. Incorporating this form as a basic concept of GST one can claim that "to exist means to be at rest or under change".

The requirement (2) is interpreted as the multitude of various objects, both material and ideal. Actually, this is the "Universe" as it appears before any systematization of its constituents, or objects. By "object" we mean any thing that can be thought about. The requirement (2) must be taken into consideration while a system is constructed: no construction is possible unless the necessary building blocks, the objects, are available.

The requirement (3), "universality", implies a property or feature which is the same for all object-systems within a given system ("the system of objects of the same kind"). Logically, it is a basis for any classification. Construction of a system involves this point as well: any given i-th system is to be built of objects belonging only to a set $\{M_i^{(0)}\}$ which is separated by reasons $a \in \{A_i^{(0)}\}$ and is called "the set of primary elements" in the following.

The requirement (4), "unity", is comprehended in twofold way. On the one hand, it is a relationship (in particular, an interaction) between the primary elements which is responsible for creation of object-systems having also some new "integral" properties which in turn can be additive, non-additive, additive-non-additive. On the other hand, it is an individual object, the object-system. The fundamental importance of the requirement for the system existence is evident.

The requirement (5), "sufficiency", is comprehended here in the same sense as in the case where one needs a sufficient amount of materials and appropriate conditions to build something. The role of this condition is obvious: without sufficient amount of primary elements or sufficient reasons neither construction, nor existence of any system is possible. In essence, the requirement (5) coincides with Leibnitz's "sufficient basis principle".

The premises (1)–(5) combined with the rules of logics enable one to obtain all the statement formulated in GST(U).

In particular, basing on the requirements (1) and (2) one can assert that "there exists a set of objects". This is equivalent to the statement of the existence of the so called universal set U, known in the set theory. From the ontological point of view, this statement coincides with the argument for the existence of Universe.

Furthermore, the premises (1)–(3) make it possible to assert: "there exists a multitude of universal objects", or equivalently, to admit the existence (in objective and subjective reality) of specific "sets of primary elements" selected by reasons $a \in \{A_i^{(0)}\}$. Such sets may be finite or infinite, may have equal or different cardinalities, they may include identical or different elements, be fuzzy or not fuzzy.

Some examples of the sets of primary elements are: (i) the set of elementary particles forming atoms: protons, neutrons, electrons which are selected by the set of features $\{A_a^{(0)}\}$ (the subscript a stands for "atom"), (ii) the set of "points", "lines" and "planes" which are elements in construction of a conceptual space and selected by features $a \in \{A_{sp}^{(0)}\}$ (sp indicates "space"), (iii) the group of reflections $\{\sigma\}$ in planes (the maximum number is 3 for finite bodies, and 4 for infinite bodies). The reflections make it possible to obtain all possible structural transformations, they are selected by features $a \in \{A_s^{(0)}\}$ ("s" indicates "symmetry").

Now we will form a combination (1)(4)(2)(3) based upon the premises (1)–(4). This combination implies that "there exists a unity of the variety of universal objects" or, equivalently, "there exists a unity of primary elements". The combination means that objects belonging to any specific set $\{M_i^{(0)}\}$, which are selected by features $a \in \{A_i^{(0)}\}$ are in certain i-th relations of unity R_i. For example, electrons, protons and neutrons may (and do) undergo some atom-creating relations which are interactions of a special type R_a; "points", "lines" and "planes" may be (and, under certain conditions, are) in relations R_{sp}: they "lie on", "are between", "are congruent", "are parallel", etc. The reflection planes may intercross under various angles, according to relations $r \in \{R_s\}$.

Because of the double meaning of the term "unity", the combination (1)(4)(2)(3) is also interpreted as "the existence of a unity of primary elements in the form of an individual object". For instance, the unity of the existing unity of protons, neutrons and electrons is an atom; the unity of "points", "lines" and "planes" is the conceptual space; the unity of the reflection planes is a symmetry transformation.

Finally, one should have in mind that the unity relations R_i, no matter where do they arise (in nature and/or in human mind), must meet requirements of certain laws. The atom-creating interactions must satisfy the laws of atomic physics, $z \in \{Z_a\}$; the space-creating relations must satisfy the axioms of union, order, congruence, continuity, parallelism, as well as the theorems which stem from the axioms, $z \in \{Z_{sp}\}$; symmetry creating relations must satisfy the axioms of group theory, $z \in \{Z_s\}$. Note that in GST(U) conditions imposed on the unity relations are called the composition laws, denoted by Z.

Basic concepts of GST(U)

Constructive derivation of the combination (1)(4)(2)(3) is, in fact, a mental construction of an object. It is of special importance that the object appears as an object-system. Consequently, here we deal, in essence, with a constructive derivation of the "object-system" concept which is one of the most fundamental concepts in GST(U).

DEFINITION 1.

The *object-system*, OS, is a unity constructed in accordance with relations (in particular, interactions) r which are elements of a set $\{R_{OS}\}$ combined with conditions z which are elements of a set $\{Z_{OS}\}$ and restrict the relations. The unity is built of primary elements m from a set $\{M_{OS}^{(0)}\}$ selected from the universal set U according to criteria a, which are elements of a set $\{A_{OS}^{(0)}\}$. The sets $\{Z_{OS}\}$, $\{Z_{OS}\}$ and $\{R_{OS}\}$, $\{Z_{OS}\}$, $\{R_{OS}\}$ and $\{M_{OS}^{(0)}\}$ may be empty, or consist of any number of elements, from one to infinity, and the elements may be identical or different.

PROPOSITION 1.

Any object O is an object-system. The proposition is true because of Definition 1, according to which an object, even if it contains a single element—itself, is also an object-system. Obviously, in this particular case the sets of relations and composition laws are empty; $\{R_{OS}\} = \emptyset$, $\{Z_{OS}\} = \emptyset$.

Furthermore, the empty system, or zero-system is an important particular case of object-system. This system contains no elements, and the set $\{A_{OS}^{(0)}\}$ is empty, as well as the sets $\{M_{OS}^{(0)}\}$, $\{Z_{OS}^{(0)}\}$, and $\{R_{OS}^{(0)}\}$. By the way, all these sets are examples of empty systems. The set itself is, of course, also an example of the object-system; in this case $\{Z_{OS}\} = \emptyset$, $\{R_{OS}\} = \emptyset$, while $\{M_{OS}\} \neq \emptyset$. In truth, "unity is a set and a set is "unity"! Probably, owing to this fact, quite different theories of finite and infinite sets have been developed in mathematics, since each type of the sets, finite and infinite, have quite different integral properties as the mathematical object-systems, as unities. For instance, in the former case (for finite sets) a part is never equivalent to the whole, while in the latter case (for infinite sets) a part can be equivalent (have equal cardinality) to the whole object.

It follows from Proposition 1 that absolutely every object can be regarded as an object-system. We will illustrate this statement with other two examples which will be employed later in analysis of the system symmetry.

Example 1. A right or left (i.e. different from its mirror image) dissymmetric object, say, an asymmetric plant leaf with its left half wider than the right half, and the right half longer than the left half. The symbol we use for such a leaf is wl' (the prime indicates that it is the right half that is longer, so the mirror image of this leaf is represented by symbol $w'l$). In this case the "primary" elements are dissymmetry factors ("dissfactors") W and L, i.e. the features, the appearance and the very existence of which make the leaf left or right. The unity relations are the leaf attribute relations; the composition law is the requirement that in free combinations of the dissfactors $W(w, w')$ and $L(l, l')$ (modifications of the dissfactors are given in parentheses) the number of dissmodifications S would be $S_2^2 = 2^2 = 4$, while with other possible combinations forbidden it is $S_2 = 2^2 - 3 = 1$.

It is noteworthy that it was just the empirical discovery of dissfactors (i.e. features the appearance and existence of which in an object make it left or right resulting in elimination of the second-kind symmetry, i.e. the mirror symmetry, from the object symmetry type) that provided the basis for the theory of left and right, the theory of dissfactors developed by the author[5]. According to the theory, only with the number of dissfactors $m = 1$ one has the Kant–Pasteur dissymmetry with the number of possible dissmodifications $N = 2$, while for m from 1 to ∞, one has a more general dissymmetry with the number of dissmodifications N from 2 to ∞. Thus the extended dissymmetry incorporates naturally the Kant–Pasteur dissymmetry as the first and simplest particular case. In the theory of dissfactors formulae are derived for calculation of the number of dissmodifications of dissobjects in the cases of total and partial combinability of dissfactors, a new formulation is given for the Pierre Curie causal principle: "if some actions reveal dissfactors, the dissfactors must be detected also in causes generating them", a possibility has been proved to develop such theories of antisymmetry, colour symmetry, colour antisymmetry, cryptosymmetry etc., which would be based upon the partial combinability of dissfactors[3,5].

Example 2. Interaction. In this case the primary elements are (i) changing and changeable objects (A and B, B and A), (ii) the action agents ("interactions") propagating from A to B and from B to A and (iii) the propagation medium. The unity relations are couplings between the primary elements. The composition law is the requirements:

$$\Delta t_{AB} < T_B, \quad \Delta t_{BA} < T_A, \quad \Delta t_{AB} \geqslant \Delta t_{min} = R_{AB}/v_{k,max}, \quad \Delta t_{BA} \geqslant \Delta t_{min} = R_{AB}/v_{k,max},$$

where Δt_{AB} and Δt_{BA} are the times of the interaction propagations from A to B and from B to A, respectively, T_A and T_B are the individual existence times for the objects A and B, R_{AB} is the distance between the objects, Δt_{min} is the minimum time necessary for the interaction agent to cover the distance R_{AB} while the agent has the highest possible finite velocity $v_k = c$ (the velocity of light in the vacuum).

Note that the first example is a static object-system and the second example is a dynamical object system, so GST(U) can be applied both to statical and dynamical object-systems.

In view of the above discussion we can admit that the combination (1)(4)(2)(3) ("there exists a unity of a set of primary elements") means also that "there exists an object-system". But "exists" means either "is at rest" or "is changing". The rest state of an object-system may be considered as a continuous transition (in time) of the object-system into itself. Logically, it may be considered as the identical transformation. Originally, this operation has been explicated as a system operation by A. V. Malikov. As for a change of an object-system, it always leads to a transition of the object-system into one or more other object-systems, which proceeds under certain laws. The resulting object-systems, in turn, undergo transformations to object-systems of the third generation, the latter are transformed to those of the fourth generation and so on. Having in mind the absolute character of motion and the relative character of rest, one concludes that such transformations are inevitable. The object-systems appearing in this way may be of the same or/and different qualitative type.

DEFINITION 2

A system of objects of a given i-th kind is, essentially, a regular set of object-systems of the same kind. The expressions we use, "of the same kind" or "of a given kind", mean that every object-system has common typical features (one and the same property); namely, each one is constructed of all (or some) primary elements m of the set $\{M_i^{(0)}\}$, in accordance with all (or some) relations r of the set $\{R_i\}$, with some (or all) composition laws z which are elements of the set $\{Z_i\}$, realized in the considered system of objects of the given kind. As for an object-system, for the system of objects of the same kind the sets $\{Z_i\}$, $\{Z_i\}$ and $\{R_i\}$, $\{Z_i\}$, $\{R_i\}$ and $\{M_i^{(0)}\}$ may be empty, or contain any number of elements, from one to infinity.

A very spectacular example of a system of objects of the same kind is the saturated hydrocarbons CH_4, C_2H_6, C_3H_8, . . . , $C_{s-1}H_{2(s-1)+2}$, C_sH_{2s+2}. All of them are built of the same primary elements, C and H, in accordance with the same relation of chemical affinity and governed by the same composition law, C_nH_{2n+2} ($n = 1, 2, . . . , s$).

There are proper systems of the same kind also for the above mentioned examples of a plant leaf and interaction. For instance, for a leaf of type wl' it is the system of 4 isomer leafs: wl', $w'l$, wl, $w'l'$. In this case the primary elements are the dissfactors W and L, the unity relation is the presence of the dissfactors in a leaf, and the composition law is the equality $S_{k_0}^{k_0} = 2^{k_0}$ ($k_0 = 2$). For the case of "interaction" one has a system of 9 action-systems (see Table 1).

The completeness of the exhaustion of the action types can be proved by means of a formula determining the number of arrangements of m elements taken k at a time with possible repetitions, i.e. $A_m^k = m^k$. In fact, looking at the action symbols one can conclude that each of them can be considered conventionally as an arrangement of 3 elements, $>$, $<$ and $=$, with repetitions, taken 2 at a time. The result is $A_3^2 = 3^2 = 9$. It should be stressed that a proof of the completeness

Table 1. Space-time system of actions

No.	Action type	Realization condition	Action symbol
1	2-action of the type $<<$	$\Delta t_{AB} < T_B,\ \Delta t_{BA} < T_A$	$<<$
2	Quasi-2-action of the type $=<$	$\Delta t_{AB} = T_B,\ \Delta t_{BA} < T_A$	$=<$
3	Quasi-2-action of the type $<=$	$\Delta t_{AB} < T_B,\ \Delta t_{BA} = T_A$	$<=$
4	1-action of the type $<>$	$\Delta t_{AB} < T_B,\ \Delta t_{BA} > T_A$	$<>$
5	Quasi-0-action of the type $==$	$\Delta t_{AB} = T_B,\ \Delta t_{BA} = T_A$	$==$
6	1-action of the type $><$	$\Delta t_{AB} > T_B,\ \Delta t_{BA} < T_A$	$><$
7	Quasi-0-action of the type $>=$	$\Delta t_{AB} > T_B,\ \Delta t_{BA} = T_A$	$>=$
8	Quasi-0-action of the type $=>$	$\Delta t_{AB} = T_B,\ \Delta t_{BA} > T_A$	$=>$
9	0-action of the type $>>$	$\Delta t_{AB} > T_B,\ \Delta t_{BA} > T_A$	$>>$

of exhaustion is an important requirement imposed by GST on every construction of a system of kind-i objects. For brevity, we omit the description of primary elements and the unity relations for each of 9 action-systems. As to the composition laws, in Table 1 we have presented only the realization conditions specific for every action kind. A restriction which is not specific here is the requirement that

$$\Delta t_{AB} \geqslant \Delta t_{\min} = R_{AB}/v_{k,\max}, \quad \Delta t_{BA} \geqslant \Delta t_{\min} = R_{AB}/v_{k,\max}\ (v_{k,\max} = c).$$

It is noteworthy that after a simple algebra the latter inequalities give rise to invariants of the Lorentz transformations of special relativity, namely, $d\tau^2$ ("the proper time of a material point") and ds^2 (the space-time interval). The invariants can be used to construct the light cone, and one is led immediately to the two-, one-, and zero-sided actions, and so to events which can or cannot be interrelated as a cause and its effect.

Other examples of systems of objects of some kinds are the systems of point-wise, linear, planar and space symmetry groups, classical or non-classical, various systems of natural numbers, Mendeleev's Table of chemical elements, homologic series in chemistry and biology, the periodic system of corollas and flowers of plants, natural and artificial systems of plants and animals, the system of social and economical structures, the linguistic system of 6 isomer words: tom, tmo, otm, omt, mto, mot.

Definition 2 and the examples we have presented suggest that a system of objects of the same kind is a regular aggregate of separate object-systems which are not, in general, enclosed in each other, being parts of a single object, like dolls in the Russian toy named *Matryoshka*.

A wide distribution of systems of similar objects in nature, society and thinking suggest an existence of a law invariant under a transition from inanimate nature to animate nature and from the letter to human society. And such a law does exist indeed!

PROPOSITION 2 (The law of systematics)

Any object is an object-system, and any object-system belongs to at least one system of objects of his kind.

The validity of this law is a direct consequence of Definitions 1 and 2, and Proposition 1.

If all the above arguments are correct, the law of systematics must be adopted as one of fundamental laws, which must be considered as an absolute system law in view of its universal applicability to the real world. Its gnosiological significance is beyond all question, as it generates a new tool for gnosiology, which is relevant to the Science as a whole and is, probably, the cardinal method of GST(U); we mean the so called S-method. The S-method enables one to study any material or ideal object not only "within the universal interrelation and interconditionality", but also as an object-system in a system of objects of the kind specific for the object concerned. The S-method is appropriate for such an approach, because it involves two special algorithms: an algorithm for representation of an object as an object-system and an algorithm for construction of a system of objects of a given kind. We have shown previously[6], taking for examples the periodical system of chemical elements and the system of cyclic carollas, that the use of the S-method provides a fundamental cognitive profit.

It goes without saying that systems of objects of the same kinds can be united in more and more large system units, "families", "classes", "types", "kingdoms", "empires", etc. Nevertheless, since Definition 2 is invariant with respect to such a unification, all the higher systems can be, in turn, interpreted as a system of objects of the same kind, but of different generalities. The sequence of systems having more and more general applicability to the real world tends finally to a limit which is the system in general. Thus we are ready for the last step, a definition of an "abstract system".

DEFINITION 3

The system S is a set of object-systems composed according to relations r from a set of relations $\{R\}$, with composition laws z from a set of composition laws $\{Z\}$. The primary elements of the systems S, m, belong to a set $\{M^{(0)}\}$ extracted from the universal set U by features a from a set of features $\{A^{(0)}\}$. In general, the sets $\{Z\}$, $\{Z\}$ and $\{R\}$, $\{Z\}$, $\{R\}$ and $\{M^{(0)}\}$ may be empty.

It is possible now to turn to an analysis of the system symmetry: the basis for the analysis is constructed.

The Symmetry of System Theory of Groups of System Transformations and Their Invariants

It is impossible to discuss a group without specifying its elements. In the case we deal with, the group elements are the system transformations.

PROPOSITION 3

There are only 4 main types of transformations of a system-object in the framework of a system of objects of the same kind, namely, identical, quantitative, qualitative and relative transformations. In other words, we mean the transformation of the object into itself, of its quantity, quality, or relations between its primary elements.

Proof. Recall that by virtue of its very existence the object-system is either resting, or moving. In the former case, it undergoes the continuous identical transformation into itself, in the latter case it is transformed into object-systems of the same (qualitatively identical) kind or of different kinds.

Evidently, in the consideration of transformations of an object-system in the framework of a system of objects of the same kind one must consider the composition laws $z \in \{Z_i\}$ in the transitions as invariable, even if only in view of the above condition. For a fixed set $\{Z\}$, however, an object-system, by definition, cannot have changed its properties, except the quantity, quality, and the unity relations of its primary elements. Thus we have only 4 transformations: identical (if the object-system is transformed into itself), quantitative, qualitative, and relative (if it is transformed into other object-systems).

We have an example of the identical transformation:

$$\text{tom} \rightleftarrows \text{tom}.$$

In this case the quantity, quality and relations of letters are invariable.
We have an example of quantitative transformations:

$$\text{tom} \underset{-e}{\overset{+e}{\rightleftarrows}} \text{tome}.$$

In this case the quality and relations (the linear order and the quality of letters) are invariable.
Some examples of qualitative transformations (interchanges of letters):

$$\frac{T \rightarrow O,\, O \rightarrow M,\, M \rightarrow T}{T \leftarrow O,\, O \leftarrow M,\, M \leftarrow T}$$

Presumably, these equilateral triangles and the letters at their vertices can be superposed by means of rotations in space. In this case the qualitative transformation of the lettes, under which "TOM" becomes "OMT" and vice versa, changes neither the quantity of the primary elements (sides, letters, angles), nor relations between them.

An example of relative transformations (permutations):

$$\text{TOM} \rightleftharpoons \text{MOT}.$$

In this case neither the quantity of the letters, nor their quality is changed.

Uniting 4 basic transformations in groups of 1, 2, 3 or 4 elements, one gets 4 basic and 11 derivative transformations, their total number being 15 (see Table 2). The completeness of the exhaustion of all variants presented in Table 2 is quite evident, because $\Sigma C_4^i = 2^4 - 1 = 15$.

Comparing the transformations 2 and 9, 3 and 10, . . . , 8 and 15, one can easily see that the differences between them are just quantitative, inessential. Having in mind the essential identity of the transformations 2–8 to the transformations 9–15 and with due account for the quantitative aspect of the problem, one draws a fundamental generalization. This generalization influences all the propositions of GST(U), so we call it "the central proposition."

CENTRAL PROPOSITION OF GST (U) (the principal law of system transformations of an object-system)

By virtue of its very existence in the framework of a system of objects of the same kind, any object-system transforms according to some laws $z \in \{Z_i\}$ either (A) to itself, by means of the identical transformation, or (B) to other object-systems, by means of one of 7 and only 7 different transformations; namely, it can change (1) quantity, (2) quality, (3) relations, (4) quantity and quality, (5) quantity and relations, (6) quality and relations or (7) quantity, quality and relations of its primary elements, all of them or only in a part of them.

It is clear from the Central Proposition that the same name, say, the Ql-transformation, must be applied both to a transformation of the quality of every primary element of the object-system and to that of some of the primary elements.

Another remarkable point of the Central Proposition is that the total set of the system transformations contains 1 identical and 7 non-identical transformations. The knowledge of their number and quality is quite important. For instance, basing on this knowledge we can assert that *there are only 7 ways in which Nature, inanimate and animate, and Society can create their object-systems*. Meanwhile, amazing as it is, philosophers have, evidently, never raised the cardinal question what are the number and types of ways in which the creation and evolution do proceed. An exception among them was Democritus of Abdera (some details on his achievements in this field have been given in [5]). Even at some occasions where the question, and the adequate answer to it, did suggest theirselves, creators of various philosophical and natural evolutionary conceptions passed the problem by. A result was a substantial omission in all such theories. For example, A. N. Severtsov[7], enumerating modes of the phylembryogenesis in his theory of the development of ontogenesis, mentioned only two (of 7 possible) variations in the stages of embryogenesis, changes in the number (prolongation or abbreviation) or in the quality (deviation). In spite of the presence of empirical material, other 5 modes of the phylembryogenesis were not separated in his theory. A similar situation takes place in the

Table 2. The list of basic and derivative transformations for an object-system within a system of objects of the same kind

Transformation type		
1—I	6—QnR	11—IR
2—Qn	7—QlR	12—IQnQl
3—Ql	8—QnQlR	13—IQnR
4—R	9—IQn	14—IQlR
5—QnQl	10—IQl	15—IQnQlR

Note. The symbols I, Qn, Ql, R are initials of the words "Identity", "Quantity", "Quality", "Relation".

modern Darwinism, a synthetic theory of evolution incorporating various conceptions of morphogenesis. In the latter case, for instance, people attempt to reduce morphogenesis ultimately to increases, or decreases, in the number and size of cells, to their differentiation or dedifferentiation, i.e. to the first and second types of generation of object-systems, while other five, Nos. 3–7, transformation modes were discarded. That is to say, all these theories have been constructed at best by 2/7 of their total volume, or equivalently, 5/7 of their construction are not yet completed. In view of this fact, it is quite natural to raise the problem of an essential, 5/7 of the whole, complement to the theories concerned.

One more remark. Having in mind the Central Proposition, one should naturally expect a dominating role played in Human Culture by two numbers, 7 and underlying 3 (as $7 = \Sigma C_3^i$. Numerous arguments in favour of this suggestion, we believe, can be easily found by the reader himself.

The above discussion concerned transformations of an individual object-system. Dealing with transformations of multitudes of object-systems, one would draw a different conclusion.

PROPOSITION 4

A multitude of object-systems within a system of objects of the same kind, by virtue of its existence, is transformed according to laws $z \in \{Z_i\}$ either to itself, by means of the identical transformation, or to other multitudes of object-systems, by means of 254 and only 254 different ways.

The increase in the number of the transformation types, from 8 to 255, has a simple reason: a transformation of a multitude of object-systems to other multitudes can occur under the action not only in a single one of 8 ways mentioned, but also by means of any 2 of 8, 3 of 8, . . . , or all 8 of 8 possible ways. Hence we have $\Sigma_{i=1}^{8} C_{u8}^{i} = 2^8 - 1 = 255$.

The calculation is valid, of course, only under the assumed conditions. If we discriminate, for instance, between the orders of transformations (as it may be essential for a study of the temporal behaviour of a process), and between multiplicities of the employed transformation types, the number of different modifications can be increased to infinity.

Thus we have described all the system transformations which are possible from the point of view of GST(U). Now we can analyse them from the point of view of theory of groups. It should be emphasized that in the consideration of sets of system transformations and antitransformations which is given below we have a single aim, to prove that these sets are symmetric, at least with respect to the selected composition laws, i.e. they are groups. Therefore we leave open the question what is the meaningful interpretation of these groups and, first of all, the composition laws associated with them.

PROPOSITION 5 (proved by A. V. Malikov)

The set of 8 transformations with the composition law T given by the Cayley scheme of these transformations is a 8-th-order group.

The scheme is given in Table 3.

Inspecting the scheme one can verify that (1) for any pair of the transformations their composition is again one of the 8 transformations, (2) the composition of any three transformations is associative, i.e. for example $(QnQlTQlR)TI = QnQlT(QlRTI) = QnR$, (3) an identity transformation I exists, its composition with any non-identical transformation is again the non-identical transformation, e.g. $ITQl = QlTI = Ql$, (4) any transformation has an inverse transformation; the composition of both is identity I (in the present case every transformation is its own inverse), (5) the composition law T is commutative, as the table is symmetric with respect to its diagonal connecting the upper left angle with the lower right angle; i.e. for any pair of transformations (a, b) one has $aTb = bTa$. (Such groups are known as *Abelian*, in honour of the Norwegian mathematician Neils Henrik Abel, who employed these groups for the first time in the theory of algebraic equations.)

Using Lagrange's theorem (1771) ("for any finite group, the order of every subgroup is a divisor of the group order") and Sylow's theorem (1872) ("a group G of order g contains a subgroup of order h if h is a divisor of g, and $h = p^n$, where p is a prime number and n is any positive integer"), one can show that there are 7 second-order subgroups, 6 fourth-order subgroups, 1 first-order subgroup and 1 eighth-order subgroup (15 subgroups in all).

Table 3. The Cayley scheme for the 8-th-order group of system transformations

T	I	Qn	Ql	R	QnQl	QnR	QlR	QnQlR
I	I	Qn	Ql	R	QnQl	QnR	QlR	QnQlR
Qn	Qn	I	QnQl	QnR	Ql	R	QnQlR	QlR
Ql	Ql	QnQl	I	QlR	Qn	QnQlR	R	QnR
R	R	QnR	QlR	I	QnQlR	Qn	Ql	QnQl
QnQl	QnQl	Ql	Qn	QnQlR	I	QlR	QnR	R
QnR	QnR	R	QnQlR	Qn	QlR	I	QnQl	Ql
QlR	QlR	QnQlR	R	Ql	QnR	QnQl	I	Qn
QnQlR	QnQlR	QlR	QnR	QnQl	R	Ql	Qn	I

The existence of 7 second-order subgroups (which are also groups) indicates that any non-identical transformation combined with the identity generates a second-order symmetry group with respect to the composition law T. This is a clear evidence that *absolutely every type of the system transformations, under certain conditions, has a harmony, a completeness and self-consistence*!

In the historic time scale, every non-evolutionary type of the system transformation looks like a cell, an embryonal form of an evolutionary system transformation. Therefore with respect to History (of inanimate or animate Nature, of Society) the identical transformation plays the role of statigenesis, the quantitative transformation—that of quantigenesis (occurring in two forms, as progress or regress), the qualitative transformation is qualigenesis, the relative transformation is the isogenesis (a one-level development), . . . , the quantitative-qualitative-relative transformation is the quanti-quali-isogenesis. Respectively, the identical and non-identical transformations generate the stati- and neo-genesis, the group and its subgroups of 8 non-evolutionary system transformations are prototypes of the (mathematically isomorphic to them) group and its subgroup of 8 evolutionary system transformations.

A new step in the study of transformations can be performed by means of a dialectic device—branching every transformation to n transformation-antitransformation pairs, i.e. n pairs of opposite types. (By the way, in every type of 4 basic transformations we have mentioned both their "+" and "−" forms.) Then from 8 transformations involved in the Central Propositions we derive 27 system antitransformations: 1 for I, 2 for Qn, Ql or R, 4 for QnQl, QnR or QlR, and 8 for QnQlR. In particular, for Qn one has $+$Qn and $-$Qn, for QnQl one has $+$Qn$+$Ql, $-$Qn$-$Ql, $+$Qn$-$Ql, and $-$Qn$+$Ql, etc. The antitransformation for I is I itself.

PROPOSITION 6

The set of 27 antitransformations with a composition law F is a 27-order Abelian group.

In order to give a schematic presentation of the action of the compsition law F, one would draw a square Cayley table, putting symbols of 27 antitransformations in its first column and the first row, while their compositions are posed in cells at intersections of the corresponding columns and rows. The result is a table containing $27 \times 27 = 729$ cells. Such a tremendous table is, however, not necessary: one can do with a representative fragment of the table (Table 4).

The fragment enables one to verify immediately the validity of all 4 axioms of the group theory. The theorems by Lagrange and Sylow indicate that there are 13 subgroups of the 3rd order, 3 subgroups of the 9th order, one subgroup of the first order and one of the 27th order (18 subgroups in all). The presence of 13 subgroups of the 3rd order is an evidence that any pair of opposite types for each of 8 transformations combined with the identical transformation and provided with the composition law F is a perfectly harmonical thriade. This is clear also from the fragment given in Table 4, which represents one of 13 subgroups of the 3rd order.

Table 4. A fragment of the Cayley table for the 27-order group of system antitransformations

P	I	+Qn	−Qn
I	I	+Qn	−Qn
+Qn	+Qn	−Qn	I
−Qn	−Qn	I	+Qn

As in the above discussion, in application to History, the group of 27 nonevolutionary system antitransformations and its subgroups turn out to be the (mathematically isomorphic to them) group of 27 evolutionary system antitransformations and its subgroups. The presence of both '' + '' and '' − '' realizations also for the evolutionary system transformations, in particular for the quantitative transformation (the quantigenesis) is confirmed in a quite convincing manner, say, by the following arguments.

In his classical monograph ''Oligomerization of homologous organs as one of the cardinal ways of evolution of animals'' (1954) V. A. Dogel summarized the immense material on various classes of animals and used it to classify processes taking place in their evolution. The processes are (1) polymerization—an increase in the number of homologous organs, (2) oligomerization—a decrease in the number of homologous organs, (3) a change of polymerization by oligomerization, and (more rarely) of oligomerization by polymerization, (4) polymerization in some organs and oligomerization in other organs and (5) a combination of polymerization with decentralization and desintegration, and a combination of oligomerization with centralization and integration of the organism, with its higher differentiation, more sophisticated organization, etc.[8].

Thus GST is a highly developed theory of evolution and development, that can be established within the higher (general-system, philosophical) synthesis in terms of two new categories, ''the form of matter evolution'' and ''the form of matter development''. However, this fact involves an opposite, complementary to them, general-system and philosophical category, namely, ''the form of the matter conservation''.

The 8 items of the Central Proposition and the respective 1 first-order subgroup and 7 second-order subgroup are in correspondence with the same number (8) of the conservation cases, taking place in the inanimate and animate nature and society, namely: (1) Qn, Ql, R, Z; (2) Ql, R, Z; (3) Qn, R, Z; (4) Qn, Ql, Z; (5) R, Z; (6) Ql, Z; (7) QN, Z; (8) Z. Here the symbols Qn, Ql, R, Z stand for 4 basic forms of the conservation, for quantity, quality, relation, and composition law of the primary elements. The examples of the first and the second forms are the conservation laws for electric, baryonic and leptonic charges which hold in quantum theory. An example of the conservation of relation is the law according to which the velocity of light in vacuum is constant. Finally, the fourth form has an example in the invariance of the physical laws with respect to the charge-space-time inversion, i.e. the Pauli–Lüders CPT transformation. The conclusion we draw is that the complementary categories ''the form of matter evolution'' and ''the form of matter conservation'' have their prototypes in quite real and fundamental properties of the world.

As we have seen, the existence of the second- and third-order subgroups in the 8-th and 27-th groups, respectively, is an evidence in favour of a symmetry in every separate system transformation which manifests itself under appropriate conditions. However, the symmetry of a separate system transformation has been investigated in the framework of GST(U) only for the case of the relative transformation. The work along this line has led us to a system theory of isomerism, and then also to an explication of the ''isomerism–symmetry'' correspondence. The detailed investigation of this correspondence has been completed with a derivation of the isomerism groups. Strange as it may seem, neither mathematicians, nor naturalists and sociologists realize the existence and significance of this correspondence to a sufficient extent. In the following, we first present principles of the system approach to the isomerism, as soon as it is necessary, and then, having got the information we need, explicate the ''isomerism–symmetry'' correspondence.

GENERAL THEORY OF ISOMERISM. ISOMERISM AND SYMMETRY

This branch of GST investigates the third main mode of generation of object-systems—replacing old relations between the primary elements by new relations.

PROPOSITION 7. (The law of isomerization)

Isomerism occurs in systems where object-systems, replacing old relations between the primary elements by new relations between the same elements, are transformed to two or more object-systems which are different in their inter-element relations.

Proof. Isomerism is such a system of objects of the same kind which consists of object-systems identical in the contents (number and types) of primary elements, but different in relations between the elements. From the mathematical point of view, an isomer is a permutation and isomerism is a set of permutations, or arrangements of *n* primary elements taken *n* at a time. So the system must have isomerism, by definition.

All forms of matter motions satisfy the conditions of the law of isomerization. Therefore isomerism *must* be present everywhere, as it was confirmed by the discoveries of isomerisms in chemistry (Woehler, Libich, Berzelius, 1822–1830), nuclear physics (Hahn, 1921), biology (Urmatsev, 1956–1957), sociology (Urmantsev, 1974), geology (Sharapov, Zabrodin, 1977–1979). It is remarkable that in geology the discovery of isomerism and its careful investigation were performed along the line initiated by predictions of the author's version of GST and owing to an extensive use of the general theory of isomerism developed in the framework of GST(*U*).

Examples of isomerisms in chemistry, nuclear physics, biology, linguistics, and sociology have been presented in the author's monograph "Symmetry of Nature and the Nature of Symmetries"[3]. V. Yu. Zabrodin has exposed numerous examples of isomerisms in geology in his monograph "System Analysis of Disjunctives"[9].

The law of isomerization governs not only the forms of matter motion, but also the forms of matter existence. A realization of this fact was the starting point which enabled us to extend substantially the conventional theory of isomerism and to conclude on the existence not only isomer structures (bodies), but also isomer spaces, isomer motions, isomer times[3]. In Table 5 we present the list of 4 fundamental and 64 fundamental and derivative isomerisms of principal form of the matter existence, and 63 isomerisms in this list are new, while 15 of them are related only to space, time and motion. (Probably, the reader has already noticed that this table,

Table 5. List of 64 possible fundamental isomerisms and symmetries (60–61 among the symmetries are new, 63 isomerisms are new)

No.	Isomerism (symmetry)	No.	Isomerism (symmetry)	No.	Isomerism (symmetry)
1	S	22	TDS	43	DSTM
2	T	23	STM	44	DTSM
3	D	24	SMT	45	TSDM
4	M	25	MST	46	TDSM
5	ST	26	MTS	47	STMD
6	TS	27	TSM	48	SMTD
7	SD	28	TMS	49	MSTD
8	DS	29	TDM	50	MTSD
9	SM	30	TMD	51	TSMD
10	MS	31	DTM	52	TMSD
11	TD	32	DMT	53	TDMS
12	DT	33	MTD	54	TMDS
13	TM	34	MDT	55	DTMS
14	MT	35	SDM	56	DMTS
15	MD	36	SMD	57	MTDS
16	DM	37	MSD	58	MDTS
17	STD	38	MDS	59	SDMT
18	SDT	39	DSM	60	SMDT
19	DST	40	DMS	61	MSDT
20	DTS	41	STDM	62	MDST
21	TSD	42	SDTM	63	DSMT
				64	DMST

Note. S: space, T: time, D: dynamic, M: matter isomerism (symmetry).

as well as the next Table 6, are also tables of the corresponding symmetries; the reason owing to which the tables have the double significance will be explained below.)

Examples of isomer spaces, isomer motions, isomer times are given in the mentioned book by the author. Here we present only examples of isomer spaces.

Evidently, in accordance with the law of isomerization, the space isomerism must be considered as the phenomenon of the existence of a variety of spaces having the same contents, but with different relations between elements. For example, there are pairs of left and right dissymetrical spaces—continuums, semicontinuums, discontinuums, the classical symmetry of which is exhausted just with the first-kind elements. Clearly, from the point of view of the theory of dissfactors[5], or, say, the theory of multiple antisymmetry[10], every isomer variety of this type may contain not only a pair of such spaces, but also more of isomer spaces. Another example is the variety of space states which are transformed into each other because of various automorphisms—one-to-one mappings of the space onto itself.

The classifications of isomerisms by the type of the operations transforming one isomer structure into the other suggested to the author a derivation of 54 structure isomerisms, 53 of which turned out to be essentially new. Those are crypto-, single and multiple anti- and/or colour isomerisms—classical, homothetic, conformal, affine, projective, topological. The results are given in Table 6.

At present, models have been constructed for many of the 54 isomerisms (cf. e.g. [3,11]). Besides, a possibility has been found for doubling, trebling etc. the number of the structure isomerisms owing to a change in the law according to which the properties (+ or −, colour, and others) are combined with themselves, as well as with basic geometrical transformations (Euclidean, homothetic, conformal and others).

An investigation of the optical (to be more exact, dissymmetrical) isomerism, a phenomenon known in stereochemistry, performed by the author[11,12] within GST, resulted in an evidence of the existence of three types of diss-isomerisms: the first type (old) with the number of isomers $S = S_{k_0}^{k_0} = 2^{k_0}$ (its examples are the isomerisms of aldohexose and lime-tree leaves), the second type (new) the isomer number for which is

$$ S_{k_0 + k_1 + \ldots + k_n}^{\rho} = \sum_{i=0}^{\rho} \left[\frac{k_0!}{(\rho - i)!(k_0 - \rho + i)!} 2^{\rho} \sum \Pi_{k_j}^{i} \right], $$

(its examples are the isomerisms of pyranohexose and isolated roots of some plants), and the third type (also new) with the isomer number $S_{0+k_1}^{1} = 2k_1$ (its examples are the isomerisms of pyranohexose with $k_0 = 0$ and cyclic corollas with an odd number of overlapping petals).

Table 6. List of 54 possible structure isomerisms and symmetries (40 among the symmetries are new, 53 isomerisms are new)

No.	Isomerism (symmetry)	No.	Isomerism (symmetry)	No.	Isomerism (symmetry)
1	*classical*	19	*affine*	37	*projective*
2	anti-	20	aff. anti-	38	proj. anti-
3	mult. anti-	21	aff. mult. anti-	39	proj. mult. anti-
4	colour	22	aff. colour	40	proj. colour
5	mult. col.	23	aff. mult. col.	41	proj. mult. col.
6	col. anti-	24	aff. col. anti-	42	proj. col. anti-
7	col. mult. anti-	25	aff. col. mult. anti-	43	proj. col. mult. anti-
8	mult. col. mult. anti-	26	aff. mult. col. mult. anti-	44	proj. mult. col. mult. anti-
9	crypto-	27	aff. crypto-	45	proj. crypto-
10	*homothetical*	28	*conformal*	46	*topological*
11	hom. anti-	29	conf. anti-	47	top. anti-
12	hom. mult. anti-	30	conf. mult. anti-	48	top. mult. anti-
13	hom. col.	31	conf. col.	49	top. col.
14	hom. mult. col.	32	conf. mult. col.	50	top. mult. col.
15	hom. col. anti-	33	conf. col. anti-	51	top. col. anti-
16	hom. col. mult. anti-	34	conf. col. mult. anti-	52	top. col. mult. anti-
17	hom. mult. col. mult. anti-	35	conf. mult. col. mult. anti-	53	top. mult. col. mult. anti-
18	hom. crypto-	36	conf. crypto-	54	top. crypto-

Note. mult.: multiple, col.: colour.

CAMWA12:1/2(B)-Z

In previous works[3,5] we have proved the existence of three types of dia-stereo-isomer-isms, as well as of dissymetrical-non-dissymmetrical and non-dissymmetrical isomerisms. We have shown that the phenomena of antipodes, dia-stereo-, cis-, trans-, ortho-, meta-, para-isomers, of tautomerism, of conformation, which were considered earlier as if they were specific for chemistry, are not restricted to any separate branch of science. It was also shown in the mentioned works that the phenomena of antipodes and dia-stereo-isomers, which were previously related only to the dissymmetrical isomerism in chemistry, are in fact present also in isomerisms of some non-dissymmetrical types.

Moreover, linguistic models enabled us to make a number of observations:

(1) For certain object-systems isomerism may be generated by any of 7 non-identical transformations.

(2) The R transformation may result in a number of transitions: (i) an isomer set transformed into another isomer set (say, the set {tom, omt} into the set {tmo, mto}), (ii) a non-isomer set transformed into a non-isomer set (say, {tom} into {omt}), (iii) a non-isomer set transformed into an isomer set and vice versa (e.g. {tom} → {tmo, mto}, or {tmo, mto} → {tom}).

(3) Each of 7 non-identical transformations can generate an isomerism from an original object with no change in its composition for instance by means of a quantitative transformation, as in the scheme

$$\text{tom} \xrightarrow{-t,\ -m} \text{o} \xrightarrow{+m,\ +t} \begin{bmatrix} \text{mot} \\ \text{tmo} \end{bmatrix}$$

the inverse transformation is also possible (e.g. as in the above scheme with reversed arrows).

(4) In accordance with Proposition 4, a non-isomer set of object-systems can be transformed into a non-isomer set and vice versa in 254 different ways.

(5) If two isomers I_1 and I_2 (or any two objects A and B) are different in at least one feature (in the case of isomers it is the structure), then they are different in infinitely many relations to other objects (in the case of isomers, in their properties).

(6) If two isomers I_1 and I_2 (or any two objects A and B) are different in at least one relation to other objects, then they must be different in at least one feature (in the case of isomers, in their structure).

(7) Isomerism is intimately related to symmetry.

Let us discuss the last point in more detail.

The relation between isomerism and symmetry is proved by means of the theory of permutation groups. Actually, this theory and the relevant mathematical ideas in general have a meaningful interpretation as the theory of isomerisms. In fact, from the mathematical point of view, isomer is a permutation, isomerism is a set of permutations, isomerization is a substitution the upper line of which means the object of the isomerization, while the lower line is the result of the process, subsequent isomerizations are the product of substitutions. The set of all substitutions with the product as the composition law is the group of permutations. Thus the set of all isomerizations with the product operation is also a group, the group of isomerism, so it reveals an "isomerism symmetry". Clearly, the latter is the invariance of the isomers in their composition with respect to the isomerism operations. Owing to these operations, i.e. isomerizations, some isomers in the set are transformed into other isomers in the same set, and the whole set is mapped to itself conserving the contents of primary elements and composition of the isomers. The arguments presented suggest the following proposition.

PROPOSITION 8

Any finite group of all isomerizations of n-th order, a group I_n, is isomorphic to the n-th order group S_n of all substitutions.

The mathematical isomorphism of the theory of groups of substitutions to the theory of groups of isomerisms permits an immediate extension of results from the former to the latter. In particular, we have the following.

THEOREM (Cayley)

Any finite group of order n is isomorphic to a subgroup of the group of substitutions of n-th degree.

The analogue of this theorem for isomerism is the following:

PROPOSITION 9

Any finite group of order n is isomorphic to a subgroup of the group of all isomerizations of n-th degree.

Hence the next proposition follows.

PROPOSITION 10

Any finite symmetry group of order n is isomorphic to a subgroup of the group of all isomerizations of n-th degree.

The isomorphism between symmetry and isomerism, established here at least for the finite groups, permits one to draw a number of conclusions, transferring the results from one branch of science to the other. The most important conclusions are as follows.

(1) The lists of 64 fundamental and 54 structure isomerisms can be considered also as lists of 64 fundamental and 54 structure symmetries. This is the reason why the Tables 5 and 6 presented above are also the tables of symmetries, both known and discovered for the first time.

(2) One can introduce the concepts of continuous and discrete isomerism transformations, finite and infinite isomerism groups.

(3) One can introduce the concept of the isomerism dimensionality. An isomerism will be considered as n-dimensional ($n = 0, 1, 2, 3$), if any isomer in the set has an n-dimensional symmetry, be it point-wise, linear, planar, or spatial. For example, the isomerism of asymmetrical aldohexoses of the composition $C_6H_{12}O_6$ or that of asymmetrical lime-tree leaves is 0-dimensional, as any of 16 isomers of the aldohexose compound or leaf has a point symmetry group (1). Isomerism of plant sprouts with left or right configurations of leaves is one-dimensional, since every sprout isomer has a one-dimensional, or linear, symmetry described by one of the "rod" symmetry groups. Sometimes, however, the isomer symmetry is changed by isomerization. For instance, depending on the ion strength and the solution temperature, RNA molecules may exist either as tangels having a point symmetry, or as threads having a one-dimensional symmetry. Respectively, such objects have not just n-dimensional symmetry, but an $n_1 - n_2 - \ldots - n_k$ -dimensional symmetry. In the case of RNA the symmetry is 0–1 - dimensional.

(4) A new idea is advanced on a possible development of a theory describing groups with $n_1 - n_2 - \ldots - n_k$ -dimensional symmetries for which the object dimensionality is not invariant under the syummetry transformations.

(5) A conclusion new for the classical theory of structure symmetries is drawn: it is possible to realise any dissymmetrical (left or right) or non-dissymmetrical object not only in two or one modifications, respectively, but also in numerous, in a limit-infinite, modifications. This is a direct consequence from, say, a possible existence of any dissymmetrical or non-dissymmetrical isomer in two or more isomer (or, more generally, polymorphic) modifications. This conclusion can be drawn also from the theory of dissfactors developed by the author[3,5].

It is not out of place to conclude this Section with an eulogy to isomerism; the phenomenon deserves it completely. Actually, though in view of the law of isomerization the isomerism is not a universal property (it is specific only for a certain type of systems), the requirements of this law are fulfilled for special cases of every form of motion and every form of the matter existence; more than that, such cases may be found in every branch of science. This fact alone is sufficient reason to adopt the concept of isomerism as a universal category for science as a whole. However, the isomerism is significant not only for natural sciences, it has a fundamental philosophical meaning; first of all, because of the fact that the relative mode of transformations of object-systems, i.e. a transition from old relations between primary elements to new ones, is not an ordinary mode of transformation, but a primary, indecomposable and irreducible, specific form of the matter evolution. Recall that from the point of view of GST(U) the relative form of the matter evolution is one of 4 basic primary types of transformation of old object-systems into new ones.

Unfortunately, the universal character and the fundamental nature of isomerism, owing to its direct relevance to genesis, symmetry, composition-structure-properties of objects in inanimate and animate nature and human society, have been realized to full extent neither by philosophers, nor by investigators in other fields, natural sciences or humanities.

Isomerism is a special case of polymorphism ("isomeric polymorphism"). Therefore one would decide that now we should turn to a consideration of polymorphism and its generating system transformations, quantitative-relative and quantitative-qualitative-relative. However, the symmetry of such transformations is still practically unknown in the framework of GST, though a number of other aspects of polymorphism have been elaborated within GST in a sufficiently complete manner. Therefore we will not dwell on polymmorphism and concern directly its necessary and tantamount complement—isomorphism. Below we start from a definition of system isomporphism, and then omitting for brevity a number of relevant system propositions we shall undertake an explication of the nature of the relation "system isomorphism—symmetry". The revelation of this relation will make it possible to describe the symmetry of a system as itself.

System Isomorphism and Equivalence

It is not a simple task to propose an adequate definition for "isomorphism". In the literature on the system theory it is taken as something primordial, in contrast to GST advocated by the author where it is introduced at some stage of the theory as an opposite to polymorphism. Most people treat isomorphism unreasonably just as the mathematical isomorphism. Meanwhile, there is a concept of isomorphism in natural sciences, which was originated from Rome del'Liesl, Le Blanc, Haüy, Bedans, and formulated in a "complete variant" by E. Mitscherlich who investigated a number of phosphates and arsenates (1819–1821). By suggestion of Jacob Berzelius, Mitscherlich named the new phenomenon "isomorphism"—the property of having the same form (cited from [13]). Afterwards the concept was adopted in mathematics, and then in other sciences, including humanities and biology, and has been reduced to just a similarity of a usually high degree and mainly in morphological features (though this is not at all obligatory).

Clearly, the mathematical definition of isomorphism cannot be adopted as basic in GST. It was our opinion that, in accordance with the spirit of GST, which is an urge towards a maximum generality, pithiness, synthesis, the appropriate definition must be satisfactory for mathematicians and scientists, but it should not coincide with their particular interpretations. Having this argument in mind, we introduced the new term "system isomorphism".

DEFINITION 4

System isomorphism is a relation R between object-systems within the same system of type-S objects, which is specified as a subset of Cartesian product of the system S, $R \subseteq S_c \times S_c$, selected by compared features $F_1, \ldots, F_k$ and having two properties: (1) reflexivity, by virtue of which any object-system $a \in S$ is in the system isomorphism to itself, aRa; (2) symmetry, by virtue of which, if an object-system a is in the system isomorphism to an object-system b, the latter is also in the system isomorphism to a, i.e. if aRb one has bRa.

We emphasize the system character, pithiness and generality of this definition.

Definition 4 has a system character, as the system isomorphism is represented as a subset which is reflexive, symmetrical and has features $F_1, F_2, \ldots, F_k$, contains pairs of similar object-systems and belongs to the system $S_c \times S_c$. Primary elements of this subsystem are all elements of object-systems of the system S, or only a part of them, the unity relations are the combination relations, the composition law is the two properties and the requirement that the elements possess some given features.

The definition is meaningful and general, since system isomorphism appears, on the hand, as a system explication of the similarity relation which is fundamental for science, arts, and practice, on the other hand, as a synthetic general category, the superlative of which is "identity" and the most common mode is "partial resemblance". An important particular case is "equivalence" having numerous special forms, among which more significant for our purpose are the relations of "equality" and "mathematical isomorphism". It is the concept of "equivalence" that makes it possible to explicate the relation "system isomorphism–symmetry".

DEFINITION 5

Equivalence is a relation R between object-systems within the same system of type-S objects defined as a subset of Cartesian product of the system S, $R \subseteq S_c \times S_c$, which has the following properties: (1) reflexivity, by virtue of which any object-system, $a \in S_c$, is equivalent to itself, aRa; (2) symmetry, by virtue of which, if an object-system a is equivalent to another object b, then b is equivalent to a, i.e. if aRb then bRa; (3) transitivity, by virtue of which if an object-system a is equivalent to b, and b is equivalent to c, then a is equivalent to c, i.e. if aRb and bRc then aRc.

The definitions of system isomorphism and equivalence are almost identical, up to point (3). This is done in order to make more manifest the fact that the second concept is a particular case of the first concept. However, this was not our main purpose; we aimed to investigate the relation "system isomorphism–symmetry" by means of the concept of equivalence.

There are at least two ways to approach the above problem: first, by means of the concept of "equality", a particular case of the equivalence relation; second, by means of a derivation of the correspondence and symmetry laws using the concept of equivalence. Let us consider these points in more detail.

Equality–symmetry

In GST(U) objects O *are considered as equal in features* F *if they can be made indistinguishable in these features after introduction of modifications* M. If we compare this definition with the definition of symmetry (cf. Introduction) and replace the word "coincidence" in the latter by "equality", it is seen that symmetry is equality or, at least, a category essentially based upon the concept of equality.

It is remarkable that each of 4 axioms of group theory (the closure axiom—indirectly, and other three—quite literally) do also state some equalities, so the group-theoretical viewpoint supports the above conclusion on symmetry.

The concept of "equality" is in a similar situation. If we replace the word "equal" in the above definition by "symmetrical", it is seen that equality is symmetry or, at least, a concept essentially based upon symmetry. This is also in agreement with the properties of the equivalence relation, as well as of the equality relation, namely, reflexivity, symmetry, transitivity, as these properties are equivalent to three group axioms, namely, on the neutral element, on inverse elements, and on the closure of group.

The double analysis, that of symmetry based on equality and of equality based on symmetry, prompts us positively to a very bold, though apparently simple, conclusion that symmetry is equality, equality is symmetry, and both concepts are identical. Respectively, asymmetry is inequality, inequality is asymmetry, and these concepts are also identical.

Independently of a conclusion one draws on the nature of the "equality–symmetry" relation, the fundamental significance of equality for symmetry, and that of symmetry for equality, is quite evident. Therefore it would be not out of place to present a brief analysis of the theory of symmetry, and the history of its development, basing on the definition of equality.

It follows from the definition that equality is a relative property. It will be shown below, taking the theory of the structure symmetry as an example, that it is a relativistic understanding of equality that was an implicit basis put by the people who developed the theory of symmetry under all the relevant constructions, both classical and non-classical schemes, elaborated during the past 60 years.

In fact, in the case of the coincidence and/or mirror equalities adopted in the classical theory of symmetry, the features to be compared are the configurations of objects, while the identifying operations are, in the first case, translations and/or rotations and, in the second case, mirror reflections (in a point, line, plane, space) and combinations of the latter ones with non-mirror motions.

As to the non-classical theories of the structure symmetry, single and multiple anti-symmetry, colour symmetry, colour single and multiple anti-symmetry, crypto-symmetry, P-symmetry, complex symmetry, dissfactors, Q-symmetry, W-symmetry, homothety, affine, conformal and curvilinear symmetries, symmetries in multi-dimensional Euclidean, non-Euclidean, pseudo-Euclidean spaces, the situation is quite similar. For example, we refer to the theory of the structure antisymmetry, because most other non-classical theories of symmetry are either its further extensions, or overlap with it substantially.

In the theory of antisymmetry, a left white glove is treated as equal (or "anti-equal") to right black one, and vice versa, because there is such a combined reflection in a plane ("anti-plane") which transfers left to right, right to left, black to white, white to black, and the whole figure of two gloves is transformed to itself.

The above arguments suggest some conclusions.

First, new theories of symmetry treat as equal also such objects (such equalities) which were considered as essentially different in previous theories (respectively, as inequalities). The unique reason why these equalities have been adopted is always the same thing, i.e. the existence of real or/and mental operations making the objects O, compared in features F, indistinguishable.

Second, because of the discovery of new identifying operations, the adoption of new types of equality is not just a simple reformulation of known facts (in terms of the theory of anti-, colour or curvilinear symmetry); it has always resulted in an advancement, since in each case the new approach enabled one to derive more completely and accurately the number, structure and type of the symmetry, possible for objects of a given class, and hence to predict the number, structure and type of all polymorphic modifications which are possible in principle. It is sufficient to recall in this connection that instead of 32 zero-dimensional, 75 one-dimensional, 80 two-dimensional, and 230 (219) three-dimensional classical crystallographic groups, in the theory of anti-symmetry there are, respectively, 122, 394, 528 and 1651 crystallographic Shubnikov groups which permit one to analyse more accurately the shapes of the crystal polyhedra, the form and structure of their vertices, edges, faces, the properties of the crystal in the whole, its physical and chemical characteristics.

A concluding remark. A generalized understanding of equality is of considerable methodological importance, since it suggests an idea that there are not just a few dozen equalities, known at present in various theories of symmetry, but an infinite sequence of equalities. Thus it gives a direction and an impact to the scientific research. Simultaneously, this approach involves in the theory a variety of unknown, peculiar equalities and symmetries, if corresponding identifying operations are introduced in a proper way. In this sense, the problem of development of new theories of symmetry is now an almost trivial task, since the most exciting aspect of symmetry, "what is equality?", is no longer a sacramental question. Meanwhile, with all this in mind, one has to admit that equal and inequal, identical and different, conserved and changed, resting and moving, symmetrical and asymmetrical, true and false under a set of transformations in one theory, may turn out and in fact are indeed, respectively, inequal and equal, different and identical, changed and conserved, moving and resting, asymmetrical and symmetrical, false and true in another theory, with respect to a different set of symmetry transformations.

Laws of correspondence and symmetry

Formally, a system of objects of kind i can be considered as a finite or infinite set of object-systems specified by means of a principle A_i which incorporates $a_i \in \{A_i^{(0)}\}$, $r_i \in \{R_i\}$, $z_i \in \{Z_i\}$. This identification permits one to extend concepts and theorems of the theory of finite and infinite, fuzzy and non-fuzzy sets directly to GST and to develop the latter also (but not exclusively) as a theory of finite and infinite, fuzzy and non-fuzzy systems.

In particular, it is by means of a simple transfer of the results that we shall prove below the existence of the laws of correspondence and symmetry which are important for GST. However, before presenting the definitions and set-theoretical schemes of the proofs, we will draw attention to some comments necessary for understanding of the matter.

As in the set theory, we assume that an infinite system of object-systems of a kind B, $S_B = \{a, b, c, \ldots\}$, has the same cardinality as an infinite system of object-systems of a kind C, $S_C = \{\alpha, \beta, \gamma, \ldots\}$, if there is a one-to-one correspondence between the object-systems belonging to these systems at least by a single law $(\alpha)f = a$, where f is a functional relation law. Consequently, it can be said that S_C has cardinality equal to that of S_B, and the notation $|S_C| \sim |S_B|$ is adequate, where the wavy line is simultaneously the symbol of equivalence, since the relation specified in this way is an equivalence relation.

Evidently, for finite systems of objects the equal cardinality is just an equal number of object-systems, so the cardinality is a generalization of the number of elements. In analogy to a pair of systems, of kinds B and C, having n_1 and n_2 elements, respectively, where a single one of three possible relations, $n_1 = n_2, n_1 > n_2$ or $n_1 < n_2$, takes place, for two infinite systems of objects, S_1 and S_2, with cardinalities m_1 and m_2, only one of three relations is possible, $m_1 = m_2, m_1 > m_2$, or $m_1 < m_2$.

PROPOSITIONS 11, 12

The laws of correspondence and symmetry. For any two systems of object-systems, S_1 and S_2, relations of only 4 types are possible:

1. S_1 and S_2 are equivalent and mutually symmetrical.
2. S_1 contains a proper part equivalent and symmetrical to S_2, and S_2 contains a proper part equivalent and symmetrical to S_1.
3. S_1 contains a proper part equivalent and symmetrical to S_2, and S_2 contains no proper part equivalent and symmetrical to S_1.
4. S_2 contains a proper part equivalent and symmetrical to S_1, and S_1 contains no proper part equivalent and symmetrical to S_2.

A relation of the fifth type, where neither S_1 contains a proper part equivalent and symmetrical to S_2, nor S_2 contains a proper part equivalent and symmetrical to S_1, is impossible.

Proposition 11, the law of correspondence, is proved by means of Zermelo's axiom of choice, as in the set theory. Besides, it is important to have in mind that, according to the Kantor–Bernstein theorem (if each of two sets (systems) is equivalent to a part of the other one, the sets are equivalent), the case 2 is, in fact, the same as the case 1. It is clear hence that the relations $m_1 = m_2, m_1 < m_2, m_1 > m_2$, where m_1 and m_2 are cardinalities of S_1 and S_2, respectively, are incompatible.

Proposition 12 is the law of symmetry; the existence of a symmetry of one of 4 types (more exactly, in view of the Kantor–Bernstein theorem, 3 types) stems, at least, from the following arguments.

First, it is a consequence of the fact that the equivalence relation (in the present case, the equal cardinality) which holds in some way between the system, does already contain the condition of a mutual symmetry, as we have seen in the above analysis of the "equality–symmetry" relation.

Second, it results from the fact that the one-to-one mappings, involved in the establishment of 4 (3) equivalence types mentioned in the correspondence law, are always sets of mappings which are mathematical groups with an appropriate mapping composition law. In fact, the set present in type 1 contains the identity mapping ϵ transforming every element $k \in S_i$ ($i = 1, 2$) into itself, in type 2 every mapping $\alpha: a \rightarrow a', a \in S_1, a' \in S_2$, has an inverse α^{-1}: $a' \rightarrow a$, in type 3 any two mappings α, β are present together with their product $\alpha \cdot \beta$.

Having in mind the problems stated in the Introduction, we shall consider the law of symmetry in more detail. The law of symmetry asserts that (i) there exists an intersystem symmetry for any two systems of kinds A and B, (ii) there is an intrasystem symmetry, a symmetry of the system as itself, if S_A and S_B are considered as subsystems of a new system S_C.

Evidently, the number of intersystem symmetries would be larger than 4 (or 3), if the systems S_A and S_B are compared in their system-generating parameters, i.e. in (1) m; (2) r; (3) z; (4) m, r; (5) m, z; (6) r, z; (7) m, r, z. Respectively, there are 7 sets for the system S_A: $\{M_A\}, \{R_A\}, \{Z_A\}, \{M_A, R_A\}, \{M_A, Z_A\}, \{R_A, Z_A\}, \{M_A, R_A, Z_A\}$, and 7 sets for the system S_B: $\{M_B\}, \{R_B\}, \{Z_B\}, \{M_B, R_B\}, \{M_B, Z_B\}, \{R_B, Z_B\}, \{M_B, R_B, Z_B\}$. In turn, various equivalence and symmetry relations can be found between any sets in the first group and any sets in the second group, $7 \times 7 = 49$ types in all (like $\{M_A\} \sim \{M_B\}, \{M_A\} \sim \{R_B\}, \ldots, \{M_A, R_A, Z_A\} \sim \{M_B, R_B, Z_B\}$), and as there are 3 basic types (mentioned in the laws of correspondence and symmetry) the total number of types is $49 \times 3 = 147$.

Similarly, we have not 4 (or 3), but 28 intrasystem symmetries, putting each of 7 sets $\{M\}, \{R\}, \{Z\}, \{M, R\}, \{M, Z\}, \{R, Z\}, \{M, R, Z\}$, relevant to systems S_A and S_B, in correspondence to itself, as well as with other 6 sets in view. With account of 3 basic types, the number of the intrasystem symmetries is not 28, but $28 \times 3 = 84$.

Thus, for arbitrary systems S_A and S_B there are $(49 + 28 \times 2) = 105$ kinds, and $105 \times 3 = 315$ types of inter- and intra-system symmetries!

A considerable progress can be attained in the investigation of the inter- and intra-system symmetries, if we take into account the fact that the conditions of the law of symmetry hold for all types of the matter motion and all forms of the matter existence: space, time, motion, and their "carrier"-substance (substratum). Taking them in various combinations and arrangements, we get, respectively, $\Sigma C_4^i = 15$, and $\Sigma A_4^i = 64$ different systems.

The first fact is responsible for the necessary existence of mechanical, physical, chemical, geological, biological and sociological symmetries, as confirmed by observations in absolutely every branch of Science and Art. It provides also with a rational explanation for the existence of all the symmetries.

The second fact is responsible for the necessary existence of inter- and intra-system symmetries in each of 15 (with account of possible orders, in each of 64) systems given in Table 5. If one would put each of 15 (or 64) systems in correspondence both with itself and with other 14 (63) systems, the number of possible symmetries (with or without an account of the presence of 3 types), the numbers of possible symmetries are, respectively, 120 and 360—for systems of 15 types, and 2080 and 6240—for systems of 64 different types. Note that the number of possible symmetries Σ_n and the completeness of the exhaustion are evaluated by means of the formula for the sum of n terms of an arithmetic progression, $\Sigma_n = (1/2)$ $(a_1 + a_n)n$, where a_1 is the first term and a_n is the n-th term of the progression. For instance, for 15 system types one has $\Sigma_{15} = (1/2)(1 + 15)15 = 120$.

The law of system similarity

In the law of correspondence, the word "equivalence" can be replaced by "system isomorphism", since the former is a particular case of the latter, which imposes on the system concerned less restrictive conditions than the former. Thus we get the law of system isomorphism—the law of system similarity, and consequently, 4 (3), 28, 49, 105, 120, 2080 system isomorphisms (according to the numbers of the symmetry types) in mechanics, physics, chemistry, geology, biology, sociology, relevant to space, time, dynamics, matter.

By virtue of the law of system similarity, the relations of the system isomorphism must take place in absolutely any pair, triplet, . . . , n-plet of systems, say, in the sequence: "substance, arrangement of stars, idea, a human fate, form, identity, beauty, life of Count Lev Tolstoy, catalase-induced desintegration of hydrogen peroxide, measure, essence, Wizard by componist G. Sviridov".

One should not think that in this arbitrary sequence we mean a universal unity in the spirit of Leibnitz's universal identity, indistinguishability. This point of view would be, at least, naive. Actually, we mean a universal system isomorphism, which may appear either as the Leibnitz identity, or as a partial similarity, or as equality, or as the mathematical or natural isomorphism, etc. That is to say, in general the system isomorphism admits a variety of realizations of the same thing by means of different primary elements or/and the unity relations, or/and the composition laws. To summarise, it reveals a variety in unity.

In the real world of systems we find a lot of evidences in favour of the system similarity. Examples of the similarity are the isomorphism between 16 isomers of lime-tree leaves and 16 isomers of aldohexoses, between 9 isomers of inosite and 9 (out of 14) isomers of barberry corolla, between cis- and trans-isomers of 1,2-dichlorethylene and cis- and trans-isomers of corolla of Viola nocturna, between right and left isomers of glyceraldehyde and right and left crab-violinists, between the general structure of genetic code, the binomial series of 2^6, icosahedron, dodecahedron, a chemical compound barena, and Radiolaria cyrcoregma dodecahedra. One can also mention similarities between homological sequences in the development of animals and plants and homological sequences of spirits and hydrocarbons, between the sequence of development of things in human material culture and the sequence in evolution of organisms, between biological evolution, biocenosis, natural selection and technical evolution, technocenosis, informational selection, between genome and language, evolutional genetics and comparative philology, between the lognormal law of distribution of galactics in Space and the same law for abundance of chemical elements in Earth's crust, distribution of structural constituents in polycrystals, distribution of animals and plants on Earth's surface, between periodical law for chemical elements and the periodical law for cyclic corollas, a rhythmic structure of chemical elements and a similar structure of the series of musical tones, etc. The number of relevant examples can be immensely increased easily. In GST(U) this must be expected, but the main point is elsewhere.

The system similarity cannot be reduced to any type of similarities known in natural sciences or sociology, in particular, to convergence or parallelism, familiar in biology. Actually, the system similarity may be not a result of a kinship or/and similar conditions of the existence. The similarity may arise just from different realizations of the same abstract system of a kind i. Such is the origin of the similarity, say, between 16 isomers of aldohexose and 16 isomers of lime-tree leaves[14], between Ohm's law of electric conductivity, Fourier's law of thermal conductivity, and d'Arsy's filtration law. The new type of similarity due only to an effect of different realizations of the same thing, has been called system community in our works.

Of course, the existence of the system community is somewhat complicating our understanding of the nature of similarity. However, neglecting it one can draw incorrect conclusions, in particular, construct wrong "life trees", as it was shown by S. V. Meyen[15] for the example of the English paleobotanist R. Melville. This reasoning suggested to the author an aphorism: "If two things are similar, the similarity is not always due to a kinship, or to identical conditions of life, or to both these reasons."

Before the appearance of GST, various correspondences, say, between qualitatively different sequences of development, or between laws discovered in different aspects of nature or society, or between number characterizing qualitatively different systems, have been never expected. Such correspondences were established empirically and, as a rule, considered by many people naively as results of purely accidental coincidences. Meanwhile, it is probably for the first time in science, such "accidental coincidences" are directly derived from a number of laws of GST, and therefore are quite expectable. It is an essential thing: such expectations combined with an algorithm of prediction of similarity developed within GST(U) give hope that it is possible, in principle, basing on known systems (say, the life of A. S. Pushkin, or Mendeleev's Periodic Table), to reconstruct e.g. texts of the 100 books by Democritus, which are supposed to be lost forever, or to predict new writings of modern or future authors. The real existence of the system isomorphism enables one, in principle, to explain the phenomenon of Jeanne d'Arc or the possibility of infinitely many performance interpretations in Art, model building in Science, versification in literature.

The law of symmetry and the whole logics of our reasoning lead us to a conclusion that absolutely every system has necessarily a symmetry in some features and under some transformations, and is asymmetrical in other respects. Such a concept of system has an important methodological and psychological significance, since it gives an investigator a hope to find symmetrical and asymmetrical properties in the system under study, prompting him to discover properties of both types.

As to the law of system isomorphism, it aims investigators to detect various similar features in different systems concerned. Both these laws have been deliberately employed by the author in the elaboration of new sections of GST, the theory of 2-, 1-, and 0-sided actions and the theory of relations of contradiction and compatibility.

Relations of 2-, 1-, 0-sided action, contradiction and compatibility

In Table 1 we have presented the space-time system of 9 actions S_a (the subscript means "action"); in Table 7 the same 9 actions are united in a group of actions with the composition law F. It is seen directly from Table 7 that (1) for any two actions a, $b \in \mathcal{G}$, their composition aFb is also an element of $\mathcal{G}$, (2) the law F is associative, since for any three actions a, b, $c \in \mathcal{G}$ one has $aF(bFc) = (aFb)Fc$. In particular, $<<F(<>F><) \rightarrow (<<F==) \rightarrow <<$, and $(<<F<>)F>< \rightarrow (>=F><) \rightarrow <<$, (3) there exists a unique F-neutral action, a quasi-0-action $==$, such that its composition with each of the 9 actions is the latter action again (see the second row and the second column in Table 7), (4) for any action a in the set there is a unique opposite action a^{-1}, such that $aFa^{-1} = a^{-1}Fa = "=="$ (see Table 8).

It is seen from Table 7 that the group is commutative, i.e. Abelian; it is of 9-th order and

Y. A. URMANTSEV

Table 7. The 9-order group of actions

F	= =	> >	< <	= <	= >	< >	> <	< =	> =
= =	==	> >	< <	= <	= >	< >	> <	< =	> =
> >	>>	< <	==	> =	> <	= <	< =	= >	< >
< <	<<	==	>>	< >	< =	> =	= >	> <	= <
= <	= <	> =	< >	= >	==	< =	> >	< <	> <
= >	= >	> <	< =	==	= <	< <	> =	< >	> >
< >	< >	= <	> =	< =	< <	> <	==	> >	= >
> <	> <	< =	= >	> >	> =	==	< >	= <	< <
< =	< =	= >	> <	< <	< >	> >	= <	> =	==
> =	> =	< >	= <	> <	> >	= >	< <	==	< =

contains, according to Lagrange's and Sylow's theorems, 6 subgroups: one of the 1st order, four of the 3rd order, and one of the 9-th order. Thus the system S_a is, in fact, symmetrical with respect to the composition law F, given in Table 7.

However, the same system S_a appears as an asymmetrical object, a groupoid (not a group) with respect to another composition law E, given in Table 9. The groupoid is constructed essentially in agreement with the common sense: in Table 9, as before, every action of Table 8 has its opposite, the action " = = " is also neutral, the E-composition of the opposite actions a and a^{-1} is the neutral element " = = ", i.e. $aEa^{-1} = a^{-1}Ea = $ " = = ". These properties are the same as in a group, yet the composition law of Table 9 is not that of a group, since it is not associative. This is seen, say, from the following example:

$$<<E(=>E<>) \to <<E<> \to <=, \quad (<<E=>)E<> \to <=E<> \to <>.$$

Besides, $aEa = a$, while $aFa = a^{-1}$. Thus the system S_a is not symmetrical with respect to the composition law E, or probably it would be better to say that it is dissymmetrical.

Table 10 is a qualitative classification of actions of A on B and of B on A, which are realized at 2-, 1-, 0-influences. This table is also a qualitative classification of relationships between the partners (A and B), so it can be considered in two aspects: as a table of actions and as a table of relationships.

It is clear from the qualitative point of view that only 9 relationships are possible. The formula for the number of arrangements with repetitions, $A_m^k = m^k$ provides us with a proof that the variants are exhausted completely. In the present case $m = 3$ (+, −, no sign), $k = 2$, so $A_3^2 = 3^2 = 9$. The system of relationships is represented by two opposite subsystems.

The first subsystem consists of pairs of objects which are in agreement with each other. This subsystem is called correlative (from Latin 'con'—'with, together'). The correlative subsystem consists of three correlative pairs: $+A+B$, $-A-B$, AB. The objects in these pairs are called the correlatives (which are in accordance, identical, similar) or *isoids* (from Greek 'isos'—'equal, identical, similar'). Examples of the correlativism are the phenomena of synergism and antagonism of ions in physiology of animals and plants, the mutual neutrality in politics, the absence of interaction, consonances in music, concordance in genetics.

Table 8. Opposite actions and their F-compositions

	+	F	−	
		= =	= =	
		> >	< <	
= = ←		= <	= >	→ = =
		< >	> <	
		< =	> =	

Table 9. Groupoid of 9 actions

E	==	<<	>>	=<	=>	<>	><	<=	>=
==	==	<<	>>	=<	=>	<>	><	<=	>=
<<	<<	<<	==	<<	<=	<=	=<	<<	=<
>>	>>	==	>>	>=	>>	=>	>=	=>	>>
=<	=<	<<	>=	=<	==	<=	><	<<	><
=>	=>	<=	>>	==	=>	<>	>=	<>	>>
<>	<>	<=	=>	<=	<>	<>	==	<>	=>
><	><	=<	>=	><	>=	==	><	=<	><
<=	<=	<<	=>	<<	<>	<>	=<	<=	==
>=	>=	=<	>>	><	>>	=>	><	==	>=

The second, alternative subsystem consists of pairs of objects which are in different and opposite relationships. We call it 'disrelative' (from Latin 'dis'—a prefix which means separation). The disrelative subsystem consists of two alternative sub-subsystems.

The first sub-subsystem consists of pairs of objects which are different and contrasting. We call it 'contra-disrelative'. The contra-disrelative sub-subsystem consists of two pairs of objects, $+A-B$ and $-A+B$. The objects in such pairs are called 'contra-disrelatives' or 'antioids' (from Greek 'anti'—'counter'). The classical example of the antioidism is the one-sided parasitism in animate nature.

The second sub-subsystem consists of pairs of objects which are different but not contrasting. We call it 'noncontra-disrelative'. The noncontra-disrelative sub-subsystem contains 4 pairs of objects, $+AB$, $-AB$, $A+B$, $A-B$. The objects in such pairs are called 'heteroids' (from Greek 'heteros'—'other'). The example of the heteroidism is the one-sided determinism in the correspondence between past and present, present and future.

Evidently, if we start from a contensive concept of actions and the requirements of the law of the system similarity and put the unique quasi-0-action " $= =$ " in correspondence to the unique conrelative relationship AB, and the 2-action "$\ll$"—in correspondence to a conrelative relationship, say, $+A+B$, then we get the mathematical isomorphism of the system 10 (see Table 10) to the system 1 (see Table 1) and consequently, to the following one-to-one correspondences:

(1) $<< \ldots +A+B,$
(2) $=< \ldots A+B,$
(3) $<= \ldots +AB,$
(4) $<> \ldots +A-B,$
(5) $= = \ldots AB,$
(6) $>< \ldots -A+B,$
(7) $>= \ldots -AB,$
(8) $=> \ldots A-B,$
(9) $>> \ldots -A-B.$

Both the systems may be considered as interpretations of each other.

Because of the mathematical isomorphism between the system 10 and the system 1 and as the system 1 is a group with respect to the composition law F and is not a group with respect to the composition law E, the system 10 can be treated both as a group and as a groupoid, as a symmetrical or asymmetrical set, with respect to the same composition laws F and E, respectively.

Below, in Table 11 we present only the group of relationships of the 9-th order with its 6 subgroups: 1 of the 1st order, 4 of the 3rd order, and 1 of the 9-th order. It is not difficult to notice that the group of relationships is completely isomorphic to the group of actions. Besides,

Table 10. Qualitative system of relationships

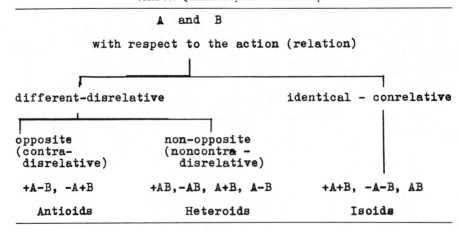

A and B

with respect to the action (relation)

different-disrelative		**identical - conrelative**
opposite (contra-disrelative)	non-opposite (noncontra-disrelative)	
+A-B, -A+B	+AB,-AB, A+B, A-B	+A+B, -A-B, AB
Antioids	**Heteroids**	**Isoids**

the group of relationships is an evidence in favour of an advantage gained from an isomorphism established between different systems, as it enables one to perform a correct and fruitful exchange of knowledge between various fields of research.

Using the isomorphism, we write down the isomorphic pairs of subgroups of actions and relationships. The first-order subgroups are " $= =$ " and AB, the 3rd-order subgroups are " $= =$ ", " $>>$ ", " $<<$ " and AB, $-A-B$, $+A+B$; " $= =$ ", " $=<$ ", " $=>$ " and AB, $A+B$, $A-B$; " $= =$ ", " $<>$ ", " $><$ " and AB, $+A-B$, $-A+B$; " $= =$ ", " $<=$ ", " $>=$ " and AB, $+AB$, $-AB$. Finally, the 9-th-order subgroups are the whole groups of actions (Table 7) and relationships (Table 11).

Evidently, the opposite forms of the actions, 2-sided, 1-sided, 0-sided ones, always in combination with the neutral action " $= =$ ", as well as the opposite forms of the relationships, conrelative, contra-disrelative, non-contra-disrelative ones, combined with the neutral relationship AB, are the third-order symmetry groups. This means, however, that absolutely all kinds of actions and absolutely all kinds of relationships under appropriate conditions have a harmony, a completeness, and self-consistence! Moreover, this harmony gets a supreme realization when one considers the complete multitudes of all possible actions and all possible relationships and tries to establish a deep parallelism between these multitudes. The statement of this fact is, on the one hand, the construction of the 9-th-order group of actions and the 9-th-order group of relationships (in the present approach no higher order groups are possible!), on the other hand, it is the exact mathematical isomorphism established between these groups.

The law of symmetry and a number of other propositions of GST(U) indicate a group nature of any system, at least in a single aspect. Hence we immediately derive contradictions and compatibility of systems.

Table 11. The 9-order group of relationships

F	AB	-A-B	+A+B	A+B	A-B	+A-B	-A+B	+AB	-AB
AB	AB	-A-B	+A+B	A+B	A-B	+A-B	-A+B	+AB	-AB
-A-B	-A-B	+A+B	AB	-AB	-A+B	A+B	+AB	A-B	+A-B
+A+B	+A+B	AB	-A-B	+A-B	+AB	-AB	A-B	-A+B	A+B
A+B	A+B	-AB	+A-B	A-B	AB	+AB	-A-B	+A+B	-A+B
A-B	A-B	-A+B	+AB	AB	A+B	+A+B	-AB	+A-B	-A-B
+A-B	+A-B	A+B	-AB	+AB	+A+B	-A+B	AB	-A-B	A-B
-A+B	-A+B	+AB	A-B	-A-B	-AB	AB	+A-B	A+B	+A+B
+AB	+AB	A-B	-A+B	+A+B	+A-B	+A-B	-A-B	A+B	-AB	AB
-AB	-AB	+A-B	A+B	-A+B	-A-B	A-B	+A+B	AB	+AB

PROPOSITION 13

Any system possesses *n* relations of contradiction and *m* relations of compatibility.

This statement is a consequence of symmetrical properties of all systems, i.e. the group properties, and the resulting presence in any system of contradictions between the opposite elements, as well as of compatibility between non-opposite elements of the group. In particular, in each of the groups, that of actions and that of relationships, 9 contradictions and 72 relations of compatibility are realized.

By virtue of the law of systematics, we get Propositions 14 and 15.

PROPOSITION 14

Any contradiction is a contradiction-system and any contradiction-system belongs to at least one system of contradictions of the same kind.

PROPOSITION 15

Any compatibility is a compatibility-system and any compatibility-system belongs to at least one system of compatibilities of the same kind.

The combination of Propositions 13, 14, 15 leads to Propositions 16 and 17.

PROPOSITION 16 (The law of system contradictoriness)

Any system contains a subsystem of contradiction-systems.

PROPOSITION 17 (The law of system compatibility)

Any system contains a subsystem of compatibility-systems.

The latter means that from the point of view of GST(*U*) the compatibility is as universal, as its opposite, the contradictoriness. Consequently, the system itself is a very harmonical aggregate—a unity of two opposite subsystems, a subsystem of contradictions and a subsystem of compatibilities.

Thus we have solved all the problems stated in the Introduction in connection with the explication of symmetry of the system; we have described all groups of non-evolutionary and evolutionary system transformations and their invariants, presented proofs of symmetry for an arbitrary system, the system as itself, analysed symmetry groups of 2-, 1-, and 0-sided actions and associated relationships. Now we are ready to analyse the system of symmetry, the symmetry as itself.

THE SYSTEM OF SYMMETRY

A direct consequence of the law of systematics is

PROPOSITION 18

Any symmetry is a symmetry-system (an object-system), and any symmetry-system belongs to at least one system of objects of the same kind.

Let us dwell on both statements of this proposition.

Symmetry as an object-system

From the ontological point of view, symmetry is a property of objects "*O*" to coincide in features "*F*" after modifications "*M*". The poetic motto by Jacob Bernoulli, "Eadem mutata resurgo" ("Being changed I am resurrected") is quite applicable to symmetry. Bernoulli quoted the motto being charmed by the symmetry of the logarithmic spiral, but his expression is wonderfully appropriate for description of the main property of any symmetry. In the framework of GST, symmetry is considered as one of realizations of an abstract system. In fact, symmetry is such an object-system, the primary elements of which are objects "*O*" and features "*F*", and the unity relations are the attribute relations of the features "*F*" to the objects "*O*", while the composition laws is the requirement that the objects must possess the features both before and after modifications "*M*".

From the gnoseological point of view, components of the real symmetry are reflected in thinking in the following way: the object is associated with the concept of a "carrier of symmetry'', it is the basis of its classification by 4 principal types (structural or crystallograph-ical, geometrical, dynamical, temporal), features ''F'' are associated with the concepts of ''invariants'', they provide with a basis for classification of symmetry of given type by subtypes. For example, in the case of crystallographical symmetries, they are classified according to the shape of the figure; the symmetries can be zero-, one-, two-, three-, . . . , n-dimensional. Modifications ''M'' are associated with the concept ''operation or transformation of symmetry''; they are a basis for exact and exhaustive classification of subtypes to classes, kinds, species. The combination of all three properties of symmetry is associated with a synthetic concept ''the theory of symmetry transformations and their invariants''.

Naturally, the group as an ideal mathematical prototype of a real symmetry, in turn, can be represented as a special object-system, a united mathematical object. All components of an object-system can be indicated in a group: the set of primary elements is the generating elements of the group, $M_{\mathscr{G}}^{(0)}$, the unity relations are the relations specified by 4 group axioms, the composition law is the multiplication law in the group.

Even if the abstract group is just an interpretation of an abstract system, it is sufficiently abstract itself, so it can be in turn interpreted in infinitely many ways in terms of quite different material or ideal objects. This statement is true also for any concrete realization of the abstract group. For instance, a 60-order group is present in a number of objects: in the mathematical alternating group of substitutions A_5, and the group of non-mirror automorphisms of geometric bodies, dodecahedron, icosahedron, and in a chemical compound barena, in a biological object Radiolaria cyrcoregma dodecahedra. A remarkable fact: such an interpretation by means of the same abstract group conserves specific properties of the objects under investigation and reveals their unity with an accuracy up to an isomorphism. The existence of infinite interpretations of the group theory and the ability of it to reveal the twofold (special and general) nature of various objects can be easily explained.

The first property is a consequence of a non-specific character of requirements stated in axioms of the theory of groups for various matter existence forms (space, time, motion) and matter evolution forms (mechanical, physical, chemical, geological, biological, social). This explains the ample applicability of the theory of groups (symmetry) which is striking even for philosophers, mathematicians, and theoretical physicists.

The second property, the possibility to represent objects in their unity and variety, results, on the one hand, from the above mentioned existence of various interpretations of the group theory, on the other hand, from the possibility to derive *all the multitude of objects of the same kind* by means of the mathematical apparatus of the theory. This enables one to explain another feature of the theory, which was numerously discussed in the literature, the power of the group they, its inherent wonderful ability to classification analysis.

Nevertheless, the above arguments explain rather the reason of the ample applicability of the group theory than the reason of its deep penetration in the very essence of objects under investigation. In this connection, one might just refer to a power of the mathematical apparatus of the theory. However, the very power is itself a result of the 4 axioms which are the basis of the theory. Meanwhile an arbitrary set of axioms would not generate an effective mathematical apparatus, nor result in a profound analysis of being: the set must reflect the most general, and at the same time sufficiently pithy, nature of things. Therefore the ultimate reason of penetrability of the group theory is, in our opinion, the objective dialectics, which is expressed in terms of its axioms, even if in part, and the corresponding dialectic character of cognition of the real world, associated first of all with the conception given by the theory of groups where every system is represented as a unity of two subsystems, consisting of contradictions and compatibilities.

Symmetry within the system of objects of the same kind

The system character of symmetry manifests itself not only in that any particular symmetry, or symmetry in general, is a special object-system, but also in the fact that all concrete symmetries, and the symmetry as itself, belong to a system of objects of the same kind.

In any special case of symmetry, for example, for the point symmetry $5m$ (one quinary axis and five reflection planes intersecting along the axis, i.e. the symmetry of Pythagorean five-pointed star) it is the fact that it belongs to the system of point symmetry groups, which are in turn members of the system of all point, linear, planar, spatial symmetry groups, while all of them are examples of the classical symmetry. The classical symmetry itself is a member of the system of 54 structure symmetries (Table 6), and the latter is included into the system of 64 fundamental symmetries (Table 5). Other relevant examples are the inter- and intra-system symmetries derived above (in the Section "Laws of correspondence and symmetry"), the numbers of which are 4(3), 28, 49, 84, 105, 120, 147, 315, 360, 2080 and 6240.

In the case of symmetry as itself, which manifests itself, in particular, as a group in general, the same property is revealed in the fact that the group belongs to the system of abstract algebraic structures of groups, rings, bodies, abstract algebras, etc.

In the framework of GST(U), the reason why symmetry (both concrete and abstract) belongs to a system of objects of the same kind coincides with the reason why an elementary particle belongs to a system of particles, atom belongs to a system of atoms, molecule belongs to a system of molecules, an organism belongs to a life tree, and a social and economic structure belongs to a system of such structures. The reason we have in mind is the existence. The existence that can be in relative rest or in absolute motion, with relative conservation or with absolute changing, persistent to any material or ideal objects, is responsible for the fact that symmetry does inevitably belong to the system of objects of the same kind. The same is the role of the very emergence of systems of objects. The emergence, existence, transformation, development of object-systems must be in an obligatory agreement with requirements of the Central Proposition of GST; in all these cases poly- and isomorphisms, symmetry and asymmetry must manifest theirselves. The future progress will show all that in concrete details for the example of emergence, existence, transformation and development of the theories dealing with symmetry and dissymmetry, harmony and disharmony.

REFERENCES

1. M. D. Mesarovic and Y. Takahara, *General Systems Theory: Mathematical Foundation*. Academic Press, New York (1975).
2. A. I. Uyemov, *System Approach and General Systems Theory* (in Russian). Mysl', Moscow (1978).
3. Yu. A. Urmantsev, *Symmetry of Nature and the Nature of Symmetry* (in Russian). Mysl', Moscow (1974).
4. Yu. A. Urmantsev, Principles of general systems theory, in *System Analysis and Scientific Knowledge* (in Russian), pp. 7–41. Nauka, Moscow (1978).
5. Yu. A. Urmantsev, On the nature of left and right (foundation of the theory of dissfactors), in *Principles of Symmetry* (in Russian), pp. 180–195. Nauka, Moscow (1978).
6. Yu. A. Urmantsev, What use can a biologist have representing an object as a system within the system of objects of the same kind? *Zhurn. obshchey biologii* **39**, 699–718 (1978).
7. A. N. Severtsov, *Morphological Regularities in Evolution* (in Russian). Academy of Sciences of the U.S.S.R., Leningrad–Moscow (1939).
8. V. A. Dogel, *Oligomerization of Homologous Organs as One of the Cardinal Ways of Evolution of Animals* (in Russian). Academy of Sciences of the U.S.S.R., Leningrad (1954).
9. V. Yu. Zabrodin, *System Analysis of Disjunctives* (in Russian). Nauka, Moscow (1981).
10. A. M. Zamorzaev, *Theory of Single and Multiple Antisymmetry* (in Russian). Shtiintsa, Kishinev (1976).
11. Yu. A. Urmantsev, An attempt of axiomatic construction of general systems theory, in *System Investigations* (in Russian), pp. 128–152. Nauka, Moscow (1972).
12. Yu. A. Urmantsev, Isomerism in animate nature. IV. Investigations of properties of biological isomers (for the example of flax corollas). *Botan. zhurn.* **58**, 769–782 (1973).
13. I. I. Shafranovskii, History of development of the theory of isomorphism. Vestnik of Leningrad University, No. 6, 62–69 (1967).
14. Yu. A. Urmantsev, On determination of signs of enanthyomorphism in non-chemical (biological) diss-isomers by means of chemical ones. *Zhurn. obshchey biologii* **40**, 351–367 (1978).
15. S. V. Meyen, Plant morphology in its nomothetical aspects. *Bot. review* **39**, 205–260 (1973).

Comp. & Maths. with Appls. Vol. 12B, Nos. 1/2, pp. 407–411, 1986
Printed in Great Britain.

0886-9561/86 $3.00 + .00

THE MOSAIC PATTERNS OF H. J. WOODS

D. W. Crowe

Department of Mathematics, University of Wisconsin—Madison, Madison, WI 53706, U.S.A.

Abstract—The mosaic patterns for the 46 two-color two-dimensional patterns, first published by H. J. Woods in 1936 are reproduced. The mosaics themselves are preceded by some comments on the historical context of Woods' contribution to the study of two-color patterned ornament. The new group/subgroup notation of H. S. M. Coxeter for these patterns is also described.

In 1856 Owen Jones' *The Grammar of Ornament*[1] was first published. Although not the only book of its kind, it has been the reference *par excellence* for mathematicians interested in the cultural aspects of patterned ornament. It is organized entirely on cultural and historical principles, not at all on stylistic or mathematical principles.

The first edition of A. H. Christie's *Pattern Design*[2] (then called *Traditional Methods of Pattern Designing*) was published in 1910. This work can be thought of in retrospect as representing the next step in the mathematization of the study of patterned ornament. Christie's organizing principle is "formal". That is, patterns are grouped together not according to the time periods or cultures in which they are found, but according to the "forms" appearing in them, and the relations among these forms. Implicit in this organization are certain geometrical principles (by no means explicitly stated) of shape and isometry.

The final step in the mathematization of the study of patterned ornament, at least as presented to the non-mathematical public, is represented by Peter Stevens' 1981 publication, *Handbook of Regular Patterns*[3]. Here the material is presented entirely on geometric principles, so that all the patterns in the book are listed according to the isometries they admit. That is, they are classified first as finite, one-dimensional, or two-dimensional. Then within these three subdivisions they are classified according to the scheme developed by crystallographers for their geometrical study of repeated patterns in space.

How did these two disciplines, the cultural-historical and the mathematical, happen to meet and interact in this particular specialty: the study of patterned ornament? Many of us have answered that the interaction was especially fostered by Andreas Speiser's early group theory textbook, *Theorie der Gruppen von endlicher Ordnung*[4]. For, in all editions beginning with the second in 1927, Speiser included a large section dealing with the symmetries of ornament, in which he introduced us to the plane crystallographic groups using the notation of his colleague, Niggli. This, combined with the thesis of his student Edith Müller on symmetries of ornaments in the Alhambra, as well as Speiser's other cultural writings, persuaded many of us that designers of ornament from ancient times had at least a subconscious understanding of basic ideas of group theory. Speiser was, in the eyes of many, the first to make this connection.

However, a few years later, there was a worker in the field of textile ornament who recognized that mathematics could be directly applied, not just to the *study* of ornamental items, but to their design and creation. This worker was H. J. Woods, the proper subject of this note.

Henry John Woods was born in 1904 and died just recently, on April 22, 1984. Although born in Brooklyn, New York, he was raised and educated in England, with a degree in mathematics from Oxford. He spent most of his working life in the Department of Textile Industries of the University of Leeds, and much of his work, including the patterns reproduced here, appeared in publications of the Textile Institute. In 1935–36 he published a four part paper with the general title, "The Geometrical Basis of Pattern Design"[5], in parts I and IV of which he describes, apparently for the first time in an explicit way, the classification of what he (along with Christie and others) calls "counterchange" patterns, that is, two-color patterns.

It is in these 1935–36 papers that Woods anticipated work that would not be done by crystallographers or mathematicians for another twenty years. Although crystallographers, a few years earlier, had recognized the *groups* of the 46 two-color patterns in the form of groups of the "two-sided plane", none had drawn them in a way which actually tiled the plane. The illustrations Woods gave are not just two color patterns in the plane, but are *mosaics* in the sense of N. V. Belov and E. N. Belova (in a 1957 paper, partially reprinted as "Mosaics for the Dichromatic Plane Groups" in [6]). That is, they cover the plane entirely, as tiles of two colors, without interstices.

Unfortunately, in contrast to the work of Speiser, Woods' work did not come to the attention of mathematicians until nearly forty years later. In particular, it was presumably completely unknown to the active Russian school of crystallographers, for Belov and Belova make no mention of Woods' work of twenty years earlier. Apparently it was Branko Grünbaum who first noticed the relevance of Woods' work to current research on colored patterns, and brought it to our attention. However, the actual patterns have not been reproduced yet, and we have taken this opportunity to do so.

Woods' original articles are well worth reading. His notation is a very reasonable one, and a modification of it was used by the archeologist Dorothy Washburn in her study of the Upper Gila[7]. However, it is not the one generally used at present, for example, by N. V. Belov and T. N. Tarkhova ("Dichromatic Plane Groups", in [6]), who call it the "rational symbol".

Table 1. The notation of H. J. Woods for the 46 two-color two-dimensional patterns, compared with the "rational symbol" of Belov and Tarkhova in [6] and the group/subgroup notation of Coxeter.

Woods' No.	Woods' Notation	"Rational Symbol"	Group/ Subgroup	Woods' No.	Woods' Notation	"Rational Symbol"	Group/ Subgroup
1.	b11	p'_b1	p1/p1	24.	$n12_1$	p'_cg	cm/pg
2.	2'11	p2'	p2/p1	25.	c112'	cm'	cm/p1
3.	2/b11	p'_b2	p2/p2	26.	$n2_12_1$	p'_cgg	cmm/pgg
4.	12'1	pm'	pm/p1	27.	n22	p'_cmm	cmm/pmm
5.	a12	p'_b1m	pm/pm(m')	28.	$n22_1$	p'_cmg	cmm/pmg
6.	$b12_1$	p'_bg	pm/pg	29.	c2'22'	cmm'	cmm/cm
7.	b12	p'_bm	pm/pm(m)	30.	c22'2'	cm'm'	cmm/p2
8.	ca12	c'm	pm/cm	31.	4'11	p4'	p4/p2
9.	$112'_1$	pg'	pg/p1	32.	4/n11	p'_c4	p4/p4
10.	$b2_11$	p'_b1g	pg/pg	33.	4'2'2	p4'm'm	p4m/cmm
11.	2'2'2	pmm'	pmm/pm	34.	4'22'	p4'mm'	p4m/pmm
12.	22'2'	pm'm'	pmm/p2	35.	42'2'	p4m'm'	p4m/p4
13.	$a2_12$	p'_bgm	pmm/pmg	36.	4/n22	p'_c4mm	p4m/p4m
14.	a22	p'_bmm	pmm/pmm	37.	$4/n2_12$	p'_c4gm	p4m/p4g
15.	ca22	c'mm	pmm/cmm	38.	$4'2_12'$	p4'gm'	p4g/pgg
16.	$2'2'_12$	pmg'	pmg/pm	39.	$4'2_12$	p4'g'm	p4g/cmm
17.	$2'2_12'$	pm'g	pmg/pg	40.	$42'_12'$	p4g'm'	p4g/p4
18.	$22'2'_1$	pm'g'	pmg/p2	41.	32'1	p31m'	p31m/p3
19.	$a2_12_1$	p'_bgg	pmg/pgg	42.	312'	p3m'	p3m1/p3
20.	$b2_12$	p'_bmg	pmg/pmg	43.	6'	p6'	p6/p3
21.	$2'2'_12_1$	pgg'	pgg/pg	44.	6'2'2	p6'mm'	p6m/p3m1
22.	$22'_12'_1$	pg'g'	pgg/p2	45.	6'22'	p6'm'm	p6m/p31m
23.	n12	p'_cm	cm/pm	46.	62'2'	p6m'm'	p6m/p6

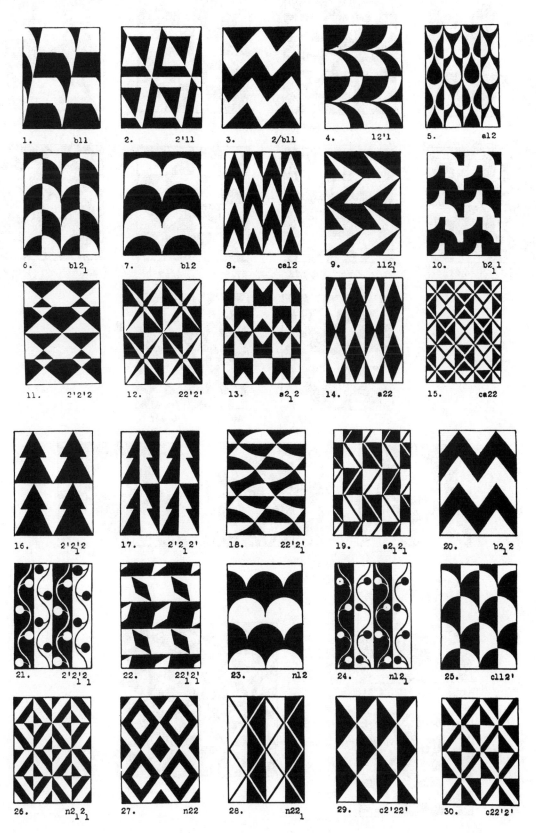

Fig. 1. (© The Textile Institute, used by permission.)

Fig. 1. (*Continued*).

From the mathematical point of view, a much more attractive notation than either of these has recently been suggested by H. S. M. Coxeter[8]. He observes that each two-color pattern has associated with it three groups. G is the symmetry group of the underlying one-color pattern, G_1 is the normal subgroup of G which preserves colors, and the quotient group G/G_1 is the group (always of order two in the two-color case) which permutes the colors. All but two of the 46 two-color patterns are completely determined by the specification of G and G_1. Hence the notation G/G_1 is appropriate for these 44 patterns. For both of the remaining two patterns, #5 and #7 on Woods' list, $G = G_1 = pm$, so a further distinction is needed for these two. It is natural to write #7 as $pm/pm(m)$ to indicate that all reflections (m for mirror) preserve color, i.e. are in the subgroup G_1; and to write #5 as $pm/pm(m')$ to indicate that some reflections reverse colors, and hence are not in G_1.

To correlate these notations, the reproduction of Woods' patterns in Fig. 1 is accompanied by Table 1 which gives the number (1–46) of Woods' drawing, his notation, the ''rational symbol'' from Belov and Tarkhova, and Coxeter's new ''group/subgroup'' symbol. Inspection of this group/subgroup notation, arranged here in Woods' order, unexpectedly reveals how aware Woods was of the significance of these color-preserving subgroups, even though it is not so obvious in his own notation. May the republication of these patterns serve as a small monument to Woods' memory.

REFERENCES

1. O. Jones, *The Grammar of Ornament*. London (1856).
2. A. H. Christie, *Pattern Design*. Dover, New York (1969). (Reprint of the second (1929) edition, first published in 1910 as *Traditional Methods of Pattern Designing*.)
3. P. S. Stevens, *Handbook of Regular Patterns*. M.I.T. Press, Cambridge (1981).
4. A. Speiser, *Theorie der Gruppen von endlicher Ordnung*, 2nd Edn. Springer Verlag, Berlin (1927).
5. H. J. Woods, The geometrical basis of pattern design, Part I—Point and line symmetry in simple figures and borders; Part II—Nets and sateens; Part III—Geometrical symmetry in plane patterns; Part IV—Counterchange symmetry in plane patterns. *Journal of the Textile Institute, Transactions* **26**, **27**, T197–210; T293–308; T341–357; T305–320 (1935, 1936).
6. A. V. Shubnikov, N. V. Belov, and others, *Colored Symmetry*. Pergamon Press, Oxford (1964).
7. D. K. Washburn, *A Symmetry Analysis of Upper Gila Area Ceramic Design*, **68**, Papers of the Peabody Museum of Archeology and Ethnology. Harvard University, Cambridge (1977).
8. H. S. M. Coxeter, *Coloured Symmetry*, presented at the Interdisciplinary Congress on M. C. Escher. University of Rome, Italy, March (1985).

Comp. & Maths. with Appls. Vol. 12B, Nos. 1/2, pp. 413–418, 1986
Printed in Great Britain.

0886-9561/86 $3.00 + .00
© 1986 Pergamon Press Ltd.

BILATERAL SYMMETRY IN INSECTS: COULD IT DERIVE FROM CIRCULAR ASYMMETRIES DURING EARLY EMBRYOGENESIS?

Klaus Sander

Institut für Biologie I (Zoologie), Albertstr. 21a, D 7800 Freiburg i.Br., F.R.G.

Abstract—The bilateral symmetry of the larval body may arise epigenetically even in animal forms with bilaterally symmetrical eggs. In insects, this can be shown by development of parallel twins from eggs or early embryos split along their midlines. The present contribution explores the possibility that epigenetic symmetrization is based on circular asymmetries pre-existing in the egg cell. This possibility is suggested by phylogenetic considerations and by data based on epimorphic regeneration of larval appendages. The author proposes an experimental test discriminating between circular and bilaterally symmetrical prepatterns in primitive insects.

Bilateral symmetry is a striking feature in the external appearance of most multicellular animals, which however may be accompanied by a wealth of hidden asymmetries[7]. The bilateral organization was reproduced faithfully and constantly through millions of generations and by myriads of individuals, which indicates that it must be encoded in the egg cell and its genome in a very reliable manner. Yet this need not mean that bilaterality is encoded as a bilaterally symmetrical "prepattern"[11] or array of singularities guiding development in mirror image fashion from the beginning. Much more likely, bilaterality is encoded as a set of prescriptions which guide interaction of the system's components among themselves and with external physical factors, and thereby *de novo* generate a bilateral pattern during early embryogenesis. This "epigenetic" origin of bilateral symmetry is strikingly evident in spiralian embryogenesis (see below) and during nematode development[12], but bilaterality appears to arise epigenetically even in those animal eggs that are themselves of bilaterally symmetrical shape and internal organization (mainly found in insects and cephalopods). At least in some insects, the evidently bilateral egg organization is not required as a prepattern on which to model the bilateral body. This is shown by the fact that the isolated left and right halves of such insect eggs (or of the embryonic rudiments developing in them) can each produce a bilaterally symmetrical larval body[4,8,9].

Several formal possibilities for epigenetically generating bilateral symmetry could be thought of. The one to be considered here, first conceived during discussions with L. Bischoff and H. O. Gutzeit[1,10], is derived from current ideas on pattern formation in regenerating insect and amphibian appendages, and might (if necessary) be defended on phylogenetic grounds. It generates bilateral symmetry on the basis of a pre-existing circular asymmetry.

The phylogenetic background of my proposition is as follows. Insects are thought to derive from annelid type ancestors with eggs developing by spiral cleavage. In this type of cleavage, the early daughter cells of the egg are arranged in polarized circular patterns expressing a definite and constant chirality or handedness while frequently lacking any indications of bilaterality (Fig. 1). Yet the adults are of strictly bilateral organization, and their median plane coincides with the rotational axis of the cleavage stages.

As for recent formal models of postembryonic patterning, these were largely influenced by the concepts of positional information[15,16] and the "clock face" or polar coordinate model[2,3], both mainly based on data from regeneration experiments. The former concept states that the individual cells or cell groups during ontogenesis somehow "sense" their relative positions in the developing system and that they "interpret" this "positional information" by embarking on position-specific developmental pathways; the overall result is that different cells perform different developmental tasks in a reproducible spatial pattern. When the interpretation process has been initiated, a cell or group of cells can be removed experimentally from its original position and yet continue its course of development; it is then thought to have acquired a "positional value", a more or less stable intrinsic property guiding its further development.

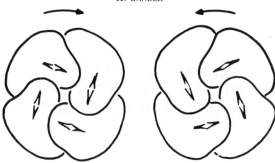

Fig. 1. Diagram of spiral cleavage at the transition from the 4-cell stage to the 8-cell stage; the equators of the mitotic spindles mark the future cleavage planes separating 4 small upper cells from 4 large lower cells. Note the two enantiomorph patterns and circular asymmetries (arrows) which in molluscs depend on the mother's genotype. In annelid worms, the future median plane coincides with the rotational axis of spiral cleavage. In many spiralians, the median plane seems predetermined already in the egg stage. However, in others it is determined by an epigenetic chance decision during early embryogenesis[13].

The "clock face model" and its more recent variants (reviewed in [14]) propose that the positional values in a row of cells form a continuum, depending perhaps on the continuous variation of some physico-chemical parameter(s) over space. A second tenet of the model is that juxtaposition of cells with strongly disparate positional values causes intercalation of the missing values (usually by intercalation of daughter cells which somehow acquire these values). In the clock face model proper, these principles are applied to a ring of cells representing, for instance, a section across the epithelial tube which forms the basic structure of an insect leg [Fig. 2(a)]. The positional values of individual cells or cell groups are numbered on this ring in analogy to the figures on a clock face, with the additional tenet that the 12/0 transition has no special status over the other positions—the hands of the clock, too, pass over this position as over any other point on the clock face. The interesting aspects of this circular array of positional values are revealed when one removes sectors from the circle of values [Fig. 2(c)] and anneals the resulting free edges [Fig. 2(d)]; experimentally this can be done by cutting bits from larval insect legs and let the wound margins fuse. This operation juxtaposes disparate positional values, and this in turn triggers intercalation of the missing values [Fig. 2(e)]. The latter process obeys the rule of "shortest route intercalation". Between any two values on the clock face there exist two possible routes of intercalation. The rule prescribes that intercalation should occur the shortest way, i.e. via the smaller number of new values; for example, when values 3 and 6 are confronted, intercalation should generate values 4 and 5 in between, and not values 7–8–9–10–11–12/0–1–2. However, this rule leaves the system with an ambiguity in situations where both possible routes are of equal length (as e.g. from 3 to 9). Such situations are credited with having dramatic consequences, namely "distal regeneration" or rather distal outgrowth[2,3] (see below). Finally, the model permits uneven spacing of positional values to occur in the intact system; for instance, the values from 3 to 8 might be crowded into one quarter of the circumference while the other values were spread over the remaining three quarters [Fig. 2(b)].

Why are these formal models relevant for our topic? Because of the fact that the embryonic rudiment, called the germ anlage, and one of its adnexes (the amnion) on principle represent an epithelial tube (albeit closed anteriorly and posteriorly). More correctly this tube can be described as a heart-shaped epithelial vesicle, the upper wall representing the bilaterally symmetrical germ anlage [Fig. 2(f)]. Each half of the germ anlage will give rise to the respective body half; the germ anlage thus forms the bilaterally symmetrical basis from which the larval body develops.

We shall now consider under what conditions a bilaterally symmetrical pattern could arise from a circular array of positional values like that shown in Fig. 2(f). On a formal level this is quite simple. The cells merely need to interpret positions on either side of a "midline position" in a symmetrical fashion. If position 6 is assumed to specify the midline [Fig. 2(f), 3(b)], cells with the positional value 5 should develop like those with positional value 7, cells with positional value 4 like those with positional value 8, and so on. The alternative possibility would be a bilaterally symmetrical array of positional values [Fig. 3(a)]. In the latter case cells from positions

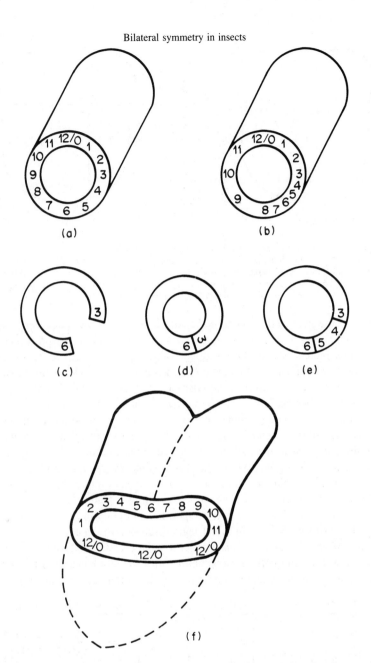

Fig. 2. Features of the clock face model[3] (a–e) and positional values postulated for the transverse plane of the embryonic rudiment in the stone cricket. (a) Clock face arrangement of positional values in a cross section through an epithelial tube (like the hypoderm of an insect leg). (b) A part of the circumference (values 4 and 5) is removed. (c) The margins have annealed, confronting positional values 3 and 6. (d) This confrontation triggers intercalation of cells with the intervening positional values (values 4 and 5). (e) Positional values may be spaced unevenly on the circumference. (f) The embryonic rudiment of the stone cricket at the time when bilateral symmetry is clearly expressed. The rudiment is a heart-shaped epithelial vesicle of which the dorsal (= upper) wall represents the germ anlage proper and the ventral (lower) wall the amnion. The transverse cut exposes circular positional values, the amnion being characterized by a single value (0/12). The "ears" of the germ anlage are the anterior rims of the head lobes.

symmetrical to the midline [e.g. position 4 on either side, Fig. 3(a)] would not recognize each other as being different when confronted experimentally, while equally located cells with differing positional values [say 4 and 8, Fig. 3(b)] would do so and should initiate intercalation between them.

A simple experiment might discriminate between these two alternatives, provided that all cells of the amnion were to share a single positional value [named 12/0 in Fig. 2(f)]—which seems tenable because the amnion throughout its temporary existence reveals no signs of regional differentiation. The crucial experiment consists in cutting the embryonic vesicle diagonally and

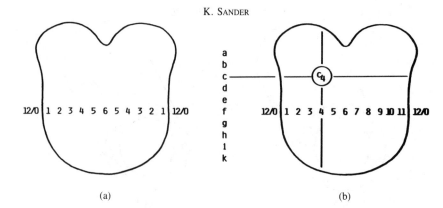

Fig. 3. Diagram showing the coordinates for alternative grids of positional values which might conceivably exist in the germ anlage of the stone cricket; the cells at grid position c_4 are marked by a circle in the right hand figure. Longitudinal values are marked by letters identical for both alternatives. Transverse values (numerals) differ in being arranged in a mirror image configuration (a) or monotonically as in Fig. 2(f) (b); in both cases, positional value 6 is the "midline value".

let the cut rim of the germ anlage heal to that of the amnion. The latter step would confront cells of positional values 1–11 or 1–6, resp., with the amnion cells of positional value 12/0, thereby triggering intercalation of the positional values missing in between. As shown in Fig. 4, this intercalation would lead to quite different grids of positional values on the two alternative assumptions. With bilaterally symmetrical positional values, intercalation would always occur "downhill", and a tapering bilaterally symmetrical pattern should result [Fig. 4(a)]. With circular positional values, shortest route intercalation would be downhill in the posterior parts (where less than half of the original germ anlage was spared), but uphill in the anterior regions where the positional value at the cut margin exceeded the midline value [Fig. 4(b)]. The net difference would seem negligible, were it not for the ambiguous situation arising in the region where the cut transects the midline. At this level both routes for intercalation are of equal length. Both together would represent a complete circle of positional values (see large numerals in Fig. 4 (b)). As stated above, such a circle is to trigger distal outgrowth (2, 3), thereby obeying the "complete perimeter theorem"[6]. In insect legs, distal outgrowth constructs a supernumerary distal part (tip) of leg; by analogy, in the germ anlage distal outgrowth should produce a supernumerary posterior end of body, containing all segments located posterior to the longitudinal position at which the complete circle formed [positions below f in Fig. 4(b)].

<u>4</u>	<u>5</u>	<u>6</u>	<u>5</u>	<u>4</u>	<u>3</u>	<u>2</u>	<u>1</u>	<u>12/0</u>	d	<u>4</u>	<u>5</u>	<u>6</u>	<u>7</u>	<u>8</u>	<u>9</u>	<u>10</u>	<u>11</u>	<u>12/0</u>
<u>4</u>	<u>5</u>	<u>6</u>	<u>5</u>	<u>4</u>	<u>3</u>	<u>2</u>	<u>1</u>	<u>12/0</u>	e	<u>4</u>	<u>5</u>	<u>6</u>	<u>7</u>	<u>8</u>	<u>9</u>	<u>10</u>	<u>11</u>	<u>12/0</u>

4 5 6 5 4 3 2 1 12/0   f   4 5 6 〈 7-8--9--10-11〉12/0
                                    〈 5-4--3--2--1 〉

<u>4</u>	<u>5</u>		<u>4</u>	<u>3</u>	<u>2</u>	<u>1</u>	<u>12/0</u>	g	<u>4</u>	<u>5</u>	<u>4</u>	3	2	1	12/0
<u>4</u>		3	2	1	<u>12/0</u>			h	<u>4</u>		3	2	1	<u>12/0</u>

(a)   (b)

Fig. 4. The patterns of intercalation expected after oblique cutting and annealing of germ anlage and amnion. The figure shows only part of the grids outlined in Fig. 3, with longitudinal positional values (letters) indicated between the two alternative grids. Transverse positional values existing before the experiment are underlined; the value on the right hand end of each line was confronted with the 12/0 value of the amnion on annealing. Seeming incongruencies in the *spacing* of intercalated values would disappear if, in agreement with the well established "rule of Barfurth", the rows of intercalated values were arranged at right angles to the cut (i.e. slanting down towards the right). No predictions are possible for the posteriormost regions since the posterior tip of the germ anlage represents a budding zone[4] where cells may not yet have acquired specific positional values. With a bilaterally symmetrical arrangement of positional values (left), intercalation leads to a bilaterally symmetrical pattern lacking an increasing number of medial elements towards the posterior end. With monotonic or clock face arrangement of positional values (right), a "complete circle" ambiguity[2] (large figures) results where the cut transects the midline. The "complete perimeter theorem"[6] would here require the outgrowth of a posterior end of body beginning with longitudinal position f. This supernumerary posterior end should stick out laterally from the body since transverse positional values 3 and 9 are located halfways between the ventral and (future) dorsal midlines represented by values 6 and 12/0, respectively.

The experiment just discussed was in fact performed some decades ago by Gerhard and Johanna Krause, using embryos of the stone cricket *Tachycines*[5]. In contrast to the predictions of Fig. 4, the "posterior" or normal abdomen did not noticeably taper. This difference can be explained by the well-proven fact that the posterior part of the germ anlage lags behind in development[4] and the cells might not yet be committed to specific positional values. A most relevant coincidence, however, is that after oblique cutting several embryos did indeed form a second tip of body projecting roughly at right angles from the cut flank (Fig. 5)—as would be predicted on the basis of circular positional values [Fig. 4(b)].

Krause and Krause[5] explained this unexpected result by assuming that part of the posterior "budding zone" had been carried anteriorly by the needle during the cutting operation. This explanation by itself is at least as tenable as the one suggested in Fig. 4(b), yet it fails to account for the fact that additional abdomens of this type were rarely if ever seen after other types of cut. At present a decision between these alternative explanations is not possible because the number of relevant published experiments is much too low, and the course of development in the obliquely cut embryos was difficult to monitor. The available experimental results thus cannot provide support for the hypothesis outlined in Fig. 4(b). However, they are compatible with this hypothesis.

In closing, some theoretical considerations pro and contra circular asymmetries as the basis for bilaterality should be discussed. Dr. J. Mittenthal pointed out to me that serious problems would arise if the transverse (or circumferential) positional values remained circular into the larval stages. In this case the clock face model would predict strong discrepancies in positional value between the legs sprouting from opposite sides of the body. Such differences did not show up in the many experiments on record which should have revealed them, for instance when a larval leg from one body side was grafted with the same dorsoventral orientation on a leg stump on the other body side[3]. However, as stressed by Curt Stern[11], ontogenesis can be viewed as a succession of spatial inhomogeneities or patterns, each serving as the prepattern for the next and more complicated pattern. With this picture in mind one might propose that the earliest bilateral pattern is based on circular positional values, but once established it serves as a prepattern generating a system of bilaterally symmetrical positional values which thereafter would guide and control further development. Among the attractions of circular primary pre-patterns I count the fact that they allow simpler formal rules (to be discussed elsewhere) for generating the type of symmetrization observed by Krause[4] after splitting the germ anlage

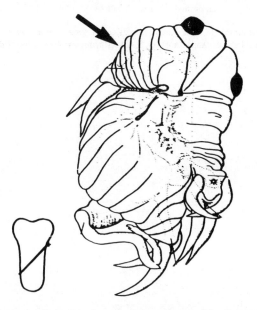

Fig. 5. Ventral view of operation performed on the embryonic rudiment of the stone cricket by Krause and Krause[5], and dorsal view of resulting larval body. Note tip of supernumerary posterior body sticking out to the left (arrow). The part oriented towards the bottom of the figure carries three sets of terminal structures instead of one; this is an additional complication not observed in other cases. (From Krause[4], used by permission.)

very early. Moreover, circular prepatterns might ultimately be linked to the chiral asymmetries characterizing so many organic molecules—asymmetries which have for long been envisioned as guiding *deviations* from symmetry in higher animals[7,16].

My speculative contribution to this issue was aimed at drawing attention to a possible formal mode of ontogenetic symmetrization overlooked so far, in the hope that it might stimulate repetition of the experiments of Krause and Krause[5] or call forth some other *experimentum crucis* in the future.

REFERENCES

1. L. Bischoff, Morphologische Untersuchungen zur larvalen Segmentierung mit normaler Symmetrie und experimentell erzeugter Spiegelbild-Symmetrie (Doppelabdomen) in *Sciara ocellaris*. Staatsexamensarbeit, Fakultät für Biologie, Freiburg i.Br. (1983).
2. S. V. Bryant, V. French and P. J. Bryant, Distal regeneration and symmetry. *Science* **212**, 991–1002 (1981).
3. V. French, P. J. Bryant and S. V. Bryant, Pattern regulation in epimorphic fields. *Science* **193**, 969–981 (1976).
4. G. Krause, Induktionssysteme in der Embryonalentwicklung von Insekten. *Ergebnisse der Biologie* **20**, 169–198 (1958).
5. G. Krause and J. Krause, Die Regulation der Embryonalanlage von *Tachycines* (Saltatoria) im Schnittversuch. *Zool. Jahrb.* **75**, 481–550 (1957).
6. J. Lewis, Continuity and discontinuity in pattern formation. In: "Developmental order: Its origin and regulation", St. Subtelny and P. B. Green (eds.), 40th Symp. Soc. Developmental Biol. Alan R. Liss, New York, pp. 512–531 (1982).
7. J. M. Oppenheimer, Asymmetry revisited. *Amer. Zoologist* **14**, 867–879 (1974).
8. K. Sander, Pattern formation in longitudinal halves of leaf hopper eggs (Homoptera) and some remarks on the definition of "embryonic regulation". *Wilh. Roux' Arch.* **167**, 336–352 (1971).
9. K. Sander, Embryonic pattern formation in insects: basic concepts and their experimental foundations. In *Pattern formation. A primer in developmental biology.* (eds. G. M. Malacinski and S. V. Bryant) pp. 245–268. Macmillan, New York (1984).
10. K. Sander and H. O. Gutzeit, *Duplicitas cruciata* and other riddles of embryonic patterning in insects explained by intercalation of pattern elements. *J. Embryol. Exp. Morph.* **82**, Suppl. 178 (1984).
11. C. Stern, Genes and developmental patterns. *Proc. Intern. Congr. Genet. 9th* (Bellagio, Italy 1953), Caryologia (suppl.) **6**, 355–369 (1954).
12. J. E. Sulston, E. Schierenberg, J. G. White and J. N. Thompson, The embryonic cell lineage of the nematode *Caenorhabditis elegans*. *Developmental Biol.* **100**, 64–119 (1983). (reviewed in *Trends in Biochemical Sciences*, Oct. 1983, pp 349–351).
13. J. A. M. van den Biggelaar, A. W. Dorresteijn, S. W. de Laat and J. G. Bluemink, The role of topographical factors in cell interaction and determination of cell lines in molluscan development. In *International Cell Biology 1980–1981*, pp. 526–538. (edited by H. G. Schweiger). Springer Verlag, Berlin (1981).
14. A. Winfree, A continuity principle for regeneration. In *Pattern formation. A primer in developmental biology.* (eds. G. M. Malacinski and S. V. Bryant) pp. 103–124. Macmillan, New York (1984).
15. L. Wolpert, Positional information and the spatial pattern of cellular differentiation. *J. theoret. Biol.* **25**, 1–47 (1969).
16. L. Wolpert, Positional information and pattern formation. In *Pattern formation. A primer in developmental biology.* (eds. G. M. Malacinski and S. V. Bryant) pp. 3–21. Macmillan, New York (1984).

Comp. & Maths. with Appls. Vol. 12B, Nos. 1/2, pp. 419–433, 1986
Printed in Great Britain.

0886–9561/86 $3.00 + .00

FORMATION OF SYMMETRIC AND ASYMMETRIC STRUCTURES DURING DEVELOPMENT OF HIGHER ORGANISMS

Hans Meinhardt

Max-Planck-Institut für Entwicklungsbiologie, Spemannstr. 35, D 74 Tübingen, F.R.G.

Abstract—Symmetric or partially symmetric structures are generated naturally during development of higher organisms. It is shown that this has its origin in the generation of positional information by local sources of morphogenetic substances and the cell geometry in which this positional information is used: The distribution of a substance in a file of cells (resembling the cross-section of the embryo before gastrulation) generated by a local source and diffusion is necessarily symmetric, it has two slopes. The mechanism of local autocatalysis and long ranging inhibition by which local sources can be generated in an initially homogeneous population of cells allows an understanding of why accidentally a second set of positional information can be generated. The result is a symmetric malformation. Partially symmetric structures are also formed during development of legs and wings of insects and vertebrates. We have proposed that the positional information for limb development is generated at the intersection of two particular borders. This leads inherently to partial symmetric patterns. The near-symmetry of insect wings as well as the perfect symmetric malformation of vertebrate limbs are discussed in terms of this model.

INTRODUCTION

During embryonic development, the complicated structure of a higher organism is generated from a single fertilized egg. In this process, symmetrical (or near-symmetrical) patterns frequently arise. An example is the two sides of a human face. The feature of symmetry in higher organisms contributes substantially to the feeling of beauty we have frequently when regarding living beings. How does this symmetry arise? What determines a normal symmetrical pattern—like the two sides of a face mentioned above—and what is the cause of symmetrical malformations such as, for instance, Siamese twins?

The esthetic appeal of symmetric patterns may be misleading since, as a rule, strongly asymmetric—or more precisely—polar patterns, are formed. This is true not only for the main axes of the body: the anteroposterior axis (head to tail) as well as for substructures, for instance, the defined polarity of a hand from the thumb to the little finger. Even structures which appear strongly symmetric such as the retina are biochemically and functionally asymmetric[1,2]. The relation of polar and symmetric pattern can be illustrated with the dorsoventral (back to belly) pattern of vertebrates. Vertebrates are bilaterally symmetric except for the heart and some other internal organs. Nevertheless the dorsoventral development is very polar in the sense that the pattern on one side is different from the other side and that a continuous transition from one side to the other exists. The reason for this is that only the dorsoventral cross-section has this polar character. The actual pattern is formed on both halves of the approximately tube-like body. Therefore, as discussed below in detail, these two halves are symmetric to each other.

Much evidence has been collected to show that the structures along the main body axes are not determined individually, but that a whole sequence of structures is generated under a common master control. The concept of "positional information" introduced by Wolpert[3] has been especially successful in explaining this feature. According to this model, graded concentration profiles of morphogenetic substances (to be called "morphogens") are generated in the developing embryo and that different concentrations are "interpreted" in different ways by the cells, i.e. they evoke particular pathways of cell differentiations. In this way, a graded concentration profile (to be called a "gradient") leads to an ordered sequence of cell differentiation in space. I will show that the hypothesis of morphogenetic gradients together with their mode of generation and the geometry of the tissue in which they act, provide an understanding of how polar and symmetric patterns can arise during embryogenesis.

INSECTS AS A MODEL SYSTEM

A butterfly with unfolded wings is a perfect example of a symmetrical pattern. The sequence of insect segments shows, on the other hand, a strongly polar pattern. The process of pattern formation in early insect embryogenesis[4] is especially convenient to clarify the relation between polar and symmetric pattern since the basic body pattern of the embryo becomes fixed before any tissue movement takes place in the course of gastrulation. The determination of the main body parts takes place in a cell sheet underlaying the egg shell. This cell sheet, the blastoderm, has thus the same ellipsoidal form as the egg.

Many experimental observations can be accounted for under the assumption that the sequence of segments arise under the control of a gradient which has its high point at the posterior egg pole[5,6]. Thus, a high concentration of this anteroposterior morphogen would cause abdominal segments while low concentrations would lead to head formation (see Fig. 5).

For the determination of the basic body anlage in a cell sheet, a second positional information system is required to organize the dorsoventral (back to belly) dimension. By this gradient, the cells within each segment become different from each other to form e.g. a wing or a leg.

While the anteroposterior gradient can arise from a point-like morphogen source, the dorsoventral gradient must be maintained along the whole egg length. This requires a stripe-like source, for instance, at the ventral side and perhaps, in addition, a stripe-like sink at the dorsal side (Fig. 1). How such stripe-like and point-like sources can be generated will be discussed below in detail. First it should be established whether the two orthogonal positional information systems we assume and the ellipsoidal shape of the cell sheet in which the determination takes place can provide a working model for the types of symmetries we observe.

ONE PEAK—TWO SLOPES: THE BASIS OF MANY NATURALLY OCCURRING SYMMETRIC PATTERNS

Let us regard first a horizontal cross-section through the blastoderm. This is an elliptical file of cells which all have obtained the same dorsoventral positional information. At each particular distance from the posterior pole, both sides obtain the same anteroposterior positional information (Fig. 1). On the other hand, if we regard a (circular) dorso-ventral cross-section, all cells have obtained a particular anteroposterior positional information. If we assume the high point of the dorsoventral gradient at the ventral side, the concentration decreases symmetrically towards the dorsal side. Each particular concentration is present twice, once on the left and once on the right side. Thus, particular structures such as legs or wings coded by a particular morphogen concentration will be present twice and in symmetric arrangement. Thus, the origin of normal symmetric pattern formation becomes clear. The cells which obtain a particular determination by one positional information system resemble a circle (or an ellipse). The second positional information system creates a symmetrical pattern within this circle since a single source plus diffusion produces in a linear array of cells necessarily two symmetrical slopes. Thus, we expect a symmetrical view if we regard an insect from the ventral, from the dorsal, from the anterior or from the posterior side despite the fact that the evoking positional information systems consist of a polar pattern.

THE GENERATION OF POSITIONAL INFORMATION BY AUTOCATALYSIS AND LONG-RANGING INHIBITION

The assumption of a gradient as the organizing agent of a sequence of structures in space contains a circular argument as long as no explanation is given how such gradient can arise. Turing[7] has shown in his pioneering paper that a reaction between two substances with different diffusiveness can lead to pattern formation. We have shown that this reaction must consist of a short range autocatalytic process coupled with a long-ranging inhibition of this autocatalysis[5,8,9]. A simple molecular realization of this scheme consists of an autocatalytic activator

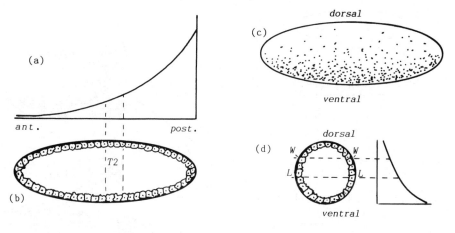

(e)

VENTRAL MIDLINE

1E	2E	3E	4E	5E	6E	7E	8E
1D	2D	3D	4D	5D	6D	7D	8D
1C	2C	3C	4C	5C	6C	7C	8C
1B	2B	3B	4B	5B	6B	7B	8B
1A	2A	DORSAL	MIDLINE			7A	8A
1B	2B	3B	4B	5B	6B	7B	8B
1C	2C	3C	4C	5C	6C	7C	8C
1D	2D	3D	4D	5D	6D	7D	8D
1E	2E	3E	4E	5E	6E	7E	8E

ANTERIOR (HEAD) POSTERIOR (TAIL)

VENTRAL MIDLINE

Fig. 1. Generation of symmetric pattern by asymmetric positional information. In insects, the determination of the basic body pattern takes place in the blastoderm—a cell sheet of ellipsoidal shape. A positional information system is assumed for the anteroposterior (a) and the dorsoventral axis (c, d). A particular combination of both positional information systems is always present twice, one on the left side, the other on the right side of the organism. (T2 = second thoracic segment in (a), W = wing, L = leg in (d)). The result are symmetrical patterns despite the fact that the evoking positional information is asymmetric. (e) Schematic view of the arrangement of differentiated cell states arising under these positional information systems. The blastoderm, approximated as a cylinder, is shown unrolled. Anteroposterior pattern is denoted 1–8, the dorsoventral A–E. The result is a symmetrical checkerboard (the periodic character of the segmental pattern in insects has been neglected). For the pairwise determination of wings and limbs (see Fig. 6 and 7) it is important that an intersection of two borders (in the example 6/7 and C/D) is always present twice.

a which catalyses in addition its own antagonist, the rapidly diffusing inhibitor h (Fig. 2). An example of this interaction is given in the following differential equations:

$$\frac{\partial a}{\partial t} = c \frac{a^2}{h} - \mu a + D_a \frac{\partial^2 a}{\partial x^2} + \rho_0 \tag{1a}$$

$$\frac{\partial h}{\partial t} = ca^2 - vh + D_h \frac{\partial^2 h}{\partial x^2}. \tag{1b}$$

Such equations are easy to read even by those not familiar with differential equations. Eq. 1(a) states that the change of the activator concentration per time unit is proportional to the (nonlinear) autocatalytic a-production, that the activator disappears proportional to the number of activator molecules present and that diffusion takes place (μ is the decay rate, D_a the diffusion constant of the activator, c is a constant). A small activator-independent production term (ρ_0)

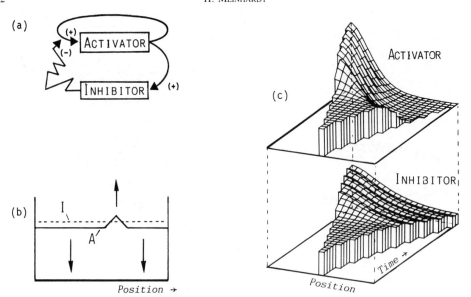

Fig. 2. Pattern formation by autocatalysis and lateral inhibition. (a) The reaction scheme. Assumed is an activator which feeds back on its own synthesis (autocatalysis). It catalyses also its antagonist, the highly diffusible inhibitor. (b) The instability: A small local elevation of the activator (A) grows since the additionally produced inhibitor (I) diffuses into the surroundings where it inhibits activator production. (c) Generation of a polar pattern. Assumed is a linear array of cells which grows at both margins. If a critical size is exceeded, pattern formation starts. A polar gradient is formed since a central activator maximum requires space for two slopes which is not available at the critical size. The polar pattern can be maintained upon further growth but the system can switch to symmetric pattern by a variety of perturbations (Fig. 3).

is able to initiate the system at very low activator concentrations. This term is very important for the transition of a normal pattern to a symmetric malformation (Fig. 3) as well for the regeneration of the pattern in tissue fragments[5].

Fig. 2 provides some intuition as to why an interaction of the type described by Eq. 1 leads to a stable pattern. It shows further that, if the total area is small (comparable with the range of the activator molecules), only one activator maximum can be formed. This maximum appears preferentially at one end of the field of cells.

Periodic structures can arise on the basis of the activator-inhibitor mechanism if the total area is much larger than the range of the inhibitor. The resulting structures can have symmetric features such as observed in the arrangement of leaves in plants or in the veins of leaves (for computer simulations and programs, see [5]).

THE DORSOVENTRAL AXIS AND THE FORMATION OF STRIPES

A somewhat more complicated molecular mechanism is required to generate a concentration profile with a gradient along one axis but with a constant concentration along the other, especially if the latter axis is more extended than the first. Such a system is—as mentioned above—required for the dorsoventral gradient. To achieve this, the area of morphogen source must not be point-like but has to have a stripe-like extension. Stripe-like patterns can be generated if two feedback loops are present which exclude each other locally but activate each other on longe-range[5,10]. In this way, the two feedback loops need each other in a symbiotic manner. Stripes in which the one or the other feedback loop is turned on are an especially stable configuration since stripe formation is necessarily connected with a long common boundary, a feature which allows an efficient mutual stabilization (Fig. 4). A cell of one type is always close to a cell of the other type. In the following Eq. 2, g_1 and g_2 describe two autocatalytic feedback loops which compete via a common repressor r but which activate each other by the highly diffusible substance s_1 and s_2:

$$\frac{\partial g_1}{\partial t} = \frac{cs_2 g_1^2}{r} - \alpha g_1 + D_g \frac{\partial^2 g_1}{\partial x^2} + \rho_0 \qquad (2a)$$

$$\frac{\partial g_2}{\partial t} = \frac{cs_1 g_2^2}{r} - \alpha g_2 + D_g \frac{\partial^2 g_2}{\partial x^2} + \rho_0 \tag{2b}$$

$$\frac{\partial r}{\partial t} = cs_2 g_1^2 + cs_1 g_2^2 - \beta r \left(+ D_r \frac{\partial^2 r}{\partial x^2} \right) \tag{2c}$$

$$\frac{\partial s_1}{\partial t} = \gamma(g_1 - s_1) + D_s \frac{\partial^2 s_1}{\partial x^2} + \rho_1 \tag{2d}$$

$$\frac{\partial s_2}{\partial t} = \gamma(g_2 - s_2) + D_s \frac{\partial^2 s_2}{\partial x^2} + \rho_1. \tag{2e}$$

Equation 3 describes a lateral activation system based on the interaction of only three substances:

$$\frac{\partial a}{\partial t} = \frac{c}{s(\kappa + b^2)} - \mu a + D_a \frac{\partial^2 a}{\partial x^2} \tag{3a}$$

$$\frac{\partial b}{\partial t} = \frac{c \cdot s}{(\kappa + a^2)} - \nu b + D_b \frac{\partial^2 b}{\partial x^2} \tag{3b}$$

$$\frac{\partial s}{\partial t} = \alpha a - \beta b \cdot s + D_s \frac{\partial^2 s}{\partial x^2}. \tag{3c}$$

In this interaction, the autocatalysis is of a hidden form: substance a inhibits substance b and vice versa. Thus, an increase of a leads to a decrease of b and, in turn, this leads, due to the decline of b-inhibition on a, to a further a increase. Thus, the mutual inhibition causes autocatalysis *and* mutual exclusiveness. The spatial pattern results from the interaction of a and b via the highly diffusible substance s; a produces s but s is inhibitory to the a production. In

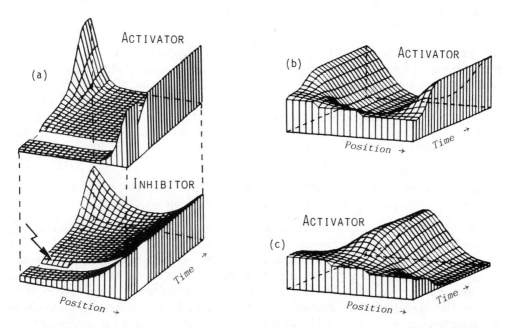

Fig. 3. Switch from polar to symmetric patterns: the origin of symmetric malformations. A polar activator-inhibitor pattern (see Fig. 2) has the tendency to switch into a symmetrical pattern, if the size of the field is larger than the critical field size (range of the activator molecules). At the non-activated side, the inhibitor concentration is naturally very low and any further lowering, caused, for instance, by UV-irradiation or leakage, can lead to the onset of autocatalysis which leads to a second activator maximum. A symmetrical pattern results (see Fig. 5). (b, c) At larger extensions of the field, symmetrical patterns are favoured even if the initial distribution is clearly asymmetric. Minor differences can decide whether one peak forms in the center or whether two peaks form at opposite ends of the field.

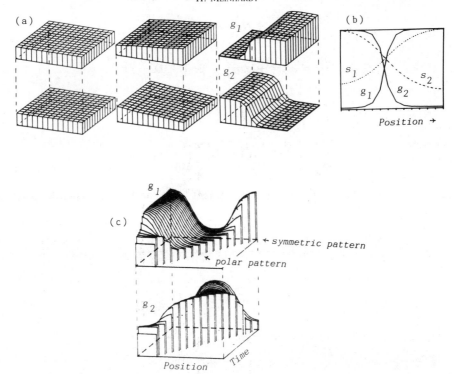

Fig. 4. Pattern formation by lateral activation. If two feedback loops (g_1 and g_2) exclude each other locally but activate each other on long range (via s_1 and s_2, Eq. 2), a stripe-like concentration profile of g_1 and g_2 emerges[10]. Such a pattern is, for instance, required to form the dorsoventral gradient of an insect along the whole anteroposterior extension of the egg (see Fig. 1(c)). (a) Stages in the g_1/g_2 pattern formation. (b) cross-section through the final pattern including the s_1 and s_2 distributions. (c) Transition of a polar into a symmetrical pattern is possible also in the lateral activation scheme. In this example, at a too large distance from g_1-cells, the g_2 state obtains little support by s_1 molecules but the high s_2 concentration induces a switch of some g_2-producing cells into g_1-producing cells (calculated with Eq. 2 in a linear array of cells, g_1 and g_2 plotted as function of time). The formation of symmetrical malformation in limb development (Figs. 8, 9) or of Siamese twins are assumed to have this origin.

contrast, b depends on s and removes s. Thus, a region in which a high a production takes place depends on the close proximity of a high b region for the removal of the "poisonous" s while the b region needs an a region for s supply. An a–b stripe (or stripes) would be the most stable configuration. The width of the stripe is determined essentially by the range of the a and b molecules. (A checkerboard-like pattern would require sharp edges and such a spatial resolution would require a low diffusion of a and b. Then, however, the system would tend to produce incoherent patches).

SYMMETRICAL MALFORMATIONS: TWO SEPARATE PEAKS PROVIDE TWO SETS OF POSITIONAL INFORMATION

In many insect species, a variety of different experimental manipulations such as egg centrifugation, UV-irradiation, punctuating or temporary ligation of the anterior egg pole can lead to a specific malformation, the double abdomen (for review, see[4]): head and thoracic segments are replaced by a second set of posterior abdominal segments. The same pattern can also arise in *Drosophila* embryos when the *bicaudal* gene is mutated in the mother fly (Fig. 5,[11]). The activator–inhibitor mechanism mentioned above provides a straightforward explanation for this instability. In normal insect development, the high point of the gradient and thus the activated region must be located at the posterior pole. From other experiments, we have to conclude that the activated region is small, confined to the posterior tenth of the egg[5,6]. Thus, the activator distribution is flat throughout the larger part of the egg. In contrast, the inhibitor with its shallow gradient is convenient to supply positional information (Fig. 5). At the anterior egg pole, the inhibitor concentration is low. This creates an instable situation since any further unspecific lowering of the inhibitor—caused, for instance, by the manipulation listed above—

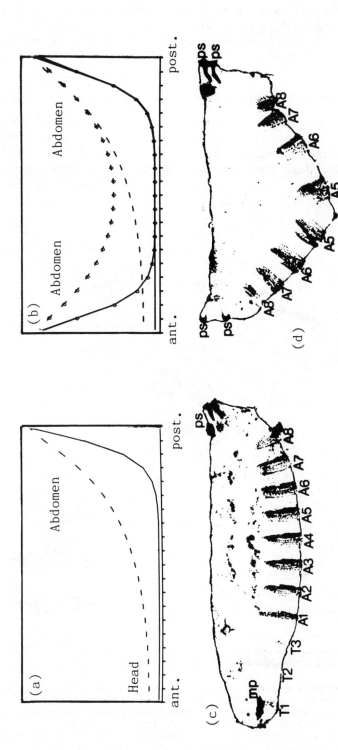

Fig. 5. Formation of symmetrical malformations in insects. (a) The normal pattern is assumed to be generated by an activator–inhibitor system. The inhibitor (———) provides the positional information proper: high inhibitor concentration determines abdominal, low concentration head structures. (b) Any lowering of the inhibitor can lead to a second activator maximum at the anterior side (see Fig. 3(a)). The positional information becomes symmetric. Abdominal structures are formed at both egg poles while head and thoracic structures are lost. (c, d) Biological example: a normal larva of *Drosophila* (c) and the symmetric malformation *bicaudal*[11], carrying two sets of terminal abdominal segments. (Photographs kindly supplied by Chr. Nüsslein-Volhard, see[11].)

can, lead to the onset of autocatalysis starting from the small activator-independent a-production (ρ_0), Eq. 2. Then, the new activator maximum grows and sharpens itself until it reaches an equilibrium with the self-produced inhibitor (see Fig. 3). The anterior egg pole is then exposed to the same positional information as the posterior egg pole. The result is a completely symmetric malformation (Fig. 5(d)). Thus, the mode of source generation by local autocatalysis and long ranging inhibition has an instability that a second peak arises at distance from the first. If this happens, a second set of positional information is generated which, together with the normal one, leads to a symmetrical malformation.

SYMMETRY AND ASYMMETRY IN LIMB FORMATION

Partial symmetry during normal development as well as strictly symmetrical malformations are also common in the development of substructures such as legs or wings. For an understanding of how both types of symmetric structures may arise, I will explain our present view of how normal limb development is initiated. The determination of limbs requires certainly a primary subdivision along the anteroposterior and the dorsoventral axes. One could imagine that a combination of a particular anteroposterior and dorsoventral determination is the signal to form

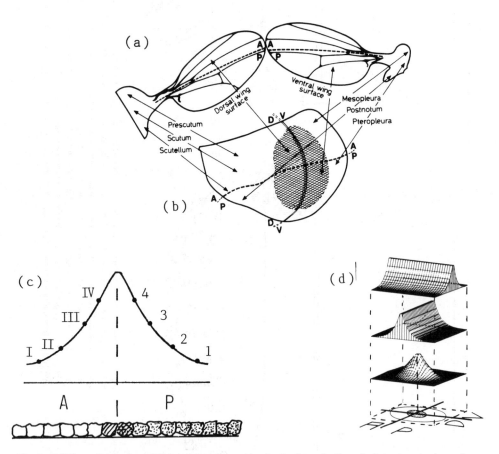

Fig. 6. Cell determination boundaries as organizing regions for the determination of substructures such as wings and legs. (a) Biological example: The wing of *Drosophila* consists of a dorsal and a ventral cell sheet. Both are symmetric to each other. (b) It is formed from an imaginal disc[17] during metamorphosis by evagination and subsequent flattening of the central part of the disc (hatched). The lines of symmetry, the D/V border as well as the A/P border, are lines of clonal restrictions and are determined very early, before the disc itself is formed[14]. (c, d) Model. Assumed is that two cell types, A (blank) and P (dotted) have to cooperate to produce a morphogen. Only in cells close to at the common border (hatched) is this cooperation possible. The result is a symmetrical pattern which can be differently interpreted in both cell types (I–IV and 1–4). The result could be a partial symmetric pattern. (d) In a cell sheet, an intersection of two borders is required. Each border produces a ridge-like positional information, and both together form a cone-shaped distribution, appropriate to determine the disc itself and the wing blade within the disc.

a limb: in terms of Fig. 1, for instance, the determination 6C. Such a field of cells would be in itself homogeneous and a type of pattern formation similar to that discussed above for the determination of the whole embryo would be required to determine the limb's fine structure.

However, many experiments on limb determination suggest another mechanism: that the borders between adjacent determinations are used to organize the surrounding cells[12,13]. Imagine that the cell of the type 6 and 7 (Fig. 1) have to collaborate to produce a limb morphogen. The cooperation would be possible only along the border (Fig. 6). The morphogen distribution would have a symmetrical ridge-like distribution centered over the boundary. The local concentration is a measure for the distance of a cell from the border.

To determine the position of a limb, a single border is certainly insufficient; an intersection of two borders is required. An intersection of two borders defines a unique point. In the example used above, this could be an intersection of the 6/7 and the C/D border (Fig. 1). Such an intersection determines not only the position of the limb but generates also the positional information for the limb's fine structure and determines thus its handedness. From the type of symmetry generated by the primary anteroposterior/dorsoventral coordinate system it is evident that a particular intersection, i.e. a particular limb, will appear in pairs (Fig. 1), both have the opposite handedness.

In insects which pass through a metamorphosis, legs and wings are "prefabricated" in imaginal discs. The cells of the discs are already determined as to which parts they form in the adult organism. This allows a rapid formation of the adult structure during metamorphosis since it requires only that the sheet evaginates and that the cells take over their predetermined functions. Borders separating anterior and posterior as well as dorsal and ventral parts of the disc are well known as the so-called compartment borders[14,15]. I have proposed that these compartment borders are formed first and the imaginal discs are formed around their intersections[5,12]. A morphogen which is produced at the intersection of two borders, i.e. produced by cooperation of three sectors or four quadrants, has the shape of a cone. This model fits well with the experimental observation. In the leg disc, for instance, the future leg segments are determined in concentric rings[16]. The most distal structures, the claws of the leg, are determined in the center of the disc. The cone-shaped morphogen gradient with its circular contour lines would account for such a pattern in a straightforward manner (see Fig. 8).

The situation is similar in the wing disc. The central part which surrounds the intersection of the borders forms the wing proper (Fig. 6,[17]). The cells at the margin form thoracic structures to which the wing is attached. The cone-shaped distribution is convenient to separate both cell populations from each other as well as the disc cells from the cells of the surrounding larval ectoderm. The two borders with their symmetrical pattern at both sides account for the partial symmetry we see in the insect wing (Fig. 6(a)). Thus, the mechanisms of cooperation of compartments provide a very natural coordinate system for the observed arrangement of pattern elements.

In arthropods or insects, leg formation is a periodic event, similar to segmentation itself. The process of segmentation can be explained under the assumption that during embryogenesis a periodic iteration of three cell states, called S, A and P is formed[5,18]. The juxtaposition of S and P cells leads to the formation of a segment border while the A–P border is, as discussed above, a prerequisite of leg and wing formation. The A–P border is thus already formed during the process of segment formation and is used a second time, for the determination of appendages.

The positional information generated by the boundary mechanism is in principle symmetrical on both sides of the border. However, the resulting pattern can be asymmetric since the cells are necessarily different on both sides of the border. The most extreme way to generate an asymmetric pattern out of the symmetrical positional information seems to be used in the anteroposterior determination of vertebrate limbs where only one of the two tissues can respond to the morphogen[19]. The competent cells are thus exposed to one half of a symmetrical, bell-shaped distribution (Fig. 7), i.e. to a monotonic gradient. This asymmetric pattern is appropriate to determine the polar sequence of the digits. The digits are formed along the line termed the apical ectodermal ridge, a structure which is very early visible. I assume that this ridge is caused by a dorsoventral border. Since both the dorsal and the ventral side of the border are used, the resulting pattern, for instance of a hand, is partially symmetric along the line of digits.

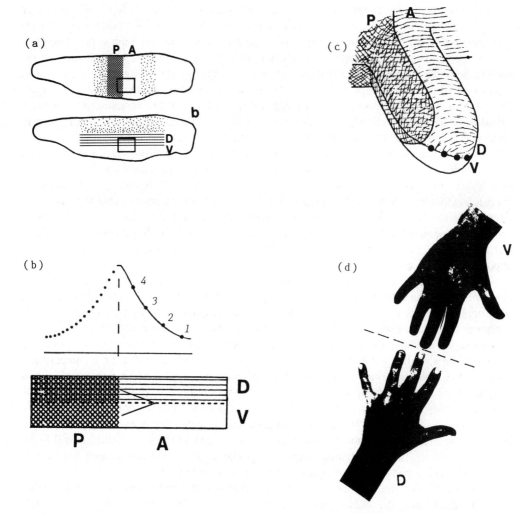

Fig. 7. Determination of vertebrate limbs. (a) The intersection of an A/P and a D/V border is assumed to generate a limb field. The morphogen resulting from the A–P cooperation is primarily symmetric (see Fig. 6). Since only the A-cells are competent to respond, the resulting pattern (the sequence of digits) is asymmetric. (c) The outgrowing limb consists thus mainly of A-tissue. The digits are assumed to be formed along a D/V border. This model accounts for many experimental observations observed after graft operations[13]. (d) Since both sides of the D/V border contribute to the limb, the dorsal and ventral side of a limb is partially symmetric.

SYMMETRICAL MALFORMATIONS OF LEGS AND WINGS

Symmetrical malformation of appendages occurs spontaneously and more frequently after certain experimental treatments. A particular genetic background can increase the frequency. Bateson[20] collected a large number of spontaneously arisen leg triplication of a variety of insect species and found that two of the three legs are mirror images of each other. A more systematic investigation was possible with the discovery of a mutation in *Drosophila* which causes leg duplication and triplication with high frequency after heat shock. According to the model, a leg duplication occurs if a second intersection of the two compartment borders is formed (Fig. 8). In the *Drosophila*-mutation just mentioned it has been shown that the heat treatment primarily causes some cell death and that this leads to a respecification of some cells of the anterior compartment into posterior cells[21,22]. As explained in detail in Fig. 8, this leads to new intersections and thus to one or two additional legs. The symmetry relations predicted by the model agree precisely with that observed by Bateson: in a triplication the three legs appear in a plane since they arise along the (linear) dorso-ventral border. The outer two legs have the same handedness as a normal leg would have at this side while the central leg has opposite handedness (Fig. 8).

Duplication and triplication are known to occur also in vertebrates. Fig. 9 provides some

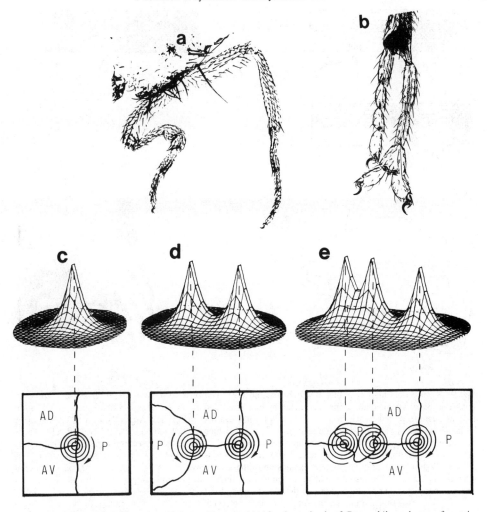

Fig. 8. Symmetrical malformations in insect legs. (a, b) After heat shock of *Drosophila* embryos of certain genetic backgrounds, supernumerary legs are formed[21,22]. (a) Two mirror-symmetric legs grow out of a single leg stump. (b) Three legs are formed instead of a single leg. (c–e) Model: The normal leg is formed at the intersection of two—the anteroposterior and the dorsoventral—compartment borders. The cone-shaped morphogen distribution, generated around this intersection, is appropriate for the circular arrangement of pattern in the disc. If some of the A-cells switch into P-cells (see Fig. 4(c)), one (d) or two (e) new intersections are formed. The resulting morphogen profile is completely (d) or partially (e) symmetric, in agreement with the observations (a, b). In agreement with the experimental observations, the model predicts that the three legs of the triplicated leg are arranged along a line, i.e. in a plane and not like the legs of a photographic tripod since the new intersections are formed on the dorsoventral compartment border. The two outer legs have the same handedness as a normal leg while the central leg has opposite handedness (arrows). Thus, the model provides a rationale for Bateson's rules[20]. (a, b after photographs kindly provided by J. Griton, see[21]).

examples. Again we expect that respecifications lead to new intersections and thus to supernumerary or symmetrical limbs. The model predicts that symmetrical limbs are possible with two little fingers but never with two thumbs at each end. To my knowledge, the observed malformations follow that prediction.

PROBLEMS IN GENERATING CARDINAL AXES ON A SPHERE

Let us go back to the primary pattern formation of the embryo. For the discussion of the coordinate system in insect embryogenesis, it was essential that the egg has a long anteroposterior extension and a small dorsoventral extension to generate the symmetrical checkerboard-like pattern as the primary grid of positional values (Fig. 1). Other eggs, for instance those of amphibians, have quite precisely a spherical shape. How can such a coordinate system be generated in a spherically shaped cell sheet? Further, how can it be achieved that both the anteroposterior and the dorsoventral axes are kept perpendicular to each other? During oogenesis,

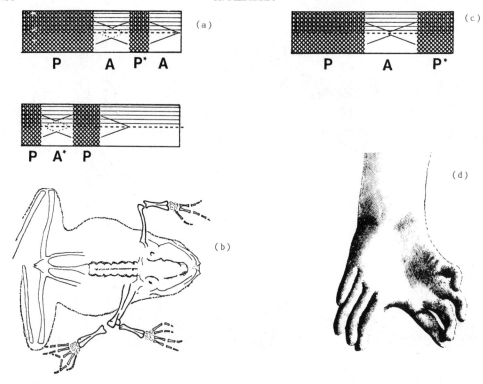

Fig. 9. Symmetrical malformations in vertebrate limbs[13]. The A and P regions which give rise to the limb field (Fig. 8) are assumed to stabilize each other by long range activation (see Fig. 4). If this mutual stabilization is insufficient, a partial switch from A to P or vice versa can occur (Fig. 4c). A patch of new P-cells in an A region (P*) or of A-cells in a P region leads to an A region which is bordered on both sides with P-cells. Thus, the model predicts posterior to the normal limb the formation of a symmetrical supernumerary limb which carries the posteriormost digit, the little finger at each end. (b) The bones of a frog with such a malformation (after[27]). (c) If an A–P respecified group of cells extends up to the anterior margin of the A region, only one additional intersection appears. The result would be a symmetrical limb with two little fingers at each end. The model predicts that a symmetrical limb with two thumbs is impossible (since an A–P–A configuration would lead to two separate limbs). (d) An observed symmetrical limb of a human. (after[20].)

the amphibian and the sea urchin eggs obtain a strong animal–vegetal polarity. The dorsoventral axis is comparatively labile and can be reoriented in both species by experimental manipulations[23,24]. However, the high dorsal (or ventral) point appears only at a certain distance from the animal and vegetal pole, i.e. on an approximately equatorial ring. This suggests how the two axes are kept perpendicular: A first pattern is created and the high point of the second positional information system has to appear at a certain intermediate level of the first. However, two high points on a sphere would not create a coordinate system like that we are familiar with on the earth consisting of parallels of latitude and meridians since a high equatorial point would create a second set of concentric circles (Fig. 10(a)) but not a symmetric checkerboard such as shown in Fig. 1.

There are two ways out of this problem. Either the dorso-ventral pattern has, similarly to insects, a long extension. This could be achieved by the formation of a dorsal and a ventral hemisphere and the border between the two runs through the animal and vegetal pole. The dorso-ventral positional information would be measured as the distance from the border between the dorsal and ventral hemisphere in a similar way as discussed above for the legs and wings. However, this would require some sort of size regulation since the circumference is much larger at the equator than near the pole. A way to avoid the size regulation problem is to use only a small part of the sphere (which is in fact a nearly flat sheet of cells). This geometry is used in chickens where the future embryo proper, the *area pellucida* occupies only a small fraction of the surface of the spherical yolk.

Another mode is suggested by the gastrulation of amphibians. At later blastula stage, the cells along a line situated somewhat below the physical equator begin to move into the hollow sphere. It could be that the cells obtain their final determination during this gastrulation. Thus,

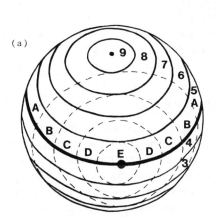
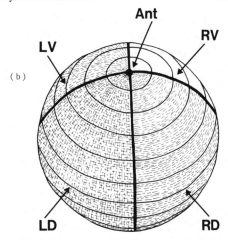

Fig. 10. Pattern formation on a sphere. (a) Two sources of morphogens on a sphere are inappropriate to generate a symmetrical checkerboard-like pattern such as required to determine a normal body plan (compare with Fig. 1(e)). (b) A subdivision into three pairs of hemispheres (A/P, D/V and L/R) provides a more convenient coordinate system which easily can be kept orthogonal by an appropriate coupling. Limbs, for instance, would be formed along the D/V border (see Fig. 7) at a certain distance from the anterior pole (Ant, a DV/LR intersection). Centrally determined structures such as the notochord, dorsal fins etc. could be determined by the L/R border. In general, the position of an organ would be determined by ratios of positional information given as the distance from the L/R and D/V borders. A switch into a symmetrical pattern (for instance L/R/L or D/V/D, see Fig. 4(c)) could lead to Siamese twins.

determination would take place in a ring, at a certain distance from the animal and vegetal pole. This would lead to the symmetrical checkerboard (Fig. 1).

TWO OR THREE CARDINAL AXES?

In principle, two cardinal axes are sufficient to organize a bilaterally symmetric organism. However, a third axis would simplify the interpretation of positional information and would facilitate the maintenance of cardinal axes orthogonal to each other. Let us assume a primary subdivision of the sphere into three pairs of hemispheres, for instance into an anterior (A) and a posterior (P), into a dorsal (D) and a ventral (V) as well as into a left (L) and a right (R) hemisphere (Fig. 10b). The mechanism of lateral activation mentioned above would be appropriate for this pattern formation. The axes of the three pairs of hemispheres would orient themselves perpendicular to each other if the condition for a high D or high V concentration would be a high R *and* a high L concentration, i.e. a R/L border. Similarly, the condition for a high R or high L would be a D/V border. To keep the AP pair perpendicular to the two other pairs, the condition for a high A or a high P concentration would be a D/V *and* a L/R border (Fig. 10(b)).

What would be the advantage of such a system? In a pure A/P–D/V system, dorsalmost and ventralmost structures must be initiated at the corresponding maximum concentrations. A system which detects the maxima would be required. The addition of a L/R system would provide the appropriate signal for all structures which have to appear along the dorsal or ventral midline, for example the notochord, dorsal fins of fishes or the ventral furrow of insect gastrulation. The somites may be specified not only by a certain D/V level but by a combination of D/V and L/R. The condition for tail formation would be an intersection of the D/V and the R/L border in the posterior hemisphere.

A direct experimental proof for the existence of a L/R pattern is not yet available. In *Drosophila*, Nüsslein-Volhard[25] found surprisingly many mutations (about 10) which abolish dorso-ventral pattern formation. Since we expect a close coupling between the D/V and the L/R pattern, it could be that some of these mutations affect primarily the L/R pattern and thus secondarily the D/V pattern. If this would be indeed the case, we would expect asymmetric left and right portions of the body (measured from the dorsal and ventral midline) in leaky mutations. Since changes in the D/V pattern are easier to detect, these mutations could be classified as pure D/V mutations. In mice, by breading and selection, strains have been obtained which carry an enlarged whisker field (a field of sensory hairs in the face)[26]. Either the right or the left field can be enlarged in different strains, indicating that the two sides are not determined just by the two symmetrical slopes of one peak. In the model, an enlarged R or L half would provide an explanation. According to the lateral activation mechanism for the L/R pattern, the genetic basis for such an enlargement would be an altered autocatalysis of the R or L system.

SIAMESE TWINS ARE A PARTICULAR TYPE OF SYMMETRIC MALFORMATIONS

If the D/V and L/R hemispherical pattern are generated by a lateral activation mechanism, they would contain an instability to switch into a symmetrical pattern (Fig. 4), similar to that discussed for the anteroposterior pattern of insects (Fig. 5). Symmetrical malformations—Siamese twins—could result. They can be fused at the dorsal or ventral side (back or belly) but also laterally. In terms of the model, the first malformation would occur if instead of a D/V pattern a D/V/D or a V/D/V pattern is formed. Laterally fused twins would suggest a L/R/L or a R/L/R instead of a L/R pattern. Indeed, it has been frequently observed that one of the siblings has a *situs inversus*.

CONCLUSION

Normal symmetrical or near-symmetrical patterns arise from the two slopes of a positional information system which is generated by a local source and diffusion to both sides in a file of cell. Strictly symmetrical malformation develops if accidentally a second source is formed. The mechanism by which local sources can be generated—short-range autocatalysis and long-range inhibition—suggests why a transition from one to two sources can occur.

REFERENCES

1. G. D. Trisler, M. D. Schneider and M. Nierenberg, A topographic gradient of molecules in the retina can be used to identify neuron position. *Proc. Nat. Acad. Sci. U.S.A.* **78**, 2145–2149 (1981).
2. R. M. Gaze, J. D. Feldman, J. Cooke and S. H. Chung, The orientation of the visuotectal map in *Xenopus*: developmental aspects. *J. Embryol. exp. Morph.* **53**, 39–66 (1979).
3. L. Wolpert, Positional information and the spatial pattern of cellular differentiation. *J. theoret. Biol.* **25**, 1–47 (1969).
4. K. Sander, Formation of the basic body pattern in insect embryogenesis. *Adv. Insect Physiol.* **12**, 125–238 (1976).
5. H. Meinhardt, *Models of Biological Pattern Formation.* Academic Press, London (1982).
6. H. Meinhardt, A model for pattern formation in insect embryogenesis. *J. Cell Sci.* **23**, 117–139 (1977).
7. A. Turing, The chemical basis of morphogenesis. *Phil. Trans. B.* **237**, 37–72 (1952).
8. A. Gierer and H. Meinhardt, A theory of biological pattern formation. *Kybernetik* **12**, 30–39 (1972).
9. A. Gierer, Generation of biological patterns and form: Some physical, mathematical, and logical aspects. *Prog. Biophys. molec. Biol.* **37**, 1–47 (1981).
10. H. Meinhardt and A. Gierer, Generation and regeneration of sequences of structures during morphogenesis. *J. theor. Biol.* **85**, 429–450 (1980).
11. C. Nüsslein-Volhard, Genetic analysis of pattern formation in the embryo of *Drosophila melanogaster. Wilhelm Roux' Arch.* **183**, 249–268 (1977).
12. H. Meinhardt, Cell determination boundaries as organizing regions for secondary embryonic fields. *Dev. Biol.* **96**, 375–385 (1983).
13. H. Meinhardt. A boundary model for pattern formation in vertebrate limbs. *J. Embryol. exp. Morph.* **76**, 115–137 (1983).
14. A. Garcia-Bellido, P. Ripoll and G. Morata, Developmental compartmentalization of the wing disk of *Drosophila. Nature New Biol.* **245**, 251–253 (1973).
15. E. Steiner, Establishment of compartments in the developing leg discs of *Drosophila melanogaster. Wilhelm Roux' Arch.* **180**, 9–30 (1976).
16. G. Schubiger, Anlageplan, Determinationszustand und Transdeterminationsleistungen der männlichen Vorderbein-scheibe von *Drosophila melanogaster. Willhelm Roux' Arch.* **160**, 9–40 (1968).
17. P. J. Bryant, Pattern formation in the imaginal wing disc of *Drosophila melanogaster:* Fate map, regeneration and duplication. *J. Exp. Zool.* **193**, 49–77 (1975).
18. H. Meinhardt, Models for positional signalling, the threefold subdivision of segments and the pigmentation pattern of molluscs. *J. Embryol. exp. Morph.* **83** (Supplement), 289–311 (1984).
19. J. M. W. Slack, Determination of polarity in the amphibian limb. *Nature* **261**, 44–46 (1976).
20. W. Bateson, *Materials for the Study of Variation.* Macmillan, London (1894).
21. J. R. Girton, Pattern triplication produced by a cell-lethal mutation in *Drosophila. Dev. Biol.* **84**, 164–172 (1981).
22. J. R. Girton and M. A. Russell, An analysis of compartmentalization in pattern duplications induced by a cell-lethal mutation in *Drosophila. Dev. Biol.* **85**, 55–64 (1981).
23. J. Gerhart, G. Übbels, S. Black, K. Hara and M. Kirschner, A reinvestigation of the role of the grey crescent in axis formation of *Xenopus leavis. Nature* **292**, 511–516 (1981).

24. G. Czihak, Entwicklungsphysiologische Untersuchungen an Echiniden. *Roux' Archiv für Entwicklungsmechanik* **154**, 29–55 (1962).
25. C. Nüsslein-Volhard, Maternal effect mutations that alter the spatial coordinates of the embryo of *Drosophila melanogaster*. In: *Determination of Spatial Organization* (S. Subtelney and I. R. Konigsberg eds.), p. 185–211. Academic Press, New York (1979).
26. H. Van der Loos, J. Dörfl and E. Welker, Variation in pattern of mystacial vibrissae in mice. *J. Heredity* **75**, 326–336 (1984).
27. H. Rahmann, W. Engels and H. J. Grüter, Doppelmissbildungen einer Vorderextremität bei *Rana temporaria*. *L. Zool. Anzeiger* **169**, 449–454 (1962).

Comp. & Maths. with Appls. Vol. 12B, Nos. 1/2, pp. 435–463, 1986
Printed in Great Britain.

0886–9561/86 $3.00 + .00
© 1986 Pergamon Press Ltd.

THE VISUALIZATION OF MUSIC: SYMMETRY AND ASYMMETRY

Roberto Donnini

Via Gramsci 353A, 50019 Sesto Fior. Firenze, Italy

Abstract—In this contribution the author attempts to explain the differences between music and its visualization (the score) and through various examples the relation between resonant result and score. In the same time he tries to show all the possible concepts of symmetry (and asymmetry) that exist inside this relation and in the graphical signs of certain scores. A new definition of symmetry is not formulated rather, a direction for a new way to conceive it is given.

Music before being played is always *something else*. That is, before music exists as an audible experience, it exists as an idea or plan in the mind of the composer.

The event of music at a deep level is never only music but always *something else*: this is why one can speak about it. This *something else* can be the score, that is to say, its visualization.

We cannot discuss music referring only to the impression or to the mood it creates. If we do so, our norms remain purely subjective, and must change every time the subject changes. Phenomenologically, music is only the playing of a score. The listener perceives the beginning of a resonant sensation: but if, at the same time, he can watch the score, a new formal and visual sensation is added to the first. The person who, instead, sees the score without listening to the music, has only a visual sensation: in this very moment he *sees* the music in a different way, that is to say, through *something else*, which can be defined as the mirror image of the music itself.

Music exists only in relationship to a score and to a musical instrument. The score and the instrument are related, and each relates to the music performed. These relationships can be seen as symmetrical. Following a score and saying that the musical writing is interesting, signifies that the formal elements of the graphical signs forming the sheet-music are in suitable relationship to that which they represent. If the signs are well-proportioned and well-balanced, and if a sort of concordance of the several parts is present by which they are integrated into a whole (H. Weyl), those who follow the score will have a positive impression.

The listener can only imagine the sound written or drawn in a score: when the air is moved by the waves produced by an instrument the ear is touched by those waves as the actual sound is portrayed by the score of music. The score represents the music through a reflected copy: between the sound and its visualisation there is the same relation as between an eagle and its prey, both having a role that can be defined as complementary and symmetrical.

Not all music has a score. The score may be replaced by a plan that the performer has clearly in his mind. In this case the performer becomes the composer. Folk music usually does not have a score: those performing or listening to a piece of folk music are familiar with the musical pattern. In folk music the feeling comes from the experience of continuous listening[1]. In general, all pieces of folk music are performed (i.e. composed) in the same way. Apart from small variants, the structure is always the following:

(i) a short introduction or a *ritornello*
(A) the first *stanza* with a certain type of music
(B) the second *stanza* with another type of music
(r) an interval (often a development of (i) or of ritornello)

and so on.

The resulting sequence (i)(A)(B)(r), (A)(B)(r), (A)(B) etc. is usually familiar to the audience. The repeated elements create a rhythm that shows the concordance of the several parts. In everyday language when all of these elements are well-proportioned and well-balanced something symmetrical is born.

Symmetry is a formal and structural pattern found in an infinite number of living and non-living things and is fundamental in human thought. There are many kinds of symmetry, but they can be reduced in general to four types: bilateral symmetry, spatial symmetry, symmetry of movement (rotational and translatory) and symmetry of color.

In music, all of these kinds of symmetry apply, but there is much controversy concerning them, although, as we can see ahead, we can find numerous examples of each. It is not difficult to find parts of scores illustrating bilateral symmetry and symmetry of movement. But in music we cannot limit the concept of symmetry to a sense of a good-proportion and a well-balanced concordance of parts. The art movements of the last seventy years have brought in to question all of the traditional aesthetic rules, so we can no longer say, "Beauty is bound up with symmetry" (H. Weyl)[2]. The established rules no longer dictate what constitutes beauty. Often in music what gives the impulse towards the discovery of the *new* is the rejection of the rules of symmetry. The symmetrical order reached through a rational construction of an idea, is forced to take a *swerve* resulting in a *difference* which causes the first asymmetry[3]: henceforth the *swerve* makes headway toward asymmetry and so on, in a manner producing, in the end a new symmetrical order. The formal inspiration of the composer balances the *swerves* and the *differences* in reference to the general idea and plan.

A fine example of this process is represented in music by the Fugues of Johann Sebastian Bach. The principle of imitation that embodies the fugue gives it symmetry in a very imaginative and flexible manner. The germ cell of the fugue is a thematic idea called the *subject* which is announced alone in one voice. Immediately it is answered in another, but not in exact repetition (the *difference*). Pitches and shapes are slightly altered in a way that can be defined as circular and the cells following it repeat each other in turn. Bach arranges the fugue inventing, repeating, breaking and continuously restoring the cells inside each bar: all this is made on the razor's edge and the inspiration regulates the process so completely that all becomes part of a great and unique idea and plan. Bach constantly weaves more symmetries, he discovers and focuses his attention on many possible formal asymmetries. In this process they become so numerous that they gradually become the contrary of the first symmetrical pattern and may be their mirror image.

The fugue was a vital form of theoretical symmetry during the Baroque period (1600–1750). But in the long run it could not endure and meet the changing requirements of evolving musical inspiration; more than that, it became a restraint.

In the history of human actions the discovery of the exception to the rule has always provided an impulse for the formulation of the *new*: "Freedom is the negative" (Hegel).

Arnold Schoenberg, the most revolutionary composer of the last seventy years, studying the rules of tonal harmony, found a new harmony. He invented another formal structure with its own symmetry. In confrontation with the classical main rule (tonality), the new rule invented by Schoenberg (atonality) appeared to be topsy-turvy, asymmetrical, an error . . . but errors often deserve the place of honour because they push beyond simple equation into fractions that never add up to one and their movement does not stop. The 25th and 40th bars of the *Toccata & Fugue in E Minor* by Bach are interesting examples of symmetry in music (Fig. 1)[4]. All the cells are formed by three notes: the central one is higher than the remaining two. The connection of the notes in the staff by a line results, therefore, in a conical shape with the point upwards:

∧ ∧ ∧ ∧ ∧ ∧ ∧

∧∧∧∧ ∧∧ ∧

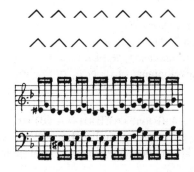

Fig. 1. Bach, *Toccata & Fugue in E Minor.*

Fig. 2.

The geometrical figure that we observe is the first symmetrical pattern of the composition identified as bilateral. The cells are repeated with a variation consisting in a one-tone change. This shows the second symmetrical pattern as a clear translatory example. But in this Fugue of Bach there is another and wider symmetry and this, again, is translatory: the progression of the bars and their forms conform to one another. In the 40th bar, on the other hand, the cells are set up by four notes forming two lines:

These points are alternately upward and downward (Fig. 2). Here too, a single note connected through the progression of the bars is represented by a zigzag line. Besides being translatory, the symmetry is rotational.

We find another example of symmetry in the next two bars (the 10th and 11th) of the first *Prelude* from the *Well-tempered Clavier* by Bach (Fig. 3). Between the first and the second group of tones (in the 10th bar) the only variation is in G clef where the F# become F. In the 11th bar the second group is equal to the first. The differences in the 10th and 11th bars consist only in the movement upwards or downwards of certain notes: the beat remains the same. When played separately the two bars can be different in sound, but the sequence of the paths of each cell presented by the whole composition produces a well-proportioned and well-balanced composition.

The bilateral symmetry controls the entire fugue. In fact all groups of notes within each bar are always the same and are repeated in this way:

$$(2 + 4) + (2 + 4) \qquad \text{(semiquavers)}$$

and

$$(1 + 1 + 1) + (1 + 1 + 1) \qquad \text{(minim + quaver + crotchet)}$$

Even Ludwig van Beethoven attempted the fugue. But his contemporaries said that he was never able to compose a faultless fugue. For example in the *Sonatas for Piano* op. 110 and in the *String Quartet* op. 113, where he adopts the fugue pattern, the formal and resonant result differs from that of the typical fugue. Beethoven, by deviating from the strict rules of the fugue, was able to produce music more congenial to his time. At the same time he infused new life into the fugue.

Another special example of symmetry is the *Passion according to St. Matthew* by Bach.

Fig. 3. Bach, "Prelude" from *The well-tempered Clavier*.

Fig. 4. Bach, *Passion according to St. Matthew.*

A passage for recitative and choir is made by three staves: the F clef maintains a constantly repeated note with a monotonous and martellato rhythm (Fig. 4). In the G clef, however, the music has a certain mobility that contrasts with the monotony of the obstinate bass. Between the static rhythm of the bass and the movement of the other voices there is a continuous attraction: the bass provides a symmetric repetition, while the other voices, in fluctuating motion, offer a clear asymmetrical pattern. This *difference* is explained as the pain of Jesus, that is to say, the torsions and the movement of the body fixed to the cross. We have here a particular instance of bilateral symmetry. When the image on the right is exactly the same as the image on the left, and the process continues on both sides a motion begins and repetition follows.

   Repetition means the exact reproduction of the thing itself, but we sometimes encounter patterns which resemble, but do not monotonously repeat, one another; *differences* are then present in the whole sequence: this contrast enables one to focus on one or the other of the elements repeated. The selection of focus in the piece of Bach is made possible by the contrast between the bass and the other voices: the former (the F clef) provides the regular symmetry, the latter (G clef) provides asymmetrical resemblance. Perception is thrown by this contrast and forced to follow one of the three musical lines given by the score that is, the visual element, or by the executed music, the resonant element. Different transcribers present the same piece of music differently: their transcriptions vary in terms of visual presentation and often in term of sound. Ferruccio Busoni, an Italian composer of the twentieth century, transcribed the music of Bach—he lived in a period when the music of Bach was studied with particular interest. Many other composers of his time like Max Reger, Albert Schweitzer, Gustav Mahler, etc. transcribed many pieces of Bach. The same piece of music transcribed by different musicians illustrates what it means to resemble but to be different and to differ but to resemble.

   We can compare the piano transcription of *Toccata & Fugue in C major* by Bach made by F. Busoni with the same bars transcribed by myself. Here, moreover, bilateral symmetry and symmetry of movement are presented with repetitions, differences and asymmetries (Fig. 5). The existence of a deep relation between the visual images of the score and the sensation produced by the sound is a real fact. This relation represents a kind of language. A similar

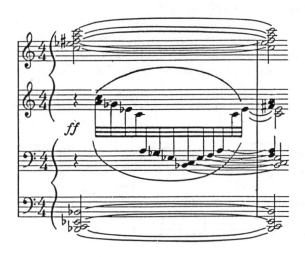

Fig. 5. Bach, *Toccata & Fugue in C major.*

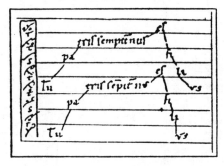

Fig. 6. Anonymous, "Tu Patris" from *Musica Enchiriadis*.

concept of language was used in music in the Middle Ages, through 12th century, that is to say, during the Ars Antiqua period. The music form was the polyphonic Organum, a way of adding (Organare) a voice to the official text of music. This addition was made in a symmetrical path: it was called *parallel Organum* if written like this:

or *oblique Organum* if written as follows:

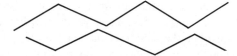

The symbols represent a new line of music moving either parallel to or in opposite direction from the given line. Many variations upon these two ways were common, but they could always lead back to the previously mentioned symmetric patterns. In Fig. 6, a neumatic score of an Organum called *Tu Patris* from *Musica Enchiriadis* by an anonymous composer (9th century), a symmetry of movement is quite clearly shown. In another piece of the thirteenth century called *Discendit de celis* both parallel and contrary motion are presented (Fig. 7). This symbolic relation in which reason sustains both music and text was quite forceful as a musical device. For example in the latin text the word *discendit* was accompanied by a group of descending notes; while the word *levavi* was accompanied by a group of ascending notes. These two are the commonest examples: a multiplicity of cases with many nuances are found in all of the music of the Ars Antiqua and often the later Ars Nova.

Fig. 7. Anonymous, *Discendit de celis*.

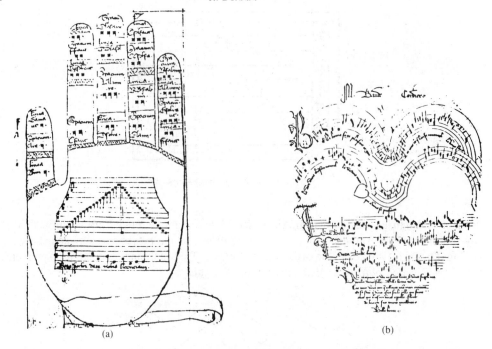

Fig. 8. (a) Diagrammatic "teaching hard". (b) Love song.

The unknown composers of these centuries tried to link the visual image of the score to the resonant image of the music—this process has never ceased. In the previous examples the median line symmetrically dividing the motion of the notes of the staff is the quintessence of the music. This graphic method represents the language binding the composer, the performer, and the listener. We recognise another connection between the look and the sound of music in the medieval composers who used red ink to set the word "blood," green for "grass," black and white for "grief," "joy," "darkness" and "light." Love-songs from the court of Avignon were written in the form of a heart: we also find circular scores and an explanation-teaching inscribed in the drawing of a hand (Fig. 8). The graphic representation of music has always been fundamental in the history of music.

Fig. 9. G. Frescobaldi.

Fig. 10. Handel, *Suite des pieces pour le clavecin.*

The graphics of certain pages of the music of Giacomo Frescobaldi is very interesting. The notes of the cells pointing upwards or downwards are joined by curved and wavy lines (Fig. 9). The alternate groups of notes are marked by an asterisk. They show a certain symmetrical pattern but in general the piece is more asymmetric than symmetric. I define these points as a *spot symmetry*, which involves neither the composition nor the main idea or plan of the composer.

A case where the symmetry seems to be the basic idea of the entire composition, is the *Suite des pieces pour le Clavecin* (second suite—1720) by George Frederick Händel (Fig. 10). Both staves G clef and F clef are developed following each other in two formal ways. In this piece, Händel is too schematic and he seems to be more interested in showing the virtuosity of the performers than in the resonant result. The two line construction (G clef is thick, F clef is thin) reflects a strong formal order and a symmetrical counterpoint.

Groups of notes set in symmetrical patterns are common in certain compositions through the centuries: the following extracts are examples taken here and there from the last three hundred years.

The symmetry in the short passage by Beethoven *Variation on a theme of Diabelli* (op. 120) is too clear to require any comment (Fig. 11). In the next extract from the *Sonata in G major* by Domenico Scarlatti the elaboration is more complicated, but still remarkable (Fig. 12). The formal pattern of the short passage from *Lotosblume* by Robert Schumann is very near to the previous extract by Beethoven (Fig. 13).

We find a symmetrical pattern also in an extract from the madrigal *Tu m'uccidi o crudele* from 5th book of Madrigali by Carlo Gesualdo da Venosa (Fig. 14). Here the melodic line hides the symmetry that remains within the musical notation. The groups of two and three notes do not show it clearly: asymmetry seems to be the rule.

Comparing the sound of a piece of music where the symmetry is very evident with another

Fig. 11. Beethoven, *Variation on a theme of Diabelli.*

Fig. 12. Scarlatti, *Sonata in G major.*

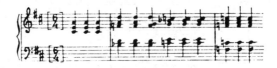

Fig. 13. Schumann, *Lotosblume.*

Fig. 14. Carlo Gesualdo da Venosa, *Tu m'uccidi o crudele.*

Fig. 15. Wagner, *Die Meistersinger.*

Fig. 16. Despres, *Victimae Pascali Laudes.*

Fig. 17. Haydn, *Symphony no.* 45.

piece where it is not, we can understand how many *differences* are produced by different ways of graphical representation. In Richard Wagner's *Meistersinger* (2nd act, 2nd scene) the strong stressing prevents the development of the melody (see Fig. 15), and influences the *pathos* of the music which at this point is remarkable principally for its symmetry.

In one of the high points of the history of music, that is to say *Victimae Pascali Laudes* by Josquin Desprès (in Fig. 16 I transcribe here the final bars), the symmetry and asymmetry are perfectly integrated. In general when the asymmetrical process rules the piece, it remains as asymmetrical as the result tends to be free. The difference between the piece by Desprès and the extract in Fig. 17 from *Symphony* (no. 45) by Franz Joseph Haydn is very clear. Through these various examples we can draw some conclusion about human perception in relation to sound and noise. In certain music the more the symmetry is evident in graphic representation, the less the melody is resonant to the ears: a melodic sequence is easier to perceive if the musical writing is asymmetrical.

Harmony is in keeping with a symmetrical representation. The mental human faculties and our thinking are governed by rhythm and principles of symmetry. Humankind reacts positively on an emotional and psychological level to the sensation of symmetry and this reaction is innate. Accordingly, it would appear that the human ear is inclined toward music containing little melody, since melody is asymmetrical. But, in reality, the contrary is true. *Easy Listening*, music with facile melody, provides the proof.

Analyzing this contrast we find many other elements. The melody is an arrangement of notes in the field of tonality. But the system of tonality is a convention and depends on experiences inherent in the structures of our culture. In the world there are many different cultures and each makes a distinct response to sound. Sound does not make music, it is music that makes sound. This article is focused on symmetry in music and not on tonality or atonality which constitute the opposite terms of a musical problem.

The composer does not compose music intending to make a symmetric or asymmetric pattern. Their visualisation is born unconsciously during the development of the score. Sometimes the composer himself arranges a particular pattern or a process that could be defined as symmetrical, by which an entire score is visualized. In the fugue, the construction follows some conventional rules. One of these, well-known, is the reflection of a principal theme, that is to say, the same notes played backwards. This is a theoretical way used by composers to force their spontaneous inspirations to follow the fugue form.

A page of the *New England Psalm Songs* by W. Billings (1770) is presented by a circular pentagram since the music should never finish (Fig. 18). Here the symmetry is not inside the bar or the cells but in the formal representation of the entire composition. This is not an exception: in the history of music we can find other examples. In a composition of 1981 by Mark Jacobs called *Analemma* the continuity between old and contemporary music is quite clearly shown. He invented a score made as an eight (see Fig. 19)[5].

In the romantic and post-romantic music there is a procedure very common for many

Fig. 18. W. Billings, *New England Psalm songs.*

composers: a group of notes goes upwards through each bar everytime. Usually these parts are simple melodies and very elemental (Fig. 20). The symmetrical result of the melodic passages might seem to contradict the observation I have made: harmony equals symmetry, melody equals asymmetry. But here the melodic result is without strong variations, widening to infinity. The compressed movement arises bar by bar in a continuous fever avoiding stops and pauses and proceeding in atonality. In fact the variations keeping this unswerving progression tend to symmetry. This way of composing was used by many musicians of the 19th and 20th century: we can find it in F. Liszt, R. Wagner, F. Chopin, A. Bruckner, G. Mahler, A. Schoenberg, A. Berg, etc.

In Chopin's *Etudes* no. 1 op. 25 the passage in Fig. 21 shows another example of unswerving progression where melodic manner and classical harmony do not clash with symmetrical path.

In another very famous piece by Chopin, the *Marche Funèbre* from op. 35, two chords are repeated constantly twice in each bar in F clef throughout the composition (Fig. 22). They are missing only in the central part and their absence produces a notably different resonant result. These two symmetrical chords carry the melodic line (G clef) fitting perfectly with the F clef. Its gentle monotony leads the listener back to the symmetrical cadenza of the F clef.

The works of Claude Debussy represent a return to *Modes* and to the scale patterns of early ecclesiastical music (Gregorian Chants, etc.). He achieves a perfect melodic declamation

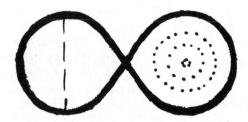

Fig. 19. This is not a score but a schematic drawing made by the author.

(a)

(b)

(c)

Fig. 20. (a) R. Wagner, *Valkyrie*—3rd act. (b) R. Wagner, *Tristan and Isolde*—3rd act. (c) F. Liszt, *Wedding*.

inspired by the *Modes*. The esaphonic (and/or penthatonic) scales and modes used by Debussy give the key for a vague and expressive sing-song not sung but whispered. The action should never be arrested: it should be continuous and uninterrupted. Melody is almost anti-lyrical and powerless to express the emotions.

Thus, the graphic visualization shifts the melody in a symmetrical path (Fig. 23). The melody is like a psalmodic chant with a poignant and penetrating quality. Structural support does not exist, chords make colors, the lighted and transparent resonances do not have fully formed melodic arches but only wisps and fleeting touches. The effect prevents the sound from being very melodic.

In the first and second bars from the first of the *Deux Arabesques* the representation reaches an extreme perfection (Fig. 24). The slur arches between groups of tones render the sensation of repetition more intense than ever. The movement in the staves is circular and then reversible: what is more symmetric than reversibility? The negation of symmetry is the irreversibility.

The essence of the instrumental technique of Debussy consists of a pure and resonant mosaic of fragmentary chords finely colored and a design of melodic cells composed mathematically—except in those short figurations where the *ostinatos* appear again with some small variation. All these elements fit into the composition with symmetric visualization. I believe the piece by Debussy for piano *Jardins sous la Pluie* from *Estampes* is the finest example containing all types of symmetry. The design of cells in each bar is repeated constantly: five fundamental types are present remaining more or less unchanged. Their order is given in Fig. 25. Here we can see spot symmetries, bilateral symmetry, translatory symmetry, symmetry of movement, *differences*, swerves and even a general symmetry visible through the entire piece. Moreover repetitions, invariances, and similes are frequently used. The above five fundamental cells rule the complete score and account for the broad, formal harmony that permeates the piece.

It is possible to find symmetrical patterns in another collection for piano solo: the *Preludes*. The collection is composed by many pieces with different titles as many distinct elements of a

Fig. 21. Chopin, *Etude*.

Fig. 22. Chopin, *Marche Funèbre*.

whole. All these *differences* are necessary to weld it into a unified whole. Each bar is the twofold image of the previous one and so on: the purpose of these differences is to create music and rhythm (see Fig. 26). The composition results in an indissoluble unity: the sequence of elements, designed to create the formal pattern and resonant sensation, is essential to complete the entire piece. From bar to bar the score appears to the eye to be the same, *something else* recomposes it each time rendering a unique sequence. This process creates a continuous intrigue: variations, *differences*, and repetitions, do not subdue the piece but give a dream-like monotony to its recurrent phrases. When Debussy spoke about Javanese and Balinese music, which strongly influenced his works, he demonstrated his understanding of its multiplayer structure and its rhythmic intricacy. Symmetry does not mean lucidity. In fact, Gamelan is based on a very simple theme but played simultaneously and at different paces. This process (which is the same as Debussy's) arrives at a complex figuration often very hard to translate into a complete visualization. The basic theme of one of the well known pieces of Gamelan from Bali is shown in Fig. 27. In certain old manuscripts Indonesian music is presented like this as in Fig. 28. The Chinese characters we can see, denote the cultural exchanges existing between all countries of Far East Asia. But the presence of Chinese characters invites us to approach symmetry at a different level.

Ideographic writing can be considered another example of symmetry but in a symbolic sense. *Music* in Chinese language is written by this symbol:

It represents five drums as the five tones of the musical scale upon a stand of wood. All transformations are apparent in the Fig. 29. But the same symbol pronounced differently means *joy* or *pleasure*, for music gladdens the heart.

Thus, symmetry can appear through a visualization, a resonant sensation, and a reflection in the human understanding. In Japanese writing (not very different from the other East Asian Countries) there are many examples of symmetry. Through the original and the transcription of the short scores in Fig. 30 we bring the music as a picture before the ear and mind. The original graphic is the symmetrical image of the transcription: both represent a language for the same music interpreted by two different cultures and the formal result is almost identical to that of the music.

In a *Noh* chant from Japan the notation is marked on the right side of each line of text (Fig. 31). The text itself and its meaning conducts the music: there is a sort of symbolic exchange between the sound of music and the words of the text: each reflects the other's symmetric image at several levels: sonorous, psychological, emotional and visual.

Fig. 23. Debussy, *Prélude à l'après-midi d'un faune*.

Fig. 24. Debussy, *Deux Arebesques*.

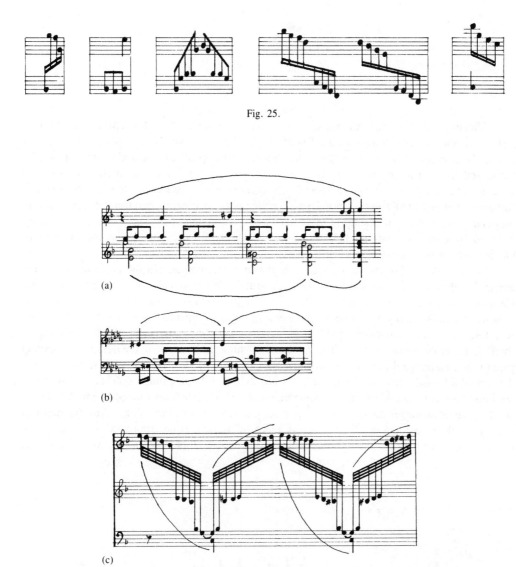

Fig. 25.

(a)

(b)

(c)

Fig. 26. Debussy, *Preludes*. (a) *Deux pas sur la neige* (5th and 6th bars). (b) *La sérénade interrompue* (98th and 99th bars). (c) Feux d'artifice (36th bar).

Fig. 27.

Fig. 28.

Tibetan notation reaches the maximum visual presentation of all Eastern music. We find distinct manners of notation varying for each instrument and from place to place. The musical writing is mnemonic with some descriptive elements. It consists of undulating lines drawn by black, red, or yellow ink (see Fig. 32). Likeness and memory are the properties of this kind of score. The line is the wavy connecting image of a hypothetic transcription on a western staff. Rhythm is missing in all these scores. Using the mnemonic manner of oral tradition, the Tibetan musicians do not need an exactly graphic representation of sign of rhythm: memory constantly renewed by everyday learning helps to complete the symmetric notation at the moment of playing: the rhythm comes from inspiration.

Thus, while not present in the score, rhythm is very important in the performance of the music. Rhythm is not one-dimensional, but represents a whole that composes the sense of music. Hermann Weyl seems to consider music under a single dimension, summarizing it as rhythm. I, however, consider rhythm to be a succession of regular, repeated intervals. Its result could resemble translatory symmetry, but we must remember that equal intervals do not produce rhythm. Common sense dictates that rhythm is symmetry because it creates a symmetrical pattern. But there is a difference between rhythm and symmetry. The first is a creative act and the second a formal result[6]. Asyncronies, accidentals, variations, omissions and syncopations are important elements of rhythm considered as music. All examples quoted in this article are short passages of longer compositions. They are extracts which I have defined as symmetrical because of their position and their form. A piece of music can be composed arranging these passages in a symmetrical or asymmetrical path or in any other way chosen by a composer.

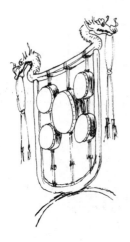
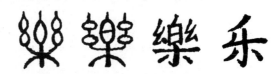

Fig. 29.

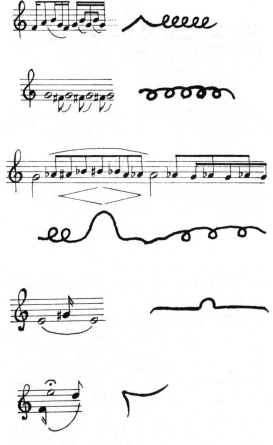

Fig. 30.

But there is a connection between these passages: on the one hand, the idea and plan, on the other, rhythm.

If we make allowance for the rhythm in the next example in Fig. 33, typical of Philippine music, we notice at once its regularity, evenness, stress and its symmetric path. But this notation is a theoretical form because all these signs are a convention that permits the score to be

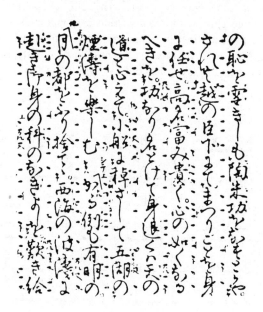

Fig. 31.

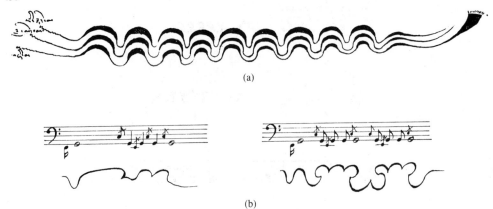

(a)

(b)

Fig. 32. (a) Conga's notation. (b) Vocal notation.

understood. I have already indicated that this convention is *something else*, something *different* from music. In reality, the sound does not have the symmetric aspect of transcription. The problem lies precisely here. We find symmetry in music but sometimes it is in the score, and not in the performed music; or else, the music produces a sensation which can be defined as symmetrical, but we cannot find graphic signs of symmetry in the score.

African rhythms are very complex because simultaneous meters with different beats are used. It is very difficult to perceive the symmetrical sensation, but through the simplification of transcription in western notation it appears very clear (Fig. 34). It is commonly recognized that in the field of music it is difficult to agree upon a unique definition of symmetry: acoustical product and score each have their own essence and consequently many specific symmetries.

As a result of the revolutionary changes in the world of Art which have taken place during this century, a new concept of notation (i.e. language) was born. Not only notes and staff but also drawings, text instructions, signs and symbols: in a word, there are no longer any conventions. Composers resort to a profusion of different notation systems with instructions in further subdivisions. Because of this, many graphic possibilities are available to a composer, and a wide area of research opens up.

Antecedents of contemporary music offer some additional examples to illustrate my point. It may be surprising to some to learn that the first bars of the 5th *Symphony* of Beethoven (Fig. 35) is an antecedent of contemporary music. The 5th Symphony can be considered as concentrated in these three chords: the continuation can appear as an appendage, that is, a development of this first special instant. The symmetry is obvious: 3 quavers, 1 minim—3 quavers, 1 minim—and so on to infinity. The sensation of infinity is reached by ceasing to listen after the first few bars. Wassily Kandinsky transcribed it with the following drawing

Fig. 33.

Fig. 34. African rhythms.

inventing a way to show its symmetrical visualization (Fig. 36). A contemporary composer and critic, Dieter Schnebell, considers the previous bars by Beethoven, together with the small piece from the *Passion according to St. Matthew* by Bach (quoted in Fig. above), and some passages from Schubert and Wagner as antecedents of a composition of 1960 by La Monte Young called *Composition* 1960 #7[7].

From the theory and concept of atonality a new notion of symmetry developed in Schoenberg's mind. Schoenberg's thesis on atonality proposed that each of the twelve tones of the western scale should be of equal importance because if any one tone predominates it becomes a tonal center in the listener's ear. The composer must work with a row of twelve tones and use them in such a way that they appear in the row or in the specific permutation arranged. Permutations of the main row are three: inversion (i), retrograde (r), retrograde inversion (r i). It is possible to see how the symmetry is shown in an atonal system in the permutation row of Alban Berg's *Violin Concerto* (1935) (Fig. 37). The tone-row is a device intended strictly for the composer' own use: it is not meant to be heard by the listener. In fact atonality is always composed using the previous three kinds of permutation: but following the score we do not find any symmetrical passage: asymmetry predominates.

In Schoenberg's *Five Piece for Orchestra* (1909) the last page of *Farben* is in my opinion the best example of asymmetry in western music. The form reaches the highest point attainable and its perfection tempts us to consider it as symmetry. Thus we have the reverse of the phenomenon: asymmetry becomes symmetrical. The visual form of the page is done by staves full of single oval signs with tails, where their stems are not joined. It seems that Schoenberg took a handful of notes and let them fall at random on an empty page. The articulation of *Farben* and especially this page was the turning point and the departure of contemporary music. A different concept of composing had been born. This handful of notes means that chords must change gently so that the entry of instruments is not emphasized: the concentration of performers upon timber becomes the most important feature in performing this piece.

A different relation with timber is achieved in Erik Satie's works. For my purpose, the best example is *Socrate* (1919). Here one finds all that must not be done in music: chords and tones which succeed each other without progress, non-development, jump-cuts, antivariations. In *Socrate*, Satie reaches a state of musical passiveness: no rhythm, no stasis, no movement, no variety, no color, no climax. Thus again, symmetry gives shape to its visualisation with another concept of music composition (Fig. 38).

John Cage and later Karlheinz Stockhausen, were the first composers properly named as contemporary. Their importance gradually grew through the years. The beginning of contemporary music is Cage's 4'33"—*tacet any instrument or group of instruments*—(1950–52). The piece's instruction is in the word "tacet:" the two numbers give the duration in minutes and seconds: music becomes *silence*. Cage carries to the extreme the concept of Anton Webern's.

Fig. 35. Beethoven, *Symphony no. 5.*

Strings and piano

Beethoven's 5th Symphony (the first measures).

Fig. 36. Beethoven, first bar of *Symphony no.* 5, as transcribed by Kandinsky.

Webern's scores are incredibly terse, silence often becoming as important as sound. Cage with 4'33" demonstrates the meaning of silence and what music can be; any sound and any noise. In these 4 minutes and 33 seconds music reaches an instant of pause, that is to say a taking of breath in *silence* after many centuries of sounds.

In 1956, Stockhausen made the first step ahead in a composition called *Klavierstuke IX*: the same chord repeated many, many times (see Fig. 39). When this obsession finally stops and the music takes form through normal atonal proceedings, the revolutionary discovery also seems to stop. The sensation of a resonant symmetry reaches its pinnacle before the start of atonal development. As in the first bars of Beethoven's 5th Symphony of in this piece by Stockhausen a sort of interrupted plan remains in the ear of the listener. He resumed the idea of *Klavierstuke IX* many years later: *Stimmung* (1968) is the natural continuation and the logical development of the same plan. Here the music reaches a complete anti-conventional representation. A section called *Kala Kasesa Ba-ú* provides a good example of a special form of symmetry: the spiral (Fig. 40).

The resonant results have the same effect of repetition. Cage again is the initiator of a new method of rendering scores completely different from conventional notation. It consists in a casual symbolic system (an aide-memoire) which synthesises all of the rules formulated in the past to record and write down music. His composition *Fontana Mix* (1956) is the inspiration for the many scores which use it as a model. Another piece called *Kontakte* (1960) by Stockhausen for electronic tape, piano and percussions is considered fundamental for contemporary music: here, values and defects, views and limits of electronic music are quite clearly shown (Fig. 41). *Treatise* (1967) by Cornelius Cardew is another very important piece of music. Its score had a great influence on contemporary composers. As the title suggests, the score is a treatise of all of the possibilities a composer has to visualize an idea or plan through music. Here no one part of the score can be considered symmetrical. The multiplicity of possibilities must be asymmetrical because only asymmetry can accommodate the infinite varieties of music; only asymmetry permits the complete discourse of which music is capable. He avoided conventionality even while adopting the conventional staff, and succeeded in visualizing a vast variety of sound.

Fig. 37. Berg. *Violin Concerto.*

Fig. 38. Satie, *Socrate*. (© Editions Max Eschig; used by permission.)

Fig. 39. Stockhausen, *Klavierstuke IX*. (© Universal Edition A. G.; used by permission.)

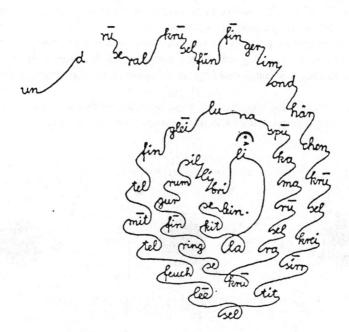

Fig. 40. Stockhausen, *Stimmung*. (© Universal Edition A. G.; used by permission.)

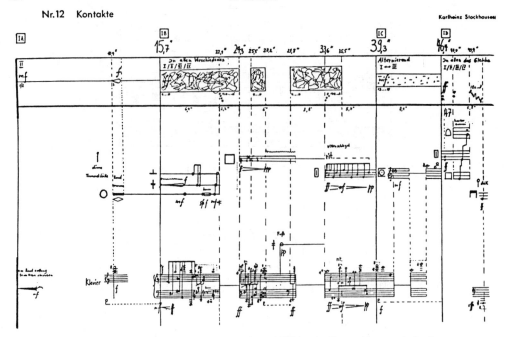

Fig. 41. Stockhausen, *Kontakte*. (© Universal Edition A. G.; used by permission.)

The symbology of its discourse attains a new level in the works of Daniele Lombardi who was greatly influenced by Cardew (Fig. 42). But Lombardi, too, needs to establish rules: graphic symbols become signs in space to help both performers and listeners who must WATCH the score during the performance. He refuses idealism in constructing his scores: sound becomes like a multiplicity of being. He calls this method *ideographic musical writing*, that is to say a method for making music comprehensible to all [8]. (See Fig. 43.) Because the listener is obliged to read the score, Lombardi gives it special attention and makes all signs very clear in order to avoid any ambiguity he uses in interpretation. In order to catch the listener's attention he uses symmetric elementary shapes such as triangles, circles, ellipses, etc.

Bill Hellermann, in a piece called *At Sea* (1977) conceived an extraordinary score: staves are not in a straight line but imitate the movement of ocean waves (Fig. 44). Their symbolic movement fills every page. The composition is meant never to stop, and every beginning is a rebirth. A kind of universal symmetry is achieved.

Today many composers continue to use the staff method with many other media and indefinite kinds of instructions. A young Italian composer Lucia Donnini[9] in a piece called *Stucklein* (1982) reaches a strange visualization of symmetry using simply staves and notes (Fig. 45).

Sylvano Bussotti, with his important composition of the 1960s *Passion selon Sade*, became the inspiration of many young composers who learned from him to arrange an original work

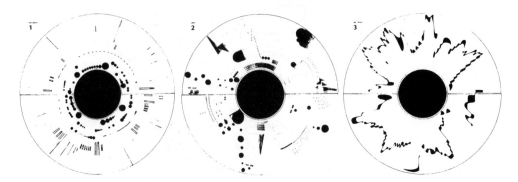

Fig. 42. D. Lombardi, *Twirls*. (© Daniele Lombardi; used by permission.)

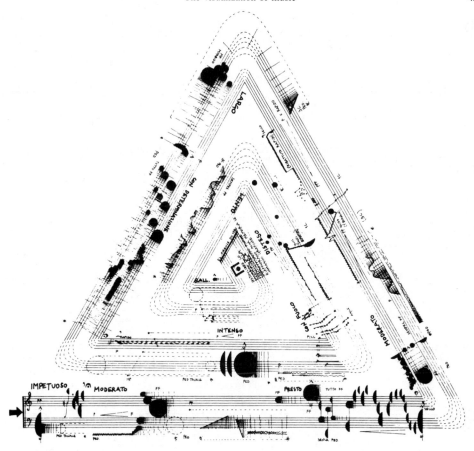

Fig. 43. D. Lombardi, *Tumbling Tumbleweed*. (© Daniele Lombardi; used by permission.)

of some magnitude with common formal structure. I wrote asking him to suggest a work which might serve as an example of symmetry in music. He answered, ''In *Passion selon Sade* there are the examples that might be suitable for your work: *Solo, Rara (Dolce)*'' (2/12/84).

I referred above to La Monte Young's *Composition* 1960 #7: the peculiarity of this piece is its extreme concentration. It represents the next step beyond Webern's and Cage's *silences*:

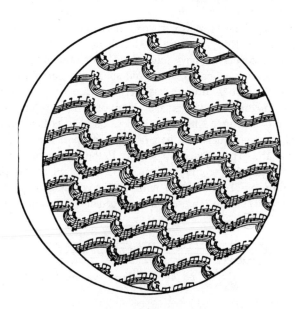

Fig. 44. B. Hellermann, *At Sea*. (© Bill Hellerman; used by permission.)

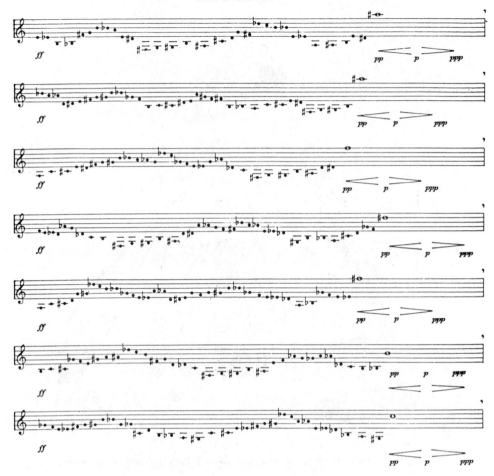

Fig. 45. L. Donnini, *Stuklein*. (© Lucia Donnini; used by permission.)

there is no longer any question of concession to the past (Fig. 46). This piece is formed by a chord of two notes with the text instruction "to be held for a long time". There have been two very famous performances: legend has it that the duration was eight hours in the previous American performance and six hours at *Darmstadter Kurse* in Germany a few years later. Here the symmetry is in sound, in resonant sensation, and in the score.

La Monte Young's piece gave rise to a group of musicians who represent the latest evolution of western music: they are T. Riley, F. Glass, S. Reich and C. Palestine. In the beginning they were rejected by other musicians and have found it difficult cultivating an audience or winning a support among conventional audiences who could apparently hear only what the music lacks in relation to the music they are used to: that is, harmonic movements and melodic perceptible forms. Their musical structure involves a steady harmonic base, often a drone and repetitive or *ostinato-like* rhythmic structure. Although all these composers have in fact received a reasonably conventional musical education, their principal support and audience has come from the world of the visual arts. Their concerts specially in the sixties and seventies took place in Art Galleries and consequently the first critical works appeared in Art Magazines. *Trio for Strings* (1958) a composition of La Monte Young, can be considered the starting point for this group of musicians (Fig. 47). *Trio* represent the antecedent of idea conceived in *Composition*

to be held for a long time

Fig. 46. La Monte Young, *Composition* 1960 #7. Reprinted from *An Anthology*. Young & MacLow, New York (1963). (© La Monte Young; used by permission.)

Fig. 47. La Monte Young, *Trio for strings*. (© La Monte Young; used by permission.)

1960 #7. This score presents a very interesting kind of symmetry: the piece has an exposition, a recapitulation and a coda; all these parts have one or more centers of symmetry. Fig. 47 is the exposition and the composer himself marked the beginning, the end and the three centers of symmetry for use in this article. All of the scores of this group of composers are really symmetrical in their visualisation: as an example I offer two extracts from Steve Reich's works (Fig. 48).

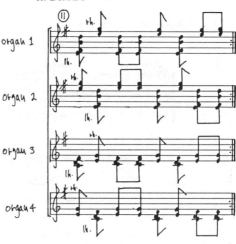

Fig. 48. S. Reich, *Phase Patterns* and *Music for Mallet Instruments, Voices and Organ.* (© Steve Reich; used by permission.)

I consider my music in its resonant result in line with the music of these composers; but my scores are different because I use more media[10]. The score (in Fig. 49) is composed of three pieces *Andar*, *Fula*, *Te-eesa* (*Suca* is another version of *Andar*) consisting of notes, staves, graphic instruction, and text instructions (not shown)[11]. The symmetry and asymmetry is quite evident both graphically and musically. Graphically, my scores are represented by formal waves of notes written in the staves displayed on an oscilloscope monitor.

In the field of music many composers have begun to use the computer to create sound. Computer music is going to replace electronic music. Computers require programs, and the score often becomes such a program, resembling a mechanical activity more than music. Music becomes numbers (frequencies, ratios, series, Fibonacci Series, Pascal triangles, etc.). In the oldest studies (specially Greek) ratios were considered to be the mathematical source of music. Because of the extensive use of computers, the theory of music in relation with mathematics and physics is once again going to be the object of research and investigation.

Stockhausen was the first composer who designed a score for electronic music: in *Studie I* (1953) the score and the music have nothing symmetrical; in explanations of the way he

Fig. 49. R. Donnini. *Andar*, *Fula*, *Te-eesa* (© Roberto Donnini).

Albert Mayr

PERIODICITÀ 1   for 5 voices and 5 stopwatches

Each of the performers – who should be situated at a certain distance from each other – chooses one of the syllables

Pe – rio – di – ci – tà

and says it at the indicated times (try to be as accurate as possible)

Articulation should be smooth, not staccato.

Pitch, timbre and intensity may vary gradually, but always within the normal speech-register of each performer .

Pe	rio	di	ci	tà
0'06"	0'02"	0'00"	0'00"	0'02"
19"	16"	15"	16"	19"
32"	30"	30"	32"	36"
45"	44"	45"	48"	53"
58"	58"			
1'11"	1'12"	1'00"	1'04"	1'10"
24"	26"	15"	20"	27"
37"	40"	30"	36"	44"
50"	54"	45"	52"	
2'03"	2'08"	2'00"	2'08"	2'01"
16"	22"	15"	24"	18"
29"	36"	30"	40"	35"
42"	50"	45"	56"	52"
55"				
3'08"	3'04"	3'00"	3'12"	3'09"
21"	18"	15"	28"	26"
34"	32"	30"	44"	43"
47"	46"	45"		
4'00"	4'00"	4'00"	4'00"	4'00"
13"	14"	15"	16"	17"
26"	28"	30"	32"	34"
39"	42"	45"	48"	51"
52"	56"			
5'05"	5'10"	5'00"	5'04"	5'08"
18"	24"	15"	20"	25"
31"	38"	30"	36"	42"
44"	52"	45"	52"	59"
57"				
6'10"	6'06"	6'00"	6'08"	6'16"
23"	20"	15"	24"	33"
36"	34"	30"	40"	50"
49"	48"	45"	56"	
7'02"	7'02"	7'00"	7'12"	7'07"
15"	16"	15"	28"	24"
28"	30"	30"	44"	41"
41"	44"	45"		58"
54"	58"			
		8'00"	8'00"	

Fig. 50. A. Mayr, *Periodicità*. (© A. Mayr; used by permission.)

Albert Mayr
NAM PLAY FOR NAMEBEARERS

Each group of players writes its own score,using the elements
given,i.e. the bitstrings resulting from the conversion of letters
to binary numbers.

The bitstrings are arranged in an array (see example, there the
bitstrings corresponding to the first 3 letters of each name were
used). The resulting score is read as follows:

each performer plays the bitstring corresponding to his name,one
bit at a time; a 1 indicates an event (acoustical,visual or other)
to be performed or triggered by the player,a 0 indicates 'tacet'.

Each (vertical) column of bits represents a group-situation within
a time-unit the length of which is not specified. The time-unit begins
when one or more active players in a situation start(s) his (their)
event(s).The other active players must begin their events before the
end of the first event in the time-unit,but the events need not be
simultaneous or have the same length. Once all active players in a
situation have brought their events to an end the respective time-
unit is considered finished.

Rests between time-units are ad libitum.

During performance the score should be projected on a screen.

example

PIETRO	0	1	0	1	1	1	0	0	1	0	0	1	0	0	0	1	0	1
GIANCARLO	0	0	0	1	1	1	0	0	1	0	0	1	0	0	0	0	0	1
FREDERIC	0	0	0	1	1	0	0	1	1	0	0	1	0	0	0	1	0	1
CHRISTIAN	0	0	0	0	1	1	0	0	1	0	0	0	0	1	1	0	0	1
PAUL	0	1	0	1	1	1	0	0	0	0	0	1	1	0	0	1	0	0
DON	0	0	0	1	0	0	0	1	0	1	1	0	0	1	0	1	0	1
BIRGID	0	0	0	0	1	0	0	0	1	0	0	1	0	1	1	0	0	1
YVES	1	0	1	0	0	0	1	0	0	1	0	1	0	0	0	1	0	1

Fig. 51. A. Mayr, *Nam Play*. (© A. Mayr; used by permission.)

composed this score, Stockhausen says that a series formed by the numbers 4 5 3 6 2 is the
basic rule: do not use any symmetric succession and avoid repeating two or more times the
same frequency in a predetermined section of music.

In computer and electronic music the frequencies and their numbers are conceived as
material for a program of music composed uniquely of intensity and rhythm. In the following

Fig. 52. M. Zazeela. Reprinted from *Selected Writings*. Heiner Friedrich (1969). (© Marian Zazeela; used by permission.)

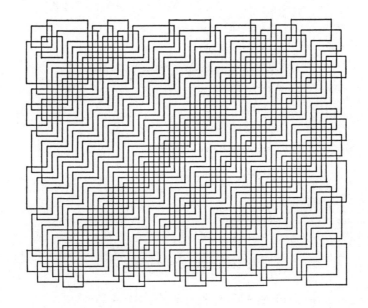

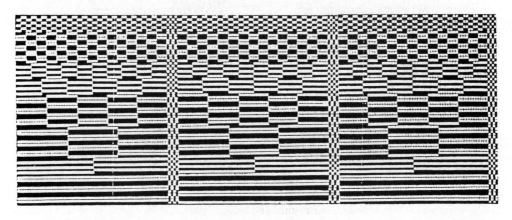

Fig. 53. J. Gibson, 15/3:5. *Primordial Relation Series* and J. Gibson and Vitalis, *Drawing for Video*. (0.9A) (© J. Gibson (1984); used by permission.)

piece by Albert Mayr called *Periodicità* (1974) the symmetry springs from the timing of the performance. Performers join simultaneously at the fourth minute the exact beat that begins again the word "periodicità" (Fig. 50). In another piece called *Nam Play* (1973) the binary system of numbers applied to the names of the players becomes the score to be played: instructions and examples are in reality the score conceived by the composer (Fig. 51).

Another branch of composers and performers uses only drawings not considered as a score, but as an aid to understanding the music when performed. Marian Zazeela's works are closely related to the music of La Monte Young (see Fig. 52). She often performs his music. During the performances slides of Zazeela's designs and drawings are projected upon the performers

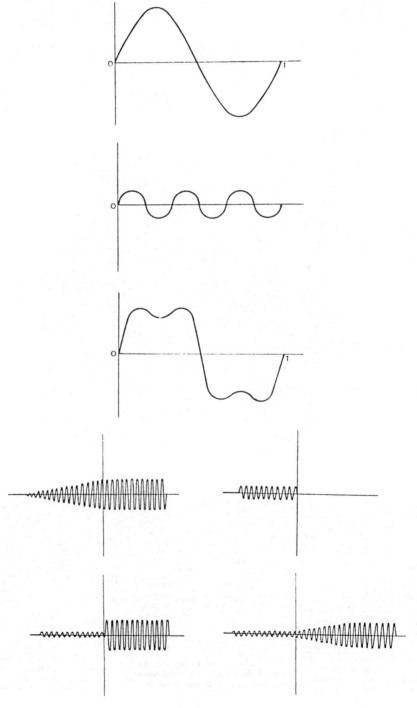

Fig. 54.

and the listeners. The sound becomes a representation of the designs and drawings, as the slides seem to be the visualisation of the music's pathos.

The drawings by Jon Gibson are different in nature but the effect on both listeners and performers is the same. But these *scores* of Gibson's have their origin in optical art, and there is apparently no symmetric intention (Fig. 53).

Music, in common with all the arts, is constantly in evolution and it is not possible to foretell the future. If it were so it would mean the certain death of the arts. It is not possible to draw conclusions about symmetry in music, because to do so would restrict its meaning. I leave the readers to their own judgement[12]. I, nevertheless, would like to finish with patterns displayed on a monitor when a frequency is visualized by a beam of electrons from a cathode-ray tube as a wave form (Fig. 54). These fluctuating lines are, perhaps, the symmetrical secret soul of music, conceived as both an artistic and a technological field of research and investigation.

REFERENCES AND NOTES

1. For example Rock music is symmetrical in a beat usually in Common Time 4/4. Rock music is the folk music of our western civilization of the last 30 years. It has no scores.
2. H. Weyll, *La simmetria*. Feltrinelli ed., Milano (1975); E. Agazzi, *La simmetria*. Soc. ed. Il Mulino, Milano (1973); Atti del Simposio, Roma 1969, *Simmetrie*. Accademia Nazionale Dei Lincei (1970); G. Rigault, *Simmetrie e Cristalli*. Loescher ed., Torino (1979).
3. G. Deleuze, *Différence et répétition*. Presse Universitaires de France, Paris (1968); E. Levinas, *Totality and infinity*. Martinus Nijhoff Publishers, The Hague (1979).
4. If not clearly mentioned, all short examples of this article are transcribed by myself.
5. The original score is in *Ear Magazine East* **8**, Number 1–2, May 1983.
6. N. Abraham, *Le temps, le rythme et l'incoscient*, in: Revue Francaise de psychanalyse, *La créativité* no. 4 Juillet 1972, Presses Universitaires de France, Paris (1972).
7. D. Schnebel, *Composition* 1960: *La Monte Young* (1971) in: Musique en Jeu no. 11 Juin 1973, Editions Du Seuil, Paris (1973).
8. D. Lombardi, *Spartito preso/La musica da vedere*. Vallecchi Ed., Firenze (1981).
9. There is no relation with the author of this article.
10. In *Leonardo*, **14**, no. 2, Spring 1981, Pergamon Press, there is an article I wrote about my scores entitled *Artistic graphic musical scores influenced by Tantric Art*.
11. A record (numbered, signed, and with hand-painted cover) of my music can be obtained from Galleria Schema, Via della Vigna Nuova 17, p.p., 50123 Florence, Italy. The record is entitled T1A, T2A (1980), Lynx Records (Ed. Lynx-xel-ha Records) (cat. no. Z 00112). Now I am preparing another record with a version of the score TUNEDLESS 2° including *Andar Fula Te-eesa Suca*: one of the instruments used in this record will be a computer.
12. I am grateful to the publishers and composers for permission to reproduce extracts from music which is their copyright. Special thanks for different reasons to Kaye Ashe, Teresa M. Brooks, Isabelle Picard, Paola Vitali, Catia Morani, Giuseppe Morrocchi, Carlo Bonomi, Aldo Golin.

Comp. & Maths. with Appls. Vol. 12B, Nos. 1/2, pp. 465–476, 1986
Printed in Great Britain.

0886–9561/86 $3.00 + .00
© 1986 Pergamon Press Ltd.

ON THE SYMMETRY OF PERIODIC STRUCTURES IN TWO DIMENSIONS

A. Gavezzotti and M. Simonetta

Dipartimento di Chimica Fisica ed Elettrochimica e Centro CNR, Università di Milano, via Golgi 19,
20133 Milano, Italy

Abstract—The symmetry aspects of periodic arrays of chemical entities with two-dimensional translation (layers) are examined. The plane groups, G_2^2, and the layer groups, G_2^3, are considered, and their mutual relationships are discussed. Examples drawn from surface chemistry in heterogeneous catalysis on regular metal surfaces are proposed. The matrix representation of the plane and layer groups is analyzed for some important cases.

1. TWO-DIMENSIONAL SYMMETRY: HYDROCARBON CHEMISORPTION AND WALL DECORATION

The arts and crafts of drawing, painting, and decoration, are no doubt strictly two-dimensional matters. A sheet of paper, a piece of canvas, a blank wall awaiting the hand of a fresco master, dispense with that precious third dimension we so much exploit in everyday life. Still, there are some craftsmen that make use of symmetry, and some who do not. Among the last, generations of painters—from Giotto to Jackson Pollock—would have considered repetition of figures by translation or rotation as the dull trick of an insipid dauber. At the other extreme, symmetry of form and colours must have been about the only concern for the anonymous arabian decorator who set out to the difficult task of pleasing the Sultan's eyes with his brightly coloured tiles on the walls of the Alhambra, in Granada. At that time, not even the most imaginative alchemist could have foreseen a connection between wall decoration and the quest for the philosophal stone. Today, gasoline is more in demand than gold, and present-day alchemists compete to find the materials that best help the conversion of crude oil into the thin liquid that propels our cars. Such materials are catalysts; and it so turns out that the best model systems available for the study of catalytic action at a molecular level are clean metal surfaces on which molecules chemisorb in a neatly ordered, symmetric, two-dimensional array. Just as the tiles that pave the Sultan's rooms at Granada: and a glance at Figs. 1(a)–1(b) should be convincing on this point.

But the striking coincidence between the two-dimensional patterns of surface chemists and the decorative imagination of craftsmen is further testified by Figs. 2(a)–2(b), where the artistic example is taken from an entirely different time and place—the exuberant Vienna of the first years of this century. Of course, the ultimate source of these coincidences lies in the group-theoretical requirement that any periodic pattern of two-dimensional figures belong to one of the 17 plane groups[1,2]. The Alhambra decorator, however, might have known more than present-day surface spectroscopists—in the case of the methylacetylene structure on the Pt(111) surface, which is strictly similar to the one shown in Fig. 1(a), one carbon monoxide molecule was found, after many efforts, to occupy the empty space on the threefold axes at the cell origin, while the arabian artist had quite naturally filled this space with a star (Fig. 1(b)).

Symmetry is pleasing to the eye, and the reason why this is so is deeply buried in the intricacies of the physiology of human perception. A suspicious mind might therefore think of a common origin for the structural motifs of Figs. 1(a–b) and 2(a–b)—namely, human bias towards the pleasantness of a symmetric drawing. Such a surmise casts a shadow on the chemical theory of chemisorption and molecular interaction; but it must be said at once that such a theory rests on a much firmer basis than just decorative whim. Once they bind to a metal surface, molecules are forced to occupy well defined positions, and are capable of mutual recognition by means, essentially, of van der Waals forces. It is, therefore, the very concrete requirement of minimum enthalpy and maximum entropy that is responsible for the pattern formation. The authors of Fig. 2(a) worked out their model on the principle of best surface-adsorbate binding,

A. GAVEZZOTTI and M. SIMONETTA

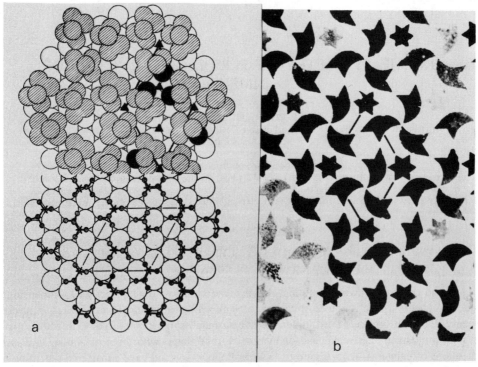

Fig. 1. (a) The surface structure of propylidyne chemisorbed on the (111) face of fcc metals[13]. H atoms are shown by dashed areas in the top part of the illustration; full lines show two different choices of the unit cell. (b) A decoration on a wall of the rooms of the Alhambra, in Granada. The similarity between (a) and (b) is more evident by observing the pattern of the black atoms in (a) and of the tips of the arrows in (b).

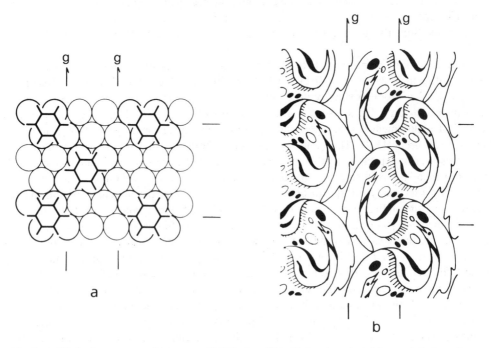

Fig. 2. (a) The surface structure of benzene on Pt(111), from [11]. (b) Drawing adapted from a cloth decoration by M. Benirschke, Wien, 1902. Glide planes are labeled g (vertical); the horizontal lines mark the vertical cell edge, while the horizontal cell edge is twice the distance between the glide planes.

and of mutual molecular avoidance, a principle that was certainly obscure to the author of Fig. 2(b).

Nonetheless, the patterns of Figs. 1b and 2b were successful in pleasing the Sultan and *belle époque* ladies, while those of Figs. 1(a) and 2(a) are less easily accepted, and are open to criticism by the surface chemistry community. There is perhaps a lesson to be learnt from this.

2. MOLECULAR CLOSE-PACKING AND SYMMETRY

It is generally true that an array of molecules—be it two- or three-dimensional—minimizes its free energy by choosing to be as compact as possible, compatibly with the impenetrability of atoms. This principle is a cornerstone of organic crystal packing theories, and, were it not for the action of the forces between the adsorbate and the metal surface, would certainly find a straightforward application to the two-dimensional case. Yet, preliminary evidence so far available[3] shows that when a monolayer of sizable organic molecules is deposited on a compact metal surface, it does obey a sort of 2-D close-packing principle.

The tendency to close-packing can be understood on an intuitive basis. Objects of any form, which attract each other, will naturally tend to come as close as possible. If one deals with spheres, one gets the most symmetrical cubic packing; with molecules, the arbitrariness of their shapes requires a more complicate choice of the packing pattern.

Symmetry, however, is not a choice, but a must. A regular and periodical juxtaposition of chemical entities entails a regular and periodical potential, and this, in turn, brings symmetry into play. The symmetry of the crystal potential function has been analyzed by P. M. Zorkii[4], who, in his so-called "superposition theorem", states that "a necessary condition for the formation of a crystal is that at least two molecules added on to the original molecule should fall in energetically equivalent special points". By "special points" is meant the lowest minima of the function $U(x, y, z)$ which is the potential of the intermolecular interaction. By applying this theorem in sequence to all molecules in the crystal, one can infer that, at the very least, there must be translational symmetry for a crystal to be formed. If one then takes into consideration also three angles, δ, ϵ, ξ, which describe the mutual orientation in space of the related molecules, and on which the U function also depends, the path is open to an extremely simple and fruitful analysis of the introduction of symmetry axes and planes relating the molecules in the crystal. The reader will be aware at this point that Zorkii's analysis has been recast here into a non rigorous, intuitive language.

As we have already mentioned, a crystal minimizes its free energy by minimization of enthalpy and maximization of entropy. As very simply and neatly stated by Kitaigorodski[5], a high packing density helps fulfilling the first requirement, and a high symmetry is coherent with the second. The best way for a crystal to achieve stability is therefore to contain symmetry elements that produce close-packing.

If one is ready to accept the view that a molecule is a real three-dimensional object made of interpenetrating atomic spheres, with a weight, a volume and a surface[6,7], and, above all, a well defined shape, then it is relatively simple to see why for example glide planes or screw axes are stability-producing symmetry elements. From this point of view, the problem looks like a jigsaw puzzle, as Fig. 3 shows. Repetition by mirroring (Fig. 3(a)) results in a bump-to-bump and hollow-to-hollow confrontation, while, most naturally, mirroring plus translation produces an efficient packing of arbitrarily shaped objects (Fig. 3(b)). Such simple reasoning explains why more than 50% of all organic structures belong to space group $P2_1/c$, which contains the simplest combination of glide planes and screw axes.

No such simplifications are possible if the view is taken that a crystal is a chemical system which, as any other, must be described quantum mechanically by a suitable wavefunction. The mere fact that a flourishing and perfectly adequate theory of molecular crystals, based on shape and size concepts, has been built and is being used, lends powerful support to the "real object" model of molecular structure. No doubt, the "steric effect" has helped in the rationalization of solution chemistry too.

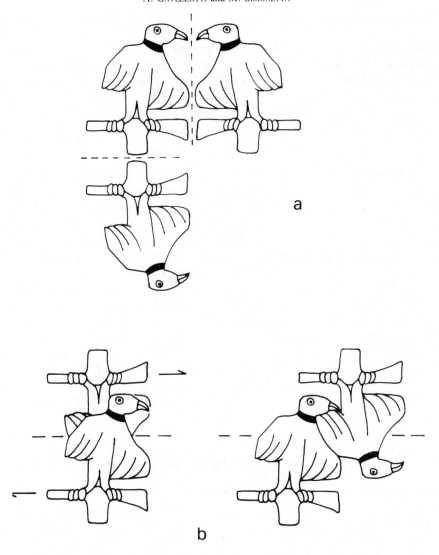

Fig. 3. (a) The effect of mirror planes (dotted lines) on a figure. (b) The effect of a glide plane on the same figure; the plane is marked by the dotted line, while the arrows indicate the direction of the displacement.

3. TWO IN TWO: THE PLANE TWICE-PERIODIC GROUPS G_2^2

There are five plane nets corresponding to two-dimensional Bravais groups, or two-dimensional lattices. Symmetry operators in two dimensions are axes of order 2, 3, 4 and 6, and mirror and glide planes; it being understood that the axes are formally perpendicular to the plane space, and that the symmetry planes are also perpendicular to this plane, on which they are represented by their trace. Putting those elements together, one obtains the 17 G_2^2 groups. Of course what has been here exposed in conversational language can be put in the formal language of group theory[8].

These groups are important because they describe the projections of three-dimensional crystalline structures, and the symmetry of sections (e.g. of electron density) through three-dimensional structures. They find a natural application in the study of molecular chemisorption, when a monolayer of molecules is deposited on the flat surface of a crystalline metal. Figure 4 shows the effect of the symmetry axes (the effect of the symmetry planes is very well illustrated in Fig. 3).

Special cases concerning the adsorption of large molecules on high-symmetry, compact metal surfaces like the fcc (111) have been treated in terms of some of the symmetry operators of Figs. 3 and 4 oriented in appropriate ways with respect to the metal atoms of the surface[9]. Figure 5 shows some (graphically attractive, but also chemically significant) results.

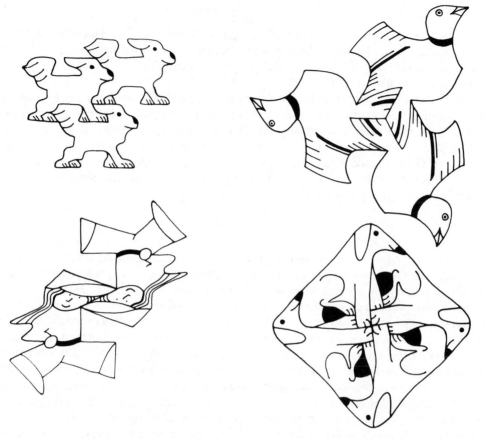

Fig. 4. The effect of pure translation, of a twofold axis, of a threefold axis, and of a fourfold axis on a plane figure.

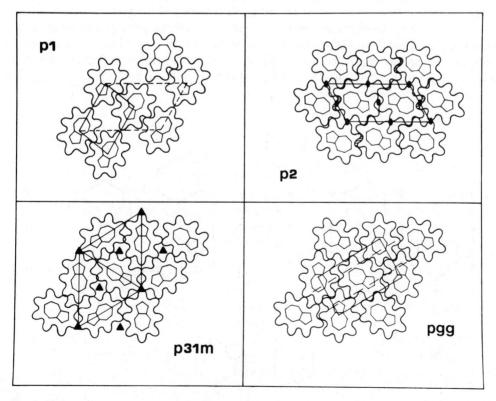

Fig. 5. Various plane groups for layers of azulene molecules. These are the most energetically favourable for adsorption on Rh(111); see [9].

4. THREE IN TWO: THE LAYER GROUPS G_2^3

A more general and adequate description of the two-dimensional packing of three-dimensional objects calls for the use of layer groups. These contain two translations which define a plane, X_1X_2, which is a singular plane; there is no periodicity along X_3, the third axis perpendicular to the singular plane.

There are 80 G_2^3 groups[8]; the symmetry elements of K, the molecular point group, are oriented so that rotations due to 3-, 4- and 6-fold axes are performed around axes perpendicular to the singular plane. Twofold symmetry axes may lie in this plane, which can also be a mirror or glide plane. Thus, the case may be that some symmetry elements interchange the X_3 coordinate, and that the space is subdivided by the singular plane into two symmetry-equivalent subspaces. All the G_2^3 groups can be derived[8] by forming the semidirect product of the two-dimensional translation group and the molecular point group:

$$G_2^3 = T_2 \,\text{\textcircled{s}}\, K.$$

17 of the G_2^3 groups do not contain symmetry elements which interchange the X_3 coordinate. No wonder, the G_2^3 groups are projected onto the G_2^2 groups along X_3. In chemisorption studies, one has to deal with a layer of molecules bound to a plane metal surface. Focussing on the overlayer only, while in general the G_2^3 groups should be used to describe its symmetry, it turns out that in most cases the G_2^2 subset is sufficient, especially when flat molecules like the aromatics are involved. In such cases, the third dimension is immaterial, in the sense that the X_3 component of the forces acting on the molecules is null, or at least, it is constant and equal to the more or less homogeneously distributed binding force between the adsorbate and the metal surface.

Of course, when the full chemisorption system is taken into account—that is, the chemisorbed overlayer plus one or more metal atom layers—there can never be two symmetry-equivalent subspaces, since the metal layer creates an indelible asymmetry. The G_2^3 groups, therefore, vanish into their G_2^2 subset, which is currently used, for example, to analyze the LEED diffraction patterns obtained from chemisorption systems. Incidentally, the symmetry of the metal layers is usually so high—since most studies are carried out on closely-packed surfaces—that the whole symmetry is limited by the adsorbate overlayer.

The analysis of the overlayer symmetry alone in terms of the layer groups is nonetheless an amusing intellectual exercise, and is not without its rewards, as will become clear later. To make just a simple example, let us consider an overlayer of pyridine molecules. One could conceive an adsorption mode in which the N atom interacts with the surface more strongly than

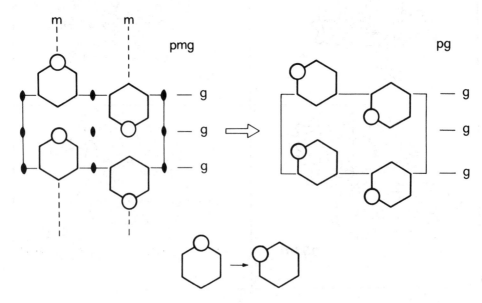

Fig. 6. A layer of pyridine molecules (the circle symbolizes the N atom), and the plane group symmetry change upon in-plane rotation.

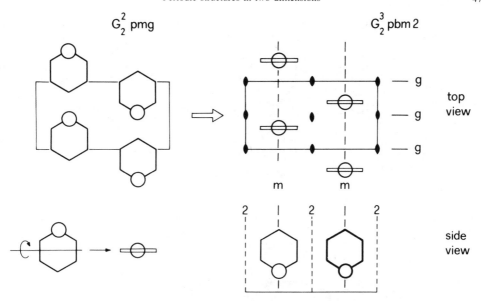

Fig. 7. Same as Fig. 6, but for out-of-plane tilting of the pyridine molecules. In the side view, the heavy contour denotes a molecule in the foreground.

the C atoms or CH groups, so that the molecule tilts out of the plane parallel to the metal surface[10], while this is not so for benzene[11].

Let us analyze the various symmetry groups that may be used to describe what happens if the pyridine molecule rotates around its main axes. We assume, as a starting point, a flat layer, as seen in Fig. 6, belonging to the plane G_2^2 group pmg–a tightly packed array. It should be noted that the mirror symmetry is introduced since it is already a molecular symmetry element; in fact, as the molecule rotates in its plane around a vertical axis, the layer mirror planes and the twofold axes disappear, and the pmg symmetry is reduced to pg (refer always to Fig. 6).

Rotation around one of the axes in the molecular plane (see Fig. 7) brings the molecule out of the layer plane. One must then use a G_2^3 group, which, in this instance, turns out to be pbm2. No symmetry elements are lost, since pbm2 projects onto pmg.

Rotation around the third axis (see Fig. 8) is more interesting, since it introduces a mirror plane that runs through the molecule, at right angles to the molecular plane. Here, a G_2^3 must be used in which the space is subdivided into two symmetry-equivalent subspaces. This is group pbmm. Of course, however, in the real case also the metal atoms plane must be taken into account, and the space group is again pbm2.

5. MATRIX REPRESENTATION

The study of regular overlayers, like that of three-dimensional crystals, must necessarily proceed through some kind of modeling. A mathematical model of a periodic, symmetric array of objects in space can be built by the use of the matrix representation of symmetry.

In any case, the first aim is the calculation of the lattice energy. In such calculations, one fundamental chemical unit—be it a molecule, or a group of ions, or a single atom, as in metals—is surrounded by other units, related to the fundamental one by the symmetry elements of the space group. The energy is then evaluated, quite often, as a lattice sum of atom–atom contributions, each of which depends explicitly on the cartesian coordinates of the atoms[12]. Therefore, starting from the atomic coordinates for the fundamental unit, $\mathbf{x}_{oj}$, one obtains the coordinates for all the surrounding units as:

$$\mathbf{x}_{ij} = \mathbf{M}_i \mathbf{x}_{oj} + \mathbf{t}_i$$

where $\mathbf{M}_i$ and $\mathbf{t}_i$ are a matrix and a vector, respectively, which express the rotations and translations brought about by symmetry. Of course, the set of $\mathbf{M}_i$–$\mathbf{t}_i$ couples must form a group.

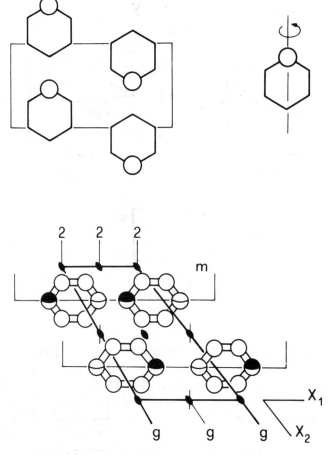

Fig. 8. Same as Figs. 6–7, but for rotation around the third axis. The N atom is denoted by a circle in the plane drawing, and by a full circle in the perspective drawing.

For G_2^2 and G_3^3 groups—the 17 plane groups and the 230 space groups, respectively—matrices and vectors are available from standard sources[2]. Starting from them, we shall now introduce a matrix representation for some interesting G_2^3 groups, see Figs. 9–10. Some considerations on the way in which plane and layer groups are related will be offered.

Essentially, the matrix representation for G_2^3 is the same as for the G_3^3 with the same label, except for a number of limitations due to the uniqueness of the z axis. For instance, there is but one p2 group in G_3^3, while there are two in G_2^3, depending on whether the 2 axis is along z or is in the xy plane. Another difference is that, while in G_3^3 the translation vector components can always be increased or decreased by an integer number of cell edges, the third component of these vectors in G_2^3 must always be zero. In fact, this component is a purely formal one, since the translation group has only two components.

The matrix representations are collected in the Table. It is understood that in this Table the third component of each of the translation vectors in G_2^3 suffers from the above limitation. In some cases, the choice of the axis orientation is not consistent with that of [2] or [8], but this was necessary in the unification of two different conventions.

Consider first Fig. 9 and Table 1. The matrix for p112 is simply obtained from that of p2, since the z coordinate is unchanged; when however one goes to p$\bar{1}$, the third coordinate changes sign, and one obtains an inversion center. If, on the other hand, the twofold rotation axis is in the xy plane, as in group p121, the coordinate along the symmetry axis (y in our case) does not change sign. It is worth noting that, the twofold axis being in the xy plane, the angle γ between the two translation unit vectors must be 90°.

More interesting is the case of glide planes. pg and p1a1 are correlated in the same way as p2 and p112 were (check the matrices in Table 1). When, however, the third dimension

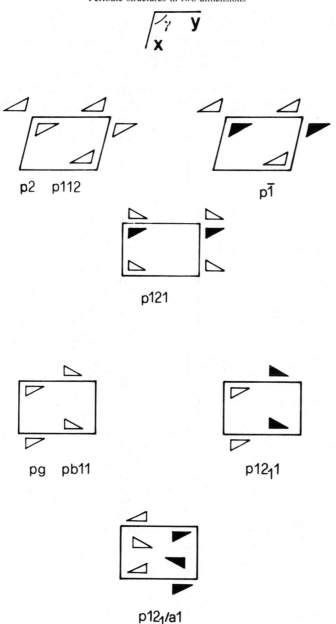

Fig. 9. Symbolic drawings of the symmetry of some G_2^3 groups (after [8]). Full triangles are above, open triangles are below the unique (xy) plane. Where two labels are given, the first refers to the corresponding G_2^2 group.

comes into play (so to speak), the glide plane may change into a screw axis, and group $p12_11$ obtains. When both glide planes and screw axes are present, at right angles, one may go to group $p12_1/a1$, which has (in Fig. 9) glide planes parallel to x and screw axes parallel to y. The coupling of these two symmetry elements results in an inversion center, which closes the symmetry group.

We may now turn to Fig. 10, where the relationship between pmg and pbm2 is seen to follow the previous examples. The coupling of a mirror and a glide plane requires a twofold axis for group closure. Group pbmm is easily obtained from pbm2, by addition of a mirror plane along the xy plane. The number of asymmetric units in the cell is doubled, and each matrix of pbm2 has a corresponding matrix in pbmm with the z coordinate changing sign.

As a final example, we propose the case of the pgg and pba2 groups, for which we need

A. Gavezzotti and M. Simonetta

Table 1. Matrix representation for G_2^2 and G_2^3 symmetry groups. Z is the number of asymmetric units in the cell. Symbols for symmetry elements are: $\bar1$, inversion center; 2, twofold rotation axis; 2_1, twofold screw axis; g, glide plane; m, mirror plane. The zeroes in the matrices are not written. Refer to Figs. 9–10

Group name	Z	Matrices and vectors	Symmetry element
G_2^2 p2	2	$\begin{bmatrix}-1&\\&-1\end{bmatrix}\begin{bmatrix}0\\0\end{bmatrix}$	2
G_2^3 p112	2	$\begin{bmatrix}-1&&\\&-1&\\&&1\end{bmatrix}\begin{bmatrix}0\\0\\0\end{bmatrix}$	2 on z
G_2^3 p$\bar1$	2	$\begin{bmatrix}-1&&\\&-1&\\&&-1\end{bmatrix}\begin{bmatrix}0\\0\\0\end{bmatrix}$	$\bar1$
G_2^3 p121	2	$\begin{bmatrix}-1&&\\&1&\\&&-1\end{bmatrix}\begin{bmatrix}0\\0\\0\end{bmatrix}$	2 on y
G_2^2 pg	2	$\begin{bmatrix}-1&\\&1\end{bmatrix}\begin{bmatrix}0\\1/2\end{bmatrix}$	g
G_2^3 p1a1	2	$\begin{bmatrix}-1&&\\&1&\\&&1\end{bmatrix}\begin{bmatrix}0\\1/2\\0\end{bmatrix}$	g yz
G_2^3 p12$_1$1	2	$\begin{bmatrix}-1&&\\&1&\\&&-1\end{bmatrix}\begin{bmatrix}0\\1/2\\0\end{bmatrix}$	2_1 on y
G_2^3 p12$_1$/a1	4	$\begin{bmatrix}-1&&\\&-1&\\&&-1\end{bmatrix}\begin{bmatrix}0\\0\\0\end{bmatrix}$	$\bar1$
		$\begin{bmatrix}1&&\\&-1&\\&&1\end{bmatrix}\begin{bmatrix}1/2\\1/2\\0\end{bmatrix}$	g xz
		$\begin{bmatrix}-1&&\\&1&\\&&-1\end{bmatrix}\begin{bmatrix}1/2\\1/2\\0\end{bmatrix}$	2_1 on y
G_2^2 pmg	4	$\begin{bmatrix}-1&\\&-1\end{bmatrix}\begin{bmatrix}0\\0\end{bmatrix}$	2
		$\begin{bmatrix}1&\\&-1\end{bmatrix}\begin{bmatrix}0\\1/2\end{bmatrix}$	m
		$\begin{bmatrix}-1&\\&1\end{bmatrix}\begin{bmatrix}0\\1/2\end{bmatrix}$	g
G_2^3 pbm2	4	$\begin{bmatrix}-1&&\\&-1&\\&&1\end{bmatrix}\begin{bmatrix}0\\0\\0\end{bmatrix}$	2 on z
		$\begin{bmatrix}1&&\\&-1&\\&&1\end{bmatrix}\begin{bmatrix}0\\1/2\\0\end{bmatrix}$	m xz
		$\begin{bmatrix}-1&&\\&1&\\&&1\end{bmatrix}\begin{bmatrix}0\\1/2\\0\end{bmatrix}$	g yz
G_2^2 pgg	4	$\begin{bmatrix}-1&\\&-1\end{bmatrix}\begin{bmatrix}0\\0\end{bmatrix}$	2
		$\begin{bmatrix}1&\\&-1\end{bmatrix}\begin{bmatrix}1/2\\1/2\end{bmatrix}$	g on x
		$\begin{bmatrix}-1&\\&1\end{bmatrix}\begin{bmatrix}1/2\\1/2\end{bmatrix}$	g on y

Table 1 (*Continued*)

Group name	Z	Matrices and vectors		Symmetry element
G_2^3 pba2	4	$\begin{bmatrix} -1 & & \\ & -1 & \\ & & 1 \end{bmatrix}$	$\begin{bmatrix} 0 \\ 0 \\ 0 \end{bmatrix}$	2 on z
		$\begin{bmatrix} 1 & & \\ & -1 & \\ & & 1 \end{bmatrix}$	$\begin{bmatrix} 1/2 \\ 1/2 \\ 0 \end{bmatrix}$	$g\ xz$
		$\begin{bmatrix} -1 & & \\ & 1 & \\ & & 1 \end{bmatrix}$	$\begin{bmatrix} 1/2 \\ 1/2 \\ 0 \end{bmatrix}$	$g\ yz$
G_2^3 p2$_1$2$_1$2	4	$\begin{bmatrix} -1 & & \\ & -1 & \\ & & 1 \end{bmatrix}$	$\begin{bmatrix} 0 \\ 0 \\ 0 \end{bmatrix}$	2 on z
		$\begin{bmatrix} 1 & & \\ & -1 & \\ & & -1 \end{bmatrix}$	$\begin{bmatrix} 1/2 \\ 1/2 \\ 0 \end{bmatrix}$	2_1 on x
		$\begin{bmatrix} -1 & & \\ & 1 & \\ & & -1 \end{bmatrix}$	$\begin{bmatrix} 1/2 \\ 1/2 \\ 0 \end{bmatrix}$	2_1 on y

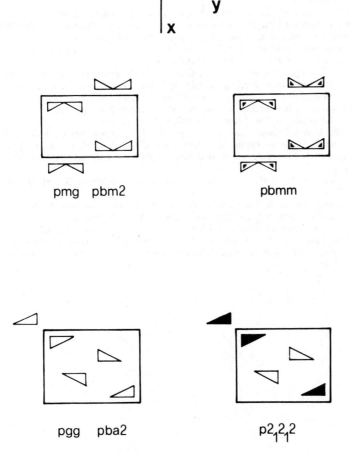

Fig. 10. Same as Fig. 9; a triangle with a dot symbolizes the superposition of a full and an empty triangle.

not bore the reader with the usual explanation. When the two glide planes turn into screw axes, one gets group p2₁2₁2.

All the groups we have illustrated contain the precious symmetry elements that we have seen to improve the chance of close packing. A similar analysis could be conducted (although by a much more tedious effort) on groups containing threefold, fourfold and sixfold axes.

6. CONCLUSIONS

The three-dimensional space groups G_3^3 are used widely by practicing chemists, and especially by crystallographers. Also, although less commonly, the corresponding two-dimensional G_2^2 groups are found to have useful applications by surface chemists, among others. As we have briefly shown, they are useful in layer packing energy calculations and, in general, for surface crystallography. For example, in some favourable cases, a plane group for the adsorbate was assigned on the basis of extinctions in the LEED patterns.

The no-man's land of G_2^3 groups is also worth exploring, though, since much useful information for the chemist can be found there. For example, they could be used to describe multilayer adsorption on regular metal surfaces, and the inclusion compounds in which one layer of organic material slips in between two layers of graphite or any other layered guest material. Some of their characteristics have been, it is hoped, illustrated by this paper.

Acknowledgement—The valuable help of Gianmario Bandera for the graphic part of this work is gratefully acknowledged.

REFERENCES

1. A. Gavezzotti, The 17 two-dimensional space groups: an illustration. *Atti Accad. Naz. Lincei, Memorie (Ser. VIII)*, **13**, 107–119 (1976).
2. *International Tables for X-Ray Crystallography*. Kynoch Press, Birmingham (1962).
3. A. Gavezzotti, M. Simonetta, M. A. Van Hove and G. A. Somorjai, Adsorbate-adsorbate interactions and the ordering of organic monolayers on metal surfaces. *Surface Sci.* **154**, 109–120 (1985).
4. P. M. Zorkii, Structure of molecular crystals. III Potential function symmetry method for asymmetric congruently equal molecules. *Sov. Phys.-Cryst.* **13**, 19–24 (1968).
5. A. I. Kitaigorodskii, *Organic Chemical Crystallography*, Consultants Bureau, New York (1961).
6. A. Gavezzotti, The calculation of molecular volumes and the use of volume analysis in the investigation of structured media and of solid-state organic reactivity. *J. Am. Chem. Soc.* **105**, 5220–5225 (1983).
7. A. Gavezzotti, Molecular free surface: A novel method of calculation and its uses in conformational studies and in organic crystal chemistry. *J. Am. Chem. Soc.*, **107**, 962–967 (1985).
8. B. K. Vainshtein, *Modern Crystallography*, vol. I, pp. 109–121. Springer-Verlag, Berlin (1981).
9. A. Gavezzotti and M. Simonetta, On the generation of trial structures and the evaluation of the formation energy for layers of chemisorbed molecules on metal surfaces: naphthalene and azulene on Rh(111). *Surface Sci.* **134**, 601–613 (1983).
10. R. M. Wexler and E. L. Muetterties, Coordination chemistry of pyridines on Ni(100). *J. Am. Chem. Soc.* **106**, 4810–4814 (1984)
11. R. F. Lin, R. J. Koestner, M. A. Van Hove and G. A. Somorjai, The adsorption of benzene and naphthalene on the Rh(111) surface: a LEED, AES and TDS Study. *Surface Sci.* **134**, 161–183 (1983).
12. A. Gavezzotti and M. Simonetta, Crystal chemistry in organic solids. *Chem. Revs.* **82**, 1–13 (1982).
13. A. Gavezzotti, M. Simonetta, M. A. Van Hove and G. A. Somorjai, Force-field calculations on the packing energy of monolayers of C_3 and C_4 hydrocarbon molecules adsorbed on single-crystal metal surfaces. *Surface Sci.* **122**, 292–306 (1982).

Comp. & Maths. with Appls. Vol. 12B, Nos. 1/2, pp. 477–485, 1986
Printed in Great Britain.

0886–9561/86 $3.00 + .00
© 1986 Pergamon Press Ltd.

A SPACE VIEW OF A SYMMETRIC OBJECT

Barbara E. Lowrey

Central Data Services Facility, Code 633, National Space Science Data Center, Space Data and
Computing Division, Goddard Space Flight Center, Greenbelt, MD 20771, U.S.A.

Abstract—Symmetric features, both natural and man-made, exist on earth and are visible in space data. This paper closely examines one circularly symmetric object, the Robert F. Kennedy stadium in Washington, DC, which is contained in a Landsat-4 Thematic Mapper image. The effect of the space measurement on the appearance of the feature varies according to the band of the image. The human perception of the data is affected by the band and also by the magnification and context of the feature. The differences in interpretability of one object demonstrate the nature of the challenges involved in achieving computer vision. The future computing systems and related developments that may enable a machine to "understand" a symmetric object are described.

1. INTRODUCTION

The earth contains many natural and man-made objects which are symmetric when viewed from space. The Landsat series of satellites has provided images of scenes of the earth below, where each scene is about 185 km by 172 km. The images are taken in several bands simultaneously, providing the opportunity to make composite colored images. One circular symmetric object, a stadium located in Washington, DC, is selected from a Landsat scene and examined in detail. Circular symmetry is readily defined in mathematical terms; however the ability of a human being to perceive the circular symmetry of a planar object is not well understood and is conditioned by the environment of the object. In space data, the digital quality of the data may alter the ability of a human to recognize an object. Circular features are particularly susceptible to distortion by digitization. The Landsat view of the stadium in Fig. 1 provides an example of the interaction of measurement and human reasoning in interpreting the image. This example demonstrates the nature of the problems to be solved in achieving "artificial intelligence" in computer vision. The prospect of future computers recognizing a digitized symmetric object is discussed in view of existing and future parallel processors.

2. IMAGES FROM THE LANDSAT THEMATIC MAPPER EXPERIMENT

The NASA Landsat series of satellites acquire digital images of earth from space and transmit the images down to ground stations on earth. The satellites Landsat 4, launched 16 July 1982, and Landsat 5, launched 1 March 1984, contain a new instrument called the Thematic Mapper. This instrument represents a significant advance in remote sensing technology over earlier multispectral scanning sensors, because of the increased number of spectral bands and of the improved spectral and spatial resolution of the reflected light received. The experiment has obtained useful observations in the disciplines of hydrology, geology, cartography, forestry, urban planning and agriculture. More information on the Thematic Mapper and the experimental results is available in [1].

The Thematic Mapper provides images taken in 7 bands (that is, wavelengths). There are 3 bands in the visible light portion of the electromagnetic spectrum: Band 1 of blue-green color (.45 to .52 μm); Band 2, green (.52 to .60 μm); and Band 3, red (.63 to .69 μm). There are 3 infrared bands: Band 4 (.76 to .90 μm); Band 5 (1.55 to 1.75 μm); and Band 7 (2.08 to 2.35 μm). There is one thermal band, Band 6 (10.3 to 12.5 μm). Each of these bands has the appearance of a black and white "picture". Any three of the bands may be combined to produce a "false-color" image.

Each image for a given band is made up of "pixels", a word derived by contracting "picture elements". A pixel from the Thematic Mapper Experiment is given a digital value obtained by measuring the quantity of light reflected from an area of ground below about 30 meters by 30 meters in size (except for the thermal band, which has a resolution of about 120 × 120 meters). The value of the pixel is a digit between 0 and 255. A value of 0 indicates an entirely "black" area of the ground; that is, no light in the measured band is reflected. A value of 255 indicates that the ground below is "white"; that is, highly reflecting in the measured band. Intermediate pixel values indicate shades that increase from dark gray to light gray, as the value of the pixel increases. The higher the value of the pixel the lighter the shade of gray will be.

Images in 3 different bands may be combined to produce a colored image. A common method is to place an infrared band on the red screen of an image analysis terminal, the red or green band on the green screen, and the blue-green band on the blue screen. This makes vegetation, which is highly reflecting in the infrared, appear as a red hue on the false-color images. Figure 1 is a photograph of a Thematic Mapper image containing Band 4 (infrared) on the red screen, Band 3 (red) on the green screen, and Band 2 (green) on the blue screen.

A digital image differs from a photograph in two respects:

(1) The smallest discernible unit is a pixel, which gives a staircase or blocky effect when the image is enlarged; while the smallest element of a photograph is grainy, which gives an enlarged photo a fuzzy look, and

(2) Digital values are available, allowing each pixel of the image to be mathematically manipulated by computers, while a photograph can not readily be resynthesized without digitization. Thus, mathematical functions can be used to recover information contained in the original image that otherwise would not have been apparent to the human observer.

The Land Analysis System (LAS)[2] at Goddard Space Flight Center is a computer system composed of a minicomputer, an array processor, image analysis terminals, photographic equipment and 240 software routines which can mathematically alter the original image values to assist human interpretation. A common enhancement is to contrast-stretch the original gray level measurements, which typically cluster in a relatively narrow range of the 0–255 domain, into the whole domain. This is analogous to increasing the contrast in a black and white photo, but the contrast is increased through selected computational alteration of digital values. This allows small but significant differences in gray levels to be resolved better by the human eye, and hence noticed and perceived by the human interpretation system. The 3 bands shown in Fig. 1 underwent contrast-stretching by the LAS system.

The application of diverse mathematical functions to images is known collectively as "image processing". The techniques of image processing are increasingly important to many endeavors, for example, medicine, materials science, meteorology and robotic vision, as well as in remote sensing. More information of the use of image processing as applied to space data is provided in [3].

3. SCENE DESCRIPTION

We shall study a symmetric object in a Landsat scene in order to provide a detailed example of the differences between images, photographs, and to examine the nature of the human interpretations of visual scenes.

A Landsat-4 scene of the greater Washington, DC, area was taken on 2 November 1982 as the satellite passed overhead on 10:14am EST. Again, Fig. 1 is a false-color photograph of 3 bands described earlier in Sec. 2. The false-color image causes the rivers to appear as a dark-blue color. The image contains the confluence of two rivers, the Potomac coming from the left, and the Anacostia, coming from the right. The area to the left of the Potomac is Virginia.

This image shows vegetation, which is highly reflective in the infra-red, as red or pink. Late fall conditions existed in the scene, so that the leaves had mostly changed color, but not fallen. The areas composed of mostly deciduous trees are bluish-red and mottled, while the grassy areas show as smooth, red areas. Urban areas containing little vegetation are blue-white to white, depending on the mix of roads, parking lots and roofs. The major urban area is downtown Washington, DC, which is in the center of the image. Urban corridors also extend along major roads into Virginia.

Figure 2 is an enlargement (×2) of a subset of this image showing the downtown area of the city in detail. The blue cloverleaf in the lower right is the Tidal Basin beside the Jefferson

Fig. 1. Thematic Mapper image of Washington DC Landsat-4. Bands 4, 3, 2.

Fig. 2. Thematic Mapper image of downtown Washington Landsat-4. Bands 4, 3, 2.

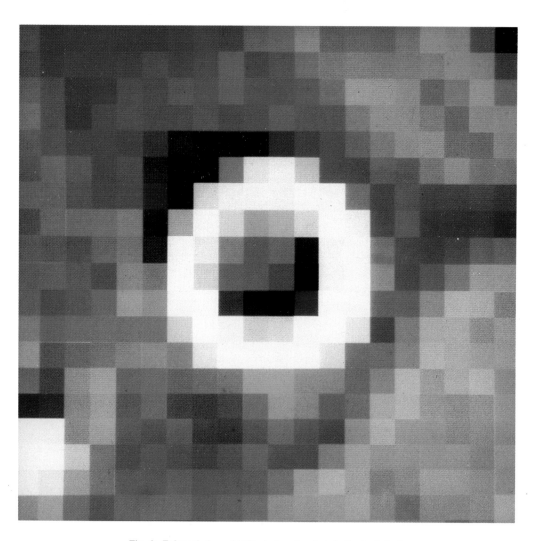

Fig. 3. Enlarged view of RFK stadium Landsat-4. Bands 4, 3, 1.

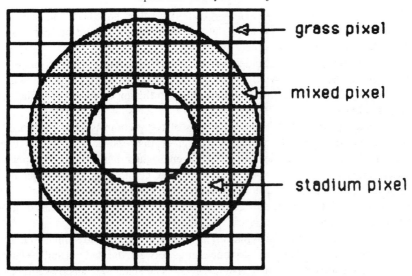

grass pixel

mixed pixel

stadium pixel

Fig. 4. Diagram showing effect of pixel resolution on measurement value.

Monument. The pink area above is the Washington Monument and White House grounds. The perpendicular, lighter pink area is the Mall, which extends eastwards toward the Capitol on the right. The Robert F. Kennedy (RFK) stadium is on a straight line east from the Mall–Capitol axis, just before the East Capitol street bridge over the Anacostia River.

The RFK stadium is a symmetric annular ring. It reflects brightly in the visible and infrared bands, thus giving it a white appearance. (The observer on the ground sees the roof of the stadium as a shiny silvery blue color). The mathematical enhancements used to increase contrast in false color images strengthen the white appearance, because the roof is a strong reflector relative to other objects in the scene. At the magnification of Fig. 2, the symmetry of the object is readily appreciated.

4. COLOR ENLARGEMENT OF THE STADIUM

The image of the RFK stadium can be enlarged digitally to the extent that the individual pixels are visible. Figure 3 shows an enlarged false color image of the stadium using Band 4 (infrared) on the red screen, Band 3 (red) on the green screen and Band 1 (blue-green) on the blue screen. The stadium has lost much of its annular aspect and instead appears to be a blocky feature. This occurs mainly as a consequence of the coarseness of the resolution of the instrument (30 meters) compared to the size of the stadium.

A schematic drawing (Fig. 4) shows how the process of digital measurement causes an annular object to appear blocky. Those pixels which contain areas of both the stadium and the surrounding grass will have values in between those solely from the stadium or from the grass.

Human perception is a major reason that the stadium looks annular in Fig. 2, and blocky in Fig. 3. At the enlarged size, visual blurring by the human visual system is much reduced. Additionally, photographic blurring due to graininess is also significantly smaller compared to an enlarged pixel size.

The interior of the stadium and the exterior area immediately surrounding the stadium are bright red in the infrared because these areas are grassy. There is a white area in the lower left corner which is the beginning of another large building. Surrounding the stadium and grass is a traffic circle and parking; the reflections from the asphalt provide the blue and blue-green pixels around the edge of the image. The deep purple to black pixels on the lower right corner of the inside of the stadium and the upper left corner of the outside of the stadium are the shadows cast by the stadium.

5. INDIVIDUAL BANDS

The color images above are constructed from individual bands. Each band may be viewed as a black and white image consisting of gray levels. Figures 5–8 show the gray levels of the enlarged stadium scene for Bands 1–4. The original pixel values have been enhanced to provide maximum contrast. The stadium appears as an extremely bright feature in every band, along with the building in the lower left corner.

Band 1, taken in blue-green light, shows the road and parking areas as more reflective than the vegetation. The general symmetry of the stadium is well preserved, allowing for the resolution of the pixels relative to the building, although the appearance of circular symmetry disappears at this enlargement.

Band 2 is taken in green light. The brightness of the stadium is less pronounced than in Band 1 because the grass reflects strongly in the green (and indeed is seen as green by the human observer). Interestingly, the asphalt reflects even brighter than the grass. (Asphalt appears gray to human because it reflects evenly in all the visible colors; it is brighter than commonly thought because "gray" tends to have a psychological connotation of "darkness".) The symmetry is less apparent in this band, due to the poorer contrast and the particular location of the pixels relative to the stadium.

Band 3 is the red band, and the symmetry is readily apparent, and similar to that of Band 1. Again, the grass is dark, so there are clear boundaries between the stadium and the surrounding grass. The asphalt surfaces are brighter than the grass. The stadium's shadows are particularly pronounced in this band.

Band 4 is an infrared band, and the grass surrounding the stadium is nearly as reflective as the stadium. The asphalt is relatively dark in the infrared. Because of the lessened constrast with the grass and the heightened effect of the shadows, the symmetry appears to be lost in this image, and the stadium itself seems to have disappeared. It may be that the observer can find the stadium by squinting at the picture or moving far away from it; this psychological effect depends on the blurring of shadows sufficiently so that the human recognizes the shadow as shadow, and then can recognize the feature.

6. CONCLUSION

The example of the image of the Robert F. Kennedy stadium taken by the Landsat-4 Thematic Mapper experiment shows that the concept of symmetry interacts with human perception and measurement technique. The more the stadium was enlarged, the more difficult it became to appreciate the circular symmetry. Further, in Band 4, human perception of shadows and knowledge that the object is present is required to pick out the stadium at all.

Mathematical techniques can be devised to recognize a stadium in a clear image, or to find a symmetric object. These techniques can be implemented readily on a computer, and demonstrated on a small subscene, containing a few hundred pixels. For example, the horizontal or vertical degree of symmetry can be calculated through the technique of flipping the image around a horizontal or vertical axis to make a second image and cross comparing the two images. Or a template of a symmetric image can be created and passed over a small scene until a good match is found. These techniques require a good deal of a priori knowledge of the selected object and, if the scene is large, considerable computer time in searching for computer matches. Similar requirements apply to a large class of problems needed to provide computer vision for robots and other purposes.

The challenge to computer vision is one of scale; how can all the "stadiums" present in this Landsat scene be found in a scene containing over 200 million pixels? The amount of computer time required by contemporary computers to "understand" images is currently infeasible. The challenge of providing automated image interpretation is expected to be met by the development of parallel computing devices, which consist of a large array of small interconnected computers. One existing parallel computer, the Massive Parallel Processor (MPP) contains over 16,000 small computers arranged in a flat array where each computer element can receive and transmit information directly to its four nearest neighbors[4]. Each computer can perform computations on an individual pixel in a 128 × 128 pixel scene and can compare it with its nearest neighbors, or with information on the overall scene which has been broadcast by a main control unit. The next generation of computers, particularly parallel processors which hold a large array in individual central processing units (CPU's) and compute on individual

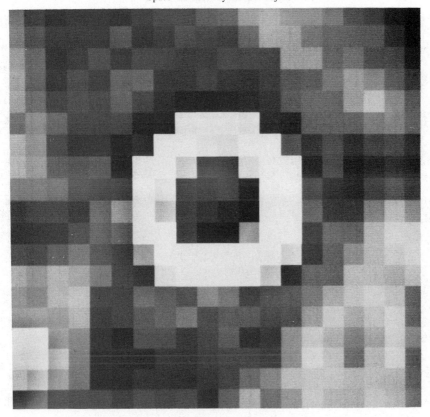

Fig. 5. Band 1 enlarged view of RFK stadium.

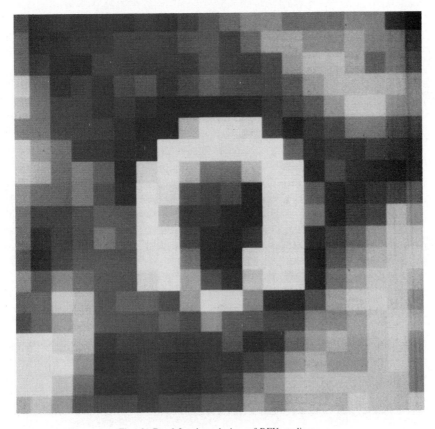

Fig. 6. Band 2 enlarged view of RFK stadium.

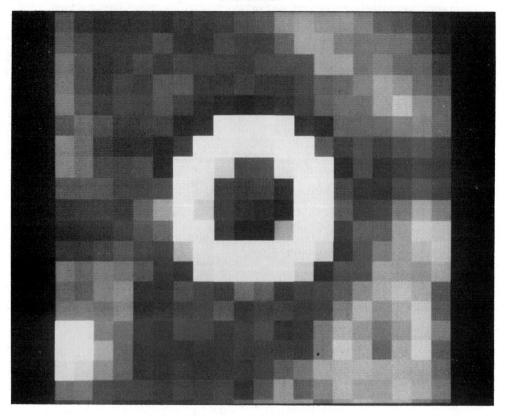

Fig. 7. Band 3 enlarged view of RFK stadium.

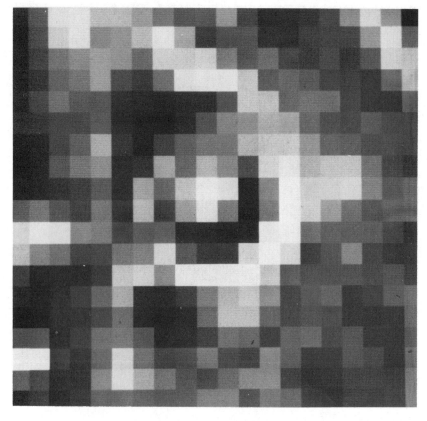

Fig. 8. Band 4 enlarged view of RFK stadium.

pixels simultaneously, will allow machines to replicate many of the sophisticated human inter-pretations necessary for computer vision.[5,6]

The next few years will see extensive development in parallel computer architecture, algorithms for parallel computation of old and new mathematical problems, programming lan-guages and systems to implement the algorithms on parallel machines. These developments will combine with better understanding of how humans interpret visual scenes and how to perform equivalent interpretations on the computer; and how humans can better interpret data presented by the computers. The development of these systems, together with the increase in capacity to store data, will provide more powerful ways of accessing and understanding information. As systems increase in power and decrease in price, major impacts will occur throughout many fields of human endeavor; art, medicine, engineering and science.

Acknowledgements—Skillful assistance from Mark Emmons and Barbara Stiles of RMS Technologies in the preparation of the photographed images is gratefully acknowledged.

REFERENCES

1. IEEE Transactions of Geosciences and Remote Sensing, *Special Issue on Landsat*-4, Vol. GE-22, no. 3, (1984).
2. D. Fischel, J. C. Lyon and P. H. Van Wie, *The capabilities of the Landsat-D Assessment System. Proceedings of ISPRS Commission* II *Symposium*, Ottawa, Ontario, Canada (1982).
3. W. B. Green, *Digital Image Processing: A Systems Approach*. Van Nostrand Reinhold Company, New York (1983).
4. K. E. Batcher, Design of a Massively Parallel Processor. *IEEE Transactions on Computers*, September (1980).
5. M. Onoe, K. Preston, Jr., and A. Rosenfeld, *Real-time Parallel Computing*. Plenum Press, New York (1981).
6. K. Preston, Jr., and L. Uhr, *Multicomputers and Image Processing*. Academic Press, New York (1982).

Comp. & Maths. with Appls. Vol. 12B, Nos. 1/2, pp. 487–510, 1986
Printed in Great Britain.

0886-9561/86 $3.00 + .00

CLASSIFICATION, SYMMETRY AND THE PERIODIC TABLE

WILLIAM B. JENSEN

Department of Chemistry, Rochester Institute of Technology, Rochester, NY, 14623, U.S.A.

Abstract—The scientific use of qualitative, descriptive class concepts versus quantitative, causal mathematical relations is first examined, both with respect to their relative importance in the fields of chemistry, physics, and biology, and with respect to the potential relevance of symmetry considerations. The graphical representation of class concepts by means of sorting maps is then introduced and illustrated with examples from the field of chemistry. When atomic number is used as one of the map coordinates, a special class of map is produced that exhibits symmetry features in the form of an approximate periodicity. This, in turn, leads both to enhanced predictive abilities for the maps in question and to chemistry's most powerful example of the use of class concepts—the periodic table. The paper concludes with a detailed analysis of the current status of the periodic table in chemistry as exemplified by the step—pyramid form of the table and with a brief commentary on the abuse of symmetry considerations in the construction and interpretation of periodic tables in general.

EXPLANATION AND CLASSIFICATION: STAMP COLLECTING RECONSIDERED

The archetypical model of the scientific method usually portrayed in books on the philosophy of science is based on the science of physics and has, since the time of Newton, pivoted around the use of fundamental theories to deduce explicit mathematical relationships between quantifiably measurable properties. In other words, it has centered around making explicit relationships of the general form:

$$u = f(x, y, z \ldots), \tag{1}$$

where u, x, y and z are properties with assignable numerical values[1–2]. Even in the absence of a fundamental theory relating the properties in question, it is possible, if a sufficiently broad range of values have been experimentally determined for the properties, to mimic this process by empirically fitting the values to some functional form—a technique that forms the basis of great deal of applied engineering. The use of such empirical relationships is generally one of interpolation since attempts to extrapolate them beyond the original range of experimental values are fraught with danger. In addition, because of the absence of a theoretical rationale for the functional form in question, such relationships are considered to be lacking scientific "explanation".

Symmetry can enter into (1) in two contexts. The first of these might be called *physical symmetry* and is the type of symmetry normally discussed in books on the subject[3–7]. Thus certain numerical values of u may periodically reappear as x, y, z, are varied, or the value of u may be invariant to certain permutations of x, y, z . . . or to changes in their signs, etc. The second kind of symmetry might be called *logical symmetry* and was first explicitly pointed out by Rosen[5], who called it the "symmetry of analogy" and defined it as ". . . the invariance of a relation or statement under changes of the elements involved in it." In other words, it is based on the observation that the same functional form can frequently be applied to a wide range of different physical systems. A good example would be the concept of a generalized work function in thermodynamics as the product of a generalized force and a generalized coordinate[8] or, as a more extreme example, the concept of a generalized systems theory for describing open systems whose laws would be invariant to the specific elements of the system, be it a reactor in a chemical plant, a living cell, an organ, an organism, a society, or an entire planetary ecosystem[9]. A variety of these analogies, commonly encountered in elementary physics, have been discussed in an interesting book by Shive and Weber[10]. The value of such analogies in stimulating research, as well as their dangers, are self-evident and will be touched on later.

A great deal of chemistry has been successfully treated in terms of the physics–engineering

488 W. B. JENSEN

model described by (1). So much so, that it is not uncommon in books on the philosophy of science to find chemistry contemptuously dismissed as a branch of applied physics with little or nothing new to teach us about the nature of the scientific method or the world. However, a closer examination of a typical chemistry text quickly raises doubts about this facile judgement. For, in addition to the quantitative, dynamic, causal concepts so typical of physics, one also encounters a large number of qualitative, static, descriptive, classificatory concepts more reminiscent of biology. Indeed, the use of such *class concepts* in chemistry is pervasive, as indicated by the partial list in Table 1, and reflects not only the nature of the subject matter but the nature of what is considered to be acceptable "chemical explanation".

Unlike the causal relationships typified by expression (1), the problem of predicting class behavior can seldom be given an explicit functional form. In the simplest possible case, the class behavior will correlate with certain ranges of a quantifiable property x of the species in question:

$$\text{Class } A = f(\text{range } A \text{ of } x) \tag{2}$$
$$\text{Class } B = f(\text{range } B \text{ of } x), \text{ etc.}$$

That is, instead of the simple one to one mapping characteristic of the relation (1), we have a many to one mapping. However, since recourse to the use of class concepts in the first place is frequently a reflection of the complexity of the phenomenon being described, the most common case is usually a good deal more complicated:

$$\text{Class } A = f(\text{range } A \text{ of } x, \text{ range } A \text{ of } y, \text{ range } A \text{ of } z \ldots), \tag{3}$$

where, although a species may have a value of x lying within the required range for class A, it will still not correlate with that class unless it stimultaneously possesses values of y and z also falling within the required ranges. Moreover, the components of relations (2) and (3) may vary widely in their precision. Classes may correspond to easily recognizable either/or situations, such as whether a series of solids are of structure type I, structure type II, etc or whether binary liquid pairs are miscible or nonmiscible in all proportions at the temperature in question, or they may correspond to some ill-defined but widely used distinction, such as whether the species in question are metallic or nonmetallic. Likewise, the properties correlating with the class behavior may range from those with assignable numerical values, such as electronegativity, radius, partial charge, or a NMR shift, to more vague concepts, such as the ability to undergo a certain type of reaction.

As suggested above, the most commonly accepted reason for the use of class concepts instead of mathematically explicit functional relationships is that the phenomena in question are too complex to treat rigorously using currently available theories and mathematical concepts. This, in turn, carries the implication that sciences making heavy use of class concepts, such as chemistry and biology, are somehow immature and still await a complete reduction of their subject matter to the rigor and precision of theoretical physics. The most extreme expression of this point of view has been attributed to Rutherford, who is reputed to have dismissed any activity not coming up to the standards of physics as stamp collecting rather than science. Though there is certainly an element of truth in this, there is also a more subtle and fundamental reason for the use of class concepts in chemistry.

The average introductory textbook frequently defines chemistry as the science of matter and the changes it undergoes, a not particularly enlighting statement since it applies with equal

Table 1. Examples of chemical class concepts

element	base	rare earth
compound	salt	chalcogen
solution	oxidant	halogen
mixture	reductant	ketone
metallic	electrophile	ester
ionic	nucleophile	ether
covalent	hard and soft ions	aldehyde
acid	alkali metal	alcohol

force to physics and even to biology. Indeed, all science deals with matter in some fashion and what is really of interest is what aspect of matter has been singled out for detailed study by the particular science. Thus physics may be naively defined as the study of the general properties of matter and biology as the study of matter when organized in living systems, but what of chemistry? The answer was provided nearly 90 years ago by Wilhelm Ostwald[11], who suggested that chemistry was the study of matter when organized from the standpoint of specific substances. In Ostwald's opinion the first and most fundamental law of chemistry lies in the observation that the multitude of material bodies around us can be classified not only in terms of size, shape, and function but in terms of substances, that is, the observation that aluminum, as defined by a certain set of specific properties, is recognizable as aluminum, be it in the form of an automobile door or a spoon. Thus on a simple Comtean hierarchy, chemistry is the first science to deal in detail with a particular organizational aspect of matter, the first science to have species, the first science to have a natural history as well as a natural philosophy component, and consequently the first science to make pervasive use of class concepts.

What is being suggested here is that class concepts play such an important roll in chemistry and biology, not because of some immaturity in these sciences, but because they have an added classificatory dimension that is largely missing in physics. In other words, stamp collecting is part of their job. Though biologists are certainly interested in understanding life in terms of the fundamental laws of physics, they are also interested in aspects of living systems for which such reductionism is of limited value, such as establishing the range of present and past living organisms, the appearance of new species via evolution, the correlation of organism size with metabolism or geographical distribution, the relative rates of limb and organ growth during development, etc. Likewise, while chemists are certainly interested in understanding chemical reactivity and structures in terms of quantum mechanics and thermodynamics, they are also interested in discovering new chemical species and in establishing simple correlations between known classes of species. Indeed these tasks are frequently more relevant to the direct solution of chemical problems. Thus the knowledge that a species is an ester, coupled with a knowledge of how substituent effects can modify ester-like behavior, may be infinitely more useful in a practical synthesis problem than a detailed quantum mechanical calculation or at least in retrospectively understanding the experimental behavior[12].

This parallelism with biology has also been noted by some historians of chemistry, who have emphasized that the 18th century chemical revolution was in many ways more Linnean than Newtonian in character[13–17], and more recently by some writers on the philosophy of science like D. W. Theobald, who has pointed out yet another interesting parallel[18]:

> Theories are rarely highly controlled by observation in chemistry, since theories are, as I have explained, generally rationalizing constructions covering vast arrays of experimental data, rather than precise mathematical formulations vulnerable to a single quantitative misfortune. . . In chemistry we are often setting out to understand what has actually occurred rather than deliberately contriving to fulfil predictions. We are, so it has been said, telling 'likely stories' rather than hazarding and testing prophecies. As we shall see, it is this difference of temporal emphasis which aligns chemistry with biology as much as with physics.

Theobald goes on to point out that this heavy use of retrodictive rationalization is closely related to the use of class concepts in both sciences as these are required to organize the factual arrays in the first place. Thus chemistry occupies a curious position in our naive Comtean hierarchy. It is bracketed on both sides by less ambiguous examples of the use of certain scientific tactics. The use of causal mathematical reductionism is better illustrated by the practice of physics, whereas the use of class concepts and retrodictive rationalization of broad factual patterns is better illustrated by the practice of biology. This may in part account for the deluge of books, both scholarly and popular, on the philosophical implications of modern physics and biology, while chemistry remains devoid of any such appeal and chemists rather philistine in their philosophical outlook[19–23].

This rather lengthy philosophical digression aside, the question arises as to what role, if any, symmetry can play in predicting and correlating the class concepts summarized by relation (3). That logical symmetry or analogy plays a role is self-evident as this is the essence of classification. The establishment of the class *alkali metals* automatically implies the existence of member-invariant class statements, such as M is an air and moisture sensitive metal forming chlorides of the type MCl and sulfides of the type M_2S. But what of the role of physical symmetry?

SORTING MAPS AND CLASSIFICATION

Answering this question requires some way of making relation (3) more explicit, and this may be done by means of an empirical procedure known as a sorting map. To construct a sorting map one requires a set of species whose class assignments are already known (e.g. structure type, miscibile–nonmiscible, reactive–nonreactive, etc.) and for which numerical values of the properties thought to determine or correlate with this class behavior are also known (e.g. electronegativity, radius, charge, etc.). Using these properties as the coordinates, the species are then plotted, a different point symbol being used for each class, and the resulting map is then examined to determine whether the properties in question have produced a significant sorting or separation of the classes. If so, the properties are considered to play a significant role in determining the class behavior and best-guess class boundaries are added to the map. These allow for interpolative prediction of class behavior for species other than those used to construct the original map provided that values for each of the property coordinates are known. Though this procedure can in principle be applied to n simultaneous properties, limitations of spatial visualization normally limit it to two or three. Some typical two-dimensional sorting maps of chemical interest are shown in Figs. 1–4.

The coordinates of the maps in Figs. 1–4 were intuitively selected based on the qualitative bonding concepts used by virtually every chemistry text to verbally rationalize the behavior in question. However, such knowledge is not a necessary prerequisite as computer programs, like the SIMCA method of Wold and coworkers[24], have been developed to systematize the search for relevant property coordinates. Here again a set of species with known class assignments (called a *learning set*) as well as a set of potentially relevant properties is required. The computer then systematically explores the usefulness of different property pairs (using one property per coordinate as well as combinations of properties) as sorting coordinates and mathematically assesses the degree of sorting obtained in each case in order to arrive at the optimal set. This set of coordinates can then be applied to interpolatively predict the class behavior of other species (called the *test set*). Wold *et al.* characterize SIMCA as method for analyzing chemical data in terms of *similarity and analogy* and point out that it is really an example of the more general subject of *pattern recognition* in chemistry.

Though sorting maps allow us to better visualize and systematize the use of class concepts, the maps in Figs. 1–4 do not appear to display any overt symmetry properties. Is this a function of our choice of coordinates or is physical symmetry simply irrelevant to the problem at hand?

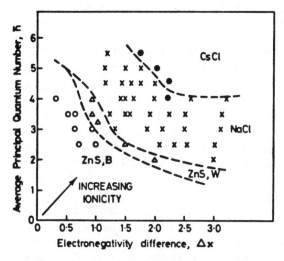

Fig. 1. Structure sorting map for binary AB octet species. Modified from [81].

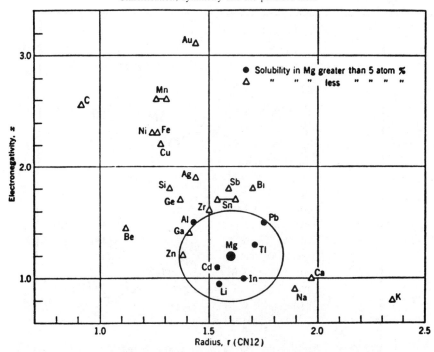

Fig. 2. Solubility sorting map for binary magnesium alloys[82].

The answer lies in the assertion of J. C. Phillips, who has pioneered the application of sorting maps to problems of solid state structures, that the sorting map procedure is in essence a form of "Mendeleevian analysis", i.e. it is identical in spirit to the procedure used by Mendeleev in deducting what is perhaps chemistry's most powerful example of the use of class concepts— the periodic table, and that, in fact, the periodic table itself is really a sorting map[25].

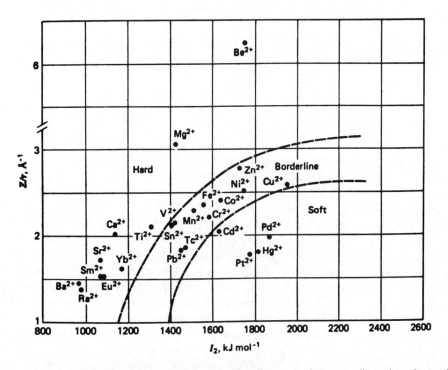

Fig. 3. Sorting map for hard and soft divalent cations. Coordinates are charge to radius ratio and second ionization potential[83].

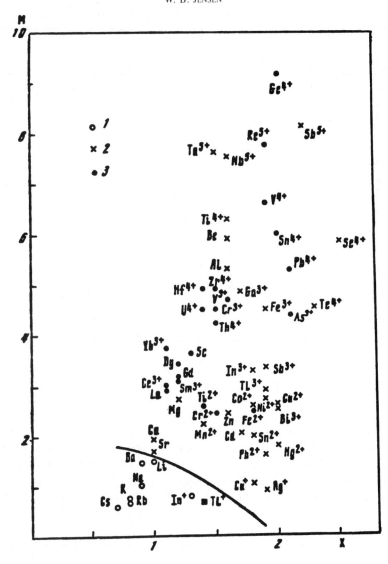

Fig. 4. Sorting map for TlCl–MCl, binary phase diagrams. Coordinates are charge to size ratio for M^{x+} and its electronegativity[84]. Type 1 points indicate eutectic formation, Type 2 indicate compound formation, and Type 3 indicate systems not yet investigated.

SYMMETRY AND CLASSIFICATION: THE PERIODIC TABLE

The simplest way to test Phillips' assertion is to plot the atomic number of an element versus the maximum number of available valence electrons per atom (i.e. maximum oxidation state) as shown in Fig. 5. Though seldom thought of in this context, this map, as Mazurs has emphasized[26], is really a type of periodic table and corresponds to a modernized version of the table (Fig. 6) proposed by Mendeleev in his 1889 Faraday Lecture based on a plot of the element's atomic weight versus its maximum oxygen valence[27–28]. As will be seen shortly, more commonly encountered versions of the periodic table may be viewed as symmetry related transformations of this more fundamental form, a step which Mendeleev appears to have taken implicitly in arriving at his original table. However, in the case of the second major originator of the periodic table, Lothar Meyer, this use of a two-dimensional property–property map as a preliminary step appears explicitly in the form of his famous atomic volume–atomic weight plot (Fig. 7)[29]. Indeed, it was not long before chemists realized that virtually any measurable property of an element, or even of the element's compounds, could be used in conjunction with the atomic weight to produce a periodicity map. Some 19th century chemists, like Thomas Carnelly, virtually based their entire careers on the production and study of these maps, and

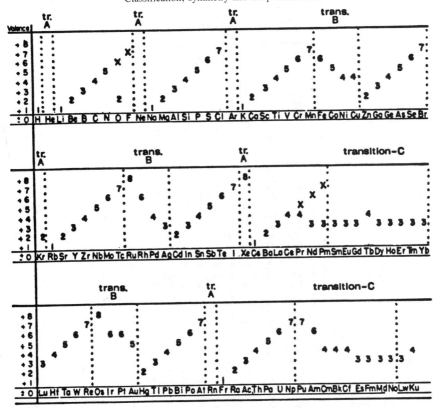

Fig. 5. Plot of maximum oxidation state versus atomic number. Except for the O and F points (see text) this is identical to a plot of the maximum number of valence electrons versus atomic number. This plot is actually continuous and has been broken into three sections solely for reasons of size.

interest in them managed to persist well into the third decade of this century[30]. Some typical examples (modernized by replacing atomic weight by atomic number) are shown in Figs. 8 and 9.

Comparison of the maps in Figs. 1–4 (Type I) with those in Figs. 5–9 (Type II) shows some significant differences. Whereas Type I maps fail to show any recognizable symmetry features and tend to clump members of the same class into the same spatial region, Type II maps show approximate symmetry in the form of an approximate periodicity and tend to place members of the same class not in the same spatial region but rather in symmetry related regions. This, in turn, leads to a significant difference in the predictive abilities of the two types of map. As noted earlier, Type I maps can be used to interpolatively predict a species' class behavior given a knowledge of the values of its sorting map coordinate properties. The reverse process of predicting the values of the coordinate properties from a knowledge of the species' class is much less useful as it can only yield a range of values, though some use of this has been made in establishing upper and lower bounds for solubility parameters[31]. Nothing can, of course, be inferred about whether or not a given class is missing any as yet to be discovered members.

The characteristic patterns arising from the symmetry features of Type II maps, however, radically alter this situation. By restricting a species to a relatively narrow portion of the map's pattern they allow not only interpolative prediction of class behavior from a knowledge of the coordinate properties, but the reverse process, as well as the interpolative prediction of missing class members and their accompanying coordinate properties. This is well illustrated by Mendeleev's famous prediction of the existence and properties of eka-silicon (germanium), eka-boron (scandium), and eka-aluminum (gallium) and by the use of the periodic table to predict both the values of as yet undetermined atomic weights and to modify incorrectly reported values[32–36]. On the other hand, extrapolative use of Type II maps has historically proved as perilous as the extrapolative use of Type I maps, as evidenced both by the difficulties of 19th century chemists in accommodating the rare earth elements and by their failure to anticipate

Symbols and Atomic Weights		The Composition of the Saline Oxides
R	A	R_2O_3
	[6]	[7]
H	1	1 = n
Li	7	1†
Be	9	— 2
B	11	— — 3
C	12	— — — 4
N	14	1 — 3° — 5°
O	16	
F	19	
Na	23	1†
Mg	24	— 2†
Al	27	— — 3
Si	28	— — 3 4
P	31	1 — 3° 4° 5°
S	32	— 2 — 4° 5° 6°
Cl	35½	1 — 3 — 5° — 7°
K	39	1†
Ca	40	— 2†
Sc	44	— — 3†
Ti	48	— — 3 4
V	51	— 2 3 4 5
Cr	52	— 2 3 — — 6°
Mn	55	— 2† 3 4 — 6° 7°
Fe	56	— 2† 3 — — 6°
Co	58½	— 2† 3 4
Ni	59	— 2† 3
Cu	63	1† 2†
Zn	65	— 2†
Ga	70	— — 3
Ge	72	— 2 — 4
As	75	— — 3 — 5°
Se	79	— — — 4 — 6°
Br	80	1 — — — 5° — 7°
Rb	85	1†
Sr	87	— 2†
Y	89	— — 3†
Zr	90	— — — 4
Nb	94	— — 3 — 5°
Mo	96	— 2 3 4 — 6°
~~(1)~~~		
Ru	103	— 2 3 4 — 6 — 8
Rh	104	— 2 3 4 — 6
Pd	106	1† 2 — 4
Ag	108	1†
Cd	112	— 2†
In	113	— 2 3
Sn	118	— 2 — 4
Sb	120	— — 3 4 5
Te	125	— — — 4 — 6°
I	127	1 — 3 — 5° — 7°
Cs	133	1†
Ba	137	— 2†
La	138	— — 3†
Ce	140	— — 3 4
Di	142	— — 3 — 5
~~(14)~~~		
Yb	173	— — 3
~~(1)~~~		
Ta	182	— — — — 5
W	184	— — — 4 — 6
~~(1)~~~		
Os	191	— — 3 4 — 6 — 8
Ir	193	— — 3 4 — 6
Pt	196	— 2 — 4
Au	198	1 — 3
Hg	200	1† 2†
Tl	204	1† — 3
Pb	206	— 2† — 4
Bi	208	— — 3 — 5
~~(5)~~~		
Th	232	— — — 4
~~(1)~~~		
U	240	— — — 4 — 6

Fig. 6. Mendeleev's plot of oxygen valence versus atomic weight. Excerpted from a larger table in [28].

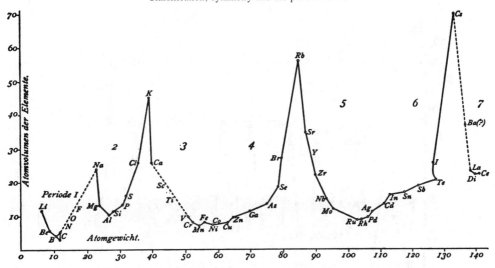

Fig. 7. L. Meyer's atomic volume–atomic weight periodicity map[73].

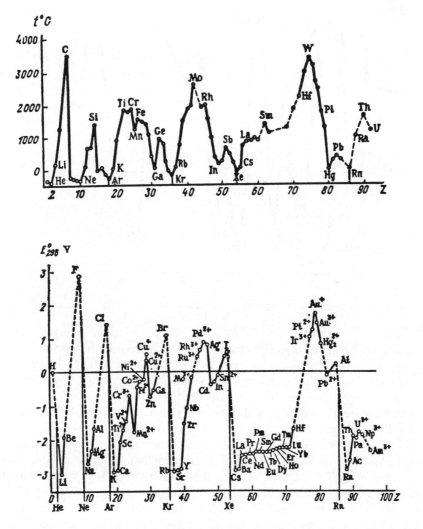

Fig. 8. Various property–atomic number periodicity maps for the elements in their standard states[85]. *Top*: melting points. *Bottom*: standard electrode potentials.

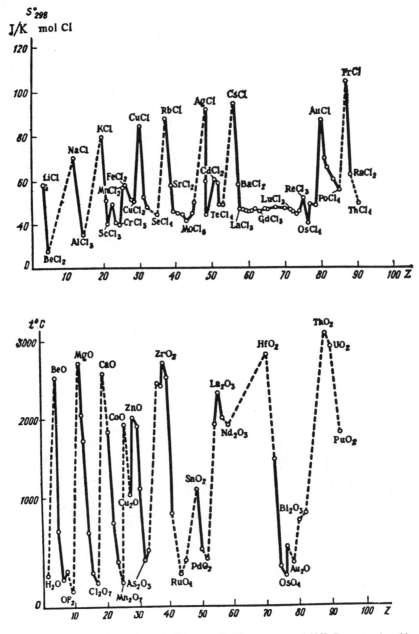

Fig. 9. Various property–atomic number periodicity maps for binary compounds[85]. *Top*: entropies of formation for chlorides. *Bottom*: Melting points of oxides.

the discovery of the noble gases[37]. Indeed, in a recent article on the subject, Karpenko estimates that about 200 "spurious" or nonexistent elements have been reported in the chemical literature since the end of the 18th century, and that the lion's share of them have been associated with attempts to extrapolate sections of the periodic table[38].

Another important aspect of Type II, versus Type I, maps lies in the 19th century chemist's conviction that any significant classification of the elements had to be based on a *natural system* of some sort, that is, it should arise naturally or spontaneously from a study of the elements' properties rather than being based on some *artificial* or arbitrarily selected criterion or theoretical preconception[17,39]. The fact that all Type II periodicity maps lead to essentially the same result and that in principle the classes spontaneously appear on the maps as elements in symmetry related positions, rather than being imposed beforehand, both suggest that the resulting classification is really a natural one. On the other hand, the assertion that the classes can be recognized solely on the basis of symmetry is probably overly ambitious as the symmetry in question is often only very approximate. In actual fact many natural classes of elements, such as the halogens, the alkali metals, and the alkaline earths, had been recognized by chemists long before the work of either Mendeleev or Meyer, thus substantially simplifying the problem of recognizing class relations on the maps. Inspection of textbooks from around the period 1865–1870 shows that the most common properties used for this purpose were valence and metallic character[40–41]. Though actual maps weren't used because of the lack of a quantitative metallicity scale, it is possible to generate a modern map that essentially reproduces these classifications by using the Herzfeld refractivity–volume ratio as the metallicity coordinate[42]. It would, of course, be historically inaccurate to imply that the development of the periodic table was motivated solely by an empirical search for a natural classification of the elements. Belief in the inherent simplicity and unity of matter, sparked by Prout's hypothesis, and a certain perverse fascination with purely numerical relationships between atomic weights, sparked by Doebereiner's triads, also played a role[32,39]. However, it is interesting that, retrospectively at least, Mendeleev emphasized only the natural classification aspect and even went so far as to deny that the periodic table implied anything about the structure of atoms or the unity of matter[17].

As noted earlier, most common periodic tables are not based directly on the property–atomic number(weight) maps themselves, but on a transformation of these maps whereby the symmetry related members of each class are brought together in a single region of the map. In addition, this transformation is both generalized and qualitative, i.e. it attempts to simultaneously represent the form of as many kinds of property–atomic number maps as possible. However, since not all of these maps display periodicity to the same degree, it is possible to prioritize them by ranking the properties relative to their importance in determining class behavior, with properties like maximum oxidation state and electronic configuration near the top of the list and properties like specific conductivity near the bottom; this, in turn, giving rise to the concept of primary, secondary, and tertiary analogies between elements (see below). The game of constructing a periodic table then becomes one of finding a transformation that will make explicit as many of these analogies as possible. This transformation may range from a simple doubling over of the map on itself at each periodic repetition, to some elaborate three-dimensional extension, but whatever the device, the task has managed to exercise the ingenuity of chemists for more than a century now and has led to literally hundreds of different suggestions.

CHOOSING A MODERN PERIODIC TABLE

Numerous books[26,32–36,39,43–48] and reviews[49–50] have been devoted to describing these suggestions, the most thorough being that of Mazurs, who lists more than 700 tables proposed between 1869 and 1970[26]. Mazurs has also developed a detailed classification of these tables, though in selecting a suitable table for modern use only three of his criteria need to be considered:

(1) three-dimensional (3D) versus two-dimensional (2D) tables,
(2) short and medium length versus long tables,
(3) chemical versus electronic configuration tables.

3D periodic tables, or—more properly—periodic models, have frequently been advanced based on the claim that they illustrate continuity more effectively than 2D or planar tables and are capable of making explicit a larger range of analogies because of the greater flexibility inherent in their 3D geometries. The forms most commonly proposed have been based on cylindrical or conical helices with occasional use of lemniscates and even spheres. Mazurs has also shown that most planar tables correspond to various 2D projections of these 3D models. In choosing a table, these 3D models have been ruled out as they are generally expensive to construct and impossible to effectively reproduce in books and on wall charts. In addition, their 3D shapes make it impossible to simultaneously view the entire table, often necessitating a fair amount of spatial manipulation in order to obtain even a general overview. Besides, the concept of continuity is understood by convention in the repeating periods of the planar tables and, as will be seen, it is possible to select a planar table capable of representing not only conventional primary *intra*-group analogies but more subtle secondary *inter*-group analogies as well.

Mazurs uses length to further divide periodic tables into three major categories:

(1) *short tables* of 8 columns, with the *d*-block interleaved with the *s* and *p*-blocks and the *f*-block set to one side,

(2) *medium tables* of 18 columns, with the *d*-block fully expanded, but the *f*-block still set to one side,

(3) *long tables* of 32 columns, with both the *d* and *f*-blocks fully incorporated and expanded.

Short tables have virtually disappeared from use in the United States, where the currently popular form is a medium length block table (which is ironically, but incorrectly, called a long table). In choosing a table, only long 32 column tables have been considered. While there may have been valid reasons for setting the *f*-block elements to one side when their electronic configurations were poorly understood and it was thought that their only oxidation state was III (thus requiring that 30 elements be crammed into two squares of group III), this is no longer the case. The only result of this convention at present is to suggest to students that these elements are somehow dull and unimportant and so can be set to one side and ignored, which is regrettably just what happens in most introductory courses and not a few advanced inorganic courses as well. The only discernable advantage of the currently popular medium block table is that it allows one to make the printing larger. In short, it is time we get rid of what D. A. Johnson has aptly termed the periodic table's "antiquated footnote".

Chemical tables are based primarily on considerations of the elements' chemical and physical properties and especially on the values of their highest oxidation states. In other words, they give precedence to the periodicity displayed by the sorting maps in figures 5 and 6. These considerations are reflected in the common conventions for numbering the classes or groups (thus group VII reflects the VII oxidation state of Cl in ClO_4^- rather than the more common I-oxidation state of Cl^-), and in the use of the A and B groups or series in the longer periods to link together elements with common maximum oxidation states but otherwise relatively distinct chemical and physical properties (e.g., Ca(II) and Zn(II)). Electron configurations tables, on the other hand, frequently dispense with family and group labels based on oxidation state, relying instead on the periodic repetition observed in both the nature (*s*, *p*, *d*, or *f*) and number of outer electrons for the isolated gas phase atoms.

Chemical tables have frequently been criticized because of the numerous exceptions to the maximum oxidation state labels. Thus F formally fails to attain the VII state of a group VII element in any of its compounds, and only Ru and Os ever attain the VIII state of group VIII. Even more serious, all of the members of group IB have maximum oxidation states greater than I (i.e., Cu(IV), Ag(III), and Au(V)[51–52]). Finally, as mentioned earlier, this approach has traditionally crammed all of the *f*-block elements into group IIIB, though many are known to exhibit higher oxidation states[53–54]. Ironically, most of these problems stem from a simple failure to update what are essentially late 19th century labels in light of recent chemical advances. As will be seen shortly, when this is done most of these problems disappear. On the other hand, it should be pointed out that similar objections can be raised concerning electronic configuration tables and labels based on idealized filling rules, as nearly 20% of the elements have ground state configurations that violate these rules. An even graver problem is He, which should be classed with the alkaline earth metals according to its electronic configuration rather than with the noble gases as indicated by its chemical and physical properties[55–56]. What all of this means is simply that the periodic table is a true natural classification based on the simultaneous consideration of as many property–atomic number maps as possible. Since none of these maps exhibits perfect periodicity, the result is the best "averaged" representation and one which is consequently imperfect with regard to any single property considered in isolation, be it maximum oxidation state or electronic configuration. In short, the table should combine the best features of both the traditional chemical table and the more recent electronic configuration tables.

Based on the above criteria—namely that the table should be a 32 column planar table that optimally combines the advantages and information content of both the traditional chemical table and the modern electronic configuration table—it was felt that the best choice, from among the multitude described by Mazurs, was the modified step-pyramid table shown in Fig. 10. This form of the table was actually proposed as early as 1882 by Bayley[57], who, however, used vertical rather than horizontal periods (Fig. 11), and again, apparently quite independently, by the Danish thermochemist J. Thomsen[58] in 1895 (Fig. 12). Though used in a modified medium length 18 column form by a variety of chemists in the early part of the century (Summarized in [59]), it did not attain any real popularity until Bohr[60–61] revived Thomsen's original long form in the early 1920s in conjunction with his work on atomic spectra (Fig. 13). As a consequence the table is frequently referred to as the Thomsen–Bohr table, though it should perhaps be more accurately called the Bayley–Thomsen–Bohr table.

ADVANTAGES OF THE STEP-PYRAMID TABLE

In addition to its overall aesthetic appeal, resulting from its symmetric shape, the step-pyramid table has a number of distinct advantages over the currently popular medium length block table, both from the electronic configuration standpoint and from the standpoint of a traditional chemical table.

Advantages as an electronic configuration table

Though the columns of the step-pyramid table are not explicitly labeled with electronic configurations, these and other structure information for isolated gas phase atoms are implicit and relatively easy to extract. The uses and advantages from this point of view can be succinctly summarized as follows:

(1) The elements are explicitly classified by electronic type as s, p, d and f-block elements. These blocks are indicated on the table by thick dividing lines and by labels at the top and bottom of each block. Coloring each block and its connecting solid diagonals a different color brings these divisions into striking relief.

(2) Because the f-block elements have been fully inserted into the table in their natural order of appearance, the ideal electronic configurations of each kind of atom can easily be obtained simply by counting the squares from H to the element in question and inserting the appropriate type of electron per square as indicated by the block labels. The required quantum number n in each case is identical to the period number of the square in question.

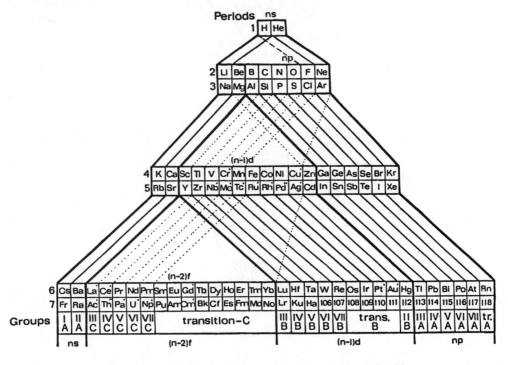

Fig. 10. The modified *Bayley–Thomsen–Bohr* step-pyramid periodic table.

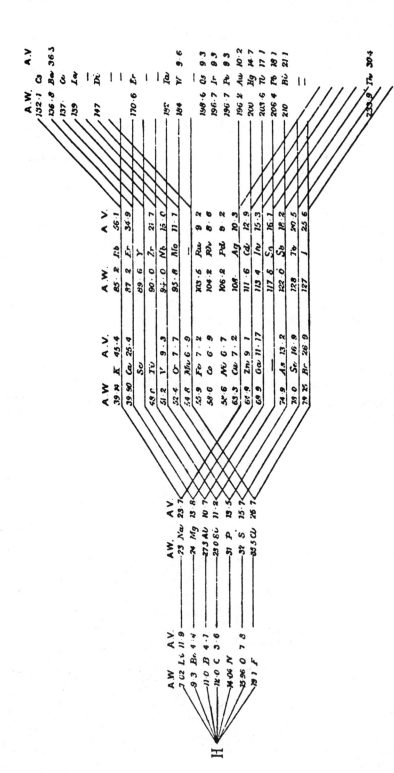

Fig. 11. Bayley's 1882 table[57].

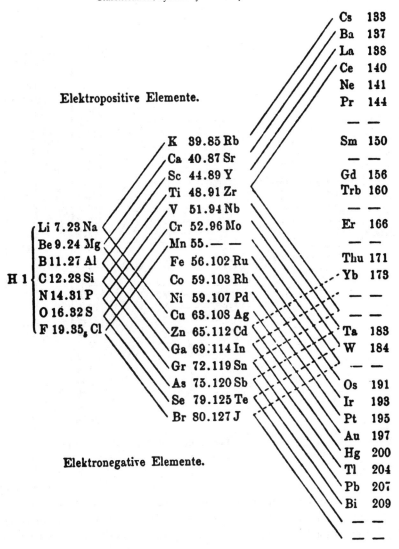

Fig. 12. Thomsen's 1895 table[58].

(3) The real spectroscopic ground state configuration of the isolated gas phase atom, when different from the ideal configuration, is obtained from the number of dots in the upper right hand corner of the element's square. In the case of d-block elements, these indicate the number of ns electrons in the ideal configuration that must be converted into $(n - 1)d$ electrons in order to obtain the real configuration. In the case of the f-block elements, they indicate the number of $(n - 2)f$ electrons that must be converted into $(n - 1)d$ electrons.

(4) In contrast the currently popular block table, the step-pyramid table incorporates the recent conclusion that La and Ac should be classified as f-block elements with irregular configurations, whereas Lu and Lr should be classified as the first members of the d-block in periods 6 and 7[62].

(5) The table not only classifies elements into groups according to their outer electronic configurations, but further subdivides the elements within each group of the p, d and f-blocks according to the electronic structures of their cores by placing each core type on a different step of the pyramid. Thus N, P, As, Sb, and Bi are all placed in group VA by virtue of their common outer ns^2np^3 configurations. However, Bi, on the bottom step, has a [Noble Gas]$(n - 2)f^{14}$ $(n - 1)d^{10}$ core, Sb and As, on the second step, have a [Noble Gas]$(n - 1)d^{10}$ core, and P and N, on the third step, have just a [Noble Gas] core. In the case of the p-block and the period 6 d-block elements, especially, these core variations lead to variations in chemical behavior associated with the d-block and f-block (or so-called lanthanide) contractions. As a result, the

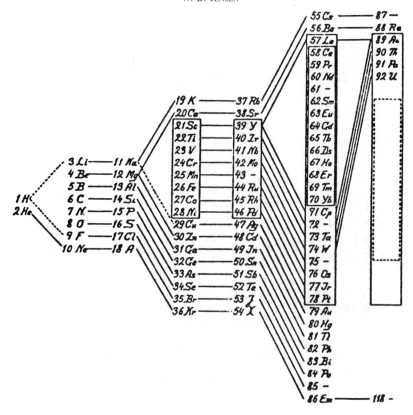

Fig. 13. Bohr's 1922 table[60].

lower member of a step tends to resemble the upper member of the step below more closely than it resembles the other member of the same step (e.g., Al resembles Ga more than it resembles B, Zr resembles Hf more than it resembles Ti, etc.). Thus the steps group elements more by similarity of total electronic configuration than by similarity of chemical properties.

Advantages as a chemical table

Following traditional chemical tables, the step-pyramid table uses both maximum oxidation state group numbers and A–B group modifiers. Its advantages over other chemical tables lie in its extension of the group concept to the f-block, resulting in an additional C series of groups to parallel the traditional A–B series; its use of an updated set of oxidation labels that eliminate most the exceptions found in the current system; and, most importantly, in its geometric ability to explicitly indicate not only primary intra-group analogies, but the secondary inter-group analogies that gave rise to the A–B modifiers in the first place, as well as certain tertiary analogies.

The original intent of the A–B group modifiers was to distinguish between elements in the long periods of the table having similar maximum oxidation states but otherwise relatively distinct chemical and physical properties (e.g. Mn and Br in period 3). As such, the order in which these labels were applied was to a certain extent arbitrary, at least as far as the 19th century chemist was concerned. With the advent of electronic configurations, however, it became apparent that this duplication resulted from two elements having the same total number of valence electrons, but otherwise distinct arrangements or configurations of these valence electrons (i.e. $ns^2(n - 1)d^5$ versus ns^2np^5). In light of this knowledge, there is little doubt that the American A–B convention, which associates A with those elements deriving their valence electrons solely from the outermost n shell and B with those elements capable of (but not necessarily) also employing electrons from the $n - 1$ shell, is superior to the European convention, which retains the arbitrary nature of the original 19th century scheme. In other words, the American convention maximizes the correspondence between the chemical and electron configuration tables, whereas the European convention doesn't. A similar logic lies behind numbering the periods 1–7 rather than 0–6 as in some older tables. Relative to a chemical table this choice is arbitrary, whereas relative to a configuration table it isn't, since the first choice allows one to associate the period number with the primary quantum number n for the atoms of the period, whereas the second choice doesn't. Consequently the first choice maximizes the correspondence between the two types of tables. Because the step-pyramid table also attempts to assign group numbers to the f-block elements, the above reasoning requires the existence of a third "C series" of groups, where C indicates an element capable of (but not necessarily) also employing $n - 2$ shell electrons as valence electrons.

The logic behind the revised oxidation-state labels is best seen by examining the maximum oxidation state–atomic number sorting map in Fig. 5. This shows a series of repeating cycles within which the maximum oxidation state progressively increases and numerically corresponds to the number of valence electrons lying outside the nearest noble gas or pseudo-noble gas core. These repeating cycles of progressive increase are, in turn, separated or bridged by shorter "transitional" series of elements within which the oxidation state variation is highly erratic and no longer shows a necessary correspondence with the number of "outer" electrons. Not all of the progressive cycles and transitional bridges are equivalent. In the A series the transitions occur with a single element (a noble gas) and the maximum oxidation state falls from a value of seven (e.g. Cl) to a value of one (e.g., K) as the cycle repeats. In the B series, on the other hand, the transitions occur more gradually, requiring four steps (e.g., Fe, Co, Ni, and Cu), and the maximum oxidation state falls from a value of seven (e.g., Mn) to a value of only two (e.g., Zn) as the cycle repeats. Finally, in the C series the transitions become even more gradual and erratic, requiring nine steps (e.g., Pu–No), and the maximum oxidation state falls from seven (e.g., Np) to a value of only three (Lr) and is not complete until the end of the f-block. Thus, although the proposed A–B–C series and transitions do not exactly parallel one another, they do form a systematic pattern of variation, the transitions becoming progressively longer (1, 4, and 9 elements respectively) and the minimum oxidation state at the end of each transition progressively greater (I, II, and III respectively).

Use of the terms transition or transitional elements in the above context actually corresponds to the historically correct usage, as these terms were originally applied solely to the group VIII triads (i.e., Fe–Co–Ni in period 4, Ru–Rh–Pd in group 5, and Os–Ir–Pt in period 6) rather than to d-block elements in general with incomplete d subshells[63]. As such, the terms were intended to suggest that these triads represented a gradual transition between the end of one cycle or period of increasing maximum oxidation states (e.g., K(I) to Mn(VII) in MnO_4) and the beginning of another (e.g., Cu(I) to Br(VII) in BrO_4^-). These elements were very similar in their atomic weights and chemical properties, exhibited variable oxidation states—reflecting their transitional natures, and frequently resisted attack by common reagents (hence the name noble metal for the heavier members). When the rare or noble gases were later discovered, it was suggested that they too were transition elements, as they also bridged the gap between successive cycles of increasing maximum oxidation states[64]. Indeed, they were considered to be more perfect examples of transitional species as the contrast between the elements at the end and beginning of successive periods (e.g., Cl(VII) and K(I)) was much sharper than that between elements at the beginning and end of successive series (e.g., Mn(VII) and Cu(I)), and the transition occurred in these cases in one step rather than three. Finally, the noble gases appeared to be chemically inert, and thus represented truly "noble" elements, in contrast to the known reactivity of the so-called noble metals. This view of the group VIII triads as imperfect "noble gases" was also used by later writers on the periodic table[65] and the observation that they should really be extended to transitional tetrads by incorporating Cu, Ag, and Au was first pointed out by Jorgensen[55–56]. Indeed, the knowledge of these elements' oxidation states was already sufficiently developed in Mendeleev's day to show the untenability of the group IB concept, and its inclusion in the periodic table is really an attempt to idealize the symmetry. Not only is I not the maximum oxidation state exhibited by these three elements, except for Ag(I), it is not even the most common oxidation state.

In terms of the modern electronic theory of atomic structure, the failure of these transitional or bridging elements to show any simple relation between maximum oxidation state and the number of electrons lying outside the nearest noble gas or pseudo-noble gas core, as well as the rapid fall in the maximum oxidation state from the value of seven observed just before the transition to the value observed after its completion, can both be interpreted as indicating that the outer electrons in these elements are undergoing a rapid transition from the status of valence electrons to the status of core electrons and consequently exhibit a certain ambiguity in their behavior. Even in the case of the f-block elements, where this transition is much more leisurely and is not really complete before the end of the block, there is evidence of this process in the increased stability of Yb(II) and No(II) relative to the more common III + oxidation state[51].

Examination of the sorting map in Fig. 5 shows only five exceptions to the revised labeling scheme (O, F, Pr, Nd, and Pm). Since our true concern with group numbers is in identifying the maximum number of accessible valence electrons and we are only using maximum oxidation state as a convenient indicator of that number, the O and F exceptions are really only formalities of our conventions for assigning oxidation states and are due to the fact that these elements are two of the most electronegative in the table. There is little doubt that in solid state oxides and fluorides, despite the II- and I-oxidation states, all six valence electrons of oxygen and all seven of fluorine are implicated in bond formation (e.g., ZnO). On the other hand, if one were to extend the series to a maximum oxidation state of VIII rather than VII, as done in the past, the number of real exceptions would increase to eleven and there would be no A transition to parallel the B and C transitions. Finally, if one were to retain Cu, Ag, and Au as group IB and add analogous groups IC(Yb, No) and IIC(Tm, Md), as done by Mazurs[26], the exceptions would increase to 18. Thus the proposed scheme extends the oxidation state labels to the f-block while simultaneously removing most of the previous exceptions from the s, p and d-blocks. (For a discussion of alternative classifications of the f-block see [66–69].)

Use of the revised labeling scheme is not, of course, limited to the step-pyramid table, as it may be used in conjunction with any suitable long form 32 column table. Rather, as mentioned earlier, the particular advantage of the step-pyramid table over other 32 column tables lies in its ability to explicitly indicate not only primary intra-group analogies like other tables, but secondary inter-group analogies and tertiary analogies as well (Table 2). A primary analogy between two elements means that the elements in question have not only the same maximum number of outer valence electrons (defined as those lying outside the first noble gas or pseudo-noble gas core) but the same idealized arrangement or configuration for these electrons and, by implication, a closely related pattern of lower oxidation states as well. Such elements fall in the same *group* as distinguished by both the group number and modifier (e.g. IA, IIIB, VC, etc.). This relationship is indicated by placing the elements in the same column or by connecting offset columns by solid diagonal tie-lines. This is the only level of analogy explicitly indicated on most tables.

A secondary analogy between two elements means that the elements in question have the same maximum number of valence electrons but differing valence electron configurations. Consequently, although the elements exhibit certain similarities and analogies when in their maximum oxidation states, these do not necessarily extend to lower oxidation states. Such elements belong to the same *family*, as indicated by identical group numbers, but to different groups of that family, as indicated by different group modifiers (e.g., In family III: groups IIIA, IIIB, and IIIC). Thus the resulting secondary intra-family analogy is inter-group rather than intra-group in nature and is explicitly indicated on the table by connecting the elements in question with a dotted diagonal tie-line.

Although other forms of the table have continued to retain the A–B group modifiers, they do not explicitly indicate the inter-group analogies that gave rise to these labels in the first

Table 2. Levels of analogy in the step-pyramid table

Order of Analogy	Nature of Analogy			Type of Tie-Line
	Same Idealized Outer Electron Configuration	Same Maximum No. of Valence Electrons†	Related Lower Oxidation States	
Primary	Yes	Yes	Similar Pattern	Solid
Secondary	No	Yes	Not Necessarily	Dotted
Tertiary	No	No	At least One	Broken

†Defined as the number of electrons lying outside of the nearest noble gas or pseudo-noble gas core.

place. As a consequence the use of these secondary analogies has become almost a form of "lost knowledge" to most users of the table, leading to an inability to decide on the best A–B convention[70] and to IUPAC's regrettable decision to abandon these labels completely[71]. In contrast, the step-pyramid table, as Bayley noted nearly 100 years ago, allows one to construct a "kinship" chart for each family of elements that explicitly traces both the primary and secondary relations, with separate branches corresponding to each primary group, and kinks within each group corresponding to different core varieties[57]. These charts may either be extracted from the table as in Fig. 14 or traced directly on the table itself by coloring in the proper squares and diagonals (both solid and dotted).

It is important to emphasize that, with few exceptions, secondary inter-group analogies are only valid for maximum oxidation states. This is the only situation for which the oxidation number is an accurate reflection of the total number of active valence electrons. For lower oxidation states the number only reflects those valence electrons committed to heteronuclear bonding and other valence electrons may be present as stereoactive lone pairs or involved in

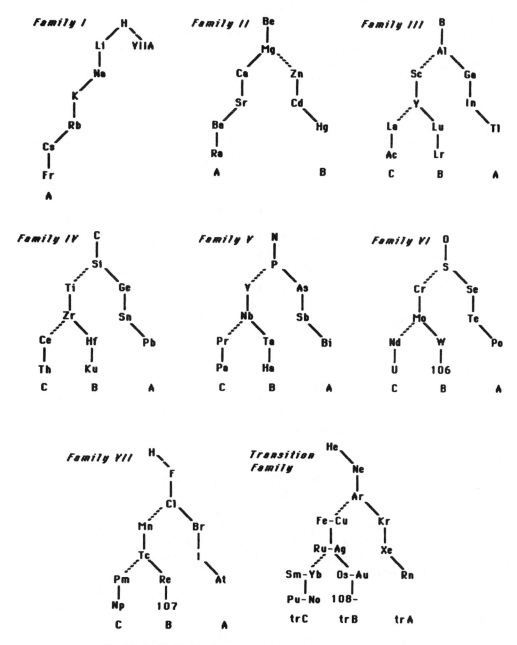

Fig. 14. Family kinship charts extracted from the step-pyramid table.

homonuclear catenation, thus leading to totally unrelated structures and reactivity patterns for the species being compared. Though, at first glance, it might seem absurd to compare Mn with Cl or Cr with S, similarities do exist for the highest oxidation states. Thus Mn_2O_7 and Cl_2O_7 have similar structures and permanganates (MnO_4^-) and perchlorates (ClO_4^-) are often isomorphous. Likewise, CrO_3 and the chain form of SO_3 have similar structures and chromates (CrO_4^{2-}) and sulfates (SO_4^{2-}) are isomorphous. It is also worth reminding oneself that large differences can occur within the same group—compare, for example, the properties of elemental N_2 with those of Bi, or C with Pb, or even those of CO_2 with SiO_2.

A tertiary analogy between two elements means that the elements in question exhibit certain similarities when at least one of the elements is in a lower oxidation state but that the elements have neither the same maximum number of valence electrons nor the same electronic configurations (e.g., Fe(III) and Al(III)). These are indicated by connecting the two elements with a broken diagonal tie-line. However, since this level of analogy is legion and highly erratic in its variation across the table, it is not for the most part explicitly indicated, though the table is capable of doing so. The sole exception is the analogy between H(I-) in its saline hydrides and the halogens X(I-) in their saline halides. This has been traditionally indicated either by floating H in space above the rest of the table or by simultaneously placing it at the heads of both groups IA and VIIA. The step-pyramid table, on the other hand, is able to indicate this analogy without removing H from the body of the table or duplicating its position.

It is also sometimes argued that H is placed above the table because it is totally unique in its properties. While it is true that H is unique relative to other IA elements, this is really a reflection of a systematic variation in the periodic table which shows that the elements in the first row of any new electronic block tend to show abnormalities relative to the elements in later rows of the same block, and that the degree of divergence decreases in the order s-block $>>$ p-block $>$ d-block $>$ f-block. Thus He, like H, is unique relative to other ns^2 elements, a fact that places it among the noble gases rather than the alkaline earth metals (Note, however, that the definition of valence electron used in Table 2 lessens this problem by defining this number as 0 for the noble gases, thus placing helium's abnormality in its core structure). Likewise, C–F are quite distinct compared to other p-block elements, and the same is true to a lesser degree for Sc–Zn relative to the heavier d-block elements. In the f-block, on the other hand, the differences are less pronounced, though all of the oxidation state exceptions occur in the first row of the block. These systematic variations are frequently implicit in the organization of topics adopted by descriptive inorganic texts based on the periodic table (see, for example, the chapter headings in reference[51]). Note also that the traditional group IB analogy is really a tertiary rather than a primary analogy and that comparison of Cu(I), Ag(I), and Au(I) with the alkali metals is no more dictated by maximum oxidation states and electronic configurations than comparison with Tl(I) or Ga(I).

In summary then, the step-pyramid table does appear to be quite successful at accomplishing the goal outlined in the last section of optimally combining the best features of both the chemical and electronic configuration tables. This, however, raises the further question of whether or not these two types of tables are redundant. Jorgensen[55–56] has in fact explicitly raised this question and has concluded that this is not the case—a conclusion supported by the continued popularity of chemical tables. The structure of the table actually reflects idealized configurations and, despite attempts at half-subshell rationales, deduction of observed gas phase spectroscopic ground states from the table requires supplementary information such as the dots used in Fig. 10. Likewise, the relation between the spectroscopic table and chemical properties is not always straightforward, at least for the transitional elements.

Mendeleev was fond of making a distinction between simple bodies and elements. By the former he meant the elements in their bonded standard states and by the latter presumably the free atoms themselves. His point was that the explanation of the periodicity displayed by "simple body" sorting maps like those in Fig. 8 or by compound sorting maps like those in Fig. 9 lay ultimately in the properties of the true "real elements" rather than in those of the simple bodies and compounds "visually known to us" (see especially footnotes 11 and 17 of Chapter 15 of [28]). Today we are, of course, able to directly measure properties of the isolated gas phase atoms and construct corresponding property—atomic number maps like those in Fig. 15. The assumption, however, that the maps in Fig. 15 "explain" those in Figs. 8 and 9, and consequently that the periodic table need only encode information about free atoms, is oversimplified. The structures of free atoms do not in fact persist in the structures of standard state simple bodies or compounds. Rather it is the valence electrons and cores that persist and which constitute our "true" elements. The value of free atom sorting maps is that free atoms are the simplest chemically interesting structures formed by our "true" elements and consequently their structures and periodicity are relatively easy to rationalize in terms of the properties of their constituent cores and valence electrons. These free atom maps can, in turn, be *correlated* with the more complex periodicity maps for standard state elements and compounds, though they do not really *explain* them in the sense outlined above. In short, we have available to us a class of periodicity maps that the 19th century chemist lacked and one that is simpler in its underlying structure (Fig. 16). However, the conclusion that the periodic table itself should reflect information about all three classes of maps remains intact and suggests that the American Chemical Society's recent proposal to replace the oxidation state labels with a simple 1–18 numbering system will result in a loss of explicit information[71]. In addition, despite claims to the contrary, such a scheme would be awkward when used in conjunction with a 32 column table[72].

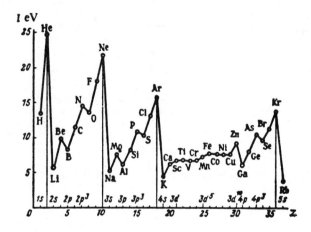

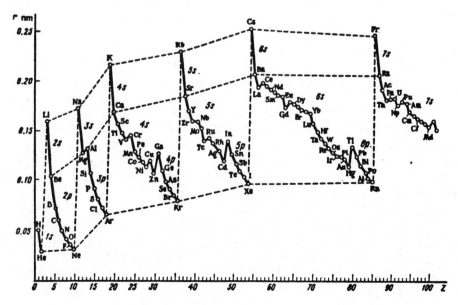

Fig. 15. Various property—atomic number periodicity maps for gas phase atoms[85]. *Top*: first ionization potential. *Bottom*: radius.

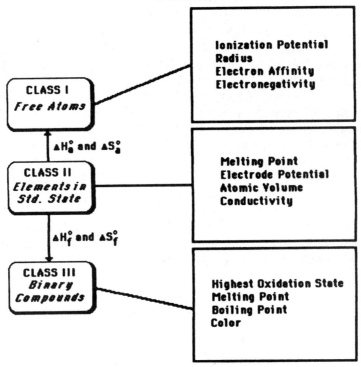

Fig. 16. The relationship between the three classes of property–atomic number periodicity maps. Maps can be constructed not only for the properties listed in each case but also for the connecting properties.

ABUSES OF SYMMETRY

In discussing the periodic table in the 1911 edition of his famous textbook, Nernst[73] commented that:

> At first the establishment of the periodic system was regarded as a discovery of the highest importance, and far-reaching conclusions as to the unity of matter were expected from it; more recently a disparaging view has been taken of these noticeable, but unfruitful, regularities. One now often finds them underrated, which is intelligible enough, since this region, which especially needs scientific tact for its development, has become the playground of dilettante speculations, and has fallen into much discredit.

Anyone who has reviewed current papers dealing with this subject or who has been the recipient of unsolicited tables knows that there is still some truth to Nernst's complaint. Indeed, one is tempted to assert that, with the demise of alchemy, the construction of revolutionary new periodic tables has become the favorite expression of eccentricity in chemistry, almost on par with perpetual motion machines in physics and trisecting angles in mathematics. The temptation to read more into the shape of the table than is really there is almost overwhelming. Even someone as great as Werner was tempted[74]. Having postulated a missing element between H and He, he decided to perfect the symmetry of his table by guaranteeing that rows of differing length always occurred in pairs. Consequently he further postulated a row of three missing elements lying above the H–X–He row. Others have been luckier. Thus Bayley correctly postulated 31 elements (sans the noble gases) for period 6 (recall Fig. 11) though his reasoning appears obscure, leading van Spronsen to comment that Bayley apparently "designed his system on primarily aesthetic grounds", i.e. on the basis of symmetry[39]. Likewise, Thomsen (recall Fig. 12), applying Werner's logic to the other end of the table, suggested that Th and U were the only known members of a second 31 column period (again sans the noble gases), a proposition that is still the subject of speculation[75–77].†

————

†Reference [39] appears to have both miscounted the length of Bayley's table and to have overlooked Thomsen's comments on the placement of Th and U.

A far more common abuse, however, involves the symmetry of logic, i.e., the abuse of analogy. In terms of periodic tables, this most often takes the form of an unsupported micro–macro analogy, leading to the belief that one's favorite form of the periodic table is a literal macroscopic scale-up of the atom, the position of each box literally representing the physical position of the corresponding differentiating electron in the atom itself. The author has personally been treated to unsolicited examples of helical atoms and planar disk atoms, both complete with s, p, d, and f electrons in total disregard for the quantum mechanical interpretation of these concepts.

The psychologist Joseph Jastrow once wrote an essay entitled "The Natural History of Analogy" in which he laid the blame for virtually every human superstition and stupidity on man's predilection for this "primitive mode of logic"[78]. There is certainly little doubt that humans exhibit a strong attraction for symmetry in their conceptual schemes as well as their wallpaper designs, and that many features, for example, of Aristotle's (and certainly Plato's) thought stem from such considerations rather than from the results of direct observation and experiment. Nevertheless, it cannot be denied that analogy plays an important role in scientific creativity. Indeed, it has been studied in some depth by philosophers of science[79]. The real danger appears not to be analogy itself, but rather the analogy that is untested or, if one is a follower of Popper, the analogy that is untestable[80].

REFERENCES

1. C. G. Hempel, *Aspects of Scientific Explanation and Other Essays in the Philosophy of Science*. Free Press, New York (1965).
2. Y. B. Zeldovitch, *Higher Mathematics for Beginners and Its Application to Physics*. Mir, Moscow (1973).
3. F. M. Jaeger, *Lectures on the Principles of Symmetry and Its Application in All Natural Sciences*. Cambridge University Press, London (1917).
4. H. Weyl, *Symmetry*. Princeton University Press, Princeton (1952).
5. J. Rosen, *Symmetry Discovered: Concepts and Applications in Nature and Science*. Cambridge University Press, Cambridge (1975).
6. J. Rosen, *A Symmetry Primer for Scientists*. Wiley-Interscience, New York (1983).
7. A. V. Shubnikov and V. A. Koptsik, *Symmetry in Science and Art*. Plenum, New York (1974).
8. V. V. Sychev, *Complex Thermodynamic Systems*. Mir, Moscow (1981).
9. L. von Bertalanffy, *General Systems Theory*. Braziller, New York (1968).
10. J. N. Shive and R. L. Weber, *Similarities in Physics*. Wiley-Interscience, New York (1982).
11. W. Ostwald, *The Principles of Inorganic Chemistry*, Chap. 1. Macmillan, London (1914).
12. C. J. Suckling, K. E. Suckling and C. W. Suckling, *Chemistry Through Models*. Cambridge University Press, Cambridge (1978).
13. W. Whewell, *History of the Inductive Sciences*, 3rd Edn., Vol. II, pp. 305–310. Appleton, New York (1873).
14. R. E. Schofield, *Mechanism and Materialism: British Natural Philosophy in an Age of Reason*. Princeton University Press, Princeton, New Jersey (1970).
15. R. E. Schofield, The Counter-Reformation in Eighteenth Century Science—Last Phase, in D. H. Roller (Ed.), *Perspectives in the History of Science*, pp. 39–66. University of Oklahoma Press, Norman (1971).
16. A. Thackray, *Atoms and Powers: An Essay on Newtonian Matter Theory and the Development of Chemistry*. Harvard University Press, Cambridge (1970).
17. D. M. Knight, *The Transcendental Part of Chemistry*, Chap. 9. Dawson, Folkestone, Kent (1978).
18. D. W. Theobald, Some Considerations on the Philosophy of Chemistry. *Chem. Soc. Rev.* **5**, 203–213 (1976).
19. M. Capek, *The Philosophical Impact of Contemporary Physics*. Van Nostrand Reinhold, New York (1961).
20. J. Powers, *Philosophy and the New Physics*. Methuen, London (1982).
21. N. Cartwright, *How the Laws of Physics Lie*. Clarendon Press, Oxford (1983).
22. D. Hull, *Philosophy of Biological Science*. Prentice Hall, Englewood Cliffs (1974).
23. M. Ruse, *The Philosophy of Biology*. Hutchinson University Press, London (1973).
24. S. Wold and M. Sjostrom, SIMCA: A Method for Analyzing Chemical Data in Terms of Similarity and Analogy, in B. R. Kowalski (Ed.), *Chemometrics: Theory and Applications*, ACS Symposium Series 52, Chap. 12. American Chemical Society, Washington D.C. (1977).
25. J. R. Chelikowsky and J. C. Phillips, Quantum Defect Theory of Heats of Formation and Structural Transition Energies of Liquid and Solid Simple Metal Alloys and Compounds, *Phys. Rev.* **B16**, 2453–2477 (1978).
26. E. G. Mazurs, *Graphic Representations of the Periodic Table During One Hundred Years*. University of Alabama, University, Alabama (1974).
27. D. Mendeleev, The Periodic Law of the Chemical Elements. *J. Chem. Soc.* **55**, 634–656 (1889).
28. D. Mendeleev, *The Principles of Chemistry*, Vol II, foldout table. Longmans, London (1897).
29. L. Meyer, *Modern Theories of Chemistry*, foldout table. Longmans, London (1888).
30. C-F. Hsueh and M-C Chiang, Periodic Properties of Elements, *J. Chinese Chem. Soc.* **5**, 263–275 (1937).
31. C. Hansen and A. Beerbower, Solubility Parameters, in A. Standen (Ed.), *Kirth-Othmer Encyclopedia of Chemical Technology*, 2nd. Edn., Suppl. Vol., pp. 889–910. New York (1971).
32. F. P. Venable, *The Development of the Periodic Law*: Chemical Publishing Co., Easton, Pennsylvania (1896).
33. G. Rudolf, *The Periodic Table and the Problem of Chemical Evolution*. Whittaker, New York (1900).
34. A. E. Garrett, *The Periodic Law*. Paul, Trench, and Trubner, London (1900).

35. C. Schmidt, *Das periodische System der chemisches Elemente*. Barth, Leipzig (1917).
36. E. Huth, *Das Periodische Gesetz der Atomgewichte und Naturlische System der Elemente*. Friedlander und Sohne, Berlin (1897).
37. R. F. Hirsh, A conflict of principles: The discovery of argon and the debate over its existence. *Ambix* **28**, 121–130 (1981).
38. V. Karpenko, The discovery of supposed new elements: two centuries of errors. *Ambix* **27**, 11–102 (1980).
39. J. W. van Spronsen, *The Periodic System of Chemical Elements: A History of the First One Hundred Years*. Elsevier, Amsterdam (1969).
40. A Naquet, *Principles of Chemistry Founded on Modern Theories*, pp. 74–75. Renshaw, London (1868).
41. L. C. Cooley, *A Textbook of Chemistry*, Chap. 3. Scribners, New York (1869).
42. P. P. Edwards and M. J. Sienko, The transition to the metallic state. *Acc. Chem. Res.* **15**, 87–93 (1982).
43. E. G. Mazurs, *Types of Graphic Representation of the Periodic System of Chemical Elements*. La Grange, Illinois (1957).
44. G. Langhammer, *Das Periodensystem*. Volk und Wissen, Berlin (1949).
45. K. Mahler, *Atombau und Periodische System der Elemente*. Salle, Berlin (1927).
46. I. Koppel, *Der Bau der Atome und das Periodische System*. Voss, Leipzig (1927).
47. D. O. Lyon, *Das Periodische System in neuer Anordung*. Deuticke, Leipzig (1928).
48. E. Rabinowitsch and E. Thileo, *Periodische System: Geschichte und Theorie*. Enke, Stuttgard (1930).
49. J. G. F. Druce, Some recent representations of the periodicity of the elements, *Chem. News* **130**, 322–326 (1925).
50. G. N. Quam and M. B. Quam, Types of graphic classifications of the elements. *J. Chem. Educ.* **11**, 27–32, 217–223, 288–297 (1934).
51. F. A. Cotton and G. Wilkinson, *Advanced Inorganic Chemistry*, 4th Edn. Wiley-Interscience, New York (1980).
52. H. Schmidbauer and K. C. Dash, Compounds of gold in unusual oxidation states. *Adv. Inorg. Chem. Radiochem.* **25**, 239–266 (1982).
53. D. A. Johnson, Recent advances in the chemistry of the less-common oxidation states of the lanthanide elements. *Adv. Inorg. Chem. Radiochem.* **20**, 1–132 (1977).
54. T. Moeller, Periodicity of the lanthanides and actinides. *J. Chem. Educ.* **47**, 417–423 (1970).
55. C. K. Jorgensen, The Periodic Table and Induction as the Basis of Chemistry. *J. Chim. Phys.* **76**, 630–635 (1979).
56. C. K. Jorgensen, The loose connection between electron configuration and the chemical behavior of the heavy elements (Transuranics), *Angew. Chem., Int. Ed. Engl.* **12**, 12–19 (1973).
57. T. Bayley, On the connection between atomic weight and the chemical and physical properties of the elements. *Phil. Mag.* **13**, 26–37 (1882).
58. J. Thomsen, Systematische Gruppierung der chemischen Elemente. *Z. Anorg. Chem.* **9**, 190–193 (1895).
59. W. B. Jensen, *The Step-Pyramid Form of the Periodic Table*. Rochester Institute of Technology, Rochester, New York (1983).
60. N. Bohr, The structure of the atom. *Nature,* **112**, 29–44 (1923).
61. N. Bohr, *The Theory of Spectra and Atomic Constitution*. p. 70. Cambridge University Press, Cambridge (1922).
62. W. B. Jensen, The positions of lanthanum (actinium) and lutetium (lawrencium) in the periodic table. *J. Chem. Educ.* **59**, 634–636 (1982).
63. W. B. Jensen, Origin of the term transition metal. *Chem. 13 News* **120**, 14 (1981).
64. J. W. Moller, *Modern Inorganic Chemistry*. p. 993. Longmans, London (1927).
65. W. Kossel, Uber Molekulbildung als Frage des Atombaus. *Ann. Phys.* **49**, 229–362 (1916).
66. G. Glockler and A. I. Popov, Valency and the periodic table, *J. Chem. Educ.* **28**, 212–213 (1951).
67. C. D. Coryell, The periodic table: the $6d$-$5f$ mixed transition group. *J. Chem. Educ.* **29**, 62–64 (1952).
68. G. Glockler and A. I. Popov, The long form of the periodic table. *J. Chem. Educ.* **29**, 358 (1952).
69. C. D. Coryell, The place of the synthetic elements in the periodic table, *Rec. Chem. Prog.* **12**, 55–64 (1951).
70. W. C. Fernelius and W. H. Powell, Confusion in the periodic table of the elements, *J. Chem. Educ.* **59**, 504–508 (1982).
71. K. J. Loening, Recommended format for the periodic table of the elements. *J. Chem. Educ.* **61**, 136 (1984).
72. G. J. Leigh, The periodic table. *Chem. Brit.* **20**, 892 (1984).
73. W. Nernst, *Theoretical Chemistry*, 3rd Edn., p. 189. Macmillan, London (1911).
74. A. Werner, Beitrag zum Aufbau des periodischen Systems. *Ber. Deut. Chem. Ges.* **38**, 914–921, 2022–2027 (1905).
75. G. T. Seaborg, Prospects for further considerable extension of the periodic table. *J. Chem. Educ.* **46**, 626–634 (1969).
76. M. Taube, The periodic system and superheavy elements. *Nukleoniks* **12**, 304–312 (1967).
77. D. Nebel, Zur Position der Actiniden und Transactinidenelemente in Periodensystem der Elemente, *Z. f. Chem.* **10**(7), 251–260 (1970).
78. J. Jastrow, *Fact and Fable in Psychology*. Houghton Mifflin, Boston (1900).
79. M. B. Hesse, *Models and Analogies in Science*. Notre Dame Press, Notre Dame, Indiana (1970).
80. K. R. Popper, *The Logic of Scientific Discovery*, Revised Edn. Hutchinson, London (1972).
81. E. Mooser and W. B. Pearson, On the crystal chemistry of normal valence compounds. *Acta Cryst.* **12**, 1015–1022 (1959).
82. L. S. Darken and R. W. Gurry, *The Physical Chemistry of Metals*. p. 87. McGraw-Hill, New York (1953).
83. W. B. Jensen, *The Lewis Acid-Base Concepts: An Overview*, p. 266. Wiley-Interscience, New York (1980).
84. P. P. Fedorov and P. I. Fedorov, The formation of compounds in binary systems comprising gallium(III), indium(III), and thallium(I) chlorides, *Zhur. Neorg. Khim.* **19**, 215–220 (1974).
85. N. S. Akhmetov, *General and Inorganic Chemistry*. Mir, Moscow (1983).

SYMMETRY
Unifying Human Understanding
(Part 2)

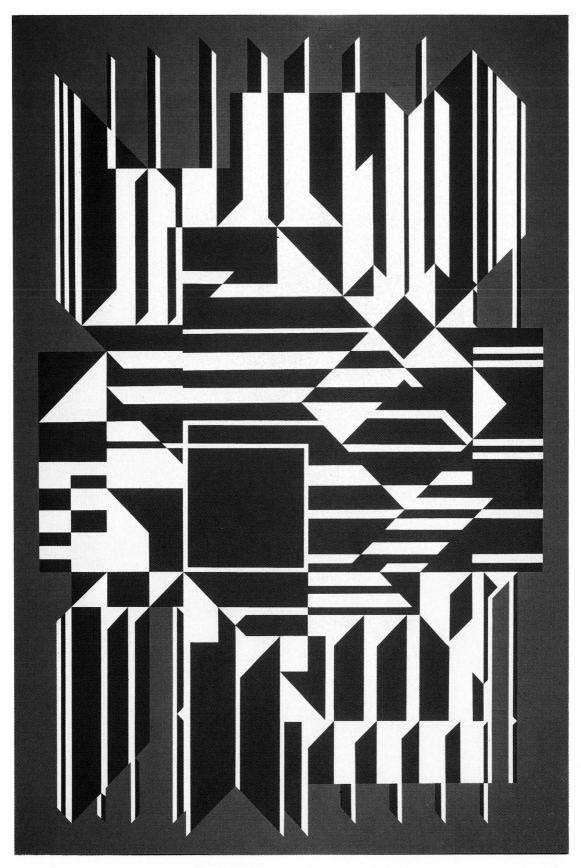

Victor Vasarely, "HORN–II" (1957), courtesy of the artist

Comp. & Maths. with Appls. Vol. 12B, Nos. 3/4, pp. 511–529, 1986
Printed in Great Britain.

CRYSTALLOGRAPHIC PATTERNS

Kh. S. Mamedov

Institute of Inorganic and Physical Chemistry, Azerbaijani Academy of Sciences, Baku 370143,
U.S.S.R.

Abstract—The primary aim of this paper is to describe the main principles of crystallographic patterns (CPs) and their distribution and connection with natural phenomena and subjects of applied art. It is shown that these principles are in many respects similar to those governing the behavior of two-dimensional crystals.

One may already find in ancient CPs the elements of colored symmetry and examples of pattern elements which have been discovered in our own times. Such CPs are very characteristic of ancient and medieval decorations of Siberia, Kazakhstan, Central Asia, Azerbaijan, and Asia Minor.

In modern times CPs are closely connected with the name of M. Escher. The growing interest in his drawings shows that CPs have considerable potential as a tool for applied art even now. It is shown that the systematic study of CPs is very useful for the understanding and description of ancient and medieval art decoration from the aforementioned countries.

In preparing this article for this special issue I have reduced and redone a larger body of material. First of all, I had to define my subject concretely, considering possible material of the same type that might appear in other articles, and excluding possible repetition. I tried to touch the common things directly related to the theme and composition of my story only in those cases when I was not sure that my own understanding is really generally accepted and that I may hold the free narrative tone to the very end without trivial considerations.

I wished to call the attention of the specialists to systematic rather than occasional review of symmetry by means of the facts given below, as well as by my own attitude and aspiration to shed light on those facts with modern scientific ideas on the laws of nature and especially on the resulting material.

Having used the *carte blanche* kindly given me by the editor (as you know, a free rein to do things is always inspirational), I have made no effort to restrict the style of my meditations. I have presented a flow of free and sincere statements, and have not attempted to impose on them a style which might conceal their individuality. A great advantage of such statements is that one's ''falsehoods'' are merely considered to be delusions, thus somehow mollifying the anger of those strict critics who feel obliged to adhere to absolute truths.

The patterns I am going to talk about are usually described in the literature as geometric patterns. It is quite enough for the pattern to be constructed from geometric elements, though it is not significant for them to be related in a geometric manner.

However, there exist patterns that may not be easily related to geometric patterns. The elements of such patterns are the images of living matter, and the characteristic of such patterns is that by their inner organization they recall double-dimensional crystals or their projections. For this reason they are called *crystallographic patterns* (CPs). The elements of such CPs may also be geometric figures.

The term ''crystallographic'' in this case calls attention to the mode of pattern organization. The pattern given in Fig. 1 is geometric, while the one given in Fig. 2 is crystallographic, though in both cases the elements are geometric. No matter what the character of the constructing element is, the CPs are constructed according to the same principle, i.e. the principle of tight packing, which results in the equivalence of design elements and the activity of the background.

Thus the ''crystallographicity'' is the determining attribute, and it is a sufficient basis for the collective term ''crystallographic patterns'' to apply.

Manuscript editor's note: Considerable effort has been made to preserve the author's original intention, while making sometimes drastic changes in his choice of English words. In some places where the original intention was particularly unclear, the author's original words remain. The reader should not be deterred by a few such passages. Apologies for any serious misinterpretations are extended to the author, who did not have an opportunity to revise the edited version. D. W. Crowe

Fig. 1. Geometrical ornaments from ancient Egypt.

My acquaintance with CPs has a very curious story. My parents and countrymen before the Revolution, and up to the Second World War, had been nomads. Until my graduation from secondary school this way of life was customary for me, and many insoluble questions arose when this type of life touched on what was discussed at school about life in general. At that time among the most agonizing and recurring questions for me were such insistent ones as these:

• Why are the homemade things that surround us in our nomadic life literally stuffed with strange geometric signs? (See Fig. 2.)

Fig. 2. The carpet ''kilim.'' Shirwan, 19th century.

• Outside our nomad tents we were living in a wonderful kingdom of various curved lines and forms. So why were the aesthetic signs not formed from them, having instead preserved these geometric patterns like some mysterious rite, transferring them to new wares?

The same questions arose in connection with folk music and poetry. Nevertheless all those signs warmed us. They could deeply move us so that we could pronounce such words as "it is fine," "it's beautiful," when we were glancing at those patterns.

In those days we could guess, after having learned from books, reproductions of artists' paintings, and the movies which were so rare and exotic for the nomads, that life in cities was somehow different. Here the aesthetic signs had been formed from the surrounding elements. We were amused by their close connection to reality. They were distinguished from reality by means of the gorgeousness of the human and nature wearings, as well as by the fact that they were motionless.

For the nomads the subjects of daily life always served as the symbol of the aesthetic sign, while in towns and cities there existed things which were specially intended to be considered as such symbols. Here also the subjects of daily life were sometimes used as the aesthetic sign symbols, but, unlike the nomads, in this case the beauty was stuck on those subjects. You couldn't conceal the patterns unless you destroyed the carpet, although you could change the painting as much as you wished without destroying the brickwork of the wall on which it was painted. Thus, one might conclude that in the cities where the straight-line geometry was predominant the aesthetic signs were formed from the elements of reality itself, with nature playing the dominating role. This reverse symmetry between the aesthetic-sign attitude of the townfold and the nomads can be clarified in the following schematic manner:

Who	The surroundings	The art objects
townfolk	geometry	nature
nomads	nature	geometry

I would call this kind of symmetry "inversy." It turns out that the view of the surroundings and the view of the aesthetic sign supplement or complement each other, i.e. what is absent in the surroundings exists in the aesthetic sign. Perhaps the famous architect and thinker Le Corbusier had this phenomenon in mind when he raised human inclination to pure geometry up to the sign of the human spareness with free will[1]. It may also be that the nomad did not need the "portrait" of an oak to be carried with him elsewhere because he could view all sorts of oaks every day and every hour, finding them surrounded with different moods at different times, while for the townfolk their inclination to nature was more a result of nostalgia. Whether it was so or not is difficult to state nowadays.

Similar questions literally pursued me till I entered the university. There they became overgrown with some other accessory questions, but worst of all was the fact that such questions were never put forward there. It was as if they did not exist at all.

In my understanding at that time, realistic art was the art that had to describe the world and the events that took place in it, and this art amazed me tremendously. It was the wonderful source of information about the world in some other connection and it concealed the "constraint" created by the carpet's "geometry." This art could be read without any difficulty and was available to everybody without any dependence on language or traditions. It was clear, and it was widely written in those days, that this trend in art was an imported trend. It was new for us, though its rudiments had been with us already. Representatives of our folk could succeed in this new trend of art, which obviously proved that we could profit from it too. But then it became incomprehensible—how could one explain the long disregard for such a necessary trend in art? It is reminiscent of the question, Why have the Chinese, the Japanese, and their cultural neighbors still not abandoned their hieroglyphs, which are extremely "noncomfortable for usage"? Surely, to answer these questions it is sufficient to grasp the concept of tradition. Tradition becomes firmer with centuries. It might be broken with great difficulty, but it is physically impossible to throw away its wreckage which prevents the construction of a new tradition, but finally is used in the construction of the new trend, etc.

But did we really have an artistic point of view as part of our own tradition? The answer is yes, we really had it. Tradition is a very important feature and a preserved characteristic of *ethnos*.

I have already mentioned what this tradition looked like for the nomads. What about the attitude of the more cultural folk layers, I mean those who became accustomed to the culture of the townfolk? It can be judged according to the patterns and drawings on subjects of their daily life, as well as the illustrated books of those days. It turns out that the inclination to geometry can be discovered here, too.

Especially interesting were the miniatures. In recent years they have become the subject of concentrated attentions by some of our art critics. For example, in one such work[2] it is said that the miniature artists of those times had generalized and emphasized the artistic fact of reality by means of the traditions of the fine arts.

From the point of view of the Renaissance the miniatures looked too primitive. They were without perspective, i.e. the paintings were flat; there was no light gradation with smooth transition from one color to another; all living beings were deformed; events which were incompatible in time and space were neighboring.

Furthermore, the intervals between the subjects were usually filled with different geometric patterns (Fig. 3).

While these works of art were being created, the scientists of the Islamic East remained the main successors of the Hellenic ideas. Why then was it necessary to deform the pictured world and to avoid the portrait likeness with nature, as was typical for the Ancient Greeks? It turns out that people long before me fell to thinking about the same problem and they came to the following conclusion. The Islamic religion forbade the portrayal of living matter. It might be even considered as a truth—why is it necessary to compete with God? I am not sure who the author of this theory was, but it is one which has had great vitality.

Thus the problem was "explained with God's help." It is evident that in such cases it is much easier for the representatives of some other tradition to invent a new explaining theory than to examine the artwork using the language of its own traditions. It is necessary to remember this fact when an outside interpretation of the classics of folklore has results which border on parody.

Besides these rather curious questions, like every other student I was busy with more direct duties such as mastering geology and adjacent sciences which were necessary for the practice of geology. Among these latter there happened to be crystallography, whose basic task was the "symmetry determination for the wooden polyhedra." It was such a boring task that the number of those wooden pieces became noticeably less by the end of our course than at the beginning. No matter how tremendous was our geological imagination we could not understand where this symmetry with its nonreal elements came from. When we had finally overcome those difficulties we found ourselves noticing them everywhere. Indeed, it was quite enough to consider the carpet patterns in order to demonstrate all necessary symmetry elements and all their functions. We found them in mathematical functions, in biological objects, in technical subjects, in music, in poetry, etc. In one word, we could find those symmetry elements in all visible things that may gladden one's eye and ear. Symmetry is explicitly or implicitly contained in the whole, which could be divided in equal or approximately equal parts; one could also find it in the result of fruitless attempts to create the whole from its parts.

We also marvelled that there existed nonsymmetrical things, possibly with a deeply hidden symmetry. Why was it so? Intuitively one could conclude that symmetrical things are more stable, but experience shows that they are much more easily noticed and with much more difficulty forgotten. Thus it is profitable for things to be symmetrical and time itself sooner or later makes them symmetrical.

In one word, the most accessible regularity in things and phenomena is their symmetry. And it is not a matter of accident that the ancient Greeks had noticed it in that faraway morning with their fresh gazes and raised it up to be the rationale of beauty, the basis of harmony. It is quite amusing that the citizens of the colonies of ancient Greece could come up with the idea of symmetry between the structure of language and things in nature. How did they learn to recognize that things and phenomena can not only be symmetrical in themselves, but also in relation to each other?

Fig. 3. From a 16th century manuscript (Archives of the Azerbaijan Academy of Sciences).

It is known that the Greeks had a very particular attitude toward their Gods; they loved and laughed at them, they abused them, and at the same time considered them to have all human virtues and shortcomings. The Gods were idle and had many problems of their own. Perhaps that is the reason the Greeks were determined to understand the laws of world creation without the help and collaboration of the Gods. To understand things through their elements, to measure them with their inner units, based on the belief that the world is atomistic, that there exists an overall harmony; that there exists symmetry between phenomena and numbers, the relation between numbers and phenomena being the most suitable language for the description of phenomena. This language was the basis for science. Every deviation from these lines always led to metaphysics.

Thus symmetry deserves to be the subject of periodic festivals[3], so that in their lofty atmosphere one can search for the relation between different manifestations of the human spirit based on symmetry. If I am not mistaken the present collection is the first correspondence festival on a truly international scale.

Symmetry has certainly not gained its present position as the basic concept of crystallography without the help of mathematics. Meanwhile, mathematics has significantly enriched itself and become more flexible and acute as the profound language for the description of natural phenomena.

Several times I have presented popular lectures before students and other audiences about the things we are trying to do in our labs during the pauses between the administrative tasks. And each time I was assured that such lectures must start with the general problems of science and its basic concepts. It is not too hard to see that in doing so we ourselves are climbing up the evolutionary stairs of knowledge. In the case being considered it is approximately as follows.

Everybody knows that there exist organizations. They are good enough when they possess the qualitatively new properties which are absent in the separate constructive elements; in that case they work steadily. Thus the organization as a whole is qualitatively more than the sum of its elements. The elements in this case must be compatible, they ought to possess the necessary qualities, which are clarified by the organization. It is also necessary to locate them in space (the spaces may even be abstract) in a certain mode, to tie them with the unique scheme and to establish the bond character in certain fragments.

One might also supplement this scheme with some new details, but it is better to say that the organization as a unified performing system is characterized by its structure. If we know the elements constructing the system and the way to construct it this means that we know its structure. The structure is the grammar of the system. The structure is the peculiarity of the system that distinguishes it as a whole on the background of a greater system; it is a property of the system. The structure is the peculiarity of the system which makes it possible to compare it with other systems no matter what the nature of their elements, etc.

How can we define symmetry itself so that one would not be obliged to describe it each time or to indicate that "that is a certain type of symmetry"? Symmetry is a property of the system, and as with every other property it is estimated as the reciprocal behaviour of the elements of the system at certain interrelations. These interrelations may be all possible rearrangements or reconstructions which leave the structures identically equal to themselves.

It is possible to measure the symmetry properties of the system. It is necessary to determine all rearrangement operations which return the structure to the initial state or, according to Shubnikov[4], who was the founder of modern crystallography in the U.S.S.R., just to determine the number of the equal parts which are formed when the structure is divided. It is obvious that the symmetry measure will depend on the meaning of "equality" and on the type of interaction which results experimentally to equality. For example, for the colorblind the patterns on carpets will have a higher symmetry. Likewise, X-rays are not sensible to isotopes and nuclear spins; for example, they will not distinguish the different hydrogen isotopes at low temperatures.

In the scientific literature symmetry is sometimes used as a synonym of invariability. Invariability and invariants are more suitable concepts when discussing the conservation laws[5]. For example, it is better to say that in translation or coding the invariant is the semantic meaning of the text.

Of great importance is the relation of symmetry to such concepts as order, entropy, microhomogeneity, complexity, and stability. It is clear why many investigators identify symmetry with order, while indeed the increase of symmetry makes the structure less regular. It is obvious from elementary considerations that for a constant total number of elementary structures the more symmetrical state requires equality, the nondistinguishable condition, of the greatest number of elements. As a result of the increase of symmetry the number of varieties of elements, as well as the complexity and stability, decrease and the (configurational) entropy correspondingly increases. All this is the consequence of determination of system information content[6].

A decrease in the number of varieties of symbols leads to a decrease of the negentropy information quantity, i.e. to the increase of entropy. As is known, the Shannon equation for information quantity determination and the Boltzmann equation for energetic-structure entropy determination are antisymmetric to each other. Symmetry also increases entropy in the latter case. Of course, the similarity between these entropies is not complete. In systems where the temperature as a concept has meaning the configuration entropy is constant, i.e. does not depend on temperature. It changes discretely only when the temperature is not furthermore utilized by the structure and the latter has to rearrange and to become able to use the energy rationally.

In such a way symmetry is, as a rule, connected with stability and steadiness. The facts which indicate the stability of the higher symmetry phases at higher temperatures, i.e. the stability of high temperature modifications, prove not the trend but the rule, which is based on the connection of symmetry with entropies. It should be remembered that in all this discussion we have silently assumed constant values of other parameters such as the character and strength of interaction, which is the main regulatory mechanism.

Complexity is the other concept which is needed in order to compare and codiscuss the problems of science and art. It is directly proportional to the information content[7].

Everybody intuitively knows that there exists a limit of complexity after which the system collapses. How shall we determine that limit? The temperature as a stimulus for the symmetry and microhomogeneity (the ratio of volume to the number of indistinguishable elements) growth is incompatible with the formation of complex chemical compounds.

Consequently the stability decreases, which is the result of the structure's complexity, must be compensated for by the simplicity of the medium structure, by introduction of the utilized energy into the structure of the elements or by giving it the machine-like dynamic structure. Such a structure is necessary for the consecutive appearances of the structure in determined states.

Thus the complexity of biologically important molecules is justified by dynamics of consequent formations and conversions and it is finally compensated with excess. Such considerations underlie an understanding of intuitive or conscious scientific and artistic constructs and the clarification of their general peculiarities.

Of significant importance is the problem of the structure and symmetry of our knowledge of reality. This is a special question. I shall only mention that we approach the whole structure of reality in two antisymmetric ways: analysis and synthesis. In the first case we start from the totality of the facts of reality and as a result the symmetry of our picture of reality is gradually increased. Meanwhile, we are sincerely happy watching the clarification which unfolds.

In the second case we start from the single fact with the most clear fragment structure and complicate it by decreasing its symmetry according to some other facts. And while doing this we are again happy about the clarification because such concepts as chaos and nothingness are equally nonavailable for our understanding. In both cases the number of facts necessary for the construction of the structure of reality is decreased.

Such an optimal structure in its possibilities for prediction contains the facts in their potential form, their number being incomparably greater than before the construction. This is obviously the reason why simplicity and symmetry are sometimes perceived as criteria for truth.

For the aesthetic action of the results of science as well as of artistic searches it is important to know from which complexity this simplicity has resulted and how complex it was. That is why for correct evaluation one needs the prehistory of the achievements. In our days the CPs are very impressive for mathematicians and crystallographers because they know pretty well what stands behind this simplicity. Surely simplicity and symmetry should be considered an advantage of the structure, all other merits remaining the same. Before the application of direct methods and computers in structural analysis it was necessary to construct the structures. And almost always the beautiful, symmetrical, simple versions appeared to be true, all other characteristics remaining the same.

In art only one method of artistic construction is recognized. Thus artists in a wide meaning of this word possess synthetic minds, and if they are lucky they manage to create a new world, a new structure, a new unity, whose elements by their nature may be abstract figures as well as fragments of observed reality.

Every time we come into contact with a new artwork we have to become analysts for a short time. What we first see are the elements the artwork is constructed from, then the symmetry, then the structure, the logic and the character of the connections between the elements, and finally the action of the artwork and the way it interacts with our concrete condition. This reminds us of the steps in decoding a text which is written in an unknown language, in a system of unknown signs and with (to us) an unknown grammar. A person may stand directly before a strange culture, or before an artwork of an abstractionist (or one who imitates him), and still remain unable to understand it. If this is so common, why are there not geniuses who code their "artworks" in such a way that after some time even they are unable to decode their work? In reality this is not the normal situation. If it were a general rule that artworks not be decodable, then they would not have social value.

Your attention is not attracted to those things which are pretty well familiar to you, about which you already have a comprehensive knowledge, the understanding of which does not require any mental efforts; nor is it attracted to those things which are absolutely unknown, which remain lost in the noise of the surroundings. Information ought to be of a certain value for you, you should be conducted to it by the form being perceived by you, and at last you must expend a certain mental effort in the process. But, of course, your admiration for an artwork and the generosity of your compliments are not sufficient. If the psychological structures and the vocabularies of the people belonging to the same culture were equal, the artist would be capable of evaluating his work himself and then the content of the information, the symmetry, would be the determining criterion of the beauty of the stereotype-minded people. But beauty and every other emotional response are born as a result of the contact of the aesthetic sign with the concrete individual.

Therefore information, as well as its structure and symmetry, gains new colors. The information content is perceived along with the corresponding associations as the content. In the same way, structure with its symmetry and corresponding associations is converted into form. Briefly speaking, content is the information that might be expressed, coded, or transported by the given form. Approximately by the same reason we relate in different ways to artworks even if they are created inside our culture within the limits of its traditions.

But why do we repeatedly return to certain aesthetic signs throughout our lives?

First of all, they have several levels, their structures being inexhaustible and organized according to the hierarchy principle like objects in nature.

Grouping the elements in a different way, or according to foreknown plans, you may see the change of the hierarchy structure, the regular transfers, their interesting interactions. The mind supplements the structure with the connections which are missed by conscious perception. So you gradually discover how the deeply realised context continually modifies the meaning of the elements and discovers their new potential possibilities. With each new level you enrich your own vocabulary and understanding with new analogs, new isomorphisms. All this can be enriched with the asymmetry of the vision. And what is very important you may always watch the increase of the whole compared with the sum of its parts. Of course, you yourself also improve with time, becoming enriched with experience and knowledge. But this can in no way add anything to the artwork that is not actually already in it.

Thus, because of the economy in the variety of elements, the resulting ambiguity of the individual effects of these elements was always one of the unwritten rules of creative art.

Thus it seems to me that with my semiphysical and semilyrical digressions I have been trying to give the key for an understanding of my personal interest in CPs and of the reasons why they were and still are perceived as aesthetic signs.

As I have mentioned above, before the direct methods in structural analysis, before the use of modern computers, the decoding of crystal structure required all the phantasy, experience and intuition of the scientist. The volume of the elementary cell, the symmetry elements, the number of molecules and atoms in the cell had all been estimated experimentally. But non-standard methods were always used in order to identify how the atoms and molecules tightly filled the volume of the elementary cell.

Thus the people working with organic and complex compounds of known discrete structures were dealing with the search for packings which would fit the minimal one considering the known limitations. In the case of inorganic compounds Pauling's rules did help significantly. It was necessary to fill space with polyhedra corresponding to the composition considering the Pauling's rules, which forbade certain modes. The results of these searches were versions and alternatives. Experiment could reject one of them, confirming the other.

Sometimes the real structure had a quite different character and was bothering and terrifying you, because you had missed such a simple structure. The question arose: Why was the structure not realised in the way you had proposed? Nowadays the analysis of the structure by automatic diffractometers, powerful computers, and well-done direct methods is carried out much faster and more easily. But such problems as why the hypothetical structures are not realised, how it is possible to evaluate the probability of their realisation, and how to choose the conditions for their realisation are not solved even now.

Briefly speaking, in those days we were dealing with the creation of crystallographic three-dimensional patterns. Such an opportunity is of interest from the point of view of the mentioned problem. Considering what was mentioned above it must be clear how great was my surprise and admiration when I discovered for the first time the CPs on the Barda mausoleum (Fig. 4) which was built in 1322. The cylindrical surface of the mausoleum is covered with a pattern whose repeating element is "Allah" written in the Arabic alphabet. Its plane group P4, which was not formally known to the artist as it was known to us in our work, came about as the result of tight packing of the word Allah. But it cannot be that the Barda mausoleum was the only one of its kind. Such a mode of creating patterns could not exist without its past, as well as without its successors.

It turns out that another good example is in Nahichevan, in the Karabaglar mausoleum (Fig. 5), which belongs to the school of the well-known architect Adjami[8]. In the Shirvanshahs Palace in Baku (15th century)[8] there is an architectural medallion with six "Ali," three of which are written in the shape of hollows in the stone, and the other three on the juts between the hollows (Fig. 6). The old masters knew well enough that it was possible to achieve equality between the background and the foreground; having done so they came up with antisymmetry, the black and white version of colored symmetry. As is known nowadays this achievement is related to that of Escher[3], who is world renowned for his CPs.

So it happened that all my "discoveries" had been long ago described by historians of our architecture, but naturally they did not pay attention to their crystallographic peculiarities.

In 1969 Dr Alan Mackay of London University kindly helped me get acquainted with illustrations of patterns from all over the world in the University library.

To my surprise, the CPs in the Middle Ages were widespread in Central Asia and Asia Minor[9,10], in Southern Azerbaijan, and in Afghanistan (Figs. 7–16). In all these countries words were used as design elements for the creation of CPs, and along with words geometric figures (polygons) were used (Figs. 8–10, 12, 15). As is seen from the figures these patterns may be considered as different examples of periodic tiling of the plane with polygons. And it is not surprising that they happened to be isomorphic with the inorganic crystals which have

Fig. 4. "Allah" in a kufic script as an element of CPs in the Tomb in Barda, Azerbaijan (1322).

Fig. 5. Example of kufic script used for decorative purposes. Karabaglar, Azerbaijan, 18th century.

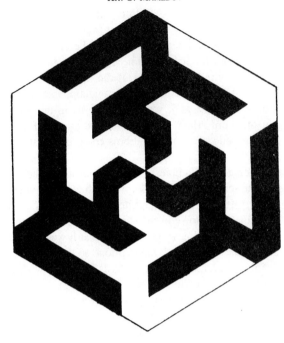

Fig. 6. Six "Ali" in kufic script. The Palace of Shirwanshahs, Baku, 14th–15th centuries.

been decoded in our own time. It should be noted that the area of distribution of patterns of this type is wider than that of the first ones.

From the fact that words and phrases were widely used as pattern elements one might conclude that indeed the inclination toward geometric patterns is connected with the Islamic religion. Here arises a natural question: If this is really so, then such a trend should not be observed before Islam. Is that so?

A review of pre-Islamic traditional fine art has shown that in Arabia, Iran, India, and partially in Afghanistan and southern parts of Central Asia, it was mainly anthropomorphic, while for the Hun and Scythian area more characteristic is the so-called scythian beast style[21],

Fig. 7. Crystallographic patterns on ceramics. South Turkmenistan (Koksur, Karatepe), 3rd millennium B.C.

Fig. 8. A wall decoration. Egypt (1094).

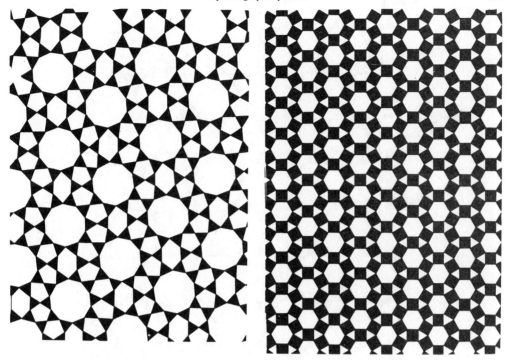

Fig. 9. A wall decoration from Jami Mosque. Isfahan, 16th century.

Fig. 10. The widespread ornament of architectural monuments of Azerbaijan and Turkey (18th century).

which has features of CPs[11] (Fig. 17). The sign in Fig. 19 was discovered in the Kazakh region of Azerbaijan[12]. Specialists believe that its age is about 4000 years. Its semantics is unknown. The same sign has also been discovered on the scythian axes in the Central Caucasus[13]. Along with this sign on the same axes some patterns were also discovered which were similar to the infinite number of patterns on the Kufic signs (Fig. 20).

But how had the masters of the Middle Ages in Azerbaijan inherited these modes of creating patterns? Please fix your attention to Fig. 21, where the ''Ali'' pattern is represented. You may notice that in its symmetry group and method of organization it repeats the Kazakh sign (Fig. 19).

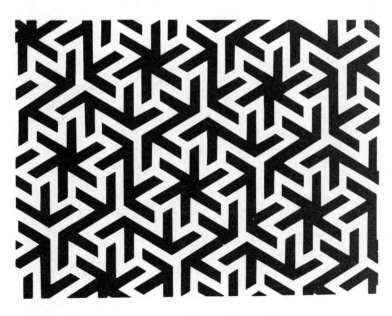

Fig. 11. Fragment of wall decoration. Sirajli madrasah, Konya, Turkey (1243).

Fig. 12. Fragment from wall decoration of Sahib Ata Mosque, Turkey (1279).

While observing the exhibits of the Novosibirsk ethnography museum I was greatly surprised to find that sign on one of the embroideries made by the Siberian people. This summer, Dr A. Mackay informed me about his discovery of a pattern in England which was constructed of the same sign. He believes that it is of Mediterranean origin. It is still incomprehensible why it has such wide dissemination.

Thus the CP had existed some thousand years before Islam. It might have been that Islam was tolerant enough about the revival of this sign in the Middle Ages. The fact that this style in the dissemination region was not imported is confirmed by numerous facts. But always keep

Fig. 13. "Muhamed" and "Ali" in a kufic script as the element of CPs. Sultaniye, Tabriz (1307–1313).

Fig. 14. Fragment from wall decoration of Bibi Hanum Mosque. Samarkand (1399–1404).

Fig. 15. Medieval CP similar to the structure of numeral biotite.

Fig. 16. The motif of the crystal structure of 1,3,5-tryphenylbenzene and the widespread ornament in the medieval Islamic East.

Fig. 17. Examples of decoration from an ancient Altay work (middle of the 1st millennium B.C.).

Fig. 18. An ornament on ceramics. Tomsk, West Siberia (end of the 4th–beginning of the 1st millennia B.C.).

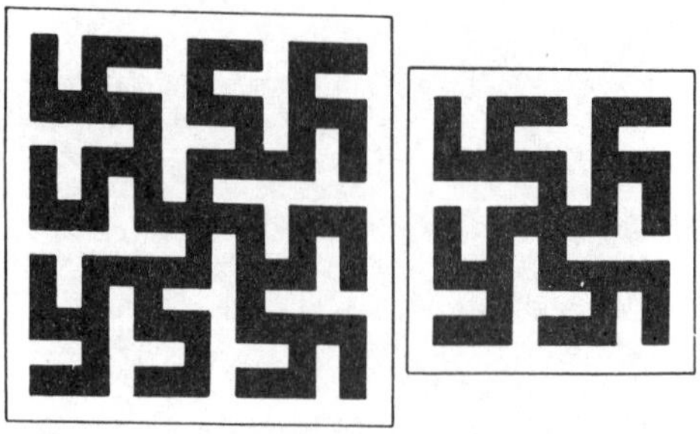

Fig. 19. Seals from Kazakh, Azerbaijan (end of the 2nd and 1st millennia B.C.).

in mind the predominant character of dissemination. Zoomorphic pictures may be found even in these regions. Such examples of crystallographic patterns can be also discovered in those world regions which are not typical for them[14]. The closest analogy exists between the patterns of the Siberian regions and the patterns of American Indians. A more careful comparison should be made of them.

    I was greatly surprised also when looking through the *UNESCO Courier*, where I found the article by Larichev[15]. There I discovered that the sculptures shown in the article were

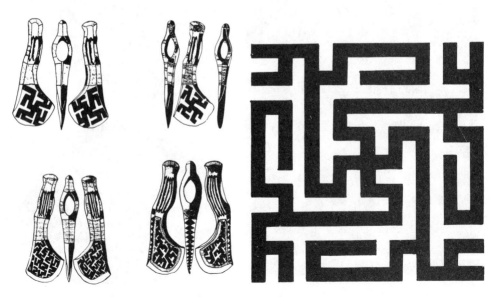

Fig. 20. Bronze axes from Tily, South Osetia (12th–10th centuries B.C.).

Fig. 21. Four "Alis" in kufic script. Isfahan, 14th–15th centuries.

compiled from complementary forms of tortoise and mammoth, mammoth and bison. According to the author these things are about 35,000 years old. It is quite amusing. It is evidence (and the author also thinks so) that the culture is not an imported one—it is of local origin. The same opinion exists about the Scythian–Siberian-type culture[21], which is in fair agreement with the character of patterns of this culture.

A systematic review of the world CPs in general and of the aforementioned regions in particular probably would be very useful for archeology. It would be interesting to observe the evolution of development of this trend and find out its psychological preconditions: Was there any concrete knowledge as a foundation for this trend? It is unbelievable that there was not. Now fix your attention on Fig. 22 with the picture of four horses by Rza Abbasi (1587–1628). Having used the symmetry (the twofold axes) the artist has achieved his final goal by picturing only two horses—you only need to turn the picture by 90° to see the other pair of horses. And you might not suppose that this result was just a matter of accident. It is interesting that using symmetry for the organization of the picture seems to result in a saving in the details which are necessary for it. It is just unbelievable that nature had not used this peculiarity of symmetry.

As already mentioned, the scientists of the Islamic East were successors of the knowledge of the Greeks. It is interesting to ask whether they could have fixed attention on such a concept as symmetry. I feel that they did not. It turns out that the Greeks who were satisfied with the latent symmetry in their artworks were the inventors of the concept of symmetry, while the peoples who had been preserving their traditions via CPs for thousands of years had not come to the symmetry concept. This means that the imitation of nature on the level of principle was not sufficient for the discovery of this principle.

Probably everything is connected with the question that the Greeks were putting themselves: "Why do things happen to be so beautiful, and how can they be constructed?" Such a question may result in such concepts as symmetry, harmony, etc.

In the poetry of the Islamic Middle Ages it was a popular opinion that God, having created the world, had then hidden the very method of its creation from the people so that they would not become the creators of new worlds. But the Greeks had decided to find out this secret against the will of the Gods and started competing with them.

In our days the restoration of CPs is connected with the name of the famous artist Escher[17]. The crystallographers consider him to be the inventor of colored symmetry. His basic contribution is considered to be the discovery of the equivalency of the background with the foreground. In Ref. [3] it is literally said that before Escher nobody could even comprehend this idea, but we can convince ourselves by numerous examples that this is not really so. The only thing to realize is that the ancient people had not known that they had really discovered something new, nor had Escher himself. He was only familiar with the Spanish versions of CPs and he did them qualitatively anew and gave us the idea of existence of an almost forgotten interesting trend. Escher is famous not only for his CPs; he made many other works which have attracted the attention of outstanding representatives of modern science. The influence of Escher's ideas is gradually spreading. CP fans have appeared in Canada[18]. We in our laboratory have also taken some interest in these ideas (Figs. 23–28).

The dissemination of CPs is favored by science. The reason is that their ideas are close to each other by their nature and the problems. The CPs are also very promising as architectural and textile patterns.

The other interesting direction of research is the combination of patterns with nonevident symmetry, i.e. nonlinear modulation of common patterns. One may find many such examples in Escher's works. If I am not mistaken such variations have also been experienced even before him.

It should be noted that in the process of constructing CPs the symmetry appears as a result of combination but not as a means of its formation, as is characteristic for other patterns. It is also interesting that the creators of CPs seem to avoid the combination with a mirror-symmetry plane. Also widely used was the antisymmetry plane. As is currently known, a mirror-symmetry plane gives a pattern a static appearance.

Of course, people with a skeptical attitude to the CPs refer to the fact that they have too severe limitations for the modern artist. In a sense this is really so. There is no prototype, and the packing of the elements is not always easily done, even being impossible for most elements.

Fig. 22. *Four horses*, by Rza Abbasi (1587–1628).

Fig. 23. "Vatan" in a kufic script as an element of CPs by Kh. S. Mamedov.

CPs are constructed almost by the principles of rhythmic musical patterns. But here the redundancy is much higher than that in music. Remember that the pattern is always viewed in context with greater systems such as large constructions, the surroundings, etc., and that we observe it from a certain angle of sight. It becomes clear that this is a sophisticated simplicity and that eventually almost everyone might understand that it is not so easy to achieve such simplicity. The structurally modular CPs in classical constructions also provide an inner scale for the whole complex. The importance of the modular agreement of the pattern and of the complex including it is quite obvious.

The CP, like music, put in order the way of thinking and direct it along a certain course. By their aesthetic action and by the construction principle they are close to the aesthetic objects

Fig. 24. "Nizami" in a kufic script as an element of CPs by H. N. Nadjafov.

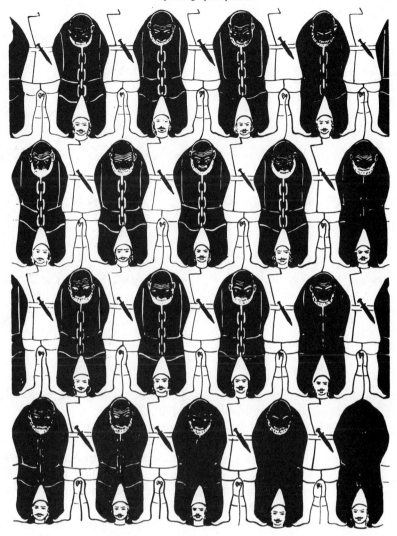

Fig. 25. *Unity*, by Kh. S. Mamedov.

of natural origin and to the best creations of such artists as Mondrian, Klee, Vasarely, and other constructivists[19].

As is known at present all basic natural laws are of a prohibiting character. Everything that is not prohibited by them is possible. Thus there exist tabus and rules of the game forbidding certain actions. But actually one can find here large possibilities. As a matter of fact, as in modern science, the whole specificity of traditional arts is connected with the character of those tabus and game rules. Nevertheless, in order to raise the CP up to the level of aesthetic sign it is still necessary to be an artist. Le Corbusier, with his words that order is the quality of geometry and a man is characterized by his attention to order, seemed to refer to the importance of the human attitude to symmetry in order to understand the archeological and architectural monuments. Indeed, besides the fact that symmetry is a significant new parameter like chemical composition and other specific features for describing and characterizing memorials, and other practical problems, it may be very useful to clarify the character of the intellectual relation of peoples to biological and aesthetic media at different stages of their development. I would like to demonstrate the practical usefulness of our knowledge about CPs with at least one example. When people restore memorials destroyed to their very foundations according to the recovered elements and their statistics they may restore the patterns and compare them with analogous monuments.

As I have said previously, our classical carpet by itself is sufficient to explain the meaning and the significance of symmetry and related concepts. In other words, these patterns may be

Fig. 26. *Falcons*, by Kh. S. Mamedov and I. R. Amiraslanov.

successfully used for explaining the plane groups of colored symmetry, the periodical packings of the plane with self-complementary elements, etc.[20].

The significance of the emotional atmosphere during the process of studying and mastering the new things and concepts is obvious. Of course, I could not embrace even superficially all problems related to CPs, though I had a great desire to do so. I hope that what I have mentioned is quite enough to admit the existence of such interesting phenomena in ornamentalistics.

Its systematic study by the methods of the exact sciences is of great interest for understanding nature and clarifying the basic principles of modern traditionalistic applied art. The importance of the latter was expressed by Paul Valéry in his letter to Le Corbusier: "One ought not even to start the search for the purest forms in art without basing it on the examples from the heritage of the past."

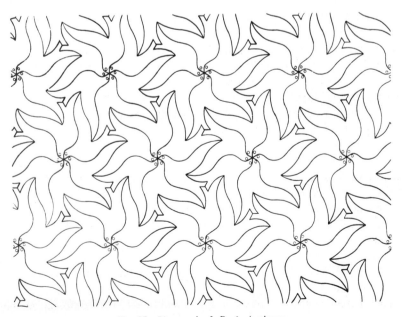

Fig. 27. *Pigeons*, by I. R. Amiraslanov.

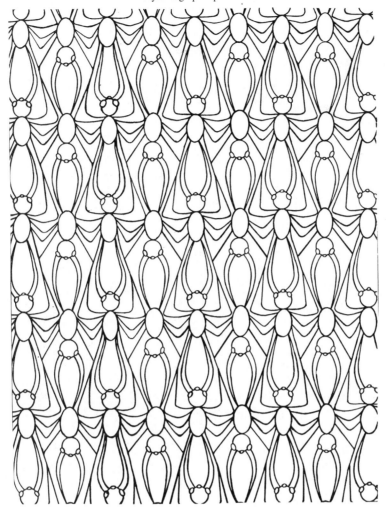

Fig. 28. *Beetles*, by Kh. S. Mamedov and I. R. Amiraslanov.

REFERENCES

1. Le Corbusier, *Architecture of the 20th Century* (translated from French). Progress (1970).
2. K. Kerimov, *Sultan Mukhmed and His School*. Iskusstvo, Moscow (1970).
3. *The Ornaments of Symmetry*. Mir, Moscow (1980).
4. A. V. Shubnikov and V. A. Koptsik, *Symmetry in Science and Art*. Nauka, Moscow (1972).
5. E. Vigner, *Studies of Symmetry*. Mir, Moscow (1971).
6. L. Brilloin, *Science and Theory of Information* (translated from French). Moscow (1960).
7. Abrahm Mol, *Theory of Information and Aesthetic Perception* (translated from French). Mir, Moscow (1966).
8. M. Uceinov, A. Bretanitski and A. Salamzade, *History of Azerbaijan Architecture*. Moscow (1963).
9. *Architectural Monuments of Central Asia*. Photographic Album of A. M. Pribitkov Planeta, Moscow (1970).
10. Selcuk Mülayim, Anadolu Türk mimarisinde geometrik süslemeler, Kültür ve turizm bakanliği yayinlari: 503, Ankara (1982).
11. Kh. S. Mamedov, I. R. Amiraslanov, G. N. Nadzhafov and A. A. Murasaliev, *Memory of Ornaments*. Azerneshr, Baku (1981).
12. I. G. Narimanov, Clay stamps from western Azerbaijan. *Material Culture of Azerbaijan*, vol. VII. Elm.
13. B. Z. Tekhov, *Central Caucasus from the XVI-X Centuries B.C.* Nauka, Moscow (1977).
14. P. Fortova-Samolova, *The Egyptian Ornament*. Artia, Prague (1963).
15. Vitalii Ye. Larichev, *Ancient People of Siberia*. Courrier, UNESCO, December (1980).
16. M. P. Griaznov, *Arzhan*. Nauka, Leningrad (1980).
17. C. H. MacGillavry, *Symmetry Aspects of M. C. Escher's Periodic Drawings*. A. Oesthoek, Utrecht (1965).
18. F. Brisse, Bidimentional symmetry and Canada. *The Canadian Minerologist* **19**, 217–224 (1981).
19. Richard P. Lohse, Standard, series, module: New problems and tasks of painting. In *Module, Symmetry, Proportion, Viston & Value Series* (Edited by Gyargy Kepes). Stadia, London (1966).
20. Emil and Milota Makovicky, Arabic geometrical patterns—a treasury for crystallographic teaching. *N. Jb. Miner. Mb. Jg.* **H.2**, 58–68 (1977).
21. S. I. Rudenko, *Art of Altai and of Near Asia*. Vostochnaia Literatura, Moscow (1981).

Comp. & Maths. with Appls. Vol. 12B, Nos. 3/4, pp. 531–545, 1986
Printed in Great Britain.

0886–9561/86 $3.00 + .00

SYMMETRY IN A NATURAL FRACTURE PATTERN: THE ORIGIN OF COLUMNAR JOINT NETWORKS

NORMAN H. GRAY

Department of Geology and Geophysics, University of Connecticut, Storrs, CT 06268, U.S.A.

Abstract—The remarkable regularity of the rock columns outlined by the cooling related contraction joints in lava flows at sites such as the Giants' Causeway, Ireland, Fingal's Cave, Scotland and the Devil's Postpile, California is well known. Columnar joints are the only system of natural fractures to approach an optimal hexagonal honeycomb-like pattern. 3-connected vertices in near-surface natural-fracture networks are almost exclusively orthogonal "T-type" junctions. Columnar joints owe their unique quasihexagonal symmetry and the prevalence of six-sided columns to the presence of "Y" junctions. Initially, the cooling-related fracture network at the upper and lower surfaces of a flow are dominated by T junctions and are similar in appearance to mudcrack patterns. However, as the tips of these cracks grow towards the center of the flow to relieve the thermal strain which develops there during cooling, the poorly positioned ones are eliminated or modified to yield a more regular network dominated by Y rather than T junctions.

INTRODUCTION

Large-scale polygonal patterns, often of striking regularity, are found in a variety of geologic settings[1]. A familiar example is the arrangement of the shrinkage cracks which form when mud puddles dry up in the summer sun. Similar, although larger, fracture networks form the "patterned ground" of permanently frozen terrain in arctic areas[2,3] and the so-called "giant desiccation polygons" found on some dry lake beds in arid regons[4,5]. Although the regularity of these patterns is on occasion impressive, the most symmetric natural fracture network is formed by the cooling related joints in surface lava flows (Fig. 1).

The perfect fit and remarkable regularity of the polygonal columns outlined by columnar joints has always attracted notice. Localities at which the symmetry of the joint network is highly developed are classic tourist attractions. The Giants' Causeway, Ireland, and Fingal's Cave on the Isle of Staffa in Scotland are the best known, but equally good examples are found at Orange, New Jersey; The Devil's Postpile, California; Devil's Tower, Wyoming; Titan's Piazza, Massachusetts; Stappi, Iceland; and at the Organ Pipes, Victoria, Australia. The names themselves suggest some of the mystery and fascination connected with this natural phenomenon.

The uniquely regular character of columnar jointing demands an explanation and many have been offered. During the 17th and 18th centuries, basalt columns were confused with crystal forms. Indeed, in the bitter arguments over the origin of basalt at the end of the 18th century, the so-called "neptunists" pointed to these "crystals" as evidence that basalt crystallized from the hot brine of a primeval ocean! In the 1800s a variety of more reasonable explanations were proposed. One seriously considered idea was that columns were "frozen" convection cells; another was that columns outlined centers around which cooling lava congealed. Although both explanations have received some attention in recent years[6], the accepted view today is that columnar joints are tensional features which relieve the thermal stresses which develop during the cooling of a lava flow[7–9]. Cracks formed on the crust of modern lava lakes have been thoroughly studied and shown to be the result of the thermal contraction accompanying cooling[10]. However, the pattern formed by these fractures looks more like mudcracks than well-developed columnar jointing (Fig. 2). The origin of the symmetry so characteristic of true columnar networks is still an open question even though the cause of the jointing itself is well understood.

THE SYMMETRY OF COLUMNAR NETWORKS

Mallet[11] and Iddings[12] recognized that a hexagonal honeycomb-like arrangement of contraction cracks would maximize the area-to-fracture-length ratio. Based on a vague idea that nature would attempt to minimize the total crack area and at the same time relieve as much of

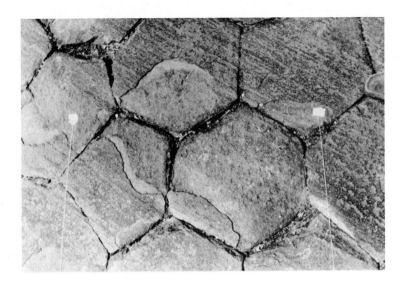

Fig. 1. Columnar joints at "The Devil's Postpile," California. The polygonal columns, seen here in cross section, are roughly 1 m across.

the thermal stress as possible, a hexagonal joint pattern would seem the ideal arrangement. However, the very uniqueness of columnar jointing is the weakest point of the argument. Mudcracks, tensional features themselves, are rarely (if ever) arranged in a hexagonal manner. To Iddings[12], the solution to the paradox lay in the uniformity of cooling in a sizable lava flow:

> . . . In a homogeneous mass the contractile force which produces cracks at certain distances will exert itself equally in all directions over a surface uniformly subjected to the cooling forces, and will at the instant of rupture act towards centers whose distance apart is dependent on the rate of cooling. If the mass is perfectly homogeneous the centers of contraction will be disposed over the surface with the greatest uniformity possible, that is, they will be equidistant throughout and the resultant fractures will be in a system of hexagons.

A version of this passage is found in almost all introductory geology texts and is often repeated in literature prepared for visitors to the popular columnar joint localities. Unfortunately, this explanation requires unbelievable coincidences. If every crack does not nucleate at precisely the correct location, in exactly the right orientation, they will miss each other and the resulting crack pattern looks more like crazed porcelain than columnar lava (Fig. 3).

Over the years other explanations have been proposed which supposedly avoid the statistical difficulties implicit in Iddings' model. Billings[13] suggested that cooling fractures nucleate around 120° Y junctions. Based on Ernsberger's[14] study of the mechanics of crack growth in heat-treated glass Spry[8] suggested that a single columnar joint might repeatedly bifurcate and produce a large number of Y junctions. Although the branching angle observed in fractured glass is rarely greater than 60° Spry argued that the homogeneity of the cooling in lavas might lead to the formation of Y junctions with interjoint angles close to the ideal 120°. These ideas are summarized schematically in Fig. 3. Neither really solve the statistical question; both simply postpone the problem to a later stage in the development of the crack network. Y junctions in Billings' model must originate in precisely the right spot with the correct orientation to produce a perfect network. Similar problems arise with a branching-crack model. The bifurcations must all occur at precisely the right location if orthogonal T-type junctions, which are absent in many columnar networks, are to be avoided.

These explanations do a disservice to the understanding of columnar jointing by over-emphasising their approach to hexagonal symmetry. In detail, even the best examples of columnar joint patterns bear only a superficial resemblance to a perfect hexagonal grid (Figs. 1 and 2). Interjoint angles are rarely 120°, the edges vary appreciably in length and the polygonal

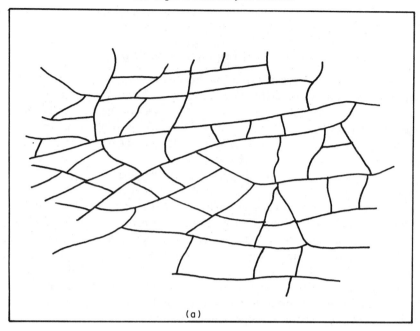

(a)

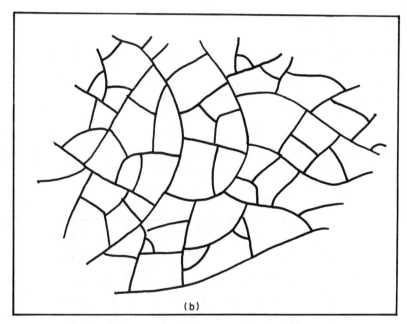

(b)

Fig. 2 (a–b). Natural fracture networks: (a). Pattern ground—thermal contraction cracks in permafrost [Traced from the cover of *Science* **66** (3901)]. Scale unknown but such polygons are typically tens of meters across. (b). Mudcracks, Mesozoic Shuttle Meadow shales, Hartford Basin, CT. The portion of the network pictured here is 5 cm across.

areas outlined by the joints are never all hexagons. Much has been made of the geometric shape of columns. Beard[15] and Spry[8] both noted that seven- and eight-sided columns are common at the better localities, whereas pentagonal columns are more prevalent at sites where columnar jointing is less well developed. The average number of sides is also correlated with the perceived "quality" of the jointing. The average at the best sites is close to six. Less spectacular localities may average only five sides per column. Networks having less than five sides per column lack the special columnar "look" that seems to have universal appeal.

Although the best developed columnar networks lack perfect hexagonal symmetry, they are regular in other respects. Every joint is perfectly straight and the individual columns are meticulously fitted together as if assembled by a master mason (Fig. 1). The symmetrical nature of a columnar network is most evident if contrasted with the genetically similar, but more irregular, patterns formed by mudcracks (Fig. 2).

JUNCTIONS IN COLUMNAR NETWORKS

All vertices in both mudcrack and columnar joint networks are 3-connected. In these near-surface fracture arrays the cracks rarely, if ever, intersect. The principal difference between the patterns is the nature of the triple junctions. In mud the cracks typically truncate each other at orthogonal "T"-like vertices. "Y"-type vertices, junctions at which none of the edges are truncated, are found in columnar networks. The better developed the jointing the more numerous the Y junctions.

The existence of two distinct vertex types is the underlying cause of the variation in the mean number of sides to the polygons in natural crack networks. As both junctions are 3-connected the number of vertices per polygonal area must average six. If however, we count vertices, or edges, by first disassembling the net a difference between T and Y junctions becomes apparent. In the original network each T junction is common to three adjacent polygons, but when separated only two of the three retain it as a vertex. The Y's, on the other hand, are shared by the same three polygons in both the assembled and disassembled states. The average shape of the separate columns in a columnar joint network is thus dependent on the relative numbers of T and Y junctions present. If J_T and J_Y are the proportions of T and Y junctions respectively, the mean number of sides per column is simply (Fig. 4 and [16])

$$2(2J_T + 3J_Y).$$

Mudcracks contain only T junctions. Hence the average mud polygon has only four sides. Columnar networks consist of anywhere from 0 to 100% Y junctions. The best developed have no T junctions and the average column is a hexagon. Columnar networks containing T junctions will, as a consequence, average less than six edges per column.

ORIGIN OF JUNCTIONS

Since the geometric form of the columns is the most characteristic feature of true columnar networks the source of the junctions, especially the Y's, is a key to understanding the origin of columnar jointing in general.

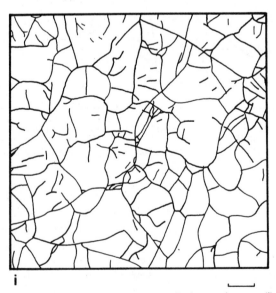

i

Fig. 2 (c). Columnar joint networks as seen in sections perpendicular to columns at (i) the crust of the March 1965 Makaopuhi lava lake (Peck and Minakami, 1968 (scale bar equals 3 m); (ii) Mt. Holyoke, MA (upper colonnade of the Holyoke flow); (iii) Columns of the Giants, Rt. 108, CA (these joints nucleated on a large block of foundered crust and are separate from the colonnades or entablature of the flow); (iv) The Devils' Postpile, CA (lower colonnade).

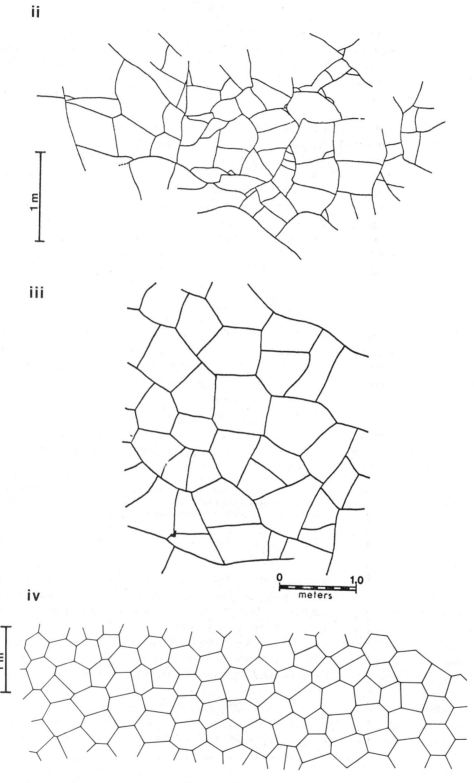

Fig. 2(c). (Continued).

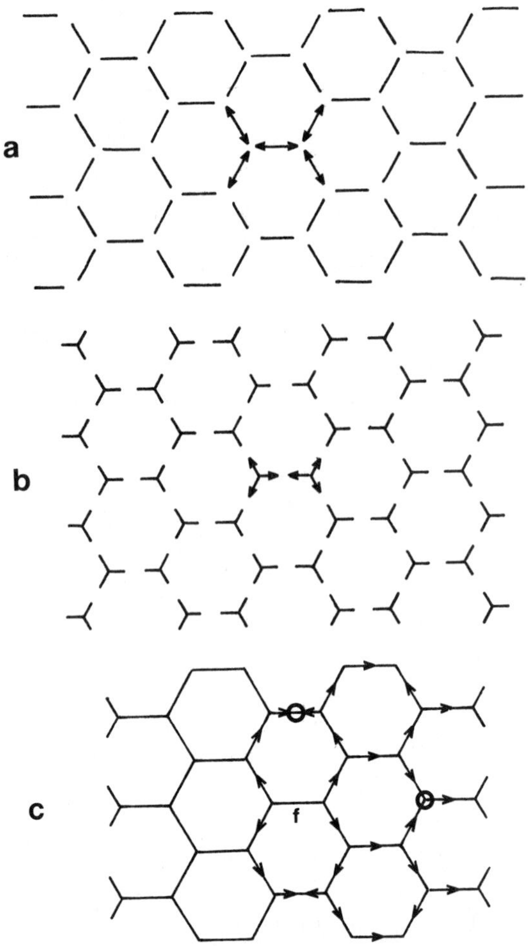

Fig. 3. Schematic illustrations of some suggested theories for the origin of columnar jointing. (a) Fractures nucleate and grow in both directions until truncated by another fracture. (b) Triple Y junctions nucleate, fracture, and then grow radially until truncated by fractures originating at neighboring sites. (c) A single original fracture (labelled f) bifurcates at regular intervals to produce a regular hexagonal pattern. Note in each case that a slight error in the location or orientation of any fracture destroys the regularity of the pattern and introduces either curved fractures or orthogonal T junctions.

T *junctions*

The origin of T junctions is no mystery. With patience one can follow the development of a T-junction network in any mud puddle after a summer shower. Individual cracks nucleate in the drying mud at roughly random sites and grow in both directions until truncated. A single crack relieves only a portion of the tensional stresses that build up in its vicinity. Stresses parallel to its length are totally unaffected. A growing crack tip orients itself to take maximum advantage of the local stress. In the vicinity of a preexisting fracture the largest residual stresses are perpendicular to its length. Cracks thus tend to truncate against preexisting fractures at near-orthogonal T-like junctions. T-type junctions, which superficially resemble Y junctions, may arise when two cracks approach each other simultaneously. Both alter direction as their stress relief fields overlap. The first crack on the scene is deflected but survives. The straggler is truncated, typically just past the point of maximum deflection of the first crack (Fig. 5).

Y *junctions*

True Y-type junctions are absent in most slowly formed two-dimensional fracture networks. Mud, spalling paint and ceramic craze are examples of material in which the cracks develop slowly as tensional stresses build up. If the fractures grow rapidly, as in quenched glass, they may well branch and produce a number of Y junctions. However, since the angle of bifurcation

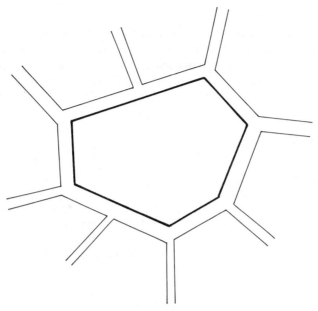

Fig. 4. Mean number of sides to the polygons in a disassembled 3-connected net as a function of the proportion of T and Y junctions. If J_T and J_Y are the proportions of T and Y junctions, G_k the number of k-gons in the *disassembled* net and V_3 the number of 3-connected vertices in the assembled state, $\Sigma kG_k = (2J_T + 3J_Y)V_3$. The factors 2 and 3 relate to the fact that each T junction is common to only two, whereas each Y junction is common to three of the disassembled polygons. Since each edge is shared by two vertices the total number of edges in the assembled net is $1.5V_3$. Euler's theorem relates the total number of polygons, edges and vertices in the assembled net: $\Sigma G_k - 1.5V_3 + V_3 = 2$. Using the first relation to eliminate V_3 the mean number of k-gons in the disassembled net is found to be $\Sigma kG_k/\Sigma G_k = 2(2J_T + 3J_Y)$.

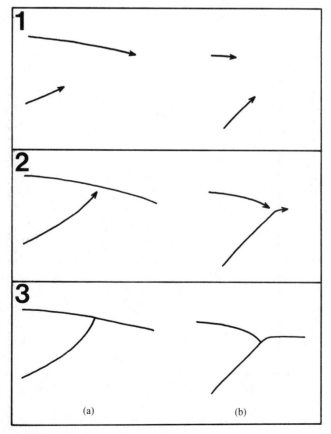

(a) (b)

Fig. 5. Origin of pseudo-Y junctions in a purely T-junction fracture array. (a) Formation of ordinary T junctions when a growing crack encounters a preexisting fracture. (b) Formation of a pseudo-Y junction when two approach each other simultaneously. Y junctions of this type are common in the upper colonnade joints on Mt. Holyoke, MA [Fig. 1(c)–i].

is rarely greater than 60° these Y junctions are not typical of those found in columnar joint networks.

How then are the columnar Y junctions formed? The answer to that question is found in the three-dimensional form of columnar joints.

Tomkieff[17] called attention to a three-tiered subdivision of jointing in flows with well-developed columnar jointing. In keeping with the neoclassic names often aplied to these sites he suggested the architectural terms "lower colonnade", "entablature" and "upper colonnade" for the three zones. The best developed jointing is found near the top of the lower colonnade. Columns in the entablature are much less regular and are typically arranged in large, spherulitic-like groupings of radiating joints (Fig. 6).

At the base of the lower colonnade orthogonal T junctions predominate. True Y junctions make their first appearance a few decimeters into the flow and progressively increase in abundance upwards relative to T junctions. At the classic localities T's are almost totally absent in the upper part of the lower colonnade. Where the exposure permits, the systematic elimination of T junctions can be followed in detail (Fig. 7). Near the base of the colonnade, junctions can be seen to have repeatedly switched from T to Y and then back to T as the fracture network grew into the cooling flow. Interjoint angles oscillate about 90° or 120° (Fig. 8). Joints can be traced from the top of the lower colonnade back to their site of origin at the lower contact, a

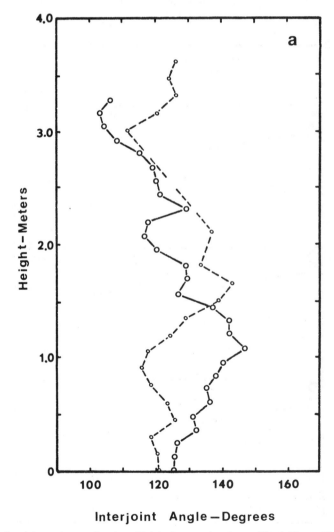

Fig. 6 (a)–(c). Interjoint angles as a function of height for some junctions at Titan's Piazza, Mt. Holyoke, MA. The origin of the height axes are arbitrary; the base of the flow is unexposed. Note the long-period oscillation around the optimal 120° interjoint angle.

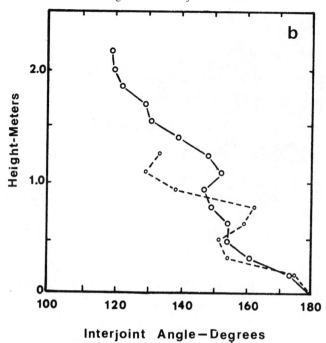

Fig. 6 (b).

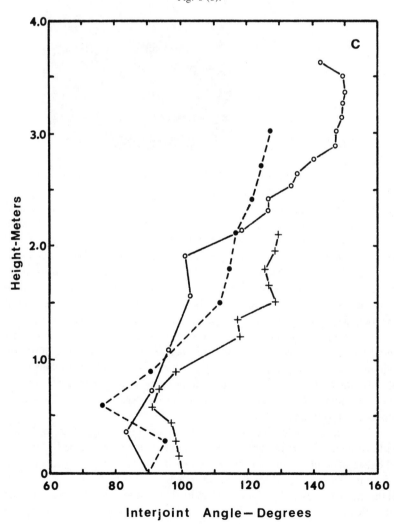

Fig. 6 (c).

Fig. 7. Y junction evolving from a T junction at Titan's Piazza, Mt. Holyoke, MA.

distance typically amounting to several tens of meters. Most joints originating at the base of the flow are truncated or die out long before reaching the upper part of the colonnade. Columns of the lower colonnade average about 1 m in diameter.

These observations suggest that an originally T-junction-dominated mudcrack-like fracture network advances slowly inward towards the center of a flow as it cools. In this way the tensional stresses are released as they develop. At any instant a joint will have progressed to the point where the energy required for continued growth is just balanced by the thermal strain stored in the adjacent rock. Since the stress fields associated with individual cracks overlap and interfere, especially in the vicinity of junctions, the joints must be in continual competition for a share of the limited energy available. Although held back by this interference, the most favorably positioned joint should be able to penetrate further than its competition and thus be in a position to take full advantage of the thermal strain accumulating in advance of the fracture network. Natural selection, akin to "survival of the fittest" in the organic world, will eliminate all but the most optimally positioned and oriented cracks.

The inward progress of the fracture network is not necessarily linear in detail. Ryan and Sammis[9] have interpreted centimeter-scale striations, sometimes termed "chisel marks"[7], found on the face of some columns as evidence of episodic advance. Cracks initially propagate by brittle fracture, but as the residual stresses wane they slow down and switch to a plastic mode until the thermal stresses build up once again. The brittle–plastic transition is accompanied by a slight deflection in the course of the fracture. The overall effect, however, is the slow inward growth of the columnar network.

Fig. 8. Junction of the entablature and lower colonnade columnar zones at Columns of the Giants, CA.

Since it uses only a portion of the thermal strain energy stored in its vicinity a "smart" T junction would find it advantageous to switch to a more efficient Y arrangement. Unfortunately, from a T junction's point of view, this option is not always available. The junctions are all interconnected and to simply change the interjoint angles at one site requires readjustment of the whole network. The competition is complex but its consequences simple. T junctions and poorly situated joints are systematically eliminated (Fig. 9). The network is eventually dominated by Y junctions. Continued competition eliminates curved joints and juggles the position of the junctions until the interjoint angles are as close as possible to 120° (Figs. 10 and 11). At this stage further evolution is energetically difficult. Natural joint networks seem unable to evolve beyond this stage to what is probably only a slightly more efficient arrangement—a perfect hexagonal net.

Columns in the entablature never develop the same perfection as those of the lower colonnade. T junctions are common throughout. As a result the average shape of a polygonal column in the entablature is considerably less than hexagonal. The typically 10–30 cm columns are also notably smaller than those of the lower colonnade.

Joints in the entablature originate in the upper colonnade and extend downward to meet the upward-growing columns of the lower colonnade (Fig. 6). In the last stage of their evolution they appear to have grown extremely rapidly to form large spherulitic-like groupings of irregular radiating columns. The almost explosive character of this growth is apparently due to the near-uniform temperature profile in the center of cooling lava flow. Once the central region is capable of cracking, any joint which finds itself slightly ahead of the rest of the fracture array will be able to quickly outdistance its competition and penetrate deeply into the unfractured portion of flow. The situation is similar to crystallites growing in a highly supersaturated solution. The center of the flow is effectively supersaturated in tensile stress and any fracture able to escape

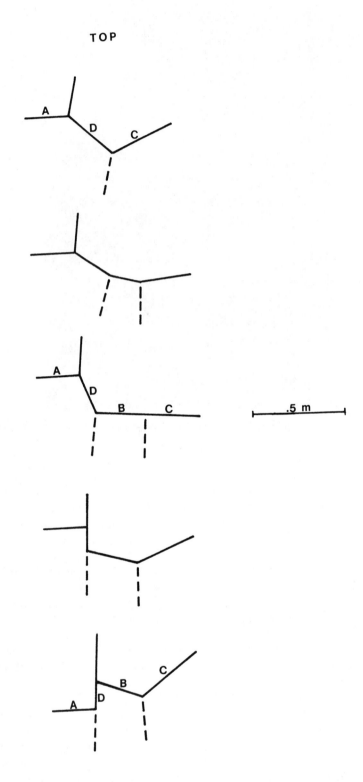

Fig. 9. Joint readjustments accompanying the elimination of a T junction at The Devil's Postpile, CA. Cross sections of a portion of the columnar network at intervals of 0.5 m. Note that Y junctions alter to T's before moving.

(a)

(b)

(c)

Fig. 10. Columnar jointing at The Devil's Postpile, CA. The lower contact of the flow is exposed at the base of photograph (a). Photograph (b) overlaps (a) to some extent but most of the area pictured in (b) lies above (a). The regular columns in (c) are close to the top of the lower colonnade, but the exact distance from the its base is unknown. Columns progressively become more regular towards the center of the flow. The pinch and swell structure displayed by columns in the upper part of (b) is a reflection of the complexity of the readjustments accompanying the wholesale elimination of the original T junctions as the fracture network grew towards the center of the flow.

Fig. 11. Details of the columnar jointing at The Devil's Postpile, CA. (a) "Puckers" in a column in the upper part of the lower colonnade. The indentations mark the site of a junction of a joint which extends into the column at the center of this photograph. This junction repeatedly oscillated between a T and Y until finally, as a T, it was eliminated. (b) Bent columns. The two column sets visible here curve into subparallelism at their mutual boundary, reflecting the interference of their strain release fields as the fracture array grew towards the center of the flow.

the interference of neighboring cracks will grow rapidly through the entablature. Since the normal stresses at the leading edge of a crack are due to the weight of the overlying material the source of entablature fractures is more likely to be the upper than the lower colonnade. Repeated bifurcations of the growing crack give rise to the spherulitic-like clusters of columns. Some of the newly formed fractures become truncated at orthogonal T junctions by earlier joints of the same branching network. As a result the entablature always contains a number of T junctions.

SUMMARY

Cooling-related contraction joints in lava flows form one of the most strikingly symmetric natural fracture patterns. Unlike most crack networks, columnar joints are characterized by Y rather than orthogonal T-type triple junctions. The average polygon outlined by fractures approaches a hexagon in the best-developed columnar networks. At the base of a flow Y junctions are absent, the average column has four sides, and the fracture pattern looks similar to mudcracks. Y junctions replace T junctions in a complex process involving modification, selection and elimination of joints during their growth towards the center of the cooling flow. The fracture network steadily evolves toward, but never achieves, the ideal hexagonal pattern which would result in the most efficient relief of thermal stresses built up during cooling.

REFERENCES

1. R. A. Nelson, Natural fracture systems: description and classification. *Bull. Am. Ass. Petrol. Geol.* **63**, 2214–2219 (1979).
2. A. H. Lachenbruch, Depth and spacing of tension cracks. *J. Geophys. Res.* **66**, 4273–4291 (1961).
3. A. H. Lachenbruch, Mechanics of thermal contraction cracks and ice-wedge polygons in permafrost. *Geol. Soc. Am. Special Paper No. 70* (1962).

4. R. Willden and D. R. Mabey, Giant desiccation fissures on the Black Rock and Smoke Creek deserts, Nevada. *Science* **133**, 1359–1360 (1961).
5. J. T. Neal, A. M. Langer and P. F. Kerr, Giant desiccation polygons of Great Basin playas. *Bull. Geol. Soc. Am.* **79**, 68–89 (1968).
6. L. H. Kantha, Basalt fingers—Origin of columnar joints? *Geol. Mag.* **118**, 251–264 (1981).
7. A. V. G. James, Factors producing columnar structure in lavas, and its occurrence near Melbourne, Australia. *J. Geol.* **28**, 458–469 (1920).
8. A. Spry, The origin of columnar jointing, particularly in basalt flows. *J. Geol. Soc. Aust.* **8**, 191–216 (1961).
9. M. P. Ryan and C. G. Sammis, Cyclic fracture mechanisms in cooling basalt. *Bull. Geol. Soc. Am.* **89**, 1259–1308 (1978).
10. D. L. Peck and T. Minakami, The formation of columnar joints in the upper part of Kilauean lava lakes, Hawaii. *Bull. Geol. Soc. Am.* **79**, 1151–1166 (1968).
11. R. Mallet, Origin and mechanism of production of prismatic (or columnar) structure of basalt. *Phil. Mag.* **4**, 201–226 (1875).
12. J. P. Iddings, Columnar structure in the igneous rocks of Orange Mountain, New Jersey. *Am. J. Science* **131**, 321–330 (1886).
13. M. P. Billings, *Structural Geology*, 2nd Edn. Prentice-Hall, New York (1954).
14. F. M. Ernsberger, Detection of strength impairing flaws in glass. *Proc. R. Soc. Lond. Ser. A* **257**, 213–223 (1960).
15. C. N. Beard, Quantitative study of columnar jointing. *Bull. Geol. Soc. Am.* **70**, 379–381 (1959).
16. N. H. Gray, J. B. Anderson, J. E. Devine and J. M. Kwasink, Topological properties of random crack networks. *Math. Geol.* **8**, 617–626 (1976).
17. S. I. Tomkieff, Basalt lavas of The Giants' Causeway. *Bull. Volcan. Napoli* **2**, 6–26 (1940).

Comp. & Maths. with Appls. Vol. 12B, Nos. 3/4, pp. 547–563, 1986
Printed in Great Britain.

0886–9561/86 $3.00 + .00

CHIRALITY IN THE WORLD OF STEREOCHEMISTRY

V. I. Sokolov
A. N. Nesmeyanov Institute of Organo-Element Compounds, U.S.S.R. Academy of Sciences,
Moscow 117813, U.S.S.R.

Abstract—Stereochemical fundamentals such as chirality, conformation and configuration are characterized. Algebraic properties of chemical sets, and in part, chiral sets are analyzed.

The stereochemical system as a whole has been consistently derived based on the definition of a molecule as the full set of all attainable conformations. A novel geometric theory of chirality is described which reveals the real sense of the elements of chirality. Finally, the mathematical foundations of stereochemistry have been thoroughly discussed in terms of the most important role played by the theorems of E. Noether, E. Ruch and K. Gödel for the problem of stereochemical configuration. For the latter, the analytical derivation as a function of the structural argument has been proposed for the first time.

INTRODUCTION

A range of ideas based on the conception of symmetry is extensively used in a variety of academic fields and general activities. In this context, one cannot help recalling a remarkable book by Hermann Weyl[1] that has paved the way for numerous subsequent works treating various aspects of symmetry in science, the arts, and life. As an example, mention should be made of a fundamental study of Shubnikov and Koptsik[2]. Prior to Weyl, contributions on symmetry had been mostly of a descriptive nature[3], whereas following the publication of his *Symmetry*, mathematical aspects occupied an important, and often prevailing, position. Crystallography and versatile applications of mathematics to physics have been and remain the most important areas where the ideas of symmetry are being efficiently employed. Symmetry approaches were long ago utilized by such sciences adjacent to physics and chemistry as spectroscopy, via which symmetry concepts first intruded into the world of molecules. The importance of symmetry-based approaches consists in that they frequently allow qualitative, conceptual results to be obtained, diverting oneself from more specific numerical data. For this reason symmetry approaches belong to nonnumerical mathematics. In this context I deem it expedient to incorporate into this paper some sections dealing with the application of nonnumerical mathematics in chemistry. It seems to me that the interdisciplinary nature of the present paper is suitable, in the best way possible, for communicating such ideas that are worth the notice of mathematicians with a view to informing them about the existence of such problems in chemistry, and in whose solution they can contribute. It is my feeling that such editions are primarily intended to strengthen links between different branches of science.

This paper will discuss the basic conceptions of stereochemistry dealing with the spatial arrangement of molecules and extensively relying on ideas related to symmetry. Stereochemistry, resting on a fundamental triad—chirality, conformation, configuration—will be interpreted basically in the spirit of my monograph[4], with incorporation of recent results.

Stereochemistry, whose emergence dates back to the early 19th century, when the optical activity of molecules was discovered, is in essence as old as organic chemistry. The first stereochemical theory put forward simultaneously by van't Hoff and Le Bel in 1874 accounted for the spatial structure of organic compounds on the basis of the tetrahedral model of the carbon atom. The geometrical approach to stereochemistry was further developed by Werner's coordination theory, which explained the structure of numerous complex compounds and their optical activities on the basis of the octahedron, which is a regular polyhedron. The study of optical isomers (enantiomers and diastereomers) formulated the idea of configuration as being a key concept of stereochemistry. Determination of the absolute configuration of molecules of optically active compounds by anomalously scattered X-rays, introduced in 1951[5], made it possible, for the first time, to identify dextrorotatory and levorotatory enantiomers in three-dimensional space. The next most essential stereochemical category, conformation, came into being in the early 1950s when conformational analysis was shaped as a branch of stereochemistry dealing with the internal mobility of molecules.

Finally, introduction of the concept "chirality" has accomplished the formulation of the fundamental triad of stereochemistry. The term chirality was coined by Lord Kelvin in his Baltimore Lectures in 1884 and 1893, but it did not become more generally known until their publication in 1904[6]. It is well known that although this term had been introduced by a physicist it was revived many years thereafter to find its utility first in stereochemistry[7], where the theory of chirality is developed at greatest length, and then it was extended to other fields of knowledge since chirality certainly has quite a general significance.

Before proceeding to the subject matter, it is to be noted that different authors have developed a variety of systems describing chemical chiral structures[7,8], differing in approach, depth of elaboration, and scope.

The most popular Cahn–Ingold–Prelog (CIP) system describing chiral molecules[7] has done great services to stereochemistry; namely, it has first systematically used the notion of chirality, introduced the concept of chiral elements and chirality patterns, and developed recipes for a consecutive naming of stereochemically complex molecules. It should be admitted, however, that it has not brought us closer to an understanding of the genuine meaning either of chirality elements or configuration in its interaction with conformation. By its very orientation the CIP system is a pragmatic tool for inventing descriptors rather than for cognizing the essence of stereochemical concepts. Characteristically, every worker concerned with the problem constructs his own system virtually without any interaction with other investigators. As far as the present author is concerned, he is an exception only to a small degree. This appears to be peculiar to the current stage, where stereochemistry is mastering the ideas of nonnumerical mathematics. It is not yet clear whether the future will see the above concepts synthesized to a practicable extent or whether someone will devise a novel system that will take into account the experience gained by earlier workers. Certain similarity can be found with the history of the formation of the theory of substitution reaction mechanisms in organic chemistry. In that period, all previous studies were generalized by the Ingold–Hughes conception[9]. Yet the important distinction is that the development of that theory was possible and was accomplished within chemistry and using its means, thereby giving rise to a new chemical branch, *viz.* physical organic chemistry. The development of the theoretical basis of chiral stereochemistry necessitates the involvement of mathematical facilities and, what is desirable, professional mathematicians.

SYMMETRY AND CHIRALITY IN STEREOCHEMISTRY

A complete set of symmetry elements is known to be intrinsic only to the sphere. Objects may be referred to as symmetrical if they retain certain elements of symmetry. For quite a long time, such objects have been called *dissymmetric*, thereby emphasizing the lack of a part of a complete set of symmetry elements. The term "asymmetric" should be applied only to those objects where not a single element of symmetry is present. All this is certainly true of molecules, but what is extremely important for them is a specific subset of the sets of symmetry elements, which is responsible for the optical activity of molecules. For a molecule to be optically active, it is essential and sufficient that it cannot be superimposed upon its flat mirror image; this property has been termed *chirality*[6]. This definition, given by Kelvin, remains the most general one; as to Prelog's wording[10], it mentions two types of motion, which is substantially redundant information: "An object is chiral if it cannot be brought into congruence with its mirror image by translation or rotation." The opposite property, that is, superimposability of an object with its mirror image, has become known as *achirality*. Chiral molecules exist as pairs of enantiomers which differ only in the sign of rotation of planes of polarized light. Achiral molecules do not possess optical activity of their own and can exhibit only the induced activity. Since chirality, as the definition implies, can be inherent to objects of any nature it is resonable to discuss chirality in the general case using sets; yet, before this is done, it seems useful to analyze special chemical sets, that is, sets of molecules, from the standpoint of algebra.

ALGEBRAIC PROPERTIES OF CHEMICAL SETS[4,11]

Let us define a complete (universal) chemical set S as the sum total of all conceivable chemical compounds. It is evident that S is infinite. One and only one of two operations to be called hereafter *chemical addition* and *chemical multiplication* is determined for each pair of the elements $a, b \in S$. The former involves mixing of unreactive molecules; the latter, a chemical reaction between molecules. The set S is closed with reference to the above operations:

$$a + b = (a + b), \qquad a' \times b' = p + q \in S$$

Let individual chemical compounds be called simple elements of the set S, and their mechanical mixtures called component elements. In the general case, multiplication is a bimolecular reaction with a component element, a product, being capable of including any number of simple elements and being concurrently the sum thereof. Special cases: monomolecular reactions ($b' = 1$, an identity element of a set) and addition reactions ($q = 0$).

   Let us explore whether the three principal algebraic laws are applicable. Chemical addition (no reaction occurs) appears to be commutative and associative:

$$a + b = b + a, \qquad a + (b + c) = (a + b) + c.$$

The first property is also intrinsic to chemical multiplication, since two reagents are equitable when interacting with each other. However, the condition of associativity is not generally fulfilled:

$$a \times b = b \times a, \qquad \text{but} \qquad a \times (b \times c) \neq (a \times b) \times c.$$

In other words, the end result of applying two consecutive reactions to a certain substrate depends, generally speaking, on the order of the sequence. It will suffice to refer to classical rules of orientation of electrophilic substitution in a benzene ring:

   Interesting results can be derived from analysis of the laws of distributivity, which relate two different operations. For the set S, the first law of distributivity which holds both for the algebra of numbers and for the algebra of sets is not obeyed (multiplication with reference to addition):

$$a \times (b + c) = a \times b + a \times c,$$
$$A \cap (B \cup C) = (A \cap B) \cup (A \cap C).$$

Indeed, if there is a mixture of two substances unreactive towards each other, a third reagent can simultaneously implicate the two species into a complex reaction, as shown below:

This law is not obeyed whenever conjugated reactions occur.

   It should be noted that the second law of distributivity, valid only for the algebra of sets rather than for the algebra of numbers, is applicable to the set S (addition with reference to multiplication):

$$a + (b \times c) \neq (a \times b) + (a \times c),$$
$$A \cup (B \cap C) = (A \cup B) \cap (A \cup C).$$

In terms of the chemical model now being investigated, this law can be interpreted as follows. Addition of A to the product of the reaction between B and C yields results which are equivalent to that of the reaction between C and C if each of them was preliminarily mixed with A. To put it differently, A is required to be chemically inert with respect to B, C and $B \times C$. But this is the condition which defines the chemical addition.

It is worthy of note that in relation to the laws of distributivity, the elements of the set S resemble sets rather than numbers. This resemblance is, however, far from being complete. For instance, of the two laws of idempotency, only an additive and not multiplicative law is obeyed, provided the corresponding operation has been defined:

$$a \cup a = a, \qquad a \cap a \neq a.$$

Considering a random element $a \in S$ as a set, let us designate all other elements of S, except a, as $\bar{a}$. It can readily be seen that the de Morgan laws are not followed:

$$\overline{a \cap b} \neq \bar{a} \cup \bar{b}, \qquad \overline{a \cup b} \neq \bar{a} \cap \bar{b}.$$

To summarize, the behavior of the elements of the complete chemical set S differs from that of both numbers and sets. S is characterized by the following properties.

1. For any pair of elements $a, b \in S$, either of the two operations chemical addition *or* chemical multiplication is defined: $a * b = p \in S$ (the law of exception).
2. Addition is commutative and associative:

$$a + b = b + a, \qquad a + (b + c) = (a + b) + c.$$

3. Multiplication is commutative and nonassociative:

$$a + b = b + a, \qquad a \times (b \times c) \neq (a \times b) \times c.$$

4. The first distributivity law (the algebra of numbers) is, generally speaking, not obeyed:

$$a \times (b + c) \neq a \times b + a \times c$$

5. For any $a,b,c \in S$, the second distributivity law (the algebra of sets) is followed:

$$a + (b \times c) = (a + b) \times (a + c).$$

6. The additive law of idempotency is obeyed, whereas the multiplicative and de Morgan laws are not.

Some other uses of this model have been dealt with elsewhere[4,12]. In conclusion, it is worthwhile to note that the complete chemical set S should be described by one of the non-associative algebras.

CHIRAL SETS[4,12]

Of subsets of S, let us single out a complete chiral set χ. To this end, let us transform each element of the set S by an operation involving mirror images. Such a procedure gives rise to reflection of the complete chemical set into itself: $R(S) \rightarrow S$. As this takes place, all the elements are divided into two nonoverlapping classes:

$$R(a) \longrightarrow a \in A \qquad \text{and} \qquad R(\chi) \longrightarrow \bar{\chi} \in X,$$
$$R(\bar{\chi}) \longrightarrow \chi \in X.$$

The elements of class *A—achiral*—are stationary points of this transformation. The elements of class *X* are *chiral* and are converted into enantiomers.

Let us call *achiral* any set whose all elements are stationary points in the transformation *R*. If, to say the least, an element is converted into the one not identical to itself then such a set will be named *chiral*. A nonempty set containing no achiral elements will be referred to as completely chiral. Using the above technique, we can break *S* into two subsets, whose combination is essentially $A \cup X = S$. Not all the elements of the subset *A* are equally related to *X*. It is expedient to single out a prochiral (see above) set *P*: $P \cup A' = A$, where A' is the achiral set per se. The relationship between these sets is schematically illustrated in Fig. 1. Prochiral elements can be regarded as "boundary" points of a chiral set, which do not belong to it. Let us use the experience of stereochemistry to provide a convenient classification for any chiral sets[4,12].

Let us call a set *G enantiomeric*, if

(1) it comprises only chiral elements χ_{ij}, and
(2) for any $\chi_i \in G$, $\overline{\chi}_i \notin G$.

If, on the contrary, for any χ_i, a set *H* contains an element $\bar{\chi}_i$ enantiomeric with respect to it and there is lack of achiral elements, then it is natural to call this set *racemic*. According to the definition, $G \cup G' = H$. In this fashion, racemic sets consist solely of pairs of enantiomers. With reference to chiral sets in general, the concept of "enantiomerism" lends itself to broader interpretation than in stereochemistry. A pair of enantiomers can imply two such elements of a set, for whom all parameters (whose nature can be accurately ascertained for specific instances) coincide except for only one parameter capable of assuming either value. Such a duality can conveniently be denoted by plus (+) and minus (−) (without any association with optical rotation). In this form, the picture of enantiomerism of chiral elements is of the greatest generality.

When for each element of a chiral set, two or more parameters can take up either value, *diastereomeric* sets result. On the other hand, the only variable parameter can assume $n > 2$ values. Consider one type of such sets. Let a set consist of groups of elements—*clusters*—for which there exist an operation *T*. Successively applying this operation $n - 1$ times to any element, it is possible to obtain all the *n* elements making up a cluster:

$$T^{(n-1)}(a_1) = a_n.$$

Such sets can be termed *cluster* sets of the rank *n*. When $n = 2$, enantiomeric pairs result; **T** stands for the operation involving the mirror image in a plane. Thus, racemic sets are essentially cluster sets of rank 2. The cluster sets of higher ranks are obviously of less concern for stereochemistry. It is noteworthy, however, that what is involved in the general case is a variety of options rather than duality.

Now divide a racemic set containing *N* pairs of elements into two subsets so as to prevent any pair of enantiomers from falling into the same set. As a result, a pair of enantiomeric subsets

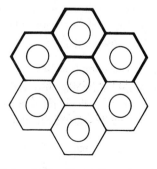

Fig. 1. Chiral molecule of hexahelicene (one enantiomer).

will be produced. Such selection of one element from each pair can be carried out by 2^{N-1} techniques. Are all these breakages equivalent to one another? Analysis of the combination of values AC† (absolute configuration), the only parameter that has the *plus* or *minus* sign, reveals the *single* breakage where all the elements belonging to the same subset are identical in sign. This, however, can be of no special consequence. The breakage referred to above does fundamentally differ from all others provided there is a genetic linkage between all the elements labeled AC rather than an accidental coincidence. The presence of such a linkage is equivalent to meeting the condition of *homochirality* introduced by Ruch. If homochirality does occur, then the above single breakage gives a *configurational series*. The foregoing discussion has no bearing on any structural model and is consequently of a general nature. A better insight into homochirality and configurational series can, however, be obtained by treating some specific chemical materials. If the chiral elements of sets involve molecules in relation to which the form of a coordination polyhedron of the chiral atom is known, then the existence of the configurational series is associated with fulfillment of the Ruch theorem[8].

THE RUCH THEOREM AND HOMOCHIRALITY

A molecule can be represented as a certain molecular skeleton with ligands attached to it. All common molecular skeletons are achiral, as are the molecules with all similar ligands. Chirality occurs provided some achiral ligands are dissimilar; chirality conditions of different-ligand molecular polyhedra are discussed elsewhere[4].

In the Ruch theory[8a], a parameter λ corresponds to each ligand and numerically characterizes certain properties related to chirality. The chirality of the whole molecule is described by a continuous function of the parameters of all ligands $\chi (\lambda_1, \ldots, \lambda_n)$. The χ specifies the chirality of a molecule. The definition suggests that for achiral molecules, the chirality function should vanish. Yet this model tells nothing of whether the χ function has zeros of different types that would correspond to chiral molecules.

It has turned out that the classes of molecules of both categories do exist, namely, chiroids a without chiral zeros and chiroids b with chiral zeros.

The Ruch theorem

A class of molecules belongs to category a only in one of the following cases:

(1) in the molecular skeleton there are only two sites intended for ligands;
(2) the number of such sites is $n > 2$, but the group of skeleton symmetry contains symmetry planes *each of which* has n-2 skeleton sites intended for ligands.

All other classes of molecules fall into category b. The concept *homochirality* relates only to chiroids a, and the arrangement of configurational series is meaningful for them; whereas for chiroids b, their meaning is of purely nomenclature and descriptor natures.

The Ruch theory has demonstrated that the concept of stereochemical absolute configuration formulated and valid for tetrahedral molecules cannot be mechanically extended to all other polyhedra. As in the case of tetrahedron T_d,‡ it has the very same meaning for chiroids of class a, which also comprise a distorted tetrahedron, disphenoid C_{2v}, and a trigonal bipyramid D_{3h}.

As regards all other molecular polyhedra, homochirality does not exist, and consequently a stereochemical nomenclature (say, for octahedral molecules) is of no greater importance than is a set of descriptors to name a molecule.

Prochirality and pseudochirality

Classification into chiral and achiral molecules turns out to be insufficient in some respects. Not all achiral molecules happen to be identical in relation to conversion into chiral species by successive structural modifications. It was found expedient to single out the submasses of achiral

---

†The reason for which this symbol has been chosen will become clear later.

‡Indicated is a point group of molecular skeleton symmetry identical to a skeleton group with the same ligands.

molecules that can be converted into chiral ones through a single replacement. In 1966 Hanson[13] defined the notion "prochirality" as follows: If the replacement of one point ligand in a restricted achiral set of point ligands by a new point ligand gives rise to a chiral set, then the original set is prochiral. Prochiral is that part of achiral that is separated by one step from chiral. More recently, the notion "the length of the chiralization way", μ, of an achiral object was introduced[14]. It stands to reason that for a prochiral molecule it is minimum, $\mu = 1$, for example, for CH_2XY, whereas for CH_3X and CH_2X_2 $\mu = 2$, and for CH_4 $\mu = 3$. Prochirality finds its principal utility in connection with the studies of enantiotopism and diastereotopism[15,16].

The presence in a molecule of more than one chiral moiety gives rise not only to diastereomers but also to some new options. The following molecules containing two chiral and enantiomeric ligands at the tetrahedral atom are achiral diastereomers because there is a symmetry plane passing through achiral ligands a and b and the central atom which is a pseudochiral center:

The problem of pseudochirality was discussed elsewhere[17,18].

Chiral and achiral point groups of symmetry

Thus the mirror image operation is employed to classify objects into chiral and achiral. In this content, it is worth recalling that as far back as the late 19th century Wulff[19] considered the symmetry plane to be the principal element of symmetry. It is apparent that the set of symmetry elements of chiral objects (molecules) cannot incorporate reflection relative to a plane. The set of all symmetry operations of a given figure makes up a point group of symmetry, whereas all symmetry elements meet at the same point, which remains undisplaced in all symmetry transformations. The action of symmetry point groups having only axes of rotation C_n and D_n gives rise to objects that cannot be superimposed upon their mirror images: these are chiral point groups as also are more complex and rare T, O and I. A special case (group C_1) involves truly asymmetric molecules devoid of other symmetry elements except identical transformation E; Fig. 2 illustrates the ratio of various molecular types to the corresponding point groups of symmetry. The action of mirror axes brings an object into congruence with its mirror image, which accounts for the achirality of point groups S_n. Since axes S_n are equivalent to a symmetry plane at $n = 1$ and to an inversion center at $n = 2$, the symmetrical criterion of chirality can be worded as follows: *a molecule is chiral if it has no mirror axes S_n*. One can suggest the "tree" of Fig. 3 for identifying the point groups of symmetry, which unlike the one suggested in [20] allows primarily for chirality.

In stereochemistry, there are two concurrent approaches to the analysis of molecular chirality. The former is based on point-group symmetry, while the latter is based on the elements of chirality[7] such as center, axis, and plane. The symmetry-group approach would seem to be quite consistent, but has substantial disadvantages from the viewpoint of practical application. Molecules differing widely in structure may fall into the same large class of molecules such as C_1, whereas structurally similar molecules frequently belong to different classes, as shown below. At the same time isomers of different topology may belong to the same point group, as exemplified by MA_2B_2 molecules:

This disadvantage has been overcome through a new symmetry-group notation proposed by Pople[21]. It provides a unique code for each molecule, but perhaps is somewhat cumbersome for routine application. The advantage of the approach based on chirality elements is that it is more intimately related to molecular structure, while its drawback is lack of definition of chirality

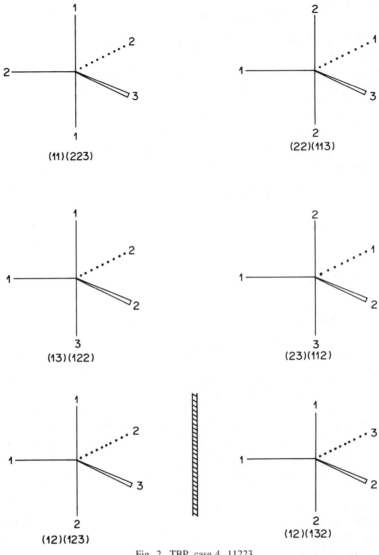

Fig. 2. TBP, case 4, 11223.

elements. A recently suggested (by the author) novel interpretation of the meaning of chirality elements, stemming from the geometric theory of chirality, will be treated below.

Conformational space of molecules

A molecule is a set of all attainable conformations or in other words, a molecule is a set whose elements are conformations[4]. Conformation is an instantaneous picture of a molecule, its mode of existence at a certain instant of time. It is assumed that the boundaries of the area wherein the molecule itself exists are available, definable, and known. This signifies that it is known how and to what extent the internal coordinates of the molecule can be changed, leaving its wholeness unaffected. It has long been known that conformations vary in reactivity; those conformations whose proportions in a population is small are frequently more active. We cannot restrict ourselves to the several most common conformations. A molecule should be thought of as being a complete sum total of possible conformations. A set of the molecule's conformations is infinite, continuous, and it represents the conformational space of the molecule. The conformational space of the molecule, with an operation assigned to it (a certain type of conformational motion) meet the requirements placed upon topological space and satisfy axioms for metric spaces. An infinite graph whose vertices are molecular conformations corresponds to conformational space, while with continuous conformational space this graph is connected.

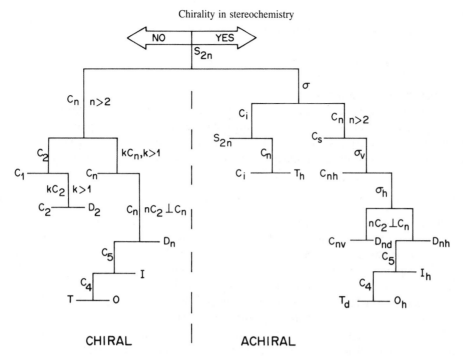

Fig. 3. Stereochemistry-oriented tree of symmetry point groups.

Clearly, various molecules, including structural isomers, form different and nonoverlapping conformational spaces. However, stereoisomers, that in contrast to structural isomers have a similar arrangement of atomic bonds, are topologically similar and should be part of the same conformational space, because, in principle, enantiomers can be interconverted by continuous conformational conversion without bond rupture.† In a sense, stereochemistry imposes more stringent requirements on objects than does topology, for, resting on chirality, it discovers an important distinction between enantiomers; yet the same conformational space is shared by a pair of enantiomers.

We shall now deal with some special cases which will emphasize the importance of the condition of conformation *attainability* in the above formulation. Conformational space, or diversity, corresponding to a cyclic molecule, represents a connected region incorporating all conformations attainable by a molecule without bond rupture. It is well known, however, that a fairly large cycle makes possible the existence of a knot molecule which is quite stable and gives a connected conformational region of its own. Relative to a simple monocycle, the knot molecule is a possible but in no way (without bond rupture) attainable conformation. From the perspective of conformational motions, there are no foregoing states for the knot molecule in the conformational region of the monocycle. In some fields of knowledge, such a situation is known as "Eden's garden" and can be formalized as follows:

$$G(m) \xrightarrow{m>P} G_1(m) \cup G_2(m), \qquad G_1(m) \cap G_2(m) = \emptyset.$$

Thus, the conformational space of a cyclic molecule, when the magnitude of the cycle exceeds some critical value, breaks into two regions lacking common points and corresponding to two *topological isomers*[23]. According to the definition given in [23], true topological isomers are invariably in "Eden's garden" with respect to each other. The following equation describes the occurrence of catenanes when there is a rise in the size of two monocycles:

$$G(m,n) \xrightarrow{m,n>q} G_1(m,n) \cup G_2(m,n), \qquad G(m,n) \cap G_2(m,n) = \emptyset.$$

Incidentally, we remark that $p \gg q$. However, for rotaxanes, which are close to, virtual analogs, of catenanes though lacking a true topological linkage, an equilibrium between isomers

†At this most general stage of analysis, we shall not be concerned with the energetic characteristics of linkages and processes.

(chain + ring $\rightleftharpoons$ rotaxane) is possible and has been observed. In this case,

$$G_1(m^r,n^c) \cap G_2(m^r,n^c) = B_m \qquad (r = \text{ring}, c = \text{chain}, B = \text{boundary}).$$

It may be assumed that other Eden's gardens of energetic rather than topological origin are likely to exist; namely, a local minimum on an energy surface, surrounded by an energy barrier whose height makes it virtually insurmountable without bond rupture. Both aspects, topological and energetic, are present in a case which has not as yet been realized, namely when an isolated atom is included into the internal region of a polyhedral hydrocarbon molecule.

CHIRALITY, CONFORMATION, CONFIGURATION, AND THEIR INTERRELATIONSHIP

Whatever has been earlier said about chiral sets is certainly applicable to the sets of a molecule's conformations, where chiral and achiral subsets can be singled out. Importantly, chiral conformations basically cannot occur for some molecules, namely, di- and triatomic molecules invariably lying in the same plane. The conformational diversity of such molecules as benzene, ferrocene, or cyclopropane virtually lacks chiral conformations because conformational motions leading toward them are energetically unfavorable. In the conformational diversity of an achiral molecule, chiral conformations may occur only as enantiomeric pairs.

The foregoing suggests that conformations should be treated as a primary notion, and chirality as a certain attribute of a set of conformations. Configuration can be defined only for this (chiral) subset. The situation is summarized in Fig. 4. In practice, what differentiates one enantiomer from another is the sign of optical rotation at a selected wavelength. Using purely chemical means, it is impossible to assign a given enantiomer (of a tetrahedral molecule) either absolute configuration (AC), that term used to mean the "handedness" of an enantiomeric molecule in three-dimensional space.† For this reason, numerous correlations of relative configurations performed before the mid-20th century had recognized the relationship of configurations of distinct molecules with the configuration of a standard chosen at random (glyceraldehyde). Determination of AC has become possible only beyond the scope of a chemical system, which vividly demonstrates the validity of the Gödel theorem in the field of stereochemistry[4].

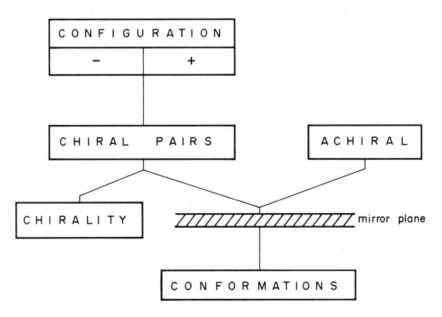

Fig. 4. The hierarchy of stereochemical fundamentals.

†The author uses the word "configuration" solely in the sense of stereochemical absolute configuration. For this reason, as the definition suggests, configuration AC is valid only for chiral, even homochiral molecules and conformations.

The problem of configuration in organic stereochemistry primarily encompasses molecules with tetrahedral chiral centers as well as with other chirality elements, such as axis and plane. Having clarified the situation with other molecular polyhedra, the Ruch theorem has convincingly demonstrated that all other polyhedra, except a trigonal bipyramid D_{3h}^5, belong among chiroids of class $\underline{b}$ for whom configurational series do not exist.[†]

This circumstance can be briefly noted in a quantor form as follows:

$$\forall M_x \{ M \in T_d^4, D_{3h}^5 \} \; \exists \; \mathcal{K},$$

to be read, *Stereochemical configuration§ exists for all chiral molecules with the skeleton of a tetrahedron or trigonal bipyramid.*

Configuration in a trigonal bipyramid

For molecules in the form of a trigonal bipyramid, configuration should therefore be determined. In a tetrahedron T_d, four different ligands form one pair of enantiomers, while all other combinations of ligands form one achiral isomer. In a trigonal bipyramid (TBP) D_{3h}, five ligands are broken into two different classes: apical (two) and equatorial (three). If all five ligands are different, then $C_5^2 = C_5^3 = 10$ pairs of enantiomers are possible; other combinations involving ligands are shown as examples below (see Fig. 2):

Case 1, 11234: four achiral diastereomers[‡] (11)(234), (23)(114), (24)(113), (34)(112), and three pairs of enantiomers: (12)(134), (13)(124), (14)(223).

Case 2, 11123: four achiral diastereomers (11)(123), (12)(113), (13)(112), (23)(111).

Case 3, 11112: two achiral diastereomers (11)(112), (12)(111).

Case 4, 11223: four achiral diastereomers (11)(223), (22)(113), (13)(122), (23)(112) and one pair of enantiomers (12)(123).

Case 5, 11122: three achiral diastereomers (11)(123), (12)(112), (22)(122). Only case 4 resembles a tetrahedron in that there is a single pair of enantiomers, though in addition to four achiral diastereomers.

THE GEOMETRICAL THEORY OF CHIRALITY[22]

Since chirality represents a general phenomenon, it would be appropriate to divert oneself for a while from molecules and focus on infinite chiral structures. In terms of methodology, there must be cases (and, indeed, they are not uncommon) when it is easier or only possible to deal with some problems if one utilizes a broader approach and addresses a wider combination (including the case where the problem to be discussed has arisen). Such an approach has been formulated in the general theory of systems and constitutes a characteristic features of what is known as "systems analysis." A theoretical foundation is furnished by the Gödel theorem of incompleteness of logical systems which we have referred to earlier in connection with the problem of determining absolute configuration.

Thus, to get more penetrating insights into chirality, let us raise a problem: how to determine chiral structures, figures of $(n - 1)$th dimensionality, which fill in an isotropic continuum $\mathbb{R}^n$. Putting it another way, we are searching for a way to chiralize isotropic straight line $\mathbb{R}^1$, plane $\mathbb{R}^2$, space $\mathbb{R}^3$, and hyperspaces $\mathbb{R}^n$,

The existence of chiral helicene molecules (Fig. 1), which are essentially fragments of a helical (spiral) surface realized in a molecular material, prompts us to complete the filling of space $\mathbb{R}^3$ with a chiral two-dimensional figure. Parametric equations for a helicoid are

$$X = a \cos \varphi, \qquad y = a \sin \varphi, \qquad z = F(a) + h\varphi \quad (-\infty < a < +\infty).$$

[†]This, however, does not imply that a question cannot be raised concerning the stereochemistry of a specific reaction. An individual reactive act can occur with or without retention of a "configuration" of a center if the entering ligand occupies or does not occupy the site of the leaving ligand.

§In Ruch's sense.

[‡]Here the definition is followed by *diastereomers are nonenantiomeric stereoisomers*, which seems to be new and is valid for TBP. The notation applied to TBP isomers is (aa)(eee).

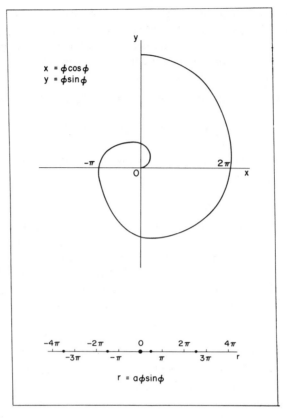

Fig. 5.

If a straight line chosen arbitrarily but perpendicular to a fixed axis slides along this axis following the above law, with $h \to 0$ and $a \to \infty$, then space $\mathbb{R}^3$ will be filled by the resulting surface of the helicoid, the meaning of chirality (the absolute configuration of a propeller) being determined by the direction of sliding, that is, by the sign.

Analogously, plane $\mathbb{R}^2$ is chirally† filled in by Archimedean spiral (Fig. 5):

$$x = a\varphi \cos \varphi, \qquad y = a\varphi \sin \varphi.$$

It is evident that when $a \to 0$ the distance between the adjacent coils of the spiral becomes an infinitesimal quantity. It is worth noting the continuity of both the surface of the helicoid and that of Archimedean spiral. Continuity, however, cannot be retained when a *chiral* sequence of points (Fig. 5) fills an infinite straight line $\mathbb{R}^1$ satisfying the law

$$r = a\varphi \sin \varphi.$$

Interestingly, the other coordinate present in the Archemedean spiral equation, namely $r = a\varphi \cos \varphi$, leads at $\mathbb{R}^1$ to an *achiral* sequence of points (if the multiplicity of the zero point is ignored) which has an inversion center at $r = -\frac{1}{2}$. This suggests that the chiral filling of continua $\mathbb{R}^n$ by the figures of $(n-1)$th dimensionality is also possible for $n > 3$ if a sinusoidal coordinate is retained.

It is to be emphasized that the chiral filling of isotropic continua $\mathbb{R}^n$ is characterised by an all-important property; namely, filling of the n-dimensional isotropic space $\mathbb{R}^n$ by a chiral $(n-1)$-dimensional figure necessitates the presence of a special $(n-2)$-dimensional figure to be "coiled" by the filling figure. For instance, a cylindrical helicoid fills in $\mathbb{R}^3$ having, as an axis, a special straight line, while Archemedes' spiral fills in $\mathbb{R}^2$ having a special point as the

---

†Here is meant the chirality on a two-dimensional plane relative to a straight line, and furthermore the chirality on a one-dimensional (straight) line relative to a point.

Table 1. Interrelationships between the dimensions of the isotropic space, the chiral filling figures and the singular achiral figures

Isotropic space to be filled	Its dim.	The chiral filling figure	Its dim.	The corresponding singular achiral figure	Its dim.
Straight line	1	sinusoidal succession of points	0	multiple zero point	less than zero†
Plane	2	Archimedean spiral	1	point (center)	0
Space	3	helicoidal surface	2	straight line (axis)	1
Hyperspace	4	helicoidal space	3	plane	2
	N		$N - 1$		$N - 2$

† Assigned because of the multiplicity of the zero point.

spiral center. The filling of $\mathbb{R}^1$ by points according to the above chirality law also requires the presence of a central *zero point* which is characterized by its multiplicity. It is quite likely that this regularity also holds for hyperspace $\mathbb{R}^4$ whose chiral filling by helical three-dimensional space requires a two-dimensional plane as a special element of the whole system. Such a situation can logically be also extended to larger n, but this has no bearing on the stereochemistry of molecules. It is understood that all these special elements of chiral systems represent *achiral* continua. The general pattern is as follows: *n-dimensional isotropic space is filled in by an (n − 1)-dimensional chiral figure having a "special" (n − 2)-dimensional achiral figure.* These relations are presented in Table 1.

The foregoing analysis leads to an understanding of the true meaning of chirality elements. A remarkable role played by achiral special figures of lower dimensionality in the chiral systems of higher dimensionality suggests that it is "chirality elements" in stereochemistry (which in themselves are well known to be achiral and hence look like semantic nonsense) that are the achiral special components of the chiral systems of higher (2 units higher exactly) dimensionality. Specifically, a "chirality plane" is an achiral element of a chiral structure in hyperspace $\mathbb{R}^4$. The epithet "chiral" as applied to center, axis, and plane is used as a sign indicating a retained relation to chiral structures of higher dimensionality, whose substantial part they constitute.

Consider now some transitions between chiral structures in spaces of varying dimensionalities. If the abscissa and ordinate of a cylindrical helix whose equations have been given above are multiplied by φ, leaving the applicate unaffected, then the equation of a conical helix results, the latter lying on the surface of a right circular cone and exhibiting a "chiral center" along with a "chiral axis." Addition of the condition $-\infty < a < +\infty$ enables one to go from a helix to a surface helicoid. Notice that the conical helical surface does not fill in all of the space, leaving unoccupied a plane orthogonal to a singular axis and passing through a singular point— the vertex of the cone, or the center.

Projection of the conical helix onto the above plane gives a Archimedean spiral, which when projected onto the ordinate axis leads to a straight line occupied by points according to the aforementioned chiral law. Since this procedure is by and large reversible, it is possible to reconstruct an *n*-dimensional picture starting from the (n − 1)-dimensional one. For example, to change from Archimedean spiral to a conical helix requires the addition of a third coordinate, the applicate, passing through the center. It is felt that a similar operation makes it possible to change from a cylindrical helical surface to helical 3-space filling in four-dimensional hyperspace $\mathbb{R}^4$.

As opposed to a conical helix, a cylindrical one, when projected onto the plane orthogonal to the cylinder axis, gives the circumference of infinite multiplicity (Fig. 6). Similarly, a cylindrical helicoid, when $h \rightarrow 0$, results in an infinitely multiple plane σ (Figs. 6, 7). An opposite limiting process, $h \rightarrow \infty$, gives an infinite set of planes σ' that contain a cylinder axis and are orthogonal to plane σ. A concurrent operation $A \rightarrow 0$ tying up plane σ to the axis will give a single point. This process can be looked upon as a consecutive reduction of virtually chiral space (that is, isotropic $\mathbb{R}^3$, chiralized as described earlier, and even $\mathbb{R}^4$) to a "chirality

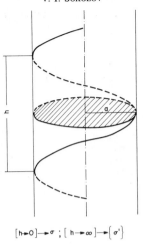

$$[h \!\rightarrow\! 0] \!\rightarrow\! \sigma \; ; \; [\,h \!\rightarrow\! \infty\,] \!\rightarrow\! \{\sigma'\}$$

Fig. 6.

plane,'' then to a ''chirality axis,'' and, finally, to a ''chirality center.'' In other words, chirality elements should be treated as the results of successive stages in the ''convolution'' of chiral space as dimensionality decreases.†

In the light of this conception, the ambiguity of identification of the chiral plane becomes understandable, for its position depends on the implicit operation of dechiralization of the molecule being analyzed. To illustrate, let us take the π-complex of a metal in whose cyclic ligand all substituents are not identical. Dechiralization can be accomplished by two basically different techniques, as shown below (Fig. 7).

(1) Removal of the π-complex-bound metal with retention of a planar chiral ligand whose plane σ plays the role of a chiral plane.

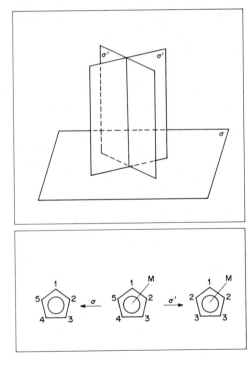

Fig. 7.

---

†A suggestion has recently been made[24] concerning a hypothetical mechanism of the formation of a chiral carbon in the form of a helicoid surface winding around some axis due to an incomplete graphitization of other metastable modifications of the carbon.

(2) Coincidence of any pair of substituents in the ligand, which results in several (five in this specific instance) planes.

In full accordance with the above analysis, all planes σ' are orthogonal to σ. The definition of chirality axis or chirality center is free from such ambiguity.†

THE NOETHER THEOREM AND CONFIGURATION

The conformational space of a molecule consists of several chiral and achiral zones; chiral zones have achiral boundaries. If an achiral zone is continuous, that is, the chirality function does not alter the sign *inside* the zone, then as a consequence of the Noether theorem, an invariant exists on this subset. The role of the function is played by chirality, which vanishes on the zone boundary. In physics, the Noether theorem is responsible for the existence of state functions. Similarly, in stereochemistry, it requires the existence of the invariant incidental to the continuous zone of chiral conformations, having an achiral boundary. If the chiral zone is not continuous, it means that inside it there occurs a jump reversal of the sign of the chirality function without going through zero, that is, the condition of homochirality is not fulfilled.

Thus, configuration turns out to be an intersection of the scope of three important theorems.

(1) The Noether theorem establishes the existence of the invariant incidental to the continuous zone of chiral conformations, that has an achiral boundary.
(2) The Ruch theorem indicates the types of molecular structures for which precisely these zones exist.
(3) The Gödel theorem states that configuration cannot be determined by purely chemical means but it can be determined via the addition of approaches employed in some other science (X-ray diffraction analysis, for instance).

Configuration as a function of a simple structural argument

It is felt that stereochemical configuration as a higher member in the above-mentioned hierarchy of fundamental notions can be derived as an analytical function of some argument which is characteristic of each conformation. It may be recalled that conformation is at the lowest level of the hierarchy.

When being considered as a function, configuration may assume three values, such as ($+$) and ($-$) for enantiomers and 0 (zero) for achiral objects. This area coincides exactly with the area of values for the well-known function *signum* (sgn) used in mathematics. This suggests its use in the analytical derivation for stereochemical configuration as follows:

$$AC = sgn[\sin \varphi],$$

where φ is the simplest structural characteristic of a given conformation that is a dihedral angle (as shown in Figs. 8 and 9) for all cases wherein the homochirality does exist, namely, planar and axial types of chirality, and chiral centers in tetrahedrons and trigonal bipyramids.

CONCLUSIONS

There are some other topics in stereochemistry that have a bearing upon the problems dealt with in the present edition. Among these, for instance, is a popular problem concerning the origin of optical activity in Nature and, to be more precise, the issue concerning the cause of the existence of natural chiral molecules in the form of unequal quantities of enantiomers and, normally, in the form of an exclusively single enantiomer. The majority of workers concerned with this problem are inclined to search for the cause in the chirality of a certain sign of various

---

†In the earlier version of the CIP system[7b], its authors insisted on the concept of chirality center that will not coincide with any atom, using an example, adamantane with four substituents at the bridgehead position. However, this example can be treated as successfully and more logically as a system comprising four chirality centers arranged as a tetrahedron. Since four points allow a single tetrahedron to be built, such an arrangement provides for the existence of only two enantiomers rather than 2^n as with independent centers.

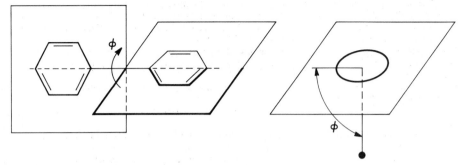

Fig. 8.

irradiations and, in the final analysis, they relate this problem to nonretention of parity in some nuclear processes. Magneto-optical effects or chiral liquid crystals can make their contribution to the discussion of the effects produced by a variety of chiral factors. In this paper, I tried to dwell on the problems I have personally been concerned with.

In his book, referred to above time and again[1], Weyl, writing about such figures as triquetras and swastikas, says that the source of the ideas of the magic power of these images seems to lie in the exciting effect of incomplete symmetry—rotations without reflections. Perhaps the point is that such a type of symmetry suggests a motion in a certain direction resembling that of a rolling wheel, and thereby "negating rest," which is intimately associated with the perception of combinations of symmetry, including a mirror plane. In any event I am willing to admit that the contemplation of a chiral molecule exhibiting incomplete symmetry and capable of rotating a polarized-light plane arouses an aesthetic feeling in me. This is one of the factors, at least for me, that makes stereochemistry so fascinating.

REFERENCES

1. H. Weyl, *Symmetry*. Princeton University Press, Princeton, New Jersey (1952); Russian translation: Moscow, Nauka (1968).
2. A. V. Shubnikov and V. A. Koptsik, *Simmetria v nauke i iskusstve*, 2nd Ed (in Russian). Moscow, Nauka (1972); *Symmetry in Science and Art*, Plenum Press, New York (1974).
3. W. D'Arcy, M. Thompson, *On Growth and Form*. Cambridge (1915).

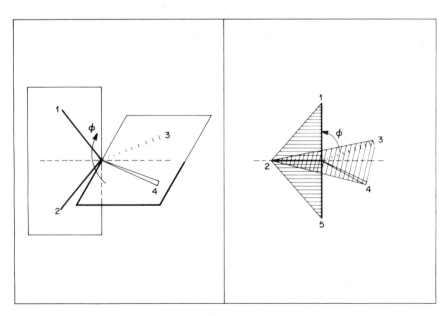

Fig. 9.

4. V. I. Sokolov, *Introduction to Theoretical Stereochemistry* (*Vvedenie v teoreticheskuyu stereokhimiyu*). Nauka, Moscow (1979,1983) (in Russian); English edition (enlarged): Gordon & Breach, New York (to appear).

5. J. M. Bijvoet, A. F. Peerdeman and A. J. Bommel, Determination of the absolute configuration of optically active compounds by means of X-rays. *Nature* **168**, 271–272 (1951).

6. Lord Kelvin, *Baltimore Lectures*, 1884 *and* 1893. C. J. Clay and Sons, London (1904).

7. (a) R. S. Cahn and C. K. Ingold, Specification of configuration about quadricovalent asymmetric atoms. *J. Chem. Soc.*, 612–623 (1951).

   (b) R. S. Cahn, C. K. Ingold and V. Prelog, The specification of asymmetric configuration in organic chemistry. *Experientia* **12**, 81–94 (1956).

   (c) R. S. Cahn, C. K. Ingold and V. Prelog, Specification of molecular chirality. *Angew. Chem. Intern. Ed.* **5**, 385–415 (1966).

   (d) V. Prelog and G. Helmchen, Basic principles of the CIP system and proposals for a revision. *Angew. Chem. Intern. Ed.* **21**, 567–583 (1982).

8. (a) E. Ruch and A. Schönhofer, Homochiralität als Klassifizierungsprinzip von Molekülen spezieller Molekülklassen. *Theoret. Chim. Acta* **10**, 91–110 (1968); E. Ruch, Die Vandermondesche Determinante als Chiralitätgemass. *Theoret. Chim. Acta* **11**, 183–192 (1972); E. Ruch, Algebraic aspects of chirality in chemistry. *Acc. Chem. Res.* **5**, 49–58 (1972).

   (b) E. Ruch, A. Schönhofer and I. Ugi, Näherungsformula für spiegelungsantimetrische Molekuleigenschaften. *Theoret. Chim. Acta* **7**, 420–432 (1967).

   (c) I. Ugi, D. Marquarding, H. Klusacek, G. Gokel and P. Gillespie, Chemistry and logical structures. *Angew. Chem. Intern. Ed.* **9**, 703–730 (1970); I. Ugi and P. Gillespie, Representation of chemical systems and interconversions by *be* matrices and their transformation properties. *Angew. Chem. Intern. Ed.* **11**, 914–918 (1971).

   (d) A. S. Dreiding and K. Wirth, Chiral arrangements of four point sets. *Math. Chem.* (*MATCH*) **8**, 341–345 (1980).

9. C. K. Ingold, *Structure and Mechanism in Organic Chemistry*, 2nd Ed. Cornell University Press, Ithaca, N.Y. (1969).

10. V. Prelog, Chirality in chemistry. *J. Molec. Cat.* **1**, **159–172 (1975/76).**

11. V. I. Sokolov, Algebraic properties of chemical sets. *Zh. str. khim.* (*Russ.*) **16**, 971–976 (1975).

12. V. I. Sokolov and I. V. Stankevich, Chiral sets. *Zh. str. khim.* (*Russ.*) **19**, 226–232 (1978).

13. K. R. Hanson, Applications of the sequence rule, I,II. *J. Amer. Chem. Soc.* **88**, 2731–2742 (1966).

14. V. I. Sokolov, On a minimum property of trigonal prism as a coordinative polyhedron. *Zh. str. khim.* (*Russ.*) **17**, 743–745 (1976).

15. K. Mislow and M. Raban, *Stereoisomeric Relationships of Groups in Molecules. Topics in Stereochemistry*, Vol. 1. Interscience, New York–London (1967).

16. V. Prelog and G. Helmchen, Pseudoasymmetrie in der Organischen Chemie. *Helv. Chim. Acta* **55**, 2581–2598 (1972).

17. H. Hirschmann and K. R. Hanson, Prochiral and pseudoasymmetric centers: Implications of recent definitions. *Tetr.* **30**, 3649–3656 (1974).

18. G. P. Schiemenz and J. Pistor, Zum Verständnis der Begriffe "Enantiomer" and "Enantiotop." *Chemica Scripta* **23**, 216–223 (1984).

19. Yu. V. Wulff, On symmetry plane as the main element of symmetry. *Proc. Warsaw Soc. of Naturalists* **7**, (6), 1–12 (1896); cited in *Selected Works on Crystallophysics and Crystallography*, (in Russian), p. 242. GITTL, Moscow–Leningrad (1952).

20. J. Donohue, A key to point group classification. *J. Chem. Educ.* **46**, 27 (1969); *Kristallographia* (*Russ.*) **26**, 908 (1981).

21. J. A. Pople, Classification of molecular symmetry by framework groups. *J. Amer. Chem. Soc.* **102**, 4615–4622 (1980).

22. V. I. Sokolov, Geometric theory of chirality; the real sense of chirality elements. *Math. Chem.* (*MATCH*), February (1985).

23. G. Schill, *Catenanes, Rotaxanes, Knots*. Academic Press, New York–London (1971).

24. V. I. Sokolov, Is chiral graphite possible? *Zh. str. khim.* (*Russ.*) **25**, 175–177 (1984).

Comp. & Maths. with Appls. Vol. 12B, Nos. 3/4, pp. 565–578, 1986
Printed in Great Britain.

0886-9561/86 $3.00 + .00

GEOMETRY AND CRYSTAL SYMMETRY

MARJORIE SENECHAL
Smith College, Northampton, MA 01063, U.S.A.

Abstract—The beautiful external forms of crystals are manifestations of their internal structures. These structures, which can be regarded as infinite periodic patterns, are determined by local forces. In this article we discuss symmetry from the "local" point of view. First we show that the symmetry of an infinite regular point set (the atomic pattern of an ideal crystal) is a consequence of the symmetry of finite configurations in the pattern. Then we briefly discuss the basic structural feature of internal crystal symmetry, the space lattice, and its relation to the crystal's external form. This form is predicted by locally defined growth rules. We conclude with an open problem, the origin of the symmetry and form of crystal twins.

Thou, silent form, dost tease us out of thought
As doth eternity.

John Keats (1795–1821)

1. INTRODUCTION

Keats was thinking of a Grecian urn when he wrote those words, but they would have been just as appropriate in an ode to a beautiful crystal (Fig. 1). For centuries the beauty of crystals has tantalized the imagination of the artist, the poet, the geometer, and the scientist. In Keats' time, the study of crystals was entering a new era. For almost two hundred years, imaginative scientists such as Kepler, Hooke, and others had tried to explain the external forms of crystals in a general way by supposing that crystals were built of minute particles (polyhedral or spherical) arranged in orderly arrays. Now such speculations on the geometry of the internal structure of crystals were being transformed into more rigorous theories which predicted definite relationships between internal structure and external form. (Had Keats been aware of this, perhaps he would have written that crystals tease us into thought . . . .)

The 19th century was the golden age of geometrical crystallography: between 1811 and 1891, the hypothesis that the internal structure of a crystal is a three-dimensional repeating pattern was developed in complete detail. The starting point for this work was the discovery in the 1780s, by the French crystallographer Romé de l'Isle, that the forms of crystals of a single mineral are always closely related. Until then, the variety of forms of crystals of the same species was considered to be an argument against any hypothesis of internal regularity. Romé's discovery suggested that on the contrary, the forms of a crystal are not arbitrary but due instead to some laws of structure and growth.

The first major breakthrough, shortly after the turn of the 19th century, was the building-block theory of crystal structure (Fig. 2) proposed by Romé's contemporary and fellow countryman Haüy. Haüy's theory not only accounted for the observed forms of many crystals, but also explained why crystal forms can have certain symmetries but not others (see Section 4). After the blocks were replaced (by Haüy's critics) with points representing their centers, symmetry became a mathematical problem and an active field of study. The possible symmetry classes of external crystal forms were enumerated in 1824, a complete classification of the simplest three-dimensional repeating patterns, the lattices, had been obtained by 1849, and finally, in 1891, the list of the 230 symmetry groups of internal crystal structures was complete. That this was a triumph of the geometrical imagination is underscored by the fact that it was not until 1912, with the discovery of the diffraction of X-rays by crystals, that this detailed and complicated theory of crystal symmetry could be verified experimentally. The symmetrical diffraction patterns recorded on photographic plates were conclusive evidence that the atoms of a crystal are arranged in symmetrical, periodic arrays.

Today, even though the symmetries of crystals have long been completely classified, our understanding of crystal symmetry is still incomplete in at least one important respect. Although

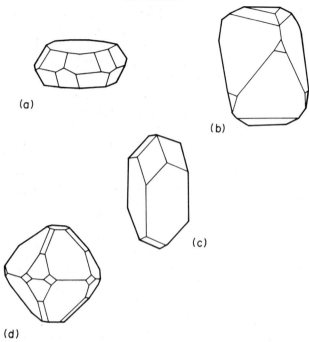

(a)

(b)

(c)

(d)

Fig. 1. Drawings of crystals, adapted from Goldschmidt's Atlas der Kristallformen: (a) hematite, (b) cinnabar, (c) harmotome, (d) chalcopyrite.

we know all the symmetries of entire systems of infinite patterns, we do not understand how these symmetrical patterns arise in a developing crystal. A crystal grows from a nucleus of a few atoms to a repeating pattern with so many atoms that it can be considered infinite. Under ideal circumstances, the growth of the pattern is orderly once it is well established: in the words of D'Arcy W. Thompson, in his classic of biology *Growth and Form*[1], "A crystal grows by deposition of new molecules, one upon one and layer by layer, superimposed or aggregated upon the solid substratum already formed." He then adds, "Each particle would seem to be

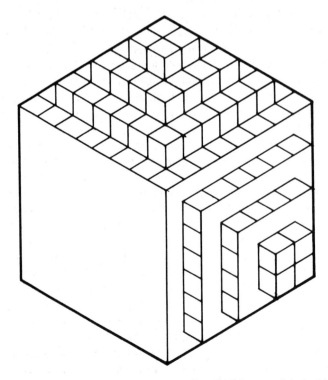

Fig. 2. Haüy's construction of a rhombic dodecahedron [see Fig. 15(c)] from parallelopiped building blocks.

influenced, practically speaking, only by the particles in its immediate neighborhood, and to be in a state of freedom and independence from the influence, either direct or indirect, of its remoter neighbors." If this is so, then the symmetry of the atomic pattern of a crystal, and also the symmetry of its external form, must be consequences of the local arrangements of particles. But how do the local arrangements come to be, and how do they work together to build a symmetric whole? This is an important problem: as the author of a recent article[2] states, "the crystal problem," that is, the problem of the origin of crystal symmetry, is a "major open question" in statistical mechanics.

A crystal is not a geometrical abstraction, and geometry alone will not solve the crystal problem. Still, it has been shown in the last eight years that geometry has a great deal to teach us about the reasons for crystal symmetry. In this paper I will try to show that recent results[3–5], together with more classical material, give us a picture which, though oversimplified, does address the crystal problem and also sheds light on the relation between internal crystal structure and external crystal form.

For the purposes of crystal geometry, we can distinguish three stages of development: the nucleation stage, in which atoms or groups of atoms begin to cluster together; the crystallite stage, in which a lattice structure is evident; and the crystal stage, in which the crystal has achieved its final external form. We will work our way backward. First, in Section 2, we show that the symmetry of an infinite regular point set (the atomic pattern of an ideal crystal) is a consequence of the symmetry of finite configurations in the pattern. Next, in Section 3, we briefly discuss the basic structural feature of internal crystal symmetry, the space lattice. In Section 4 we discuss the relation between the external form of a crystal and its space lattice, and note that this form is predicted by locally defined growth rules for crystallites. Finally, in Section 5 we reach the nucleus and conclude with a discussion of an open problem, the origin of the symmetry and form of crystal twins.

2. LOCAL SYMMETRY AND GLOBAL STRUCTURE

A crystal is a structure in which a vast number of atoms (or molecules) are arranged in a more or less orderly three-dimensional pattern. In order to study its geometry and symmetry, we must replace this imperfect and complicated real object by an abstract one which is perfectly orderly and much simpler, but shares certain of its key characteristics. Thus, instead of a real crystal we consider a set of points in three-dimensional space which represents the centers of its atoms. Then we place requirements on the set that model those that nature places on the crystal. First, since atoms cannot be placed arbitrarily close together, we assume that the point set is discrete, that is, (1) there is a minimum distance r between any pair of points in the set. And next, since the number of atoms in a crystal is enormously large, and since these atoms are arranged in a pattern that continues in all directions, we assume that our set has infinitely many points, homogeneously distributed at least in a statistical sense; that is, (2) there are no arbitrarily large "holes" in space—any sphere of radius greater than some number R has points of the set interior to it. This set will be a geometrically satisfactory model of a simple crystal structure (one in which all the atoms are alike and play equivalent roles) if all the points in the set are equivalent; such a set is called a regular system of points (Fig. 3).

Hilbert and Cohn–Vossen explained regularity in these words[6]: "Let us draw the lines connecting some fixed point of a regular system with all the other points of the system, and then do the same for a second fixed point. Then . . . the two configurations of straight line segments obtained in this way are congruent, i.e., there is a well-defined motion . . . that brings one of these figures into coincidence with the other." Thus, to each point of the set we associate an infinite "spider," and it is the congruence of these spiders that defines the regularity of the system of points.

Translated into point–set geometry, the crystal problem becomes: Is the regularity of the entire system built up from local configurations of points? Can we show that the congruence of the infinite spiders is the result of some less-restrictive conditions?

The answers to these questions are yes, as Delone, Dolbilin, Shtogrin and Galiulin showed in 1976[3]. Since this important result is not widely known, we will outline a proof.

Assume that we have a set of points in three-dimensional space satisfying conditions (1)

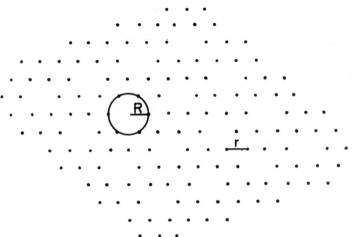

Fig. 3. A portion of a point set satisfying conditions (1) and (2). For any such set, $r < 2R$; in this example,
$r = R$.

and (2) above, but with no other restrictions placed on them (for the moment). Let x be any point of the set and let $S_x(\rho)$ be the finite spider joining x to all points of the set within a sphere of some positive radius ρ (Fig. 4). If $S_x(\rho)$ is congruent to the spider $S_y(\rho)$ about some other point y, then there is a motion—for example, a reflection in the dotted line—which brings $S_x(\rho)$ into coincidence with $S_y(\rho)$. [In fact, the reflection interchanges $S_x(\rho)$ and $S_y(\rho)$.] The idea of the proof is that if ρ is properly chosen, and if the spiders of radius ρ about all the points of the set are congruent, then any motion which brings one of these finite spiders into coincidence with another brings their infinite spiders into coincidence as well, and so the system is regular.

Let us look again at Fig. 4. The reflection which interchanges $S_x(\rho)$ and $S_y(\rho)$ is not a symmetry of the set of points. As Fig. 5 shows, the reflection does not even bring the next nearest neighbors about x into coincidence with the next nearest neighbors about y; the larger spiders are brought into coincidence only by 180° rotation. But, as we can see by careful inspection, this rotation brings not only the second nearest neighbors about x, but also the third, the fourth, and so on, onto the corresponding neighbors about y. Can we always choose the radius of the spider so that this must happen? We will show that if the points of the set satisfy a certain local condition then the answer is yes.

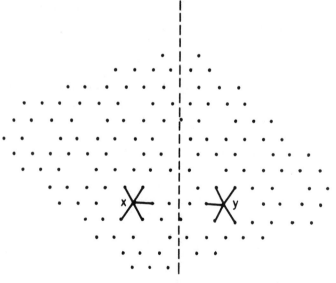

Fig. 4. The finite spiders joining points x and y to their nearest neighbors in the set. The spiders are congruent. They can be interchanged by reflection across the dotted line or 180° rotation about the midpoint of the line segment joining x and y (not shown).

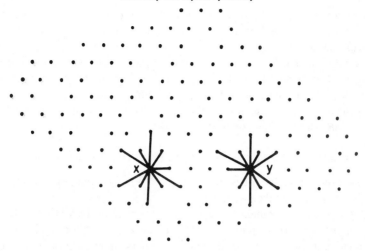

Fig. 5. The finite spiders joining *x* and *y* to their nearest and next nearest neighbors. Only 180° rotation about the midpoint of the line segment joining *x* and *y* can bring one of these spiders into coincidence with the other.

First, we note that the symmetry group of any spider is finite, and if the spider is enlarged, its symmetry group cannot increase: either it remains the same, or it decreases. But even if it decreases, eventually the symmetry must stabilize, since in a decreasing chain of symmetry groups the number of symmetries at each stage is less than the number in the preceeding stage. (In our example, the symmetry of the spiders which include both first and second nearest neighbors has stabilized, since these spiders have no symmetry at all.)

Next, we impose a third condition on our point set (3): the spiders of radius $\rho_0 = \rho + 2R$ about all the points of the set are congruent, where ρ is the smallest radius such that the symmetry groups of the spiders of radius ρ and radius $\rho + 2R$ are the same.

Now we can complete the proof. Let *x* and *y* be any two points of the set. Condition (2) ensures that every point of the system has a neighboring point less than distance $2R$ away. So, if we enlarge the radius of a spider by $2R$, we necessarily increase the number of arms. Condition (3) ensures that the motion that brings $S_x(\rho)$ into coincidence with $S_y(\rho)$ also carries $S_x(\rho + 2R)$ onto $S_y(\rho + 2R)$. If *w* is a point of the set which is within distance $2R$ of *x*, then $S_w(\rho)$ is entirely contained in $S_x(\rho + 2R)$ and so $S_w(\rho)$ is carried to a spider $S_z(\rho)$, where *z* lies within distance $2R$ of *y* (Fig. 6). But then, since $S_w(\rho + 2R)$ has the same symmetry as $S_w(\rho)$ and is congruent to $S_z(\rho + 2R)$, we see that the same motion must also carry $S_w(\rho + 2R)$ to $S_z(\rho + 2R)$. Any point of the set can be joined to any other by a chain of line segments each of length

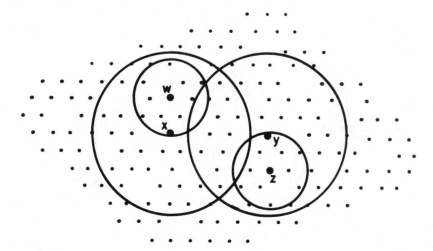

Fig. 6. Since $S_x(\rho + 2R)$ contains $S_w(\rho)$, the motion which carries $S_x(\rho + 2R)$ to $S_y(\rho + 2R)$ carries $S_w(\rho)$ to $S_z(\rho)$. (The arms of the spiders are not shown.)

less than 2R [condition (2)], so by repeating this argument, the same motion carries any point of the set onto some other point of the set, that is, the infinite spider about x is brought into coincidence with the infinite spider about y, and hence the system of points is regular.

This theorem shows that Thompson's hypothesis (quoted in the introduction) has a structural basis in geometry. It does not tell us, however, what is near and what is remote, that is, what the radius actually is. It has been shown that in two dimensions ρ can always be chosen to be 4R[7], and recent results concerning an analogous problem for tilings[8] suggest that in three dimensions ρ = 6R. But this has not yet been proved.

3. THE STRUCTURE OF A REGULAR SYSTEM OF POINTS

What does a regular system of points look like? What are the key features that allow us to classify these point sets by their symmetries?

Figure 3 is a typical two-dimensional regular system. If we look at it carefully, and draw the (finite) spiders about some more of the points, we see that they occur in six different orientations: relative to any one of them, the other spiders either have the same orientation, or else they are rotated 60°, 120°, 180°, 240°, or 300° (Fig. 7). In fact, it follows from the discreteness of the system that in any two-dimensional system the number of orientations of the spiders will be finite. We can see this with the help of a little analytic geometry.

In the plane every rotation keeps a single point fixed, the center of rotation. If the center is a point of the system, then the angle of rotation must be at least 60°: let p be the center, and q one of its neighbors at the minimal distance r. Then q is rotated about p to q', also at distance r from p. Since the distance from q to q' must be at least r, the angle of rotation must be at least 60° [Fig. 8(a)]. On the other hand, if the center of rotation is o, a point which does not belong to the system, then there is a point of the system, say q, whose distance d from o is not greater than R. When the system is rotated about o, q is carried to some point of the system q', a distance w ⩾ r away. Applying the law of cosines to the triangle qoq' [Fig. 8(b)], we have

$$2R^2(1 - \cos \theta) \geqslant 2d^2(1 - \cos \theta) = w^2 \geqslant r^2,$$

so θ cannot be arbitrarily small. It follows that in either case only a finite number of angles of rotation are possible, and therefore the spiders can have only finitely many orientations.

Now since there are infinitely many spiders and only finitely many orientations, we conclude that there are at least two spiders of the same orientation, either one of which can be brought into coincidence with the other by translation. Since this translation is a symmetry of the entire system, there must be infinitely many spiders of this same orientation in the row defined by the first two. In fact, there must be a spider of this orientation which is not in that row, so we have translation in a second direction as well. For if the centers of all the spiders of this

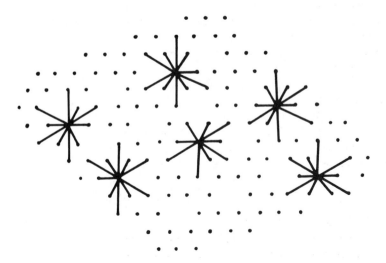

Fig. 7. The spiders in this regular system of points occur in six orientations.

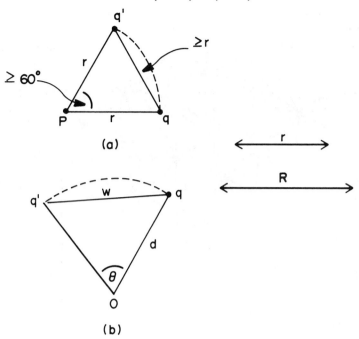

Fig. 8. The angle of rotational symmetry of a two-dimensional regular system of points cannot be arbitrarily small, whether the center of rotation is (a) or is not (b) a point of the system.

orientation lay in a single row, then the centers of all the spiders of any other orientation would also have to lie in a single row, and thus the points of the system would lie in a finite number of rows. But then the system would have holes of arbitrarily large size. The two translations generate a lattice (Fig. 9), whose nodes are the centers of the spiders of the same orientation.

In fact, the centers of the spiders of each orientation form a lattice, and so the regular system of points consists of a finite number (in this example, six) of interpenetrating lattices. The lattice is the principle structural element of a regular system of points: the different symmetry groups of regular point systems are obtained by considering which symmetries are compatible with the lattice structure. The details for the two-dimensional case are lucidly explained in [9].

More generally, a regular system of points in n dimensions contains an n-dimensional translation lattice, but if n is greater than 2 this is not easy to prove. The difficulty in three dimensions arises from the fact that spiders of different orientations may be related by a screw rotation, which is a combination of rotation and translation. No matter how small the angle of rotation, a screw rotation still moves a point a minimum distance because of the translation component. This creates considerable complications in the proof. The theorem can still be proved by elementary methods, however[10]. (See also the original proof by Schoenflies[11]).

Elementary methods cannot be used in higher dimensions, because when n is greater than 3 our simple picture of a rotation breaks down. The existence of lattices in regular systems of points in four and higher dimensions was established by Bieberbach in 1910 as part of his affirmative answer to the first part of Hilbert's 18th problem[12]. (At an international congress of mathematicians in 1900, Hilbert presented a list of 23 important problems as a challenge to the mathematicians of the 20th century[13]. The first part of the 18th problem asked whether the number of symmetry groups of regular systems of points in n-dimensional space is finite.)

We now return to three-dimensional space to consider the influence of the lattice on the form and symmetry of crystals.

4. LATTICES AND POLYHEDRA

In 1849 the French crystallographer Bravais proved that there are essentially 14 types of point lattices in three-dimensional space[14]. He then applied his lattice theory to some fundamental crystallographic problems, including the relation between internal structure and external form. In particular, he showed that we can determine, from the lattice of a crystal, the

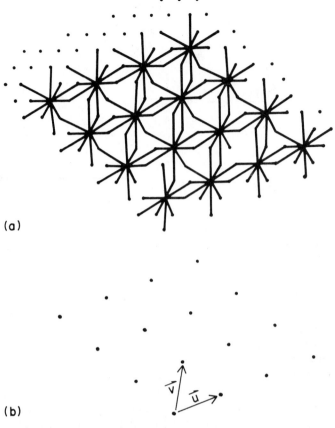

(a)

(b)

Fig. 9. In a regular system of points, the centers of the spiders of a single orientation (a) lie at the points of a lattice (b). This lattice can be generated by the two translations indicated by vectors **u** and **v**.

planes that are likely to appear as its faces[15]. In the intervening 137 years, Bravais' ideas have stood the test of time rather well. Reality is much more complicated, of course, but the faces of many crystals with simple structures are those which are predicted by Bravais' theory[16]. This theory shows how far a little geometry (together with a few additional assumptions) can take us toward an understanding of crystal form.

In outline, Bravais' theory is this: Let us assume that in a crystal the attractive forces between atoms decrease with increasing distance between them, and that if an atom has several near neighbors, its attractive force is distributed among them. Consider now an infinite set of atoms whose centers lie at the nodes of a three-dimensional lattice (Fig. 10). The points of the lattice lie equally spaced in rows, the rows lie equally spaced in planes, and the planes are stacked at equal distances to form the lattice. There are infinitely many ways to dissect the lattice into rows and planes, and the planes normal to different directions may differ greatly in the spacing of the points that they contain. The more densely the points are arranged in a lattice plane, the greater is the distance to the next parallel plane in the stack. In the language of forces, this says that layers of densely packed lattice planes are weakly bonded, while stacks of relatively open planes have strong interplanar attractions.

Crystal growth ("one upon one, layer by layer") can be expected to respect this argument. The "rate-determining step" in growth is thought to be the formation of successive layers. If this is so, then the shorter the distance between parallel planes, the faster the crystal will grow in the perpendicular direction. But the faster a crystal face grows, the smaller it will be relative to faces which grow more slowly (Fig. 11). Thus, as general rule, a crystal will be bounded by its faces of slowest growth; that is, the surface layers can be expected to be those lattice planes which are most densely packed. This is Bravais' "Law of Reticular Density."

The Russian crystallographer Wulff pointed out in 1908[17] that Bravais' law provides the basis for predicting a crystal's external symmetry and form. Wulff had already published his famous construction for the so-called equilibrium form of a crystal, which is the polyhedron

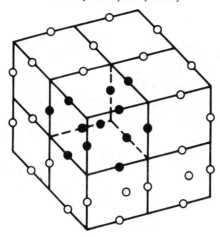

Fig. 10. A portion of a three-dimensional lattice. In this lattice, the points are located at the centers and midpoints of the edges of cubes which are stacked together to fill all the space.

bounded by the faces of lowest free surface energy. This construction does not presuppose a crystal lattice; it is only necessary to know the free surface energies of the possible crystal faces. The construction is very simple: from a point P (inside a hypothetical crystal), draw vectors perpendicular to all possible crystal faces, whose lengths are proportional to the corresponding free surface energies. At the tip of each vector, construct the plane perpendicular to it. The lower the energy, the closer the plane will be to P. Each plane divides space into two half-spaces, one which contains P and one which does not. The intersection of the half-spaces containing P is a closed convex polyhedron about P; this is the Wulff equilibrium form.

Now Wulff applied his ideas to a lattice L. Since large interplanar spacings in L correspond to low free energies of lattice planes, he constructed a second lattice L* whose interplanar spacings are reciprocal to those of the original lattice L (Fig. 12). (The reciprocal lattice is a well-known construction which plays an important role in X-ray crystallography.) For every set of parallel lattice planes in L, there is a vector in L* perpendicular to them whose length is reciprocal to the distances between parallel planes. The shorter the vector, the lower the free energy of the plane. Thus the Wulff construction, applied to all the vectors of L* issuing from some reciprocal lattice point o, will give us the equilibrium form of the crystal with lattice L (Fig. 13). It should be noted that this polyhedron is, except for a scaling factor, the "Dirichlet domain" of the point o of the reciprocal lattice, which is the closure of the region of space whose points are closer to o than to any other point of L*. There are only five possible topological

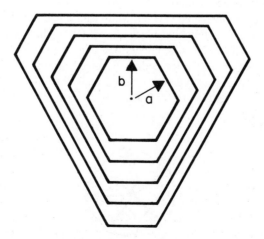

Fig. 11. The largest faces of a crystal are those with the slowest growth rates. In this two-dimensional example, the growth rate per unit time in direction a is twice the rate of growth in direction b. After four units of time, the b faces are clearly predominant.

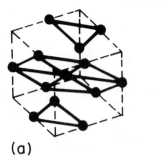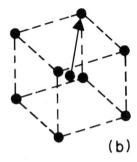

(a)   (b)

Fig. 12. (a) A set of parallel planes in the lattice of Fig. 10. (b) A portion of the reciprocal lattice. Its points are located at the centers and vertices of cubes. The reciprocal lattice vector perpendicular to the planes in (a) is shown.

types of Dirichlet domains for three-dimensional lattices; they are Fedorov's five parallelohedra (Fig. 14).

The polyhedron just constructed is the Wulff form of a crystal with lattice L; it is the form predicted by Bravais' law. The crystal forms found in this way may vary in symmetry (depending on the symmetry of L), but in every case their symmetries must be compatible with the topology of the parallelohedron of the lattice. Thus by inspecting Fig. 14 we conclude immediately that the rotational symmetry of a crystal can only be twofold, threefold, fourfold or sixfold. (This fundamental result was known to Haüy; the argument presented here makes the dependence of external symmetry on internal structure especially clear.)

Of course the entire atomic structure, not just the abstract lattice, helps to determine the morphology of a real crystal. Furthermore, the forms of real crystals are highly sensitive to the environmental conditions in which they grow. But even so, many crystals do have a Wulff shape[18]. An historical review of the problem of the relation between crystal structure and morphology can be found in [19].

The theoretical Wulff form is fundamental, in the sense that it says everything that can be said on the basis of lattice geometry. Surprisingly, Wulff's construction also turns out to be the form that a crystal achieves when its growth is studied as a local process. We conclude this section with a brief discussion of some interesting work on cellular automata by Willson[4,5].

Instead of an infinite set of points, we now consider a finite set which occupies some of the nodes of an infinite lattice. This configuration can be regarded as a fledgling "crystallite," which grows as more of the nodes become occupied. We assume that the rule that governs growth is a local one, so that whether or not a node becomes occupied at a given time depends only on some rule which is a function of the presence or absence of points at neighboring nodes. What can be said about the shape of the configuration as it grows?

To model the growth of the crystallite, Willson has studied the growth of configurations which are governed by ordered transition rules. He begins with a fledgling crystallite w, which has already reached a certain threshold size but whose shape is otherwise arbitrary. The growth of w is governed by a rule F, a function which specifies how the occupancy of a node at one moment of time is determined by its neighbors one unit of time earlier. F is assumed to be ordered, which means that once a lattice node is occupied, it remains that way and set–subset relations among configurations are preserved during transition. Thus Fw represents the crystal after one unit of time, F^2w after two units of time, and so forth. The growth of the configuration is represented by the sequence of configurations w, Fw, F^2w, . . . , F^pw, ... Remarkably,

Fig. 13. According to Wulff, the form of a crystal with the lattice of Fig. 10 is the truncated octahedron, which is the smallest polyhedron enclosed by planes perpendicular to the vectors of the reciprocal lattice [Fig. 12(b)].

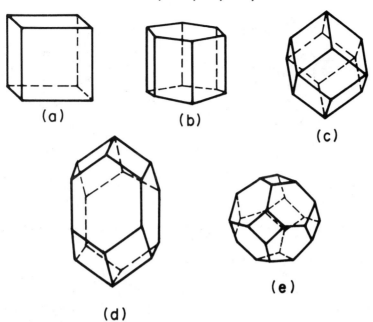

Fig. 14. Fedorov's five parallelohedra are the Wulff forms of crystals whose atomic patterns are three-dimensional lattices. No lattice can have rotational symmetry which is not a symmetry of a parallelohedron; thus the rotation can be only one-, two-, three-, four-, or sixfold.

it turns out that as p increases, $F^p w$ approaches a polyhedron W whose shape depends only on F. Thus the crystallite w has a polyhedral destiny which is completely determined by its transition rule. Moreover, W can only have those angles and symmetries which are permissible in crystals. In fact, W turns out to be the intersection of certain half-spaces; it is closely related to the Wulff construction.

Willson's results complement the theorem of Sec. 2 in a very pleasing way: together they assert that local geometry determines both the structure and form of an abstract crystal.

We now proceed from the crystallite to the very first stage of crystallization, the nucleus.

5. CRYSTAL TWINS

Not all crystals have convex polyhedral forms. Some crystals have curved faces; others, like snow crystals, are not polyhedral but dendritic. Perhaps most fascinating of all are the forms of crystal twins, two crystals of the same kind which are joined together along a face or even interpenetrate (Fig. 15).

The mutual arrangement of the individual crystals which constitute a twin is not arbitrary, but always obeys a "twin law," a symmetry operation which interchanges the positions of the individuals. For example, in Fig. 15(a), the twin law is reflection in a plane, in Fig. 15(b) it is 180° rotation about an axis, and in Fig. 15(c) it is 60° rotation followed by reflection in a plane perpendicular to the rotation axis.

As we have seen, the external symmetries of single crystals are due to their internal regularity, so we may also suppose that the twin laws indicate a well-defined relation between the orientations of the atomic patterns of the individuals of a twin. (X-ray studies confirm this.) It is also plausible to assume that the boundary between the individual crystals has a structure compatible with both orientations; if so, the boundary structure is likely to be closely related to the normal one. Very few experiments have been carried out to test this hypothesis, however.

The "twin problem" is to develop a theory of twinning which explains the form and symmetry of twins on the basis of what is known (or can reasonably be assumed) about their genesis, development, and internal structure.

What do we know about the genesis of crystal twins? Some kinds of twins are formed after the crystal is grown (by mechanical stress, or in a phase transition to a modified structure). Others are formed in the course of growth. Growth twins are not well understood: the widespread

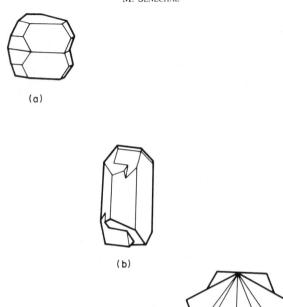

(a)

(b)

(c)

Fig. 15. Twinned crystals, adapted from Goldschmidt's Atlas der Kristallformen: (a) gold, (b) orthoclase, (c) fluorite.

assumption that they are created when two crystallites coalesce seems to be incompatible with the observed twin laws and with experimental studies of the conditions under which twinning occurs[20].

There is, however, an alternative theory of the genesis of growth twins, which assumes that twinning begins in the crystal nucleus. This idea has been mentioned in the literature from time to time, but it has never been worked out in detail. In this concluding section I will describe some of the interesting geometric problems which it entails.

Every crystal begins its existence as a nucleus. In our simplified model, we can think of the nucleus as a small number of equal spheres closely packed together in a sort of knobby ball. The ball grows as more spheres are added to it. At some point, somehow, the configuration begins to look less like a ball and more like a portion of the lattice characteristic of a crystal, that is, it becomes a crystallite. To understand how this happens, we need to study the geometry and growth of clusters of spheres.

It is easy to see that in the plane, a maximum of six circles of a given radius can be arranged around another circle of the same radius, and the pattern thus established can be continued forever. But in dimensions greater than two, sphere-packing problems are notoriously difficult. It took almost 200 years to prove that, in three dimensions, 12 and not 13 is the maximum number of equal spheres which can be placed in contact with another one. These twelve spheres can be arranged in many different ways; the three best-known arrangements are shown in Fig. 16. If we continue the first arrangement [Fig. 16(a)] we obtain the lattice of Fig. 10. The second [Fig. 16(b)] can also be extended to a regular system of points (this one is not a lattice). In the third [16(c)] the 12 spheres lies at the vertices of an icosahedron, which has fivefold rotational symmetry; this configuration cannot be extended to a regular system of points.

It is tempting to suppose that the first two arrangements in Fig. 16 model the nuclei of normal crystals, since the patterns they establish can be continued indefinitely. It has also been suggested that the frequent twinning of the many metals whose atomic structure is the lattice of Fig. 10 can be accounted for by a nucleus of the type shown in Fig. 16(b). This configuration has a median reflection plane; the layers of spheres above it and below it could, however, be continued in the pattern of Fig. 10. (Thus a nucleation twin would arise from an already twinned nucleus.)

The puzzling suggestion that this simple picture may not be correct comes to us from recent work on the geometry of small clusters by physicists and other scientists who are interested in them. Small clusters of spheres in three-dimensional space are an important model for phenomena in many fields (see [21] for a broad review). They have been studied intensively by various methods, including computer simulations and experimental studies of atomic clusters in metals and rare gases. One fact persistently emerges from these studies: if the number of spheres is relatively small, a stable (low-energy) configuration is likely to have icosahedral symmetry.

This has important implications for crystal growth. We noted above that the icosahedron has fivefold rotational symmetry, which cannot occur in a regular system of points, so if we continue to add equal spheres to an icosahedral cluster, maintaining its fivefold symmetry, we cannot form a crystal. On the other hand, the icosahedral symmetry itself cannot be maintained beyond several additional layers of spheres. This suggests that the structure of any crystal, whether it will eventually be single or twinned, undergoes a major transition as it changes from a nucleus to a crystallite. It also suggests a new model for the formation of nucleation twins.

Let us suppose that the transition from nucleus to crystallite begins locally, perhaps at several different places in the growing cluster. If the growth of the nucleus is normal, these transitions will "cooperate" and a single normal crystal will result. On the other hand, if the growth of the nucleus is chaotic, then during the transition it may simply fall apart. Twinning may be an intermediate case, in which the transition begins in several locations, which then develop independently into domains of normal growth, possibly with different orientations. In order for growth to continue, these domains would have to cooperate to the extent of sharing boundary structures (the boundary surface might be planar in some cases, or quite complicated in others). This process could account for crystal triplets, quadruplets, and so forth, as well as twins.

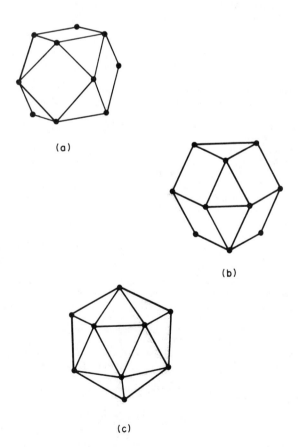

(a)

(b)

(c)

Fig. 16. Polyhedra representing three arrangements of 12 equal spheres about a central sphere of the same size. The centers of the outer spheres are located at the vertices of the polyhedra.

The theory just outlined, though quite vague, raises some challenging problems. Since clusters may grow to as large as 1000 atoms before their symmetry changes, we will need to study the growth and transition of complex multilayer structures. This problem has been considered by Mackay[22] and also by Farges and his colleagues[23] and others, but we do not yet have a clear picture of the sequences of the transition. Among the unsolved problems presenting themselves for consideration are the following.

How large can a multiilayer icosahedral cluster of equal spheres grow before its symmetry necessarily breaks down?

Under what conditions does the breakdown of this symmetry bring about a transition to a configuration with crystallographic symmetry? What are the small-scale structural changes involved and how do they cooperate to bring about a new overall pattern?

Under what conditions does the transition lead to a multiple structure? How are the symmetries of the multiple structures related to the known twin laws?

And finally, can we predict the forms of crystal twins[24]?

The answers to these questions may make an important contribution to the solution of one of the most fascinating problems in crystal geometry.

Acknowledgments—I would like to thank Doris Schattschneider, Stephen J. Willson, Branko Grünbaum, and the students in my applied mathematics course "The Geometry of the Solid State" for very helpful comments on an earlier draft of this paper. I am also grateful to Piet Hartman, N. N. Sheftal', Ravil Galiulin, Jean Farges and Gerard Torchet for stimulating discussions of crystal twins.

REFERENCES

1. D'Arcy W. Thompson, *Growth and Form*. Cambridge University Press (1917).
2. C. Radin, Tiling, Periodicity, and Crystals. Preprint, University of Texas.
3. B. N. Delone, N. P. Dolbilin, M. I. Shtogrin and R. V. Galiulin, A local criterion for regularity of a system of points. *Doklady Akademii Nauk SSSR* **227** (1976) (in Russian). English translation in *Soviet Math. Doklady* **17**, 319–322 (1976).
4. S. J. Willson, Limiting shapes for configurations. *Journal of Computer and System Sciences* **15**, 243–261 (1977).
5. S. J. Willson, On coherent growth of configurations. *SIAM Journal on Mathematical Analysis* (to appear).
6. D. Hilbert and S. Cohn–Vossen, *Geometry and the Imagination*. Chelsea, New York (1953).
7. M. I. Shtogrin, unpublished result.
8. P. Engel, On the regularity of discrete point sets (to be published).
9. D. Schattschneider, The plane symmetry groups, their recognition and notation. *American Mathematical Monthly* **85**, 439–450 (1978).
10. B. N. Delone and M. I. Shtogrin, A simplified proof of Schoenflies' Theorem. *Doklady Akademii Nauk SSSR* **219**, 95–98 (1974) (in Russian).
11. A. Schoenflies, *Krystallsysteme und Kristallstruktur*. Leipzig, Teubner (1891). Reprinted 1984 by Springer-Verlag.
12. L. Bieberbach, Über die Bewegungsgruppen der *n*-dimensionalen Euklidischen Räume mit einem endlichen Fundamentalbereich. *Göttinger Nachrichten*, 75–84 (1910).
13. D. Hilbert, Mathematische Probleme. *Göttinger Nachrichten*, 253–297 (1900).
14. A. Bravais, Memoire sur les systèmes formés par des points distribués regulierement sur un plan ou dans l'espace. *Journal de l'Ecole Polytechnique* **33**, 1–128 (1850).
15. A. Bravais, Etudes crystallographiques. *Journal de l'Ecole Polytechnique* **20**, 101–276 (1851).
16. G. Friedel, Etudes sur la loi de Bravais. *Bulletin de la Societé française de Minèralogie* **20**, 326–455 (1907).
17. G. Wulff, Zür Theorie der Kristallhabitus. *Zeitschrift für Kristallographie* **45**, 433–473 (1908).
18. N. I. Leonyuk, R. V. Galiulin, L. I. Al'shinskaya and B. N. Delone, The practical determination of perfect habits of crystals. *Zeitschrift für Kristallographie* **151**, 263–269 (1980).
19. C. J. Schneer, Morphology and crystal growth, in *Crystallography in North America* (Edited by D. McLachlan and J. P. Glusker). American Crystallographic Association (1983).
20. M. Senechal, The genesis of growth twins in crystals. *Kristallografia* **25**, 908–915 (1980) (in Russian). English translation in *Soviet Physics—Crystallography* **25**, 520–524 (1980).
21. M. R. Hoare, Structure and dynamics of simple microclusters, in *Advances in Chemical Physics* (Edited by I. Prigogine), Vol. 40. John Wiley & Sons (1979).
22. A. L. Mackay, A dense noncrystallographic packing of equal spheres. *Acta Crystallographica* **15**, 916–918 (1962).
23. J. Farges, M. F. de Feraudy, B. Raoult and G. Torchet, Relaxation of Mackay Icosahedra. *Acta Crystallographica* **A38**, 656–663 (1982).
24. M. Senechal, The mechanism of formation of certain growth twins of the penetration type. *Neues Jahrbuch für Mineralogie* **11**, 518–524 (1976).

Comp. & Maths. with Appls. Vol. 12B, Nos. 3/4, pp. 579–583, 1986
Printed in Great Britain.

0886-9561/86 $3.00 + .00

THE GENERALIZED PETERSEN GRAPH G(24, 5)

H. S. M. Coxeter

Department of Mathematics, University of Toronto, Toronto, Canada M55 1A1

Abstract—The symmetric trivalent graph with 48 vertices has a group of automorphisms of order 288 with the presentation

$$\delta^2 = \sigma^3 = (\delta\rho)^2 = (\delta\sigma)^2 = (\sigma\rho)^2 = 1, \quad \rho^4 \rightleftarrows \sigma.$$

The three involutory elements $R_1 = \sigma\rho$, $R_2 = \sigma^{-1}\rho\sigma^{-1}$, $R_3 = \rho\sigma$ generate one subgroup of order 48, while

$$T_1 = \delta\rho, \quad T_2 = \sigma^{-1}\rho\delta\sigma, \quad T_3 = \rho\delta$$

generate another. In either case the generators can be represented as colored edges of the graph.

1. INTRODUCTION

The Petersen graph, which has 10 vertices, is sometimes denoted by G(5, 2) because it can be drawn as a regular star-pentagon {5/2} inside an ordinary pentagon {5}, with corresponding vertices joined. Frucht, Graver and Watkins[5, p. 212] generalized this to G(n, d), whose 2n vertices belong to polygons {n} and {n/d}. If this graph is to be vertex-transitive and edge-transitive, there are just five possible values for n/d:

$$5/2, \quad 8/3, \quad 10/3, \quad 12/5, \quad 24/5.$$

Thus the most complicated graph of this kind is G(24, 5). When suitably edge-colored, it serves as the Cayley diagram for the unimodular group $GL(2, 3)$ of order 48[2, p. 111], generated by involutions R_1, R_2, R_3 such that

$$R_1R_2R_3R_1 = R_2R_3R_1R_2 = R_3R_1R_2R_3.$$

The same graph, differently colored, serves as the Cayley diagram for a less familiar group of the same order 48, whose three generating involutions T_1, T_2, T_3 satisfy

$$(T_2T_3)^2 = (T_3T_1)^2 = (T_1T_2)^2.$$

Both these presentations were used by R. M. Foster in 1966, as alternative ways to describe the symmetrical trivalent graph with 48 vertices, whose group of automorphisms has order 288.

2. THE PARTIAL PLANE 8₃

The unimodular group $GL(2, 3)$, of order 48, is the group of 2 by 2 matrices of determinant ± 1 over the field $GF[3]$, that is, the group of projective and antiprojective collineations in the partial plane 8_3 which is derived from the affine plane $AG(2, 3)$ by omitting the origin and the 4 incident lines[3, p. 230]. One way to generate it is by the 3 involutions

$$R_1 = \begin{bmatrix} 0 & 1 \\ 1 & 0 \end{bmatrix}, \quad R_2 = \begin{bmatrix} -1 & 1 \\ 0 & 1 \end{bmatrix}, \quad R_3 = \begin{bmatrix} 1 & 1 \\ 0 & -1 \end{bmatrix} \quad (\text{mod } 3),$$

which satisfy the presentation

$$R_\nu^2 = 1, \quad R_1R_2R_3R_1 = R_2R_3R_1R_2 = R_3R_1R_2R_3. \tag{2.1}$$

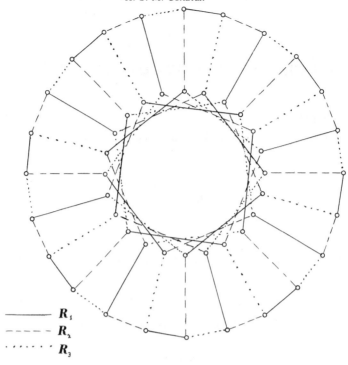

$$
\begin{aligned}
&\underline{\hspace{2.5cm}} \quad \boldsymbol{R}_1 \\
&\text{-----} \quad \boldsymbol{R}_2 \\
&\cdots\cdots \quad \boldsymbol{R}_3
\end{aligned}
$$

Fig. 1.

With this choice of generators, $GL(2, 3)$ has the Cayley diagram shown in Fig. 1, which is based on the graph G(24, 5) in the notation of Frucht, Graver and Watkins. (Starting at any vertex of the graph, we can trace the paths $R_1R_2R_3R_1$, $R_2R_3R_1R_2$ and see that they both take us to the same other vertex.) As the three R's all behave alike, it is natural to seek a matrix S, of period 3, that will transform each R_ν into the preceding one, in cyclic order, so that $R_2 = SR_1S^{-1}$ and $R_3 = S^{-1}R_1S$. The extended group that includes S is denoted by 2[8]3 or 3[8]2, because it has the presentation

$$R^2 = S^3 = 1, \quad (RS)^4 = (SR)^4, \tag{2.2}$$

where R is any one of the R_ν. These relations imply $(RS)^h = 1$, where $h = 24$[2, pp. 95, 106, 158]. From 2.1 we easily deduce

$$
S = \begin{bmatrix} 1 & 0 \\ 1 & 1 \end{bmatrix}.
$$

Since this matrix belongs to $GL(2, 3)$, it transforms that group by an *inner* automorphism, and the extended group 2.2, of order $3 \times 48 = 144$, is

$$2[8]3 \cong \mathcal{C}_3 \times GL(2, 3) \tag{2.3}$$

[2, p. 107]. Thus $GL(2, 3)$ itself has the alternative presentation

$$R^2 = S^3 = 1, \quad (RS)^4 = (SR)^4 = Z, \quad Z^6 = 1 \tag{2.4}$$

in terms of

$$
R = \begin{bmatrix} 0 & 1 \\ 1 & 0 \end{bmatrix} \quad \text{and} \quad S = \begin{bmatrix} 1 & 0 \\ 1 & 1 \end{bmatrix}.
$$

3. ROTATION GROUPS IN 4 DIMENSIONS

When each point (x_0, x_1, x_2, x_3) of Euclidean 4-space is represented by a quaternion

$$x = x_0 + x_1 i + x_2 j + x_3 k,$$

any motion keeping the origin fixed appears as a quaternion transformation $x \to \bar{a}xb$ or, more concisely,

$$\bar{a}xb$$

where a and b are fixed quaternions of unit norm[2, pp. 67–78] and $\bar{a}$ is the quaternion conjugate of a. Any finite group of direct orthogonal transformations can be described by the combination

$$(\mathbf{L}/\mathbf{L}_K; \mathbf{R}/\mathbf{R}_K)$$

of four finite groups of quaternions (cyclic, dicyclic or binary polyhedral)[4, pp. 50–57] such that $\mathbf{L}/\mathbf{L}_K \cong \mathbf{R}/\mathbf{R}_K$.

One such group of motions, of order 144, is

$$2[8]3 \cong (\mathscr{C}_{12}/\mathscr{C}_6; \langle 4, 3, 2\rangle/\langle 3, 3, 2\rangle) \tag{3.1}$$

[2, p. 101 (§10.8)], which is Du Val's No. 8 with $m = 3$. In the form 2.2, it is generated by motions R and S (the R_3 and R_2 of [2, (10.71)]) which are

$$ix(i + k)/\sqrt{2} \quad \text{and} \quad e^{\pi i/3} x(1 + i + j + k)/2.$$

The subgroup $GL(2, 3)$ is generated by R and its transforms by $S^{\pm 1}$; thus the R_1, R_2, R_3 of 2.1 can be represented by the half-turns

$$ix(j + k)/\sqrt{2}, \quad ix(k + i)/\sqrt{2}, \quad ix(i + j)/\sqrt{2}. \tag{3.2}$$

It is interesting to compare these with the half-turns

$$ixi, \quad ixj, \quad ixk$$

which generate the 'Pauli' group $\langle 2, 2, 2\rangle_2$, of order sixteen[3, p. 221].

4. A NEW ROLE FOR THE GRAPH G(24/5)

Figure 2 shows a different coloring for the same graph G(24/5). Tracing paths along its edges, we see that the 3 colors represent generators T_1, T_2, T_3[2, pp. 93 (Ex. 4), 173] which satisfy the relations

$$T_\nu^2 = 1, \quad (T_2 T_3)^2 = (T_3 T_1)^2 = (T_1 T_2)^2. \tag{4.1}$$

To verify that no further relations are needed, we must ensure that the group so presented has order 48. Accordingly, we investigate the subgroup of index 2 generated by the products

$$I = T_2 T_3, \quad J = T_3 T_1, \quad K = T_1 T_2,$$

which satisfy

$$I^2 = J^2 = K^2, \quad IJK = 1. \tag{4.2}$$

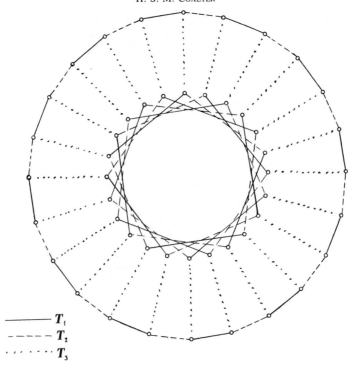

$$\underline{\quad\quad}\; T_1$$
$$-\;-\;-\;T_2$$
$$\cdots\cdots\;T_3$$

Fig. 2.

Since these relations hold for

$$I = -\omega i, \quad J = -\omega j, \quad K = -\omega k,$$

where ω, of period 3, commutes with i, j, k. This[1, p. 377, §11] is the group

$$\langle -2, -2, 2\rangle \cong \mathscr{C}_3 \times \langle 2, 2, 2\rangle \qquad (4.3)$$

where $\langle 2, 2, 2\rangle$ is the quaternion group of order 8 whose elements are ± 1, $\pm i$, $\pm j$, $\pm k$. Thus the order of 4.2 is 24, and that of 4.1 is 48, in agreement with the number of vertices of the graph.

We shall soon see that, when represented as motions in 4-space, T_1, T_2, T_3 are the half-turns

$$ix(j - k)/\sqrt{2}, \quad jx(k - i)/\sqrt{2}, \quad kx(i - j)/\sqrt{2}, \qquad (4.4)$$

whose products in pairs are $-ixa$, $-jxa$, $-kxa$, where

$$a = (1 + i + j + k)/2, \qquad (4.5)$$

as in [2, p. 76]. For, since $-kjixa^3 = (-1)x(-1) = x$, these products satisfy the relations 4.2. Hence the group 4.1 of order 48 is

$$(\langle 2, 2, 2\rangle/\langle 2, 2, 2\rangle; \langle 3, 2, 2\rangle/\langle 3, 2, 2\rangle),$$

the 'central product' of $\langle 2, 2, 2\rangle$ and the dicyclic group $\langle 3, 2, 2\rangle$ of order 12 (with an unusual choice of generators). This is Du Val's No. 10 with $m = 3$ and $n = 2$.

Since the three T's can be permuted, we may introduce an extra motion S, of period 3, which transforms each T_ν into the preceding one in cyclic order, so that $T_2 = ST_1 S^{-1}$ and $T_3 = S^{-1}T_1 S$. The extended group that includes S has the presentation

$$S^3 = T^2 = (S^{-1}TST)^2(STS^{-1}T)^2 = 1, \tag{4.6}$$

where S is $\bar{a}xa$ and T is any one of the T_ν, say $ix(j - k)/\sqrt{2}$. This group, of order $3 \times 48 = 144$, is thus seen to be

$$(\langle 3, 3, 2\rangle/\langle 3, 3, 2\rangle; \langle 3, 2, 2\rangle/\langle 3, 2, 2\rangle),$$

the central product of the binary tetrahedral group $\langle 3, 3, 2\rangle \cong SL(2, 3)$ and the dicyclic group $\langle 3, 2, 2\rangle$[2, p. 80]. This is Du Val's No. 14 with $m = 3$.

Since 4.6 has an automorphism of period 2 which interchanges S and S^{-1} while leaving T invariant, we can derive a group of 288 by adjoining an involuntary element R, which yields the presentation

$$R^2 = S^3 = T^2 = (RS)^2 = (RT)^2 = (S^{-1}TST)^2(STS^{-1}T)^2 = 1$$

or, since

$$(S^{-1}TST)^2(STS^{-1}T)^2 = (RSRTST)^2(STRSRT)^2 = (RST)^4 S(TS^{-1}R)^4 S^{-1},$$
$$R^2 = S^3 = T^2 = (RS)^2 = (RT)^2 = 1, \quad (RST)^4 \rightleftarrows S. \tag{4.7}$$

In terms of new generators $B = RST$ and $C = (ST)^{-1}$, we have

$$R = BC, \quad S = B^2C^2, \quad T = CB^2C^2$$

and the presentation becomes

$$(BC)^2 = (B^3C^2)^2 = (B^2C^3)^2 = 1, \quad B^4 \rightleftarrows C^2, \tag{4.8}$$

implying $B^{24} = C^{12} = 1$.

It is interesting to compare this with the group $[4 + 4, 3]$ of order 96, in which $B^4 \rightleftarrows C$ and $B^8 = 1[3, p. 243 (8.1)]$.

Since the graph G(24, 5) (without colors) has 72 edges and is 2-regular, its group of automorphisms has order $2^2 \times 72 = 288$, and can thus be indentified with the above group 4.8. Frucht, Graver and Watkins[5, pp. 217–218] derived the equivalent presentation

$$\delta^2 = \sigma^3 = (\delta\rho)^2 = (\delta\sigma)^2 = (\sigma\rho)^2 = 1, \quad \rho^4 \rightleftarrows \sigma \tag{4.9}$$

from 2.2 by defining $\rho = S^{-1}R = (RS)^{-1}$, $\sigma = S$ and adjoining an involutory element δ which transforms ρ and σ into their inverses. This can be reconciled with 4.7 by putting

$$\rho = RST, \; \delta = RS, \; \sigma = S; \quad R = \sigma\delta, \; S = \sigma, \; T = \delta\rho.$$

REFERENCES

1. H. S. M. Coxeter, The binary polyhedral groups, and other generalizations of the quaternion group. *Duke Math. J.* **7**, 367–379 (1940).
2. H. S. M. Coxeter, *Regular Complex Polytopes.* Cambridge University Press (1974).
3. H. S. M. Coxeter, The group of genus two. *Rend. Semin. Mat. Brescia* **7**, 219–248 (1984).
4. Patrick Du Val, *Homographies, Quaternions and Rotations.* Oxford University Press, London (1964).
5. R. Frucht, J. E. Graver and M. E. Watkins, The groups of the generalized Petersen graphs. *Proc. Camb. Philos. Soc.* **70**, 211–218 (1971).

Comp. & Maths. with Appls. Vol. 12B, Nos. 3/4, pp. 585–616, 1986
Printed in Great Britain.

0886–9561/86 $3.00 + .00
© 1986 Pergamon Press Ltd.

AN INTERPLAY
BETWEEN THE PHENOMENON
OF CHEMICAL ISOMERISM AND SYMMETRY
REQUIREMENTS: A PERENNIAL SOURCE OF STIMULI
FOR MOLECULAR-STRUCTURE CONCEPTS, AS WELL
AS FOR ALGEBRAIC AND COMPUTATIONAL CHEMISTRY†

ZDENĚK SLANINA

The J. Heyrovský Institute of Physical Chemistry and Electrochemistry, Czechoslovak Academy of
Sciences, Máchova 7, CS-121 38 Prague 2, Czechoslovakia

Abstract—One of the most general chemical phenomena, isomerism, can be fruitfully described by various theoretical approaches. Moreover, in some situations theoretical techniques are preferable or even the only source of information. All methodological approaches of theoretical chemistry (i.e., quantum-chemical within the Born–Oppenheimer approximation, rigorous quantum-mechanical, molecular-mechanical, algebraical or statistical-mechanical) can be productively combined with symmetry considerations. This article concentrates on symmetry properties of stationary points of potential energy hypersurfaces (for both rigid and nonrigid species), and especially symmetry rules for transition states including their symmetry numbers, and also discusses the role of symmetry in isomer enumerations, topological reduction of hypersurfaces and the graph-like state of molecules, as well as the thermodynamical and kinetical interplay of isomers of different symmetries. It is concluded that the symmetry approach to the problems of isomeric chemistry continues to be a productive complement of purely computational means, enabling direct insight into the logical structure behind chemical problems.

1. INTRODUCTION

The concept of isomerism, introduced into chemistry under this name by Berzelius[1], is continually being tested and broadened. Recently, it has been accepted that a definition of isomerism cannot[2,3] be given in the absence of defined experimental criteria of the conditions and methods of observation. Isomers are thus considered[4–6] as individual chemical species with identical molecular formulae that display at least some differing physicochemical properties, and that are stable for long periods of time in comparison with those during which measurements of their properties are made. New types of isomerism are being recognized (e.g. [7–13]) and even very simple compounds (e.g. [14–18], ClO_2, N_2O_4, H_2SO_4 or the smallest amino acid, glycine[19]) can be observed (or expected) to exhibit isomerism. This procedure is assisted considerably by the theoretical approaches that at present enable the discovery of new isomers (e.g. the closed form of ozone[20,21]) or the prediction of hitherto unforeseen reactions within an isomeric system (e.g. [22,23]). A number of chemically bizarre isomers of small organic molecules has recently been characterized in the study of interstellar species (for a review, see [24]).

Algebraic methods represented the first theoretical approach to yield useful results in the study of isomerism. As the use of these techniques does not necessarily require the application of quantum theory, it is not surprising that, according to Ref. [25], the first serious works date from as early as 1874, when three papers were published on the problem of isomer enumeration by Cayley[26], Körner[27] and van't Hoff[28]. It should, however, be noted for the sake of completeness that the very first application of algebraic methods in a chemical context involved the introduction of graphs for the presentation of molecules. This key step in the history of chemistry was taken[29] in the middle of the last century and is connected with works published by Couper, Crum Brown, Franklad, Kekulé, etc. Nonetheless, Dalton should probably be considered as the first chemical theoretician[30], as the concept of molecular structure began to develop in his atomic hypothesis formulated in 1807.

In addition to enumeration, algebraic techniques (set theory[31,32], group theory[33–35], graph theory[36,37], and information theory[37,38]) are frequently used for classification and characterization purposes. In spite of their abstract character, these algebraic methods permit

†Dedicated to the 50th anniversary of the publishing of Pólya's Hauptsatz[234].

the rationalization or systemization of the relationships between isomers (see, for example, [39–51]). Graph theory represents a basis for the analysis of structure–activity relationships[52,53] in terms of topological indices (see [54–58]), carried out in the framework of the additivity principle (cf. [59,60]). Moreover, as already mentioned, algebraic procedures can lead to predictions of new reactions—automerizations proceeding through six-membered transition states (studied by Balaban[23]) are a classic example. Although, in contrast to quantum-chemical techniques, the algebraic approach does not lead to a description of the energetics, nonetheless it permits direct insight into the intristic logical structure behind chemical problems and, in addition, its use is not limited by the size of the system studied (or at least not as strictly as with quantum-chemical methods). The algebraic generalization of the notion of isomerism was formulated for the computer-assisted design of syntheses[32,61–67]. Later on, the theoretical techniques were enriched by the addition of quantum- and statistical-mechanical, quantum-chemical and molecular mechanics methods. In the two latter cases, the individual isomers and the relations between them are treated as minima and their interrelations on the corresponding potential-energy hypersurface(s)[68–81].

In all these approaches to isomeric chemistry symmetry considerations have played important and fruitful role. This article will be concerned with the classical as well as recent theoretical results in the study of symmetry properties of isomeric species and particularly with those lying on the frontiers of algebraic, quantum and computational chemistry.

2. THE CONCEPT OF POTENTIAL-ENERGY HYPERSURFACE

The concept of a potential-energy hypersurface is a consequence of the separation[82–85] of the nuclear and electronic motions as proposed by Born and Oppenheimer[82] in 1927. Adiabatic approximation underlies many of our chemical concepts; for example, the notion of molecular structure and energy barriers. The phenomenon of isomerism can be interpreted straightforwardly in these classical terms as the occurrence of more than one local minimum within one hypersurface or within the hypersurfaces of potential energy corresponding to various electronic states. The feasibility of transitions among single minima is determined by the barrier heights separating the adjacent minima and by temperature (at lower temperatures also by the possibility of quantum-mechanical tunnelling), and possibly also by transition probabilities between energy hypersurfaces. The same factors determine in principle the possibility of distinguishing between two configurations, some qualitative criteria being outlined for this purpose[86,87]. Ideally, isomeric structures (especially in the case of stereochemically nonrigid systems[88–94]) should be treated by means of the whole potential-energy hypersurface(s) comprising all possible configurations[68,69,95]. This level is, however, nowadays unattainable; nevertheless the question of the experimental distinguishing of structures corresponding to the individual local minima is a cardinal question of the concept of isomerism itself[2,3].

In the oldest nonrigid or fluxional molecules described (ammonia with respect to inversion[96], cyclopentane with respect to pseudorotation[97], trigonal bipyramids with respect to pseudorotation[98], rearrangements of bullvalene[88]—see Fig. 1), relatively fast transitions occurred between identical structures. A classical example is bullvalene (tricyclo $3.3.2.0^{4.6}$) deca-2.7.9-triene) exhibiting fast transitions of the position of the cyclopropane ring and double bonds, as depicted in Fig. 1. Theoretical considerations have proved[88,99] that there are a total of 1,209,600 ways in which this transition can occur. Averaging of the positions of the hydrogen atoms can be made possible by these transitions.

The adiabatic approach is an approximation only and therefore the same holds for the conventional quantum-chemical picture of isomerism. The classical concept of molecular structure is not consistent with requirements of quantum theory and therefore it has recently been criticized[100–106]. A rigorous quantum-mechanical interpretation of isomerism, i.e. a description in terms of the eigenfunctions of the total molecular Hamiltonian (which is the same for all species having a given summary formula) has not yet been completed. The identification of certain eigenfunctions of the molecular Hamiltonian with particular chemical species is an approximation[107]. In other words, the uniqueness of molecular species, which consists of the same kind and number of nuclei along with the same number of electrons, is only an approximation (which is good in most cases). Wilson[108] pointed out that chemists prepared substances by methods which select particular structures rather than exact energy states. Thus, the resulting

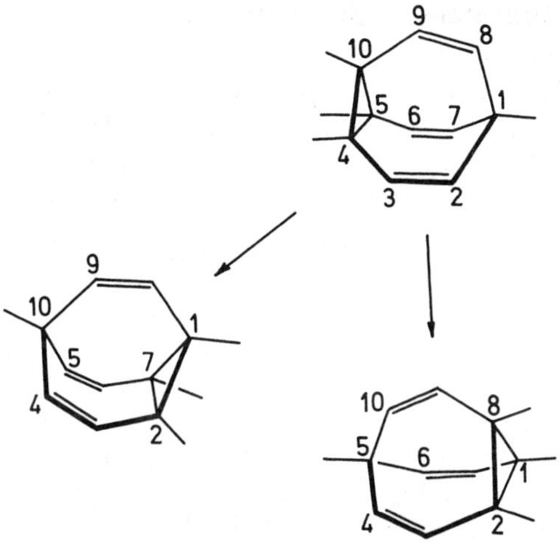

Fig. 1. An example of a fluxional molecule (bullvalene); 4 of 1,209,600 equivalent structures are depicted together with interconversions (according to Ref. [88]).

states need not be eigenfunctions of energy, i.e. stationary states. In fact, the uniqueness of isomers is well founded in terms of the nonstationary states provided the change with time is sufficiently slow.

It is well known that the present state of numerical quantum chemistry does not permit the carrying out of a routine evaluation of potential-energy hypersurfaces for systems of chemical interest. Low-number atomic systems consisting of nuclei at the very beginning of the periodic system represent an exception[73,80,109–111]. However, instead of the construction of the whole hypersurfaces quantum chemistry is able to localize and characterize the hypersurface stationary points, i.e. points having zero first derivatives of energy with respect to the nuclear coordinates. Thus, also in the field of isomeric chemistry, description of an isomeric system in terms of local energy minima and saddle points represents the present state-of-the-art limit in theoretical studies, in both quantum chemistry[74–76] and molecular mechanics[112–114] (see Fig. 2 for an illustrative example[115]). Identification of the stationary points is possible by an analysis of the eigenvalues of the force constant matrix F at these points[70–81]. Three kinds can be distinguished, namely local minimum, transition state or activated complex, and the higher types of stationary point; their F matrixes have zero, one, and two or more negative eigenvalues, respectively. There are three[116,117] significant levels of stationary-point location and identification: (i) double numerical differentiation, (ii) analytical differentiation followed by numerical differentiation, and (iii) double analytical differentiation. The pioneer works of Pulay[118–120] and of McIver and Komornicki[121] showed the way of mastering technique (ii). At present the effective treatment (ii) is used[122] within various quantum-chemical methods including the *ab initio* correlated wavefunctions[123–130]. Moreover, technique (iii) has recently been mastered[127] for *ab initio* SCF wavefunctions, which is promising[124–126] for the multiconfiguration SCF level to be completely settled soon. The F matrix serving the identification of stationary points can also be used for the carrying out of normal coordinate analysis[131,132]. This treatment enables us to describe the motion of nuclei around the stationary point. We can thus summarize that now in practically all areas of computational chemistry the methodology has reached the level to enable the study of chemical reactivity phenomena in terms of energy hypersurface critical points and their characteristics. However, the harmonic description represents the upper limit; the calculations of higher derivatives of energy are still quite scarce[120]. That is why it is necessary so far to describe the motion of nuclei in terms of the rigid rotator and harmonic oscillator (RRHO). Let us note that within this simple approach the motions of individual stationary points are independent.

3. SYMMETRY PROPERTIES OF STATIONARY POINTS

Symmetry has long played an essential role in the study of molecular or electronic structures[133]. Typical uses of symmetry in molecular physics often leads to a Boolean structure

1,3 - DIOXOCANE CONFORMATIONS

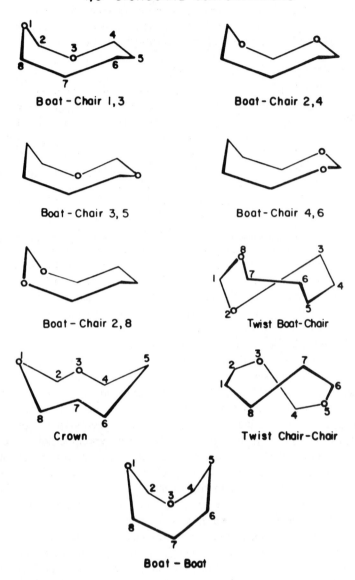

Fig. 2. Low-energy minima of 1,3-dioxacene (according to Ref. [115]).

of answers ("yes/no"). Although symmetry considerations cannot replace quantum-mechanical calculations, there are broad regions where they can point out calculations useful for the given problem[134,135]. Two distinct levels of utilization of the symmetry of chemical species can be distinguished (in the framework of the Born–Oppenheimer approximation). The first involves the symmetry of the electronic wave function (see, for example, [136,141]) of the relevant state, while the second is connected with the symmetry of the nuclear wave function, i.e. with the symmetry of the normal vibrational modes. A classic example of the utilization of symmetry rules for the correlation of electronic states is contained in the results of Wigner and Witmer[141], which determine the molecular states of diatomic molecules from separate atomic states. The best known results in the field of symmetry control of chemical reactions are certainly the Woodward–Hoffmann rules[142,143], while further important results have been developed by other workers (see, e.g. [144–151]). The Woodward–Hoffmann rules are concerned with concerted processes and postulate the conservation of electronic orbital symmetry. They do not introduce any other limits to the size or shape of the reaction system except for the requirement of the existence of at least one symmetry element which is common to both the reactants and products. The orbital correlation diagram is then constructed considering this symmetry. If the occupied orbitals of the reactants are correlated only with the occupied orbitals of the products, the reaction is considered to be allowed. When, however, any of the occupied orbitals of the reactants is correlated with an unoccupied orbital of the products (and vice versa), the reaction is considered to be forbidden. Among other things, the number of isomerization reactions has been treated[142,143,152] in terms of the Woodward–Hoffmann rules.

3.1 *Symmetry properties of rigid and nonrigid molecules*

The simplest approach to molecular species conceived of them as rigid or quasirigid aggregates. It is assumed that atomic nuclei vibrate rapidly with small amplitude around a fixed equilibrium position. The symmetry properties are conventionally described using a point group reflecting the geometrical symmetry of the equilibrium nuclear configuration. It was, however, soon realized that the point group of symmetry is not the only symmetry present in a molecule and that the point group is not completely utilizable for the study of rotating molecules[154,155]. Alternatively, a transition can be made from the symmetry elements to the permutations of identical nuclei, as the complete molecular Hamiltonian is invariant under all such permutations (in contrast to the point group of symmetry). The partial results of Wilson[154,155] and Hougen[156,157] were completed by Longuet–Higgins[158], who entirely demonstrated explicitly that the permutation–inversion group is connected with the fundamental symmetries of the complete molecular Hamiltonian (molecular symmetry group) and introduced the key concept of a feasible operation as a transformation which can be achieved without passing over an insuperable barrier, i.e. within the time scale of the experimental observation. This classical analysis of Longuet–Higgins[158] provided a basis for the subsequent broad development of the study of nonrigid molecules[159,163]. The molecular symmetry group was introduced[158] as the set of all feasible permutations P (under the given conditions) of the space and spin coordinates of the identical atomic nuclei in the molecule (including the identity E) and all the feasible permutation inversions $P^* = PE^*$ (where E^* is the operation of inversion through the molecular centre of mass of the space coordinates of all the particles). The particular group of molecular symmetry of a given nonrigid molecule is a subgroup of the complete permutation–inversion group. Berry[164] interpreted (cf. [162]) the introduction of the permutation–inversion group as a process of induction from the point group to an empirically determined supergroup generated from the point group by a finite set of feasible generators.

Study of the symmetry properties of individual nonrigid molecules soon revealed that the corresponding permutation–inversion groups are quite large (typically[162] $10–10^3$ elements—cf. Fig. 3) and also are not very common (isomorphous with the usual groups). This led to attempts to formulate groups of nonrigid molecules as semidirect products. An example is the technique of isodynamic operations introduced by Altmann[166] or the isometric group approach developed by Günthard *et al.*[167–169]. It should be noted that the relationship between permutation–inversion and the point groups is not strictly hierarchial, but rather complementary[161]. The point group of each nonlinear molecule is isomorphous with the subgroup of the complete permutation–inversion group formed by the completely feasible elements[159–163]. Provided that the discussion is limited to classification of the vibronic (rather than the rovibronic) wave functions, both types of groups lead to the same classification results. This permits, for example, the work with point groups of symmetry in discussion of the properties of the energy hypersurface related to the normal vibrational modes.

Another approach toward a more complete specification of molecular symmetry than that attained so far using point groups of symmetry alone involves the concept of a framework group proposed by Pople[170], which is especially useful for computer handling with structure–symmetry relations. It has been demonstrated[171] that the framework group for nonrigid molecules can readily be extended to the molecular symmetry group. The formalism of the framework group was simplified[172] in terms of the site symmetry groups[173–175].

Introduction of the concept of feasible permutations is closely connected to the understanding of the isomerism of nonrigid molecules. The existence of a larger number of geometrically nonequivalent minima on the hypersurface represents a more complicated situation[176] than that originally conceived by Longuet–Higgins[158]. Of all types of symmetry operations[158] of the complete Hamiltonian, only two (simultaneous inversion of the positions of all the particles in the centre of mass and permutation of the positions and spins of any set of identical nuclei) can undergo changes during transitions between isomers. This leads to the concept of the internal

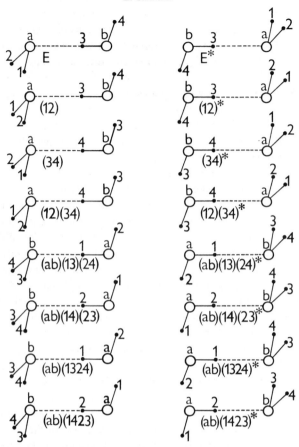

Fig. 3. Effect of each permutation–inversion group operation on the water dimer molecule (according to Ref. [165]).

symmetry structure[177,178] of the complete Hamiltonian, that is entirely described by the group of all nuclear permutations and of the spatial inversion. However, the isomers are not associated with the irreducible representation of this group, as a particular isomer is not described by the complete Hamiltonian but by a local Hamiltonian partially localized in the corresponding domain of the configurational space. In every such domain, the complete Hamiltonian is approximated by the localized Hamiltonian invariant for permutations that are feasible without leaving that domain. Thus, operations with respect to which the complete Hamiltonian is invariant are split into operations leaving a localized Hamiltonian invariant and operations mutually transforming the localized Hamiltonians. Beginning from the complete permutation–inversion group (i.e. from completely flexible molecule), the symmetry can be reduced by gradually forbidding some of the permutations of identical nuclei and/or the space inversion. Each step in this process yields a set of nonrigid isomers. Gradual carrying out of this forbidding process yields a set of groups embedded in one another that, at some step in this process, become isomorphous with the usual groups describing the symmetry of rigid species. It is apparent that every nonrigid isomer is characterized by its own Longuet–Higgins group of feasible permutations (i.e. allowed in our restriction scheme).

The symmetry properties of isomers, or of degenerate rearrangements, were studied in a number of works (see, e.g. [47,62,168,178–183]). Gilles and Philippot[178] studied the structure of the Longuet–Higgins group, proposed a method for mapping all types of nonrigidity in the given system and utilized the introduction of the technique of generators for the classification of isomerization processes. Utilization of the groups of nonrigid molecules permitted direct rationalizations, for example, of knowledge concerning thermal nitroxide formation[179] or conditions during the rearrangements of $C_2H_5^+$ and $C_5H_5^+$ ions[181].

A situation described by the Jahn–Teller theorem[184], stating that the electronic wave function in an energy minimum cannot exhibit spatial degeneracy (except for linear molecular configurations), can be considered to be a special case of symmetry-conditioned nonrigidity. Here the potential energy hypersurface must exhibit a nonzero gradient for a degenerate electronic state, resulting in distortion of the configuration exhibiting this degeneracy (see, e.g. [73,185,186]). The Jahn–Teller distortion leading to stabilization of the species can lead to the production of an isomeric system[187–189] for which the name Jahn–Teller isomers has been proposed[187]. It should be noted, for the sake of completeness, that this isomerism can be considered to be a special case of orbital isomerism[190] based on alternative occupation of distinguishable orbitals by electrons.

3.2 Transition states and symmetry

A number of rigorous results have been obtained concerning the symmetry of transition states. This is a result of a certain exceptional position of transition states among other types of stationary points. In simple chemical reactions, the saddle points on the hypersurface can be connected directly with the reactants and products, e.g. by the steepest descent path; the matrix of the second derivatives of the energy with respect to the coordinates in saddle points exhibit only one negative eigenvalue. It is thus intuitively clear that the symmetry of the transition state should reflect these factors in some way and should be related to the symmetry of the reactants and products (cf. [73,191,192]).

The Murrell–Laidler theorem[193] can be considered as an extremely important result in this connection; this theorem states that the conventional vibrational FG matrix analysis of the transition state can lead to only one negative (imaginary, if the level of frequencies is concerned) eigenvalue. To retain logical consistency, however, a transition state must be defined here in energy terms alone (the lowest energy barrier separating the reactants and products) and its differential properties should be obtained deductively. Alternatively, the Murrell–Laidler theorem can be formulated as stating that only one reaction path can lead through the transition state (provided that the possibility of its splitting in the region between the transition state and the reactants or products is not considered). Proof through contradiction is readily carried out: assume that the given transition state contains two normal vibrational coordinates Q_r and Q_s, both with negative eigenvalues. Consider points $[Q_r^0 \pm \delta Q_r, Q_s^0 \pm \delta Q_s]$ lying in the region around transition state Q_r^0, Q_s^0 (all the other coordinates kept constant). The Taylor expansion (regardless of the choice of signs) yields

$$U(Q_r^0 \pm \delta Q_r, Q_s^0 \pm \delta Q_s)$$

$$\approx U(Q_r^0, Q_s^0) + \frac{1}{2}\left(\frac{\partial^2 U}{\partial Q_r^2}\right)_0 \delta Q_r^2 + \frac{1}{2}\left(\frac{\partial^2 U}{\partial Q_s^2}\right)_0 \delta Q_s^2 < U(Q_r^0, Q_s^0). \quad (1)$$

Naturally, points $[Q_r^0 \pm \delta Q_r, Q_s^0]$, $[Q_r^0, Q_s^0 \pm \delta Q_s]$ also lie energetically lower than the transition state. It is then apparent if the pathway through points $[Q_r^0 - \delta Q_r, Q_s^0]$; $[Q_r^0, Q_s^0]$; $[Q_r^0 + \delta Q_r, Q_s^0]$ is replaced by pathway $[Q_r^0 - \delta Q_r, Q_s^0]$; $[Q_r^0 - \delta Q_r, Q_s^0 - \delta Q_s]$; $[Q_r^0, Q_s^0 - \delta Q_s]$; $[Q_r^0 + \delta Q_r, Q_s^0 - \delta Q_s]$; $[Q_r^0 + \delta Q_r, Q_s^0]$, avoiding the transition state, the latter pathway is the lower of the two. This contradicts the definition of a transition state and consequently only one reaction path can actually pass through the given transition state. An elementary application of the Murrell–Laidler theorem can be demonstrated on solution of the symmetry[191] of the activated complex in the autoisomerization of ozone:

$$O_1 - O_2 - O_3 \rightleftharpoons O_2 - O_1 - O_3 \rightleftharpoons O_2 - O_3 - O_1. \quad (2)$$

The stable form of ozone is represented by an open C_{2v} structure with a valence angle[194] close to $2/3\pi$. A structure with symmetry D_{3h} can be suggested as the activated complex taking part in this interconversion. The symmetry structure of the vibrational modes of this species[191] is $\Gamma_{D_{3h}} = A_1' + E'$ (see Fig. 4). However, the totally symmetric mode A_1' cannot be suggested as a possible reaction coordinate, as it represents dissociation into three oxygen atoms. From this point of view, both degenerate modes with symmetry E' are suitable, but they are not compatible with the Murrell–Laidler theorem as they would represent two orthogonal directions with negative eigenvalues. Thus the structure with symmetry D_{3h} is excluded as a possible

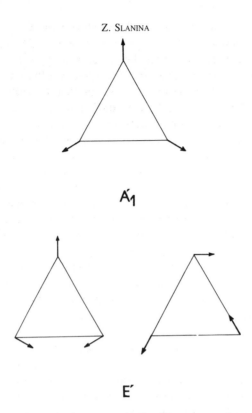

$$\acute{A}_1$$

$$E'$$

Fig. 4. Normal vibrational modes for an A_3 molecule of D_{3h} symmetry (according to Ref. [191]).

transition state but can be considered as a local minimum and/or higher type of saddle point or as a maximum. In actual fact, a quantum-chemical study has demonstrated[20,21,195] the existence of a local minimum with D_{3h} symmetry and thus isomerism of ozone. As is apparent from symmetry considerations, the search for a transition state with the latter point group is bound to be unsuccessful; the activated complex must exhibit lower symmetry. It should be noted that the transition state for reaction (2) has been shown[196] to be a structure with C_{2v} symmetry. However, when passing to a transition state with one or more zero eigenvalues, some peculiarities can appear in connection with the contribution of cubic or higher terms to the Taylor expansion (1). For example, when both eigenvalues vanish for a two-dimensional system, the corresponding hypersurface can be formed[197] by three valleys meeting at a point (called a "monkey saddle"). For example, this situation was indicated[198] for the isomerization in the $C_5H_5^-$ system. However, detailed reanalysis of the results[198] has demonstrated that the assumed region of a transition state with symmetry D_{3h} is formed by a maximum with this symmetry and three equivalent, usual transition states, each corresponding to one reaction path and sufficiently clearly displaced from the central maximum.

Stanton and McIver[197,199–201] further developed, formalized and illustrated the Murrell–Laidler theorem. For formulation of the symmetry rules they used the concept of a transition vector, designating the normal vibrational coordinate of the transition state with a negative eigenvalue. In terms of a transition vector, their rules are[197] (i) the transition vector cannot belong to a degenerate representation of the transition-state point group; (ii) the transition vector is antisymmetric under every transition-state symmetry operation which converts the reactants to the products; and (iii) the transition vector is symmetric with respect to every operation that leaves either the reactants or the products unchanged, or changes the reactants into equivalent reactants or products into equivalent products. A special consequence of rule (ii) is, for example, the fact that a structure that would exhibit a threefold (or higher odd) axis of rotation, converting the reactants into the products, cannot act as a transition state. For example, of the three following possible activated complexes in the four-centre reaction of $H_2 + H_2$, i.e. structures with symmetries D_{3h}, T_d and D_{4h}, the first two can immediately be excluded after analysis of the corresponding diagrams (so that the calculation need not be carried out for these two species[202]). Similarly, rule (ii) demonstrates that, in narcissistic reactions[203,204] (i.e. in processes where the reactant or product are interconverted through reflection or S_n rotation) the transition vector must be antisymmetric with respect to this operation, provided it is a transition-state symmetry operation. It should be noted that symmetry rules are useful not only for the selection of structures that could act as transition states, but that they can also simplify the numerical side of localization of stationary points on the hypersurface (cf. [121,205,206]). For example[197], it follows from rule (ii) that, if the group of the transition state includes a symmetry operation that converts the reactants to products, then the energy of the transition state will be a local minimum in the subspace of coordinates belonging to the totally symmetric representation of the group (conversion of the search for a transition state into energy minimization).

Pechukas[207] used a somewhat more specific assumption, i.e. he analyzed a situation in which the transition state is linked directly to the reactants and products by paths of steepest descent, and obtained stricter rules, i.e. the rules for conservation of nuclear symmetry along the path of steepest descent. He demonstrated that the symmetry of the transition state is the symmetry of the entire path from the reactants through the transition state to the products, namely that the symmetry groups of the reactants and products must contain the symmetry group of the transition state linking them. Processes in which the reactant and product are physically indistinguishable, when the transition state can exhibit additional symmetries involving interchange of the reactant and product configurations, are an exception to this rule. Notably, it follows from the requirement that the symmetry of the transition state between physically distinguishable reactants and products must be included in their own symmetry, that, for example[207], (a) linear transition states must yield linear reactants and products, (b) planar transition states must yield planar reactants and products, (c) optically inactive transition states must yield optically inactive reactants and products, and (d) the point group of symmetry of the transition state can be no larger than the largest subgroup common to the point groups of both the reactants and products, etc. It should be noted that, even though the above rules for the symmetry of the activated complexes were originally formulated in terms of point groups of symmetry, conversion into the terms of the Longuett–Higgins groups has also been given[181].

The Jahn–Teller theorem[184] can also be used for selection among possible transition-state structures. The possible use of this theorem in connection with transition states need not be trivially apparent (it could appear that the Jahn–Teller distortion occurs along the transition vector); nonetheless, it is readily justifiable[208] and follows from the existence of a direction in which the energy decrease is linear in terms of the distortion. Use of the Jahn–Teller theorem permits (independent of the above arguments) exclusion, for example, of the T_d structure from possible participation as an activated complex in the four-centre reaction $H_2 + H_2$, as this symmetry leads to a degenerate electronic configuration. Mixing of close but nondegenerate electronic states leads to a pseudo-Jahn–Teller effect[209], in which the change in energy is not a linear but a quadratic function of the distortion. Utilization of the pseudo-Jahn–Teller effect to predict transition states is important for structures with low-lying electronic excited states[210–215].

It should be noted for completeness that the use of symmetry rules assumes a certain nontrivial symmetry of the transition state, and is not useful for evaluation of asymmetric structures. As was pointed out by McIver[201], there is no a priori reason why activated complexes should be symmetric. Consequently, the possibility of an asymmetric structure should always be considered. It is especially necessary for a symmetry-forbidden reaction to test whether the barrier cannot be lowered by considering a path of lower symmetry. This is true of the process

$$H_2 + CH_2 \longrightarrow CH_4, \qquad (3)$$

which the Woodward–Hoffmann rules state to be symmetry-forbidden if the reaction path retains C_{2v} symmetry. However, the barrier can be lowered considerably[216] if, for example, only C_s symmetry is required, as indicated in Fig. 5.

Formulation of the Murrell–Laidler theorem was originally inspired[193] by the once-important problem of the relationships and usefulness of symmetry numbers and statistical factors for the construction of rotational partition functions. The statistical factors can be used as an alternative means for exclusion of some activated complex structures[193,217]. Recently,

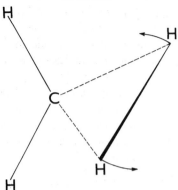

Fig. 5. Preferred distortion from C_{2v} to C_s symmetry of the reaction path for reaction (3) (according to Ref. [191]).

a great deal of attention has been devoted[193,217–227] to the question. Even though this extensive discussion eventually demonstrated[226,227] that the conventional symmetry numbers are consistent with the formulation of the theory of absolute reaction rates (and thus that the symmetry numbers, and not the statistical factors, are the correct objects for the calculation of the one-way rate passage through a transition state), analysis of this classical concept of rate processes was very fruitful. At the beginning of this debate, two reasons were given for not using symmetry numbers. It appeared[218,220] that they can lead to erroneous rate expressions for symmetric reactions and also that complications can occur when optically active species are present. It has been suggested that these difficulties can be overcome by replacing symmetry numbers by statistical factors[218–222]. For example, Schlag[218] recommended replacing ratios of conventional symmetry numbers by reaction-path degeneracy factors derived from the group-theory approach. This technique was then simplified by replacing this latter procedure by the direct-count method[219]. A variation of this method involves the distinguishing of statistical factors for the forward and backward reactions[220]; other modifications have been suggested[221,222]. Special attention has been paid to the use of statistical factors for optically isomeric pathways and optically isomeric activated complexes[223,224]. A further stage in the development of the problem is concerned with the discovery[193,217,225] that statistical factors need not always be an appropriate tool, as it has been found that the resultant ratio of forward and backward rate constants leads to erroneous equilibrium constant values. This apparent contradiction was finally removed[226,227] by demonstrating that the symmetry number method used in a certain, unambiguous, "proper" way leads to adequate expressions for the rate formulae.

   Particular care is necessary in considerations of processes involving optically active species and for symmetric reactions, i.e. reactions where the reactants and products would be indistinguishable if atomic labelling was not used. A prototype bimolecular symmetric reaction is the process $H + H_2 \rightarrow H_2 + H$; simultaneous consideration of the corresponding isotopically labelled process with formal utilization of symmetry numbers σ leads to the following scheme:

$$H + D_2 \longrightarrow H - D \ - \ D \longrightarrow HD + D,$$
$$\sigma = 2 \qquad\qquad \sigma = 1 \qquad\qquad \sigma = 1 \qquad\qquad (4)$$

$$H + H_2 \longrightarrow H - H \ - \ H \longrightarrow H_2 + H,$$
$$\sigma = 2 \qquad\qquad \sigma = 2 \qquad\qquad \sigma = 2 \qquad\qquad (5)$$

indicating that (neglecting differences in the partition functions resulting from mass differences) reaction (4) is clearly favoured over reaction (5) by a factor of 2, although it would be expected that the rates should be identical under these conditions. This error results from routine application of the theory of absolute reaction rates for reaction (5). This theory determines the rate of passage through the transition state in one direction; however, for reaction (5), passage in either direction leads to the (required) product. It thus follows from the identical nature of the reactants and products in reaction (5) that twice the rate constant for passage through the transition state in a single direction must be used for rate considerations (e.g., in determination of the lifetime of an H_2 molecule against H atom exchange). This rule then leads to the elimination of the reported paradox in reactions (4) and (5). The whole problem was based on the proper understanding of the quantities produced by the theory of absolute reaction rates.

4. ENUMERATION OF ISOMERIC STRUCTURES

If the structural information on a given chemical system is reduced to the form of a graph, then the tools of graph theory developed for enumeration of nonisomorphic graphs[228] can be equally well employed for the determination of the number of possible isomers. Because of symmetry elements frequently present in chemical species, the isomer enumerations cannot, however, be reduced to pure combinatorics. The modern era in graphical enumeration dates[229] from 1935 when Pólya published his now classic theorem. (Results obtained before the appearance of Pólya's theorem are reviewed in Refs. [5,25,29,229].) In the solution of the general enumeration problem, Pólya fruitfully combined the classical method of generating functions with basic results from the theory of permutation groups. The method of generating functions consists of replacing the sequence $a_0, a_1, a_2, \ldots$ (determining the numbers of figures of certain types) by the function

$$S(x) = a_0 + a_1 x + a_2 x^2 + \cdots . \tag{6}$$

In fact, Cayley[230] has already used the method of generating functions in connection with enumerations. The usefulness of the mathematical theory of permutation groups for the consideration of the symmetry of enumeration problems was already pointed out in the works by Redfield[231] and Lunn and Senior[232].

Consider p points and the permutation group H of order h permuting these points. Let permutation $\alpha \in H$ exhibit $j_r(\alpha)$ cycles[233] of order r $(r = 1, \ldots, p)$. The cyclic structure of group H can be described in terms of its cycle index $Z(H)$, defined as a polynomial in variables $s_1, s_2, \ldots, s_p$:

$$Z(H) = \frac{1}{h} \sum_{\alpha \in H} s_1^{j_1(\alpha)} s_2^{j_2(\alpha)} \ldots s_p^{j_p(\alpha)}, \tag{7}$$

or alternatively

$$Z(H) = \frac{1}{h} \sum H_{j_1, j_2, \ldots, j_p} s_1^{j_1(\alpha)} s_2^{j_2} \ldots s_p^{j_p}, \tag{8}$$

where $H_{j_1, j_2, \ldots, j_p}$ designates the number of permutations of type $[j_1, j_2, \ldots, j_p]$ (i.e. the number of permutations with identical sequence $j_1, j_2, \ldots, j_p$) and summation in (8) ranges over all possible choices of the types $[j_1, j_2, \ldots, j_p]$. Using the cycle index of the permutation group of the problem, Pólya[234] demonstrated how the enumeration of all combinatorically possible configurations can be reduced to enumeration of their equivalence classes. Consider different figures containing different sorts of objects (say, three) with the generating function of figures $f(x, y, z) = f_1$, and the derived functions $f_r = f(x^r, y^r, z^r)$. Then according to Pólya's theorem the generating function for enumeration of nonequivalent configurations, $F(x, y, z)$, is obtained by replacing variables s_r in the cycle index of the permutation group of the problem by the corresponding functions derived from the generating function of figures f_r:

$$F(x, y, z) = \frac{1}{h} \sum H_{j_1, j_2, \ldots, jp} f_1^{j_1} f_2^{j_2} \cdots f_p^{j_p}, \tag{9}$$

where the summation ranges over the types $[j_1, j_2, \ldots, j_p]$.

The enumeration theorem is considered to be a milestone, not only in graph theory but in mathematics as a whole[235]. For completeness, however, a few comments should be made. Mention was made previously of Redfield's work[231] which was long neglected and which contains many of the ideas later rediscovered independently by Pólya. Redfield's paper was recently analyzed[236] and it was shown that it includes, replaces and simplifies many of the results for molecular combinatorics of recent years. It is interesting to note that this work was the only one Redfield ever published (cf. [235]). Another recent and interesting reanalysis[237,238] of early original results related to chemical enumeration is concerned with Körner's proof of the homotopy of the hydrogens in benzene, published in 1869, pointing out the potential usefulness of analogous arguments in the discussion of the structure of fluxional molecules.

Pólya himself treated the problem of enumeration of chemical isomers in his later works[239,240]. Figure 6 serves two illustrative examples. Before the derivation of Pólya's theorem, there was a great deal of scepticism about the possibility of a general solution, or even of obtaining an analytical formula, for an arbitrary enumeration problem. This led to the study of the possibility of the solution of a given enumeration problem iteratively, using recursion formulae, the first contribution of that type being carried out by Henze and Blair[241,242]. Table 1 presents results of enumeration in selected acyclic series.

4.1 *Enumeration for nonrigid molecules*

Suitable selection of the elements of the permutation group appearing in the enumeration theorem can lead to a description of some types of stereoisomerism (see, for example, [243,246]) and even molecular nonrigidity[238,246–251] and thus one can approach more closely the actual physical conditions existing within the given system. When considering enantiomers as different isomers, the corresponding permutations must be eliminated in the enumeration and a lower-order permutation group must be used. In the same way, cis and trans isomers can be included in the enumeration, permitting only permutations that change the positions of all the participating substituents simultaneously. This method was used[243–246] for expanding the enumeration of structural isomers to include some types of stereoisomerism. Pólya already had consid-

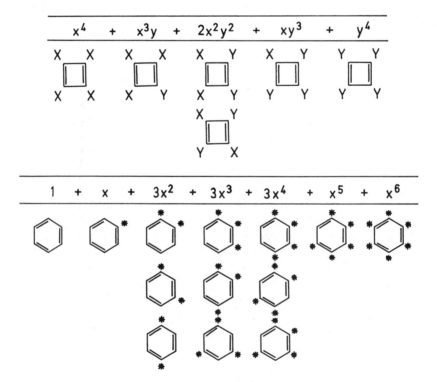

Fig. 6. Representation of the solution of an enumeration problem by a generating function for cyclobutadiene fully substituted by two kinds of univalent substitutents (X, Y), and for benzene successively substituted by one kind of univalent group (*).

Table 1. Isomer enumeration results in some acyclic series[a,b]

n (carbon atom content)	1	2	3	4	5	10	15	20	25
$C_nH_{2n} + 1^-$	1	1	2	4	8	507	48 865	5 622 109	712 566 567
	1	1	2	5	11	1553	328 092	82 300 275	22 688 455 980
$C_nH_{2n} + 2$	1	1	1	2	3	75	4 347	366 319	36 797 588
	1	1	1	2	3	136	18 127	3 396 844	749 329 719
RCH_2OH	1	1	1	2	4	211	19 241	2 156 010	269 010 485
		1	1	2	5	551	110 500	27 012 286	7 333 282 754
R_1R_2CHOH			1	1	3	194	19 181	2 216 862	281 593 237
			1	2	5	768	162 672	40 807 290	11 247 841 224
$R_1R_2R_3COH$				1	1	102	10 443	1 249 237	161 962 845
				1	1	234	54 920	14 480 699	4 107 332 002
$R_1C{\overset{O}{\underset{OR_2}{\diagup\!\!\!\diagdown}}}$		1	2	4	9	599	57 564	6 589 734	832 193 902
		1	2	4	10	1319	273 172	67 819 576	18 581 123 978
$R_1R_2C{=}CR_3R_4$		1	1	3	5	377	36 564	4 224 993	536 113 477
		1	1	4	6	895	185 310	46 244 031	12 704 949 506

[a]According to Refs [239, 241–243]; the numbers of structural isomers are given in the upper line, the numbers of stereoisomers in the lower line; R_1, R_2, R_3, for secondary and tertiary alcohols, and R_2, for esters, are alkyl radicals.
[b]R, R_i denote an alkyl radical or hydrogen atom.

ered[240] three types of permutation groups: a group of spatial formulae leading to the number of stereoisomers, an extended group of spatial formulae leading to the number of stereoisomers minus the number of pairs of optical isomers, and, finally, a group of structural formulae generating the number of structural isomers. For example, for cyclopropane the order of the group of spatial formulae is 6, of the extended group of spatial formulae it is 12, and of the group of structural formulae it is 48. (Table 1 lists[239,241–243] numerical illustrations of enumeration of stereoisomers in acyclic series and comparison with the values from enumeration distinguishing only structural isomers.)

It was already mentioned that, for nonrigid molecules, the point group of symmetry corresponding to an idealized rigid structure does not permit[158] the correct description of the symmetry of the problem. For this purpose it must be extended by the inclusion of symmetry elements realized by intramolecular motions. Thus, in employing Pólya's theorem for enumeration of the isomers of nonrigid molecules, it is necessary to utilize[246] a permutation group including both the operations of the corresponding point group as well as every permutation feasible through the internal degrees of freedom. A permutation is considered feasible in the sense introduced by Longuet–Higgins[158] as a permutation realizable without transition over an insurmountable barrier (under the given conditions). Pólya's theorem has been extended[246] for enumeration of nonrigid compounds on the basis of a generalized wreath-product method[252,253]. The generalized wreath-product group represents an extension of the Longuet–Higgins results[158] for symmetry group of nonrigid systems. This concept has been applied[249], for example, to reactions generated by twofold and threefold internal rotations. While the use of conventional point groups of symmetry leads, for example, to a total of 18 isomers for $C_2H_2Br_2Cl_2$, the wreath-product method yields only 6 isomers. These types of considerations are important for the rationalization and prediction[246,249] of the number of signals in NMR spectra at low and high temperatures.

A further interesting example of enumeration for a nonrigid skeleton is provided by the results obtained[247,248] for cyclohexane. The most stable structure of cyclohexane belongs to the D_{3d} point group of symmetry. The use of the permutation group corresponding to this point group in the enumeration theorem leads to a number of isomers that is contrary[247,248] to experimental results. Simultaneously, it has been found empirically that the use of Haworth's (planar) projection formulae leads[247,248] to correct results, although they ignore the reality of the cyclohexane conformation. In the use of planar formulae, enumeration is carried out in terms of the D_{6h} point group of symmetry. Leonard et al.[247] provided a rationalization of these facts. The D_{3d} point group does not include operations corresponding to the nonrigidity of the cyclohexane molecular skeleton. It has been demonstrated[247] that the symmetry of cyclohexane can be adequately represented by the introduction of a new symmetry element (termed R_6), which combines a sixfold rotation with a flip motion of the ring interconnecting one chair form with another. The effect of the R_6 operation is to permute all the positions on the same side of the ring. Group D_{3d} is replaced by group $D_{3d}R_6$ for construction of the permutation group; this group is obtained by expanding group D_{3d} by operation R_6. Group $D_{3d}R_6$ leads to the same cycle index as group D_{6h}, and thus also to the same enumeration results. In fact, this is a result of the circumstance that the D_{6h} and $D_{3d}R_6$ groups[254] are isomorphic. This example indicates[247,248,254] that the utilization of the symmetry groups of nonrigid molecules[158,246] should make a significant contribution towards making the results of the enumeration theorem closer to physical reality.

5. TOPOLOGICAL REDUCTION AND CHARACTERIZATION OF POTENTIAL HYPERSURFACES

The contemporary literature contains a relatively large number of enumeration results based on Pólya's theorem and/or on the iterative approach (e.g. enumeration in acyclic series[243], cyclic compounds[255], or other classes of organic and inorganic substances[5,256–261] as well as enumeration of rearrangement pathways[262–268] or the exhaustive generation of isomers[269–273], especially in connection with the search for structure(s) consistent with experimental information). From the point of view of the theory of chemical reactivity based on stationary points of potential energy hypersurface there is a key question to determine the relationship between the numbers and types of stationary points and the results of isomer enumeration. The correspondence is, of course, not unambiguous. Every structure included in the enumeration need not necessarily correspond to a stationary point on the hypersurface. A single configuration in the enumeration may be represented by more than one single point on the hypersurface. In principle, however, the results of enumeration yield a guess for the determination of the order of the number of stationary points on the given hypersurface. An alternative approach to the problem has recently been supplied by topological reduction and characterization of hypersurfaces. The topological approach can be understood as a generalization or abstraction of the geometrical model, where the topology of the hypersurface suppresses particular geometric characteristics retaining certain general (and chemically important) properties.

A basis for a topological representation of the potential-energy hypersurfaces was formed by the works of Krivoshey et al.[274–277]. Using a suitably introduced topology, a one-dimensional object can be assigned to every hypersurface in such a manner that the differential–geometrical properties of the corresponding hypersurface are ignored, but its topological properties (number and mutual arrangement of stationary points) are preserved. A mathematical background for this procedure is provided by the results of Kronrod[278].

An equipotential level of the hypersurface consists of one or more components. Each separate contour of an equipotential level is ascribed a point in the one-dimensional representation, and the points are ordered according to the energy values. It can be demonstrated[274,276,278] that, regardless of the dimension of the hypersurface, the topological space always leads to a tree-type graph. This tree-type graph is a geometric picture of the mapping of the given hypersurface in terms of the constant function-value contours.

The tree representing the topological characteristics of the potential-energy hypersurface is a union of points and the edges connecting them. Two types of points will be considered: end points and branch points at which graph branching occurs (the branch point coincides with at least three edges). While the end points correspond[276] to local energy minima or maxima, branch points correspond to saddle points. It is apparent that every path on the given potential-energy hypersurface can be interpreted as a path on its tree[276,279].

Figure 7(a) depicts schematically the classical LEP potential-energy hypersurface[280] for the linear H_3 molecule. On the surface, domains 1 and 2 are designated, corresponding to the reactants and products, respectively, and activated complex 3. Domain 4 corresponds to a shallow minimum close to the saddle point, which is an artefact of the LEP method. Domain 5 represents a local energy maximum. Figure 7(b) depicts the corresponding tree: minima 1, 2 and 4 and maximum 5 are its end points and saddle point 3 is the branch point.

Another topological approach to potential hypersurfaces was recently proposed by Mezey[281,282], who suggested two different schemes for the partitioning of nuclear configuration space R with dimension n into mutually exclusive subsets. The first scheme employs the curvature properties of the given hypersurface and leads to subsets (coordinate domains)

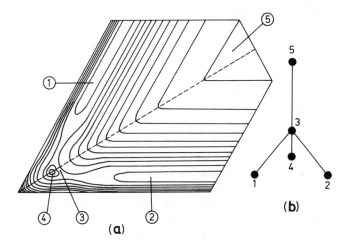

Fig. 7. Tree representation (b) of the potential energy surface [280] of the linear H_3 molecule (a).

D_μ^i characterized by the number μ of negative eigenvalues of the Hessian matrix constructed in a subspace with dimension $n - 1$. All the local minima and transition states belong to domains of the D_0^i type. Provided that the excluded domain D_{excl} is also considered (i.e. the domain where the distance of at least two nuclei attains zero value and/or where there are no continuous second derivatives of the energy) the following complete partitioning of space R is obtained:

$$R = \bigcup_{\mu,i} D_\mu^i \cup D_{excl}. \tag{10}$$

The second partitioning of R is based on the number and types of stationary points on the hypersurface and employs one of their most important properties. The extremity of every steepest descent path P_r on the hypersurface is either a stationary point $\mathbf{r}_s^{(l)}$ or a point in the excluded domain D_{excl}. Consequently, points $\mathbf{r} \in R$ can be classified according to the stationary point towards which the steepest descent path starting from these points is directed. Thus, the concept of the catchment region $C^{\mathbf{r}_s(l)}$ of stationary point $\mathbf{r}_s^{(l)}$ is introduced. A complete partitioning of space R is then given as

$$R = \bigcup_{l} C^{\mathbf{r}_s(l)} \cup C^{\overline{D}_{excl}} \cup \overline{D}_{excl}, \tag{11}$$

where $C^{\overline{D}_{excl}}$ designates the catchment region of $\overline{D}_{excl}$, and $\overline{D}_{excl}$ represents the closure of the excluded domain D_{excl}.

Both partitionings (10) and (11) of space R consist of mutually exclusive subsets and both permit chemically significant topologization of R in a natural manner. Both of the resulting topological spaces are now used for the topological analysis of selected chemical properties and concepts[283–288]; for example, molecular structure and reaction mechanisms have been interpreted[287,288] as open sets in a topological space. Such an interpretation of the molecular structure is close to the concept of the structure derived from the topological properties of the molecular charge density[30,289–292], as can be demonstrated[290] by the conjectured homeomorphism between the charge density and the potential energy.

Finally, bounds for the number of stationary points on energy hypersurfaces should be mentioned in this section. A theory of critical points of functions of many variables has been developed on the basis of differential and algebraic topology leading to Morse inequalities for the lower bounds for the number of critical points[293,294]. These inequalities were adapted by Mezey[295,296] for the study of potential-energy hypersurfaces.

6. THE GRAPH-LIKE STATE OF MOLECULES

Reduction of a real molecular system to a chemical graph or a stereograph essentially lies at the basis of any chemical enumeration. During the transfer to chemical graphs, only information on the existence of chemical bonds is retained; all information relevant to the geometrical arrangement is omitted. If it is assumed that all the bonds are flexible (which is manifested in the permutation group of the graph), the ''graph-like'' concept of the state of molecules is obtained (Fig. 8); this concept was developed systematically by Gordon *et al.* (see, e.g. [60,297–303]). This completely flexible state is hypothetical (it requires more flexibility than the real state of a molecule possesses); nonetheless, a number of molecules have a similar nonrigidity. The best-known example is the HCNO, or ammonia molecule because of its inversion motion. Although this degree of abstraction of real molecules may appear unsuitable for chemical purposes, the simplicity of the graph-like state permits rigorous solution of some basic problems, especially in statistical mechanics[304,305]. A number of properties, including the entropy estimates, kinetic aspects or phase transitions, especially for polymers, can be treated in these terms, often representing a reduction of the problem to one of an enumerative nature.

The graph-like state of molecules is connected[306] with the topological entropy S_t linked by a simple relationship with order $|G|$ of the symmetry group of its graph (R is the gas constant):

$$S_t = -R \ln|G|. \tag{12}$$

Useful rules have been derived for the calculation of order $|G|$ when the graph is of the tree type[306]. During the transition from the graph-like state of a molecule to its real state (Fig. 8), the integral degeneracy factor g of all the energy states of the real molecule appears, which leads to a contribution of S_m to the metric entropy with magnitude

$$S_m = R \ln g. \tag{13}$$

The sum of the topological and metric entropy defines the combinatorial entropy S_c:

$$S_c = S_t + S_m. \tag{14}$$

For suitably selected processes, this combinatorial entropy can make a decisive contribution towards the value of the overall entropy. Calculation of the combinatorial entropy does not require more information than the molecular graph and the manner of its correspondence to the real state, i.e. whether, and how many, bonds of the real state of the molecule permit free rotation. Table 2 gives an example of the calculation of the combinatorial entropy for n-octane and 2,2,3,3-tetramethylbutane. It follows from the data in Table 2 that the contribution of the combinatorial entropy to the entropy term of the isomerization of octanes equals the value

$$\Delta S_c = 6R \ln 3 = 54.8 \text{ J/mol K}, \tag{15}$$

which represents[306] a total of 74% of the standard overall entropy change found experimentally at a temperature of 298 K. The concept of the combinatorial entropy was also utilized by Gordon[304] in discussion of the third law of thermodynamics.

The graph-like state of molecules or matter might play a role in the description of polymers similar to the ideal gas state in thermodynamics and physics of gases and liquids[297]. The concept of the graph-like state of matter was used, for example, in the study of substitution effects and polymer distributions[297], the kinetics of linear polymerization reactions[299], and Rayleigh scattering[300]. An interesting relationship was found in the study of glass transitions of polymers[302]. The authors of the work[302] demonstrated that a limiting case in the statistical theory of these transitions is a famous problem in graph theory. They managed to find a connection between the number of configurations of the system and the number of Hamiltonian walks on a lattice graph. A closed Hamiltonian walk on the given graph is understood to be a cycle passing through each vertex of the graph just once. Here, somewhat less stringent conditions are placed on Hamiltonian walks—their initial and final points need not be identical. In the theory of glass transitions of polymers, the value of the limit

$$\lim_{N \to \infty} (H_N)^{1/N} = n_H \tag{16}$$

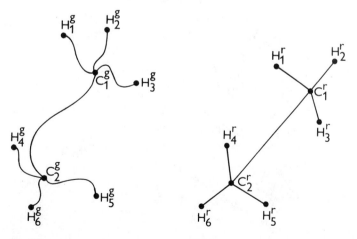

Fig. 8. Conversion of graph-like (absolutely flexible) ethane into its three-dimensional real (rigid) state (according to Ref. [306]).

Table 2. An example of the combinatorial entropy S_c with two isomeric octanes[a]

Combinatorial characteristic	n-octane	2,2,3,3-tetramethylbutane
g	2^8	2^8
IGI	$2^9 3^2$	$2^9 3^8$
S_c[b]	-Rln (2×3^2)	-Rln (2×3^8)

[a]According to Ref. [306].
[b]R = 8.31434 J/K/mol.

is important, where H_N is the number of Hamiltonian walks on a lattice graph with N vertices. Graph theory has been employed to obtain characteristic (16) for a number of various lattices[302]. The topological approach to statistical thermodynamic theory, especially for nonlinear polymers (including biopolymers[307,308]) is promising for solution of a number of further problems in this field[303,307–309].

7. INTERPLAY OF ISOMERS OF DIFFERENT SYMMETRIES

Systematic investigation of a potential-energy hypersurface (regardless of the degree of sophistication of quantum-mechanical calculations) often reveals several different local energy minima all represented by one species in an experiment and/or several different saddle points corresponding to activated complexes in a single rate process, these points frequently possessing different symmetries. Potential-energy criteria can sometimes prove that only one structure to play an important role. However, it may also happen that two or more isomeric structures of comparable stability coexist and are indistinguishable under experimental conditions. Then any structure-dependent observable can be considered as an average value resulting from contributions of all the isomers in question[76,81,310].

Carbon atom aggregates, which are present[311] at high temperatures in the gas phase above graphite, have recently been studied[312] by means of the MINDO/2 method. Comprehensive search for stationary points on potential-energy hypersurfaces of the C_n ($4 \leq n \leq 7$) was carried out[312]. Four or more stationary points were found[312] on each hypersurface, at least two of them being minima (Fig. 9). For example, three isomers (D_{2h}, D_2, T_d) were found

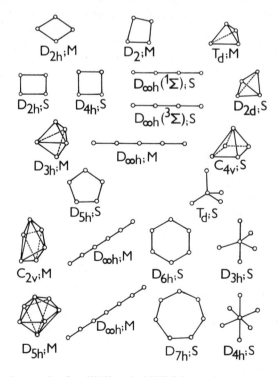

Fig. 9. Schemes of stationary points found[312] on the MINDO/2 potential energy hypersurface of C_n ($n = 4$–7); M - minimum, S - saddle point.

Table 3. Standard MINDO/2 enthalpy ΔH_T° and entropy ΔS_T° terms for partial (D_{2h}, D_2, and T_d) and overall C_4 aggregate formations at a temperature of $T = 2400$ K[a]

Process	ΔH_T° (kJ/mol)	ΔS_T° (J/mol/K)	log $K_p^{\,b}$
$4C(s) \rightleftharpoons C_4(g; D_{2h})$	904.1	210.2	−8.700
$4C(s) \rightleftharpoons C_4(g; D_2)$	952.2	213.6	−9.568
$4C(s) \rightleftharpoons C_4(g; T_d)$	1301.5	205.5	−17.593
$4C(s) \rightleftharpoons C_4(g)^c$	909.9	213.6	−8.645

[a]According to Ref. [312].
[b]K_p in atm; 1 atm = 101 325 Pa. Standard state—ideal gas at 101 325 Pa.
[c]Overall process and its (total) characteristics.

in the case of the C_4 aggregate. Accordingly, three partial equilibria [(17)–(19)] take place in the gas phase above graphite:

$$4C(s) \rightleftharpoons C_4(g; D_{2h}), \tag{17}$$

$$4C(s) \rightleftharpoons C_4(g; D_2), \tag{18}$$

$$4C(s) \rightleftharpoons C_4(g; T_d). \tag{19}$$

However, the experimental (mass-spectrometrical) values[311] of reaction characteristics do not correspond to any of the individual processes (17)–(19) but to the overall equilibrium (20):

$$4C(s) \rightleftharpoons C_4(g). \tag{20}$$

It is clearly evident that it is necessary for comparison with the experimental data to sum up weighted contributions of the isomers.

The weighting treatment in the case of C_4 formation is illustrated in Table 3. It is evident that mere approximation of the characteristics of the overall process [Eq. (20)] using those of the most stable structure (D_{2h}) may be quite misleading. This is especially true with the entropy term: the total entropy term concides circumstantially with that of D_2 isomer formation.

Another example of reaction component isomerism in the field of usual chemical equilibria is difluoroamino radical dimerization studied[313] using the CNDO/2 method and exhibiting isomerism of the dimer (gauche- and trans-N_2F_4; see Fig. 10). The effects of weighting treatment are demonstrated in Fig. 11. While the total ΔH_T^0 term is predominantly connected with the formation of gauche-N_2F_4, in case of the ΔS_T^0 term the effect of the isomerism is more marked— the difference between the summary and any partial terms being significant in the whole presented

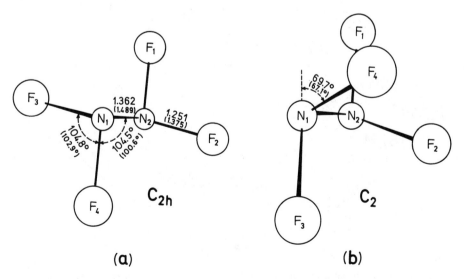

(a)   (b)

Fig. 10. CNDO/2 structures[313] of trans- (a) and gauche-N_2F_4 (b) (bond lengths in 10^{-10} m, experimental values in parentheses).

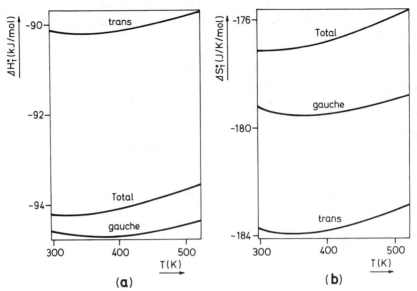

Fig. 11. Temperature dependences of CNDO/2 standard enthalpy ΔH_T^o (a) and entropy ΔS_T^o (b) terms[131] for the partial (to trans- or gauche-N_2F_4) and overall (total) dimerizations of NF_2.

temperature range. A third example of this type can be found in the recent study of the ethanal–ethenol system[314].

The phenomenon of reaction-component isomerism should not be considered only with equilibrium processes. The original concept of the activated-complex theory simply presumes that the transformation of reactants into products goes through a single activated complex. However, recent systematic investigations (e.g. [315–319]) of stationary points by various quantum-chemical methods have indicated the possibility of the existence of more than one stationary point meeting the requirements set to an activated complex of the given rate process (i.e. parallel transition states [319]). A common example is cis–trans interconversion around —N═N— bond[315–318] which is realized by two mechanisms, *viz.* the inversion and the rotation mechanisms, both being connected with a comparable activation energy. Figure 12

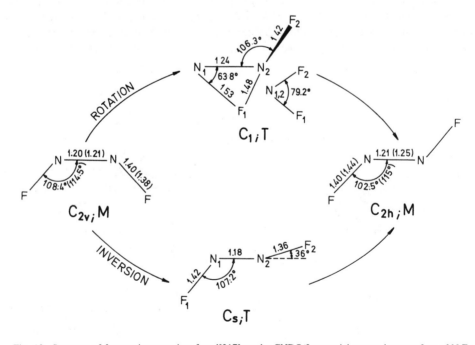

Fig. 12. Structure of four stationary points found[317] on the CNDO/2 potential energy hypersurface of N_2F_2; *M* - minimum, *T* - transition state, bond lengths in 10^{-10} m; experimental values in parentheses.

illustrates this fact, giving the schemes of stationary points on an energy hypersurface of N_2F_2 which are important in the rate process:

$$\text{cis-}N_2F_2 \text{ (g)} \longrightarrow \text{trans-}N_2F_2 \text{ (g)}. \tag{21}$$

On the one hand, it is clear that an experiment will generally give only the overall rate characterization involving the contributions of all partial paths from reactants to products through individual activated complexes. Differentiation to separate partial-activation processes hardly seems possible in any experiment. However, on the other hand, a theoretical description of such rate processes in terms of activated complex theory primarily yields the partial-activation parameters corresponding to transitions through individual activated complexes, i.e. C_s and C_1 transition states realizing[317,320] rate process (21) (Fig. 13). Thus, with confrontation of theoretical and experimental data the problem arises of what is the contribution of each partial path through the particular activated complex to the total effective values of rate characteristics obtained in the experiment. The equilibrium hypothesis (on which the activated complex theory is based) enables a straightforward application of the results of the general equilibrium isomeric problem yielding the required weighting scheme[321].

Properties of the weighting treatment were thoroughly studied[320] in the case of kinetics of isomerization[(21), Fig. 13]. For both $\Delta H_T^{\neq}$ and $\Delta S_T^{\neq}$ terms there are temperature ranges where the total value differs markedly from any of the corresponding partial ones. The results indicate that the summary $\Delta S_T^{\neq}$ term can generally be expected to be (in accordance with its physical interpretation) more influenced by the phenomenon of activated-complex isomerism than the summary $\Delta H_T^{\neq}$ term. This can be seen in a more rigorous form from the inequalities[322] describing the limit behaviour of the summary characteristics: while the total $\Delta H_T^{\neq}$ term can never get out of the interval defined by the lowest and the highest partial-enthalpic term, the total $\Delta S_T^{\neq}$ term can exceed the highest partial entropic term even by the value of $R \ln n^{\neq}$ (where $n^{\neq}$ denotes the number of different activated complexes in play). This weighting was also carried out[321] for the chair (D_{3d}) to boat (D_2) isomerization of cyclohexane (C_s- and C_2-activated complexes).

One field of quantum-chemical research is of special interest in connection with the presented conception of the reaction-component isomerism—the field of weak intermolecular interactions. Ample isomerism of multimolecular clusters is met with very often. Comparable stability of single isomers may frequently be supposed. Temperature dependences of the weight factors of individual multimolecular clusters may be quite varied[322] due to interchanges in the stability order. Clearly enough, the present experiment can generally yield only the total

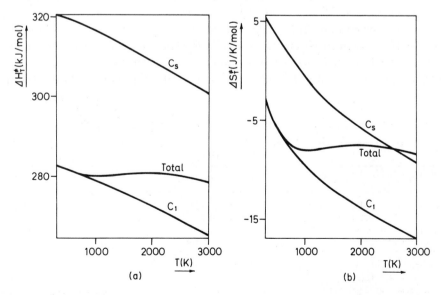

Fig. 13. Temperature dependences of CNDO/2[317] activation enthalpy $\Delta H_T^{\neq}$ (a) and entropy $\Delta S_T^{\neq}$ (b) of the cis–trans isomerization of N_2F_2 for C_s and C_1 activated complexes and for the overall activation process (total).

Z. SLANINA

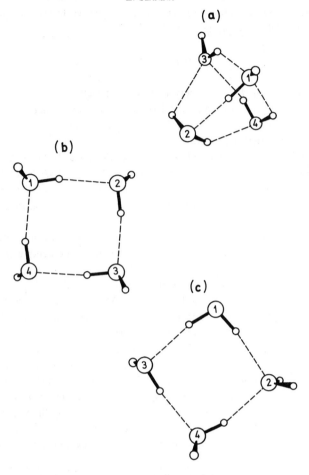

Fig. 14. Minimum-energy water tetramers: (a) pyramid, (b) S_4 cyclic, (c) asymmetric cyclic (according to Ref. [323]).

effective values of characteristics of the multimolecular clusters and the easy establishing of full interisomeric thermodynamic equilibrium is to be expected. Thus, the carrying out of theoretical studies of multimolecular clusters up to the level of multiconfiguration thermodynamic functions is quite urgent. The thorough study by Owicki *et al.*[323] supplied information, for example, on the isomerism of $(H_2O)_4$ (Fig. 14). The interplay[322] of partial characteristics of three (known[323]) tetramers is demonstrated in Table 4. Both the overall enthalpy and entropy terms differ markedly from those of the energetically most stable isomer (pyramid[323]). Similar results were also recently obtained for other mixtures of isomeric multimolecular clusters[324–333].

Table 4. Standard enthalpy ΔH_T° and entropy ΔS_T° terms for the partial[a] and overall[b] water tetramerization at $T = 400$ K

Process	ΔH_T° (kJ/mol)[c]	ΔS_T° (J/K/mol)[c]
$4H_2O(g) \rightleftharpoons (H_2O)_4(g; P)$[d]	−77.36	−335.1
$4H_2O(g) \rightleftharpoons (H_2O)_4(g; S_4)$[d]	−73.01	−330.1
$4H_2O(g) \rightleftharpoons (H_2O)_4(g; A)$[d]	−59.79	−308.4
$4H_2O(g) \rightleftharpoons (H_2O)_4(g)$[e]	−74.69	−324.4

[a]According to Ref. [323].
[b]According to Ref. [322].
[c]Standard state—ideal gas at 101 325 Pa.
[d]P—pyramid. S_4—S_4 cyclic. A—asymmetric cyclic tetramer; see Fig. 14.
[e]Overall process and its (total) characteristics.

Table 5. Temperature dependences[a] of weights w_P, w_{S_4}, and w_A of pyramid (P), S_4 cyclic (S_4), and asymmetric cyclic (A) water tetramers

T (K)	w_P	w_{S_4}	w_A
100	0.997	0.003	1×10^{-8}
400	0.624	0.297	0.078
700	0.334	0.267	0.399

[a]According to Refs [322, 323].

For multimolecular clusters, as well as for all other isomeric mixtures, it is instructive to consider explicitly their weight factors. The weight w_i of individual isomers can be expressed[76,81,310] in the molecular terms as follows:

$$w_i = \frac{q_i \exp\left[-(e_0^{(i)} - e_0)/kT\right]}{\sum\limits_{j=1}^{m} q_j \exp\left[-(e_0^{(j)} - e_0)/kT\right]}, \tag{22}$$

where q_k is the partition function of the kth isomer in the set of m cluster isomers:

$$q_k = \sum_j \exp\left[-\frac{e_j^{(k)} - e_0^{(k)}}{kT}\right], \tag{23}$$

where the sum is taken over all the eigenstates of the kth isomer, and the $e_j^{(k)}$ denote their energies; $e_0^{(k)}$ is the ground-state energy (the lowest value of the $e_0^{(k)}$ set is designated e_0).

Table 5 presents the temperature dependences of the weights of the water tetramers (Fig. 14). Whereas at the beginning of this temperature range the weight of the asymmetric cyclic tetramer is least important, at the end of the range it becomes the highest of all the three weight values[322,323].

The relative stability interchange is apparently an important event for an isomeric set. Quite recently, theoretical study has supplied us[329,330] with a very pronounced example of this kind, *viz.* the $(CO_2)_2$ system. The *ab initio* calculations of Brigot *et al.*[328] proved existence of two local minima (called parallel (P) and T form—see Fig. 15) on the potential-energy hypersurface of $(CO_2)_2$ and enabled construction[329,330] of w_i (Fig. 16). Whereas in the region of the lowest temperatures studied the structure P is entirely dominant, a relatively mild temperature increase causes, at first, distinct mutual approaching of the relative stabilities up to the temperature of about 230 K, when the equilibrium mixture is equimolar, the T structure being the more stable component above this temperature, although increase in its weight is slower here than below the temperature of inversion of the relative stabilities. It is particularly important that at the normal sublimation point of CO_2 the two isomeric forms will coexist at comparable concentrations (59% P and 41% T form). In the light of the finding, any possible future experiment in gas phase near the normal sublimation point of CO_2 should be organized with respect to the possibility of simultaneous presence of the both isomers.

Similarly to the gas-phase clusters, interplay of several isomeric clusters of different symmetries modelling gas–solid interactions has been studied recently[331–333].

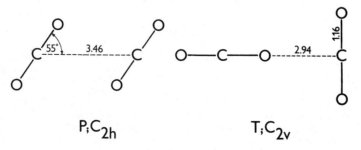

P;C_{2h} T;C_{2v}

Fig. 15. Structures[328] of the parallel (P) and T isomers of $(CO_2)_2$ (distances in 10^{-10} m).

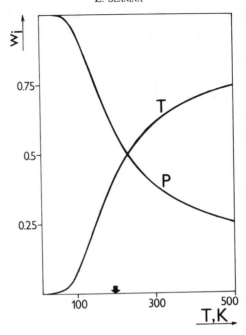

Fig. 16. Temperature dependences[329] of the weight factors w_i of the parallel (P) and T isomers of $(CO_2)_2$ (the arrow indicates the normal sublimation point of CO_2).

8. CONCLUSIONS

The recent development of symmetry considerations convincingly demonstrates that the use of symmetry properties has become a useful complement to other methods for studying potential-energy hypersurfaces. This is especially true of the representation of the hypersurfaces in terms of stationary points, which is, at present, the most frequently used method for quantum-chemical description of chemical systems. Algebraic chemistry is gradually improving the procedures for determining the numbers of stationary points on hypersurfaces, their classification and mapping of relationships between them. The individual algebraic techniques differ in their effectivenesses for individual types of systems and problems (e.g. while graph theory is more universally applicable, group theory leads to useful results primarily in situations with sufficiently high symmetry). In the latter case symmetry considerations can remarkably simplify computational procedures and offer a direct insight into a logical structure of chemical problems, as exemplified by symmetry rules for transition states. However, no simple relationships can so far be given for the interplay of isomers of different symmetries in equilibrium mixtures. Combined algebraic and computational approaches also shed new light on the problem of the observation distinguishability or nondistinguishability of isomeric structures as a key question in modern theoretical and experimental studies in isomeric chemistry.

Acknowledgments—The author is grateful to Profs. V. Kvasnička, J. N. Murrell, D. Papoušek, L. Valko, and R. G. Woolley for their constructive criticism at various stages of the work. The author acknowledges the kindness of the following organizations for the use of copyrighted material: Academic Press, Inc., The American Chemical Society, The American Institute of Physics, The Chemical Society, D. Reidel Publishing Company, Elsevier Scientific Publishing Company, North-Holland Publishing Company, Springer-Verlag and Verlag Chemie.

REFERENCES

1. J. J. Berzelius, Ueber die Zusammensetzung der Weinsäure und Traubensäure (John's Säure aus den Vogesen), über das Atomgewicht des Bleioxyds, nebst allgemeinen Bemerkungen über solche Körper, die gleiche Zusammensetzung, aber ungleiche Eigenschaften besitzen. *Ann. Phys. Chem.* **19** (der ganzen Folge **95**), 305–335 (1830).
2. E. L. Eliel, On the concept of isomerism. *Isr. J. Chem.* **15**, 7–11 (1976/77).
3. K. Mislow and P. Bickart, An epistomological note on chirality. *Isr. J. Chem.* **15**, 1–6 (1976/77).
4. K. Mislow, *Introduction to Stereochemistry*. W. A. Benjamin, New York (1966).
5. D. H. Rouvray, Isomer enumeration methods. *Chem. Soc. Rev.* **3**, 355–372 (1974).
6. J. Dugundji, R. Kopp, D. Marquarding and I. Ugi, A quantitative measure of chemical chirality and its application to asymmetric synthesis. *Top. Curr. Chem.* **75**, 165–180 (1978).

7. J. Gažo, Structure and properties of some copper (II) coordination compounds. Distortion isomerism of copper (II) compounds. *Pure Appl. Chem.* **38**, 279–301 (1974).

8. K. Mislow, Stereochemical consequences of correlated rotation in molecular propellers. *Acc. Chem. Res.* **9**, 26–33 (1976).

9. H. Manohar, D. Schwarzenbach, W. Iff and G. Schwarzenbach, A new type of isomerism among the chelates of a novel stereoselective ligand: The chemistry and crystal structures of three palladium (II) complexes with *cis*-3,5-diaminopiperidine. *J. Coord. Chem.* **8**, 213–221 (1979).

10. F. Cozzi, A. Guenzi, C. A. Johnson, K. Mislow, W. D. Hounshell and J. F. Blount, Stereoisomerism and correlated rotation in molecular gear systems. Residual diastereomers of bis(2,3-dimethyl-9-triptycyl)methane. *J. Am. Chem. Soc.* **103**, 957–958 (1981).

11. Y. Kawada and H. Iwamura, Bis(4-chloro-1-triptycyl) ether. Separation of a pair of phase isomers of labeled bevel gears. *J. Am. Chem. Soc.* **103**, 958–960 (1981).

12. Y. Kawada and H. Iwamura, Phase isomerism in gear-shaped molecules. *Tetrahedron Lett.* **22**, 1533–1536 (1981).

13. C. A. Johnson, A. Guenzi and K. Mislow, Restricted gearing and residual stereoisomerism in bis(1,4-dimethyl-9-triptycyl)methane. *J. Am. Chem. Soc.* **103**, 6240–6242 (1981).

14. S. W. Benson and J. H. Buss, Halogen-catalyzed decomposition of N_2O and the role of hypohalite radical. *J. Chem. Phys.* **27**, 1382–1384 (1957).

15. I. C. Hisatsune, J. P. Devlin and Y. Wada, Infrared spectra of some unstable isomers of N_2O_4 and N_2O_3. *J. Chem. Phys.* **33**, 714–719 (1960).

16. H. A. Bent, Dimers of nitrogen dioxide. II. Structure and bonding. *Inorg. Chem.* **2**, 747–752 (1963).

17. F. Bolduan and H. J. Jodl, Raman spectroscopy on matrix-isolated NO^+, NO_3^-, and N_2O_4 in Ne. *Chem. Phys. Lett.* **85**, 283–286 (1982).

18. R. L. Kuczkowski, R. D. Suenram and F. J. Lovas, Microwave spectrum, structure, and dipole moment of sulfuric acid. *J. Am. Chem. Soc.* **103**, 2561–2566 (1981).

19. R. D. Suenram and F. J. Lovas, Millimeter wave spectrum of glycine. A new conformer. *J. Am. Chem. Soc.* **102**, 7180–7184 (1980).

20. J. S. Wright, Theoretical evidence for a stable form of cyclic ozone, and its chemical consequences. *Can. J. Chem.* **51**, 139–146 (1973).

21. G. Karlström, S. Engström and B. Jönsson, Multiconfigurational SCF and CI calculations on the open and closed forms of the ozone molecule. *Chem. Phys. Lett.* **57**, 390–394 (1978).

22. A. T. Balaban, D. Fărcaşiu and R. Bănică, Graphs of multiple 1,2-shifts in carbonium ions and related systems. *Rev. Roum. Chim.* **11**, 1205–1227 (1966).

23. A. T. Balaban, Chemical graphs. III. Reactions with cyclic six-membered transition states. *Rev. Roum. Chim.* **12**, 875–898 (1967).

24. S. Green and E. Herbst, Metastable isomers: a new class of interstellar molecules. *Astrophys. J.* **229**, 121–131 (1979).

25. D. H. Rouvray, The pioneers of isomer enumeration. *Endeavour* **34**, 28–33 (1975).

26. A. Cayley, On the mathematical theory of isomers. *Phil. Mag.* **47**, 444–447 (1874).

27. W. Koerner, Studj sull'isomeria delle cosi dette sostanze aromatiche a sei atomi di carbonio. *Gazz. Chim. Ital.* **4**, 305–446 (1874).

28. J. H. van't Hoff, Voorstel tot Uitbreiding der Tegenwoordig in de Scheikunde gebruikte Structuur-Formules in de Ruimte, benevens een daarmeê Samenhangende Opmerking omtrent het Verband tusschen Optisch Actief Vermogen en Chemische Constitutie van Organische Verbindingen. Greven, Utrecht (1874).

29. D. H. Rouvray, Uses of graph theory. *Chem. Brit.* **10**, 11–18 (1974).

30. R. F. W. Bader and T. T. Nguyen-Dang, Quantum theory of atoms in molecules—Dalton revisited. *Advan. Quantum Chem.* **14**, 63–124 (1981).

31. G. Ege, Isomerien in der Chemie. Graphen- und mengentheoretische Betrachtung. *Naturwissenschaften* **58**, 247–257 (1971).

32. I. Ugi, P. Gillespie and C. Gillespie, Chemistry, a finite metric topology—synthetic planning, an exercise in algebra. *Trans. N. Y. Acad. Sci.* **34**, 416–432 (1972).

33. E. Ruch, Homochiralität als klassifizierungsprinzip von molekülen spezieller molekülklassen. *Theor. Chim. Acta* **11**, 183–192 (1968).

34. E. Ruch, W. Hässelbarth and B. Richter, Doppelnebenklassen als klassenbegriff und nomenklaturprinzip für isomere und ihre abzählung. *Theor. Chim. Acta* **19**, 288–300 (1970).

35. W. Hässelbarth and E. Ruch, Classifications of rearrangement mechanisms by means of double cosets and counting formulas for the numbers of classes. *Theor. Chim. Acta* **29**, 259–268 (1973).

36. K. Mislow, On the classification of pairwise relations between isomeric structures. *Bull. Soc. Chim. Belg.* **86**, 595–601 (1977).

37. N. Trinajstić, *Chemical Graph Theory, Vols. I, II.* CRC Press, Boca Raton, FL (1983).

38. D. Bonchev, O. Mekenyan and N. Trinajstić, Isomer discrimination by topological information approach. *J. Comput. Chem.* **2**, 127–148 (1981).

39. J. Brocas, Some formal properties of the kinetics of pentacoordinate stereoisomerizations. *Fortschr. Chem. Forsch.* **32**, 43–61 (1972).

40. A. T. Balaban, Chemical graphs. XVI. Intramolecular isomerizations of octahedral complexes with six different ligands. *Rev. Roum. Chim.* **18**, 841–854 (1973).

41. J. G. Nourse and K. Mislow, Dynamic stereochemistry of tetraarylmethanes and cognate systems. The role of the permutation subgroup lattice. *J. Am. Chem. Soc.* **97**, 4571–4578 (1975).

42. J. G. Nourse, Generalized stereoisomerization modes. *J. Am. Chem. Soc.* **99**, 2063–2069 (1977).

43. M. Randić, A systematic study of symmetry properties of graphs. I. Petersen graph. *Croat. Chem. Acta* **49**, 643–655 (1977).

44. D. J. Klein and A. H. Cowley, Permutational isomerism with bidentate ligands and other constraints. *J. Am. Chem. Soc.* **100**, 2593–2599 (1978).

45. M. Randić, Symmetry properties of graphs of interest in chemistry. II. Desargues–Levi graph. *Int. J. Quantum Chem.* **25**, 663–682 (1979).

46. E. K. Lloyd, Graphs corresponding to intramolecular rearrangements. *Match* **7**, 255–271 (1979).
47. J. G. Nourse, Applications of the permutation group in dynamic stereochemistry, in *The Permutation Group in Physics and Chemistry* (Edited by J. Hinze), pp. 28–37. Springer-Verlag, Berlin (1979).
48. J. Brocas, R. Willem, J. Buschen and D. Fastenakel, Isomerization modes for idealized and distorted skeleta. (II). Polytopal isomerization reactions. *Bull. Soc. Chim. Belg.* **88**, 415–434 (1979).
49. M. Randić, Symmetry properties of graphs of interest in chemistry. III. Homotetrahedryl rearrangement. *Int. J. Quantum Chem., Quantum Chem. Symp.* **14**, 557–577 (1980).
50. J. G. Nourse, Self-inverse and non-self-inverse degenerate isomerizations. *J. Am. Chem. Soc.* **102**, 4883–4889 (1980).
51. R. B. King, Chemical applications of group theory and topology. X. Topological representations of hyperocta-hedrally restricted eight-coordinate polyhedral rearrangements. *Theor. Chim. Acta* **59**, 25–45 (1981).
52. L. B. Kier and L. H. Hall, *Molecular Connectivity in Chemistry and Drug Research.* Academic Press, New York (1976).
53. A. T. Balaban, A. Chiriac, I. Motoc and Z. Simon, *Steric Fit in Quantitative Structure–Activity Relations.* Springer-Verlag, Berlin (1980).
54. D. H. Rouvray, The search for useful topological indices in chemistry. *Am. Scientist* **61**, 729–735 (1973).
55. A. Graovac, I. Gutman and N. Trinajstić, *Topological Approach to the Chemistry of Conjugated Molecules.* Springer-Verlag, Berlin (1977).
56. D. Bonchev and N. Trinajstić, Information theory, distance matrix, and molecular branching. *J. Chem. Phys.* **67**, 4517–4533 (1977).
57. D. Bonchev, Information indices for atoms and molecules. *Match* **7**, 65–112 (1979).
58. O. Mekenyan, D. Bonchev and N. Trinajstić, Chemical graph theory: Modeling the thermodynamic properties of molecules. *Int. J. Quantum Chem.* **18**, 369–380 (1980).
59. D. H. Rouvray, The additivity principle from a topological view-point. *Chem. Tech.* **3**, 379–384 (1973).
60. J. W. Essam, J. W. Kennedy, M. Gordon and P. Whittle, The graph-like state of matter. Part 8. LCGI schemes and the statistical analysis of experimental data. *J. Chem. Soc. Faraday Trans.* II **73**, 1289–1307 (1977).
61. J. Dugundji and I. Ugi, An algebraic model of constitutional chemistry as a basis for chemical computer programs. *Fortschr. Chem. Forsch.* **39**, 19–64 (1973).
62. J. Dugundji, P. Gillespie, D. Marquarding, I. Ugi and F. Ramirez, Metric spaces and graphs representing the logical structure of chemistry, in *Chemical Applications of Graph Theory* (Edited by A. T. Balaban), pp. 107–174. Academic Press, London (1976).
63. J. Brandt, J. Friedrich, J. Gasteiger, C. Jochum, W. Schubert and I. Ugi, Computer programs for the deductive solution of chemical problems on the basis of a mathematical model of chemistry, in *Computer-Assisted Organic Synthesis* (Edited by W. T. Wipke and W. J. Howe), pp. 33–59. ACS Symp. Ser. 61, American Chemical Society, Washington, D.C. (1977).
64. I. Ugi, J. Bauer, J. Brandt, J. Friedrich, J. Gasteiger, C. Jochum and W. Schubert, New applications of computers in chemistry. *Angew. Chem. Int. Ed. Engl.* **18**, 111–123 (1979).
65. V. Kvasnička, Graph theory approach. *Collect. Czech. Chem. Commun.* **48**, 2097–2117 (1983).
66. V. Kvasnička, Synthon approach and graph grammar. *Collect. Czech. Chem. Commun.* **48**, 2118–2129 (1983).
67. V. Kvasnička, M. Kratochvíl and J. Koča, Reactions graphs. *Collect. Czech. Chem. Commun.* **48**, 2284–2304 (1983).
68. A. D. Liehr, Topological aspects of the conformational stability problem. Part I. Degenerate electronic states. *J. Phys. Chem.* **67**, 389–471 (1963).
69. A. D. Liehr, Topological aspects of the conformational stability problem. Part II. Non-degenerate electronic states. *J. Phys. Chem.* **67**, 471–494 (1963).
70. J. W. McIver, Jr. and A. Komornicki, Structure of transition states in organic reactions. General theory and an application to the cyclobutene–butadiene isomerization using a semiempirical molecular orbital method. *J. Am. Chem. Soc.* **94**, 2625–2633 (1972).
71. V. G. Dashevsky, *Conformations of Organic Molecules.* Khimija, Moscow (1974) (in Russian).
72. O. Ermer, Calculation of molecular properties using force fields. Applications in organic chemistry. *Struct. Bond.* **27**, 161–217 (1976).
73. J. N. Murrell, The potential energy surfaces of polyatomic molecules. *Struct. Bond.* **32**, 93–146 (1977).
74. P. G. Mezey, Analysis of conformational energy hypersurfaces. *Progr. Theor. Org. Chem.* **2**, 127–161 (1977).
75. P. G. Mezey, Optimization and analysis of energy hypersurfaces, in *Computational Theoretical Organic Chemistry* (Edited by I. G. Csizmadia and R. Daudel), pp. 101–128. D. Reidel, Dordrecht (1981).
76. Z. Slanina, Chemical isomerism and its contemporary theoretical description. *Advan. Quantum Chem.* **13**, 89–153 (1981).
77. E. Ōsawa and H. Musso, Application of molecular mechanics calculations to organic chemistry. *Top. Stereochem.* **13**, 117–193 (1982).
78. V. G. Dashevsky, *Conformational Analysis of Organic Molecules.* Khimija, Moscow (1982) (in Russian).
79. E. Ōsawa and H. Musso, Molecular mechanics calculations in organic chemistry: examples of the usefulness of this simple non-quantum mechanical model. *Angew. Chem. Int. Ed. Engl.* **22**, 1–12 (1983).
80. J. N. Murrell, S. Carter, S. C. Farantos, P. Huxley and A. J. C. Varandas, *Molecular Potential Energy Functions.* Wiley, New York (1984).
81. Z. Slanina, *Contemporary Theory of Chemical Isomerism.* D. Reidel, & Academia, Dordrecht and Prague (1986).
82. M. Born and R. Oppenheimer, Zur quantentheorie der molekeln. *Ann. Phys. 4 Volge* **84** (der ganzen Reihe **389**), 457–484 (1927).
83. M. Born, Kopplung der elektronen- und kernbewegung in molekeln und kristallen. *Nachr. Akad. Wiss. Göttingen, Math. Phys. Kl*(6), 1–3 (1951).
84. M. Born and K. Huang, *Dynamical Theory of Crystal Lattices*, pp. 402–407. Oxford University Press, London (1954).
85. İ. Özkan and L. Goodman, Coupling of electronic and vibrational motions in molecules. *Chem. Rev.* **79**, 275–285 (1979).
86. E. L. Muetterties, Stereochemically nonrigid structures. *Inorg. Chem.* **4**, 769–771 (1965).

87. I. B. Bersuker, *Structure and Properties of Coordination Compounds*, p. 192. Khimija, Leningrad (1971) (in Russian).
88. W. v. E. Doering and W. R. Roth, Thermische unlagerungsreaktionen. *Angew. Chem.* **75**, 27–35 (1963).
89. F. A. Cotton, Fluxional organometallic molecules. *Acc. Chem. Res.* **1**, 257–265 (1968).
90. I. Ugi, D. Marquarding, H. Klusacek and P. Gillespie, Berry pseudorotation and turnstile rotation. *Acc. Chem. Res.* **4**, 288–296 (1971).
91. O. Bastiansen, K. Kveseth and H. Møllendal, Structure of molecules with large amplitude motion as determined from electron-diffraction studies in the gas phase. *Top. Curr. Chem.* **81**, 99–172 (1979).
92. G. O. Sørensen, A new approach to the Hamiltonian of nonrigid molecules. *Top. Curr. Chem.* **82**, 97–175 (1979).
93. M. Hargittai and I. Hargittai, *The Molecular Geometries of Coordination Compounds in the Vapour Phase*. Akadémiai Kiadó, Budapest (1977).
94. I. Hargittai, Gas electron diffraction. A tool of structural chemistry in perspectives. *Top. Curr. Chem.* **96**, 43–78 (1981).
95. S. H. Bauer, The variety of structures which interest chemists, in *Physical Chemistry, An Advanced Treatise*, Vol. IV (Edited by D. Henderson), pp. 1–17. Academic Press, New York (1970).
96. D. M. Dennison and G. E. Uhlenbeck, The two-minima problem and the ammonia molecule. *Phys. Rev.* **41**, 313–321 (1932).
97. J. E. Kilpatrick, K. S. Pitzer and R. Spitzer, The thermodynamics and molecular structure of cyclopentane. *J. Am. Chem. Soc.* **69**, 2483–2488 (1947).
98. R. S. Berry, Correlation of rates of intramolecular tunneling processes, with application to some group V compounds. *J. Chem. Phys.* **32**, 933–938 (1960).
99. N. M. Sergeev, Dynamic nuclear magnetic resonance. *Usp. Khim.* **42**, 769–798 (1973).
100. R. G. Woolley, Quantum theory and molecular structure. *Advan. Phys.* **25**, 27–52 (1976).
101. R. G. Woolley, On the description of high-resolution experiments in molecular physics. *Chem. Phys. Lett.* **44**, 73–75 (1976).
102. R. G. Woolley and B. T. Sutcliffe, Molecular structure and the Born–Oppenheimer approximation. *Chem. Phys. Lett.* **45**, 393–398 (1977).
103. R. G. Woolley, The quantum interpretation of molecular structure. *Int. J. Quantum Chem.* **12**(Suppl. 1), 307–313 (1977).
104. H. Essén, The physics of the Born–Oppenheimer approximation. *Int. J. Quantum Chem.* **12**, 721–735 (1977).
105. R. G. Woolley, Must a molecule have a shape? *J. Am. Chem. Soc.* **100**, 1073–1079 (1978).
106. R. G. Woolley, Further remarks on molecular structure in quantum theory. *Chem. Phys. Lett.* **55**, 443–446 (1978).
107. S. Aronowitz, Intermolecular correlation in a new approximation scheme. *Int. J. Quantum Chem.* **14**, 253–269 (1978).
108. E. B. Wilson, On the definition of molecular structure in quantum mechanics. *Int. J. Quantum Chem., Quantum Chem. Symp.* **13**, 5–14 (1979).
109. R. F. W. Bader and R. A. Gangi, *Ab initio* calculation of potential energy surfaces. *Theor. Chem.* **2**, 1–65 (1975).
110. A. Devaquet, Potential energy surfaces. *Fortschr. Chem. Forsch.* **54**, 1–73 (1975).
111. E. E. Nikitin and L. Zülicke, *Selected Topics of the Theory of Chemical Elementary Processes*. Springer-Verlag, Berlin (1978).
112. S. R. Niketić and K. Rasmussen, *The Consistent Force Field: A Documentation*. Springer-Verlag, Berlin (1977).
113. U. Burkert and N. L. Allinger, *Molecular Mechanics*, ACS Monograph 177. American Chemical Society, Washington (1982).
114. P. M. Ivanov and E. Ōsawa, Remarks on the analysis of torsional energy surfaces of cycloheptane and cyclooctane by molecular mechanics. *J. Comput. Chem.* **5**, 307–313 (1984).
115. P. W. Pakes, T. C. Rounds and H. L. Strauss, Conformations of cyclooctane and some related oxocanes. *J. Phys. Chem.* **85**, 2469–2475 (1981).
116. C. J. H. Schutte, The *ab initio* calculation of molecular vibrational frequencies and force constants. *Struct. Bond.* **9**, 213–263 (1971).
117. M. Kanakavel, J. Chandrasekhar, S. Subramanian and S. Singh, Calculation of optimum geometries and force fields by the CNDO/force method. *Theor. Chim. Acta* **43**, 185–196 (1976).
118. P. Pulay, *Ab initio* calculation of force constants and equilibrium geometries in polyatomic molecules. I. Theory. *Mol. Phys.* **17**, 197–204 (1969).
119. P. Pulay, *Ab initio* calculation of force constants and equilibrium geometries in polyatomic molecules. II. Force constants of water. *Mol. Phys.* **18**, 473–480 (1970).
120. P. Pulay, *Ab initio* calculation of force constants and equilibrium geometries. III. Second-row hybrides. *Mol. Phys.* **21**, 329–339 (1971).
121. J. W. McIver, Jr. and A. Komornicki, Rapid geometry optimization for semi-empirical molecular orbital methods. *Chem. Phys. Lett.* **10**, 303–306 (1971).
122. P. Pulay, Direct use of the gradient for investigating molecular energy surfaces. *Mod. Theor. Chem.* **4**, 153–185 (1977).
123. S. Kato and K. Morokuma, Energy gradient in a multi-configurational SCF formalism and its application to geometry optimization of trimethylene diradicals. *Chem. Phys. Lett.* **65**, 19–25 (1979).
124. J. D. Goddard, N. C. Handy and H. F. Schaefer, III, Gradient techniques for open-shell restricted Hartree–Fock and multiconfiguration self-consistent-field methods. *J. Chem. Phys.* **71**, 1525–1530 (1979).
125. R. N. Camp, H. F. King, J. W. McIver, Jr. and D. Mullally, Analytical force constants for MCSCF wave functions. *J. Chem. Phys.* **79**, 1088–1089 (1983).
126. D. J. Fox, Y. Osamura, M. R. Hoffmann, J. F. Gaw, G. Fitzgerald, Y. Yamaguchi and H. F. Schaefer III, Analytic energy second derivatives for general correlated wavefunctions, including a solution of the first-order coupled-perturbed configuration–interaction equations. *Chem. Phys. Lett.* **102**, 17–19 (1983).
127. J. A. Pople, R. Krishnan, H. B. Schlegel and J. S. Binkley, Derivative studies in Hartree–Fock and Møller–Plesset theories. *Int. J. Quantum Chem., Quantum Chem. Symp.* **13**, 225–241 (1979).

128. K. Yamashita, A. Tachibana, T. Yamabe and K. Fukui, An aspect of molecular dynamics on the CI potential energy surface. *Chem. Phys. Lett.* **69**, 413–416 (1980).

129. B. R. Brooks, W. D. Laidig, P. Saxe, J. D. Goddard, Y. Yamaguchi and H. F. Schaefer, III, Analytic gradients from correlated wave functions via the two-particle density matrix and the unitary group approach. *J. Chem. Phys.* **72**, 4652–4653 (1980).

130. R. Krishnan, H. B. Schlegel and J. A. Pople, Derivative studies in configuration–interaction theory. *J. Chem. Phys.* **72**, 4654–4655 (1980).

131. E. B. Wilson, Jr., J. C. Decius and P. C. Cross, *Molecular Vibrations.* McGraw-Hill, New York (1955).

132. M. A. Elyashevich, A simple method of calculation of vibrational frequencies of polyatomic molecules. *Dokl. Akad. Nauk SSSR* **28**, 605–609 (1940) (in Russian).

133. D. H. Duffey, *A Development of Quantum Mechanics Based on Symmetry Considerations.* D. Reidel, Dordrecht (1984).

134. J. Maruani and J. Serre (Eds), *Symmetries and Properties of Non-Rigid Molecules, A Comprehensive Survey.* Elsevier, Amsterdam (1983).

135. R. B. King (Ed), *Chemical Applications of Topology and Graph Theory.* Elsevier, Amsterdam (1983).

136. J. Serre, Group theory and the molecular orbital method, in *Molecular Orbitals in Chemistry, Physics, and Biology* (Edited by B. Pullman and P.-O. Löwdin), pp. 133–149. Academic Press, New York (1964).

137. P.-O. Löwdin, Group algebra, convolution algebra, and applications to quantum mechanics. *Rev. Mod. Phys.* **39**, 259–287 (1967).

138. H. H. Jaffé and R. L. Ellis, A treatment of symmetry in MO calculations. *J. Comput. Phys.* **16**, 20–31 (1974).

139. M. Orchin and H. H. Jaffé, *Symmetry, Orbitals, and Spectra.* Wiley-Interscience, New York (1971).

140. I. G. Kaplan, Applications of group-theory methods in quantum-chemical calculations. *Usp. Khim.* **48**, 1027–1053 (1979) (in Russian).

141. E. Wigner and E. E. Witmer, Über die struktur der zweiatomigen molekelspektren nach der guantemechanik. *Z. Phys.* **51**, 859–886 (1928).

142. R. B. Woodward and R. Hoffmann, Die erhaltung der orbitalsymmetrie. *Angew. Chem.* **81**, 797–869 (1969).

143. R. B. Woodward and R. Hoffmann, *Die Erhaltung der Orbitalsymmetrie.* Verlag Chemie, Weinheim (1970).

144. H. C. Longuet-Higgins and E. W. Abrahamson, The Electronic mechanism of electrocyclic reactions. *J. Am. Chem. Soc.* **87**, 2045–2046 (1965).

145. T. F. George and J. Ross, Analysis of symmetry in chemical reactions. *J. Chem. Phys.* **55**, 3851–3866 (1971).

146. K. Fukui, Recognition of stereochemical paths by orbital interaction. *Acc. Chem. Res.* **4**, 57–64 (1971).

147. R. G. Pearson, Orbital symmetry rules and the mechanism of inorganic reactions. *Pure Appl. Chem.* **27**, 145–160 (1971).

148. J. A. Berson and L. Salem, Subjacent orbital control. An electronic factor favoring concertedness in Woodward–Hoffmann "forbidden" reactions. *J. Am. Chem. Soc.* **94**, 8917–8918 (1972).

149. W. T. Borden and L. Salem, Electron repulsion in pericyclic transition states. *J. Am. Chem. Soc.* **95**, 932–933 (1973).

150. D. M. Silver, Hierarchy of symmetry conservation rules governing chemical reaction systems. *J. Am. Chem. Soc.* **96**, 5959–5967 (1974).

151. A. Rassat, A graph system equivalent to Woodward–Hoffmann rules. *Tetrahedron Lett.* **46**, 4081–4084 (1975).

152. J. J. Gajewski, *Hydrocarbon Thermal Isomerizations.* Academic Press, New York (1981).

153. E. B. Wilson, Jr., The degeneracy, selection rules, and other properties of the normal vibrations of certain polyatomic molecules. *J. Chem. Phys.* **2**, 432–439 (1934).

154. E. B. Wilson, Jr., The statistical weights of the rotational levels of polyatomic molecules, including methane, ammonia, benzene, cyclopropane and ethylene. *J. Chem. Phys.* **3**, 276–285 (1935).

155. E. B. Wilson, Jr., Symmetry considerations concerning the splitting of vibration–rotation levels in polyatomic molecules. *J. Chem. Phys.* **3**, 818–821 (1935).

156. J. T. Hougen, Classification of rotational energy levels for symmetric-top molecules. *J. Chem. Phys.* **37**, 1433–1441 (1962).

157. J. T. Hougen, Classification of rotational energy levels. II. *J. Chem. Phys.* **39**, 358–365 (1963).

158. H. C. Longuet-Higgins, The symmetry of nonrigid molecules. *Mol. Phys.* **6**, 445–460 (1963).

159. R. G. Jones, The use of symmetry in nuclear magnetic resonance. *NMR Bas. Princip. Progr.* **1**, 97–174 (1969).

160. J. Serre, Symmetry groups of nonrigid molecules. *Advan. Quantum Chem.* **8**, 1–36 (1974).

161. P. R. Bunker, *Molecular Symmetry and Spectroscopy.* Academic Press, New York (1979).

162. G. S. Ezra, *Symmetry Properties of Molecules.* Springer-Verlag, Berlin (1982).

163. D. Papoušek and M. R. Aliev, *Molecular Vibrational–Rotational Spectra.* Academia, Prague (1982).

164. R. S. Berry, A general phenomenology for small clusters, however floppy, in *Quantum Dynamics of Molecules: The New Experimental Challenge to Theorists* (Edited by R. G. Woolley), pp. 143–195. Plenum Press, New York (1980).

165. T. R. Dyke, Group theoretical classification of the tunneling–rotational energy levels of water dimer. *J. Chem. Phys.* **66**, 492–497 (1977).

166. S. L. Altmann, The symmetry of nonrigid molecules: the Schrödinger supergroup. *Proc. Roy. Soc.* **298A**, 184–203 (1967).

167. H. Frei, P. Groner, A. Bauder and Hs. H. Günthard, Isometric group of semi-rigid models: relation to the permutation–inversion group and extension of the concept to non-rigid molecules. *Mol. Phys.* **36**, 1469–1495 (1978).

168. H. Frei, A. Bauder and Hs. H. Günthard, The isometric group of non-rigid molecules. *Top. Curr. Chem.* **81**, 1–97 (1979).

169. H. Frei, A. Bauder and Hs. H. Günthard, Isometric groups of non-rigid molecules: homomorphisms and reformulation of a theorem on semi-direct product factorization. *Mol. Phys.* **43**, 785–797 (1981).

170. J. A. Pople, Classification of molecular symmetry by framework groups. *J. Am. Chem. Soc.* **102**, 4615–4622 (1980).

171. D. H. McDaniel, Conversion between framework group notation and permutation group notation for molecules. *J. Phys. Chem.* **85**, 479–481 (1981).

172. R. L. Flurry, Jr., Site symmetry and the framework group. *J. Am. Chem. Soc.* **103**, 2901–2902 (1981).

173. R. L. Flurry, Jr., Site symmetry in molecular point groups. *Int. J. Quantum Chem.*, *Quantum Chem. Symp.* **6**, 455–458 (1972).
174. R. L. Flurry, Jr., The use of site symmetry in constructing symmetry adapted functions. *Theor. Chim. Acta* **31**, 221–230 (1973).
175. R. L. Flurry, Jr., *Symmetry Groups: Theory and Chemical Applications*. Prentice-Hall, Englewood Cliffs, NJ (1980).
176. B. J. Dalton, Classification of the states of non-rigid molecules. *Mol. Phys.* **11**, 265–285 (1966).
177. J.-M. F. Gilles and J. Philippot, Internal symmetry groups of non-rigid molecules. *Int. J. Quantum Chem.* **6**, 225–261 (1972).
178. J.-M. F. Gilles and J. Philippot, Symmetry of non-rigid molecules and isomerization processes. *Int. J. Quantum Chem.* **14**, 299–311 (1978).
179. Y. Ellinger and J. Serre, An application of the non-rigid molecule group theory to a problem of chemical reactivity. *Int. J. Quantum Chem.*, *Quantum Chem. Symp.* **7**, 217–221 (1973).
180. M. Randić, On discerning symmetry properties of graphs. *Chem. Phys. Lett.* **42**, 283–287 (1976).
181. T. D. Bouman, C. D. Duncan and C. Trindle, Group theory and reaction mechanisms: permutation theoretic prediction and computational support for pseudorotation modes in $C_2H_5^+$ and $C_5H_5^+$ rearrangements. *Int. J. Quantum Chem.* **11**, 399–413 (1977).
182. R. S. Berry, Symmetry and thermodynamics from structured molecules to liquid drops, in *The Permutation Group in Physics and Chemistry* (Edited by J. Hinze), pp. 92–120. Springer-Verlag, Berlin (1979).
183. H. B. Bürgi, Space group-like symmetries in the conformational analysis of flexible molecules. *Match* **9**, 13–14 (1980).
184. H. A. Jahn and E. Teller, Stability of polyatomic molecules in degenerate electronic states. I. Orbital degeneracy. *Proc. Roy. Soc.* **161A**, 220–235 (1937).
185. R. Englman, *The Jahn–Teller Effect in Molecules and Crystals*. Wiley-Interscience, London (1972).
186. I. B. Bersuker, *Electronic Structure and Properties of Coordination Compounds*. Khimija, Leningrad (1976) (in Russian).
187. M. J. S. Dewar, S. Kirschner, H. W. Kollmar and L. E. Wade, Orbital isomerism in biradical processes. *J. Am. Chem. Soc.* **96**, 5242–5244 (1974).
188. A. Komornicki and J. W. McIver, Jr., Structure of transition states. 4. MINDO/2 study of rearrangements in the C_6H_{10} system. *J. Am. Chem. Soc.* **98**, 4553–4561 (1976).
189. M. J. S. Dewar, G. P. Ford, M. L. McKee, H. S. Rzepa and L. E. Wade, The Cope rearrangement. MINDO/3 Studies of the rearrangements of 1,5-hexadiene and bicyclo[2.2.0]hexane. *J. Am. Chem. Soc.* **99**, 5069–5073 (1977).
190. M. J. S. Dewar, S. Kirschner and H. W. Kollmar, Orbital isomerism as a controlling factor in chemical reactivity. *J. Am. Chem. Soc.* **96**, 5240–5242 (1974).
191. J. N. Murrell, Symmetry and the transition state, in *Quantum Theory of Chemical Reactions*, Vol. 1 (Edited by R. Daudel, A. Pullman, L. Salem and A. Veillard), pp. 161–176. D. Reidel, Dordrecht (1980).
192. P. Pechukas, Transition state theory. *Ann. Rev. Phys. Chem.* **32**, 159–177 (1981).
193. J. N. Murrell and K. J. Laidler, Symmetries of activated complexes. *Trans. Faraday Soc.* **64**, 371–377 (1968).
194. T. Tanaka and Y. Morino, Coriolis interaction and anharmonic potential function of ozone from the microwave spectra in the excited vibrational states. *J. Mol. Spectrosc.* **33**, 538–551 (1970).
195. R. R. Lucchese and H. F. Schaefer, III, Energy separation between the open (C_{2v}) and closed (D_{3h}) forms of ozone. *J. Chem. Phys.* **67**, 848–849 (1977).
196. J. N. Murrell, K. S. Sorbie and A. J. C. Varandas, Analytical potentials for triatomic molecules from spectroscopic data. II. Application to ozone. *Mol. Phys.* **32**, 1359–1372 (1976).
197. R. E. Stanton and J. W. McIver, Jr., Group theoretical selection rules for the transition states of chemical reactions. *J. Am. Chem. Soc.* **97**, 3632–3646 (1975).
198. W.-D. Stohrer and R. Hoffmann, Bond-stretch isomerism and polytopal rearrangements in $(CH)_5^+$, $(CH)_5^-$, and $(CH)_4CO$. *J. Am. Chem. Soc.* **94**, 1661–1668 (1972).
199. J. W. McIver, Jr., On the existence of symmetric transition states for cycloaddition reactions. *J. Am. Chem. Soc.* **94**, 4782–4783 (1972).
200. J. W. McIver, Jr., and R. E. Stanton, Symmetry selection rules for transition states. *J. Am. Chem. Soc.* **94**, 8618–8620 (1972).
201. J. W. McIver, Jr., The structure of transition states: are they symmetric? *Acc. Chem. Res.* **7**, 72–77 (1974).
202. C. W. Wilson, Jr. and W. A. Goddard, III, *Ab initio* calculations on the $H_2 + D_2 = 2HD$ four-center exchange reaction. I. Elements of the reaction surface. *J. Chem. Phys.* **51**, 716–731 (1969).
203. L. Salem, J. Durup, G. Bergeron, D. Cazes, X. Chapuisat and H. Kagan, Narcissistic reactions. *J. Am. Chem. Soc.* **92**, 4472–4474 (1970).
204. L. Salem, Narcissistic reactions: synchronism *vs.* nonsynchronism in automerizations and enantiomerizations. *Acc. Chem. Res.* **4**, 322–328 (1971).
205. H. Metiu, J. Ross, R. Silbey and T. F. George, On symmetry properties of reaction coordinates. *J. Chem. Phys.* **61**, 3200–3209 (1974).
206. P. Scharfenberg, Invariance criteria and symmetry conservation rules for geometry optimizations. *Theor. Chim. Acta* **53**, 279–292 (1979).
207. P. Pechukas, On simple saddle points of a potential surface, the conservation of nuclear symmetry along paths of steepest descent, and the symmetry of transition states. *J. Chem. Phys.* **64**, 1516–1521 (1976).
208. J. N. Murrell, The Jahn–Teller theorem applies to transition states. *Chem. Commun.* 1044–1045 (1972).
209. U. Öpik and M. H. L. Pryce, Studies of the Jahn–Teller effect. I. A survey of the static problem. *Proc. Roy. Soc.* **238A**, 425–447 (1957).
210. R. F. W. Bader, Vibrationally induced perturbations in molecular electron distributions. *Can. J. Chem.* **40**, 1164–1175 (1962).
211. L. S. Bartell, Molecular geometry. Bonded *versus* nonbonded interactions. *J. Chem. Educ.* **45**, 754–767 (1968).
212. R. G. Pearson, A symmetry rule for predicting molecular structure and reactivity. *J. Am. Chem. Soc.* **91**, 1252–1254 (1969).
213. R. G. Pearson, A symmetry rule for predicting molecular structures. *J. Am. Chem. Soc.* **91**, 4947–4955 (1969).

214. L. Salem, Conditions for favorable unimolecular reaction paths. *Chem. Phys. Lett.* **3**, 99–101 (1969).
215. L. Salem and J. S. Wright, Vibrational modes, orbital symmetries, and unimolecular reaction paths. *J. Am. Chem. Soc.* **91**, 5947–5955 (1969).
216. J. N. Murrell, J. B. Pedley and S. Durmaz, Potential energy surfaces for the reaction $CH_2 + H_2 \rightarrow CH_4$. *J. Chem. Soc., Faraday Trans.* II **69**, 1370–1380 (1973).
217. J. N. Murrell and G. L. Pratt, Statistical factors and the symmetry of transition states. *Trans. Faraday Soc.* **66**, 1680–1684 (1970).
218. E. W. Schlag, Symmetry numbers and reaction rates. *J. Chem. Phys.* **38**, 2480–2482 (1963).
219. E. W. Schlag and G. L. Haller, Symmetry numbers and reaction rates. II. The computation of the reaction-path degeneracy for bimolecular reactions. *J. Chem. Phys.* **42**, 584–587 (1965).
220. D. M. Bishop and K. J. Laidler, Symmetry numbers and statistical factors in rate theory. *J. Chem. Phys.* **42**, 1688–1691 (1965).
221. H. S. Johnston, *Gas Phase Reaction Rate Theory*. Ronald Press, New York (1966).
222. C. S. Elliott and H. M. Frey, Reaction of methylene with cyclobutene. Part 1. *Trans. Faraday Soc.* **64**, 2352–2368 (1968).
223. V. Gold, Statistical factors in the Brönsted catalysis law and other free energy correlations. *Trans. Faraday Soc.* **60**, 738–739 (1964).
224. R. A. Marcus, Dissociation and isomerization of vibrationally excited species. III. *J. Chem. Phys.* **43**, 2658–2661 (1965).
225. D. M. Bishop and K. J. Laidler, Statistical factors for chemical reactions. *Trans. Faraday Soc.* **66**, 1685–1687 (1970).
226. E. Pollak and P. Pechukas, Symmetry numbers, not statistical factors, should be used in absolute rate theory and in Brønsted relations. *J. Am. Chem. Soc.* **100**, 2984–2991 (1978).
227. D. R. Coulson, Statistical factors in reaction rate theories. *J. Am. Chem. Soc.* **100**, 2992–2996 (1978).
228. F. Harary and E. M. Palmer, *Graphical Enumeration*. Academic Press, New York (1973).
229. A. T. Balaban and F. Harary, Early history of the interplay between graph theory and chemistry, in *Chemical Applications of Graph Theory* (Edited by A. T. Balaban), pp. 1–4. Academic Press, London (976).
230. A. Cayley, On the theory of the analytical forms called trees. *Phil. Mag.* **13**, 172–176 (1857).
231. J. H. Redfield, The theory of group-reduced distributions. *Am. J. Math.* **49**, 433–455 (1927).
232. A. C. Lunn and J. K. Senior, Isomerism and configuration. *J. Phys. Chem.* **33**, 1027–1079 (1929).
233. M. Hall, Jr., *The Theory of Groups*. Macmillan, New York (1959).
234. G. Pólya, Un problème combinatoire général sur les groupes de permutations et le calcul du nombre des isomères des composés organiques. *C. R. Acad. Sci.* **201**, 1167–1169 (1935).
235. N. L. Biggs, E. K. Lloyd and R. J. Wilson, *Graph Theory*, pp. 1736–1936. Clarendon Press, Oxford (1976).
236. R. A. Davidson, Isomers and isomerizations: elements of Redfield's combinatorial theory. *J. Am. Chem. Soc.* **103**, 312–314 (1981).
237. J. M. McBride, Completion of Koerner's proof that the hydrogens of benzene are homotopic. An application of group theory. *J. Am. Chem. Soc.* **102**, 4134–4137 (1980).
238. K. Balasubramanian, Comments on McBride's completion of Kroner's proof that hydrogens of benzene are homotopic. *Theor. Chim. Acta* **59**, 91–93 (1981).
239. G. Pólya, Algebraische berechnung der anzahl der isomeren einiger organischer verbindungen. *Z. Kristallogr., Kristallgeometrie, Kristallphys., Kristallchem.* **93**, 415–443 (1936).
240. G. Pólya, Kombinatorische anzahlbestimmungen für gruppen, graphen und chemische verbindungen. *Acta Math.* **68**, 145–254 (1937).
241. H. R. Henze and C. M. Blair, The number of structurally isomeric alcohols of the methanol series. *J. Am. Chem. Soc.* **53**, 3042–3046 (1931).
242. H. R. Henze and C. M. Blair, The number of isomeric hydrocarbons of the methane series. *J. Am. Chem. Soc.* **53**, 3077–3085 (1931).
243. R. C. Read, The enumeration of acylic chemical compounds, in *Chemical Applications of Graph Theory* (Edited by A. T. Balaban), pp. 25–61. Academic Press, London (1976).
244. R. W. Robinson, F. Harary and A. T. Balaban, The numbers of chiral and achiral alkanes and monosubstituted alkanes. *Tetrahedron* **32**, 355–361 (1976).
245. K. Balasubramanian, Combinatorial enumeration of chemical isomers. *Indian J. Chem.* **16B**, 1094–1096 (1978).
246. K. Balasubramanian, A generalized wreath product method for the enumeration of stereo and position isomers of polysubstituted organic compounds. *Theor. Chim. Acta* **51**, 37–54 (1979).
247. J. E. Leonard, G. S. Hammond and H. E. Simons, The apparent symmetry of cyclohexane. *J. Am. Chem. Soc.* **97**, 5052–5054 (1975).
248. J. E. Leonard, Isomer number of nonrigid molecules. The cyclohexane case. *J. Phys. Chem.* **81**, 2212–2214 (1977).
249. K. Balasubramanian, Enumeration of internal rotation reactions and their reaction graphs. *Theor. Chim. Acta* **53**, 129–146 (1979).
250. I. Hargittai, Degas' dancers: an illustration for rotational isomers. *J. Chem. Educ.* **60**, 94 (1983).
251. R. L. Flurry, Jr., Isomer counting for fluctional molecules. *J. Chem. Educ.* **61**, 663–665 (1984).
252. K. Balasubramanian, The symmetry of groups of nonrigid molecules as generalized wreath products and their representations. *J. Chem. Phys.* **72**, 665–677 (1980).
253. K. Balasubramanian, Graph theoretical characterization of NMR groups, nonrigid nuclear spin species and the construction of symmetry adapted NMR spin functions. *J. Chem. Phys.* **73**, 3321–3337 (1980).
254. R. L. Flurry, Jr., On the apparent symmetry of cyclohexane. *J. Phys. Chem.* **80**, 777–778 (1976).
255. A. T. Balaban, Enumeration of cyclic graphs, in *Chemical Applications of Graph Theory* (Edited by A. T. Balaban), pp. 63–105. Academic Press, London (1976).
256. B. Alspach and S. Aronoff, Enumeration of structural isomers in alicyclic hydrocarbons and in porphyrins. *Can. J. Chem.* **55**, 2773–2777 (1977).
257. R. B. King and D. H. Rouvray, The enumeration of electron-rich and electron-poor polyhedral compounds. *Theor. Chim. Acta* **48**, 207–214 (1978).

258. A. T. Balaban and V. Baciu, Chemical graphs. XXXV. Application of Pólya's theorem to catamantanes. *Match* **4**, 131–159 (1978).

259. M. Randić, Characterization of atoms, molecules, and classes of molecules based on paths enumerations. *Match* **7**, 5–64 (1979).

260. C.-S. Chung, Computation of the number of isomers of coordination compounds containing different monodentate ligands. *J. Chem. Educ.* **56**, 398 (1979).

261. M. Randić, Graphical enumeration of conformations of chains. *Int. J. Quantum Chem.*, *Quantum Biol. Symp.* **7**, 187–197 (1980).

262. W. G. Klemperer, Enumeration of permutational isomerization reactions. *J. Chem. Phys.* **56**, 5478–5489 (1972).

263. W. G. Klemperer, Enumeration of permutational isomerization reactions. II. *Inorg. Chem.* **11**, 2668–2678 (1972).

264. W. G. Klemperer, Topological representation of permutational isomerization reactions. *J. Am. Chem. Soc.* **94**, 6940–6944 (1972).

265. W. G. Klemperer, Dynamic stereochemistry of polytopal isomerization reactions. *J. Am. Chem. Soc.* **94**, 8360–8371 (1972).

266. W. G. Klemperer, The steric courses of chemical reactions. I. *J. Am. Chem. Soc.* **95**, 380–396 (1973).

267. W. G. Klemperer, Steric courses of chemical reactions. II. *J. Am. Chem. Soc.* **95**, 2105–2120 (1973).

268. M. G. Hutchings, J. B. Johnson, W. G. Klemperer and R. R. Knight, III, The steric courses of chemical reactions. 3. Computer generation of product distributions, steric courses, and permutational isomers. *J. Am. Chem. Soc.* **99**, 7126–7132 (1977).

269. R. K. Lindsay, B. G. Buchanan, E. A. Feigenbaum and J. Lederberg, *Applications of Artificial Intelligence for Organic Chemistry: The DENDRAL Project.* McGraw-Hill, New York (1980).

270. J. Lederberg, G. L. Sutherland, B. G. Buchanan, E. A. Feigenbaum, A. V. Robertson, A. M. Duffield and C. Djerassi, Applications of artificial intelligence for chemical inference. I. The number of possible organic compounds. Acyclic structures containing C, H, O, and N. *J. Am. Chem. Soc.* **91**, 2973–2976 (1969).

271. L. M. Masinter, N. S. Sridharan, J. Lederberg and D. H. Smith, Applications of artificial intelligence for chemical inference. XII. Exhaustive generation of cyclic and acyclic isomers. *J. Am. Chem. Soc.* **96**, 7702–7714 (1974).

272. C. C. Davis, K. Cross and M. Ebel, Computer calculation of alkane isomers. *J. Chem. Educ.* **48**, 675 (1971).

273. W. E. Bennett, Computation of the number of isomers and their structures in coordination compounds. *Inorg. Chem.* **8**, 1325–1328 (1969).

274. I. V. Krivoshey, General properties of multi-dimensional potential surfaces of chemical reactions. 1. Models of complex adiabatic and non-adiabatic reactions. *Zh. Strukt. Khim.* **10**, 316–319 (1969) (in Russian).

275. V. I. Zhuravlev, I. V. Krivoshey and L. A. Sleta, General properties of multi-dimensional potential-energy surfaces of chemical reactions. V. An algorithm for construction of tree of chemical reactions from tabulated values of potential-energy function. *Zh. Struct. Khim.* **16**, 951–955 (1975) (in Russian).

276. I. V. Krivoshey and L. A. Sleta, A topological stochastic approach to the study of multidimensional potential energy surfaces of chemical reactions. *Theor. Chim. Acta* **43**, 165–174 (1976).

277. I. V. Krivoshey, General properties of multi-dimensional potential-energy surfaces of chemical reactions. VI. Character type of activated states in the case of multi-dimensional potential-energy surfaces. *Zh. Strukt. Khim.* **17**, 227–229 (1976) (in Russian).

278. A. S. Kronrod, On two-variables functions. *Usp. Matem. Nauk* **5**, 24–134 (1950) (in Russian).

279. S. Fraga, K. M. S. Saxena and M. Torres, *Biomolecular Information Theory*, pp. 76–78. Elsevier, Amsterdam (1978).

280. H. Eyring, H. Gershinowitz and C. E. Sun, The absolute rate of homogeneous atomic reactions. *J. Chem. Phys.* **3**, 786–796 (1935).

281. P. G. Mezey, Reactive domains of energy hypersurfaces and the stability of minimum energy reaction paths. *Theor. Chim. Acta* **54**, 95–111 (1980).

282. P. G. Mezey, Catchment region partitioning of energy hypersurfaces, I. *Theor. Chim. Acta* **58**, 309–330 (1981).

283. P. G. Mezey, Critical level topology of energy hypersurfaces. *Theor. Chim. Acta* **60**, 97–110 (1981).

284. P. G. Mezey, Manifold theory of multidimensional potential surfaces. *Int. J. Quantum Chem.*, *Quantum Biol. Symp.* **8**, 185–196 (1981).

285. P. G. Mezey, Level set topology of the nuclear charge space and the electronic energy functional. *Int. J. Quantum Chem.* **22**, 101–114 (1982).

286. P. G. Mezey, Quantum chemical raction networks, reaction graphs and the structure of potential energy hypersurfaces. *Theor. Chim. Acta* **60**, 409–428 (1982).

287. P. G. Mezey, The topology of energy hypersurfaces. II. Reaction topology in Euclidean spaces. *Theor. Chim. Acta* **63**, 9–33 (1983).

288. P. G. Mezey, Reaction topology: manifold theory of potential surfaces and quantum chemical synthesis design, in Ref. [135], pp. 75–98.

289. R. F. W. Bader, T. T. Nguyen-Dang and Y. Tal, Quantum topology of molecular charge distributions. II. Molecular structure and its change. *J. Chem. Phys.* **70**, 4316–4329 (1979).

290. Y. Tal, R. F. W. Bader and J. Erkku, Structural homeomorphism between the electronic charge density and the nuclear potential of a molecular system. *Phys. Rev.* **21A**, 1–11 (1980).

291. R. F. W. Bader, Quantum topology of molecular charge distributions. III. The mechanics of an atom in a molecule. *J. Chem. Phys.* **73**, 2871–2833 (1980).

292. Y. Tal, R. F. W. Bader, T. T. Nguyen-Dang, M. Ojha and S. A. Anderson, Quantum topology IV. Relation between the topological and energetic stabilities of molecular structures. *J. Chem. Phys.* **74**, 5162–5167 (1981).

293. M. Morse and S. S. Cairns, *Critical Point Theory in Global Analysis and Differential Topology.* Academic Press, New York (1969).

294. J. Milnor, *Morse Theory.* Princeton University Press, Princeton (1963).

295. P. G. Mezey, Lower and upper bounds for the number of critical points on energy hypersurfaces. *Chem. Phys. Lett.* **82**, 100–104 (1981).

296. P. G. Mezey, Errata, *Chem. Phys. Lett.* **86**, 562 (1982).

297. M. Gordon and T. G. Parker, The graph-like state of matter. I. Statistical effects of correlations due to substitution effects, including steric hindrance, on polymer distributions. *Proc. Roy. Soc. Edinburgh* **69A**, 181–197 (1971).

298. M. Gordon and J. W. Kennedy, The graph-like state of matter. Part 2. LCGI Schemes for the thermodynamics of alkanes and the theory of inductive inference. *J. Chem. Soc., Faraday Trans.* II **69**, 484–504 (1973).

299. M. Gordon and W. B. Temple, The graph-like state of molecules. III. Ring-chain competition kinetics in linear polymerization reactions. *Makromol. Chem.* **160**, 263–276 (1972).

300. K. Kajiwara and M. Gordon, Graphlike state of matter. V. Theory of Rayleigh scattering from randomly cross-linked chains of arbitrary primary distributions. *J. Chem. Phys.* **59**, 3623–3632 (1973).

301. M. Gordon, S. B. Roos-Murphy and H. Suzuki, The graph-like state of matter—VI. Combinatorial approach to the excluded-volume perturbation. *Eur. Polym. J.* **12**, 733–740 (1976).

302. M. Gordon, P. Kapadia and A. Malakis, The graph-like state of matter. VII. The glass transition of polymers and Hamiltonian walks. *J. Phys.* **9A**, 751–769 (1976).

303. M. Gordon and W. B. Temple, The graph-like state of matter and polymer science, in *Chemical Applications of Graph Theory* (Edited by A. T. Balaban), pp. 300–332. Academic Press, London (1976).

304. M. Gordon, Chemical combinatorics. Part II. On the third law of thermodynamics. *J. Chem. Soc.* A, 737–740 (1970).

305. M. Gordon and W. B. Temple, Chemical combinatorics. Part 3. Stereochemical invariance law and the statistical mechanics of flexible molecules. *J. Chem. Soc., Faraday Trans.* II **69**, 282–297 (1973).

306. M. Gordon and W. B. Temple, Chemical combinatorics. Part I. Chemical kinetics, graph theory, and combinatorial entropy. *J. Chem. Soc.* A, 729–737 (1970).

307. M. D. Frank-Kamenetskii, A. V. Lukashin and A. V. Vologodskii, Statistical mechanics and topology of polymer chains. *Nature* **258**, 398–402 (1975).

308. M. D. Frank-Kamenetskii and A. V. Vologodskii, Topological aspects of polymer physics: theory and its bio-physical applications. *Usp. Fiz. Nauk* **134**, 641–673 (1981) (in Russian).

309. W. C. Forsman, Graph theory and the statistics and dynamics of polymer chains. *J. Chem. Phys.* **65**, 4111–4115 (1976).

310. Z. Slanina, *Theoretical Aspects of the Phenomenon of Isomerism in Chemistry*. Mir, Moscow (1984) (in Russian).

311. J. Drowart, R. P. Burns, G. DeMaria and M. G. Inghram, Mass spectrometric study of carbon vapor. *J. Chem. Phys.* **31**, 1131–1132 (1959).

312. Z. Slanina and R. Zahradník, MINDO/2 study of equilibrium carbon vapor. *J. Phys. Chem.* **81**, 2252–2257 (1977).

313. Z. Slanina, J. Schlanger and R. Zahradník, CNDO/2 study of the difluoroamino radical dimerization. *Collect. Czech. Chem. Commun.* **41**, 1864–1874 (1976).

314. Z. Slanina, C_p^o of equilibrium isomeric mixtures: the ethanal–ethenol system. *Thermochim. Acta* **78**, 47–54 (1984).

315. M. S. Gordon and H. Fischer, A molecular orbital study of the isomerization mechanism of diazacumulenes. *J. Am. Chem. Soc.* **90**, 2471–2476 (1968).

316. J. M. Howell and L. J. Kirschenbaum, Substituent effects and the cis–trans isomerization of diazenes. *J. Am. Chem. Soc.* **98**, 877–885 (1976).

317. Z. Slanina, CNDO/2 study of the mechanism and kinetics of the cis–trans isomerization of N_2F_2 (g) using the activated-complex theory. *Chem. Phys. Lett.* **50**, 418–422 (1977).

318. G. Favini and R. Todeschini, The molecular structure of vinyl azide. *J. Mol. Struct.* **50**, 191–193 (1978).

319. K. Jug, Quantum chemical methods and their applications to chemical reactions. *Theor. Chim. Acta* **54**, 263–300 (1980).

320. Z. Slanina, Isomerism of activated complexes and tunnel effect and their illustration with an example of *cis–trans* isomerization kinetics of N_2F_2(g). *Collect. Czech. Chem. Commun.* **43**, 2358–2367 (1978).

321. Z. Slanina, Rate processes with isomerism of the activated complex. *Collect. Czech. Chem. Commun.* **42**, 1914–1921 (1977).

322. Z. Slanina, Isomeric structures, weights and effective characteristics of clusters. *Advan. Mol. Relax. Interact. Process.* **14**, 133–148 (1979).

323. J. C. Owicki, L. L. Shipman and H. A. Scheraga, Structure, energetics, and dynamics of small water clusters. *J. Phys. Chem.* **79**, 1794–1811 (1975).

324. M. R. Hoare, Structure and dynamics of simple microclusters. *Advan. Chem. Phys.* **40**, 49–135 (1979).

325. Z. Slanina, Isomerism of the activated complex in water dimer interconversion. *Advan. Mol. Relax. Interact. Process.* **19**, 117–128 (1981).

326. Z. Slanina, The role of the "less stable minimum-energy structure" in evaluation of the characteristics of the HF–HCl van der Waals system. *Chem. Phys. Lett.* **82**, 33–35 (1981).

327. Z. Slanina, HF–ClF isomerism: the consequences regarding system thermodynamics. *Chem. Phys. Lett.* **83**, 418–422 (1981).

328. N. Brigot, S. Odiot and S. H. Walmsley, *A priori* and empirical calculations of external vibration frequencies of the dimer of carbon dioxide. *Chem. Phys. Lett.* **88**, 543–546 (1982).

329. Z. Slanina, On the temperature dependence of the relative stability of two forms of $(CO_2)_2$. *J. Mol. Struct.* **94**, 401–405 (1983).

330. Z. Slanina, Cluster isomerism—theoretical treatments and thermodynamic consequences: a case study of C_p^o with CO_2 clusters. *Surf. Sci.* **157**, 371–379 (1985).

331. Z. Slanina, The role of adsorption-complex isomerism in confrontation of quantum-chemical and experimental quantities describing an adsorption process. *Zh. Fiz. Khim.* **57**, 1164–1167 (1983) (in Russian).

332. Z. Slanina, Adsorption-complex isomerism and quantum chemical study of gas–solid interactions: a model example. *Theor. Chim. Acta* **60**, 589–598 (1982).

333. Z. Slanina, Adsorption-complex isomerism and quantum-chemical studies in heterogeneous catalysis. *Int. J. Quantum Chem.* **23**, 1563–1570 (1983).

Comp. & Maths. with Appls. Vol. 12B, Nos. 3/4, pp. 617–627, 1986
Printed in Great Britain.

0886–9561/86 $3.00 + .00

MILTON'S MATHEMATICAL SYMBOL OF THEODICY

LEE M. JOHNSON
Department of English, Buchanan Tower #504, University of British Columbia, Vancouver, B. C.,
Canada V6T 1W5

Abstract—Symmetry in literature usually calls attention to thematic relationships, and at times it even has a distinctively mathematical character, as in the works of John Milton. In particular, Milton uses the Fibonacci series to construct golden sections in the epic, *Paradise Lost*, and in the pastoral elegy, *Lycidas*. Golden sections in these works symbolize the conjunction of heavenly and earthly conditions for the purpose of explaining the poet's theodicy or divine imperatives in relation to possibilities of human destiny.

The application of symmetry, proportion, and principles of similitude to descriptions of the natural world establishes one of our longest cultural traditions and is essential to current inquiries into the design of nature.[1] With respect to the products of human nature, especially in relation to architecture, music, and the visual arts, a similarly venerable study of symmetry extends from Pythagoras and Vitruvius to the present.[2] Our concern here will be to consider the role of symmetry in literature and, more particularly, in poetry, which, like verse drama, has long invited attention to the interrelationships among its lines, stanzas, verse paragraphs, and larger architectural elements. A sense of symmetry may attend, for example, the arrangement of parts of speech within a line or phrase, as in the golden line and the tri-colon crescendo.[3] The same sense accompanies the placement of a caesura within a line of verse or the appreciation of delicate internal balances within a highly developed form such as the Petrarchan sonnet. The resulting sense of order could be rather casual, however, should it merely gratify an aesthetic impulse or call attention to interrelationshps among literary motifs and themes. Symmetry in literature does have such functions, of course, but there are also times when it has a strictly mathematical basis which should not be overlooked. For this reason, a discussion of the complex significance of symmetry in verse should probably begin with those works whose aesthetic properties clearly depend upon the fulfilment of a distinctive mathematical pattern.

It so happens that most scholarly discussions of strictly mathematical symmetry in verse are directed to poems written well before the present century. Why this is so is a question worth taking up at the end of the essay. Even in the older poems, the mere presence of interesting numbers may not suit our argument: discussions of symmetry cannot include most forms of number symbolism, for example, because a symbolic number by itself does not establish a symmetrical relationship, which requires the participation of two or more elements. The kind of inquiry that concerns us, then, is represented by several classical scholars, who claim to have discovered the structures of geometrical rings or golden sections in the works of ancient poets.[4] Such scholarship, despite considerable difficulties in method, often meets the issue of mathematical symmetry head-on and strongly suggests that principles of mathematical composition and poetical imagination are not incompatible. Such findings, regardless of shortcomings and controversies that surround them, help us, at the very least, to reconsider our assumptions about an art. In my case, I was encouraged to look anew at the poetry of William Wordsworth, who asserted that the two main sources of his inspiration were geometrical thought and the history of poetry, and whose philosophical verse employs the golden section for his most comprehensive statements.[5] The most important poet to Wordsworth was probably his English predecessor, John Milton, whose works have long been examined for their formal qualities. Although many scholars have commented on various large-scale symmetrical interrelationships among the twelve books of *Paradise Lost*, James Whaler has written on the purely mathematical symmetries of rhythms in Milton's verse paragraphs.[6] Whaler's main argument is that Milton sometimes constructs phrases in accordance with mathematical sequences: for example, a phrase of ten beats grouped as one plus two plus three plus four is an ascending series of symbolic value, and it can be expanded arithmetically (2, 4, 6, 8) or geometrically (2, 4, 8, 16) to account for a large portion of the rhythms in a verse paragraph. Although this is not the place to test the soundness of Whaler's arguments, which I think are often wrong, his attitudes towards Milton's art are based on instincts that are surely right. The same instincts, after all, were partly responsible for bringing the mathematically educated Wordsworth back to Milton's works time and again. Accordingly, I have developed an interest in the geometrical symmetries of Milton's verse and would here like to present one of them—the golden section—and to consider its thematic significance.

In *Paradise Lost*, there are several different kinds of mathematically symmetrical patterns; but the golden section is reserved for the special purpose of proclaiming Milton's theodicy, his justification of divine providence while weighing the existence of evil with the goodness and sovereignty of God. In other words, the golden section is a symbol associated with the most fundamental concept in the epic. As such, it encircles the poem, appearing in the opening invocation and again near the end of Book XII, in which the archangel Michael finally reveals to Adam how the life of Christ promises the restoration of mankind. Now, what is assumed in the preceding remarks is that mathematical symmetry in a literary work should be at one with the work's thematic significance. In literary scholarship, the discovery of a symmetrical pattern is not in itself particularly interesting; for the sorts of patterns one encounters are not mathematically complicated and, without a clear relationship to the verbal content of a work, are either of no value to the reader or give the impression of having been arbitrarily or even irresponsibly imposed.[7] In these matters, then, the realization of poetical purposes in mathematical symmetry is our aim as we turn our attention to the details of Milton's verse.

In the opening lines of *Paradise Lost*, the key repeated terms concern man, God, and creation on literary and cosmic scales:

> Of Man's First Disobedience, and the Fruit
> Of that Forbidden Tree, whose mortal taste
> Brought Death into the World, and all our woe,
> With loss of *Eden*, till one greater Man
> Restore us, and regain the blissful Seat,   5
> Sing Heav'nly Muse, that on the secret top
> Of *Oreb*, or of *Sinai* didst inspire
> That Shepherd, who first taught the chosen Seed,
> In the Beginning how the Heav'ns and Earth
> Rose out of *Chaos*: Or if *Sion* Hill   10
> Delight thee more, and *Siloa's* Brook that flow'd
> Fast by the Oracle of God; I thence
> Invoke thy aid to my advent'rous Song,
> That with no middle flight intends to soar
> Above th' *Aonian* Mount, while it pursues   15
> Things unattempted yet in Prose or Rhyme. $\frac{16}{10}$
> And chiefly Thou O Spirit, that dost prefer
> Before all Temples th' upright heart and pure,
> Instruct me, for Thou know'st; Thou from the first
> Wast present, and with mighty wings outspread   20
> Dove-like satst brooding on the vast Abyss
> And mad'st it pregnant: What in me is dark
> Illumine, what is low raise and support;
> That to the highth of this great Argument
> I may assert Eternal Providence,   25
> And justify the ways of God to men.[8]

The first of the two sentences (1–16) in this verse paragraph presents the boundaries of human history as marked by the first and second Adams: "Man's First Disobedience" and the "greater Man" who eventually presides over the closing of time. Literary creation is also an issue, as Milton recalls how his muse helped Moses teach the genesis of things and will now help him to soar above the inspiration given to classical poets. In other words, the fame and creative actions of the poet are set in a human context of biblical and classical traditions. By contrast, the second sentence (17–26) shifts the point of view to the divine context: God's creative actions in cosmic history are recounted and lead the poet to ask that he be able to link the divine and human orders by explaining "the ways of God to men"—by attempting a theodicy.

The relationship between human and divine history, between this verse paragraph's first sentence of sixteen lines and second sentence of ten lines, is, in mathematical terms, a golden section. Working back from the minor to the major parts of the ratio, we see that $10:16::16:26$ or $5:8::8:13$, which expresses arithmetical approximations to the geometrical concept

$$a:b::b:a + b$$

by means of numbers in the Fibonacci series. Why Milton associates the golden section with the divine order requires some careful speculation. He would not have known of the term golden section, of course, for that is a nineteenth-century appellation. He probably knew the proportion, first, as the division of a form into extreme and mean ratio; for this is how Euclid describes it in II.11 and VI.30. He could also have known of the medieval Fibonacci numbers. What is more pertinent, however, is that the ratio was known in the Renaissance as the *divine proportion*. Kepler so termed it, as did the Italian mathematician Luca Pacioli, who wrote a treatise on the subject and who, as a friend of Leonardo da Vinci and Piero della Francesca, was concerned about its application to the arts.[9]

Given that attributions of value to the golden section in the Renaissance are germane to Milton's use of the ratio, let us briefly consider Pacioli's *Divina Proportione*. In Chapter V, Pacioli discusses the suitability of the title for his treatise. In short, he calls the proportion divine because of its similarities to traditional descriptions of God, as he illustrates by calling attention to five of its properties.

(1) the divine proportion is a unique form of unity, which unity is the supreme epithet of God;

(2) just as one and the same substance is found in the three persons of the Holy Trinity, so one and the same proportion is always appropriately found in the three terms of the ratio;

(3) just as God cannot be properly defined and made comprehensible to us through words, so the proportion cannot be expressed through rational numbers;

(4) just as God is always unchanging in every part, so the proportion is always continuous in all its parts;

(5) just as God's quintessence is conferred on the four elements of earth, water, air, and fire and through these on all other things in nature, so the proportion is inherent in the pentagonal dodecahedron and thereby accounts for the other regular solids and their four elements which cannot be made proportional to one another without the ratio.[10]

Although most of Pacioli's lengthy treatise is a purely mathematical description of the golden section's properties, there is much in the preceding summary which could elicit a smile of incredulity from the modern reader. And yet, Pacioli's third and fifth properties, which concern the ratio's status as a concept and its application to the natural world, represent principles that are central to modern discussions of the golden section. It is true, for example, that the ratio cannot be defined exactly by numbers: its algebraic formulation—take away one from the square root of five and divide the remainder by two—yields an infinitely non-repeating decimal fraction. There is reason to doubt that knowledge of the golden section is derived from empirical experience, but, as our bibliographical footnotes indicate, the proportion is also thought to be one of the mathematical mainsprings in the symmetry and design of nature. The golden section, in other words, is one of the areas of mathematics that continually provokes philosophical speculation on the nature of mathematical reality and the theory of knowledge. Thus, in the twentieth-century's successor to Pacioli's work—namely, *The Divine Proportion: A Study in Mathematical Beauty*—the British physicist and mathematician H. E. Huntley reflects on the properties of the *sectio divina* and its occurrence in nature before aligning his views with those mathematicians and philosophers from Plato onwards who think that mathematical reality is discovered rather than invented:

> These were facts a million years before Fibonacci was born. No mathematician created them. Someone discovered them and expressed them in mathematical symbols, again "thinking God's thoughts after Him."[11]

Enough has probably been said to indicate why, in general, the divine proportion is a suitable mathematical symbol for Milton's attempt to construct a theodicy. Milton lived at a time of cultural history in which the mathematical properties of the ratio were associated with the Deity. More particularly, as the opening invocation in Book I and the further examples in this essay demonstrate, a theodicy establishes a relationship between divine and human viewpoints, just as, in the golden section, there is a relationship between the geometrically exact concept and the necessarily imperfect arithmetic used to approximate it. Yet, the geometrical limit confers on the numbers whatever meaning they have, which metaphorically expresses precisely what a religious poet must do in the explanation of "God's ways to men."

The special status of the divine proportion as a symbol of theodicy has larger implications for the nature of Milton's art and the other kinds of symmetry to be found in it. Milton does not use the ratio as a general principle of composition, and the unusual form of the invocation in Book I becomes evident in comparison with the other invocations in Books III, VII and IX, all of which employ different configurations but are much more characteristic of the overall style of *Paradise Lost*. The poet's celebrated meditation on light and blindness at the beginning of Book III, for example, occupies a verse paragraph of 55 lines. The first 50 lines divide in half after the phrase "dim suffusion veil'd": the opening 25 lines are a direct address to light and concern the presence of light in the extraterrestrial realms, from Heaven to Hell and Chaos; in the next 25 lines, the poet considers his blindness in the terrestrial context, as he recalls how he is separated from the expressions of human nature and the imagery of external nature. Lines 51–55 contain a renewal of the direct address to light and a prayer in which the poet asks for the strength of inward sight. What is remarkable about this closing section is that it is encircled by one of the rare instances of rhyme in the generally unrhymed verse of the epic: the rhyme of "Celestial Light" (51) with "mortal sight" (55). The rhyme, moreover, circles back to the beginning of the verse paragraph—"Hail holy Light" (1) and "since God is Light" (3), thereby enclosing symbolically, in the unending line of a circle, the entire scale of creation which has supplied the paragraph with its imagery. Were we to remove the circle of light from the basically two-part structure of this paragraph, we would be close to the form of the invocation in Book VII. Here, addressing his muse as Urania, Milton writes a verse paragraph of 39 lines which divide into a simple binary symmetry at the phrase "Half yet remains unsung" (21), which not only calls attention to the half-way point of the paragraph but of the entire epic as well. Similarly, the invocation in Book IX is in two balanced parts, its 47 lines dividing in half at the end of line 24. Now, however, the invocation must introduce the sad events of the Fall; and the formal qualities of the paragraph include a binary division indicated by a colon instead of the stronger full stop, an unusually large number of enjambed lines, and a lowering of tone, as Milton augments the growing distance between God and mankind by referring to his "Celestial Patroness" not in a direct address but in the third person. A look back at all four invocations reveals, then, that on the level of symmetry there is a descending sequence in accordance with the divine proportion, a circle, a strong instance of simple binary division, and finally a weaker one. On the thematic level, direct addresses to the spirit of God, to its metaphorical expression as light, and to the muse under the traditional name of Urania give way to the third-person mention of a "Celestial Patroness." The analysis of symmetry in the four invocations corroborates Michael Fixler's argument that Plato's "four furors" or levels of inspiration are invoked in descending order in Milton's epic.[12]

The range of symmetry exemplified in the invocations anticipates the large-scale shape of the epic, and the arrangement of the books has received a good deal of attention. In fact, that arrangement is a vexed question because *Paradise Lost*, as first published in 1667, had ten books and did not show its present form until two of the earlier books were split in 1674. The ambiguity in the two editions is, in my opinion, related to the issue of whether the Fall is "fortunate" or "unfortunate" and so will be discussed later. For the present, let us briefly touch on the overall symmetry in the twelve-book form of the poem. First, it is clear that the phrase "Half yet remains unsung" at the beginning of Book VII calls attention to the epic's division into halves. Arranging Books VII to XII below Books I to VI, we see that the four invocations are paired: I with VII and III with IX, and further contrasts emerge from each pair of books.[13] *Paradise Lost* also supports the arrangement of its books as a ring-symmetry: I/XII, II/XI, III/X, IV/IX, V/VIII, and VI/VII.[14] Satan's acceptance of Hell in Book I, for example, is thus related to Adam's and Eve's acceptance of going down into the fallen world below Eden in Book XII; or, to cite another instance, the depiction of Eden in its innocence (IV) is paired with Eden during the Fall (IX). In these and other schemes, the arrangements of the books have a thematic rather than a mathematical basis: symmetrical relationships depend on parallel or contrasting episodes in the narrative but not on having, say, the same number of lines or fixed ratios between books.

What large-scale thematic symmetry can do, however, is to suggest those places in which mathematical symmetry is likely to occur. On the basis of the divine proportion in the invocation of Book I, then, we might expect another golden section to appear in Book VII, by means of the poem's arrangement in halves, or in Book XII, which is the counterpart in the poem's ring-symmetry. Book VII, however, details the creation of the universe and does not directly deal with the theme of theodicy; nor do its verse paragraphs, despite their various kinds of symmetry, display the golden section. Near the end of Book XII, on the other hand, theodicy becomes the central subject, as the archangel Michael tells Adam how God will provide the possibility of redemption from the effects of original sin in human history. The climax of Michael's argument is the demonstration of God's ways to mankind through the life of Christ, and the archangel's account of that life occupies a verse paragraph of eighty lines (XII, 386–465) designed as a divine proportion. Adam's immediate response to Michael's good news is also cast as a divine proportion, though on the smaller scale of a verse paragraph with sixteen lines (XII, 469–84). These two passages, which we should now examine in some detail, therefore bring to a close the promised theme and symbolic form announced at the beginning of the poem.

The divine proportion at the outset of Book I distinguishes between creative actions in the divine and human spheres; in his speech from Book XII, Michael concentrates on the moral dimensions of actions as contrasts between death and life. His verse paragraph has, as was mentioned, eighty lines: the first fifty concern the life of Christ from His ministry to His death and resurrection (386–435); the remaining thirty lines describe Christ's appearances to the disciples, ascension, and eventual role at the Last Judgment (436–65). The divine proportion thus divides the life of Christ into earthly and heavenly careers, occurring precisely at the description of the Resurrection. Again, as in the invocation from Book I, Fibonacci numbers approximate the ratio $3(30):5(50)::5(50):8(80)$.

The first 50 lines, which deal with Christ's earthly ministry, are further divided in half and contain an interesting verbal battle between the words "death" and "life." The opening 25 lines (386–410) clearly tip the balance on the side of "death":

> To whom thus *Michael*. Dream not of thir fight,
> As of a Duel, or the local wounds
> Of head or heel: not therefore joins the Son
> Manhood to Godhead, with more strength to foil
> Thy enemy; nor so is overcome   390
> *Satan*, whose fall from Heav'n, a deadlier bruise,
> Disabl'd not to give thee thy death's wound:
> Which hee, who comes thy Saviour, shall recure,
> Not by destroying *Satan*, but his works
> In thee and in thy Seed: nor can this be,   395
> But by fulfilling that which thou didst want,
> Obedience to the Law of God, impos'd
> On penalty of death, and suffering death,
> The penalty to thy transgression due,
> And due to theirs which out of thine will grow: 400

So only can high Justice rest appaid.
The Law of God exact he shall fulfil
Both by obedience and by love, though love
Alone fulfil the Law; thy punishment
He shall endure by coming in the Flesh   405
To a reproachful life and cursed death,
Proclaiming Life to all who shall believe
In his redemption, and that his obedience
Imputed becomes theirs by Faith, his merits
To save them, not thir own, though legal works.   410

Christ's reward for exemplifying "the Law of God" (397, 402) is death (392, 398, 406). Milton makes the word "death" more prominent by placing it at the ends of lines and phrases. By contrast, in the next 25 lines (411–35), "life" gains the upper hand:

For this he shall live hated, be blasphem'd,
Seiz'd on by force, judg'd, and to death condemn'd
A shameful and accurst, nail'd to the Cross
By his own Nation, slain for bringing Life;
But to the Cross he nails thy Enemies,   415
The Law that is against thee, and the sins
Of all mankind, with him there crucifi'd,
Never to hurt them more who rightly trust
In this his satisfaction; so he dies,
But soon revives, Death over him no power   420
Shall long usurp; ere the third dawning light
Return, the Stars of Morn shall see him rise
Out of his grave, fresh as the dawning light,
Thy ransom paid, which Man from death redeems,
His death for Man, as many as offer'd Life   425
Neglect not, and the benefit embrace
By Faith not void of works; this God-like act
Annuls thy doom, the death thou shouldst have di'd,
In sin for ever lost from life; this act
Shall bruise the head of *Satan*, crush his strength   430
Defeating Sin and Death, his two main arms,
And fix far deeper in his head thir stings
Than temporal death shall bruise the Victor's heel,
Or theirs whom he redeems, a death like sleep,
A gentle wafting to immortal Life.   435

Now the word "death" is buried in the middle of lines and phrases whereas "life" has the final word at their endings (414, 425, 429, 435). The paragraph appropriately resolves its verbal contest as it reaches its divine proportion on the phrase "immortal Life."

The final 30 lines are directed to Christ's spiritual career and, like their counterpart in the ratio's major section, divide into two equal parts. The first 15 lines (436–50) restate the triumph of "Life" (438, 443):

Nor after resurrection shall he stay
Longer on Earth than certain times to appear
To his Disciples, Men who in his Life
Still follow'd him; to them shall leave in charge
To teach all nations what of him they learn'd   440

And his Salvation, them who shall believe
Baptizing in the profluent stream, the sign
Of washing them from guilt of sin to Life
Pure, and in mind prepar'd, if so befall,
For death, like that which the redeemer di'd.   445
All Nations they shall teach; for from that day
Not only to the Sons of *Abraham's* Loins
Salvation shall be Preacht, but to the Sons
Of *Abraham's* Faith wherever through the world;
So in his seed all Nations shall be blest.   450

From the preceding statement on the universality of Christ's teachings, the final fifteen lines (451–65) depict the Son of God in spiritual glory:

Then to the Heav'n of Heav'ns he shall ascend
With victory, triumphing through the air
Over his foes and thine; there shall surprise
The Serpent, Prince of air, and drag in Chains
Through all his Realm, and there confounded leave;   455
Then enter into glory, and resume
His Seat at God's right hand, exalted high
Above all names in Heav'n; and thence shall come,
When this world's dissolution shall be ripe,
With glory and power to judge both quick and dead,   460
To judge th' unfaithful dead, but to reward
His faithful, and receive them into bliss,
Whether in Heav'n or Earth, for then the Earth
Shall all be Paradise, far happier place
Than this of *Eden*, and far happier days.   465

The words "Heav'n" and "Earth" are repeated throughout this final passage (451, 458, 463), for the defeat of evil and death leaves the earth in a condition which is superior to the original paradise in Eden. In the words of Milton's *Christian Doctrine*: "*The restoration of Man* is the act whereby man, being delivered from sin and death by God the Father through Jesus Christ, is raised to a far more excellent state of grace and glory than that from which he had fallen."[15] With the conclusion of Michael's speech, Milton's theodicy is, in principle, complete.

Proclaimed by the poet at the beginning of the epic, the "greater Man" who is to "regain the blissful Seat" of Eden has now been identified and described at length by a representative of Heaven; all that remains is for Adam, on behalf of earth and the human race, to express his understanding of God's ways:

O goodness infinite, goodness immense!
That all this good of evil shall produce,   470
And evil turn to good; more wonderful
Than that which by creation first brought forth
Light out of darkness! full of doubt I stand,
Whether I should repent me now of sin
By mee done and occasion'd, or rejoice   475
Much more, that much more good thereof shall spring,
To God more glory, more good will to Men
From God, and over wrath grace shall abound. $\dfrac{10}{6}$
But say, if our deliverer up to Heav'n
Must reascend, what will betide the few   480

> His faithful, left among th' unfaithful herd,
> The enemies of truth; who then shall guide
> His people, who defend? will they not deal
> Worse with his followers than with him they dealt?

The association of theodicy with the divine proportion, which distinguishes the verse paragraphs of the poet and the archangel, also extends to Adam's—6:10::10:16 or, in terms of the Fibonacci series, 3:5::5:8. The fact that Adam's speech echoes Michael's, not only in theme but in geometric design, is important to the dramatic relationship between the two characters and to the difficult question of whether or not the Fall is fortunate. With respect to the dramatic relationship, Michael has instructed Adam throughout Books XI and XII on the nature of human history; typically, the archangel presents a scene or situation which Adam then misinterprets and which thus requires a correction and a further example. What is evident in their exchange on the life of Christ, however, is their agreement: no correction or reproof is needed, for Adam's education has come to a successful close.

What Adam has learned, though, creates problems for many a Miltonist. In expressing his understanding of God's ways, Adam paraphrases the medieval hymn *O Felix Culpa* and suggests that the Fall is fortunate: sin occasions grace, and mankind appears to attain a happier state than in the original Eden. The implication that God has therefore been working for the Fall all along leads some scholars to question God's goodness, and there are those who, in defense of God, insist that the Fall must be seen as utterly unfortunate—that mankind would have been happier in a state of obedient innocence.[16] Conversely, there are others who think that a legalistic God, justly insisting on a seemingly unending punishment of a sinful and unredeemable humanity, would be deficient in love and compassion and should therefore also have His goodness questioned. Either way, from the human point of view, God's goodness seems to be imperiled. It is important to note Milton's sensitivity to the problem. In his ensuing speech (485–551), Michael emphasizes that even religion itself will not escape the unfortunate consequences of the Fall, which becomes fortunate only at the end of time. Although this verse paragraph and its eighty-line predecessor on the life of Christ end in the same way by announcing the final bliss to be enjoyed by the faithful, the two speeches could not differ more in their composition: the divine proportion, lucid subdivisions, and stylistic grandeur of the first utterance give way to a paragraph of no clearly discernible internal form, heavy enjambement, and a prosaic diction. The two sides of the Fall are thus poignantly captured by Milton's two different approaches to poetic architecture: perfect symmetry in contrast to apparent formlessness. How may these differences be resolved? Freely to obey God's will, which is to choose the transcendental limit symbolized by the divine proportion, rather than to follow mankind's wilfulness, which eschews limits and produces disorder, presupposes respect for both implications of the Fall and resolves into the spiritual condition of the "paradise within," the theme that closes the unrhymed love-sonnet in Michael's final speech (574–87) and brings the potential of the fortunate Fall to life before its actual occurrence at the Last Judgment.[17] What has not changed is that, innocent or fallen, mankind is still presented with a choice and an exercise of free will. In the fallen state, individuals still must choose whether or not the Fall is to be fortunate. In the light of that choice, the notion that the Fall can be defined without reference to free will and can be merely classified as either fortunate or unfortunate becomes simplistic, if not meaningless. The difference between the choices given to innocent and fallen conditions is even implicit in the divine proportion itself, which, by juxtaposing perfect geometrical symmetry with imperfect arithmetical approximations, coordinates the differences between the divine and human wills. As a symbol of Milton's theodicy, then, the divine proportion depends on contrasts, applies providential actions to a fallen world, and would not be found in a state of innocence.

It is also possible that Milton's own choices of attitudes towards the Fall are reflected in the contrasting numbers of books in the 1667 and 1674 editions of *Paradise Lost*. Why he changed the poem from ten to twelve books has received much speculation, which need not concern us here, but we should consider the relative prominence of his theodicy and its divine proportion in the two editions. Books VII and VIII, which present Raphael's account of the creation and Adam's account of his own, were combined in 1667, as were Books XI and XII on the history of the human race. In 1667, Book VII (now VII and VIII) and X (now XI and

XII) were the longest in the poem. Book X, in particular, was the most massive, standing at nearly 1,550 lines. The effect of Milton's theodicy and golden sections in the speeches of Michael and Adam near the close of the original edition was proportionately small and buried in context, perhaps as a way of underpinning the unfortunate dimensions of the Fall. In 1674, however, Book XII, at 649 lines, becomes one of the shortest in the epic, and the speeches of Michael and Adam now dominate the tone of the book's latter half. The new book gives ample space to God's love, which before trailed in the wake of a lengthy history of human sinfulness. In particular, the life of Christ becomes the central interest in the new book and is thereby more likely to be associated with the "greater Man" at the poem's outset. Not only does the twelve-book form of *Paradise Lost* more clearly align itself with the classical epic tradition, but the sharper focus on Christ as the true type of epic heroism clarifies the choices the reader can make about a spiritual destiny.

Milton's preferences in mathematical symmetry and balance reflect his thematic preoccupations, and his use of the golden section as a symbol for the divine rescue of an afflicted humanity appears to have been deeply considered throughout his poetical career. Besides *Paradise Lost*, there is one other work in which the relationship of death of life is resolved with the help of the divine proportion: the pastoral elegy *Lycidas*, written three decades earlier. Throughout its first nine stanzas, the power of death over Lycidas and life in general renders useless the attempt to live in accordance with any lasting values. It is the task of stanza ten to offer a contrast to the disproportionately long lament and to rescue Lycidas, the poet, the poem, and the world from adversity:

> Weep no more, woeful Shepherds weep no more,   165
> For *Lycidas* your sorrow is not dead,
> Sunk though he be beneath the wat'ry floor,
> So sinks the day-star in the Ocean bed,
> And yet anon repairs his drooping head,
> And tricks his beams, and with new-spangled Ore,   170
> Flames in the forehead of the morning sky:
> So *Lycidas*, sunk low, but mounted high,
> Through the dear might of him that walk'd the waves,
> Where other groves, and other streams along,
> With *Nectar* pure his oozy Locks he laves,   175
> And hears the unexpressive nuptial Song,
> In the blest Kingdoms meek of joy and love. $\dfrac{13}{8}$
> There entertain him all the Saints above,
> In solemn troops, and sweet Societies
> That sing, and singing in their glory move,   180
> And wipe the tears for ever from his eyes.
> Now *Lycidas*, the Shepherds weep no more;
> Henceforth thou art the Genius of the shore,
> In thy large recompense, and shalt be good
> To all that wander in that perilous flood.   185

Conventional stylistic analysis of *Lycidas* turns on two main points: its form as a canzone or madrigal and the nature of its rhyming pattern.[18] With respect to the form of the stanza as a canzone, the clear subdivision of the fronte at the end of line 171 and the equally clear division of the fronte from the sirima between lines 177–78 are notable; for the first nine stanzas are, in comparison, irregularly designed. The effect of a normative textbook example of the stanzaic form is thereby reinforced in the same manner as a theme in music would be were it heard at the end of the variations based on it. Moreover, the rhyme on "more"(165) and "floor"(167), reappearing in "more"(182) and "shore"(183), virtually encloses the stanza in a circle of rhyme, the symbolism of which establishes a context for other instances of this key rhyme in the poem, particularly for the opening line—"Yet once more, O ye Laurels, and once more"— which is unrhymed in its own stanza and therefore strikes one as an instance of irregularity. To the preceding observations on the crucial importance of stanza ten, one may add that the principal division of its content between the fronte and the sirima is a divine proportion with Fibonacci numbers in the sequence of $8:13::13:21$. Attention to the divine proportion is useful, for it explains, as the canzone and the rhymes do not, what kinds of content and theme will appear and how they will be treated. The divine proportion, as Milton's employment of it in *Paradise Lost* suggests, requires the conjunction of earth and heaven on the question of life's purpose and destiny. Here, Lycidas is transported from death on earth to lasting life in a pastoral heaven governed by Christ as the good shepherd. At that point, the stanza's divine proportion occurs and symbolizes the role of the heavenly realm as a pattern by means of which the vicissitudes of earthly life may be measured and imbued with value. Thereafter, the apotheosis of Lycidas as ''the Genius of the shore'' becomes the literary equivalent of the mathematical symbolism: that a presiding transcendental order watches over a world of affliction and mutability.

Throughout this essay, I have assumed that Milton's use of mathematical symmetry is deliberate and therefore justifies a close examination of its artistic and thematic significance. In the case of the divine proportion, we have been concerned with only four passages—three from *Paradise Lost* and one from *Lycidas*—which is a bit like finding a handful of needles in the haystack of the Miltonic canon. Despite the importance of the passages in which the divine proportion occurs, its rarity is undoubtedly the main reason why, even in a climate of literary study that shows an interest in symmetry, it has been overlooked. Were Milton's use of the golden section an unconscious activity, the fortuitous occurrence of the ratio would be widespread and would not be exclusive to passages which share the same kind of literary purpose and theme. Of course, the sheer complexity and range of Milton's verse and patterns of symmetry would discourage almost any attempt to identify specific patterns. Nearly all other verse reflects much more limited artistic resources, is repeatedly expressed in a relatively small range of forms, and is therefore much more likely to have all its salient qualities and characteristics appreciated in a shorter interval of time. Such, at least, is the supposition one might entertain in relation to the discovery of something new in Milton's verse, which still has its secrets despite being read by so many for so long.

The appearance of accidental symmetry in verse is nevertheless an interesting question, although it would seem to have little application to an older poet as self-conscious as Milton. Many poets from the past century or so, on the other hand, have been suspicious of mathematical and rational thought, as if it were inimical to the creative imagination. Why this is so reflects many factors, one of the most important being a profound change in how people feel about the universe. In Milton's time, as was true back to antiquity, the cosmos was usually thought to be full of beautiful proportions and harmonies which a poet quite naturally expected to be incarnated in the poetical universe of words. Recent attitudes towards the universe, by contrast, often assume or express a sense of fear and alienation in the face of indifferent, hostile, or disordered natural forces. It would be important, in my view, to find out whether modern poets who have abandoned metrical verse and profess a dislike for mathematical order nevertheless unconsciously betray a sense of mathematical symmetry in their works. In particular, arguments about the golden section as a natural expression of human order could be tested, no doubt with the facilitating help of computers. Whatever the results, one must not forget that there are still modern poets who know that mathematical and poetical symmetry are mutually supportive activities of the same human brain, which, however divided into different hemispheres or assigned to different intellectual cultures, is responsible for ensuring that the future of poetry will be commensurate with its past.

ENDNOTES

1. On symmetry in nature, György Doczi's *The Power of Limits: Proportional Harmonies in Nature, Art, and Architecture* (Boulder & London: Shambala, 1981) is notable among modern books on the subject for its vast historical range of examples and useful bibliography. See also D. W. Thompson's *On Growth and Form*, 2nd edn. (Cambridge: Cambridge University Press, 1959).
2. In addition to Doczi, see Paul H. Scholfield, *The Theory of Proportion in Architecture* (Cambridge: Cambridge University Press, 1958); Rudolph Wittkower, ''The Changing Concept of Proportion,'' *Daedalus*, 89 (Winter, 1960), 199–215; Jay Hambidge, *The Elements of Dynamic Symmetry* (New Haven: Yale University Press, 1926); Ernö Lendvai, *Bela Bartók: An Analysis of His Music*, translated by T. Ungar (London: Kahn & Averill, 1971);

and J. A. Rothwell's dissertation, "The Phi Factor: Mathematical Proportions in Musical Forms" (The University of Missouri, 1977).

3. The golden line shows bilateral symmetry in its disposition of parts of speech: for example, from Gray's *Elegy*, "The plowman homeward plods his weary way" (3) has a sequence of noun-modifier-verb-modifier-noun. Other linquistic elements in a line may also have bilateral symmetry: in Wordsworth's "St. Paul's," the line "Deep, hollow, unobstructed, vacant, smooth" (17) reveals a syllabic sequence of one, two, four, two, and one—which gives a precise mathematical definition to the pattern. The tri-colon crescendo is an arrangement of three terms or clauses: from Milton's *Paradise Lost*, the line "My word, my wisdom, and effectual might" (III, 170) places three nouns in a semantic sequence which moves from a creative principle ("my word") to its knowledge ("my wisdom") and its power ("effectual might"). Amplifying the verbal sequence is an incremental mathematical pattern: one syllable in "word," two syllables in "wisdom," and three syllables in "effectual" (treated as three syllables for the sake of the meter)—which is a simple mathematical way of ensuring that the rhythms of key words will reinforce the verbal sense of a crescendo.

4. Strictly mathematical symmetry in ancient poetry is examined by Guy Le Grelle, S. J., "Le premier livre des Gèorgiques, poéme pythagoricien," *Les Etudes Classiques*, 17 (1949), 139–235; George E. Duckworth, *Structural Patterns and Proportions in Virgil's Aeneid: A Study in Mathematical Composition* (Ann Arbor: University of Michigan Press, 1962); Edwin L. Brown, *Numeri Vergiliani: Studies in Eclogues and Georgics (Collection Latomus)*, 63 (1963); and Otto Skutsch, "Symmetry and Sense in the *Eclogues*," *Harvard Studies in Classical Philology*, 73 (1969), 156 ff. Thematic and structural symmetries of a more general character are discussed by Cedric H. Whitman in his chapter "The Geometric Structure of the Iliad" from *Homer and the Heroic Tradition* (Cambridge, Mass.: Harvard University Press, 1958); see also Dieter Lohmann, *Die Komposition der Reden in der Ilias* (Berlin: de Gruyter, 1970); Brooks Otis, *Virgil: A Study in Civilized Poetry* (Oxford: Oxford University Press, 1963); and Gilbert Highet, *The Speeches in Vergil's Aeneid* (Princeton: Princeton University Press, 1972).

5. See my *Wordsworth's Metaphysical Verse: Geometry, Nature, and Form* (Toronto: University of Toronto Press, 1982).

6. James Whaler, *Counterpoint and Symbol: An Inquiry into the Rhythms of Milton's Epic Style. Anglistica*, 6 (Copenhagen: Rosenkilde & Bagger, 1956).

7. As described by Duckworth, for example, the golden sections which organize the entirety of the *Aeneid* have nothing to do with the verbal content of Virgil's epic but are purely descriptive of a method of composition. One is left with the question of what significance, if any, the geometrical patterns have for the literary character of the poem. Whereas Duckworth modestly confines his study to matters of compositional technique and does not venture into questions of literary interpretation, John T. Shawcross attempts both, but with unhappy results, in his recent book *With Mortal Voice: The Creation of Paradise Lost* (Lexington: The University Press of Kentucky, 1982). Shawcross, taking the whole of the ten-book 1667 edition of Milton's poem as his basis, imposes a golden section on line 477 in Book VIII but does not do so for the other point of golden section in Book V (p. 62). In the longer 1674 twelve-book form of the epic, however, the two points of golden section would be different; but Shawcross passes silently over this, presumably because the selected lines at those points are not of any particular interest. Does this mean that Milton used one golden section in 1667 and later threw out geometrical form in 1674? Of course not, for Shawcross's attempt to impose a geometrical pattern does not have any reason for being sought in the first place. Shawcross then compounds his difficulties by pointing out that Book VIII is 0.616 of Books VIII plus III and that Book VII is 0.613 of Books I plus II (pp. 62, 63). But what about proportions between other combinations of books which do not approximate the golden ratio? Nothing of literary value emerges to distinguish certain combinations of books from others, and we have not even considered that such combinations change utterly between the ten-book and twelve-book editions of the poem. Shawcross's results are thus arbitrary and unconvincing shots in the dark which, after his further confusion of symmetry with his primary interest in number symbolism, produce the sort of work that leads to an understandable dismissal of the mathematical analysis of literature.

8. John Milton, *Paradise Lost*, I, 1–26, in *John Milton: Complete Poems and Major Prose*, edited by Merritt Y. Hughes (New York: The Odyssey Press, 1957), pp. 211–12. All subsequent citations of Milton's poetry are from this edition.

9. For Kepler, see H. E. Huntley, *The Divine Proportion: A Study in Mathematical Beauty* (Dover Publications, 1970), p. 25; Luca Pacioli, *Divina Proportione*, edited and translated into German by Constantin Winterberg from the Venetian edition of 1509 printed by Antonio Capella (Vienna: Carl Graeser, 1889); reprinted in *Quellenschriften für Kunstgeschichte und Kunsttechnik*, N. F. II (Hildesheim & New York: Georg Olms, 1974).

10. Pacioli, pp. 43–44. I have paraphrased Chapter V of Pacioli's treatise at some length because, not existing in an English translation, its qualities of thought and argument would be otherwise difficult to appreciate.

11. Huntley, p. 154; see also p. 38 for G. H. Hardy's thoughts on mathematical reality in *A Mathematician's Apology*.

12. Michael Fixler, "Plato's Four Furors and the Real Structure of *Paradise Lost*," *Publications of the Modern Language Association*, 92 (1977), 952–62.

13. Douglas A. Northrop, "The Double Structure of *Paradise Lost*," *Milton Studies*, 12 (1978), 75–90.

14. J. R. Watson, "Divine Providence and the Structure of *Paradise Lost*," *Essays in Criticism*, 14 (1964), 148–55.

15. John Milton, *The Christian Doctrine*, Book I, Chapter XIV, in *The Student's Milton*, edited by Frank Allen Patterson (New York: F. S. Crofts & Co., 1930), p. 1005.

16. The best analysis of Milton's views on theodicy, free will, and predestination is by Dennis Danielson, *Milton's Good God: A Study in Literary Theodicy* (Cambridge: Cambridge University Press, 1982). In the final chapter, "*Paradise Lost* and the Unfortunate Fall" (pp. 202–27), however, Danielson states the case against the *felix culpa* by forgetting that, to Milton, love and free will take precedence over all other theological concepts, including different definitions of the Fall.

17. See my "Milton's Blank Verse Sonnets," *Milton Studies*, 5 (1973), 129–53.

18. F. T. Prince, *The Italian Element in Milton's Verse* (Oxford: Oxford University Press, 1954); Clay Hunt, *Lycidas and the Italian Critics* (New Haven: Yale University Press, 1979); J. A. Wittreich, Jr., "Milton's 'Destin'd Urn': The Art of *Lycidas*," *Publications of the Modern Language Association*, 84 (1969), 60–70.

Comp. & Maths. with Appls. Vol. 12B, Nos. 3/4, pp. 629–639, 1986
Printed in Great Britain.

0886-9561/86 $3.00 + .00

SYMMETRY IN COURT AND COUNTRY DANCE

Arthur L. Loeb

Department of Visual and Environmental Studies, Carpenter Center for the Visual Arts,
Harvard University, Cambridge, MA 02138, U.S.A.

Abstract—Court and Country (Contra-) Dances constitute highly formalized and organized patterns in time and space. Although the floor patterns of these dances at any given time can be analyzed in terms of reflection and rotation symmetries, their transformations as the dances proceed are not so easily classified, as the dancers must disturb mirror symmetry in order not to pass through each other. Certain rhythms are analyzed, and a reason for medieval preference for triple time rather than quadruple time is advanced. The distinction between configurations of even and of odd numbers of dancers or couples is demonstrated, as is the symbolic value of the symmetry of some renaissance court dances. The structural differences between traditional English and Scottish Country Dances are presented in the light of historical relations between the two countries.

INTRODUCTION

Motion is one of the most basic characteristics of life. We admire the graceful motion of a gazelle, a galloping horse, even of a snake. The harmonious interaction of all parts of the body is necessary to provide effective locomotion. This harmonious interaction in turn leads to what we call rhythm: a periodic repetition of the same contractions and extensions of muscles and limbs. If it is true that play and games are the way in which we learn and hone our skills needed for survival, then the Dance certainly provides us with basic training in the harmonious use of our body. It is therefore not surprising that Dance is common to all civilizations, and that it changes style as civilizations change and succeed each other.

Most work on symmetry is done on static patterns; in view of the importance of dance as an orderly and periodically repeating phenomenon it would seem to be in order for this volume to include a representation of dance as a symmetrical pattern in both space and time. There is, to begin with, the phenomenon of rhythm; at the next level there is the repetition of the thematic pattern of the music, which conforms with the structure of the dance. Then there is the configuration of the dancers on the floor, which is not just a static two-dimensional pattern, but also needs a dimension to indicate the direction in which each dancer is moving.

It is impossible to do justice to these fundamental parameters in a single article; this author does not feel qualified to cover every one of them adequately in depth anyway. Yet, just as the study of magnetic phenomena opened the field of geometric symmetry again by the introduction of Color Symmetry, so the consideration of structures periodic in both space and time may lead to new aspects in symmetry theory. For this reason we present here some of the principal aspects of symmetry in Court and Country Dance.

RHYTHM

For simplicity's sake let us consider a very basic musical pattern, although by no means a simple one, namely a drumbeat. To understand rhythm we must assume that the beat will be sounded at least three different levels of intensity, which we shall call piano, mezzo and forte. Let us begin by hearing a sequence of drumbeats sounded at a constant level, say forte, at constant time intervals. These drumbeats provide the *pulse* and the *tempo* of the musical pattern established by the drum. Short time intervals produce a fast tempo, longer ones a slower tempo.

Now let the drummer play a softer beat, say at mezzo level, between each pair of forte beats. Although the time-interval between successive beats is now shortened, we would not say that the music has speeded up: the tempo is still determined by the time lapse between the forte beats, which still provide the pulse of the music. The mezzo-level beats could be spaced exactly half-way between the forte beats, or off-center:

$$f.m.f.m.f.m.f.m.f.m.f \ldots$$

or

$$f.mf.mf.mf.mf.mf.mf.m \ldots$$

or

$$f..mf..mf..mf..mf..mf \ldots$$

or

$$fm..fm..fm..fm..fm..f \ldots$$

Here the dots indicate absence of a beat; a *piano* beat could equally well have been placed here. The important principle is that we have here a periodically repeating pattern of sounds emitted at three different intensities, f, m, and p or .. Many other sequences could be devised; the sequence of beats characteristic of a sound pattern is called its *rhythm*.

Each of the four examples given has translational symmetry. The repeat unit, or unit cell, is called a *measure*. Since the seventeenth century it has become customary to place a vertical bar in front of the leading forte beat of each measure. Modern editions of earlier music have frequently forced the notation into this bar-line straightjacket, a procedure which obscures the rhythmical flexibility and complexity of this earlier music. The bar-line notation is a historical symptom of the fact that music and dance had become rhythmically more regular and less complex, as examplified by the driving regular pulse of the Italian baroque concerti of Vivaldi, Bach and contemporaries.

Of the four examples of rhythm, only the first has mirror symmetry. It is represented by the *March*, a walking or professional dance, in which one leg, either left or right, always falls on a strong beat. Not all Marches, however, have so symmetrical a rhythmical pattern. Some Marches follow the second rhythm example, usually by means of a hesitation on the forte beat, the other leg moving through rapidly on the weaker beat.

The sixteenth-century *Pavane* is an example of a march-like processional dance, usually danced by all at the beginning of a formal ball. The *Polonaise* is thought to be a variant having the rhythm of the second example; it is reputed to have been first danced at the inauguration of Henri de Valois, later to be Henri III of France, as King of Poland in 1573.

We have noted that the forte beats ordinarily mark the beginning of a measure. The measure is subdivided into as many subdivisions as are necessary to mark the occurrence of weaker beats. In the first and the last of the four rhythm examples each measure is subdivided into four units: the music is said to be in quadruple time. Similarly, the second and third example is said to be in triple time. It is remarkable that, whereas we tend to think of quadruple time as being the norm, being capable of having mirror symmetry, it was triple time, being intrinsically less symmetrical, which was the *tempus perfectum* of the Middle Ages. Whereas the quadruple-time Pavane was the most common processional dance in the Renaissance, the common Medieval processional dance was the *Basse Dance* in triple time.

The predilection for triple time in the Middle Ages may be understood in the light of the flexibility and flowing nature of the music, dance and costume this period, in contrast with the rather rigid forms which characterize the Renaissance. Two measures in triple time may be combined, the second receiving a subsidiary first beat, so that the following sextuple pattern in three dynamic levels results:

$$fppmppfppmppfppmpp \ldots$$

Without a change in measure length, this pattern may be transformed into

$$fpmpmpfpmpmpfpmpmp \ldots$$

In modern notation the former example corresponds to 6/8 time, the latter to 3/4 time; in either case the measure is now subdivided into six units, but modern notation tends to use the so-called quarter note as the norm wherever such is possible, as in the second example. These two patterns were often used in combination; the substitution of one in a structure constituted mainly of the other is called a *Hemiola*. An example of a Hemiola would be the following pattern:

$$fppmppfppmppfpmpmpfpmpmpfppmppfppmpp \ldots$$

Although the Hemiola was characteristic of the Middle Ages, it has not disappeared from music and dance. A dance mentioned by Shakespeare is the *Sinkapace,* a name derived from the French *cinque pas,* a different name for the Galliard. The basic rhythm of the Galliard is:

$$fppmpp \ldots,$$

but the five steps are distributed over the measure as follows:

$$fppmpp \ldots$$
$$1234.5$$

On the fourth beat the dancer leaps up high, and does not return to earth till the sixth beat, thus causing a stress in the middle of the measure, a syncopation whose name is suspiciously homophonous with the work *sinkapace*! The Galliard is a showdance, usually danced by one couple at the time. A particular form of the Galliard is La Volta, in which the gentleman lifts his partner up high on the fourth beat, supporting her on his knee. In most galliards the hemiola is used by the dancer to perform three leaps on alternate feet:

$$fppmpp \ldots$$
$$1.2.3.$$

In the nineteenth century many composers, notably Brahms and Dvorak, made frequent use of the Hemiola.

Triple time thus lends itself to subdivision into two subgroups of three as well as into three subgroups of two. It is similar to the planar symmetry groups having twofold, threefold and sixfold symmetry, which accomodate motifs having twofold symmetry in three distinct orientations as well as threefold-symmetrical motifs in two distinct orientations. Examples of dances in triple time or in the related sextuple time are, in addition to the basse-danses and galliards already mentioned, the *Menuet, Waltz* and *Jig* or *Gigue*.

FLOOR PLANS

Figure 1 is page 241 of Fabritio Caroso's *Raccolta di varij Balli,* a 1630 reprint of his *Della nobiltà di Dame* of 1600. The beauty of its design as well as the reference to mathematics are notable, and witness to the importance attached to the symmetry of the dance illustrated, in this instance *Contrapasso Nuovo*. This dance is a round dance for exactly three couples. Round dances have always been quite common, as they permit dancers to mix, and because all couples are equivalent, as distinct from processional or line dances, where there is a distinction between leading and last couples. The sixteenth-century *Bransles* (in England called *Brawls*) were social mixers for any number of couples arranged around a circle. *Contrapasso Nuovo,* however, is a *Ballo,* a dance composed of distinct movements in which every dancer traverses a characteristic pattern. In this instance, the three couples each change partners three times, so that at the conclusion of the dance all finish with their original partner. In contrast to the Bransles, therefore, *Contrapasso Nuovo* must be danced by exactly three couples, otherwise the dancers would not finish with their original partner. The symmetry is therefore threefold rotational, as indicated by the illustration.

There is mirror symmetry as well: three mirrors intersect at the center of threefold rotational symmetry. The author[1] has shown that the mirrors intersecting at odd-fold centers of rotational symmetry are *polar,* that is to say, their context is different in one direction from that in the opposite direction. In this particular instance of *Contrapasso Nuovo,* each mirror line passes

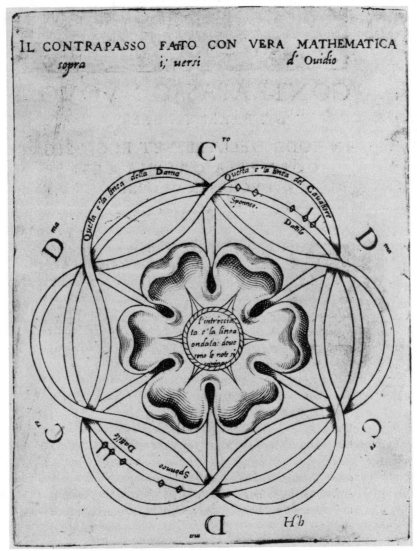

Fig. 1. *Contrapasso nuovo* (Fabritio Caroso). (Reproduced with the kind permission of the Theater Collection
in the Pusey Library at Harvard University, Dr. Jeanne Newlin, Curator.)

between the two partners of the same couple on one side, between separate couples on the other
(see Fig. 2). In this and following illustrations the symbols "O" and "x" indicate respectively
a lady and a gentleman. Observe that the mirror image of each dancer in every one of the
mirrors is a dancer of the opposite gender. The mirrors are therefore color-active: the pattern
is invariant to reflection accompanied by gender reversal.

The static configuration of Fig. 2 does not give us any information about the direction in
which the dancers are moving. In Fig. 1 we note ribbons woven around each other, which
disturb the mirror symmetry of the pattern, and are symbolic of the way in which the dancers
weave around each other. A simple but very fundamental fact of life is that dancers cannot
move through each other, even for the sake of preserving mirror symmetry. The question whether
a pair of dancers pass each other giving right or left hands, or passing right or left shoulders,
poses major problems to the dance historian, the choreographer as well as the dancer. We propose
to use here the following notation for describing two dancers passing each other, regardless
whether they give hands or simply pass shoulders:

O
O//x or \\
x

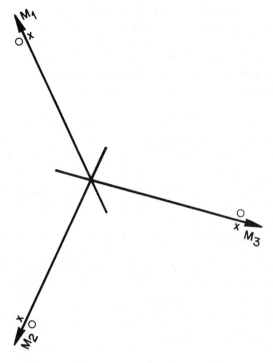

Fig. 2. Mirror symmetry in *Contrapasso Nuovo*.

means dancers pass right shoulders or give right hands, in other words, move to their left diagonally to pass.

O

O\\X or //

X

means dancers pass left shoulders etc.

Thus one figure in *Contrapasso Nuovo,* in which the dancers weave around the circle alternately giving each other right and left hands (as in the *Grand rights and lefts* in modern American folk dancing), would start as follows:

X\\⁰

O//X

X//ₒ

As the dancers proceed around the circle, each encounters a dancer of opposite gender at sixty degree intervals, at which point the triangular configuration of the three couples is inverted compared to that of the previous encounter, and the symbol for passing shoulders is reversed as well:

X\|ₒ

O/X

ₒ\X

The mirrors also reverse their polarity. In summary, then, the *Grand right and left* motion transforms the configuration such that the entire configuration, including the polarity of the mirrors and the passing symbols, is reversed.

   In the *Quadrille* or *Square Dance* the symmetry is fourfold rotational; there are four non-polar mirrors of two different kinds[2]. The dancers encounter each other at forty-five degree intervals; the square formation of the dancers now rotates forty-five degrees as well, the mirrors, being non-polar, do not change polarity, but the passing symbols do. In general, then we can say that in a round dance for n couples the grand rights and lefts figure transforms the configuration by a rotation through $(180/n)$ degrees, accompanied by a reversal of mirror polarity if present, and a reversal of passing symbols. That a symmetry analysis of this dance is indeed relevant to the intentions of its designer, is confirmed by the original manuscript and by Julia Sutton's analysis of the symmetry of *Contrapasso Nuovo*[3].

SYMMETRY RELATIONS BETWEEN DANCERS

In the preceding section we discussed the color-active mirror symmetry between partners. There are many interactions between dancers, and the symmetry of a dance frequently symbolizes these interactions. Beside the interaction between partners, there are flirtations with other dancers' partners, and rivalries between members of the same gender. Particularly charming are the *Balli* of the fifteenth century, many of which tell a simple story, and which may be considered forerunners of the romantic story ballet.

The first of these to be considered here is *La Gelosia* (Jealousy) by Domenico da Piacenza (Paris, Bibliothéque Nationale, fds.it.972, ca.1425)[4]. In this dance a number of couples start in processional position, i.e. in couples behind each other. The number of couples in this dance is flexible, but the repeat pattern in the music is determined by the exact number of dancers participating, so that the number of couples may not be said to be arbitrary, as would be the case, say, for most *basse danses*. Ordinarily the dance is described for three couples.

The pattern of the dance is performed as many times as there are couples, ordinarily three times. According to modern usage we shall call each traversal a *Round*. Each Round consists of four sections, designated by Brainard respectively A, B, C, D. A is a processional section, in which all couples remain together. Jealousy sets in the second section, when the first man deserts his lady for a brief flirtation with the second lady whose partner dances forward to join first lady. The fickle first man soon deserts second lady for the third lady, whose partner in turn dances forward to join the deserted second lady. As a result the ladies have not moved relative to the original configuration of the dance, but their partners have executed a cyclic permutation $1 \rightarrow 2 \rightarrow 3 \rightarrow 1$, indicating that first lady became partnered by second man, second lady by third man, and third lady by first man. With more than three couples the same cyclic permutation pertains. The third section of *La Gelosia* again is a procession, this time danced by the couples newly formed in the previous section. Finally, all couples utilize the color mirror which runs down the middle of the set: all change places with partners by the right hand, and back again by the left. Each Round was characterized by the cyclic permutation of the second section, and after dancing a number of Rounds equal to the number of couples in the set, all are reunited with their original partners. Accordingly, *La Gelosia* is a progressive dance in which the men only progress by a cyclic permutation.

Not all dances of the fifteenth century are couple dances: *Vercepe*, the next example[5], is danced by five dancers who start behind each other, the men in first, third and fifth position, the ladies in the even-numbered positions. The choreographer was again Domenico da Piacenza (Paris, BN 972 fol.13–14). In this configuration the third dancer occupies a central position. The symmetry of this dance, which probably denotes a skirmish in the *Commedia dell'arte* tradition, is twofold rotational around this central figure, the *Capitano*. The men and the ladies alternately circle around the center, each to music and steps characterizing movements proper to their gender.

A Ballo by Guglielmo Ebreo entitled *La Ligiadra* (Paris, BN 973 fol.33–34)[6], is a notable example where the symmetry of the dance symbolizes the relationships between the dancers. It is written for two couples who start behind each other; the first section is processional. Following the procession first man and first lady turn halfway around their own axis, with the result that first man now faces second man, while first lady faces second lady (arrows indicate the direction in which the dancers face):

First Couple: Second Couple:
○ →   ← ○
x →   ← x

Both couples then exercise a number of ceremonial movements and pass through each other, after which a flirtation ensues. First man and second lady approach each other, and dance a pattern resembling a mirror-image letter "S":

$$\text{First } \circ \to \qquad \leftarrow \text{x First}$$
$$\text{Second } \circ \to \qquad \leftarrow \text{x Second}$$

We would expect the first lady and second man to dance the mirror image of this figure on the other diagonal, i.e. a real figure "S." Instead, these two dance the same figures as their respective partners had done, that is to say, also a mirror-image "S." To understand the subtlety of this maneuver, observe that, when the first man and second lady changed places in their flirtation, they arrived in their new positions turning their backs to their rivals. However, when first lady and second man change places describing an inverted "S" on their diagonal, they turn in toward their original partners, whereas, if they had described a real "S," they would have turned their backs on their original partners. In *La Ligiadra*, Guglielmo therefore resisted mirror symmetry to indicate that the flirtations were only momentary and harmless: the dance finishes with the original partners dancing together again, finally curtseying to their colleagues.

CONTRA- AND COUNTRY DANCES

There is little doubt that these two names are related, but there is no agreement as to which of the two is derived from the other. The *Contradance* is a line dance in which the dancers form two rows, partners facing each other across the way. Ordinarily the gentlemen are on one side, the ladies opposite them. However, there are contradances in which men and ladies alternate in each row, still facing partners. The two arrangements are as follows:

```
        ○  ○  ○  ○  ○  ○  ○  ○  ○  ○  ...  ○  ○
TOP                                                  BOTTOM
        x  x  x  x  x  x  x  x  x  x  ...  x  x
```

and

```
        ○  x  ○  x  ○  x  ○  x  ○  x  ...  ○  x
TOP                                                  BOTTOM
        x  ○  x  ○  x  ○  x  ○  x  ○  ...  x  ○
```

In either case there is a color-active mirror down the center of the dance. (The top of the dance is the direction toward which the aligned dancers turn in order to have the ladies on the right of their partners.) Perpendicular to this mirror are mirror lines which in the former case are all color-inactive, while in the latter case they are alternately color-active (those passing between adjacent couples), and color-inactive (those passing through the dancers).

It is probable that the English word *Country Dance* is simply a derivation of *Contra-Dance,* for the Country Dances are not necessarily danced only in the country, but they are usually danced in contra-dance formation, in distinction from *Square Dances* or *Quadrilles,* and *Round Dances*. Most of the information which we possess about fashionable dances during the decades following the re-publication of Fabritio Caroso's treatise in 1630 are the various editions of *John Playford's The English Dancing Master,* which were published in the middle and second half of the seventeenth century. English and Scottish Country Dances are historically closely linked, but have developed differently; at present the distinctions are in the style of dancing and the footwork. For our purpose the principal difference between English and Scottish Country Dancing is in the formation of *sets*: whereas the English dances are danced in an arbitrarily long row, the Scottish divide into sets of ordinarily four couples. In the English dances everyone eventually dances with everyone else, whereas the Scots never dance outside their own set.

In either style of country dancing the couples progress, that is to say, they move up or down the line of the dance, partners commonly progressing together. In English Country Dancing the odd-numbered couples (as counted from the top of the dance) proceed down the dance, while the even-numbered ones progress toward the top[7]. This progression is accomplished by the execution of some figure by adjacent couples, which causes these couples to change places. The result is as follows:

```
time   ↓    1   2   3   4   5   6   7   8  . . .
axis        2   1   4   3   6   5   8   7  . . .
```

If the same procedure were followed again, couples would simply shuttle back and forth. However, the rule is that each couple, on reaching top or bottom of the dance, must wait out one round before reversing direction. (The even-numbered couples, who are moving up the dance are called *passive,* the odd-numbered ones *active.* On reaching one end of the dance one changes from active to passive or vice-versa.) The result is the following progression: (dancing couples are shown **bold,** the ones waiting a round out are plain)

1	**2**	**3**	**4**	**5**	**6**	**7**	**8**	**9**	...
2	**1**	**4**	**3**	**6**	**5**	**8**	**7**	**10**	...
2	**4**	**1**	**6**	**3**	**8**	**5**	**10**	7	...
4	2	**6**	**1**	**8**	**3**	**10**			...

We note that the odd numbers indeed do move down, the even ones up the dance. The progression is analogous to the progressions in bell ringing called plain bob[8]. The same permutations also occur in dance configurations where all dancers are in single file, usually starting with ladies and gentlemen alternating. In this instance the dancers change places by passing shoulders, and the parameter determining which shoulder should be passed first becomes relevant again. Also important is whether the total number of dancers or couples is even or odd. The following pattern illustrates the sequence for even numbers of couples or dancers in a row:

1	**2**	**3**	**4**	**5**	**6**
2	**1**	**4**	**3**	**6**	5
2	**4**	**1**	**6**	**3**	5
4	**2**	**6**	**1**	**5**	**3**
4	**6**	**2**	**5**	**1**	3
6	**4**	**5**	**2**	**3**	**1**
6	**5**	**4**	**3**	**2**	**1**

At this point the original configuration has inverted, and from here on the progression continues as follows:

6	**5**	**4**	**3**	**2**	**1**
5	**6**	**3**	**4**	**1**	2
5	**3**	**6**	**1**	**4**	2
3	**5**	**1**	**6**	**2**	**4**
3	**1**	**5**	**2**	**6**	4
1	**3**	**2**	**5**	**4**	**6**
1	**2**	**3**	**4**	**5**	**6**

The symmetry of inversion between the two halves of the dance is evident. Note furthermore that the even-numbered rounds are characterized by the fact that the end dancers remain in place, while in the odd-numbered rounds all dancers progress. Interesting also is that in line dances where the gentlemen start in the odd, the ladies in the even places, two instances occur where the genders are completely segregated.

Now consider the following example having an odd number of couples or dancers in the line:

1	**2**	**3**	**4**	**5**
2	**1**	**4**	**3**	5
2	**4**	**1**	**5**	**3**
4	**2**	**5**	**1**	**3**
4	**5**	**2**	**3**	1
5	**4**	**3**	**2**	**1**

Again we note that the pattern has inverted. It continues as follows:

5	4	3	2	1
5	3	4	1	2
3	5	1	4	2
3	1	5	2	4
1	3	2	5	4
1	2	3	4	5

In contrast to the case of even numbers, there is always one couple or dancer who is immobile, alternately at the top or bottom of the dance.

This progression pattern is so rich in possibilities of mixing and introducing dancers, that it is known by many different names. We noted that in English Country Dances (and New England Contras) it is the standard progression. In renaissance line dances it was known as *la haye*[9], a name which survives in English dances as *the hay,* while in Scottish country dances it is known as the *reel,* a designation not to be confused with one of the three basic rhythms of Scottish dances, which bears the same name. In round dances, where there is no end position, it becomes the *Grand Rights and Lefts* referred to above, known in Scottish dances as the *Grand Chain.*

SCOTTISH COUNTRY DANCES

When Marie de Guise married James V of Scotland in 1537, she brought with her the music and dances of the French court. Her daughter Mary was briefly Queen of France, but after the death of her husband the King, she returned to Scotland, and has generally become known as Mary Queen of Scots. After her son, James VI King of Scotland, also became James I King of England, the court at Edinborough lost much of its lustre, but at the same time maintained much of the French court traditions. In the seventeenth century the Country dance developed in Scotland much as in England, but the French influence on Scottish Country Dances is evident in such terms as *Allemande, Poussette,* and *Pas de Basque.* After the Roman Catholic descendants of James II were defeated by the Hanoverian Kings of England, at Culloden in 1745, all evidence of Scots national pride was suppressed. Nevertheless, we find on Scottish ball programs of the second half of the eighteenth century alternation of Menuets and Country Dances: this tradition went on continuously from the renaissance court of France till in the nineteenth century a rage of Celtic fashion spread throughout Europe, with Robert Schumann composing songs on German translations of Thomas More and Robert Burns, and Sir Walter Scott being widely read. Queen Victoria and Prince Albert frequently wore the Tartan, and Country Dances were danced in Scottish castles as well as on the village greens. In distinction from the Highland dances, the Country Dances hail primarily from the Scottish Lowlands, the region of Scotland closest to England.

Since in the early twentieth century the tradition threatened once more to disappear, the Royal Scottish Country Dance Society was founded in 1923 to research and record the traditional Country Dances. Some dance historians have doubts about the historical validity of the dances so canonized, but there can be no doubt about the fact that the definition and circumscription by the Society has assured a twentieth-century Scottish Country dance having a recognizable idiom of its own, based on careful folkloric and historical investigation. Scottish Country Dances are being composed and danced on special occasions such as wedding receptions, in all corners of the earth, by Scots, Dutch, Hawaiians and Chinese, but especially by mathematicians.

As stated above, Scottish Country Dances are danced in rows, as are their English cousins, but the lines are subdivided into discrete sets, most commonly comprising four couples, although some dances exist for five-couple sets. Like English Country Dances, some Scottish ones are danced by two couples together; the first couple progresses down the line of the dance as long as there is a couple below them to dance with. Each couple on reaching the end of the set, waits one round until a couple becomes available to dance with. The progression for two-couple dances in a four-couple set is therefore as follows (first couple leads off with second, while third and fourth couples wait till approached by first couple; cf. Ref. 7):

$$
\begin{array}{cccc}
\mathbf{1} & \mathbf{2} & 3 & 4 \\
2 & \mathbf{1} & \mathbf{3} & 4 \\
\mathbf{2} & 3 & \mathbf{1} & \mathbf{4} \\
3 & 2 & 4 & 1 \\
\mathbf{3} & \mathbf{4} & 2 & 1 \\
4 & \mathbf{3} & 1 & \mathbf{2} \\
\mathbf{4} & 1 & 3 & \mathbf{2} \\
1 & \mathbf{4} & 2 & \mathbf{3} \\
1 & 2 & \mathbf{4} & \mathbf{3}
\end{array}
$$

Notice that, in order to return to the original configuration the third and fourth couples would need to change places with each other; this is rarely done at the end of the dance.

For three-couple dances in a four-couple set the same rules apply: the leading couple dances as long as there are two couples below, then go to the bottom, and all others join as soon as there are enough couples above and below to dance with. The progression is:

$$
\begin{array}{cccc}
\mathbf{1} & \mathbf{2} & 3 & 4 \\
2 & \mathbf{1} & \mathbf{3} & 4 \\
\mathbf{2} & \mathbf{3} & \mathbf{4} & 1 \\
3 & 2 & 4 & 1 \\
\mathbf{3} & \mathbf{4} & \mathbf{1} & 2 \\
4 & 3 & 1 & 2 \\
\mathbf{4} & \mathbf{1} & \mathbf{2} & 3 \\
1 & 4 & 2 & 3 \\
\mathbf{1} & \mathbf{2} & \mathbf{3} & 4
\end{array}
$$

Each Scottish Country Dance is danced in one of three rhythms: in fast quadruple time (*reel*), slow quadruple time (*strathspey*) or fast triple time (*jig*). Scottish Country Dances still are essentially ballroom dances, and the footwork, which is beyond the scope of this chapter, is quite stylish. We shall, however, review a few of the symmetrical configurations typical of Scottish Country Dances and symbolic of Scotland or for the events for which they were written.

Particularly common is for the first couple to change places with their partner across the dance from them, then *casting down* one place, which means passing behind the second couple, while the second couple steps up into first place. First couple then crosses back by the right hand; the resulting configuration is: (the subscripts indicate original first, second, third and fourth couples respectively)

$$
\begin{array}{cccc}
\mathrm{O}_2 & \mathrm{O}_1 & \mathrm{O}_3 & \mathrm{O}_4 \\
\mathrm{X}_2 & \mathrm{X}_1 & \mathrm{X}_3 & \mathrm{X}_4
\end{array}
$$

There is a name for some of the relationships between the dancers in this configuration. The first couple is now between second and third couples; the dancers directly opposite them are, of course, their partners. The person to the right of the partner is *first corner,* the one to the left of the partner is *second corner.* Third lady is first man's first corner, third man is first lady's second corner. The diagonals of this rectangular arrangement form a St. Andrew's cross, named for Scotland's patron saint. Accordingly, many dances are danced along these diagonals by having first couple face, then dance with first corners, and then repeat with second corners:

$$
\begin{array}{cccc}
\mathrm{O}_2 & & \mathrm{O}_3 & \mathrm{O}_4 \\
 & \mathbf{X}_1 & & \\
 & \mathrm{O}_1 & & \\
\mathrm{X}_2 & & \mathbf{X}_3 & \mathbf{X}_4
\end{array}
$$

and

A particularly imaginative use of this St. Andrew's cross is made in the dance called *Mairi's Wedding*. Here the four dancers on the diagonal dance *half reels;* we saw above that at the halfway point in a reel (hay) the order is reversed. The dancers begin the figure by facing first corners, as diagrammed above. By initiating the half reel passing right shoulders, the first couple describes a loop around the vertices of the rectangle, then meets partners in the middle, and pass partners by the left shoulders to face second corners. Meanwhile the first corners, by dancing the half reel from the end positions, will have changed places with each other. Repeating the maneuver with second corners on the other diagonal, they will cause the original rectangle to become inverted, with each corner finishing diagonally opposite starting position, and first couple once more facing their (original) first corners. All now dance half reels once more on each diagonal, with the result that all finish again in their original position. Whereas this figure in *Mairi's Wedding* (there is considerably more to this dance) uses only the vocabulary of the Scottish Country Dance, a very pretty symbolic pattern was created: while their corners shuttle back and forth along the arms of the St. Andrew's cross, the first couple, in looping around the vertices of the rectangle, describe a four-leaf clover, symbolic of good luck.

CONCLUSIONS

We have demonstrated by examples ranging over a span of five centuries of dance in the Western European tradition, that symmetry constitutes an essential component of these dances, and can serve a symbolic function. Furthermore, as these dances are symmetrical in space as well as in time, there are some aspects to the transformations and the symmetries of these patterns which are not covered by the classical theories of the symmetry of space.

Acknowledgements—The author acknowledges gratefully the assistance of the staff of the Theater Collection in the Pusey Library, Harvard University, and in particular of Dr. Jeanne Newlin, its Curator, in having a page of the original Caroso publication reproduced in Fig. 1.

The author is also pleased to acknowledge the guidance of Dr. Ingrid Brainard, Dr. Julia Sutton, and of the teachers of the Boston Branch of the Royal Scottish Country Dance Society, in particular of Dr. Barbara Little, as he learned to dance each of the dances discussed here, and many more.

REFERENCES

1. A. L. Loeb, *Color and Symmetry*. John Wiley, N.Y., 1971, reprint Robert Krieger (1978), p. 60.
2. A. L. Loeb, op. cit.
3. Julia Sutton (translator and editor) and F. Marian Walker (musical transcriber) *Nobiltá di Dame by Fabritio Caroso*, Chapter IV. Oxford University Press, to be published.
4. I. G. Brainard, *The Art of Courtly Dancing in the Early Renaissance*, pp. 71–74. West Newton, Massachusetts (1981).
5. Ingrid Brainard, op. cit. 77–81.
6. Ingrid Brainard, op. cit. 68–70.
7. For the sake of clarifying the symmetry of the dances, we are here letting all couples start together; this is the way in which the dances are usually practiced. In authentic performance, however, the top two couples only begin the dance, and the lower couples join in gradually as the first couple progresses down the dance. When all couples have joined the dance, it proceeds as described herein.
8. Alice Dickinson, Change Ringing: Theory and Practice in *Patterns of Symmetry* (ed. Marjorie Senechal and George Fleck). University of Massachusetts Press, Amherst (1977), pp. 44–49.
9. M. S. Evans (transl.) and J. Sutton (ed.), *T. Arbeau, Orchesography*, p. 170. Tablature de la dance de la haye.

Comp. & Maths. with Appls. Vol. 12B, Nos. 3/4, pp. 641–653, 1986
Printed in Great Britain.

0886–9561/86 $3.00 + .00
© 1986 Pergamon Press Ltd.

SYMMETRY IN MOORISH AND OTHER ORNAMENTS†

Branko Grünbaum and Zdenka Grünbaum
University of Washington, Seattle, WA 98195, U.S.A.

and

G. C. Shephard
University of East Anglia, Norwich NR4 7TJ, England

Abstract—An investigation of the Moorish ornaments from the Alhambra (in Granada, Spain) shows that their symmetry groups belong to 13 different crystallographic (wallpaper) classes; this corrects several earlier enumerations and claims. The four classes of wallpaper groups missing in Alhambra (pg, p2, pgg, p3m1) have not been found in other Moorish ornaments, either. But the classification of repeating patterns by their symmetry groups is in many cases not really appropriate—account should be taken of the coloring of the patterns, of their interlace characteristics, etc. This leads to a variety of "symmetry groups", not all of which have been fully investigated. Moreover, the "global" approach to repeating ornaments is only of limited applicability, since it does not correspond to the way of thinking of the artisans involved, and does not cover all the possibilities of "local" order. The proper mathematical tools for the study of such structures which are only "locally orderly" remain to be developed.

The idea of investigating the ornaments and decorations of various cultures by consideration of their symmetry groups appears to have originated with Pólya[22]. This mathematically motivated approach, which was in sharp contrast to the earlier descriptive methods, gained wider recognition through the influential book of Speiser[27]. The earlier methods, which are exemplified by such works as those of Jones[18], Bourgoin[4], Day[6], Grasset[9], and many others, relied largely on the analysis of ornaments by considering the character of their motifs. By contrast, the mathematical approach depends on the symmetries of the design considered as a whole, and is to be found in the publications of Müller[21], Shepard[25], Weyl[29], Garrido[8], Shubnikov & Koptsik[26], Washburn[28] and others. Among mathematically inclined investigators the newer, quantitative way of looking at the physical evidence has become universally adopted; however, many practitioners of the decorative arts and their analysis for ethnographic, anthropological, archeological and other purposes still use almost exclusively the descriptive method.

The present investigation arose from a desire to clarify and settle contradictory statements regarding the symmetry groups of the Moorish ornaments that are to be found in the Alhambra, in Granada (Spain). The background facts are the following. It is well known that there are 17 classes of symmetry groups of planar ornaments which repeat in at least two nonparallel directions; these are known as the (classes of) wallpaper (or crystallographic plane) groups. In Fig. 1 are shown examples of patterns, with a very simple motif, of each of the 17 classes of wallpaper groups. In an early application of group-theoretic methods to the analysis of historic ornaments, Müller[21] examined the patterns and tilings in the Alhambra and found that there are 11 different groups present. In contrast to her findings, Coxeter[5] states that 13 wallpaper groups are represented there,‡ while the number 17 was claimed by others (see, for example, Belov[2], Fejes Tóth[7, p. 43], Martin[20, p. 111]). So when an opportunity to visit the Alhambra was provided by a Guggenheim Fellowship, an on-the-spot investigation to settle these conflicting claims was undertaken. The results of this investigation, as well as comments and observations which arose in this connection, form the topics of the present paper.

The bottom line is relatively easy to draw. After a reasonably thorough examination of the Alhambra, 13 different wallpaper groups were identified among the symmetry groups of

†The research reported in this paper has been supported by National Science Foundation grants MCS 8001570 and MCS 8301971, and by a Fellowship from the John Simon Guggenheim Memorial Foundation. One version of this material was presented in an address to the Pacific Northwest Section meeting of the Mathematical Association of America in Moscow, Idaho, on June 18, 1983.

‡In correspondence during 1981 between Professors H. S. M. Coxeter, J. J. Burckhardt and one of the authors (B.G.), the source of the number 13 was traced to a misunderstanding of some statements in Müller[21]. The same conclusion was reached independently by Professor D. W. Crowe.

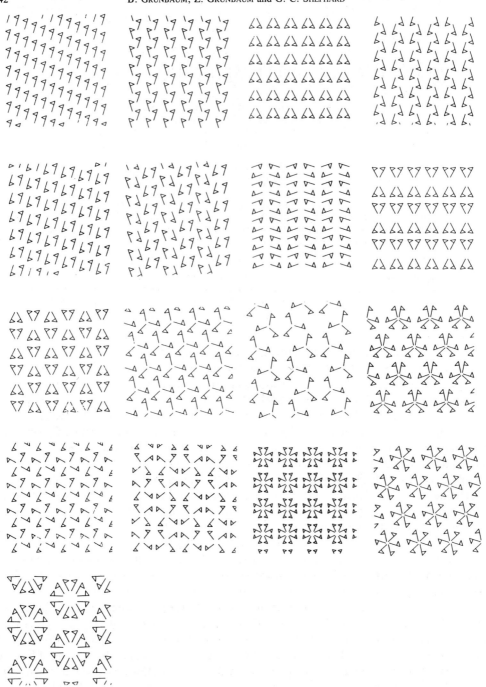

Fig. 1. Patterns in which the common motif is a small "flag", which can exemplify the 17 classes of wallpaper groups.

the ornaments found there. Actually, Müller seems to have missed only one group present in the Alhambra palace proper (namely the group pm, of which the example is given in Fig. 6 below). The other group that she missed seems not to be represented in the palace, but only in the Museum of Alhambra, and it is not clear whether this museum was accessible to her. The four groups which have not been found in the Alhambra (pg, p2, pgg, p3m1) do not appear to be represented in other Moorish artifacts either (though naturally, a really systematic examination of the enormous volume of extant materials may turn up some or all of them). It is of interest to note that two of these four groups have been located in Toledo (Spain) in buildings approximately contemporaneous with some of those in the Alhambra—one (p2, see Fig. 7) in a church, and the other (p3m1, see Fig. 8) in a synagogue. On the other hand, it seems that the groups pg and pgg fail to be represented not only in Moorish decorations, but in Islamic ornaments in general.

The attempt to determine which of the wallpaper groups are present in the Alhambra turned out to be rather interesting regardless of the numerical answer obtained, since the obvious (and even the correct) answer is not necessarily the best or the most appropriate. The material forces one to consider mathematical questions which would probably not have arisen otherwise.

Among the first serious difficulties that one encounters is to decide what is it that one is counting; for many of the ornaments there are several different yet reasonable ways in which they can be considered symmetric, and these lead to different values for the numbers of groups found. Some examples will clarify this.

(i) To begin with, there is the symmetry group in the most immediate sense: we look at the ornament, exactly as it is (except that we imagine it to be continued indefinitely in all directions and we ignore minor variations that are due to practical considerations). We then ask what isometries (rigid motions) map the ornament precisely onto itself, under preservation of all its properties. For example, in the tilings from the Alhambra shown in Fig. 2, the symmetries include reflections in the vertical and in the horizontal lines through the centers of the colored (brown or green) tiles, but not reflections in lines through the centers of the black tiles because such reflections would map brown tiles onto green ones, and so fail to be symmetries; this symmetry group is pmm.

(ii) Next, there is the symmetry group of the *underlying uncolored* ornament—the coloring of the tiles is disregarded, and the isometries that map the resulting uncolored ornament onto itself are considered. In the example shown in Fig. 2—where the underlying tiling consists of the horizontal and vertical "dogbones"—this leads to many symmetries additional to those in (i); among them are reflections in vertical and in horizontal lines through the centers of all the tiles, 90° rotations about the meeting points of quadruplets of tiles, etc. (these form the group p4g).

As additional examples we consider the tilings shown in Figs. 3 and 4. In the first the coloring destroys most of the symmetries and the symmetry group is p1; however, the underlying uncolored tiling has many rotational symmetries and its symmetry group is p3. Figure 4 shows an analogous situation, except that here the modifications to the underlying tiling consist not in coloring the tiles, but in adding inscriptions or other designs to some of them. The tiling with inscriptions has symmetry group p1; ignoring the inscriptions we find that the underlying tiling has symmetry group p6m.

(iii) Then there are the *color-symmetry* groups: instead of ignoring the colors in multicolor ornaments, or of considering the colors as unrelated to each other, we consider *color symmetries*, that is, isometries which map the underlying ornament onto itself *coupled* with *consistent* permutations of colors, so that the combination maps the colored pattern onto itself.

For example, the tiling with asymmetric trefoils shown in Fig. 5 admits as rotational symmetries only 120° turns about the centers of the trefoils and the points where six trefoils meet. The underlying tiling admits 60° turns about the latter points; these can be made into color symmetries by agreeing to accompany each 60° rotation by an interchange of colors. In the tiling shown in Fig. 6 some—but not all—symmetries of the underlying uncolored tiling can be made into color symmetries by suitable choices of permutations of the colors. By contrast, every symmetry of the uncolored tiling underlying the three-colored tiling shown in Fig. 7 can be made into a color symmetry by coupling it with a suitable permutation of colors. (Colored patterns with this property are called "perfectly colored"; see Grünbaum & Shephard[12,14] or Senechal[24] for details of the known results.) In the four-color tiling in the center of Fig. 8 blue, gray and black tiles are equivalent under color symmetries of the tiling, but the white tiles are not equivalent to them.

(iv) Many ornaments in Moorish art (as in the art of many other cultures) are *interlace patterns* (for example, Figs. 9 and 10 can be interpreted as showing interlaced polygons). Then one can consider either the group of symmetries [as in (i) above] or that of the "underlying" pattern formed by the "overlapping polygons," in which we disregard the fact that certain portions of the polygons (which may be reasonably assumed to exist) are "hidden" by parts of other polygons. For purposes of discussion, we consider the simpler interlace pattern of squares shown in Fig. 11(a), and its underlying pattern shown in Fig. 11(b). Due to the way

LEGENDS FOR COLOR FIGURES ON PAGES 645–646

Fig. 2. A colored tiling from the Alhambra. Its symmetry group is pmm, and the symmetry group of the underlying tiling is p4g.

Fig. 3. Another colored tiling from the Alhambra. All symmetries of the underlying tiling (except some of the translational symmetries) are eliminated by the coloring; the colored tiling has symmetry group p1, while the underlying tiling has symmetry group p3.

Fig. 4. A stucco wall decoration from the Alhambra. With the inscriptions and designs in the star-shaped regions, the only symmetries are translations and the symmetry group is p1. The underlying ornament, in which the inscriptions and designs are disregarded, has symmetry group p6m.

Fig. 5. A two-color tiling from the Alcazar in Sevilla (Spain) in which the symmetry group is p3 and the underlying uncolored tiling has symmetry group p6. Every 60° rotational symmetry of the underlying tiling can be made into a color symmetry of the colored tiling by coupling it with an interchange of the two colors.

Fig. 6. A multicolored tiling from the Alhambra; for purposes of the current discussion we disregard the fact that the tiling is in part wrapped about a cylindrical column. The colored tiling has symmetry group pm (the only symmetries other than translations are reflections in vertical lines bisecting the white tiles), and the underlying uncolored tiling has symmetry group p4g. Among color symmetries of the tiling are reflections in the horizontal lines that bisect the tiles, as well as appropriate vertical translations coupled with cyclic permutations of the colors brown, green and blue; the 90° rotational symmetries of the underlying tiling cannot be made into color symmetries.

Fig. 7. A three-colored tiling on the floor of a church in Toledo (Spain) which has symmetry group p2—one of the wallpaper groups that seems not to have been used in Moorish ornaments. The underlying uncolored tiling has group p6m.

Fig. 8. A wall decoration from a synagogue in Toledo (Spain) with symmetry group p3m1, which is another of the wallpaper groups apparently missing in Moorish ornaments. The underlying tiling is the same as in Fig. 7.

Fig. 9. A polygonal interlace pattern from the Alcazaba in Malaga (Spain). The reflective symmetries of the underlying pattern are not symmetries of the interlace, but can be combined with layering interchanges to form layered symmetries.

Fig. 10. A pattern of interlaced regular octagons from the Alhambra; variants of this ornament occur frequently in Islamic art from many different countries. The symmetry group is p4, but reflections coupled with layering interchange are symmetries of the layered pattern.

Fig. 11. (a) A simple interlace formed by squares, which has no reflective symmetries (symmetry group is p4). (b) The underlying pattern of overlapping squares, which admits reflective symmetries.

Fig. 12. A colored interlace pattern from the Alhambra. Variants of this pattern are frequently found in Islamic art from many countries.

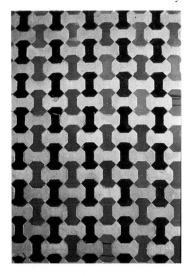

Fig. 2.

Fig. 3.

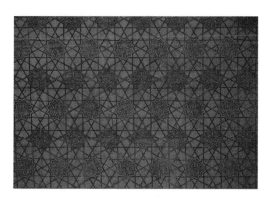

Fig. 4.

Fig. 5.

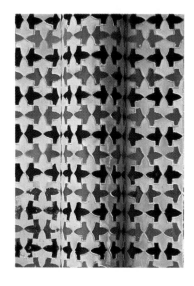

Fig. 6.

Fig. 7.

Fig. 8.

Fig. 9.

Fig. 10.

(a)

(b)

Fig. 11.

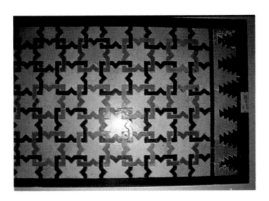

Fig. 12.

in which the squares are interlaced, the pattern in Fig. 11(a) clearly admits no reflections as symmetries (its symmetry group is p4), while the pattern in Fig. 11(b) has such symmetries (its symmetry group is p4m). However, just as with colored patterns, there is a third way of looking at this situation. A *layered symmetry* is an ordinary symmetry possibly combined with systematic interchanges of top and bottom layers at crossings, which maps the ornament onto itself. By using these the interlace pattern in Fig. 11(a) again acquires reflective symmetries—layered symmetries in which reflections are coupled with layer interchange. Similar considerations apply to the interlace patterns shown in Figs. 9 and 10.

(v) Finally, some ornaments are both colored and layered. For example, for the tiling from the Alhambra shown in Fig. 12, the complete set of symmetries arises only if we agree to consider isometries coupled with both color changes and interchanges of layers.

The above remarks show that in order to analyse Moorish patterns fully we have to consider not only the wallpaper groups [which are needed for the symmetry groups as described in (i) and (ii) above] but also groups which involve colors, or interlacing, or both. Some (but not all) of these kinds of groups have been studied. For example, it has been known for more than 50 years that, in analogy to the 17 classes of wallpaper groups, there are 46 classes of color-symmetry groups for two-colored patterns. (Here, as throughout the paper, we are discussing only the groups of ornaments periodic in at least two nonparallel directions; the "frieze groups" and finite groups can be treated similarly, but we do not consider these here.) During the last few years it has been shown that there are 23 classes of color groups for three colors, and that the corresponding numbers for four, five and six colors are 96, 14, and 90. (For the three-color result see Grünbaum[10] and Grünbaum & Shephard[13,14]; for the number of n-color groups for all n ≤ 60 see Wieting[30]; a general survey with many references is given in Schwarzenberger[23].) Bearing these facts in mind it becomes obvious that not only are some (uncolored) symmetry groups missing in the Alhambra, but most of the color groups (even for only few colors), layered groups, and color-layered groups are also not represented. In fact, the more complicated of these groups have not even been enumerated so far! So the statement that "all symmetry groups were used by ancients", which is often repeated in connection with Moorish ornaments (as well as for those of the Egyptians) is no more than a myth! For further discussion and illustrations of this aspect see Grünbaum[11].

Yet another statement that gained wide acceptance through frequent repetition concerns the wealth of decorations present in the Alhambra—or in Islamic art in general, or in the art of ancient Egyptians, or Cretans, etc. Many authors state or imply that the variety of designs found in each of these cultures is overwhelming and boundless. But detailed study shows that no such assertions can be taken seriously. While the Morrish ornaments clearly exhibit a large number of designs, what is really surprising is how few of the many possibilities were utilized. Some examples should serve to illustrate how the number of possible designs of each kind mentioned above is astronomically large, even within bounds set by practical considerations, and that in consequence it was impossible to actually use even a moderate part of their number. One source of the tremendous abundance of possibilities lies in changing parameters which do not affect any of the symmetries of the pattern under consideration. For example, the nine interlace patterns shown in Fig. 13 are all the "same" in that they have the same layered symmetry group as the pattern in Fig. 11(a)—but their aesthetic and decorative effects are very distinct.

In many analyses of Moorish patterns it is mentioned with considerable awe that even just with regular octagons, or with regular dodecagons, their artisans knew how to create three or four distinct interesting patterns. Without wishing to detract from the skill involved, it is a fact that just expanding regular polygons centered at fixed positions leads to a theoretically infinite and practically very large number of designs that appear to be quite different (see the examples in Figs. 14 and 15). Actually, with almost any motif huge numbers of decorative patterns can be created, even if one insists that every copy of the motif plays the same role (that is, the symmetry group of the pattern acts transitively on the copies of the motif). In most cases there is still no complete enumeration of the possibilities, but it is known that in a reasonably detailed classification of patterns formed by nonoverlapping congruent circular disks there are 131 classes; for patterns formed by nonoverlapping congruent line segments there are more than 200 classes, and so on. In each case many of the patterns depend on several parameters which can be

Fig. 13. A variety of interlace patterns formed by square motifs which differ only in relative distances and widths, and do not differ in their symmetry groups or their layered symmetry groups.

continuously varied. For results and references on this and related topics see Grünbaum and Shephard[13–17]. In view of these facts, it is clearly inappropriate to anticipate that any significant part of these patterns has been used in practice.

But even where it could be reasonably expected that inventiveness would produce appreciable variability, the actual material in existence presents near-monotony. An example, mentioned above, is that most of the color symmetry groups possible with even a few colors have never been used. An illustration of unused possibilities appears in Fig. 16, which shows three colorings which we devised for a relatively simple interlace with square motifs. We must also note the remarkable paucity evident in the layerings used in interlace patterns. Among thousands of interlace patterns from historical artifacts that we have seen, only two have crossings which depart from the simplest one-over-one-under variety—and even for these it is not clear whether they were really meant to exhibit a different crossing sequence or whether they arose through a mistake of the artisan or the recorder. (One of the two is a variant from Turkey of the ornament shown in Fig. 10; it was recorded by Aslanapa[1], and has also been reproduced in Grünbaum & Shephard[16]. In this variant the crossing pattern is not regular, and the symmetry group does not act transitively on the octagons.) An illustration in [16] shows a dozen variants of the crossings (from among the hundreds possible, even if equivalence of the motifs under the symmetry group is required) of the pattern of interlaced octagons shown in Fig. 10. No systematic

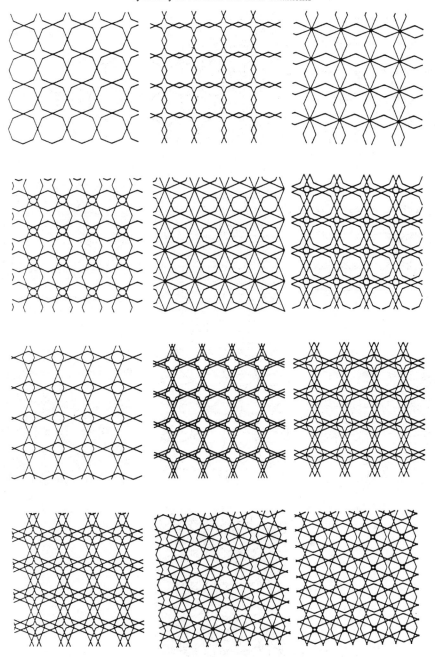

Fig. 14. A number of patterns formed by overlapping regular octagons; again the differences arise from different relative sizes, and not from changes in symmetry properties.

investigation of the possibilities seems to have been carried out so far. The lack of variety in crossing sequences is even more surprising in view of the fact that many different ones arise very naturally in the weaving of fabrics.

It should also be noted that in a certain sense all the preceding may be misdirecting us. We—mathematicians and some other scientists—may find it convenient and useful to interpret regularity of a pattern in terms of its group of symmetry (or color symmetry, etc.). In this way we can apply the results of algebra and other mathematical disciplines to the study of such patterns. However, it could be argued that this is not the concept of regularity that artisans (Moorish or any other) had in mind as they were creating their art. In fact, until a century or so ago, even to mathematicians regularity of mathematical objects had a completely different meaning. The difference between the two approaches is to a large degree the contrast of the *global* and *local* points of view. Mathematicians used to define regularity of objects such as

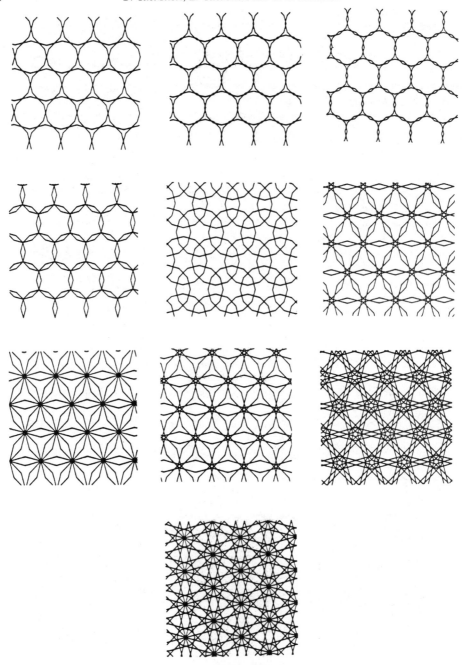

Fig. 15. A selection of patterns formed by regular dodecagons, which differ only in their sizes.

the Platonic polyhedra by requirements of congruent faces, equal angles, and other local properties; now it is customary to define regularity by the transitivity of the symmetry group on the set of flags. In the same way, it seems likely that the artisans meant to create ornaments in which each part is related to its immediate neighbors in some specific way (and not by attempting to obtain global symmetries of the infinitely extended design). We illustrate this remark with the simple tiling of squares and rectangles shown in Fig. 17. Every square tile touches two other square tiles and two rectangular tiles—one on its end and one on its side. Every rectangular tile touches four square tiles (two at its ends and two at its sides) and two rectangular tiles. From designs such as that shown in Fig. 2 it seems likely that a Moorish artist would consider this to be a perfectly legitimate pattern, yet it may be extended in such a way that its symmetry group contains only translations parallel to one direction! Of course "local" uniformity frequently leads to "global" symmetry, but it may well be that the former was the main objective

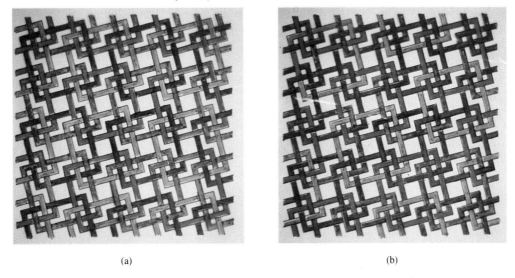

(a)   (b)

Fig. 16. (a)–(c) A four-coloring and two five-colorings of an interlace pattern of squares.

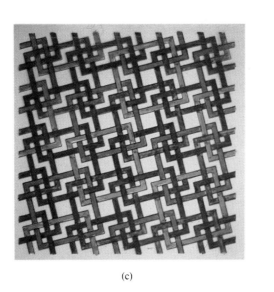

(c)

Fig. 16.

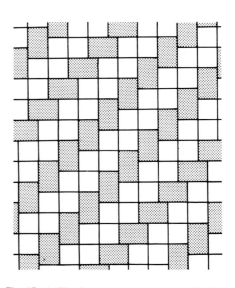

Fig. 17. A tiling by square and rectangular tiles; it is "locally" regular but is not regular in the "global" sense.

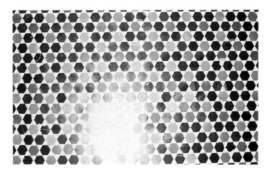

Fig. 18. A tiling from the Alcazar in Sevilla (Spain), in which the colors of the hexagons are apportioned in ratio 6:2:1.

while the latter was an accidental consequence. The interested reader may find more on this topic in [11]; the concept of "local regularity" or "orderliness" appears to be well worth investigating as a generalization of the approach to symmetry via groups. The latter may be needlessly restrictive and inappropriate in many contexts.

One last remark concerning colored patterns appears to be called for here. As mentioned earlier, very few of the groups of color symmetries are represented among Moorish and other ornaments. One reason for this situation seems to be that in many of the historical (and contemporary) multicolor decorations the various colors are not meant to play the same role. The different colors are apportioned to the copies of the motifs in unequal numbers; very frequently the proportions are $2:1:1$, $4:2;1:1$, $6:2:1$, $6:3:1:1:1$ or some similar ratios. For example, in Fig. 2 we have four white tiles for every two black ones and for each brown and green tile. In Fig. 6 the ratios are $6:3:1:1:1$, while among the hexagons in Fig. 18 six black ones go with two blue and one brown. The mathematical theory of such colorings still awaits development.

We can summarize our conclusions as follows.

1. Only 13 of the 17 wallpaper groups occur as symmetry groups of ornaments in the Alhambra. The other four groups seem not to have been used by Moorish artists, or their use was very infrequent. In this context it is worth recalling that according to Makovicky & Makovicky[19], 6 of the 17 groups account for about 98% of the designs in the rich collection of Bourgoin[3]. But the importance of determining the number of groups used is greatly reduced when we appreciate that the mode of thinking in terms of symmetry groups was totally alien to the artisans (and mathematicians) of antiquity and the Middle Ages.

2. Despite the richness of the ornaments created by Moorish craftsmen, the designs actually used represent only an infinitesimally small fraction of those that are possible, even if allowance is made for the practical difficulties of realizing certain designs in ancient times. This applies to plain designs as well as to colored or interlaced ones.

3. The study of interlace patterns from a mathematical point of view is still in a most rudimentary stage, and needs considerable development before it can be applied to the description and classification of historical artifacts.

4. The various kinds of symmetry groups are useful in the description of many of the artifacts, but more general approaches (based on "adjacency relations" or other "local" criteria) are necessary for a better understanding of the ornaments and artwork, and of the ways their creators thought about them.

5. The approach to multicolor ornaments via color symmetries appears to be inadequate in many respects and inappropriate to describe practical examples. It seems that new mathematical tools for the understanding and classification of such patterns will have to be developed.

REFERENCES

1. O. Aslanapa, *Türkische Fliesen und Keramik in Anatolien*. Baha Matbaasi, Istanbul (1965).
2. N. V. Belov, Moorish patterns of the Middle Ages and the symmetry groups [in Russian]. *Kristallografiya* **1**, 474–476 (1956). English translation: *Soviet Physics—Crystallography* **1**, 482–483 (1956).
3. J. Bourgoin, *Les éléments de l'art arabe: le trait des entrelacs*. Firmin-Didot et Cie., Paris (1879). Plates reprinted in *Arabic Geometrical Patterns and Designs*. Dover, New York (1973).
4. J. Bourgoin, *Grammaire élémentaire de l'ornement, pour servir à l'histoire, à la theorie et à la pratique des arts et à l'enseignement*. Delagrave, Paris (1880). New edition: *L'ornement*. Editions d'aujourd'hui, Paris (1978).
5. H. S. M. Coxeter, Review of *N-dimensional Crystallography*, by R. L. E. Schwarzenberger. *Bull. London Math. Soc.* **13**, 245–248 (1981).
6. L. F. Day, *Pattern Design*. Batsford, London (1903). Paperback reprint, Taplinger, New York (1979).
7. L. Fejes Tóth, *Reguläre Figuren*. Akadémiai Kiadó, Budapest (1965). English translation: *Regular Figures*. Pergamon, New York (1964).
8. J. Garrido, Les groupes de symetrie des ornements employes par les anciennes civilisations du Mexique. *C. R. Acad. Sci. Paris* **235**, 1184–1186 (1952).
9. E. Grasset, *Methode de composition ornamentale*. Librairie Centrale des Beaux-Arts, Paris (1905).
10. B. Grünbaum, Color symmetries and colored patterns. Mimeographed notes, University of Washington, January 1976.
11. B. Grünbaum, The Emperor's new clothes: full regalia, G-string, or nothing? *Math. Intelligencer* **6**, 47–53 (1984).
12. B. Grünbaum and G. C. Shephard, Perfect colorings of transitive tilings and patterns in the plane. *Discrete Math.* **20**, 235–247 (1977).

13. B. Grünbaum and G. C. Shephard, Classification of plane patterns. Mimeographed notes distributed at the special session on Tilings, Patterns and Symmetries, Summer meeting of the American Math. Society, Seattle, August 1977, 66 pp.
14. B. Grünbaum and G. C. Shephard, Incidence symbols and their applications. *Proc. Symp. Pure Math.* **34**, 199–244 (1979).
15. B. Grünbaum and G. C. Shephard, A hierarchy of classification methods for patterns. *Zeitschrift für Kristallographie* **154**, 163–187 (1981).
16. B. Grünbaum and G. C. Shephard, Tilings, patterns, fabrics and related topics in discrete geometry. *Jahresber. Deutsch. Math.-Verein.* **85**, 1–32 (1983).
17. B. Grünbaum and G. C. Shephard, *Tilings and Patterns*. Freeman, New York (1986).
18. O. Jones, *The Grammar of Ornament*. Quartich, London (1856); reprinted 1910, 1928.
19. E. Makovicky and M. Makovicky, Arabic geometrical patterns—a treasury for crystallographic teaching. *Jahrbuch für Mineralogie Monatshefte* 1977, No. 2, 58–68 (1977).
20. G. E. Martin, *Transformation Geometry. An Introduction to Symmetry*. Springer, New York (1982).
21. E. Müller, Gruppentheoretische and Strukturanalytische Untersuchung der Maurischen Ornamente aus der Alhambra in Granada. Ph.D. thesis, University of Zürich. Baublatt, Rüschlikon (1944).
22. G. Pólya, Über die Analogie der Kristallsymmetrie in der Ebene. *Zeitschrift für Kristallographie* **60**, 278–282 (1924).
23. R. L. E. Schwarzenberger, Colour symmetry. *Bull. London Math. Soc.* **16**, 209–240 (1984).
24. M. Senechal, Color groups. *Discrete Appl. Math.* **1**, 51–73 (1979).
25. A. O. Shepard, *The Symmetry of Abstract Design with Special Reference to Ceramic Decoration*. Publication 574, Carnegie Institution of Washington, Washington, D.C. (1948).
26. A. V. Shubnikov and V. A. Koptsik, *Symmetry in Science and Art*. Plenum Press, New York (1974). Russian original published by Naùka Press, Moscow (1972).
27. A. Speiser, *Die Theorie der Gruppen von endlicher Ordnung*, Second Edn. Springer, Berlin (1927); Third Edn, Springer, Berlin (1937) [= Dover, New York (1943)]; Fourth Edn, Birkhäuser, Basel (1956).
28. D. K. Washburn, A symmetry analysis of upper Gila area ceramic design. *Papers of the Peabody Museum of Archeology and Ethnology* **68** (1977).
29. H. Weyl, *Symmetry*. Princeton University Press, Princeton (1952). Paperback reprint, 1982.
30. T. H. Wieting, *The Mathematical Theory of Plane Chromatic Ornaments*. Dekker, New York (1981).

Comp. & Maths. with Appls. Vol. 12B, Nos. 3/4, pp. 655–671, 1986
Printed in Great Britain.

0886–9561/86 $3.00 + .00
© 1986 Pergamon Press Ltd.

THE GEOMETRY OF COASTLINES:
A STUDY IN FRACTALS

JAY KAPPRAFF

New Jersey Institute of Technology, Newark, NJ 07102, U.S.A.

Abstract—The geometry of coastlines, based on an empirical study by Lewis Richardson, is presented as a way of introducing the subject of fractals developed by Benoit Mandelbrot. It is shown how the statistically self-similar nature of coastlines can be generalized to an interesting class of point sets, curves and surfaces with the same property. Brownian and fractional Brownian motion are introduced as ways of generating statistically self-similar curves with the appearance of coastlines and mountain ranges.

1. INTRODUCTION

There can be little doubt that Euclidean geometry has had a large effect on the cultural history of the world. Not only mathematics but art, architecture and the natural sciences have created their structures either from the elements of Euclidean geometry or its generalizations to projective and non-Euclidean geometries. However, by its nature, Euclidean geometry appears to be more suitable to deal with the "ordered" aspects of phenomena or as a way to describe the artifacts of civilization than as a tool to describe the "chaotic" forms that occur in nature. For example, the concepts of point, line and plane, which serve as the primary elements of Euclidean geometry, are acceptable as models of the featureless particles of physics, the horizon line of a painting or the facade of a building. On the other hand, the usual geometries are inadequate tools with which to render geometrical expressions of a cloud formation, the turbulence of a flowing stream, the pattern of lightning, the branching of trees and alveoli of the lungs or the configuration of coastlines.

In the early 1950s, a mathematician, Benoit Mandelbrot, discovered new geometrical structures suitable for describing these irregular sets of points, curves and surfaces from the natural world. He coined the word "fractals" for these entities and invented a new branch of mathematics to deal with them, an amalgam of geometry, probability and statistics. Although there is a strong theoretical foundation to this subject, it can best be studied through the medium of the computer. In fact, fractal geometry is a subject in which the mathematical objects are generally too complex to be described analytically, but it is an area in which computer experiments can lead to theoretical formulations.

Mandelbrot created his geometry in 1974 after observing fractal patterns arise in many diverse areas of research such as the structure of noise in telephone communications, the fluctuation of prices in the options market and a statistical study of the structure of language. In 1961 Mandelbrot brought his attention to an empirical study of the geometry of coastlines carried out by a British meteorologist, Lewis Richardson. Mandelbrot was able to comprehend the theoretical structure behind Richardson's data and see how this structure could be generalized and abstracted. This article is devoted to a discussion of how Richardson's work on the geometry of coastlines led Mandelbrot to formulate his fractal geometry, and it is meant to serve as an introduction to Mandelbrot's work. Mandelbrot's recent book, *The Fractal Geometry of Nature*[1] is the primary reference for this article, and several of its figures have been reproduced.

2. HISTORICAL PERSPECTIVE

Before beginning a discussion of Mandelbrot's analysis of Richardson's data, it is useful to place the subject of fractals in historical perspective.

Calculus was invented in the latter part of the 17th century by Newton and Leibnitz in order to deal mathematically with the variation and changes observed in dynamic systems, such as the movement of the planets and mechanical devices. Through calculus, the concepts of "continuity" and "smoothness" of curves were quantified. However, since the approach to this subject was intuitive, "continuity" and "smoothness" were limited in their range of

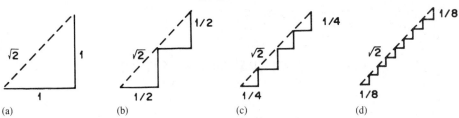

Fig. 1. Four stages in the generation of a nowhere-smooth curve.

possibilities to "ordered" or "tame" motions. It was only when the logical foundation to the subject was completed by the middle of the 19th century that mathematicians were motivated to search for extreme examples of variability with which to test the now rigorous definitions of continuity and smoothness. Much to the surprise of the mathematical community, Weirstrass, Georg Cantor and Giuseppe Peano were able to create "pathological" or "chaotic" curves and point sets that confounded intuition. Weirstrass constructed a curve spanning finite distance that was infinite in length, continuous but nowhere smooth. Peano created a curve that could fill up a square without crossovers (although segments of the curve touched). Cantor constructed a set of points as numerous as all the points within a unit interval but was so sparse as to take up negligible space or, in mathematical terms, "be of measure zero," justifying its characterization by Mandelbrot as a "dust."

Most mathematicians considered these "pathological" creations to be interesting curiosities but of little importance. However, Mandelbrot, following on the footsteps of his teacher, Paul Levy, the father of modern probability theory, saw these irregular curves and sets as models for the shapes and forms of the natural world, and he formulated his fractal geometry in order to "tame" the "monstrous" curves.

A simple example demonstrates how pathological curves can easily be constructed to confound intuition and show that basing mathematical conclusions on visual perceptions can lead to error. Compare the lengths of the sequence of zig–zag curves shown in Fig. 1. Clearly the segment lengths of each curve sum to 2 units. It is easy to see that the steps can be made small enough so that, to appearance, they are indistinguishable from the linear segment between initial and final points with length $\sqrt{2}$.

3. COASTLINES

Let us now consider the geometry of coastlines. Most people's concept of a coastline derives from two sources, observations of coastlines on a map and a vacation visit to the seashore. However, both of these experiences yield different impressions. A coastline shown in an atlas gives the impression of the coast as a comparatively smooth and "ordered" curve of finite length easily determined from the scale of the map. However, a visit to the coast generally reveals rugged and "chaotic" terrain with rocks jutting out to the sea and an undulating shoreline dynamically varying over time and space.

Look again at maps illustrating a coastline at an increasing sequence of scales. At a scale of 1 in. to 100,000 ft, the scale of a good roadmap, the coast appears quite smooth and indicates a bay by means of a small indenture. At a scale of 1 in. to 10,000 ft we notice within the bay a small inlet not indicated on the previous map. At 1 in. to 1000 ft a small cove is demarcated within the inlet. Finally, at 1 in. to 100 ft, the scale of a highly detailed map, the cove is seen to have almost as much detail as the original map of the entire coastline. In fact, Richardson discovered that the coastline has effectively unbounded length and is nowhere smooth; this ever increasing detail is taken into account at decreasing scales. Scales below 1 in. to 50 ft are not considered since they bring into focus irrelevant details, whereas scales above an upper cutoff are also not considered since crucial details of the coastline would be deleted.

Even more remarkable than having infinite length, Mandelbrot discovered that any segment of the coastline is "statistically self-similar" to the whole coastline. By a self-similar curve, we mean a curve such that any segment is mathematically similar to the whole, i.e. a magni-

PLATE XXX
Logarithmic Spiral and Shell-Growth

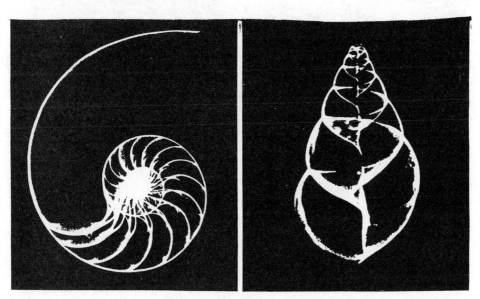

PLATE XXXV
Shell, Logarithmic Spiral and Gnomonic Growth
(Photographs: Kodak Limited)

Fig. 2. Self-similar growth in nature.

fication or contraction of the whole. It is well known that among smooth curves, only the logarithmic spiral is self-similar and this curve serves as the configuration of Nautilus shells, seashells and the horns or horned animals shown in Fig. 2.

In order to be statistically similar, two curves need only have the same statistical distribution of their features, such as ins and outs, under a magnification or contraction in their geometric scale, as will be described later in some detail. Statistical self-similarity can never be empirically verified for naturally occurring curves such as coastlines, since there are an infinite number of features associated with such curves. However, the distribution of enough aspects of these curves can be examined to conclude with a high degree of certainty whether or not two curves are statistically self-similar. Also, as we shall see, mathematical models can be formulated with exact statistical self-similarity.

The notion of self-similarity also pervades the realm of art, as shown in the hierarchal patterns of the paintings *The Great Wave* by Katsushika Hokusai and *The Deluge* by Leonardo Da Vinci, shown in Fig. 3 along with a short doggerel by Richardson reminiscent to me of the latter painting.

The Deluge, by Leonardo Da Vinci.

Big Whorls have little whorls,
which feed on their velocity;
And little whorls have lesser whorls,
And so on to viscosity.
                    by L.F. Richardson

The Great Wave, by Katsushika Hokusai.

Fig. 3. Hierarchies of self-similar curves in art.

4. A GEOMETRIC MODEL OF A COASTLINE

In this section we present a mathematical model of a coastline that is geometrically self-similar at a sequence of scales and infinite in length. In the next section, we will present a more sophisticated model that also embodies statistical self-similarity. In order to describe these mathematical models, we will give operational definitions to the following basic concepts: self-similarity, scale, length of a curve, dimension, geometrical fractals and random fractals.

4.1 *Length and scale of a coastline*

Viewing a curve at a given scale and the definition of its length are two intimately connected notions. There are many different ways to represent a curve at a given scale. One method is illustrated in Fig. 3, where the curve on the left, spanning the unit interval [0, 1], is shown on the right at scales of 1, $\frac{1}{3}$ and $\frac{1}{9}$ in Figs. 3(a), (b) and (c) respectively. The scaled curves are derived from the actual curve by subdividing the curve with dividers set to intervals of length

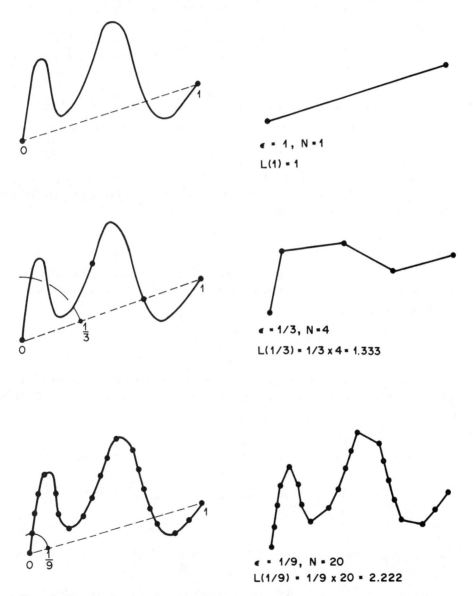

Fig. 4. Determination of the length L of a curve spanning [0, 1] by approximating the curve with N linear line segments. (a) Representation of curve at scale of $\epsilon = 1$; (b) representation of curve at scale of $\epsilon = \frac{1}{3}$; (c) representation of curve at scale of $\epsilon = \frac{1}{9}$.

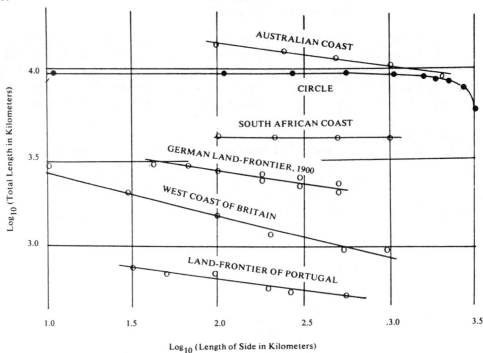

Fig. 5. Richardson's empirical data concerning the rate of increase of coastline's lengths at decreasing scales.

equal to one-third and one-ninth of the unit, starting at the beginning of the curve as illustrated by the arcs in Fig. 4. The marked points are then connected with line segments. The length of the curve, $L(\epsilon)$, at scale ϵ is then defined by,

$$L(\epsilon) = \epsilon \times N(\epsilon), \tag{1}$$

where $N(\epsilon)$ are the number of segments of length ϵ that span the curve. The total length, L, of the curve is then defined as the limiting value that $L(\epsilon)$ approaches as ϵ approaches zero or, mathematically,

$$L = \lim_{\epsilon \to 0} L(\epsilon) \tag{2}$$

Richardson applied this definition to determine the coastal length of many different countries, and he discovered that, for each of them, the number of segments at scale ϵ satisfied the empirical law

$$N(\epsilon) = K\epsilon^{-D}, \tag{3}$$

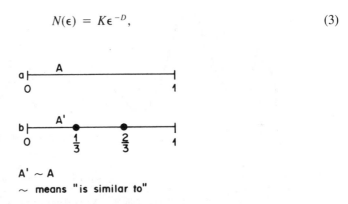

Fig. 6. The unit interval, a trivial example of a self-similar curve with dimension $D = 1$.

where K and D are constants depending on the country. Inserting eqn (3) in (1),

$$L(\epsilon) = K\epsilon^{1-D}. \tag{4}$$

The results of Richardson's analysis are shown in Fig. 5, where (4) yields a straight line when $\log L(\epsilon)$ is plotted against $\log \epsilon$. Also, as a result of eqns (2) and (4), the total length of the coastline L is effectively infinite.

Richardson's data indicates that the configuration of coastlines is derived from a general law of nature, and Mandelbrot's analysis of Richardson's data led to the following expression of that law:

> Each segment of a coastline is statistically similar to the whole, i.e. the coastline is statistically self-similar.

4.2 Geometrically self-similar curves

In his book *The Fractal Geometry of Nature*[1], Mandelbrot presents a procedure for constructing curves that are geometrically self-similar. To understand how self-similar curves relate to Richardson's law, it is sufficient to set $K = 1$ and rewrite (4) as

$$L(\epsilon) = \epsilon \, 1/\epsilon^D. \tag{5}$$

First, consider a trivial example of a self-similar curve, the straight-line segment of unit length shown in Fig. 6(a). This segment is self-similar at any scale. For example, at the scale $\frac{1}{3}$, Fig. 6(b) shows that three similar editions of the segment replicates the original. Thus, from eqn (1),

$$L(\epsilon) = \tfrac{1}{3} \times 3$$

or

$$L(\epsilon) = \tfrac{1}{3} \times 1/\tfrac{1}{3}^1,$$

and consequently $D = 1$ in eqn (5).

Now consider a less trivial example of a curve, self-similar at a sequence of scales $(\frac{1}{3})^n$, $n = 0, 1, 2, 3, \ldots$ known as the Koch snowflake. Since the curve is infinite in length, continuous and nowhere smooth, it cannot be drawn, as we explained for the example shown in Fig. 2. However, it can be generated by an infinite process, each stage of which represents the curve as seen at one of the scales in the above sequence. Figures 7(a), (b) and (c) show views of the Koch snowflake at scales of 1, $\frac{1}{3}$ and $\frac{1}{9}$, respectively both as linear segments on the left and incorporated into triangular snowflakes on the right. The snowflake is generated iteratively by replacing each segment of one stage by four identical segments of length one-third the original in the next stage. Thus, whereas for stage 1

$$L(1) = 1,$$

for stage two

$$L(\tfrac{1}{3}) = \tfrac{1}{3} \times 4$$

or

$$L(\tfrac{1}{3}) = \tfrac{1}{3} \times 1/(\tfrac{1}{3})^D,$$

where D is easily determined as

$$D = \frac{\log 4}{\log 3} = 1.2618. \ldots$$

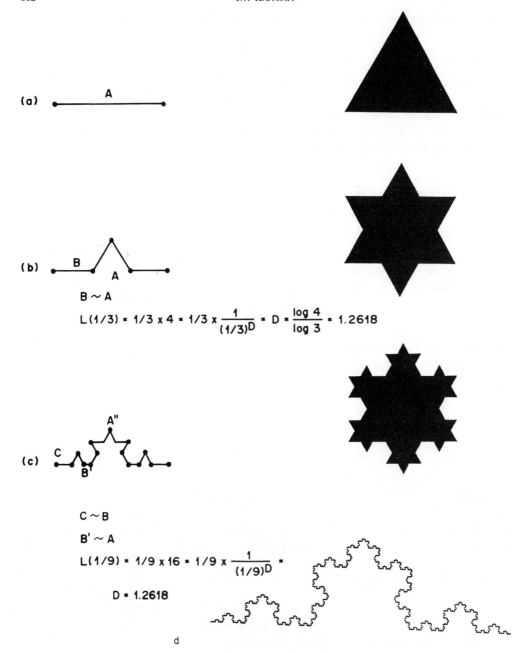

(a)

(b)

$B \sim A$

$$L(1/3) = 1/3 \times 4 = 1/3 \times \frac{1}{(1/3)^D} = D = \frac{\log 4}{\log 3} = 1.2618$$

(c)

$C \sim B$

$B' \sim A$

$$L(1/9) = 1/9 \times 16 = 1/9 \times \frac{1}{(1/9)^D} =$$

$$D = 1.2618$$

d

Fig. 7. The Koch snowflake, a nontrivial example of a self-similar curve with dimension $D = 1.2618$. (a) Koch snowflake at scale of $\epsilon = 1$; (b) Koch snowflake at scale of $\epsilon = \frac{1}{3}$; (c) Koch snowflake at scale of $\epsilon = \frac{1}{9}$; (d) Koch snowflake at an advanced stage in its generation.

For each successive stage,

$$L(\epsilon) = \epsilon \times 1/\epsilon^D,$$

with the same value of D. Each segment of a given stage is seen to be similar to a segment three times as large in the previous stage, as illustrated in Fig. 7. Thus, in the limit, each segment of length $(\frac{1}{3})^n$ of the Koch snowflake must be geometrically similar to the whole, satisfying both Richardson's data and Mandelbrot's interpretation of it. This property of self-similarity at a sequence of scales is more evident in Fig. 7(d) of a Koch snowflake at an advanced stage in its development.

Mandelbrot shows that, as for the Koch snowflake, any geometrically self-similar curve

satisfies

$$D = \frac{\log N}{\log (1/r)}, \tag{6}$$

where N is the number of congruent segments of length r that replace the unit interval in the initial stage of the iteration. Thus, for the Koch snowflake $N = 4$ and $r = \frac{1}{3}$. Mandelbrot refers to D as the dimension of the curve, and he shows that for curves of infinite length, spanning the finite distance,

$$1 < D \leq 2,$$

The magnitude of D is a measure of the "roughness" of the curve.

Fig. 8. Another fractal curve with dimension $D = \frac{3}{2}$.

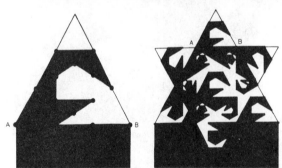

The first two steps in constructing Benoit Mandelbrot's Peano-snowflake curve

Fig. 9. The third stage in the generation of a space-filling Peano curve filling the interior of a Koch snowflake.

The relationship between N and r, expressed by eqn (6), is quite general and is illustrated for other geometrically self-similar structures in Figs. 8, 9, and 10. Figure 8 is an analogous curve to the Koch snowflake with dimension $D = \frac{3}{2}$, while Fig. 9 is the third stage of a space-filling Peano curve of dimension 2 that fills up the interior of the Koch snowflake. The point set in Fig. 11 represents six stages in the generation of what Mandelbrot calls a Cantor "dust" and has dimension $D = .6309$. Starting with the unit interval, each stage is generated from the preceding stage by removing the middle third of each remaining subinterval. What remains after an infinity of stages is an infinite set of points interspersed within empty space. The dimension is determined in the same manner as for the Koch snowflake, from eqn (6), and is

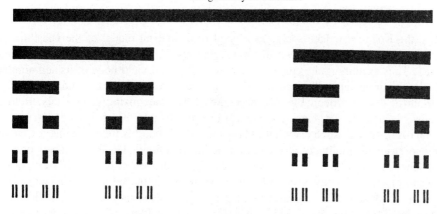

The Cantor dust uses [0,1] as initiator, and its generator is

□ CANTORIAN TRIADIC BAR AND CAKE (HORIZONTAL SECTION DIMENSION D=log 2/log 3=0.6309).

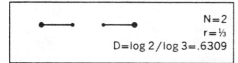

$$N = 2$$
$$r = \tfrac{1}{3}$$
$$D = \log 2 / \log 3 = .6309$$

Fig. 10. Six stages in the generation of a ''cantor triadic bar,'' an approximation to a Cantor ''dust'' with dimension $D = .6309$.

shown in Fig. 10 to be .6309. However, since it would be difficult to illustrate even a few stages in the generation of a Cantor dust, Fig. 11 shows an analogous sequence of bars with finite thickness.

Finally, Mandelbrot coined the term fractal curves to refer to curves with dimension $D >$ 1, the term fractal surfaces to refer to surfaces with dimension $D > 2$, and the term fractal point sets for point sets with $D > 0$. Although, according to this definition, fractals need not be self-similar, the class of self-similar fractals generated by Mandelbrot's recursive procedure appear to be the ones of greatest interest and I refer to them as ''geometrical fractals'' in order to distinguish them from the statistically self-similar curves discussed in the next section, which I refer to as ''random fractals.''

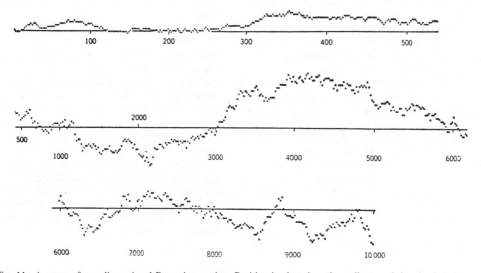

Fig. 11. A curve of one-dimensional Brownian motion. Position is plotted on the ordinate and time is plotted on the abscissa.

5. A STATISTICAL MODEL OF A COASTLINE

Although the Koch snowflake serves as a good mathematical model of the coastline, it fails to represent actual coastlines in two important respects. Its sequence of scales are bound to powers of $\frac{1}{3}$. Thus examining the curve at intervals of $\frac{1}{4}$ would yield none of its self-similar properties. Also, as irregular as the snowflake is, its structure is completely ordered, unlike coastlines. Both of these shortcomings can be removed by randomizing the fractals. Before describing how Mandelbrot introduced the element of statistics into the analysis of fractals, it is useful to revisit the linear Koch snowflake shown in Fig. 7, but this time imagine the curve to represent a spectrum of amplitudes of sound over an interval of time. Generally, a phonograph recording of sound, such as the sound of a violin, changes if the record is played fast or slow. In fact, a record of whale sounds is inaudible until the record is played at a sufficiently high speed. However, Koch snowflake music would clearly sound the same played at one-third the speed and then amplified three times. More precisely, if the amplitude at time t is represented by the function $B(t)$, the scaling property of the snowflake music is a statement about the identity of the functions $B(t)$ and $B(rt)/r$, where r is the self-similar scale. Such sounds are called scaling noises and have been studied by a colleague of Mandelbrot's, Richard Voss, at IBM Watson Research Center[2].

5.1 Brownian motion as a generator of random fractals

Curves that are statistically self-similar at every scale yield scaling noises that sound alike on a record played at any speed. One such curve can be simulated by playing a game in which a marker is moved forward or backwards along a line of equally spaced squares with equal

JEAN PERRIN'S CLASSIC DRAWINGS OF PHYSICAL BROWNIAN MOTION

Physical Brownian motion is described in Perrin 1909 as follows: "In a fluid mass in equilibrium, such as water in a glass, all the parts appear completely motionless. If we put into it an object of greater density, it falls. The fall, it is true, is the slower the smaller the object; but a visible object always ends at the bottom of the vessel and does not tend again to rise. However, it would be difficult to examine for long a preparation of very fine particles in a liquid without observing a perfectly irregular motion. They go, stop, start again, *mount*, descend, *mount again*, without in the least tending toward immobility."

The present plate, the only one in this book to picture a natural phenomenon, is reproduced from Perrin's *Atoms*. We see four separate tracings of the motion of a colloidal particle of radius 0.53μ, as seen under the microscope. The successive positions were marked every 30 seconds (the grid size being 3.2μ), then joined by straight intervals having no physical reality whatsoever.

To resume our free translation from Perrin 1909, "One may be tempted to define an 'average velocity of agitation' by following a particle as accurately as possible. But such evaluations are *grossly wrong*. The apparent average velocity varies crazily in magnitude and direction. This plate gives only a weak

idea of the prodigious entanglement of the real trajectory. If indeed this particle's positions were marked down 100 times more frequently, each interval would be replaced by a polygon smaller than the whole drawing but just as complicated, and so on. It is easy to see that in practice the notion of tangent is meaningless for such curves."

This Essay shares Perrin's concern, but attacks irregularity from a different angle. We stress the fact that when a Brownian trajectory is examined increasingly closely, Chapter 25, its length increases without bound.

Furthermore, the trail left behind by Brownian motion ends up by nearly filling the whole plane. Is it not tempting to conclude that in some sense still to be defined, this peculiar curve has the same dimension as the plane? Indeed, it does. A principal aim of this Essay will be to show that the loose notion of dimension splits into several distinct components. The Brownian motion's trail is *topologically* a curve, of dimension 1. However, being practically plane filling, it is *fractally* of dimension 2. The discrepancy between these two values will, in the terminology introduced in this Essay, qualify Brownian motion as being a fractal. ■

(a)

Fig. 12. Jean Perrin's drawings of physical Brownian motion. (a) Text from *The Fractal Geometry of Nature* by Mandelbrot. (b) Perrin's data.

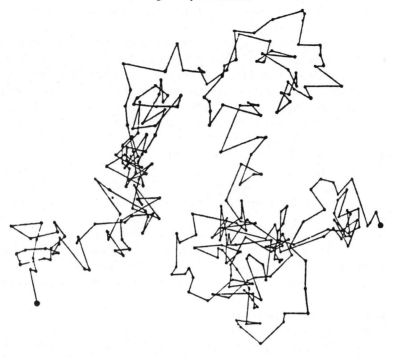

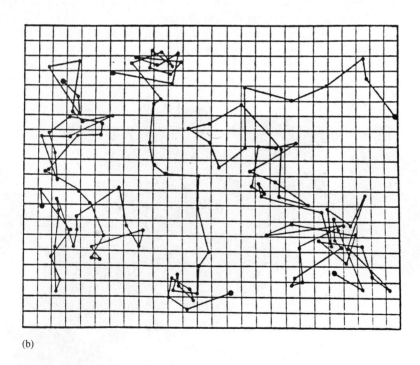

(b)

Fig. 12. (*Continued*)

probability (by flipping a coin). If each move is considered to take place during one unit of time, the movement of the marker in time, also known as a random walk, is shown in Fig. 11 for 10,000 time units or moves. In fact, this motion follows the scaling law that $B(t)$ and $B(rt)/\sqrt{r}$ have the same probability distribution for all scales r. Following a plausibility argument of Mandelbrot's, we show in the Appendix that this curve has dimension $D = \frac{3}{2}$. Also, Mandelbrot has shown that the points where the curve $B(t)$ intersects the time axis forms a set analogous to the Cantor dust shown in Fig. 10. The set has dimension $D = \frac{1}{2}$ and any interval of the time axis has the same statistical distribution of gaps between successive zeros of $B(t)$ when the gaps are adjusted according to the scale.

Actually, the previous motion approximates a one dimensional Brownian motion of a particle. One of the first scientists to observe Brownian motion of a particle suspended in a fluid was Jean Perrin. Figure 12 and the accompanying text from the *The Fractal Geometry of Nature* is a good introduction to Brownian motion along with its property of statistical self-similarity at every scale. Thus, Brownian music is an example of a scaling noise at any scale. A corollary of the self-similarity is the fact that there is no way to statistically distinguish segments of a Brownian curve over equal intervals of time from each other. Thus, in the language of statistics, Brownian motion is a "stationary" process. Another feature of Brownian motion exhibited by these curves is the complete independence of past and future positions of the particle with respect to the present position. In other words, there is no "correlation" between where the particle has come from during the interval $[-t, 0]$ and where it is headed in the interval $[0, t]$.

In terms of statistics, the change of position of a particle undergoing Brownian motion over time, $\Delta B(t) = B(t + \Delta t) - B(t)$, is distributed with a Gaussian distribution (the famous normal distribution) with mean $\langle \Delta B(t) \rangle = 0$ and variance (the square of the standard deviation) $\langle \Delta B(t)^2 \rangle = |\Delta t|$. This information permits Brownian motion to be generated continuously in time.

Brownian motion in the plane can also be generated approximately by a random walk through a lattice of squares (graph paper) in which over each interval of time Δt there is an equal chance for the marker to move up–down and right–left. The path in the plane generated by this motion has dimension $D = 2$, which indicates that eventually every point in the plane will be crossed by the curve and the motion is plane filling. Such a path is shown in Fig. 13. It will have many self-intersections and is far too erratic to serve as a model for coastlines. As a matter of fact, Brownian motion along the line illustrated in Fig. 11 with $D = \frac{3}{2}$ is also too irregular to serve as the model of a coastline which, according to Richardson and Fig. 5, tends to have a dimension $D \simeq 1.2$ with no self-intersections.

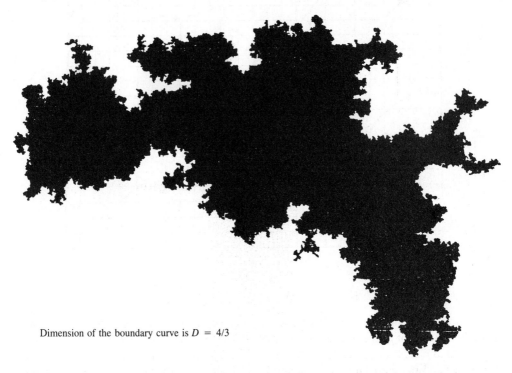

Dimension of the boundary curve is $D = 4/3$

Fig. 13. A curve drawn by two-dimensional Brownian motion. The outer boundary of the curve constitutes a "random fractal" curve of dimension $D = \frac{4}{3}$. The random walk continues inside the outer boundary but is not visible in the diagram.

5.2 *Fractional Brownian motion: an improved model of coastlines*

In search of better statistically self-similar models for coastlines, Mandelbrot rediscovered a generalization of Brownian motion which he calls fractional Brownian motion, where

$$\langle \Delta B(t) \rangle = 0, \qquad \langle \Delta B(t)^2 \rangle = \Delta t^{2H}$$

and ΔB again follows a Gaussian distribution, while the resulting traces in the plane have dimension $1/H$ for $0 < H < 1$ and are statistically self-similar with $B(t) = B(rt)/r^H$. It follows that $H = \frac{1}{2}$ reproduces ordinary Brownian motion in which a particle at any given time has an equal chance to move up–down or left–right. On the other hand, the movement of a particle under fractional Brownian motion tends to persist in the direction the particle is moving when $\frac{1}{2} < H < 1$. As a result, past and present are "correlated" with respect to the present position of the particle. This motion results in few self-intersections and fractal curves of greater regularity reminiscent of coastlines. Values of H such that $0 < H < \frac{1}{2}$ result in antipersistent motions in which the curve tends to reverse its direction from moment to moment.

D = 2.5

(a)

D = 2.1

(b)

Fig. 14. Two Brownian landscapes. (a) Rough landscape is generated by Brownian motion ($D = 2.5$). (b) Smooth landscape is generated by fractional Brownian motion ($D = 2.1$).

Mandelbrot calls the tendency of a motion to "persist" in time, as persistent fractional Brownian motion does, the "Joseph effect." After all it was Joseph, during biblical times in ancient Egypt, who enhanced his position and fortunes by realizing the propensity for feast and famine to persist over seven-year cycles. Much later, in 1906, a British civil servant Harold Hurst discovered empirically, contrary to the conventional wisdom of the time, that the fluctuations in the discharges of the river Nile were not random from year to year. Hurst applied his data to the operations of the newly built Aswan dam. Mandelbrot made the important discovery that Hurst's data followed a curve of fractional Brownian motion. Since these motions were self-similar at every scale, Mandelbrot refers to noises generated with amplitude spectra that follows a fractional Brownian curve as Hurst noises.

Finally, Fig. 14 depicts mathematical models of two Brownian landscapes. The upper landscape is an ordinary Brownian model while the lower one is fractional Brownian. Both landscapes represent wrinkled surfaces with fractional dimensions between 2 and 3. As for Brownian motion on a line and in the plane, the ordinary Brownian landscape is quite rough, more like a typical mountain scene. Analogous to the way the line and plane Brownian motion is generated, the ordinary Brownian scene is generated by starting with a plane. On one side of a randomly placed line in the plane, the plane is raised or lowered one unit with respect to the other with equal probability. Another line is randomly placed on the plane, and again one side is raised or lowered with respect to the other. This morphology is repeated iteratively with each successive elevation or depression diminished in scale by a factor of $\sqrt{k}$, where k is the number of the iteration. Generation of the fractional Brownian landscape is quite technical and is not described.

Brownian curves and surfaces are fractals, self-similar at every scale, or what Mandelbrot refers to as "creaseless." However, Mandelbrot's book is filled with many practical ways to construct statistically random fractals self-similar at a sequence of scales as for the Koch snowflake.

6. CONCLUSION

In certain ways, fractal geometry has already found useful application in describing certain aspects of the natural world. It has served as a tool for geographers to study the branching of rivers and the geometry of river basins. It is being considered by land surveyors as a way to characterize patterns of vegetation and land quality. It has given some insight to astronomers as to the distribution of stars and galaxies. It is being used as a tool for the study of the turbulence of fluids. There is potential for medical researchers to begin characterizing mammalian tumors through their fractal dimensions. Realistic scenery is now being generated by computers as backdrops to movies. Most recently, a conference on fractals was convened to discuss applications of fractals to statistical physics[3]. As the ideas become more familiar, important applications to every field are expected to emerge.

Acknowledgments—I wish to thank Dr. Mandelbrot for reading this manuscript, making several corrections and helping to place the subject in the correct historical perspective. I would also like to express my appreciation to him for the extraordinary work that he has done to create a new branch of mathematics that sheds light on so much in our world.

The following figures were reprinted from *The Fractal Geometry of Nature* with the permission of W. H. Freeman and Company, New York. Figure 3—Plate C3 and Plate C16; Fig. 5—Plate 33; Fig. 7—Plates 42 and 44; Fig. 8—Plate 51; Fig. 10—Plate 80; Fig. 11—Plate 241; Fig. 12—Plate 13; Fig. 13—Plate 243; Fig. 14—Plate 265. Figure 9 was reprinted from the April 1978 *Scientific American*. Figures 1(a)–(c) was reprinted from *The Geometry of Art and Life* by Matila Ghyka, published by Dover Press, New York (1977). Figure 1(d) was reprinted from *Patterns in Nature* by Peter Stevens, © 1974 by Peter Stevens. Reprinted by permission of Little, Brown and Company.

REFERENCES

1. B. B. Mandelbrot, *The Fractal Geometry of Nature*. W. H. Freeman, New York (1983).
2. R. F. Voss, $1/f$ noise in music; music from $1/f$ noise. *J. Accoust. Soc. Am.* **63**, 258–263 (1978).
3. M. Shlesinger (Ed.), Proceedings of the Gaithesburg Conference on Fractals. *J. Statist. Phys.* **36** (1984).

APPENDIX

A plausibility proof that the dimension of Brownian motion on a line $B(t)$ has dimension $D = \frac{3}{2}$.

Proof: The unit interval, $[0, 1]$, is covered by $1/\Delta t$ squares of length Δt, where $r = \Delta t$ is the scale of the curve. Since the standard deviation of a Brownian motion over a unit of time Δt was given by

$$\langle \Delta B^2 \rangle^{1/2} = \sqrt{\Delta t},$$

the expected variation of the curve $B(t)$ during Δt is

$$B_{\max} - B_{\min} \simeq \sqrt{\Delta t}.$$

But the number of scale lengths equivalent to $\sqrt{\Delta t}$ is

$$\frac{\sqrt{\Delta t}}{\Delta t} = \frac{1}{\sqrt{\Delta t}}.$$

Since there are $1/\Delta t$ scale lengths,

$$N = \frac{1}{\Delta t} \quad \frac{1}{\sqrt{\Delta t}} = \Delta t^{-3/2}.$$

Thus,

$$D = \frac{\log N}{\log (1/r)} = \frac{\log \Delta t^{-3/2}}{\log \Delta t^{-1}} = 3/2.$$

Comp. & Maths. with Appls. Vol. 12B, Nos. 3/4, pp. 673–695, 1986
Printed in Great Britain.

0886–9561/86 $3.00 + .00
© 1986 Pergamon Press Ltd.

IN BLACK AND WHITE: HOW TO CREATE PERFECTLY COLORED SYMMETRIC PATTERNS

Doris Schattschneider
Moravian College, Bethlehem, PA 18018, U.S.A.

Abstract—The use of isometries to create and "perfectly color" symmetric tilings and patterns is explained. The reader is not presumed to have knowledge of isometries, group theory, or computer science. A student, designer, teacher, or any other person interested in the interplay of geometry and art (particularly geometric symmetry), and the possibility of implementation using computer graphics, can learn from this paper.

The title of this article is intended to be read both literally and figuratively. We will provide, "in black and white," directions for a designer to follow in order to create a symmetric pattern and then "perfectly color" it, and we will restrict the discussion primarily to coloring symmetric patterns with two colors (black and white). There is no lack of mathematical literature on this topic; indeed, there is a confusing array of such literature since the topic of color symmetry has been explored and developed only in the last 25 years (Ref. [20] lists over 100 papers related to the subject). A few of the most recent papers (see [20] and [21]) are helpful in pointing out some of the restrictive assumptions or differing interpretations of "color symmetry" made by various authors, which are certain to confuse a reader who naively seeks information on the topic. In this article, we rely on the definitions and theory of color symmetry which have emerged as the current "standard," and which have provided the most logical means of analysing, classifying, and creating colored symmetric patterns.

We begin with a discussion intended to help a designer to understand and use the geometric machinery called isometries to create symmetric patterns and tilings of the plane. Once this is done, we can explain how to "perfectly color" such designs. Those readers familiar with isometries and the generation of periodic patterns may skip the first section.

1. ISOMETRIES AND THE GENERATION OF DESIGNS AND TILINGS

Decorative art and graphic design have been an important part of every culture in every period of history, and continue to be important today. We have included in the references just a few collections of such designs; there is a wealth of literature available which attests to the variety of styles and applications of such art. (The Dover Pictorial Archives is an excellent source of such illustrations.)

Many (perhaps most) of the graphic designs found in commercial and decorative art possess symmetry: a part of the design (a motif or a single tile) is repeated regularly to create the whole design. There are just four different ways in which a motif or tile can be related to a congruent copy of itself if this relationship is described in terms of geometric transformations which preserve shape. Mathematicians call these transformations *isometries* (*isos* = equal, *métron* = measure), and the four distinct types of transformations are translations, rotations, reflections, and glide reflections.

1.1 *Translation*

A *translation* of points in the plane shifts all points the same distance in the same direction. A vector (arrow) shows the direction of the shift, and its length is the distance the points are shifted. When the same translation T is repeatedly applied to a motif, it creates a row of equally spaced images of the motif, all facing in the same direction. Reversing the direction of the vector, but not its length, we obtain a translation, denoted by $-T$, which can repeat the motif *ad infinitum* in the opposite direction (see Fig. 1).

The border, or frieze design so created (which theoretically extends infinitely), is said to be *generated* by the translation T, and is called a *periodic border design*. The "period" of the

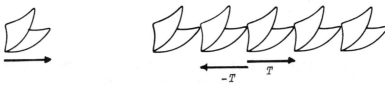

Fig. 1. Repeatedly applying the translation T to a motif creates a periodic border design.

design is just the length of the vector T. The whole border design has the property that if it is shifted by the translation T or $-T$, then motifs are superimposed on their congruent copies— to a viewer, the border would look exactly the same as before the shift.

A *periodic border tiling* is created in the same manner: begin with a single tile, and shift it repeatedly by a single translation. Since such a tiling is expected to have the tiles interlock with each other (no gaps between and no overlaps of tiles), the tile must fulfill simple conditions which are imposed by the translation. Assuming that we are translating the tile horizontally to its adjacent copy, the two tiles will interlock, or match, only if the left side of each tile has the same boundary shape as its right side. We can describe precisely how such a tile can be created (see Fig. 2). Begin with any parallelogram, and join the bottom left corner A to the top left corner B with any curve which does not intersect itself, and then translate a copy of that curve so that the copy joins the lower right corner D to the upper right corner C of the parallelogram. The bottom edge AD of the parallelogram can be taken as the vector of the translation T which carries this out. If the border is to be just one tile "high," then join B to C with any curve that doesn't intersect itself and which, after translation by T or $-T$, does not intersect the left or right edge of the tile (except, of course, at the corners). Do the same to create the bottom edge of the tile, joining A and D. The completed tile, translated repeatedly by T and $-T$, will produce a periodic border tiling.

Given a periodic border design generated by a translation T_1, if we take a second translation T_2 whose vector is not parallel to T_1 and repeatedly apply T_2 and $-T_2$ to the border, then a "wallpaper" pattern is created in which the motif repeats regularly in two directions, and the design extends throughout the plane. Such a design is called a *two-dimensional (or planar) periodic design, generated by* T_1 *and* T_2 [see Fig. 3(a)].

If we choose a point P on the original motif, and record all of the copies of P as it is repeatedly translated by T_1 and $-T_1$, T_2 and $-T_2$, then the images form an infinite array of dots, a *lattice* of points in the plane. Any two adjacent dots in this lattice represent the endpoints of a translation vector which, when applied to the whole pattern, superimposes it on itself. In fact, any vector whose endpoints are any two dots in the lattice of a periodic planar pattern will correspond to a translation which will superimpose the whole pattern on itself. This is a consequence of vector addition: the result of first applying translation T_1 and then applying translation T_2 is the same as applying a translation called $T_1 + T_2$ where the three translation vectors are related by the triangle diagram in Fig. 3(b). Similarly, $2T_1$ is a vector in the same direction as T_1, having twice the length of T_1. Successive application of several translations can produce a translation having any desired vector which joins two lattice points.

A *periodic planar tiling*, in which a single tile is repeated by translations to fill the plane, can be thought of as a periodic border tiling in which copies of the border are "carefully" stacked to fill the plane. The qualifer "carefully" refers to the exact procedure of stacking: the infinite border is translated repeatedly by the same vectors (T_2 and $-T_2$ above) to fill the plane. Of course, as with the tile border, we require that the tiles fit against each other exactly; this requires that the "top" and "bottom" of the border tiling match, as well as "left" and "right" edges of the tile. So to create such a tile, begin as before with any parallelogram and create

Fig. 2. The creation of a tile which fills out a periodic border tiling by translation.

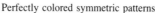

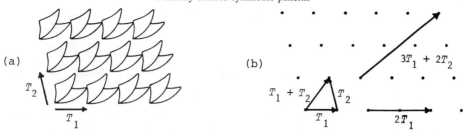

Fig. 3. (a) Independent translations T_1 and T_2, repeatedly applied to a motif to generate a wallpaper pattern. (b) The lattice of the pattern in (a), together with illustrations of the "vector addition" of translations T_1 and T_2.

left and right matching edges, using the bottom of the parallelogram as the vector T_1 to match these edges. Then, in the same manner, create matching bottom and top edges of the tile, using the left side of the parallelogram as the vector T_2 which matches these edges. As before, take care that edges of the tile (and their translates) do not actually cross. Figure 4 shows the creation of two such tiles; note that part of the left edge of the tile is used as part of the bottom edge of the tile in Fig. 4(b).

1.2 *Rotation*

A *rotation* of points in the plane moves points by turning the plane about a fixed point (called the *center of rotation*). When the angle of the turn is an integral divisor of 360°, say $360°/n$, then the rotation is called an *n-fold rotation*. This terminology emphasizes the fact that if a motif in the plane is repeated by successive rotations through an angle of $360°/n$ about a fixed center, then the motif is returned to its original position after n such turns. A design so created is said to have *n-fold rotational symmetry* (see Fig. 5).

A rotation of 360° ($n = 1$) sends each point in the plane to its original position; this isometry has the same effect as leaving each point fixed, and is called the *identity* isometry. In the special case $n = 2$ (the angle of rotation is 180°), the rotation is often called a *half-turn*, and a design created by a motif and its image under a half-turn is said to have *point* symmetry, or *central symmetry*. This terminology emphasizes the fact that in such a design, each point P on the motif has a unique corresponding point P' on the image of the motif: the two points P and P' are endpoints of a line segment having as its midpoint the fixed center of the half-turn. Figure 6 illustrates the concept of central symmetry; the "letter-form S" of Fig. 6(b) was designed by Wallace Walker.

Just as a tile which fills a border by translation is created by altering the sides of a parallelogram, a tile which fills out a "rosette" by n-fold rotations can be created by altering the sides of a circular sector (wedge) AOB whose central angle is $360°/n$. Join the vertex O of the sector to endpoint A of the arc of the sector, using any curve which does not intersect itself. Rotate the curve about O through an angle of $360°/n$, so that the image of the curve

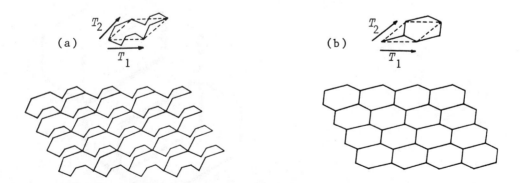

Fig. 4. By altering a parallelogram, a tile is created which fills the plane by repeated application of translations T_1 and T_2. The tile in (a) has two pairs of "parallel edges," and may be thought of as a "generalized parallelogram." The tile in (b) has three pairs of parallel edges and is called a "par-hexagon"; a tile created in a similar manner, with bent or curved edges, may be thought of as a "generalized par-hexagon."

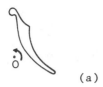 (a)  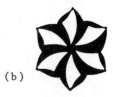 (b)

Fig. 5. (a) Rotating a motif 120° about a fixed center O creates a design with threefold rotational symmetry. (b) A design with sixfold rotational symmetry.

joins O to B. (If the image of the curve crosses the original curve, adjust the original curve and repeat.) The tile is completed by joining A to B with any simple curve. Rotating this tile $n - 1$ times about O will fill out the rosette tiling (see Fig. 7).

1.3 *Reflection*

A *reflection* of points in the plane is determined by a fixed line, called the *mirror line*, or *reflection axis*: every point not on the line is sent to its mirror image with respect to the line, and every point on the line is left fixed. (This is a "two-sided" mirror; that is, points and their mirror images literally exchange positions.) Geometrically, this means that every point P not on the mirror line is sent to a point P' (its mirror image) where P and P' are endpoints of a line segment whose perpendicular bisector is the mirror line. If the same reflection is repeated twice, then every point in the plane is sent to its original position. Figure 8(a) illustrates these remarks; the mirror line here and in all subsequent illustrations is indicated by a double line, as is cumstomary in the literature.

The repeated action of two reflections, each having a different mirror line, is more interesting. If R_1 is a reflection in mirror line m_1 and R_2 is a reflection in mirror line m_2, and m_1 is parallel to m_2, then the reflection R_1 followed by the reflection R_2 will have the same effect on points in the plane as a translation whose vector is perpendicular to the mirror lines and has length twice the distance between them. Most readers have witnessed the infinite repetition of images and their reflections which occurs when an object is placed between two parallel mirrors. Thus the design generated by repeatedly applying R_1 and R_2 to a motif is a periodic border design, in which adjacent motifs are mirror images, and every other motif is a translated image [see Fig. 8(b)].

If the mirror lines m_1 and m_2 intersect in point O, then the reflection R_1 followed by R_2 will have the same effect on points in the plane as a rotation about the point O. This phenomenon is familiar to all in the action of a kaleidoscope: when colored objects are placed between two mirrors which form an angle of $180°/n$, a beautiful symmetric repetition of the objects and their reflections fills out a circular design. The viewer sees through the peephole of the kaleidoscope n copies of matched left and right images of the original objects. Thus to create a design having this kaleidoscopic symmetry, simply place a motif between two mirror lines which intersect

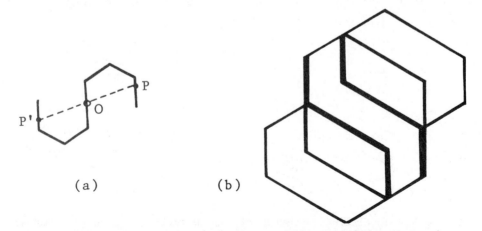

(a)   (b)

Fig. 6. (a) In a design with central, or point, symmetry, each point P can be paired with its image P': P and P' interchange places when the design is rotated 180° about its center O. (b) *Letterform S*, by Wallace Walker, exemplifies central symmetry.

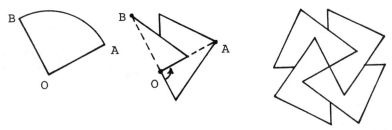

Fig. 7. A tile which fills out a rosette tiling with fourfold rotational symmetry can be created from a circular wedge with central angle 90°.

at an angle of 180°/n and repeatedly reflect the object and its reflections in the mirror lines (Fig. 9).

A single tile which fills out a rosette in this manner must necessarily be a wedge with straight sides: the sides of the tile are mirror lines of the completed rosette. The only freedom in creating such a tile is the choice of the "outside" boundary segment [see Fig. 9(c)].

1.4 Glide reflection

A *glide reflection*, as its name suggests, is a transformation of points in the plane which combines a translation (glide) and a reflection. It may be obtained by a reflection followed nonstop by a translation which is parallel to the mirror line, or by a translation followed by a reflection in a mirror line parallel to the translation vector. A single vector, called the *glide vector* (dashed in our illustrations), is often used to denote both the mirror line and the translation vector. When a single glide reflection is repeatedly applied to a motif, it creates a row of equispaced images of the motif, in which reflected and direct images alternate. The pattern is much like footprints in the sand, which are equispaced and alternate: left foot, right foot. We can observe that the effect of a glide reflection applied twice successively to a motif is the same as a translation T through twice the distance of the glide. If $-G$ denotes the glide reflection with glide vector opposite to that of G, then G and $-G$ generate a periodic border design whose translation period is $2G = T$ (see Fig. 10).

One way to create a tile which will fill out a border by successive applications of a glide reflection is to begin with a rectangle ABCD having M and N as midpoints of sides AB and CD, respectively [Fig. 11(a)]. Join corners A and B with a nonintersecting curve. Translate the curve to CD and then reflect the translated curve in the mirror line MN. The glide-reflected image of the curve will join C and D, but in the "opposite" manner: the endpoint at A will have its image at C and the endpoint at B will have its image at D. (If the glide-reflected image intersects the curve from A to B, adjust the original curve and repeat.) Join A to D and B to C with curves that do not intersect the left and right "edges," except at A, B, C, D. This tile will fill out a border by repeated application of the glide reflection with glide vector MN.

Another way to create a tile that will fill out a border by glide reflection is to begin with a rectangle ABCD having center O, with M the midpoint of AB and P the midpoint of AD [see Fig. 11(b)]. Join P to B and join B to C with nonintersecting curves. Glide-reflect these curves using as glide vector MO; the image of the curve joining P and B will join C and P, and the image of the curve joining B and C will become the bottom edge of the next tile created by repeating the glide reflection. Now fill out the border tiling by repeated application of the glide vector MO.

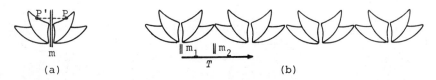

Fig. 8. (a) A reflection in mirror line m sends each point in the plane to its mirror image; if P and P' are mirror images of each other with respect to m, then the reflection in m causes them to interchange positions. (b) Repeated reflection of a motif in two parallel mirrors creates a periodic border design. Reflecting the motif first in mirror m_1 and then reflecting its image in m_2 has the same effect as translating the motif by the vector T.

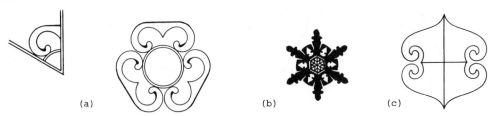

Fig. 9. (a) Repeated reflection of a motif in two mirrors which intersect in an angle of 60° creates a rosette design with threefold kaleidoscopic symmetry. (b) A snowflake has sixfold kaleidoscopic symmetry. (c) A tile created by altering the ''outside'' boundary of a circular wedge with central angle 90° can be reflected in the two straight sides of the wedge to create a rosette tiling having twofold kaleidoscopic symmetry.

Our purpose here is not to thoroughly explore the four types of isometries and their properties, but rather to give the designer an idea of how they are defined, how some of them interact, and how periodic designs and tilings can be created using the isometries. Transformation geometry is the study of isometries and their interactions; it must be thoroughly investigated to fully understand why (from a mathematical point of view) such a limited number of kinds of symmetric designs are possible. (The next section makes clear how limited the possibilities are.) There are no artistic fetters imposed *a priori* on a designer by mathematicians, but if the designer wishes to use isometries to create certain kinds of symmetry in a design, then the geometry will impose certain restrictions. Reference [27] gives an elementary introduction to transformation geometry; Ref. [3] also contains a discussion of isometries.

Designers with access to computer graphics have a powerful aid in the creation of designs using isometries. Some computer systems (even microcomputers) have ''built in'' the actions of translation, reflection and rotation. For those that do not, each of the four isometries can be represented very easily by a 3 × 3 matrix which acts on points in the plane, represented as vectors in the form $(x, y, 1)$ (the first two coordinates are the usual cartesian coordinates of the point; see, for example, [16]). Thus, utilizing ''hardware'' or ''software,'' or both, a designer can enter a motif into the computer, have the computer perform transformations on the motif to create a design, and then scrutinize the result and keep it (print or store), adjust it, or destroy it. Using the computer as a scratch pad can allow for a great deal more experimentation than might be possible producing various trial designs by hand.

2. SYMMETRY GROUPS AND THE CLASSIFICATION OF DESIGNS

Mathematicians (and many scientists) classify a design or tiling according to its *symmetry group*: this is the collection of all isometries which, when applied to the design or tiling, create an image which is superimposed exactly on the original so that, to the eye, it seems as though no transformation has taken place. (The design is said to be *invariant* under these isometries.) Isometries which leave a design invariant are often called symmetries of the design; thus the mathematicians have claimed the descriptive adjective ''symmetry'' as the noun which names the action (transformation) which creates the symmetric properties of the design. The word ''group'' is not only descriptive of a certain collection of isometries, but is also used in its technical mathematical sense. Although the details of group structure are not essential for our descriptive purpose, it is this algebraic structure which underlies the classification theory.

Classifying designs by symmetry groups is not especially useful from an artistic point of view—this method of identification tells us little about the artistic merit or richness of the

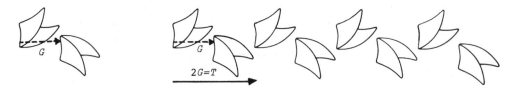

Fig. 10. Repeated application of a glide reflection G to a motif creates a periodic border design. Two successive applications of G to the motif has the same effect as translating the motif by the vector T.

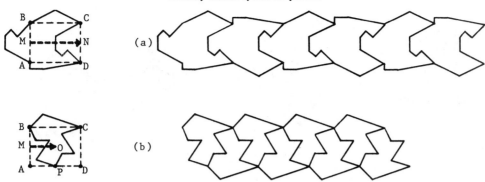

Fig. 11. Two ways of creating a tile which fills out a periodic border by repeated application of a glide reflection. Note that although the same rectangle ABCD guides the formation of the tiles in both (a) and (b), the glide vector in (b) is half the length of the glide vector in (a).

design. For example, the symmetry group of a simple asterisk ∗ is the same as that of the snowflake rosette design in Fig. 9(b), and, even worse, every asymmetric design (one which is not invariant under any isometry except the identity) has the same symmetry group, namely, the group consisting only of the identity isometry. To be fair in our criticism, we should note that when this system of classification is used in a fairly limited context of design, such as designs on pottery in a stratum of an archeological excavation, it can be very useful in grouping similar designs (for example, see Ref. [25]).

Our emphasis here is on the creation of designs through the use of isometries, rather than the classification of completed designs. The designer controls the artistic "input" by creating the motif or tile and then controls the symmetry of the design or tiling by choosing particular isometries to repeat the motif or tile. However, by using isometries to create designs and tilings, the end product will have its symmetry group generated by the same isometries that generate the design—the classification is free. We need to make the word "generate" (used in two ways in the previous sentence) more precise in order to carefully describe the different symmetry groups which are possible.

Every designer wants to use a minimal number of "ingredients" to create a design which will be made up of repeated images of a single motif or tile. There are three distinct categories of designs which were introduced in Section 1: rosette designs, periodic border designs, and periodic planar designs. The corresponding tilings of each type were also introduced. For each of these categories, we want to define the concept of *generating region* for a design, and *minimal set of generators* for a design.

Rosette designs

In order to be able to refer unambiguously to the area occupied by a rosette design, we think of such a design as being inscribed in a circle, centered at O, where the radius of the circle can be as large as needed. If the design is to be created by rotating a motif n times about O, then a wedge (circular sector) having angle $360°/n$ at O is a minimal area in which to place the motif. (Think of the motif as carved in a wooden block for a circular print: the finished print must fill out the circular area, with n equispaced images of the motif swirling about O.) The single rotation R of $360°/n$ about O, acting repeatedly on the wedge, will produce the completed rosette design [Fig. 12(a)]. We also assume that the motif chosen is "minimal"; that is, there is not a smaller portion of the motif which, when acted on by rotations about O, will produce the whole design. The wedge will be called a generating region for the design (it is a smallest area of the plane whose images fill out the whole circle containing the design). The single rotation R is a generator of this design: repeatedly applying R to the generating region containing the motif creates the whole design. The symmetry group of this design consists of the n rotations which are multiples of R; the notation for the symmetry group is C_n (the cyclic group of n elements).

The wedge (with straight sides) is not the only generating region possible for such a design; in fact, if the motif is very bent or curved, a better generating region would be a "wedge" with bent or curved sides, created in a manner similar to that shown in Fig. 7 [see Fig. 12(b)].

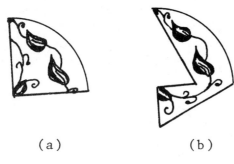

(a)   (b)

Fig. 12. A rosette design with fourfold rotational symmetry has symmetry group C_4. The wedge in (a) is a generating region for the design; the altered wedge in (b) is another generating region for the same design [note that the generating region (b) "fits" the motif better].

Even though the shapes of the two generating regions shown in Fig. 12 is different, note that their area is the same. This will be true in general: any two generating regions for a design will have the same area.

If the rosette design is to have n-fold kaleidoscopic symmetry, a wedge having angle $180°/n$ at O is the most natural to choose as the generating region in which to place the motif. The two reflections R_1 and R_2, whose mirror lines are the sides of the wedge, will generate the design; that is, repeated reflection of the motif and its images in these two mirror lines will fill out the whole rosette design. If we assume that the motif does not have mirror symmetry (if it did, it could be split along its symmetry axis and just half of it used) then two isometries are necessary to create the whole design. Thus for this type of design, a minimal set of generators will contain two isometries. Although the *number* of isometries in a minimal set of generators for a design is unique, the *choice* of these isometries is not always unique. For example, two other isometries which generate the design in Fig. 9(a) are the reflection R_1 and a rotation of 120° about O.

In general, the symmetry group of a design having n-fold kaleidoscopic symmetry consists of the n rotations in C_n, together with n reflections, where the angle between adjacent mirror lines is $180°/n$. The group is denoted D_n, and is called the dihedral group with n mirrors. Each of the $2n$ isometries in D_n is a "product" (a finite combination) of the two isometries in a minimal set of generators for the design. Thus we say that these two isometries generate the symmetry group D_n. By means of these definitions, the generators of the design and the generators of the symmetry group can be considered the same.

Rosette tilings

For rosette tilings, we need not consider an (artificial) circle surrounding the tiling; the edge of such a tiling provides its own well-defined encircling boundary. Thus a generating region for a tiling with rotational symmetry or with kaleidoscopic symmetry will be a smallest tile which, when acted on repeatedly by the generating isometries, fills out the whole tiling. The tiles whose congruent copies fill out the rosette tilings shown in Figs. 7 and 9(c) are generating regions for those tilings. Minimal sets of generators for rosette tilings are the same as those for rosette designs having the same symmetry group.

Periodic border designs

We will consider these designs as enclosed between two parallel lines (the edges of the border), that is, enclosed in a strip of finite width and infinite length, and having centerline L which is equidistant from the edges. To be called "periodic," such a border design must be generated by a finite set of isometries which includes a translation, and all of which superimpose the strip on itself. It is easy to discover that the only isometries that can be used to generate a periodic border design are translations (with vector parallel to L), glide reflections (with glide vector on L), reflections (with mirror line L, or with mirror lines perpendicular to L), and half-turns (about centers which lie on L). Although it might seem that there could be a great variety of combinations of these isometries to create different border designs, the interactions are actually quite limited: there are just seven distinct symmetry groups which are generated by combinations of these isometries. [By distinct, we mean (roughly) that the groups contain different collections of isometries.]

The simplest periodic border design is that created by repeatedly applying a translation T (and $-T$) to a motif, as in Fig. 1. If we think of stamping out such a design with the motif engraved in a linoleum block, then a generating region (the block) can be taken as a parallelogram with base equal to the vector of T and height the width of the strip. We will call this particular generating region a *translation unit*, to indicate that it is a smallest region which, when translated repeatedly by T and $-T$, produces the whole border design. (This assumes that the motif is "minimal" with respect to translation symmetry; that is, no smaller portion of the motif could produce the whole design when acted on repeatedly by some shorter translation vector.) We say that T generates the design, and T generates the symmetry group of the design, which will consist of all translations of the form mT, m an integer. Of course we could also take as a generating region an "altered" translation unit: let a curved boundary replace the left edge of a translation unit, and translate the boundary, using the vector of T, to a matching right edge. The shapes of the generating regions are different, but areas remain equal.

Every periodic border design will have regular repetitions of its motif by translations; the shortest translation vector which leaves the design invariant will be taken as the base of the translation unit for that design. For example, the design generated by a single glide reflection G shown in Fig. 10 has the translation vector of $2G$ as its shortest translation vector. A generating region for a periodic border design will be a smallest region of the border which, when acted upon repeatedly by the generators of the design, will create the whole border design. For example, a generating region for the design just mentioned, generated by G, can be taken as half a translation unit, since G acting on this generating region containing the motif will create the whole design.

We have organized the information which describes each of the seven types of periodic border designs in Table 1. There is no completely standard notation for these seven symmetry groups, but we have used notation found in several of the references. For each type of design we give the size of a generating region (relative to a translation unit), the number of isometries in a minimal set of generators (and the possible choices of isometries for these sets). Two of the seven borders are generated by a single isometry, four borders have a minimal set of generators containing two isometries, and one border requires three isometries as generators. Table 1 also contains a description of the isometries in each symmetry group, and shows a motif placed in a typical generating region and a portion of the border design with that motif. The placement of half-turn centers and mirror lines and lengths of glide vectors is important, and is evidence of the fact that the "product" of two isometries is equivalent to another isometry. In Table 1, the length of a minimal translation vector is the same for all seven borders, and all translation units have equal area. Using this table, a designer can choose a motif, decide the symmetry type of design to create, place the motif in a suitable generating region and use a minimal set of generators (properly positioned) to create the desired border design.

Border tilings

For periodic border tilings, we need not enclose the tiling in an artificial strip of finite width: the edges of the tiling outline a well-defined area of the plane which is covered by the tiling. Thus a generating region for such a tiling will be a tile of minimal area, which, when acted on repeatedly by a set of isometries, fills out the whole border. A translation unit in such a tiling is a minimal block of tiles which fills out the whole border by translations alone. Table 2 parallels Table 1, and gives information on the creation of periodic border tilings, with an illustrative example for each border type.

Periodic planar designs

In Sec. 1, we described how two independent translations T_1 and T_2, applied repeatedly to a motif, created a wallpaper pattern, and how the action of these translations on a single point of the motif created a lattice of points in the plane (see Fig. 3). The parallelogram having as its sides the vectors T_1 and T_2 will be called a *lattice unit*; this is a generating region for this design which has only translation symmetry. The translations T_1 and T_2 can be taken as a minimal

Table 1. The seven distinct types of periodic border design.

Border type	Symmetries	Translation Unit with generating region (shaded)	Area gen. reg.	Minimal number generators	Minimal sets of generators	Generating region with motif	Border
11	translations		1	1	T		
1g	translations, glide-reflections		1/2	1	G		
12	translations, half-turns		1/2	2	H_1, H_2 H_2, T		
m1	translations, reflections perpendicular to edge		1/2	2	R_1, R_2 R_2, T		
1m	translations, reflection in L, glide-reflections		1/2	2	L, T $T, G = T+L$ $L, G = T+L$		
mg	translations, half-turns, glide-reflections, reflections perpendicular to edge		1/4	2	R, G H, G R, H		
mm	translations, half-turns, glide-reflections, reflection in L, reflections perpendicular to edge		1/4	3	L, R_1, R_2 L, H, R_2 L, G, H^2 L, T, H $L, H, L+R_1$ L, G, H^2 T, G, R_2 T, G, H^2		

Table 2. The seven symmetry types of periodic border tilings. The left column contains notation for the symmetry group and the center column shows a generating region together with a minimal set of generators for each tiling.

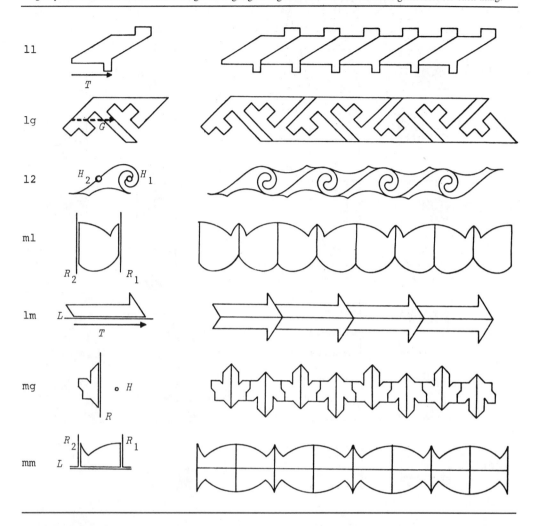

set of generators for the design (we assume that the motif is "minimal," that is, no smaller portion of the motif will generate the whole design when acted on by some set of isometries).

A wallpaper design having only translation symmetry is the simplest periodic planar design. Additional isometries can be used to create periodic planar designs having more complex symmetry, but the choices for such isometries, both in type and in number, are again very limited. Only n-fold rotations, where $n = 2, 3, 4$, or 6, and selected reflections and glide reflections can also be symmetries of periodic planar designs, and there are just 17 distinct symmetry groups of such designs. (The proof that there are 17 distinct symmetry groups utilizes both geometry and group theory; see, for instance, Refs. [4] and [6].) Every periodic planar design must have in its symmetry group two "shortest" independent translations (these correspond to the periodic nature of the design): a *lattice unit* for such a design is a parallelogram having as its sides the vectors of these two translations (many authors refer to this parallelogram as a unit cell). A *translation unit* for the design will be a minimum area of the plane, which when acted on repeatedly by these two translations, fills out the whole design. The shape of a translation unit can vary, but for a given design, any translation unit has its area equal to the area of a lattice unit. For periodic planar designs, a generating region will be a minimal portion of the plane which, when acted on repeatedly by the symmetries of the design, fills out the whole design. For each of the 17 types of design, all generating regions for that design will have the same area (but not necessarily the same shape), and this will be a fraction of the area of a lattice unit for that design. (Many of the mathematical references use the term fundamental domain for what we call generating region.)

Table 3. For each symmetry group, the left figure shows a lattice unit containing a generating region (shaded) and a minimal set of generators. The right figure shows a lattice unit with all symmetries that occur within and on the boundary of that lattice unit. The translations that are symmetries of the pattern are generated by vectors which form the sides of the lattice unit.

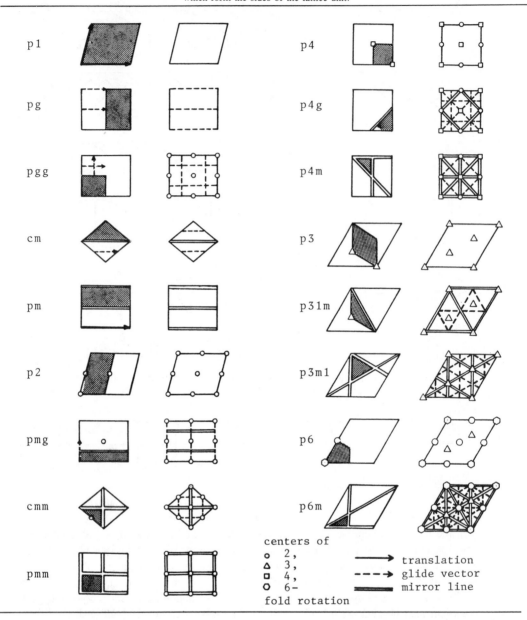

centers of
- ○ 2,
- △ 3,
- □ 4,
- ◯ 6-fold rotation

- ⟶ translation
- ----→ glide vector
- ━━━ mirror line

Reference [18] explains the notation for the plane symmetry groups; we summarize here, in Table 3, information which illustrates the concepts introduced above. Table 3 shows, for each of the 17 types of design, a lattice unit containing a shaded generating region and a minimal set of generators for that design. The full set of symmetries relative to a lattice unit for each design is also shown. (Of course, the minimal sets of generators shown in Table 3 are not unique; finding all of the different minimal sets of generators for each symmetry group, as we have done for the border designs in Table 1, is an interesting and educational exercise.) By placing a motif in a generating region and acting repeatedly on the region by the generating isometries, a designer can create a periodic planar design having the specified symmetry group.

Periodic plane tilings

A simple way to create a tile which fills the plane by successive translations in two directions was described in Sec. 1: carefully alter a parallelogram. Tiles with simple geometric shapes (triangles, rectangles, diamonds, hexagons) can be used to create tilings having more symmetries than just translations. But much more variety is possible: the fanciful animate tiles of M. C. Escher are, in most cases, generating regions of periodic plane tilings (see Ref. [14]). Reference [19] contains a recipe for creating a single tile (with enormous freedom in creating its boundary) which will fill the plane using as generators three half-turns. A great variety of tilings is displayed in the papers by Grünbaum and Shephard; a single illustrative tiling for each of the 17 symmetry types is contained in [18]. (A word of explanation for these references: an isohedral, or tile-transitive, tiling of the plane is one in which a single tile, acted on repeatedly by a set of generators for the symmetry group of the tiling, fills out the whole tiling. In some cases, this single tile may be larger than a generating region; that is, part of the tile may be sufficient to create the whole tiling.)

The minimal "ingredients" for a periodic plane tiling are the same as for a periodic planar design. A generating region will be a tile of minimal area whose images fill out the whole plane when acted on repeatedly by generators of the symmetry group. A translation unit will be a minimal block of tiles which fills out the whole plane tiling using only translations. To illustrate the variety of shapes which can be chosen for generating regions, we show, in Table 4, for each of the 17 symmetry groups a generating region together with generating isometries and a translation unit for the corresponding tiling. The boundaries of the generating regions are created by "altering" the simple shaded generating regions shown in Table 3, using the generating isometries to create boundaries of the tile that are compatible with the action of these isometries. For all but three symmetry groups (the exceptions are pmm, p3m1 and p4m) we have been able to choose as generating regions tiles which have no inherent symmetry, and hence the tilings they create have as symmetries only those isometries which are products of the generating isometries.

3. PERFECT COLORING IN BLACK AND WHITE

Many of the most striking periodic designs in decorative art are colored in black and white. There are many likely reasons for this: aesthetic (stark contrast), psychological (emphasize figure and ground), economic (black ink, white paper), and others. A balance of black and white is sought if the design is to have the two colors reflect its periodic nature. Thus, if we think of a periodic tiling generated by a single tile (rosette, border, or plane-covering), and wish to color each tile either black or white, a balanced 2-coloring would have equal numbers of black and white tiles. If, in addition, the symmetries of the (uncolored) tiling are to be emphasized by the coloring, then the individual tiles should be colored white or black in such a way that each symmetry of the uncolored tiling either:

(i) transforms all white tiles to white tiles and all black tiles to black tiles, or
(ii) transforms all black tiles to white tiles and all white tiles to black tiles.

A symmetry of the colored tiling which satisfies (i) or (ii) is called a *two-color symmetry* of the tiling. When every symmetry of the uncolored tiling is also a two-color symmetry of the colored tiling, we say that the coloring is *perfect*, or *compatible*. (Mathematically, each symmetry of the uncolored tiling induces a permutation of the colors in the colored tiling.)

To illustrate these ideas, we have colored the border tiling of type 1m from Table 2 in black and white in three different ways (see Fig. 13). Each of these coloring is "balanced"; there are equal numbers of black and white tiles (assume that the color scheme repeats in the same manner *ad infinitum*, as the border extends to the left and right). Of these three colorings, only Fig. 13(a) is perfect: the reflection L and translation T interchange black and white tiles and the glide reflection $L + T$ preserves colors of tiles, and all products of these isometries satisfy (i) or (ii). In coloring 13(b), while L is a two-color symmetry (interchanging black and white), both $L + T$ and T are not color symmetries—both preserve some colors and interchange some colors. In coloring Fig. 13(c), all three symmetries L, T, and $L + T$ of the uncolored tiling are not color symmetries—all preserve some colors and interchange others.

Looking at coloring 13(b), we can see that the translation with vector $2T$ and a glide reflection with glide vector twice the length of the vector of $L + T$ are two-color symmetries, and so 13(b) could be a perfect coloring for the uncolored tiling whose generating tile is a block

D. SCHATTSCHNEIDER

Table 4. Column 1: symmetry group; column 2: minimal number of generators; column 3: a generating region and one minimal set of generators; column 4: area of a generating region relative to a lattice unit; column 5: a block of tiles which forms a translation unit; column 6: a lattice unit for the tiling.

Symmetry Group	Generators	Gen. Region	Area	Trans. Unit	Lattice Unit
p1	2		1		
pg	2		1/2		
pgg	2		1/4		
cm	2		1/2		
pm	3		1/2		
p2	3		1/2		
pmg	3		1/4		
cmm	3		1/4		
pmm	4		1/4		

of two consecutive "half-arrow" tiles. Thus in order to decide whether a coloring is perfect or not, one must first decide what will be considered as the symmetry group of the uncolored tiling. (A mathematical observation: enlarging the generating region of a tiling amounts to taking as the symmetry group of the uncolored tiling a proper subgroup H of the full symmetry group of the uncolored tiling. A coloring of the original tiling which is not compatible with the full symmetry group of the tiling may be compatible with the smaller symmetry group H.) No reinterpretation of a generating region for the uncolored tiling can make 13(c) a perfect coloring, since the reflection L is a symmetry of any uncolored "block" of the arrow tiles, and L "mixes colors" in Fig. 13(c)—keeps some colors the same and interchanges others.

Creating a periodic tiling and perfectly coloring it in black and white is a very easy task, if a designer utilizes the "minimal ingredients" we have defined in Sec. 2. First, create a tile which is a generating region for the tiling, and choose a minimal set of generators which will produce the whole tiling. Decide the color of the generating tile and also decide, for each isometry in the minimal set of generators, if it is to preserve colors [it satisfies (i)] or if it is to reverse colors [it satisfies (ii)]. This makes each generating isometry a color symmetry of the tiling. Then act repeatedly on the single tile by the color isometries in the minimal set of generators—the whole tiling, with its perfect coloring, results. The fact that we have chosen a *minimal* set of generators to create the tiling and induce the coloring guarantees that every symmetry of the uncolored tiling will be a color symmetry—no isometry which is a symmetry of the uncolored tiling can "mix" colors in the colored tiling. This is because every symmetry of the uncolored tiling is a product of isometries in a minimal set of generators, and none of

Table 4. (*Continued*)

Symmetry Group	Generators	Gen. Region	Area	Trans. Unit	Lattice Unit
p4	2		1/4		
p4g	2		1/8		
p4m	3		1/8		
p3	2		1/3		
p31m	2		1/6		
p3m1	3		1/6		
p6	2		1/6		
p6m	3		1/12		

the isometries in a minimal set of generators can be written as a product of other isometries in the same minimal set.

This technique of creating perfectly colored black and white periodic tilings is especially adaptable to computer graphics implementation. Either using hardware or software, a computer graphics program can specify that a region be "filled" in black, or left uncolored. Thus, in having the computer produce the images of a single tile by transforming it by generating

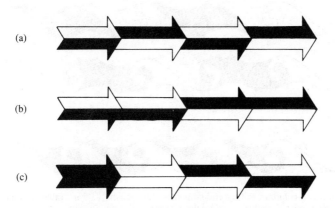

Fig. 13. Three two-colorings of periodic border tiling 1m from Table 2. Only coloring (a) is perfect.

isometries, one can attach to each generating isometry a command that will "color" the tile black or white.

For two given perfect colorings of the same (uncolored) periodic tiling, interpretation of the phrase "have the same two-coloring" has different meanings—certainly among designers, and even among mathematicians! Both designers and mathematicians can agree that if a complete interchange of colors (black ↔ white) of one coloring produces the other coloring, then the two colorings are "the same." After all, one is a positive and the other a negative of the same coloring. But if a complete interchange of colors will not produce the same coloring, most designers and some mathematicians do not want to consider the colorings as the same. Most mathematicians consider two perfect colorings to be the same if the color symmetry groups are "the same," i.e. they are generated by the same types of isometries. To illustrate what this means, let us adopt the convention of putting a prime after the symbol for an isometry if that isometry is to interchange colors [it satisfies (ii)]. Thus the symbol H_1 for the half-turn which is one of the two generators for the border tiling 12 in Table 2 remains H_1 if the color of the image of a tile when H_1 acts on it is the same as the original tile (H_1 preserves color), and the symbol H_1' signals the fact that the color of the image of a tile when H_1 acts on it is the reverse of the color of the original tile (H_1' sends a black tile to a white tile and a white tile to a black tile). In some literature on color symmetry, H_1 would be called a symmetry of the colored tiling and H_1' called an antisymmetry of the colored tiling. The two half-turns H_1 and H_2 generate the border tiling 12 in Table 2. We can choose to have H_1 preserve or reverse colors, and we can (independently) choose to have H_2 preserve or reverse colors. If both H_1 and H_2 preserve colors, then all tiles in the border they generate will have the same color—we have the single color (or uncolored) tiling as in Table 2. If H_1 preserves and H_2 reverses colors, then we get the two-coloring shown in Fig. 14(a) (generators H_1 and H_2'); if H_2 preserves and H_1 reverses colors, we get the two-coloring shown in Fig. 14(b) (generators H_2 and H_1').

A mere interchange of black and white will not make these two colorings the same. To a designer, the black blocks of two tiles and white blocks of two tiles in the two two-colorings are distinct. But to the mathematician, these two two-colorings are "the same" in the sense that their color symmetry groups are "the same": each group is generated by one color-reversing and one color-preserving half-turn. A designer can see that each of these colored tilings can be described as consisting of alternating black and white blocks (of two tiles) where the blocks have central symmetry. So even from a design point of view, one could accept the classification of these two colored designs as "the same."

The fourth possible coloring of the tiling is generated by H_1' and H_2', and is shown in Fig. 14(c). This coloring is distinct, both artistically and mathematically, from the others. So we can summarize the "different" perfect colorings (using two colors) of this single tiling as follows: there is one single-color (or uncolored) coloring; artistically, there are three colorings using both colors, and mathematically, there are two colorings using both colors (there are two distinct two-color symmetry groups).

Fig. 14. Three perfect two-colorings of the periodic border tiling of type 12 from Table 2. Although visually all three appear different, mathematicians consider the colorings in (a) and (b) to be "the same"; that is, they have the "same" two-color symmetry group.

The different perfect colorings for rosette tilings are easily described. For a tiling having only n-fold rotational symmetry (symmetry group C_n), no perfect coloring using two colors is possible if n is odd, and there is just one perfect coloring using two colors if n is even. For a tiling having n-fold kaleidoscopic symmetry (symmetry group D_n), for $n > 1$, there are three perfect colorings using two colors, but only two distinct two-color symmetry groups. (Recall that D_n is generated by two reflections R_1 and R_2 which intersect at an angle of $180°/n$; each of these can be chosen to be color-preserving or color-reversing.) We illustrate these remarks in Fig. 15, which shows two rosette tilings and their perfect two-colorings. Colored tilings which have the same color symmetry group are bracketed.

In Table 5, we illustrate the different perfect two-colorings of border tilings from Table 2, indicating their generators. When two colorings have the same two-color symmetry group, we have bracketed them. There are 17 distinct two-color symmetry groups of the periodic border tilings; if we add the 7 single-color tilings, we see that (from the point of view of symmetry groups), there are 24 distinct ways to perfectly color a periodic tiling using at most two colors.

Perfect two-colorings of periodic plane tilings can be created in the same manner as we have described for border tilings, utilizing information on generating regions and minimal sets of generators for these tilings (Table 4). We illustrate in Fig. 16 the three possible perfect two-colorings of the p1 tiling in Fig. 4(a). The colorings in Fig. 16 are a good illustration of how vastly different are the artistic and mathematical interpretations of the concept of "same coloring." The colorings of Figs. 16(b) and 16(c) give the visual effect of alternating black and white strips of tiles which fill the plane—if direction is disregarded, they could be considered "the same type" of coloring. However, coloring 16(a) is distinctly different from those in (b) and (c), a checkerboard pattern of black and white tiles in which the outline of each tile is clearly visible. Artistically, coloring (a) must be considered as distinct from those in (b) and (c). But all three colorings have the "same" color symmetry group—a group generated by one color-preserving and one color-reversing translation. (The color symmetry group of coloring 16(a) is not only generated by T_1' and T_2', but also generated by T_1' and $T_3 = T_1' + T_2'$, and T_3 is a color-preserving translation.) Thus the mathematical literature will indicate that there is just one two-color symmetry group for a p1 tiling.

Fig. 15. (a) The only perfect two-coloring of the C_4 rosette tiling in Fig. 7. (b) Three perfect two-colorings of the D_2 rosette tiling in Fig. 9(c). The two colorings which are bracketed have the same two-color symmetry group; visually, these two colorings have the same property of having a mirror line divide the tiling into a black half and a matching white half.

Table 5. All possible perfect two-colorings of the tilings in Table 2. Generators for each coloring are given, and colorings having the same two-color symmetry group are bracketed.

Border Type	Distinct 2-color groups	Generators of 2-coloring	
11	1	T'	
1g	1	G'	
12	2	$\begin{cases} H_1,\ H_2' \\ \\ H_1',\ H_2 \end{cases}$ $H_1',\ H_2'$	
m1	2	$\begin{cases} R_1,\ R_2' \\ \\ R_1',\ R_2 \end{cases}$ $R_1',\ R_2'$	
1m	3	$L,\ T'$ $L',\ T$ $L',\ T'$	
mg	3	$R,\ H'$ $R',\ H$ $R',\ H'$	

Table 5. (*Continued*)

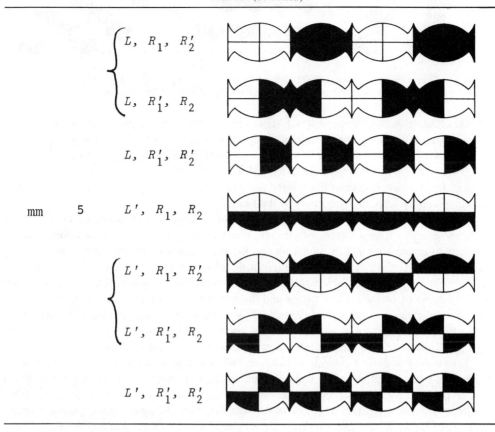

		$L, \ R_1, \ R_2'$
		$L, \ R_1', \ R_2$
		$L, \ R_1', \ R_2'$
mm	5	$L', \ R_1, \ R_2$
		$L', \ R_1, \ R_2'$
		$L', \ R_1', \ R_2$
		$L', \ R_1', \ R_2'$

The shape of a generating tile, and the way in which that tile is surrounded by adjacent tiles, are extremely important factors in determining the visual effect of a two-coloring. The tiling in Fig. 4(b) can be generated by exactly the same translations T_1 and T_2 which generate the tiling in Fig. 4(a), but the hexagon tile is shaped so that three, rather than four, tiles meet at a point. Three perfect two-colorings for this tiling are shown in Fig. 17; all give the visual impression of alternating strips of black tiles and white tiles. Thus to say that these three colorings have the "same two-color symmetry group" is in better agreement with the visual effect of the tilings.

These examples point out the need for a classification of perfectly two-colored tilings which gives consistent information about the visual impression of colored tilings of the "same type." For uncolored tilings, Grünbaum and Shephard have provided a classification of tilings by "isohedral type" which is much finer than the classification by symmetry group, and takes into account the manner in which each tile is surrounded by congruent copies of itself[7]. A finer classification of colored tilings is contained in their forthcoming book, *Patterns and Tilings* (W. H. Freeman).

Another artistic concern in coloring with two (or more) colors is ignored by the mathematical technique of perfectly coloring periodic tilings. Most artists (and many mathematicians) wish to have the shape of each individual tile visible; thus no two tiles having the same color should share a boundary. (This is often termed the *map-coloring restriction*: map makers must color

(a) (b) (c)

Fig. 16. The three two-colorings of the p1 tiling in Fig. 4(a). Coloring (a) is generated by T_1' and T_2'; coloring (b) is generated by T_1 and T_2'; coloring (c) is generated by T_1' and T_2.

 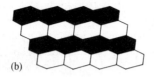 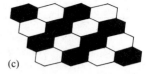

(a)   (b)   (c)

Fig. 17. Three perfect two-colorings of the tiling in Fig. 4(b). Colorings (a), (b) and (c) are generated by the same translations which generate the corresponding colorings of the tilings in Fig. 16.

adjacent countries with different colors.) Many perfect two-colorings do not satisfy this restriction; note that of the six perfectly colored tilings shown in Figs. 16 and 17, only one of these is "map-colored." [A minimum of three colors is required for any map-coloring of the tiling in Fig. 4(b).] In coloring his periodic tilings, M. C. Escher insisted that this restriction be met; his own classification system of his colored tilings emphasizes this property to a greater extent than compatibility with all symmetries of the tiling.

We do not provide here an illustration for each of the possible two-colorings and two-color symmetry groups of the 17 types of periodic plane tilings. However, our description of the technique to create a perfect two-coloring and the information in Table 4 should make this a relatively easy exercise. A word of caution: a threefold rotation cannot be a color-reversing isometry. We summarize in Table 6 the number of distinct two-color symmetry groups for each of the 17 types of periodic plane tilings and the number of different ways to create a perfect two-coloring using minimal sets of generators for these tilings. Although there are 46 (mathematically distinct) two-color symmetry groups, the number of "artistically distinct" colorings is far greater than this (and this number is open to a variety of interpretations).

The concept of perfect coloring in black and white can also be applied to periodic designs (rather than tilings), but there is a degree of ambiguity as to how the design is colored which must be resolved by the designer in order to apply the coloring technique we have described for tilings. The ambiguity arises because of the figure-and-ground aspect of these designs which is not present in the tilings, in which each tile is colored a single color. The motif is a "figure," placed on a (back)ground, and the designer may or may not choose to consider the ground as part of the design to be perfectly colored. If only the motif is colored, and the ground remains "neutral," then color-reversing isometries will only change the color of the motif, and not affect the ground at all (in effect, the ground is treated as a third color, and is not changed by any isometry of the design). If the ground is considered as part of the design, it is normally colored in contrast to the motif—the figure is against the ground. In this case, color-reversing isometries will change the color of both figure and ground: white against black becomes black against white. (Some of the most confusing illustrations that occur in the literature on two-colored designs and tilings is a result of not making clear the role of the white vs the black

Table 6.

Symmetry Group	Number of Two-Color Groups	Number of Ways of Creating a Perfect Two-Coloring Using a Minimal Set of Generators
p1	1	3
pg	2	3
pgg	2	3
cm	3	3
pm	5	7
p2	2	7
pmg	5	7
cmm	5	7
pmm	5	7
p4	2	3
p4g	3	3
p4m	5	7
p3	0	0
p31m	1	1
p3m1	1	1
p6	1	1
p6m	3	3

portions of a design—is the white merely background which does not change color, or is some or all of the white a part of the design which will change to black under a color-reversing isometry?) There can even be a "mixed" interpretation of black and white in a two-color design; for example, the motif will contain portions of white and portions of black, and will be outlined in black against a white ground. A color-reversing isometry will leave the ground white and the outline black, but reverse the coloring of the black and white portions of the motif. There is no ambiguity in creating such a design, since (presumably) the designer has a clear idea of which portions of the motif and ground that the color-reversing isometries are to affect, and which they are to ignore. However, classification of the resulting two-color design by color symmetry group (without the knowledge of the assumptions made by the designer) is an ambiguous task.

We illustrate these comments in Fig. 18 with three periodic border designs, each of which has been perfectly colored, using the same set of generating isometries, but with differing interpretations of figure and ground.

4. PERFECTLY COLORING DESIGNS AND TILINGS WITH THREE OR MORE COLORS

This section is necessarily brief—in a word, one cannot avoid learning a bit of group theory in order to understand how what we have described for two colors can be extended in a very nice way to create periodic tilings (and designs) colored with several colors, in which the coloring is compatible with the symmetries of the uncolored tiling. If we begin with a periodic tiling of the plane, which has been generated in the manner described in Sec. 2, and the tiles are each colored with a single color and k colors are used in the coloring, such a k-coloring is called *perfect* (or *compatible*) if each symmetry of the uncolored tiling induces a permutation of the k colors. By this, we mean that all tiles having the same color (for instance, red) are transformed by a symmetry of the tiling to tiles all having the same color (for instance, blue). (Although a different symmetry may send the red tiles to tiles of a different color, say green, no symmetry of the uncolored tiling may "mix" colors, sending some red tiles to blue and other red tiles to green.)

Although simple perfect colorings of some tilings can be discovered by experimentation (or using some "obvious" colorings), group theory is the key to finding all such perfect colorings. Each subgroup S of the symmetry group G of the uncolored tiling, where the index of S in G is k, determines a perfect k-coloring. The symmetries in S act on one chosen tile which is colored a particular color (say white) and all of the images of that tile under the isometries in S are also colored white. Each of the k cosets of S in G is identified with one of the k colors, and the isometries in that coset are used to transform the original (white) tile to

Fig. 18. Three perfectly two-colored border designs of type p1 which differ in interpretation of figure and ground.

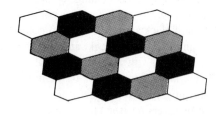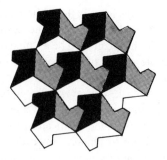

Fig. 19. A perfect three-coloring of the tiling in Fig. 4(b), and a perfect three-coloring of the p3 tiling in Table 4.

images having the color attached to the coset. In this way, the tiling is colored with the k colors, and the technique of coloring guarantees that the k-coloring is perfect.

We can illustrate these ideas by examining the three different perfect two-colorings of the border of type mg shown in Table 5. To be concrete, we specify the generating tile of the border as the first tile at the left side of the portion of the border depicted in the table. In the first coloring (generated by R, H'), the subgroup S of symmetries which sends the generating tile to white tiles is generated by the reflection R and the translation T; the index of S in the symmetry group G of the tiling is 2. The coset of S in G which contains the half-turn H contains all the symmetries which send the generating tile to black tiles (this coset consists of all of the half-turns and glide reflections which are symmetries of the border). In the second coloring (generated by R', H), the subgroup S to which the color white is attached is generated by the half-turn H and translation T. This subgroup S also has index 2 in G and the coset of S in G to which the color black is attached consists of all of the reflections and glide reflections which are symmetries of the border. In the third coloring (generated by R', H'), the subgroup S to which the color white is attached is generated by the glide reflection $R + H$; again, S has index 2 in G. The coset of S in G to which the color black is attached consists of all half-turns and reflections which are symmetries of the border.

Reference [2] gives an elementary introduction to group theory; Refs. [8], [21] and [22] are perhaps the most readable of the many papers on color symmetry. Reference [20] gives an excellent overview of the evolution of the concepts and techniques put forth by many investigators as the theory of color symmetry developed. Reference [26] is a fairly technical (mathematical) work, but the most systematic treatment available, and indicates how the process of describing and enumerating the k-color groups can be carried out with the aid of a computer.

Our last illustration, Fig. 19, shows two perfect three-colorings: a coloring of the tiling for which the perfect two-colorings in Fig. 17 were not map-colorings, and one of our p3 tilings from Table 4, for which no perfect two-coloring exists.

REFERENCES

1. W. A. Bentley and W. J. Humphreys, *Snow Crystals*. McGraw-Hill, New York (1931); Dover, New York (1962).
2. F. J. Budden, *The Fascination of Groups*. Cambridge Univ. Press (1972).
3. H. S. M. Coxeter, *Introduction to Geometry*, 2nd Edn. Wiley, New York (1969).
4. H. S. M. Coxeter and W. O. Moser, *Generators and Relations for Discrete Groups*, 2nd Edn. Springer-Verlag, Berlin (1964).
5. J. Encisco, *Design Motifs of Ancient Mexico*. Dover, New York (1953).
6. L. Fejes Tóth, *Regular Figures*. Pergamon Press, New York (1964).
7. B. Grünbaum and G. Shephard, The eighty-one isohedral tilings of the plane. *Math. Proc. Cambridge Phil. Soc.* **82**, 177–196 (1977).
8. B. Grünbaum and G. Shephard, Perfect colorings of transitive tilings in the plane. *Discrete Math.* **20**, 235–247 (1977).
9. *Handbook of Designs and Motifs*. Tudor, New York (1950).
10. J. D. Jarratt and R. L. E. Schwarzenberger, Coloured frieze groups. *Utilitas Math.* **19**, 295–303 (1981).
11. O. Jones, *The Grammar of Ornament* (orig. 1856). Van Nostrand Reinhold, New York (1972).
12. E. H. Lockwood and R. H. MacMillan, *Geometric Symmetry*. Cambridge Univ. Press (1978).
13. A. L. Loeb, *Color and Symmetry*. Wiley, New York (1971).
14. C. H. MacGillavry, *Symmetry Aspects of M. C. Escher's Periodic Drawings*. Oosthoek, Utrecht (1965); reprinted as *Fantasy and Symmetry, The Periodic Drawings of M. C. Escher*. Harry Abrams, New York (1976).

15. A. C. Racinet, *Handbook of Ornaments in Color*, 4 vols. (orig. 1879, 1883). Van Nostrand Reinhold, New York (1977-78).
16. D. F. Rogers and J. A. Adams, *Mathematical Elements for Computer Graphics*. McGraw-Hill, New York (1976).
17. R. Roth, Color symmetry and group theory. *Discrete Math.* **38**, 273-296 (1982).
18. D. Schattschneider, The plane symmetry groups: Their recognition and notation. *Am. Math. Monthly* **85**, 439-450 (1978).
19. D. Schattschneider, Will it tile? Try the Conway criterion! *Math. Mag.* **53**, 224-233 (1980).
20. R. L. E. Schwarzenberger, Colour symmetry. *Bull. London Math. Soc.* **16**, 209-240 (1984).
21. M. Senechal, Point groups and color symmetry. *Z. Kristall.* **142**, 1-23 (1975).
22. M. Senechal, Color groups. *Discrete Appl. Math.* **1**, 51-73 (1979).
23. P. S. Stevens, *A Handbook of Regular Patterns*. MIT Press, Cambridge (1980).
24. A. V. Shubnikov and V. A. Koptsik, *Symmetry in Science and Art* (trans. by D. Harker). Plenum Press, New York (1974).
25. D. Washburn, *A Symmetry Analysis of Upper Gila Area Ceramic Design*. Peabody Museum Papers, Vol. 68. Harvard University (1977).
26. T. W. Wieting, *The Mathematical Theory of Chromatic Plane Ornaments*. Dekker, New York (1982).
27. I. M. Yaglom, *Transformation Geometry I*. Math. Assn. of America (1962).

Comp. & Maths. with Appls. Vol. 12B, Nos. 3/4, pp. 697–723, 1986
Printed in Great Britain.

0886–9561/86 $3.00 + .00
© 1986 Pergamon Press Ltd.

SYMMETRY, HYBRIDIZATION AND BONDING IN MOLECULES†

Zvonimir B. Maksić

Theoretical Chemistry Group, The "Rudjer Bošković" Institute, 41001 Zagreb, Yugoslavia,
and The Faculty of Natural Sciences and Mathematics, The University of Zagreb,
Marulićev trg 19, 41000 Zagreb, Yugoslavia

Abstract—Brief historical review of the applications of symmetry arguments in interpreting the electronic and geometric structures of molecules is given. The hybridization of atomic orbitals is thoroughly discussed. It was shown that hybridization model describes directional features of covalent bonding and a number of other properties which are not directly related otherwise. This versatility in rationalizing a large number of experimental data of different kinds and a high interpretative power give to the model some semblance of truth. The model's apparently good reflection of molecular properties could be traced to the fact that hybrids conform with the local symmetry of an atom in a molecular environment. An analogy between the hybridization model and crystal field theory is found. The epistemological role of the hybridization is stressed. It provides a bridge between the most elementary first principles of quantum mechanics and the concept of a chemical bond, which is a basic tenet of the phenomenological theory of molecules.

Symmetry, as wide or as narrow as you may define its meaning, is one idea by which man through ages has tried to comprehend and create order, beauty and perfection.

H. Weyl

INTRODUCTION

Symmetry is one of the basic principles which helps a great deal in our efforts to understand the universe. It underlies, for example, the conservation laws of linear and angular momenta, which are a consequence of the homogeneity and isotropy of space. Similarly, the conservation of energy can be derived from the invariance of the equations of motion to the shift in time[1]. Elementary particles follow specific symmetry rules. Hence, it is not surprising that symmetry gives a deep insight into the structure of matter. It has strongly influenced the development of the theory of the electronic structure of molecules and crystals, particularly after the advent of quantum mechanics. Indeed, symmetry has revealed some of the very fundamental facets of chemical bonding. It found especially wide and important applications in spectroscopy by determining the so-called selection rules. Wigner writes, ". . . The actual solution of quantum mechanical equations is, in general, so difficult that one obtains by direct calculations only crude approximations to the real solutions. It is gratifying, therefore, that a large part of the relevant results can be deduced by considering the fundamental symmetry operations"[2]. He mentioned additionally that almost all rules of spectroscopy follow from the symmetry of the problem.

Symmetry arguments were an important guide even in the prequantum era. It is well known that Pasteur, van't Hoff and Le Bel recognized in the last century a close connection between the optical activity and asymmetry. It is also noteworthy that the mathematical tools necessary for a description of crystal symmetry (230 three-dimensional space groups, 32 crystallographic point-groups and 14 Bravais lattices) were coined before the end of the 19th century, that is to say, well ahead of the discovery of the X-ray method[3]. It is also interesting to mention that in 1883, some 30 years before the X-ray diffraction technique was actually invented, an English scientist by the name of Barlow assigned quite correctly the structure of a number of crystals by using abstract group theory[4]. However, the paramount importance of symmetry in gripping the basic features of the microworld was fully realized only after the rise of quantum theory, as noted above. The role of the newly developed spectroscopic techniques and of molecular (and crystal) structure determination methods in the 1950s should not be underestimated in this respect.

---

†Dedicated to Professor Linus Pauling—Prometheus of the modern chemistry—on the occasion of his 85th birthday.

Symmetry entered quantum chemistry via the classical paper of Bethe[5] on the term splitting of atoms and ions imbedded in crystals. That was the beginning of crystal field theory, which was subsequently developed by Van Vleck[6,7] and others[8,9]. In spite of the fact that the influence of ligands was mimicked by point charges forming the electrostatic field exerted on central atomic states, crystal field theory was highly effective because of its inherent symmetry content. Therefore, the following citation of Moffitt and Ballhausen[10] seems to be in place here: "It will be a long time before a method is developed to surpass in simplicity, elegance and power that of crystal field theory. Within its extensive domain it has provided at least a deep qualitative insight into the behaviour of a many electron system." The crystal field theory was later replaced by the ligand-field molecular orbital theory, which properly takes into account covalency (overlap) effects[11–13]. Its simplified version named the angular overlap method (AOM), is particularly useful due to its great applicability[14]. Molecular orbital theory was put forward predominantly by Mulliken[15,16]. The molecular orbitals (MO) wavefunctions, which describe the one-electron charge distribution over the whole domain of a molecule, enjoy wide popularity due to the ease of their computational implementation. They belong to the irreducible representations of the respective point-symmetry group[16] and are a convenient vehicle for the interpretation of molecular spectra. The MO theory has faithfully served chemistry for more than 50 years. It is far beyond the scope of this article to review all fields of its application. We shall mention instead their use in rationalizing concerted chemical reactions, which resulted in the well-known orbital symmetry conservation rules[17], thus establishing the intimate relationship between the symmetry properties of reactants and products.

The complicated vibrations of nuclei in molecular systems can be simplified and systematized by the use of symmetry and its mathematical tool-group theory. The pioneering work in this field was done by Wigner[18] and Bright–Wilson[19,20]. The motion of nonrigid molecules is even more intricate, but the use of symmetry is illuminating, as shown by the inspired paper of Longuet–Higgins[21] and by other researchers[22,23]. The rigorous calculation of wavefunctions in moderately large molecular systems is an extremely difficult task despite the exploitation of powerful high-speed computers. This applies particularly if the electron correlation is estimated as accurately as possible. Significant breakthrough in the computation of energy matrix elements was recently achieved by the use of unitary group representations[24,25]. Last but not least, submolecular particles like electrons are identical, thus exhibiting permutational symmetry properties[26,27].

The rough and incomplete historical sketch presented above conclusively shows that symmetry is of tantamount importance in tackling problems of the electronic structure of molecules. It is obviously an invaluable tool in treating properties of crystals where the number of particles, i.e. the complexity of problems, is immensely increased[28]. Thus, symmetry and group theory will always remain one of the most powerful weapons in the theoretical arsenal.

There is one aspect which was not discussed so far. The building blocks of molecules and crystals are atoms. This point of view is close to chemical intuition because structural formulas of compounds are composed of atomic symbols. We usually say that a water molecule is built from two hydrogen atoms and one oxygen atom. We rarely describe this molecular system as an ensemble of two protons, a nucleus with charge $+8|e|$ (involving 16 nucleons) and 10 electrons. There must be a good reason for this attitude. Indeed, there is a comprehensive evidence that atoms retain their identity after the formation of chemical bonds. They are not identical to atoms *in statu nascendi*, but are somewhat perturbed by their chemical environment instead. Atoms in molecules are best described as *modified atoms*. One of the apparent changes is a descent of symmetry. Free atoms are spherically symmetric, while their local symmetry, dictated by the nearest neighbours, is significantly lower in the hierarchy. It is clear that the local atomic wavefunctions (or densities) will be given as a linear combination of the free-atom eigenfunctions. This was ingeniously recognized by Pauling as early as 1931 and published in a paper[29] which is another milestone in the history of the theory of chemical bonding. Pauling's hybrid orbitals paved a large avenue called the valence bond (VB) theory, which dominated early quantum chemistry. To be fair, one should mention that hybrids were almost simultaneously invented by Slater also[30]. However, it was Pauling's work which conclusively showed their efficacy in rationalizing molecular and crystal structures. It turned out that hybridization is a simple and intuitively appealing model possessing at the same time a very high content of chemical information providing an important link between the fundamental quantum concepts, and empirical knowledge gathered by experimental chemists. Since the renewed interest might well lead to a renaissance of hybridization and VB theory[31–56], we shall dwell on the former in some more detail in this article, which is a review in a sense that it covers a rather extensive field of applications of hybridization. However, inclusion of all the literature on this topic is not intended. We shall see that the hybridization concept yields a rich harvest of chemically relevant results and penetrating conclusions. Theory will be presented on the elementary level, which permits easy digestion by a nonspecialist. In particular, only spatial wavefunctions will be considered. The spin parts can be obtained by using the formalism of the permutation symmetry group[26,27].

HYBRIDIZATION OF ATOMIC ORBITALS

If you know a thing, it is simple; if it is not simple, you don't know it.

Oriental proverb

Atomic orbitals

Description of the hydrogen atom and the related hydrogen-like ions requires solution of the corresponding Schrödinger equation involving the spherically symmetric central potential $V(r) = -e^2 Z/r$. The symmetry of the problem is best exploited by the use of the polar coordinate system (Fig. 1). Then the wavefunction can be written as a product of the radial and angular parts:

$$\Phi_{nlm} = R_{nl}(r)Y_{lm}(\vartheta, \varphi), \tag{1}$$

where n, l, m denote the quantum numbers[57]. This is a remarkable result because the well-known spherical harmonics $Y_{lm}(\vartheta, \varphi)$ depend only on the symmetry of the central field and not on its details. The latter are stored in the radial wavefunction $R_{nl}(r)$. Consequently, the angular $Y_{lm}(\vartheta, \varphi)$ functions can be determined once and for all and in fact that was done a long time ago. A few lowest spherical harmonics are given in Table 1. The one-electron function of the form (1) which describes average behaviour of an electron in the atom is called an atomic orbital. It has a strict meaning only in one-electron atoms (ions). In many-electron atoms the electrons do not move independently. They permanently interact and concomitantly the total wavefunction Ψ depends on their relative positions. It is gratifying, however, that a fairly realistic picture is obtained by the independent-electron approximation in the familiar Hartree–

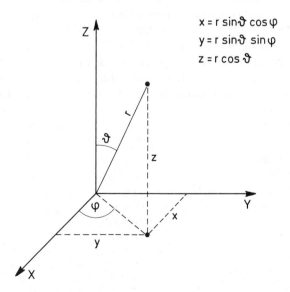

$$x = r \sin\vartheta \cos\varphi$$
$$y = r \sin\vartheta \sin\varphi$$
$$z = r \cos\vartheta$$

Fig. 1. Relation between Cartesian and polar coordinates.

Table 1. The angular parts of the hydrogen-like atom wavefunctions

Symbol	Spherical harmonic
$Y_{00} = s$	$\frac{1}{2}\sqrt{\pi}$
$Y_{10} = p_0$	$(\sqrt{3}/2\sqrt{\pi})\cos\vartheta$
$Y_{11} = p_1$	$(\sqrt{3}/2\sqrt{2\pi})\sin\vartheta e^{i\varphi}$
$Y_{1-1} = p_{-1}$	$(\sqrt{3}/2\sqrt{2\pi})\sin\vartheta e^{-i\varphi}$
$Y_{20} = d_0$	$(\sqrt{5}/4\sqrt{\pi})(3\cos^2\vartheta - 1)$
$Y_{22} = d_2$	$(\sqrt{15}/4\sqrt{2\pi})\sin^2\vartheta e^{i2\varphi}$
$Y_{2-2} = d_{-2}$	$(\sqrt{15}/4\sqrt{2\pi})\sin^2\vartheta e^{-i2\varphi}$
$Y_{21} = d_1$	$(\sqrt{15}/2\sqrt{2\pi})\sin\vartheta\cos\vartheta e^{i\varphi}$
$Y_{2-1} = d_{-1}$	$(\sqrt{15}/2\sqrt{2\pi})\sin\vartheta\cos\vartheta e^{-i\varphi}$

Fock method. The single electron moves in the average electrostatic field of all the remaining particles and the total wavefunction is obtained in the iterative self-consistent fashion[57]. It should be recalled also that Ψ satisfies the antisymmetry condition and its simplest form is given by the single determinant

$$\Psi(1 \ldots n) = \frac{1}{\sqrt{n!}} \begin{vmatrix} \Phi_1(1) & \cdots & \Phi_1(n) \\ \vdots & & \vdots \\ \Phi_n(1) & \cdots & \Phi_n(n) \end{vmatrix}. \tag{2}$$

Hence, the orbital concept is preserved albeit it holds only to a certain limit. It is important to emphasize also that all electrons are equally distributed over all orbitals in the total wavefunction $\Psi(1 \ldots n)$ (2). In spite of their approximate nature, atomic orbitals (AOs) have an important heuristic role in chemistry and provide a convenient basis for quantum-chemical calculations on molecules. While the best atomic orbitals are obtained by the self-consistent field approach, a reasonable representation is offered by the concept of nuclear screening. The interelectronic repulsion can be approximated by the scaling of the nuclear charge Z, which assumes then the $Z - S$ form. Here S denotes screening. It can be determined either variationally or roughly, for certain groups of electrons, by using the well-known Slater rules[57]. We shall not be concerned with the specific form of the radial $R_{nl}(r)$ parts of AOs because the symmetry properties are incorporated in their angular parts. It should be noted in passing, however, that AOs presented in Table 1 have defined well both the angular momentum and its projection to the Z axis. On the other hand, the most salient feature of covalent bonds is their directionality in space. If chemical bonding is to be interpreted on an orbital basis, then the directional nature of the AOs should be emphasized as much as possible. Is there an equivalent basis set which is better in this respect? The answer to this question is positive because (a) the AOs belonging to the same subshell (Y_{lm}, where $m = -l, \ldots, +l$) are degenerate, and (b) the total wavefunction Ψ (2) is invariant to all unitary transformations of orbitals Φ_i. It turns out that by making simple linear combinations of Y_{lm} and sacrificing the magnetic quantum number m, one obtains AOs with favourable directional properties. They are listed in Table 2 and depicted in Fig. 2. The shaded lobes in the latter correspond to domains where AOs assume positive values. It should be

Table 2. Directional atomic orbitals

Symbol	Orbital
s	$\frac{1}{2}\sqrt{\pi}$
p_X	$\frac{1}{2}(p_1 + p_{-1}) = (\sqrt{3}/2\sqrt{\pi})\sin\vartheta\cos\varphi$
p_Y	$-\frac{1}{2}(p_1 - p_{-1}) = (\sqrt{3}/2\sqrt{\pi})\sin\vartheta\sin\varphi$
p_Z	$p_0 = (\sqrt{3}/2\sqrt{\pi})\cos\vartheta$
$d_{X^2-y^2}$	$\frac{1}{2}(d_2 + d_{-2}) = (\sqrt{15}/4\sqrt{\pi})\sin^2\vartheta\cos 2\varphi$
d_{XY}	$-\frac{1}{2}(d_2 - d_{-2}) = (\sqrt{15}/4\sqrt{\pi})\sin^2\vartheta\sin 2\varphi$
d_{XZ}	$\frac{1}{2}(d_1 + d_{-1}) = (\sqrt{15}/2\sqrt{\pi})\sin\vartheta\cos\vartheta\cos\varphi$
d_{YZ}	$-\frac{1}{2}(d_1 - d_{-1}) = (\sqrt{15}/2\sqrt{\pi})\sin\vartheta\cos\vartheta\sin\varphi$
d_{Z^2}	$d_0 = (\sqrt{5}/4\sqrt{\pi})(3\cos^2\vartheta - 1)$

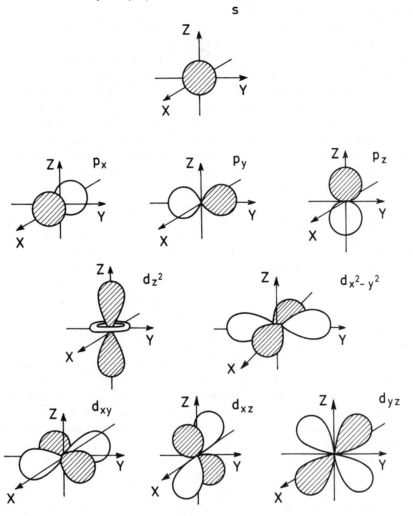

Fig. 2. Schematic representation of the angular parts of a central potential one-electron wavefunction.

mentioned that subscripts in the spatial representation of AOs, e.g. p_X, p_Y, p_Z, etc., yield immediately the functional dependence of the orbital on Cartesian coordinates. The simple unitary transformations indicated in Table 2 illustrate very nicely the process of adaptation of the basis set for the description of the chemical bonding. It will appear that the basis of AOs presented in Table 2 is very useful for this purpose, but it does not suffice, particularly for tetracoordinate atoms. This will be discussed in some more detail in the next section.

Hybrid orbitals

One of the most remarkable facets of molecules (apart from some exceptional cases) is the relative rigidity of their geometric structures. It is this rigidity which permits the widespread use of molecular "ball-and-stick" models in chemistry instead of, for example, the liquid-drop model, which would be completely inadequate. One of the main aims of theoretical chemistry is to rationalize the directional character of covalent bonds. The most attractive and intuitively appealing way to do this is the use of the atomic orbital concept. The underlying idea is that atoms retain their identities to a considerable extent within the molecule, and is close to chemical experience. Indeed, notions of atoms and bonds are of central importance for phenomenological descriptions of properties of myriads of molecules. It is empirically known that bonds between a given pair of atoms have similar properties. For instance, the bond energy in H_2O is 118 kcal/mole, which can be compared with the corresponding O—H value of 112 kcal/mole in CH_3OOH. Hence, the bond energies exhibit the characteristic additivity at least to a first approximation. Furthermore, certain stretching vibrational frequencies are fairly constant for a given bond type (e.g. the CH frequency is ~ 2900 cm^{-1} and the OH frequency is ~ 3600 cm^{-1}) which is used for analytical purposes to identify the characteristic groupings of atoms within complex molecules. The bond distances between the atoms, which are characterized by their coordination numbers, do not vary much as a rule (for example, the C—H bond attached to the tetracoordinated carbon atom is roughly ~ 1.10Å in a large number of organic compounds). Is there any relation between the relative constancy of a variety of bond properties and their transferability across the families of related molecules and directional features of covalent bonds? Anticipating conclusions of the forthcoming discussion we can say that there is a close connection between these properties although they seem to be unrelated at the first sight.

Empirical knowledge shows that many properties of covalent bonds are nearly constant in similar moieties, suggesting that the total electronic charge can be approximately partitioned into pairs of electrons which are almost completely localized in the region of each bond. The idea that bonds are formed by shared electron pairs was first put forward by Lewis[58] as early as 1916. The quantum-mechanical basis and proper interpretation of the Lewis postulate was given by Heitler and London in their benchmark paper[59], which represents the birthday of the valence bond (VB) theory and of quantum chemistry at the same time. A pair of bonding electrons in the bond A—B is described in this approach by the spatial wavefunction

$$\Psi_{AB}(1, 2) \sim [\psi_A(1)\psi_B(2) + \psi_B(1)\psi_A(2)], \qquad (3)$$

where ψ_A and ψ_B are atomic orbitals centered on the nuclei A and B, respectively. The spins of electrons placed on ψ_A and ψ_B are opposite and the spin function of the form $[\alpha(1)\beta(2) - \alpha(2)\beta(1)]$ is omitted for the sake of simplicity. Consider the simple VB treatment of bonding in the H_2S molecule. The electron configuration of sulfur is $(1s)^2(2s)^2(2p)^6(3s)^2(3p)^4$. Neglecting the core formed by the nucleus and full electron shells one obtains the configuration of valence electrons $(3s)^2(3p_Z)^2(3p_X)(3p_Y)$, where the AOs $3p_X$ and $3p_Y$ are populated by an electron each possessing the same spin (Hund's rule). Since the number of unpaired spins determines valency of the atom in question[59], sulfur can bind two hydrogens. The bond strength is roughly given by the exchange integral, which in turn is approximately proportional to the corresponding overlap integral between the ψ_A and ψ_B AOs[59]. Hence, according to the *maximum overlap criterion*[29,30] the hydrogen $1s_H$ functions should be placed on the X and Y axes at some optimal distance, implying that the H—S—H angle is 90° (Fig. 3). This is in reasonable accordance with the experimental value of 92.2°, indicating that the simple VB picture offered by the spatial part of the total wavefunction

$$[\psi_{3p_X}(1)\psi_{1sH_1}(2) + \psi_{3p_X}(2)\psi_{1sH_1}(1)][\psi_{3p_Y}(3)\psi_{1sH_2}(4) + \psi_{3p_Y}(4)\psi_{1sH_2}(3)]$$

is essentially correct. The bond angles in H_2Se (91°) and H_2Te (89.5°) can be explained by the same token. By reversing the argument, one can say that two hydrogen atoms approaching a sulfur atom diminish its symmetry. The Z axis, unspecified otherwise, now assumes a direction perpendicular to the plane of the HSH atoms. The $3p$ subshell is split by the descent of symmetry. The $3p_Z$ orbital is populated by the lone pair, while $3p_X$ and $3p_Y$ functions, remaining degenerate, can be used for chemical bonding with hydrogens forming two equivalent S—H bonds. It is important to emphasize that good agreement with experiment was obtained by

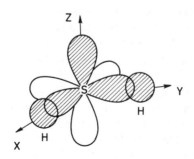

Fig. 3. Simple bonding scheme in the H_2S molecule.

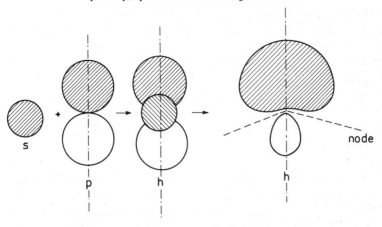

Fig. 4. Formation of a hybrid by superposition of one s- and one p-type orbital.

utilizing the maximum overlapping principle in addition to the symmetry arguments. This holds quite generally because symmetry yields only qualitative information about systems under study. Its strength, however, can be considerably enhanced by using additional criteria dictated by the physics of the situation. The simple interpretation of the shapes of H_2S, H_2Se and H_2Te is attractive, but it has limitations. For example, the bond angle in the H_2O molecule is $104.5°$, indicating that other effects are also of importance. However, they have nothing to do with symmetry and concomitantly will not be discussed here.

Stereochemistry of the carbon atom requires a new concept. Its electron configuration $(1s)^2(2s)^2(2p)^2$ suggests that carbon is divalent. In spite of the fact that unusual molecules like CH_2 and CF_2 do exist, the whole of organic chemistry rests on the tetravalency of carbon atoms. The latter is most easily explained by the promotion of one electron from the $2s$ level to the empty $2p$ orbital. Thus, the so-obtained four uncoupled spins are responsible for the carbon quadrivalency. They do not yield desired directional characteristics, though, because the configuration $(1s)^2(2s)(2p_X)(2p_Y)(2p_Z)$ involves only three orbitals projected into X, Y and Z directions. The CH_4 molecule, on the other hand, has beautiful tetrahedral structure. The problem was elegantly solved by Pauling[29] and somewhat later by Slater[30]. They introduced hybrid-orbital linear combinations of free-atom eigenfunctions

$$h_i = c_{i1}(2s) + c_{i2}(2p_X) + c_{i3}(2p_Y) + c_{i4}(2p_Z). \tag{4}$$

Hence by sacrificing the angular momentum quantum number l, it is possible to construct four equivalent orthonormal hybrids of the form (4), which are directed toward the corners of a tetrahedron. Furthermore, these hybrid orbitals have the axial symmetry required for the description of four local σ-bonds. They have a number of favourable properties. First, hybrids have a characteristic mushroom-like form (Fig. 4), which is a consequence of the reinforcement of the s and p orbitals in the domain of their positive signs, and destructive interference and partial cancellation of the s and p amplitudes in the region of their opposite signs. Hence, the hybrid orbital is strongly polarized toward its nearest neighbour. It was shown by Pauling that tetrahedral hybrids not only have proper directional features but also considerable bond strength, which overcompensates the price in energy paid by the s–p promotion[29]. Finally, by maximizing the overlapping power in one particular direction the hybrid's amplitudes are minimized in all others. Consequently, the nonbonded repulsions are usually minimized at the same time.

It is easy to see that symmetry operations of the T_d point group transform tetrahedral orbitals between themselves. Therefore, they form a basis for the representation of the T_d group. This representation is reducible because it is four dimensional, while the highest dimension of the irreducible representations of the T_d group is 3 (Table 3). Two points of importance will be considered here: (a) selection of the AOs of the central atom suitable for the construction of the general hybrid orbitals belonging to the T_d symmetry and (b) determination of the coefficients $c_{ij}(4)$ by group-theoretical arguments. For this purpose we shall make use of group represen-

Table 3. Characters of the (ir)reducible representations of the T_d group and the corresponding basis functions

T_d	I	$8C_3$	$3C_2$	$6S_4$	$6\sigma_d$	Basis sets
A_1	1	1	1	1	1	$s,\ f_{xyz}$
A_2	1	1	1	-1	-1	
E	2	-1	2	0	0	$(d_{z^2},\ d_{x^2-y^2})$
T_1	3	0	-1	1	-1	$(f_{x(z^2-y^2)},\ f_{y(z^2-x^2)},\ f_{z(x^2-y^2)})$
T_2	3	0	-1	-1	1	$(p_x,\ p_y,\ p_z)(d_{xy},\ d_{xz},\ d_{yz})(f_{x^3},\ f_{y^3},\ f_{z^3})$
Γ_h	4	1	0	0	2	$(h_1,\ h_2,\ h_3,\ h_4)$

tations. The characters of the reducible representation Γ_h generated by four equivalent tetrahedral hybrids are easily obtained by inspection. They are simply equal to the numbers of the hybrid AOs which retain their positions after the symmetry operation was performed. The permutations of hybrids by symmetry operations are easily examined with the aid of Fig. 5, where the hybrids h_i directed to the hydrogens H_i are represented by arrows. Knowing the character of the Γ_h representation (Table 3), its irreducible components are found by employing the great orthogonality theorem[60]. The result reads

$$\Gamma_h = A_1 \oplus T_2. \tag{5}$$

Hence, the hybrid orbitals can be composed of AOs which belong to A_1 and T_2 irreducible representations, respectively. The spherical harmonics of the given l span the $2l + 1$ dimensional irreducible representation of the group of three-dimensional rotations K_h. They are generally split by the descent of symmetry $K_h \rightarrow T_d$. A simple way to examine the behaviour of the spherical harmonics in the molecular field of the T_d symmetry is to take their Cartesian representation (Table 2) and execute the required coordinate transformation corresponding to the I, C_3, C_2, S_4 and σ_d symmetry operations[60,61]. One finds out that the p subshell remains degenerate while d and f subshells are split into two and three levels, respectively (Table 3). It appears that functions s and f_{xyz} belong to the irreducible representation A_1. On the other hand, three sets of the three independent functions (p_x, p_y, p_z), (d_{xy}, d_{xz}, d_{yz}) and $(f_{x^3}, f_{y^3}, f_{z^3})$ span the irreducible representation T_2. Obviously, the hybridization scheme is not unique and several combinations of AOs which belong to A_1 and T_2 representations are possible. One can combine one s orbital and three p_x, p_y and p_z orbitals to form the most conventional sp^3 tetrahedral hybrids. Alternatively, the sd^3, sf^3, fd^3 and f^4 schemes may take place or any combinations thereof. We shall postpone their discussion for the moment, in order to obtain

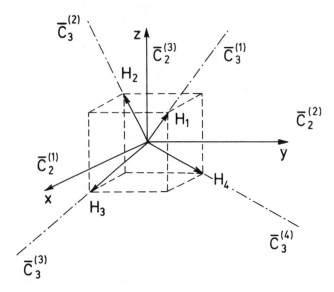

Fig. 5. Tetrahedral directions and some symmetry elements of the T_d point group.

explicit form of the sp^3 hybrids $h_i(4)$. This can be achieved by using a very practical device called projection operator[60,61]:

$$\hat{P}_\mu = \sum_R \chi_\mu(R)^* \hat{O}_R, \tag{6}$$

where $\chi_\mu(R)$ is the character of the μth irreducible representation corresponding to the symmetry operation R and the summation is extended over all symmetry operations of the point group in question. The operator denoted by $\hat{O}_R$ is associated with the transformation of the basis set functions induced by the symmetry operator $\hat{R}$ and defined by

$$\hat{O}_R f(\mathbf{r}) = f(\hat{R}^{-1}\mathbf{r}). \tag{7}$$

The operator $\hat{P}_\mu$ has a property that when applied to an arbitrary function, the component of that function belonging to the μth irreducible representation is projected out while the rest of it is abolished. The effect of the $\hat{O}_R$ operators on the tetrahedral hybrids h_i are summarized in Table 4. We can arbitrarily single out the h_1 hybrid and apply the $\hat{P}_{A_1}$ operator to extract the A_1 contribution of the hybrid. Straightforward application of the formula (6) and the data given in Table 4 yields

$$\hat{P}_{A_1} h_1 = 6(h_1 + h_2 + h_3 + h_4). \tag{8}$$

Since the hybrids h_i are orthogonal by assumption, the normalized form of $\hat{P}_{A_1} h_1$ should be equal to the s orbital which will be used for hybridization. Hence

$$(h_1 + h_2 + h_3 + h_4)/2 = s. \tag{9}$$

Analogously, $\hat{P}_{T_2}$ operator applied to h_1, h_2 and h_3 hybrids gives after normalization three linearly independent functions belonging to the T_2 irreducible representation. They are generally equal

Table 4. Transformation of the T_d hybrids under $\hat{O}_R$ for all symmetry operations R of the T_d group

R	h_1	h_2	h_3	h_4
I	h_1	h_2	h_3	h_4
$C_3^{(1)}$	h_1	h_4	h_2	h_3
$(C_3^{(1)})^2$	h_1	h_3	h_4	h_2
$C_3^{(2)}$	h_3	h_2	h_4	h_1
$(C_3^{(2)})^2$	h_4	h_2	h_1	h_3
$C_3^{(3)}$	h_4	h_1	h_3	h_2
$(C_3^{(3)})^2$	h_2	h_4	h_3	h_1
$C_3^{(4)}$	h_2	h_3	h_1	h_4
$(C_3^{(4)})^2$	h_3	h_1	h_3	h_4
$C_2^{(1)}$	h_3	h_4	h_1	h_2
$C_2^{(2)}$	h_4	h_3	h_2	h_1
$C_2^{(3)}$	h_2	h_1	h_4	h_3
$S_4^{(1)}$	h_4	h_1	h_2	h_3
$(S_4^{(1)})^3$	h_2	h_3	h_4	h_1
$S_4^{(2)}$	h_2	h_4	h_1	h_3
$(S_4^{(2)})^3$	h_3	h_1	h_4	h_2
$S_4^{(3)}$	h_3	h_4	h_2	h_1
$(S_4^{(3)})^3$	h_4	h_3	h_1	h_2
$\sigma_d^{(1)}$	h_1	h_2	h_4	h_3
$\sigma_d^{(2)}$	h_1	h_4	h_3	h_2
$\sigma_d^{(3)}$	h_1	h_3	h_2	h_4
$\sigma_d^{(4)}$	h_4	h_2	h_3	h_1
$\sigma_d^{(5)}$	h_3	h_2	h_1	h_4
$\sigma_d^{(6)}$	h_2	h_1	h_3	h_4

to some orthonormalized linear combinations of the p_X, p_Y and p_Z AOs:

$$\hat{P}_{T_2}h_1 \sim (3h_1 - h_2 - h_3 - h_4)/\sqrt{12}$$

$$= \pm(a_{11}p_X + a_{12}p_Y + a_{13}p_Z)/(a_{11}^2 + a_{12}^2 + a_{13}^2)^{1/2}, \quad (10a)$$

$$\hat{P}_{T_2}h_2 \sim (-h_1 + 3h_2 - h_3 - h_4)/\sqrt{12}$$

$$= \pm(a_{21}p_X + a_{22}p_Y + a_{23}p_Z)/(a_{21}^2 + a_{22}^2 + a_{23}^2)^{1/2}, \quad (10b)$$

$$\hat{P}_{T_2}h_3 \sim (-h_1 - h_2 + 3h_3 - h_4)/\sqrt{12}$$

$$= \pm(a_{31}p_X + a_{32}p_Y + a_{33}p_Z)/(a_{31}^2 + a_{32}^2 + a_{33}^2)^{1/2}. \quad (10c)$$

Applying the operator $\hat{O}_{C_3^{(1)}}$ to eqn (10a) one obtains

$$(3h_1 - h_2 - h_3 - h_4)/\sqrt{12} = \pm(p_X + p_Y + p_Z)/\sqrt{3}, \quad (11)$$

which follows from the fact that the left side is invariant and

$$\hat{O}_{C_3^{(1)}}p_X = p_Z, \quad \hat{O}_{C_3^{(1)}}p_Y = p_X, \quad \hat{O}_{C_3^{(1)}}p_Z = p_Y.$$

Consequently, $a_{11} = a_{13}$, $a_{12} = a_{11}$, $a_{12} = a_{13}$, and the only dilemma left is the sign of the right-hand side in (11). Taking into account that p_X, p_Y and p_Z AOs behave like components of the position vector, the linear combination $(p_X + p_Y + p_Z)/\sqrt{3}$ is colinear with the h_1. Hence, the positive sign is appropriate:

$$(3h_1 - h_2 - h_3 - h_4)/\sqrt{12} = (p_X + p_Y + p_Z)/\sqrt{3}. \quad (12a)$$

Similarly, by acting with $\hat{O}_{C_3^{(2)}}$ and $\hat{O}_{C_3^{(3)}}$ operators on eqns (10b) and (10c), respectively, one readily obtains

$$(-h_1 + 3h_2 - h_3 - h_4)/\sqrt{12} = -(p_X + p_Y - p_Z)/\sqrt{3} \quad (12b)$$

and

$$(-h_1 - h_2 + 3h_3 - h_4)/\sqrt{12} = (p_X - p_Y - p_Z)/\sqrt{3}. \quad (12c)$$

Simple algebra shows that the system of eqns (12a–c) can be transformed to an equivalent form:

$$p_X = (h_1 - h_2 + h_3 - h_4)/2,$$

$$p_Y = (h_1 - h_2 - h_3 + h_4)/2, \quad (13)$$

$$p_Z = (h_1 + h_2 - h_3 - h_4)/2.$$

Equations (9) and (13) can be succintly written as

$$\begin{Vmatrix} s \\ p_X \\ p_Y \\ p_Z \end{Vmatrix} = \frac{1}{2} \begin{Vmatrix} 1 & 1 & 1 & 1 \\ 1 & -1 & 1 & -1 \\ 1 & -1 & -1 & 1 \\ 1 & 1 & -1 & -1 \end{Vmatrix} \cdot \begin{Vmatrix} h_1 \\ h_2 \\ h_3 \\ h_4 \end{Vmatrix}. \quad (14)$$

The inverse matrix transformation would yield the desired composition of the hybrids $h_i(i = 1, \ldots, 4)$ in terms of s, p_X, p_Y and p_Z AOs. Noting that the two orthogonal basis sets (14) are related by the orthogonal matrix, one immediately obtains

$$
\begin{Vmatrix} h_1 \\ h_2 \\ h_3 \\ h_4 \end{Vmatrix} = \frac{1}{2} \begin{Vmatrix} 1 & 1 & 1 & 1 \\ 1 & -1 & -1 & 1 \\ 1 & 1 & -1 & -1 \\ 1 & -1 & 1 & -1 \end{Vmatrix} \cdot \begin{Vmatrix} s \\ p_X \\ p_Y \\ p_Z \end{Vmatrix}.
\tag{15}
$$

Thus the explicit form of the sp^3 hybrids (4) reads

$$
\begin{aligned}
h_1 &= (s + p_X + p_Y + p_Z)/2, \\
h_2 &= (s - p_X - p_Y + p_Z)/2, \\
h_3 &= (s + p_X - p_Y - p_Z)/2, \\
h_4 &= (s - p_X + p_Y - p_Z)/2.
\end{aligned}
\tag{16}
$$

Let us focus attention to the alternative sd^3 and sf^3 hybridization schemes. The crucial point of the procedure above was the behaviour of the p_X, p_Y and p_Z AOs under the action of the $\hat{O}_{C_3^{(i)}}$ ($i = 1, 2, 3$) operators. It can be easily shown that the vectors

$$
\begin{Vmatrix} p_X \\ p_Y \\ p_Z \end{Vmatrix}, \quad \begin{Vmatrix} d_{YZ} \\ d_{XZ} \\ d_{XY} \end{Vmatrix} \quad \text{and} \quad \begin{Vmatrix} f_{X^3} \\ f_{Y^3} \\ f_{Z^3} \end{Vmatrix}
$$

have the same transformation properties when the $\hat{O}_{C_3^{(1)}}$, $\hat{O}_{C_3^{(2)}}$ and $\hat{O}_{C_3^{(3)}}$ are applied. Hence, the sd^3 and sf^3 hybrids are straightforwardly obtained from eqns (16) by substitutions based on formal equivalence:

$$
\begin{aligned}
p_X &\longleftrightarrow d_{YZ} \longleftrightarrow f_{X^3}, \\
p_Y &\longleftrightarrow d_{XZ} \longleftrightarrow f_{Y^3}, \\
p_Z &\longleftrightarrow d_{XY} \longleftrightarrow f_{Z^3}.
\end{aligned}
\tag{17}
$$

The sp^3, sd^3, sf^3 and the earlier-mentioned tetrahedral hybridization compositions like fp^3, fd^3 and f^4 are all equivalent from the point of view of group theory. Apparently, some more physical insight is needed here to determine the optimal hybridization. Consider the sd^3 scheme for the C atom. Radial parts of the AOs were tacitly assumed to be independent on the quantum number l so far. The matter of truth is that the chain of inequalities for the effective nuclear charge

$$
Z_s > Z_p > Z_d > Z_f
$$

usually holds, implying that

$$
E_s < E_p < E_d < E_f.
$$

To be more specific, the $2p$ level lies by some 207 kcal/mole above the $2s$ level. The $3d$ level is about 230 kcal/mole less stable than the $2p$ level, etc. Therefore, it is obvious that the sp^3 scheme is much more important in describing the charge distribution of the tetracoordinate carbon atom than the sd^3 one. Another difference is given by the number of angular nodes of p and d orbitals. The d_{z^2} orbital (Fig. 2) has two positive lobes along the Z axis and mixing in a certain amount of the s orbital will reinforce the hybrid in two diametraly opposite directions. Hence, sd_{z^2} will be less polarized toward a specific neighbour than the sp_z hybrid. Magnitudes of the corresponding overlap integrals clearly show that the latter is more favourable. It is also easy to see that the f_{z^3} orbital behaves like the p_z orbital as far as the Z axis is concerned. The sf_{z^3} hybrid, however, has less efficient overlapping power. Nonetheless, it is plausible to assume that an optimal hybrid will have some admixture of d and f orbitals. Indeed, a small amount of the d orbital would be beneficial. The sp^3 hybrid has a small negative lobe which can be further diminished by the contribution of the properly oriented d orbital. The large positive lobe is enhanced at the same time. Neglecting the influence of the f_{XYZ} orbital, which belongs to the A_1 irreducible representation like the s orbital but is obviously less effective, we conclude that the general hybrid of the tetrahedral symmetry takes a form

$$h_1 = a(2s) + b(2p_X + 2p_Y + 2p_Z) + c(3d_{YZ} + 3d_{XZ} + 3d_{XY})$$
$$+ d(4f_{X^3} + 4f_{Y^3} + 4f_{Z^3}). \quad (18)$$

The coefficients in formula (18) cannot be estimated by group theory. The most accurate procedure would be the use of the variation theorem. However, some idea about the magnitude of the coefficients could be obtained by the simple bond strength criterion. According to Pauling, the optimal hybrid is the one with the maximum angular amplitude in the bond direction[29]. Taking into account the promotion energies, Pauling[62] found that the hybrid

$$h_1 = 0.50(2s) + 0.83(2p_Z) + 0.20(3d_{Z^2}) + 0.14(4f_{Z^3}) \quad (19)$$

has a pronounced bonding power. Hence, the contributions of the d_{Z^2} and f_{Z^3} AOs of 4% and 2%, respectively, are small but significant. Generally speaking, the efficiency in the mixing of AOs depends very much on their sizes, which should be taken into account in the semiquantitative calculations.

The hybridization of the trigonal (ethylene) and digonal (acetylene) carbon atoms can be analyzed along the same lines. It should be pointed out in this respect that the widely used sp^2 and sp^1 canonical hybridization schemes are more exception than rule (*vide infra*). Two approaches are possible in the treatment of multiple bonding: (a) σ–π separation of orbitals which in turn differ in symmetry characteristics relative to the bond axis and (b) deformed tetrahedron picture, which leads to the bent-bond representation of the multiple bonding. These two approaches differ in the choice of the basis set. They are equivalent in the MO theory if the calculations are pursued far enough. However, the bent-bond picture is to be preferred in the VB method if the calculation involves only the most important valence structures, particularly if d and f AOs are employed[62].

The hybridization is most effective in carbon atoms. This is not surprising because the loss in energy by the $2s$–$2p$ promotion is highly rewarded by the formation of the two additional covalent bonds. This is not the case, for example, in oxygen and nitrogen, where the increase in s character of hybrids improves overlapping, but does not yield new bonds. Hence, the role of hybridization is less pronounced in this type of atom. Concomitantly, the hybridization in atoms other than carbon was less extensively studied. It should be noted that the general $s^k p^l d^m$ hybridization schemes were investigated by Kimball[63,64] and were extended later to $s^k p^l d^m f^n$ phenotypes by the group-theoretical study of Eisenstein[65]. It is also noteworthy that hybridization of the central atom suitable for the π-bonding with appropriate ligand AOs can be determined on the same footing[60,61]. The most relevant hybrids for inorganic compounds are described in Pauling's classical book *The Nature of the Chemical Bond*[62]. This armamentarium was considerably enlarged by a series of papers based on the maximum of bond strength criterion which appeared over the last ten years[34,36,37,39,41,43,53,54]. Some recent advances were discussed in depth by Herman[52].

HYBRIDIZATION AND MOLECULAR PROPERTIES

The charge distribution of an atom in a molecule is anisotropic. Consequently, it is advantageous to use some sort of chemically adapted AOs, which conform themselves to the local site symmetry dictated by the immediate environment, if a simple and economical description of molecular properties is desired. This type of AO provides hybridization. Hence, hybrids can be considered as local wavefunctions of the zeroth order. It follows as a corollary that hybrids are very helpful in rationalizing those molecular properties which can be ascribed to chemical bonds. This is indeed the case[47,51,52,62,66], as shall be seen shortly.

Molecular shape and size

Hybrid orbitals were designed to explain directional properties of homopolar bonds. Therefore, they should give some insight into the bond angles in molecules. In spite of the fact that hybrids do not always coincide with straight lines passing through the neighbouring nuclei, they do give some useful information about bond angles and regularities in their changes. This is based on the intimate relation between the hybrid composition and the interhybrid angle ϑ. Consider for simplicity two equivalent and orthogonal sp^n hybrids:

$$h_i = a_i(2s) + (1 - a_i^2)^{1/2}(2p)_i \quad (i = 1, 2), \tag{20}$$

where $(2p)_i$ is the properly oriented p orbital $(2p)_i = \cos \gamma_X(2p)_X + \cos \gamma_Y(2p)_Y + \cos \gamma_Z(2p)_Z$ and $\cos \gamma_\alpha(\alpha = X, Y, Z)$ are the direction cosines. The orthogonality requirement yields

$$\cos \vartheta_{12} = -1/n_i \quad (i = 1, 2), \tag{21}$$

where the hybridization parameter $n_i = (1 - a_i^2)/a_i^2$ is by definition a positive number. It gives a ratio between the p and s populations in the hybrid. Hence, it immediately follows that two equivalent hybrids can form an angle which must be larger than $90°$, which leads to bent bonds in small rings. Furthermore, the orthogonality condition yields for the C_{2v} local symmetry case a relation

$$\cos \vartheta_{34} = (\cos \vartheta_{12} + 1)/(3 \cos \vartheta_{12} - 1), \tag{22}$$

ϑ_{34} denoting the angle between the second pair of the hybrids placed on the same nucleus. It is easy to see that an increase in one angle is accompanied by a decrease in the other. This is empirically known as the Thorpe–Ingold effect[67]. Therefore, a functional relationship between the two structurally independent parameters [(22)] is established through the hybridization concept. It should be kept in mind that the relation (22) leans rather heavily on the orthogonality requirement, which is plausible but by no means necessary, because hybrid AOs are not atomic eigenfunctions. Nevertheless, it is expected that overlapping of the single-center hybrids is very low because the electrons usually assume a state of the maximum spin multiplicity according to Hund's rule. Hence their spatial wavefunctions will try to avoid each other as much as possible to satisfy the Pauli principle. Even if the electron spins are random, the overlapping should be small because the valencies would otherwise be internally saturated. On the other hand, one can suppose on intuitive grounds that hybrids emanating from the same atom are orthogonal (or nearly so) because the interaction between bonds should be at a minimum if their individuality is to be retained. A concept of a localized covalent bond, on the other hand, is a well-founded empirical fact. Study of the hybridization by the semiempirical MO methods showed that the deviations from orthogonality are small for carbon atoms[68]. Relation (22) holds, of course, only approximately due to the neglect of higher AOs, steric effects, etc.[69,70]. It explains nicely, however, the opening up of the bond angle in CH_2 groups attached to small carbocycles[71,72]. The same feature exhibits the SiH_2 group in silacarbocycles[73]. This is easily explained by the increase in s content of the C—H (or Si—H) hybrids caused by the concomitant increase of the p character in hybrids describing a small ring. This rehybridization is a consequence of the tendency of the molecule to reduce the angular strain. The opposite is the case of the CH_2 group in the ethylene moiety. A multiple bond is always stronger in competing for the s character. Therefore, the HCH angle decreases to $117.5°$[71], which is an experimental fact. It should be mentioned that in these calculations[71–73] the maximum overlap criterion is employed for the prediction of interhybrid angles. This approach also gives fair estimates of dihedral angles in polycyclic systems and explains puckering of some six-, seven-, and eight-membered carbocycles[74,75]. Pauling's maximum hybrid orbital strength model is simple and attractive because it is free of parametrization. It gives purely theoretical estimates of bond angles. Stereochemistry is determined in this model exclusively by the central atom. This is a strong point because it gives conclusions which are perfectly general. The weakness, however, is complete neglect of the ligand influence. The model was applied to interpret and predict the structural properties of inorganic compounds, particularly transition-metal complexes[34,36,37,41,43,53,76–78] employing spd and $spdf$ basis sets. The fundamental importance of enneacovalence (nine covalent bonds) in determining structures of transition-metal compounds was revealed. Analysis of the sp^3d^5 hybrids has led to some remarkable conclusions: (a) there are two nodal cones of an sp^3d^5 hybrid which make angles of 73.2° and 133.6°, respectively, with the hybrid orbital axis. These angles determine domains in space where other hybrids can take place. It appears that the nodal cone is exactly the direction in which the other hybrid(s) assume the maximum bonding strength. (b) Two optimal polyhedra for enneacovalent transition metal atoms were determined. They correspond to hybrids with maximum bonding power and have the following forms: (1) trigonal prism with equatorial caps on the three rectangular faces and (2) tetragonal antiprism with a polar cap on one of the two bases. These polyhedra seem to play the same crucial role in transition-metal chemistry as the tetrahedron in organic chemistry. Furthermore, the best bonding set of eight sp^3d^4 hybrids form either a Archimedean (square) antiprism or a tetragonal dodecahedron. A large number of transition-metal structures was analysed along these lines and reasonable agreement with the experimental data, when available, was obtained. Admixtures of f and g orbitals in describing bonding in transition-metal clusters was discussed[39]. Finally, optimal hybrid AOs for pentacovalent bonding situations were described[54].

    Hybridization not only gives reasonable descriptions of molecular shapes, but also provides fair estimates of their sizes. Results of the iterative maximum overlap (IMO) method[71,75] should be emphasized. The underlying idea that bond radius directly depends on the hybrid's composition was pointed out first by Coulson[79]. Additionally, Dewar and Schmeising have shown that CC and CH bond lengths can be classified according to canonical hybridization states of the constituent carbon atoms[80]. This has been generalized in the IMO approach by including more flexible sp^n (n is any real number) hybrids within the framework of the IMO approach and good estimates of the structural characteristics of hydrocarbons were obtained. They can be favourably compared with the data obtained by much more intricate computations[72]. The IMO method seems to enjoy a remarkable predictive value. For example, the shortening of the interatomic distances belonging to central rings in rotanes was predicted on the basis of the IMO calculations[81]. This was subsequently confirmed by experimental measurements[82]. The estimated structure of the parent tetrahedrane[71a] (not yet synthetized) seems to be essentially correct[71b]. A compression of the double bonds emanating from small rings due to the redistribution of the s characters is predicted[83,84]. The structural properties of the disiloxy group and its derivatives were recently interpreted by hybrid orbitals distilled from *ab initio* wavefunctions[85]. The relevance of the hybridization concept in rationalizing molecular structures was recently discussed by Allen *et al.*[86]. To conclude, hybridization provides relations not only between the bond angles themselves but also serves as a guide in interpreting a number of relationships which do exist between bond lengths and angles.

Charge density distributions

    Electron isopycnic (isodensity) maps reflect the changes in charge distributions accompanying the formation of molecules[87,88]. Particularly informative for this purpose are deformation density contours. They are defined as a difference between the total molecular electron density and the superposition of atomic densities obtained by placing neutral atoms on the equilibrium positions and setting all their interactions equal to zero. This ensemble of atoms is called a promolecule. Its density distribution can be found only by accurate *ab initio* calculations. Hence, strictly speaking the deformation density maps are not observable. Nevertheless, they visualize the changes caused by bonding interactions in a very transparent way. One observes as a rule humps of the electron density in the region between the bonded atoms if they have moderate electronegativities. There is usually a depletion of charge in the rest of the molecules, particularly in their peripheral parts. Two important points should be stressed: (a) The changes in density are small as compared with the total molecular density, thus supporting the idea of perturbed atoms; (b) humps of the density nicely reflect a decrease in symmetry and distortion of atoms in chemical environments. This is exemplified by the X-ray deformation density map

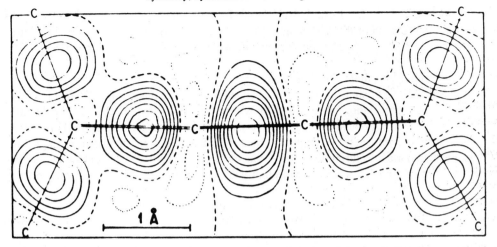

Fig. 6. Deformation density contours in butatriene part of the tetraphenylbutatriene.

taken in the butatriene plane of the tetraphenyl-substituted derivative (Fig. 6). One can notice distributions which are characteristic for sp^2 and sp^1 phenotypes of hybridization. The depicted map does not prove the "existence" of hybrids in this and other molecules, because for one thing the hybrid AOs are one-electron wavefunctions. Consequently, they are no more and no less real than the other atomic orbitals, the hydrogen wavefunction being an exception since it is a true solution. Figure 6 does show, however, that hybrid AOs are well adapted for a description of the atomic charge anisotropy.

A completely new feature is found in small strained rings. Theoretical isopycnic and deformation density maps based on local hybrid AOs exhibit considerable shift of the bond density off the geometric straight line joining the neighbouring carbon atoms[90]. Therefore the well-known bent bonds appear, discussed first in cyclopropane by Förster[91] and elaborated on later by Coulson and Moffitt in their famous paper[92]. It is interesting to mention that bent bonds were theoretically predicted some 25 years before they were experimentally found by X-ray measurements[87]. There are also numerous *ab initio* studies of strained systems by this point[93]. A number of unusual properties of small cyclic and polycyclic compounds can be rationalized in terms of bent bonds, to mention only Baeyer angular strain and pronounced chemical reactivity, for example. A bent chemical bond is not axially symmetric any more. It provides another example which shows that new facets can be expected if the symmetry is destroyed.

Energetic properties and hybridization

The relation between the directional properties of bonds and the total molecular energy is something which can be anticipated. If the postulate of the chemical bond formed by a localized pair of electrons is adopted, then it is plausible to assume that their spatial distributions will minimize the interbond repulsion, leading to the stable geometrical structure. The position of a hybrid and its direction in space depend on its s, p, etc. composition, but so does its bonding power and overlapping ability. This yields an additional link between the energy and directional features. It is difficult to delineate bonding and nonbonding effects because they are strongly interlocked. However, the former prevail at any rate since molecules are stable species. Therefore we shall focus our attention on stabilizing interactions. As mentioned earlier, the maximum hybrid strength yields a qualitative index of bond energies[62]. Good estimates of instantaneous bond dissociation energies (IBDE) in hydrocarbons were obtained by scaling the corresponding bond overlaps[94]. The former were defined as the energies necessary to break the bond, leaving the formed fragment radicals unrelaxed. A sum of weighted bond overlap integrals was correlated with the observed enthalpies of formation ΔH_f, supposing that the bond energy follows an additivity scheme. Quite reasonable results were obtained for a variety of hydrocarbons[95]. They are reliable enough to predict ΔH_f values in unknown compounds and to offer useful estimates of the heats of hydrogenation ΔH_h[95]. By reversing the argument, one can say that hybridization and its characteristic transferability between similar molecular fragments gives the simplest explanation of the ΔH_f additivity. The angular strain caused by deformed electron densities can be related to hybrid bending and a concomitant defect in overlapping[95].

Spin–spin coupling constants across a bond

Indirect coupling of spins of the directly bonded nuclei takes place via the coupled electron pair. Hence the interpretation by localized MOs seems to be an obvious choice. The situation is not simple, however, because the coupling constant is a sum of several terms[96]. Fortunately, the Fermi contact term (FC) is a dominant effect in hydrocarbons and thus can be singled out as the most important contribution. Since the FC term depends on the extent of electron penetration into the nuclei, a direct connection with s characters of the corresponding hybrids follows:

$$J(AB) = k_{AB}c_A^2c_B^2 + l_{AB}, \tag{23}$$

where A and B stands for C and H atoms, c_A^2 and c_B^2 are s-orbital populations of the hybrids, and the optimal empirical parameters are denoted by k_{AB} and l_{AB}. Relation (23) is of course an approximate one. It involves the well-known Muller–Pritchard (A=C, B=H) and the Frei–Bernstein (A=C, B=C) formulas if the adjusting factors k_{AB} and l_{AB} assume the prescribed values[96]. The average excitation energy ΔE appearing in the denominator of the first term was included in the k_{AB} parameter. Furthermore, the normalization constant of the VB function describing the AB fragment is neglected here. Its inclusion gives slight quantitative improvement[97] but it is not essential for our purpose here. Extensive semiempirical calculations have conclusively shown that relation (23) holds to a reasonable degree of accuracy[96–98]. Some care has to be exercised when multiple bonds are involved because non-Fermi terms cannot be omitted anymore[96,99]. Nevertheless, despite some reservations[100], it is beyond doubt now that $J(CC)$ and $J(CH)$ couplings are intimately related to hybridization states of the carbon atoms in question. If this is accepted, then the empirically observed relations between the $J(CH)$ coupling constants, $d(C—H)$ bond distances[101] and C—C—C bond angles in cyclic compounds[102] become perfectly understandable. The same conclusion holds for the connection between $J(C—H)$ and $J(CC)$ couplings involving the common carbon atom[103]. Furthermore, a linear relation between the $\nu(C—H)$ stretching frequencies and $J(C—H)$ spin–spin couplings was proposed on theoretical grounds[104]. Finally, the experimental CC stretching force constants were linearly related to the corresponding $J(CC)$ coupling constants[105]. Needless to say, the unifying concept underlying all these interrelations between different observables is hybridization.

Photoelectron spectroscopy and hybridization

Photoelectron spectroscopy (PES) is a playground for the MO theory. Simple MO schemes are usually very effective in interpreting PES spectra and provide useful aids in their assignments[106]. Hybrid orbitals seem to be inappropriate for this purpose at first glance. However, this is not always the case. We shall discuss it in some more detail because the symmetry arguments are particularly useful in this respect. Consider, for example, the CH_4 molecule. The molecular orbitals can be constructed consecutively by forming first the localized two-center orbitals:

$$\lambda_i = \frac{1}{\sqrt{2}}(h_i + H_i), \tag{24}$$

where h_i are sp^3 hybrids (16) and H_i stands for the $1s_{H_i}$ hydrogen functions. It is tacitly assumed that the C—H bonds are purely covalent and the overlap is neglected in the normalization constant for simplicity. The delocalized MOs of A_1 and T_2 symmetry species are readily produced by using projection operator (6) technique (*vide supra*). They read as follows:

$$A_1: \quad \psi_1 = \frac{1}{2\sqrt{2}}[(h_1 + h_2 + h_3 + h_4) + (H_1 + H_2 + H_3 + H_4)], \tag{25a}$$

$$\psi_2' = \frac{1}{2\sqrt{6}} [(3h_1 - h_2 - h_3 - h_4) + (3H_1 - H_2 - H_3 - H_4)]; \quad \text{(25b)}$$

$$T_2: \quad \psi_3' = \frac{1}{2\sqrt{6}} [(-h_1 + 3h_2 - h_3 - h_4) + (-H_1 + 3H_2 - H_3 - H_4)], \quad \text{(25c)}$$

$$\psi_4' = \frac{1}{2\sqrt{6}} [(-h_1 - h_2 + 3h_3 - h_4) + (-H_1 - H_2 + 3H_3 - H_4)]. \quad \text{(25d)}$$

By using formula (9) one straightforwardly obtains

$$\psi_1 = \frac{1}{\sqrt{2}} \left[s + \frac{1}{2} (H_1 + H_2 + H_3 + H_4) \right]. \quad \text{(26a)}$$

One can take advantage of the fact that the molecular orbitals belonging to the T_2 irreducible representation are degenerate. They can therefore be combined to simplify the formulas (25b–d). Utilizing eqns (12a–c) one gets

$$\psi_2 = \frac{1}{\sqrt{2}} \left[p_X + \frac{1}{2} (H_1 - H_2 + H_3 - H_4) \right], \quad \text{(26b)}$$

$$\psi_3 = \frac{1}{\sqrt{2}} \left[p_Y + \frac{1}{2} (H_1 - H_2 - H_3 + H_4) \right], \quad \text{(26c)}$$

$$\psi_4 = \frac{1}{\sqrt{2}} \left[p_Z + \frac{1}{2} (H_1 + H_2 - H_3 - H_4) \right]. \quad \text{(26d)}$$

Hence, the hybrid AOs completely disappeared. They are replaced by the conventional s, p_X, p_Y and p_Z free-atom AOs. The set of formulas (26) is a source of a widespread misconception that hybrids do not "exist." In fact, it shows only that s and p orbitals are a more sensible choice for the MO scheme in the highly symmetric T_d system. The hybrid basis set also yields MOs of A_1 and T_2 symmetries, but the corresponding formulas (25) are more complicated. The opposite case is found in distorted tetrahedra occurring in highly strained systems. Let us consider a prototype of the angularly strained molecule cyclopropane. Two hybridization schemes were put forward to describe the strained three-membered ring. They are depicted in Fig. 7. The Coulson–Moffitt bent-bond model describes the cyclopropyl ring by hybrids which are somewhat accommodated to the small internuclear angle of 60°. Their interorbital C—C—C angle is therefore smaller than 109.5°, but of course larger than 90°, as required by eqn (22), thus forming bent bonds (BB). The Walsh model describes bond bending by tangential p_t orbitals.

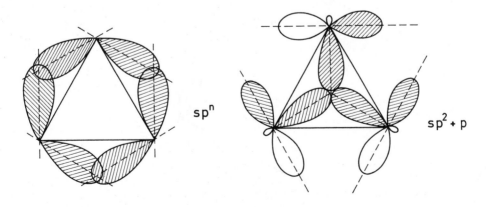

Coulson - Moffitt picture   Walsh picture

Fig. 7. Coulson–Moffitt and Walsh models for the cyclopropyl ring.

Each carbon donates in addition one sp^2 hybrid directed to the center of the ring. This model is widely applied in interpreting PES ionization potentials in strained cyclic and polycyclic compounds involving three-membered rings. The reader probably shares the writer's uneasy feeling caused by the unfavourable overlapping of p_t orbitals, which is always a warning for a danger. It should be strongly emphasized that free-atom AOs, Coulson–Moffitt and Walsh basis sets are completely equivalent if the calculations are carried out far enough, because they are interrelated by orthogonal transformations. This, however, is not the issue. The question we would like to answer is which of the three basis sets leads most directly to the satisfactory result, or in other words, which starting point is the best in the sense of the Occam's razor principle. Heilbronner *et al.*[107] have conclusively shown by using a simple Hückel-type approach that the bent bond (BB) model yields in cyclopropane the final result in a smooth way. The final step of taking into account the configuration interaction (CI) could be safely skipped. On the contrary, the Walsh (W) basis set is considerably less convenient and only the explicit CI treatment gives acceptable results. A somewhat more complex situation appears in bicyclo(1.1.0)butane, which we shall consider now because it is a nice example of the use of symmetry arguments in discussing the electronic structure of molecules. The Walsh scheme representing the generalization of the cyclopropyl case (Fig. 8) is not compatible with the geometric structure of the molecule[108]. A refined W model consisting of the sp hybrids on the bridgehead carbons accompanied by a pair of p_r radial and p_t tangential AOs on each center is more appropriate and it will be utilized in what follows. The terminal carbon atoms are described by the $sp^2 + p_t$ basis as in cyclopropane (Fig. 8). The best BB model of bicy-

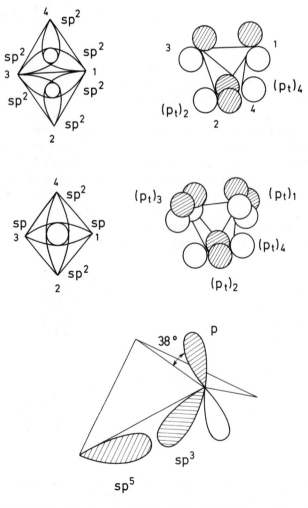

Fig. 8. Walsh and bent-bond models for the bicyclo(1.1.0)butane skeleton.

clo(1.1.0)butane is that emerging from the *ab initio* calculations of Newton and Schulman[109]. The weak central bond is described by *p* orbitals inclined by 38° to the internuclear line, while the perimetric C—C bonds are formed by the sp^3–sp^5 hybrids (Fig. 8). We shall try to describe the carbon skeleton by the semilocalized molecular orbitals (SLMO) belonging to irreducible representation of the C_{2v} point-symmetry group† by using BB and W basis sets. In order to make a fair comparison between these two basis sets, we shall slightly adjust the W model and use the $sp^{0.5}$ hybrid at the bridgeheads. Then the adopted BB model and W basis correspond to identical hybridization states[108]. For example, the bridge bond is given by the same *p* orbital inclined by 38°, etc. The SLMOs are readily obtained by inspection of the table of characters for the C_{2v} group[62]. The orbital composition of SLMOs is schematically displayed in Fig. 9. Several comments are in place here. The energy scale is not the same for the BB- and W-SLMO levels. It can be easily shown that $1'a_1$, $1a_2$ and $2'a_1$ levels of A symmetry are equal for both models because the corresponding SLMO functions are identical. The $1'a_1$ and $2'a_1$ wavefunctions are of the A_1 symmetry and can mix together. Their + and − combinations yield the most stable and the highest occupied MO (HOMO) orbitals, respectively. It is obvious by using a simple overlapping argument that the $1'a_1$ and $2'a_1$ orbital schemes depicted in Fig. 9 give the dominant contribution to the corresponding correct SLMO. The former are given for

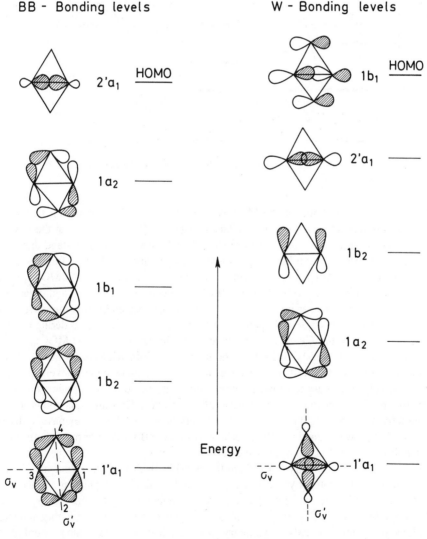

Fig. 9. Semilocalized molecular orbitals of the bicyclo(1.1.0)butane skeleton obtained by using BB and W models.

†These are symmetry-adapted linear combination of basis set functions, or SALCs in Cotton's terminology[60].

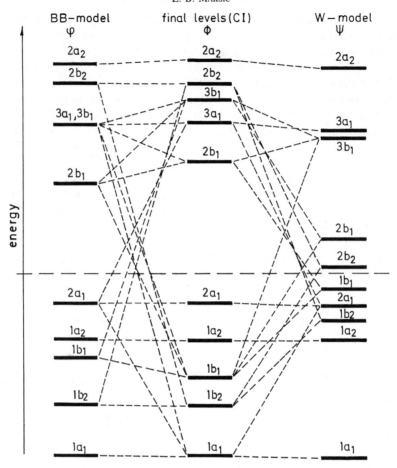

Fig. 10. One-electron CC levels in bicyclo(1.1.0)butane.

better inspection. The ordering of the SLMOs is easily obtained by assuming that intraring overlap is more important than the transannular overlapping. One observes that the BB and W schemes predict different symmetries for HOMOs. It will appear in a later stage that the BB-SLMO ordering is correct. A simple and plausible parametrization within the framework of the model Hamiltonian of the Hückel-type provides quantitative estimates of the SLMO levels[108]. Furthermore, CI with virtual levels yields the final one-electron eigenvalues and eigenvectors. They are the same (Fig. 10) for both BB and W models, as expected since the corresponding basis sets are related by a simple orthogonal transformation. However, the ordering of the BB-SLMOs is preserved by the CI, while considerable reordering of the W-SLMOs is necessary to obtain the final result. The W-SLMOs of the b_1 and b_2 symmetries are by far too unstable. This is a consequence of the shift of s character to the virtual (unoccupied) levels of the same symmetries, which in turn are too low. A serious difficulty faced by the application of the W basis is caused by the symmetry characteristics of the p_t orbitals. They are antisymmetric relative to the symmetry plane passing through the terminal carbon atoms. Hence they have to disappear in the $1b_2$ W-SLMO wavefunction. The hybrid AOs in the BB model are much more flexible. A pair of hybrids emanating from the same carbon can either assume the same sign or adopt the opposite signs. Concomitantly, apart from the HOMO, B-SLMOs have four nice intrabond overlaps, and are thus more physically acceptable and aesthetically pleasing. Since the BB scheme is a better and more realistic starting point, the final CI may be omitted for qualitative purposes. Importance of this finding lies in the fact that hybrid AOs are highly transferable between the similar moieties, and the symmetry reasoning displayed above can give enlightening insight into the ordering of the one-electron levels. The moral of the story is that a shrewd choice of the basis set accompanied by symmetry arguments can save a lot of computational efforts.

Miscellaneous molecular properties and hybridization

Compton profiles depend on the momentum distribution in a molecule[110]. The latter exhibits additivity in well-localized systems, which can be interpreted in terms of LMOs and in *ultima linea* by hybridization[111–113]. It appears that the average momentum of, for example, the C—H bond increases with the *p*-orbital population in a hybrid[110,111].

Bond stretching frequencies and force constants show hybridization dependence[47,79,104,105]. Hybrid orbitals seem to be a natural choice for a basis set in force-field methods, but some care has to be exercised due to the incomplete orbital following of the nuclear motions[114]. Molecular total (and bond) dipole moments depend dramatically on hybridization[79,115], but the charge migration contribution is also very important. The same conjecture is valid for electric field gradients and the corresponding asymmetry parameters[116,117]. Magnetic properties like diamagnetic shielding (σ^d) of the nuclei and diamagnetic susceptibility of molecules (χ^d) are indirectly dependent on hybridization via molecular geometry. Simple approximate formulas involving only interatomic distances were suggested for the calculation of these entities. The results of the IMO method based on local hybrid orbitals for σ^d and χ^d in hydrocarbons are in good accordance with observed values and/or *ab initio* data[51].

SYMMETRY, HYBRID ORBITALS AND MORE INVOLVED MOLECULAR WAVEFUNCTIONS

The exploitation of symmetry in complex quantum-chemical calculations of molecular wavefunctions is of utmost importance. The reader is referred to the illuminating discussion of the factorization of secular equations by Byers Brown[118], and a review article on symmetry adaptation and applications of Wigner–Racah algebras in quantum chemistry by Kibler[119]. We shall confine our discussion here to the choice of the symmetry-adapted functions at the atomic level. It is intuitively clear that the use of the chemically adjusted hybrid AOs must be advantageous in incomplete (simple) VB and MO calculations. The preceding discussion of the BB basis in describing carbon skeletons in cyclopropane and bicyclo(1.1.0)butane was very illustrative in this respect. Indeed, if we employ incomplete sets and the interactions are treated only to a certain degree of accuracy, then the starting point becomes extremely important. The atomic basis set should in this case reflect the salient features of the molecular environment, the local symmetry being the most important one. The first use of the polarized (i.e. hybrid) AOs can be traced to the 1930s. Dickinson[120] and Rosen[121] employed the ($1s_H + 2p_H$) basis for the calculation of ground states of H_2^+ and H_2 systems, respectively. The *ab initio* MO studies based on the hybrid AO sets are rather sparse. The early work of Hoyland[122] on hydrocarbons and of Petke and Whitten[123] on small heteroatomic molecules should be mentioned. The simulated *ab initio* scheme[124] relies on the transferability of the matrix elements from smaller fragment molecules to larger systems. It is much easier to keep track of transferable matrix elements in hybrid basis sets, because they are independent of the choice of the coordinate systems. The role of hybrid AOs in VB and GVB[32,38,42] methods was already mentioned. It should be pointed out, however, that the use of hybrid orbitals within the VB function gives a fair approximation to the linear combination of several VB configurations[125]. The group function method of McWeeny and Klessinger[126] involves full CI within the limited basis of two local hybrid AOs belonging to a given bond geminal. The total electronic wavefunction is written then as an antisymmetrized product of pair functions describing inner core, bond and lone pairs of electrons[126,127]. It is a pity that this approach did not enjoy more applications.

Properly adjusted hybrid orbitals are local wavefunctions of the zeroth order. Notwithstanding their remarkable properties, hybrid basis sets did not find widespread use in semiempirical theories of chemical bonding. They are, however, pivotal in the simple maximum overlap[128,129] or iterative maximum overlap[46,51,71,73,75,81,83,84] procedures. The former can be employed only if the geometry of molecules is known, which is a serious drawback. The hybrid-based MO schemes were used by Fukui and coworkers[130] in discussing molecular properties and reactivity. A conceptually important semiempirical method was devised by French researchers[131,132]. It is called PCILO because the localized MOs formed by hybrid orbitals undergo a perturbational procedure in order to take into account configuration interaction.

Unfortunately, the integral approximation scheme is borrowed from the CNDO method, which should be considered as a "*Schönheits Fehler.*" The structure of the method indicates that PCILO will be effective in well-localized systems, and not unexpectedly in this type of compounds it gives useful results. Current semiempirical MO approaches disregard the local symmetry-adapted basis sets, which seems to be a source of considerable difficulties. There are arguments which indicate that various zero-differential overlap (ZDO) approximation schemes can be justified, at least to a large extent, if a basis set of uniformly (or Löwdin) orthogonalized hybrid AOs is utilized[133,134]. Finally, it should be mentioned that hybrid basis sets proved very useful in approximate calculations of the electronic structure and properties of covalent solids[135,136].

We shall note in passing that hybridization is an important concept in interpreting intricate molecular wavefunctions. The hybridization indices can be extracted from the first-order density matrix elements[44,68,137–139], and compared with results obtained by more elementary procedures involving hybrid orbitals which are built in the model. This is a remarkable result because the one-determinantal wavefunction (2) is invariant to all orthogonal transformations of the basis set and yet the directional properties of bonds are stored in its density matrix. It is a rather fortunate feature, too, because otherwise the single determinants would not be acceptable wavefunctions for a description of chemical bonding.

FINAL REMARKS AND CONCLUSIONS

Hybridization is a very old concept created by men to describe a new property by using the well-known entities. In ancient Greece, a Centaur (Fig. 11)[140] was coined using an unmutilated human torso and the body of a horse, in an artistic attempt to represent a hillbilly. Even older examples of nowadays unusual hybrid creatures can be found in Egyptian mythology. It is likely that the idea of hybridization emerged by observing hybrid animals and plants in nature. It is also not surprising that the hybridization is used today in the biosciences; among these we mention only hybridoma cells obtained by a fusion of lymphocytes with plasma-cytoma cells[141,142]. The former produce a specific antibody, while the latter are immortal. Their hybrids are capable of producing monoclonal antibodies *ad infinitum*. This discovery was awarded a Nobel Prize for Medicine in 1984.

Hybridization in chemistry is an intellectual model designed to describe directional features of chemical bonds. A hybrid orbital formed by s, p, . . . AOs has certain characteristics of

Fig. 11. Centaur in fight.

each component and yet exhibits remarkable new properties which are not inherent to pure AOs. Hybrids not only rationalize a large part of stereochemistry, but provide simple explanations for a number of otherwise unrelated observables, thus suggesting some underlying authenticity of the model. This "grain of truth" is to be ascribed, at least partly, to the symmetry content of hybrids. They are adapted to conform with the local symmetry of an atom in a molecular environment. The bonding power of the hybrids is increased by the use of additional criteria like maximum bond strength or overlapping. These simple criteria and slight empirical adjustments significantly augment the performance of the model. One can find full analogy with the crystal field theory, where the symmetry represents a backbone of the approach and the semiquantitative agreement with experiment is achieved after some parametrization. Hybridization and localized molecular orbitals have considerable epistemological importance because they provide one of the cornerstones of the quantum theory of chemical bonding and valence. They describe and explain the most important axiom of the experimental chemistry—the chemical bond, which is a firm and well-established empirical concept. It is impossible to derive the concept of a chemical bond from the first principles. Hence, the hybridization is an important link between the rigorous quantum theory of molecules and phenomenology. It is difficult to find a simpler concept which is so rich in chemical information and has at the same time such a pervasive interpretative power. In my opinion, there are still nuggets waiting to be picked up in a gold field called hybridization, particularly within the framework of the semi-empirical theories of molecules and crystals.

Acknowledgments—This manuscript was completed in part within the contract for scientific cooperation between the Universities of Hamburg and Zagreb. Vehement and highly stimulating discussions with Professor M. Grodzicki are gratefully acknowledged.

REFERENCES

1. E. P. Wigner, *Symmetries and Reflections*. Indiana University Press, Bloomington (1970).
2. E. P. Wigner, *Group Theory*. Academic Press, New York (1960).
3. M. von Laue, in *International Tables for X-ray Crystallography*. Lynch Press, Breminthaven (1965).
4. L. Pauling, in *Structure and Bonding in Crystals* (Edited by M. O'Keeffe and A. Navrotsky), Vol. 1. Academic Press, New York (1981).
5. H. Bethe, Termaufspaltung in Kristallen. *Ann. Phys.* **5**, 133–208 (1929).
6. J. H. Van Vleck, Theory of the variations in paramagnetic anisotropy among different salts of the iron group. *Phys. Rev.* **41**, 208–215 (1932).
7. J. H. Van Vleck, *The Theory of Electric and Magnetic Susceptibilities*. Oxford University Press, Oxford (1932).
8. R. Schlapp and W. G. Penney, Influence of crystalline fields on the susceptibilities of paramagnetic ions. II. The iron group, especially Ni, Cr and Co. *Phys. Rev.* **42**, 666–686 (1982).
9. C. J. Gorter, Note on the electric field in paramagnetic crystals. *Phys. Rev.* **42**, 437–438 (1932).
10. W. Moffitt and C. J. Ballhausen, Quantum theory. *Ann. Rev. Phys. Chem.* **7**, 107–136 (1956).
11. C. J. Ballhausen, *Introduction to Ligand Field Theory*. McGraw-Hill, New York (1962).
12. B. N. Figgis, *Introduction to Ligand Fields*. John Wiley & Sons, New York (1966).
13. W. Haberditzl, *Quantenchemie. Band 4. Komplexverbindungen*. Dr. A. Hüthig Verlag, Heidelberg (1979).
14. C. E. Schäffer, Two symmetry parametrizations of the angular overlap model of the ligand field. Relation to the crystal field model. *Structure and Bonding* **14**, 69–110 (1973).
15. R. S. Mulliken, Electronic structures of polyatomic molecules and valence. *Phys. Rev.* **40**, 55–62 (1932).
16. R. S. Mulliken, Electronic structures of polyatomic molecules and valence. IV. Electronic states, quantum theory of the double bond. *Phys. Rev.* **43**, 279–301 (1933).
17. R. B. Woodward and R. Hoffmann, *The Conservation of Orbital Symmetry*. Verlag Chemie, Weinheim (1971).
18. E. Wigner, The elastic characteristic vibrations of symmetrical systems. English translation of the paper written in German and published in *Göttingen Nachrichten* 133–146 (1930). Reprinted in A. P. Cracknell, *Applied Group Theory*. Pergamon Press, Oxford (1968).
19. E. B. Wilson, Jr., The degeneracy, selection rules and other properties of the normal vibrations of certain polyatomic molecules. *J. Chem. Phys.* **2**, 432–439 (1934).
20. E. B. Wilson, Jr., Symmetry considerations concerning the splitting of vibration–rotation levels in polyatomic molecules. *J. Chem. Phys.* **3**, 818–821 (1935).
21. H. C. Longuet–Higgins, The symmetry groups on nonrigid molecules. *Mol. Phys.* **6**, 445–460 (1963).
22. P. R. Bunker, *Molecular Symmetry and Spectroscopy*. Academic Press, New York (1979).
23. G. S. Ezra, *Symmetry Properties of Molecules. Lecture Notes in Chemistry*, Vol. 28. Springer Verlag, Berlin (1982).
24. J. Hinze (Ed.), *The Unitary Group. Lecture Notes in Chemistry*, Vol. 22. Springer Verlag, Berlin (1981).
25. M. D. Gould and G. S. Chandler, Unitary group approach to many-electron problem, Parts 1–3. *Int. J. Quant. Chem.* **25**, 553–601, 603–633, 1089–1109 (1984).
26. C. D. H. Chisholm, *Group Theoretical Techniques in Quantum Chemistry*. Academic Press, London (1976).
27. I. G. Kaplan, *Simmetrija Mnogoelektronih Sistem*. Nauka, Moskva (1969).
28. R. S. Knox and A. Gold, *Symmetry in the Solid State*. W. A. Benjamin, New York (1964).
29. L. Pauling, The nature of the chemical bond. Applications of results obtained from the quantum mechanics and

from a theory of paramagnetic susceptibility to the structure of molecules. *J. Am. Chem. Soc.* **53**, 1367–1400 (1931).

30. J. C. Slater, Directed valence in polyatomic molecules. *Phys. Rev.* **37**, 481–489 (1931).

31. K. A. R. Mitchell and T. Thirunamachandran, Valence-bond calculations on the ground state of BeH_2. *Chem. Phys. Letters* **6**, 407–408 (1970).

32. W. A. Goddard III, T. H. Dunning, Jr., W. J. Hunt and P. J. Hay, Generalized valence bond description of bonding in low-lying states of molecules. *Acc. Chem. Res.* **6**, 368–376 (1973).

33. D. M. Silver and M. Karplus, Valence-bond approach to conservation of symmetry in concerted reactions. *J. Am. Chem. Soc.* **97**, 2645–2654 (1975).

34. L. Pauling, Valence-bond theory of compounds of transition metals. *Proc. Natl. Acad. Sci. USA* **72**, 4200–4202 (1975).

35. D. M. Chipman, B. Kirtman and W. E. Palke, The perfect-pairing valence bond model for water molecule. *J. Chem. Phys.* **65**, 2556–2561 (1976).

36. L. Pauling, Correlation of nonorthogonality of best hybrid bond orbitals with bond strength of orthogonal orbitals. *Proc. Natl. Acad. Sci. USA* **73**, 274–275 (1976).

37. L. Pauling, Angles between orthogonal *spd* bond orbitals with maximum strength. *Proc. Natl. Acad. Sci. USA* **73**, 1403–1405 (1976).

38. F. W. Bobrowicz and W. A. Goddard III, The self-consistent field equations for generalized valence bond and open-shell Hartree–Fock wave functions, in *Methods of Electronic Structure Theory* (Edited by H. F. Schaefer III), pp. 79–127. Plenum Press, New York (1977).

39. L. Pauling, Structure of transition-metal cluster compounds: Use of an additional orbital resulting from the f, g character of *spd* bond orbitals. *Proc. Natl. Acad. Sci. USA* **74**, 5235–5238 (1977).

40. R. G. A. R. Maclagan and G. W. Schnuelle, Valence-bond studies of the AH_2 molecule. *Theoret. Chim. Acta* **46**, 165–172 (1977).

41. L. Pauling, The nature of the bonds by the transition metals with hydrogen, carbon and phosphorus. *Acta Cryst.* **B34**, 746–752 (1978).

42. W. A. Goddard III and L. B. Harding, The description of chemical bonding from *ab initio* calculations. *Ann. Rev. Phys. Chem.* **29**, 363–396 (1978).

43. L. Pauling, The nature of the bonds formed by transition metals in bioorganic compounds and other compounds, in *Frontiers in Bioorganic Chemistry and Molecular Biology* (Edited by Yu. A. Ovchinnikov and M. N. Kolosov), pp. 1–20. Elsevier, Amsterdam (1979).

44. J. P. Foster and F. Weinhold, Natural hybrid orbitals. *J. Am. Chem. Soc.* **102**, 7211–7218 (1980).

45. B. Kirtman, W. E. Palke and D. M. Chipman, The valence bond orbital model as an interpretative framework for understanding electronic structure. *Isr. J. Chem.* **19**, 82–87 (1980).

46. Z. B. Maksić, K. Kovačević and A. Moguš, Investigation of the hybridization in small-ring hydrocarbons by the IMOA method. *J. Mol. Structure Theor. Chem.* **85**, 9–24 (1981).

47. W. A. Bingel und W. Lüttke, Hybridorbitale und ihre Anwendungen in der Strukturchemie. *Angew. Chem.* **93**, 944–956 (1981).

48. K. H. Aufderheide, Generalized localized atomic orbitals. *J. Chem. Phys.* **76**, 1897–1909 (1982).

49. G. A. Gallup, R. L. Vance, J. R. Collins and J. M. Norbeck, Practical valence-bond calculations. *Adv. Quant. Chem.* **12**, 229–272 (1982).

50. J. H. van Lenthe and G. G. Balint-Kurti, The valence-bond self-consistent field method (VB-SCF): Theory and test calculations. *J. Chem. Phys.* **78**, 5699–5713 (1983).

51. Z. B. Maksić, Variable hybridization—a simple model of covalent bonding. *Pure & Appl. Chem.* **55**, 307–314 (1983).

52. Z. S. Herman, Recent advances in simple valence-bond theory and theory of hybrid bond orbitals. *Int. J. Quant. Chem.* **23**, 921–943 (1983).

53. L. Pauling and Z. S. Herman, Valence-bond concepts in coordination chemistry and the nature of metal–metal bonds. *J. Chem. Ed.* **61**, 582–587 (1984).

54. Z. S. Herman and L. Pauling, Hybrid bond orbitals and bond strengths for pentacovalent bonding. *Croat. Chem. Acta* **57**, 765–778 (1984).

55. W. E. Palke, On determining orbital hybridization. *Croat. Chem. Acta* **57**, 779–786 (1984).

56. G. Del Re and C. Barbier, *In situ* atomic orbitals and extended basis molecular calculations. *Croat. Chem. Acta* **57**, 787–800 (1984).

57. M. A. Morrison, T. L. Estle and N. F. Lane, *Quantum States of Atoms, Molecules and Solids*. Prentice-Hall, New Jersey (1976).

58. G. N. Lewis, *Valence and the Structure of Atoms and Molecules*. The Chemical Catalog Company, New York (1923).

59. W. Heitler and F. London, Wechselwirkung neutraler Atome und homöopolare Bindung nach der Quantenmechanik. *Zeits. Physik* **44**, 455–472 (1927).

60. F. A. Cotton, *Chemical Applications of Group Theory*, 2nd. Edn. Wiley-Interscience, New York (1971).

61. D. M. Bishop, *Group Theory and Chemistry*. Clarendon Press, Oxford (1973).

62. L. Pauling, *The Nature of the Chemical Bond*, 3rd Edn. Cornell University Press, Ithaca, New York (1960).

63. G. Kimball, Directed valence. *J. Chem. Phys.* **8**, 188–198 (1940).

64. H. Eyring, J. Walter and G. E. Kimball, *Quantum Chemistry*. John Wiley & Sons, New York (1944).

65. J. C. Eisenstein, Use of f orbital in covalent bonding. *J. Chem. Phys.* **25**, 142–147 (1956).

66. H. A. Bent, An appraisal of valence-bond structures and hybridization in compounds of the first-row elements. *Chem. Rev.* **61**, 276–311 (1960).

67. B. Testa, *Principles of Organic Stereochemistry*. M. Dekker, New York (1979).

68. Z. B. Maksić and M. Randić, Comparative study of hybridization in hydrocarbons. *J. Am. Chem. Soc.* **95**, 6522–6530 (1973).

69. M. J. S. Dewar, H. Kollmar and W. K. Li, Valence angles and hybridization indices in "sp^3 hybridized" AX_2Y_2 systems. *J. Chem. Ed.* **52**, 305–306 (1975).

70. B. Klahn, The relations between the valence angles of sp^3-hybridized central atoms for all possible local sym-

metries. *J. Mol. Struct. Theor. Chem.* **104**, 49–77 (1983).

71. (a) K. Kovačević and Z. B. Maksić, Calculation of bond lengths and angles of hydrocarbons by the iterative MOA method. *J. Org. Chem.* **39**, 539–545 (1974).
   (b) H. Irngartinger, A. Goldman, R. Jahn, M. Nixdorf, H. Rodewald, G. Maier, K. D. Malsch and R. Emrich, Tetra-tert-butyltetrahedrane—crystal and molecular structure. *Angew. Chem. Int. Ed. Engl.* **23**, 993–994 (1984).

72. Z. B. Maksić, K. Kovačević and A. Moguš, Semiempirical versus *ab initio* calculations of molecular properties. II. Comparative study of interatomic distances and bond angles in some strained medium size hydrocarbons as obtained by the STO-3G, MINDO/3 and IMOA methods, *Theoret. Chim. Acta* **55**, 127–132 (1980).

73. M. Eckert-Maksić, K. Kovačević and Z. B. Maksić, The electronic structure of organosilicon compounds. III. Iterative maximum overlap calculations on some cyclic and polycyclic silanes. *J. Organomet. Chem.* **168**, 295–310 (1979).

74. Z. Meić and M. Randić, Hybridization in 1,3,5-cycloheptatriene and some related molecules by the method of maximum overlap. *Croat. Chem. Acta* **40**, 43–48 (1968).

75. Z. B. Maksić and A. Rubčić, Geometry of molecules. III. Iterative maximum overlap calculations of bond lengths in some conjugated polyenes and their alkylated derivatives. *J. Am. Chem. Soc.* **99**, 4233–4241 (1977).

76. L. Pauling, Bond angles in transition-metal tricarbonyl compounds: A test of the theory of hybrid bond orbitals. *Proc. Natl. Acad. Sci. USA* **75**, 12–15 (1978).

77. L. Pauling, Bond angles in transition-metal tetracarbonyl compounds: A further test of the theory of hybrid bond orbitals. *Proc. Natl. Acad. Sci. USA* **75**, 569–572 (1978).

78. L. Pauling, Evidence from bond lengths and bond angles for enneacovalence of cobalt, rhodium, iridium, iron, rhutenium and osmium in compounds with elements of medium electronegativity. *Proc. Natl. Acad. Sci. USA* **81**, 1918–1921 (1984).

79. C. A. Coulson, *Valence.* Oxford University Press, Fair Lawn, New Jersey (1961).

80. M. J. S. Dewar and H. N. Schmeising, A reevaluation of conjugation and hyperconjugation: the effects of changes in hybridisation on carbon bonds. *Tetrahedron* **5**, 166–178 (1959).

81. K. Kovačević, Z. B. Maksić and A. Moguš, Geometry of molecules. Part 4. Iterative maximum overlap calculations of interatomic distances, bond angles and strain energies in some rotanes and related spirocompounds. *Croat. Chem. Acta* **52**, 249–263 (1979).

82. A. Almeningen, O. Bastiansen, B. N. Cyvin, S. Cyvin, L. Fernholt and C. Rømming, The molecular structure of [4]-rotane. *Acta Chem. Scand. A* **38**, 31–39 (1984).

83. M. Eckert–Maksić and Z. B. Maksić, Geometry of molecules. Part 5. Interatomic distances and electronic structures of some alkyl-subsituted cyclopropanes and cyclopropenes by the IMOA method. *J. Mol. Struct. Theor. Chem.* **86**, 325–340 (1982).

84. Z. B. Maksić and M. Eckert–Maksić, Geometry of molecules. Part VI. Interatomic distances and electronic structures of some alkyl-substituted four- and five-membered cyclic hydrocarbons, *J. Mol. Struct. Theor. Chem.* **91**, 295–311 (1983).

85. M. D. Newton, Theoretical probes of bonding in disiloxy groups, in *Structure and Bonding in Crystals* (Edited by M. O'Keeffe and A. Navrotsky), Vol. I, pp. 175–193. Academic Press, New York (1981).

86. F. H. Allen, O. Kennard and R. Taylor, Systematic analysis of structural data as a research technique in organic chemistry. *Acc. Chem. Res.* **16**, 146–153 (1983).

87. P. Coppens and E. D. Stevens, Accurate X-ray diffraction and quantum chemistry: The study of charge density distributions. *Adv. Quant. Chem.* **10**, 1–35 (1977).

88. V. H. Smith, Jr., P. F. Price and I. Absar, Representation of the electron density and its topographical features. *Isr. J. Chem.* **16**, 187–197 (1977).

89. Z. Berkovitch–Yellin and L. Leiserowitz, Electron density distribution in cumulenes. A low temperature X-ray study of tetraphenylbutatriene. *J. Am. Chem. Soc.* **97**, 5627–5628 (1975).

90. Lj. Vujisić, D. Lj. Vučković and Z. B. Maksić, Charge density distribution in small strained rings. A local hybrid orbital study. *J. Mol. Struct. Theor. Chem.* **106**, 323–332 (1984).

91. Th. Förster, Die gegenseitige Beeinflussung der Valenzen im Kohlenstoffatom. *Z. Phys. Chem. B* **43**, 58–78 (1939).

92. C. A. Coulson and W. E. Moffitt, The properties of certain strained hydrocarbons. *Phil. Mag.* **40**, 1–35 (1949).

93. M. D. Newton, in *Applications of Electronic Structure Theory* (Edited by H. F. Schaefer, III), pp. 223–275. Plenum Press, New York (1977).

94. Lj. Vujisić and Z. B. Maksić, Hybridization in 2,5-dimethyl-7,7-dicyanonorcaradiene by the maximum overlap approximation. *J. Mol. Struct.* **7**, 431–436 (1971).

95. K. Kovačević, M. Eckert–Maksić and Z. B. Maksić, The calculation of the heats of formation, heats of hydrogenation and strain energies in nonconjugated hydrocarbons by the maximum overlap approximation. *Croat. Chem. Acta* **46**, 249–259 (1974).

96. J. Kowalewski, in *Ann. Rep. NMR Spectroscopy* (Edited by G. A. Webb), Vol. 12, pp. 82–176. AC, London (1982), and the references cited therein.

97. Z. B. Maksić, Calculation of *J*(CC) and *J*(CH) coupling constants in hydrocarbons by the maximum overlap method. *Int. J. Quant. Chem.* **5**, 301–306 (1971); Z. B. Maksić, M. Eckert–Maksić and M. Randić, Correlation between C—H and C—C spin–spin coupling constants and *s* character of hybrids calculated by the maximum overlap method. *Theoret. Chim. Acta* **22**, 70–79 (1971).

98. C. Van Alsenoy, H. P. Figeys and P. Geerlings, A CDOE/INDO LMO study of the nuclear spin–spin coupling constants between directly bonded C—H and C—C atoms. *Theoret. Chim. Acta* **55**, 87–101 (1980).

99. A. Laaksonen, J. Kowalewski and V. R. Saunders, Finite perturbation MCSCF and CI calculations of nuclear spin–spin coupling constants for some molecules with multiple bonds. *Chem. Phys.* **80**, 221–227 (1983).

100. V. M. S. Gil and C. F. G. C. Geraldes, in *Nuclear Magnetic Resonance of Nuclei Other Than Protons* (Edited by T. Axenrod and G. A. Webb), pp. 219–231. Wiley-Interscience, New York (1974).

101. J. B. Stothers, *Carbon-13 NMR Spectroscopy, Organic Chemistry*, Vol. 24. Academic Press, New York (1974).

102. K. Mislow, Correlation of C—H coupling constants and internuclear angles in cyclic molecules. *Tetrahedron Lett.* 1415–1420 (1964).

103. P. E. Hansen, in *Progress in Nuclear Magnetic Resonance Spectroscopy* (Edited by J. W. Emsley, J. Feeney and

L. H. Sutcliffe), Vol. 14, pp. 175–296. Pergamon Press, Oxford (1981).

104. Z. B. Maksić, Z. Meić and M. Randić, Correlation between C—H stretching frequencies and hybridization in hydrocarbons. *J. Mol. Struct.* **12**, 482–485 (1972).

105. K. Kamienska–Trela, Correlation of the CC spin–spin coupling constants with the stretching force constants of single and double carbon–carbon bonds. *Spectr. Acta* **36A**, 239–244 (1980).

106. A. D. Baker and C. R. Brundle, *Electron Spectroscopy Theory, Techniques and Applications*, Vols. 1, 2, 3. Academic Press, London (1977–1979); E. Heilbronner and H. Bock, *The Hückel MO Model and its Applications*, Vols. 1 and 2. Verlag Chemie, Weinheim, and John Wiley, Chichester (1976).

107. E. Honegger, E. Heilbronner and A. Schmelzer, Do Walsh-orbitals "exist?", *Nouv. J. Chim.* **6**, 519–526 (1982); E. Honegger, E. Heilbronner, A. Schmelzer and W. Jian-Qi, A reexamination of the Walsh- and Förster–Coulson–Moffitt (F.C.M.) orbital concept, and its relevance for the interpretation of PE spectra. *Isr. J. Chem.* **22**, 3–10 (1982).

108. M. Eckert–Maksić, Z. B. Maksić and R. Gleiter, A comparative study of bent-bond *vs.* Walsh model in strained systems. Bicyclo(1.1.0)butane. *Theoret. Chim. Acta* **66**, 193–205 (1984).

109. M. D. Newton and J. M. Schulman, Theoretical studies of bicyclobutane. *J. Am. Chem. Soc.* **94**, 767–773 (1972).

110. I. R. Epstein, Compton scattering and the chemistry of momentum space. *Acc. Chem. Res.* **6**, 145–152 (1973), and the references cited therein.

111. C. A. Coulson, Momentum distribution in molecular systems. Part I. The simple bond. *Proc. Camb. Phil. Soc.* **37**, 55–66 (1941); C. A. Coulson and W. E. Duncanson, Momentum distribution in molecular systems. Part II. Carbon and the C—H bond, *Proc. Camb. Phil. Soc.* **37**, 67–73 (1941); C. A. Coulson, Bond energies and the Compton profile for molecules. *Mol. Phys.* **26**, 507–508 (1973).

112. I. R. Epstein, Molecular momentum distribution and Compton profiles. II. Localized orbital transferability and hydrocarbons. *J. Chem. Phys.* **53**, 4425–4436 (1970).

113. A. Rozendaal and E. J. Baerends, Electron momentum density distribution in homonuclear diatomic molecules. *Chem. Phys.* **87**, 263–272 (1984).

114. D. M. Chipman, W. E. Palke and B. Kirtman, Are bonds bent? To what extent do bond orbitals follow nuclear motions? *J. Am. Chem. Soc.* **102**, 3377–3383 (1980); W. E. Palke and B. Kirtman, *J. Mol. Struct. Theor. Chem.* **104**, 207–213 (1983).

115. C. A. Coulson, The dipole moment of the C—H bond. *Trans. Farad. Soc.* **38**, 433–444 (1942).

116. E. A. C. Lucken, *Nuclear Quadrupole Coupling Constants*. Academic Press, London (1969).

117. S. Vega, Nuclear quadrupole resonance in solids. *Isr. J. Chem.* **16**, 213–219 (1977).

118. W. Byers Brown, in *Quantum Theory of Atoms, Molecules and the Solid State* (Edited by P. O. Löwdin), pp. 123–132. Academic Press, London (1966).

119. M. Kibler, Symmetry adaptation and Wigner–Racah algebras in quantum chemistry. *Croat. Chem. Acta* **57**, 1075–1095 (1984).

120. B. N. Dickinson, The normal state of the hydrogen molecule-ion. *J. Chem. Phys.* **1**, 317–318 (1933).

121. N. Rosen, The normal state of the hydrogen molecule. *Phys. Rev.* **38**, 2099–2114 (1931).

122. J. R. Hoyland, *Ab initio* bond-orbital calculations. I. Application to methane, ethane, propane and propylene. *J. Am. Chem. Soc.* **90**, 2227–2232 (1968); J. R. Hoyland, *Ab initio* bond-orbital calculations. II. An improved procedure for saturated hydrocarbons. *J. Chem. Phys.* **50**, 473–478 (1969).

123. J. D. Petke and J. L. Whitten, *Ab initio* studies of orbital hybridization in polyatomic molecules. *J. Chem. Phys.* **51**, 3166–3174 (1969).

124. J. E. Eilers and D. R. Whitman, Simulated *ab initio* molecular orbital technique. I. Method. *J. Am. Chem. Soc.* **95**, 2067–2073 (1973); J. E. Eilers, B. O'Leary, A. Liberles and D. R. Whitman, Simulated *ab initio* molecular orbital technique. II. Benzenoid aromatic hydrocarbons. *J. Am. Chem. Soc.* **97**, 5679–5985 (1975); J. E. Eilers, B. O'Leary, B. J. Duke, A. Liberles and D. R. Whitman, Simulated *ab initio* molecular orbital techniques. IV. Cyclohexanes. *J. Am. Chem. Soc.* **97**, 1319–1326 (1975).

125. J. Gerratt, Valence bond theory. *Spec. Period. Rep., Theoret. Chem.*, Vol. 1, *Quantum Chemistry*. The Chem. Soc., Burlington House, London (1974).

126. M. Klessinger and R. McWeeny, Self-consistent group calculation on polyatomic molecules. I. Basic theory with an application to methane. *J. Chem. Phys.* **42**, 3343–3354 (1965).

127. M. Klessinger, Self-consistent group calculations on polyatomic molecules. II. Hybridization and optimum orbitals in water. *J. Chem. Phys.* **43**, S117–119 (1965).

128. I. Hubač, V. Laurinc and V. Kvasnička, The generalized method for construction of hybrid orbitals by the maximum overlap method. *Chem. Phys. Lett.* **13**, 357–360 (1972); V. Kvasnička, V. Laurinc and I. Hubač, Generalized theory of maximum overlap. *Coll. Czech. Chem. Comm.* **37**, 2490–2496 (1972).

129. R. Boča, P. Pelikan, L. Valko and S. Miertuš, Maximum overlap approximation calculations on polyatomic molecules. I. EMOA method. *Chem. Phys.* **11**, 229–236 (1975).

130. K. Fukui, Hybrid-based molecular orbitals and their chemical applications, in *Sigma Molecular Orbital Theory* (Edited by O. Sinanoğlu and K. B. Wiberg), pp. 121–129. Yale University Press, New Haven (1970), and the references given therein.

131. S. Diner, J. P. Malrieu, P. Claverie and F. Jordan, Fully localized bond orbitals and the correlation problem. *Chem. Phys. Lett.* **2**, 319–323 (1968).

132. S. Diner, J. P. Malrieu, F. Jordan and M. Gilbert, Localized bond orbitals and the correlation problem. III. Energy up to the third order in the zero-differential overlap approximation. Application to σ-electron systems. *Theoret. Chim. Acta* **15**, 100–110 (1969).

133. D. B. Cook, The "invariance principle" in approximate molecular orbital theories. *Theoret. Chim. Acta* **40**, 297–302 (1975).

134. D. B. Cook, *Structures and Approximations for Electrons in Molecules*. Ellis Horwood Ltd., Chichester (1978).

135. A. A. Levin, *Solid State Quantum Chemistry*. McGraw-Hill, New York (1977).

136. W. A. Harrison, *Electronic Structure and the Properties of Solids*, W. H. Freeman and Co., San Francisco (1980).

137. C. Trindle and O. Sinanoğlu, Local orbital and bond index characterization of hybridization. *J. Am. Chem. Soc.* **91**, 853–858 (1969).

138. M. S. Gopinathan and K. Jug, Valency. I. A quantum chemical definition and properties. *Theoret. Chim. Acta* **63**, 497–509 (1983).
139. M. S. Gopinathan and K. Jug. Valency. II. Applications to molecules with first row atoms. *Theoret. Chim. Acta* **63**, 511–527 (1983).
140. G. Hafner, *Kreta und Hellas*. Holle Verlag, Baden-Baden (1967).
141. G. Köhler and C. Milstein, Continuous culture of fused cells secreting antibody of predefined specificity. *Nature* **256**, 495–497 (1975).
142. C. Milstein and G. Köhler, Cell fusion and the derivation of cell lines producing specific antibody, in *Antibodies in Human Diagnosis and Therapy* (Edited by E. Haber and R. M. Krause). Raven Press, New York (1977).

Comp. & Maths. with Appls. Vol. 12B, Nos. 3/4, pp. 725–750, 1986
Printed in Great Britain.

0886–9561/86 $3.00 + .00

SYMMETRY IN MUSLIM ARTS

ERZSÉBET RÓZSA
Mandula utca 24, Budapest, H-1025, Hungary

Abstract—The present article gives a brief summary of the manifestations of symmetry in Muslim arts. The material was collected in different Muslim countries and was selected in such a way as to show the existence of the basic patterns of symmetry in every branch of art and at the same time in every aspect of everyday life of the Muslim community.

Fig. 1. In the name of God, the Merciful, the Compassionate.

When one enters a mosque for the first time, what surely strikes the eye is the lack of paintings and sculptures so natural in our Christian culture. Instead, we find decorations of various kinds floating and curving along the walls, which are usually referred to as "geometric". Certainly the reason of this great difference is not the inability of Muslim artists to describe living beings— but then what?

Nowadays, when—for political and economic reasons—so much attention is paid to the Arabic and Muslim world, it is common knowledge that the holy book of Islam, the *Quran*, forbids the description of beings having a soul. However, vast as the territory of the Muslim world once was and still is, the Muslims could not free themselves from the effects of the previous cultures flourishing in the parts of the world conquered by them, which fact resulted in several counterexamples to the above statement.

Consider the year 622 (the year of the emigration of the followers of Muhammad from Mecca to Medina), the beginning of Islam. We find that within a relatively short period—about 100 years—the Muslims conquered the Arabic peninsula, North Africa, Spain, Southern France (for a while), Sicily, the whole territory called the Middle East today, and penetrated into Central Asia and India. Several significant cultures were flourishing in these areas at the time, and their effects on the Muslim world and thought gave birth to such works of art as the beautiful Persian and Turkish miniatures and carpets, which are also full of descriptions of human beings. Nevertheless, firmly and powerfully as religion governed and still governs the life of the Muslim community, the prohibition of the *Quran* influenced Muslim arts very strongly.

What were the sources of this prohibition?

Muslim religious thought was greatly effected by the other two great religions of the territory, Judaism and Christianity, so much so that the 5 "pillars" of Islam have close connections with their regulations. The first "pillar", the testimony (*shahādah*), is considered to be the most important, as it is the manifestation of Islam being a monotheistic religion, saying "There is no other god than God, and Muhammad is God's prophet" (*lā ilāha illa-llāha wa muhammad rasūlu-llāh*). The other "pillars", the fasting (*ṣaum*); the prayer 5 times a day (*salāh*); the poor tax (zakāh) and the pilgrimage to Mecca (ḥajj) which every Muslim must perform at least once in his life; are also clearly related to the basic concepts of Judaism and Christianity.

Prohibition of description has the closest relationship with the testimony, which is also

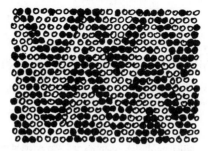

Fig. 2. Cone mosaics from Warka, Mesopotamia.

called *"tawḥīd"*, the profession of the unity of God. God being One and having no compan-
ion[1], it is only He, who has the power and ability of creation, the perfection of which Man
can never reach. *"Al-muṣawwir"*, one of His 99 attributes, means the Bestower of Forms and
Colours and is derived from the same root as the word *"ṣūrah"* (picture)[2]. In the Arabic
translation of the Old Testament this very word is used to denote the subject of prohibition[3],
proving the relations and the interconnections of the great monotheistic religions referred to
above.

Religion was not the only field where Arabs were influenced by other cultures. Their quick
expansion brought them into conflict with the two great political powers of the time. Byzantium
and the Sassanid Iran, both having a well-developed, flourishing culture: the former based on
Hellenistic heritage and Christianity, the latter fostered and preserved the heritage of ancient
Assyria and the Achaemenian empire. Muslim arts were influenced and inspired by both.

> Sassanian art must be credited with the creation of a new style of abstract, pseudo-floral
> ornament, based on traditions of Assyrian and Achaemenian art, in which rhythmic repetition
> and symmetry are the main principles.[4]

This statement is illustrated by the examples below. The famous conic mosaics of Warka
represent ever-returning patterns, while the row of flowers above the heads of the guards in the
Persepolis Palace of Darius can be found in the oldest mosque of Cairo built by Ibn Tulūn
(Figs. 2–5).

On the other hand, "Christian art of Egypt, Syria and Mesopotamia furnished models for
a number of decorative schemes found in early Islamic monuments"[5]. On the walls of the
Armenian Temple of Jerusalem the same motifs (Fig. 6) can be found as on the outer surface
of the Dome of the Rock in Jerusalem (Fig. 7), which is a unique monument in Islam, erected
by Caliph Abd al-Malik (685–705). The outline of the Dome is octagonal (Fig. 8). It was built
on the site where the Temple of Solomon had stood, atop the rock, which—according to
tradition—was that of the sacrifice of Abraham, whom the Arabs consider their ancestor. Besides
its religious significance it is important from the point of view of the arts as well. It incorporates
elements of Byzantine tradition, such as wooden stays connecting the capitals, the small dome,

Fig. 3. The cupola of the Mu'ayyad Mosque, Cairo, Egypt.

Fig. 4. The way leading to the audience hall of Darius I, Persepolis, Iran.

Fig. 5. Decoration from the al-Qatai Mosque built by Ibn Tulūn, Cairo, Egypt (876–79).

Fig. 6. Fragment of the Armenian Temple, Jerusalem, Israel.

Fig. 7. Outer decorations of the Dome of the Rock, Jerusalem, built by Caliph Abd al-Malik (691).

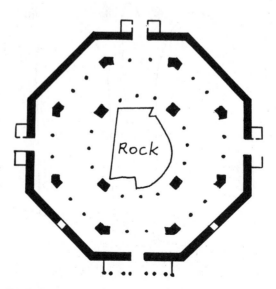

Fig. 8. The Dome of the Rock, built by Caliph Abd al-Malik.

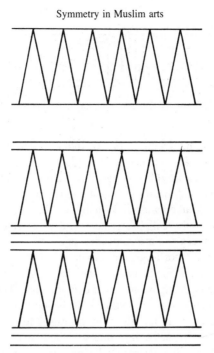

Fig. 9. Decoration from the golden lyre with the bull head, Ur, Mesopotamia.

the widespread use of mosaic ornamentation, and of Sassanid arts, as the cornice with small arches, twin colonettes and the royal symbols of crowns and jewels, which are characteristic of both traditions. All these borrowed marks are used together with the inscriptions, which are among the most characteristic elements of Muslim arts. In Oleg Grabar's opinion "these mosaics proclaim the military victory over Islam's two early enemies, Byzantium and the Sassanid Iran" but also the victory over Judaism and Christianity and "project the spread of the new universal Religion."[6]

Although, when speaking about arts in ancient times, we usually refer to temples and palaces, we must not forget the humble and common objects of everyday life, such as pots, tools and musical instruments, found in excavations. The motifs decorating them are still very popular and widely used in present day folk art (Figs. 9–12).

The examples of "rhythmic repetition and symmetry" underlying Sassanid art and its predecessors, some of the geometrically arranged patterns of Christian art, and finally the extricate schemes to be found all over the Muslim world expose the question of the role of mathematics in Muslim arts. When we come across the manifestations of such a "geometric" art, we cannot help realising that an arranging principle, mathematics must have been used consciously there. It was

> a means . . . by which man can arrive at the knowledge of the Essence; his knowledge of the basic structure of the material world, by which he is surrounded, makes it possible for him to arrive at the Essence of the secret of God Almighty's creative power.[7]
>
> The spiritual world was reflected in the sensible world not through various iconic forms, but through geometry and rhythm, through arabesques and calligraphy which reflect directly the worlds above and ultimately the supernal sun of Divine Unity.[8]

Such a conscious usage of mathematics in works of art may surprise the laymen, who usually have not even the faintest idea how much present day sciences owe to the Arabs: from the so-

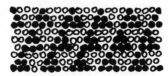

Fig. 10. Cone mosaics from Warka, Mesopotamia.

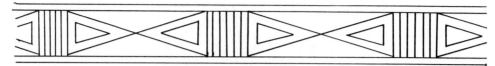

Fig. 11. Bracelet made of beads, Libya.

called "Arabic" numerals, (which were in fact of Indian origin but were transmitted to us by the Arabs), to the heritage of many a Greek scholar and scientist, whose work would have been lost to us, had not the Arabs translated them. And they did not stop at translating, they developed the knowledge thus acquired, and made significant contributions to several sciences, especially in the fields of Astronomy, Mathematics and Medicine. Sciences and arts were cultivated on a high level at the courts of the caliphs, princes and petty rulers, who found pleasure in patronising the leading scientists and artists of their age.

At the same time we must not forgo the important role that the representatives of Mathematics, the numbers, must have played outside the courts, among less or uneducated people. They were and still are the constant companions of everyday life. They were given magic values and powers, and as such were considered something beyond human reach. For example the number 5 has a lot of occurences in the life of an average Muslim. Islam has 5 "pillars"—a Muslim has to say his prayers 5 times a day—the so-called "Fatima hand" has 5 fingers (Fig. 13), imitating a real hand, but having magical significance: it defends its holder from the bewitching of the evil eye. If one feels that somebody wants to cast an evil eye upon him, he must say *"khamsah fī ʿaynak"* meaning "5 into your eye", thus breaking the force of the evil.

Having summed up all these effects and phenomena, now we examine them in their manifestations in the works of art, looking for "symmetry, or the series of ways in which a single motif can be repeated an exact number of times within a circle," which "is the most fundamental manifest aspect of Islamic geometric art."[9] As Muslim art can be geometric in its every branch, examples of symmetric nature will be presented from the simplest patterns to the most elaborated ones and they will be grouped according to the field of art they belong to.

Architecture—partly science, partly art—has a lot of symmetrical features even to the eye of a layman. They are so natural and we are so accustomed to them that walking in the street we usually do not even realise that the buildings we see are symmetrical. As early as ancient times there were efforts to reach "perfect" forms, the most well-known of which are the

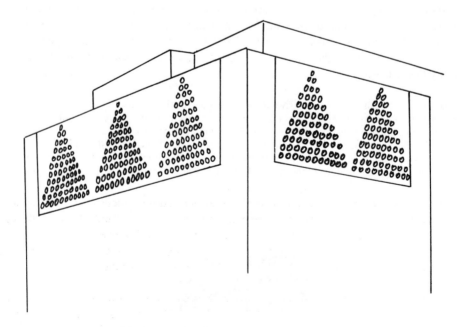

Fig. 12. Fragment of a common living house in Jerusalem.

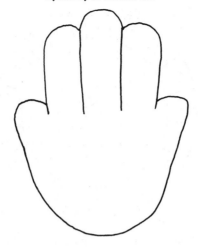

Fig. 13. Outline of a Fatima-hand made of copper, Libya.

Pyramids of Gizah in Egypt. In the Middle East the circular design had great traditions, because of the cosmological and magical significance attached to it. In 762, when Caliph al-Manṣūr founded his new capital, Baghdad, it was also of this circular design (Fig. 14). However, the designs are often rectangular whose mirror planes are perhaps the most frequent manifestations of symmetry, clearly recognisable even for the man in the street. So much so that in everyday speech when using the word "symmetry", we usually refer to this type (Fig. 15).

To our minds Muslim architecture usually means mosques and minarets; and though we usually do not realise how much an Egyptian minaret may differ in shape and form from a Turkish one—walking in the Fatimid part of Cairo we find minarets of angular type, while the Turkish ones look like well sharpened pencils, the shape best known here in Hungary, too, as we also have some left from the days of the Turkish occupation. We will agree on one thing: the general concept of the minaret is that it should be tall and slim. At least that is how we would describe it, were we to sum up its main characteristics in words. However, the minaret of the Great Mosque of al-Mutawakkil at Samarra is of a different shape; it looks like a conical helix (Fig. 16), wide at the bottom and narrowing towards the top. This pattern, though described in the plane, has earlier parallels on a pot from Hacılar in Anatolia (Fig. 17) and in the beard of a statue from Assur-nasir-apli's Northwest Palace in Nimrud (Fig. 18). The design of the minaret in the Topkapi Sarayi, Istanbul, follows the spiral pattern: the diameter of the circles is unchanged all along (Fig. 19).

But even minarets "simple" in shape could be decorated in such a way as to give us examples of various classes of symmetry. This is the case of the minaret of the Selimiye Mosque in Edirne (Fig. 20), which represents the same class of symmetry as the carved window of the cupola of the Shah Mosque at Isfahan (Fig. 21).

Cupolas, as we have seen above, could be equipped with windows looking like lace of stone—a common feature even with living houses, but there they were usually made of wood. The cupolas, however, were more often tiled or carved (Figs. 22–23). These structures are usually mounted over the central prayer hall, which may have several aisles separated by rows of columns sometimes connected by nicely shaped arches (Fig. 24).

Walking in the Fatimid part of Cairo, one cannot help recognising the ever-returning ridge ornaments. They vary from the relatively simple ones like that of the Azhar Mosque (Fig. 25), which has an ancient parallel on the top of the Ishtar Gate originating from Babylon (Fig. 26) to more sophisticated, floral ones, as that of the Ghuriyya (Fig. 27). Niches built into the sides of the mosques were also a characteristic feature (Fig. 28) having parallels from the Sassanid Iran as well (Fig. 29).

As we have already mentioned, tiles were frequently used in decorations both inside and outside of buildings, especially in eastern Islam, Iran and Central Asia. Turkey was famous first for her Iznik, then for her Kütahya tiles, which were even taken to Cairo to decorate the

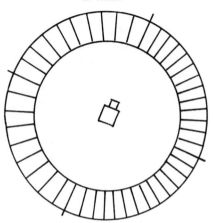

Fig. 14. Design of the new capital of Caliph al-Manṣūr, Baghdad, Iraq, founded in 762.

Blue Mosque there, which has the same sort of tiles as the Sultanahmet Mosque in Istanbul—usually referred to as *the* Blue Mosque, because of the colour of its tiles. But palaces were also often covered with tiles; perhaps the most famous example is the Topkapı Sarayı, in Istanbul, which has a great variety of tiles (Fig. 30).

The mosaics from Iran presented below are of a somewhat different kind, though they are also ornamented by floral patterns, our sketches can only indicate their main outlines (Figs. 31–32).

Beside the tiles and mosaics, carving was also a popular and widely used device. Muslim artists often created so sophisticated patterns that sometimes it is difficult to follow all the turns of the lines. In a given decoration several interwoven patterns may be combined, which are highly symmetrical within themselves (Fig. 33).

The ceramics used in contemporary life such as cups, plates and even ash-trays may have decorations originating from the early times (Fig. 34).

There are numerous examples of symmetries characterised by periodicity in two directions. They were applied to decorations covering extended surfaces of walls, floors and columns. We have already seen some examples above (e.g. Figs. 30–32) and more are given below (Figs. 35–37).

The designs were not carved always into stone: woodwork was also highly developed. A very special sort of decoration was that of the *Quran* stands, used to hold the copy of the holy

Fig. 15. Design of the tomb of the Tāj Mahal, "the crown of the palace", Agra, India (1632–1648).

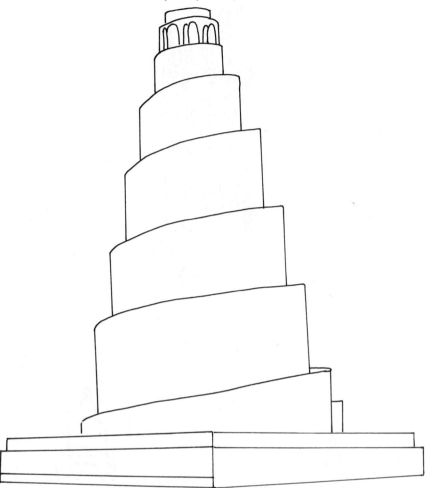

Fig. 16. Minaret of the Great Mosque at Samarra.

Fig. 17. Terracotta pot from Hacılar, Anatoly, 6th c.B.C.

E. RÓZSA

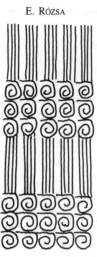

Fig. 18. Beard of a giant guardian from Assur-nasir-apli's Northwest Palace, Nimrud, Assyria, 9th c.B.C.

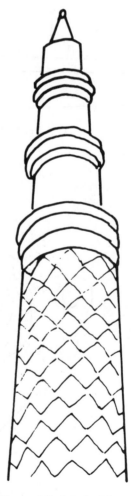

Fig. 19. A minaret from the Topkapı Sarayı, Istanbul, Turkey.

Fig. 20. Minaret of the Selimiye Mosque, Edirne, Turkey.

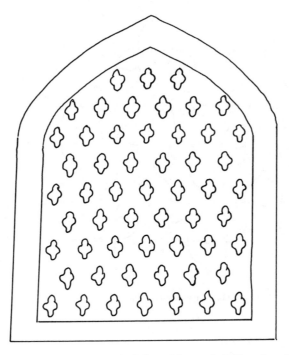

Fig. 21. The Shah Mosque, carved window of the cupola, Isfahan, Iran, 17th c.

Fig. 22. The tomb of Amir Sulayman, Cairo, Egypt, 1544.

E. RÓZSA

Fig. 23. Decoration of the cupola of the tomb of Azrumuk, Cairo, Egypt.

Fig. 24. View of the Moti Masjid's prayer hall, Agra, India, 17th c.

Fig. 25. Ridge of the Azhar Mosque, Cairo, Egypt.

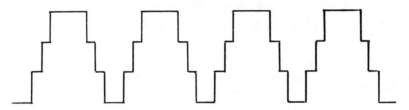

Fig. 26. Ridge of the Ishtar Gate from Babylon.

Fig. 27. Ridge of the Ghuriyya, Cairo, Egypt, 1500–1516.

Fig. 28. Masjid-i Shah, Isfahan, Iran (1611–16).

Fig. 29. Taq-i Kisra, the Hall of Khosrou, Ctesiphon.

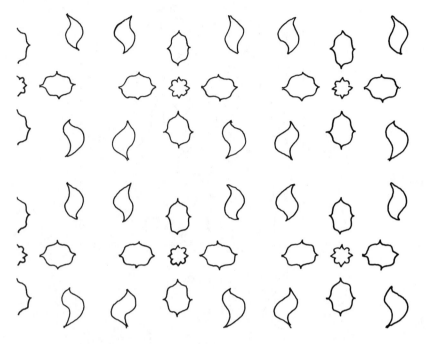

Fig. 30. Tiles from the Topkapı Sarayı, Istanbul, Turkey.

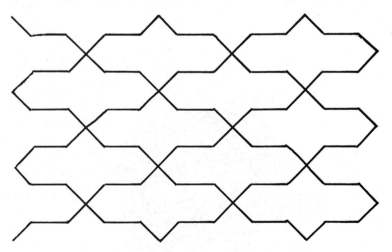

Fig. 31. Mosaics from the Gulistan Palace, Tehran, Iran.

Fig. 32. A corner of the Blue Mosque, faience mosaic decoration, Tabriz, Iran, 1462–65.

Fig. 33. Carving from the well of the Mu'ayyad Mosque, Cairo, Egypt, 1416–1420.

Painted Cup

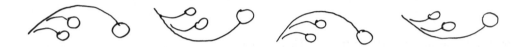

Painted Ash-tray

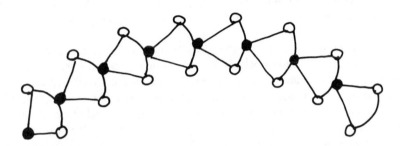

Painted Plate

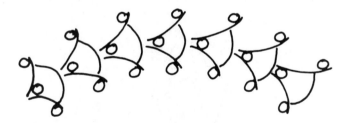

Painted Plate

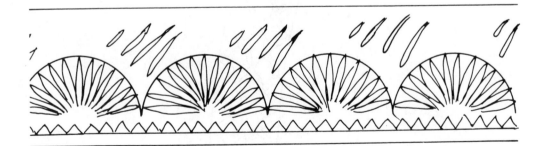

Carved Cup

Fig. 34. Decorations on present-day cups, plates and ash-trays, Tunisia.

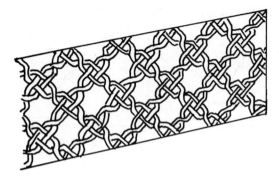

Fig. 35. Carved decoration from the Aḥmad Pasha Mosque, as-Saray al-Ḥamra, Tripoli, Libya.

book while reading or reciting from it. They usually bear the name of God or some quotations from the *Quran*. The example here (Fig. 38) has the name of God carved onto it.

Another device is inlay work, which is frequently used with wood. Walking in the bazaars of Egypt and Syria, we can find a great variety of handmade boxes, plates or small tables inlaid with mother-of-pearl (Figs. 39–40). Inlay work was also used with several materials other than wood (Figs. 41–42). In India it was used to decorate outer surfaces of buildings as well (Figs. 43–44).

Another common thing to be found in every bazaar, which is also inseparable from our concept of the Middle East, is carpets. They are manufactured all over the Muslim world. Carpets of Bukhara or Iran are as famous as the Berber ones of North Africa, the Maghreb. They are woven by hand in small "workshops" usually open to the visitor. Several types can be distinguished not only according to place, but also to patterns, technique and material. Here I would like to present carpets of more or less geometric design as it is easier for the eye to grasp their symmetry than it would be in the case of the meticuously floral patterns (Figs. 45–51).

Besides symmetry, there is another special characteristic feature of Muslim art, penetrating into every branch, and that is Arabic script, used any time and any place. Entering a mosque

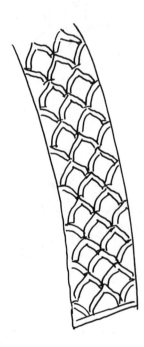

Fig. 36. A carved curve from the Aḥmad Pasha Mosque, as-Saray al-Ḥamra, Tripoli, Libya.

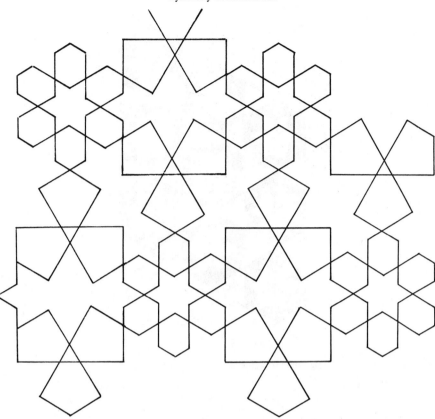

Fig. 37. Pattern from the mausoleum of Sayyida Rukhaiyah, Cairo, Egypt, 1154–1160.

or a palace, a public building or a common living house, even a person unable to read the Arabic script realises that the elaborated and artistic ribbons between the geometric and floral decorations represent some sort of writing (Fig. 52). Sometimes they are so richly decorated that it is difficult even to an eye accustomed to these letters to read them. We may even recognise that the forms of letters and their decorating elements are different in different mosques or places.

Calligraphy has developed into a much more meaningful and sophisticated art in the Muslim world than it has ever been in Europe. The famous calligraphers were always highly respected members of the courts and of the whole community. The reasons may be simple: illiterate people often render magic qualities to letters; and the Arabs, being a nomadic people, considered letters to be magical signs, which even after the rise of Islam preserved this magic significance—

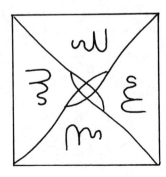

Fig. 38. Carved wooden Quran stand, Iran, 1360.

Fig. 39. Decoration of an inlaid box from Syria.

Fig. 40. Wooden box from Egypt, inlaid with mother-of-pearl.

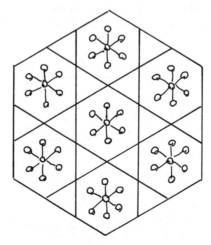

Fig. 41. Faience inlay work from the mausoleum of Gauhar Shād, wife of Shād Rokh, Herat, Afghanistan,
1432.

Fig. 42. Decoration of the back of a throne in the Topkapı Sarayı, Istanbul, Turkey.

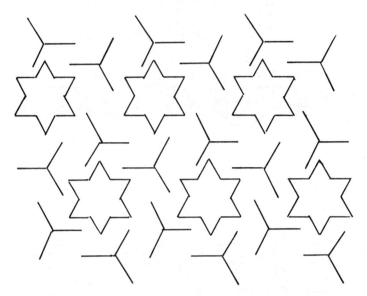

Fig. 43. Inlay work from one of the corner towers of the tomb of Itimād ad-Daula, Agra, India, 1628.

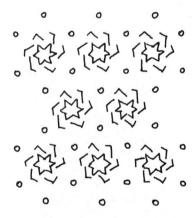

Fig. 44. Inlay work from the mausoleum of Itimād ad-Daula, Agra, India, 1628.

Fig. 45. Yağcıbedir carpet from Turkey.

Fig. 46. Milas carpet, Turkey.

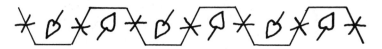

Fig. 47. Kula carpet, Turkey.

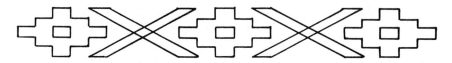

Fig. 48. Kula carpet from Turkey.

Fig. 49. Yağcıbedir carpet from the mountain villages of the Aegean region of Turkey.

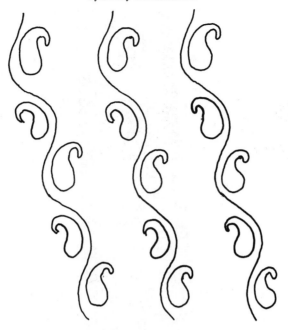

Fig. 50. A Persian rug from Khorasan.

traces of this can be found in the *Quran*. This respect, however, underwent a change, acquired a new meaning: the letters came to "embody the word of God as revealed in the sacred book, the Holy *Quran*."[10] The rise of Islam urged the need to record the *Quran* precisely and to beautify it so that it became worthy of the divine revelation. These efforts resulted in several variants of Arabic script and are widely used in buildings and carvings, just as well as in books or any other branch of art. A tradition developed with the copies of the *Quran:* the first two pages of any copy are beautifully decorated, frequently displaying patterns of symmetry (Fig. 53).

The Arab calligraphers did not stop at the several variants of script. They formed them into different patterns and lines, sometimes interweaving several lines, sometimes playfully arranging the letters in symmetrical forms (Figs. 54–56). Sometimes a figure was formed from the letters of a name or a sentence, called "ṭughrā". It was the calligraphic emblem used first by the Oghuz, then by the Seljuq rulers. It was later adopted by the Ottoman sultans, who developed it into a distinctive calligraphic art form (Fig. 57).

In these few pages I have tried to give a brief sketch of the importance of symmetry in Muslim art, to show that it is a basic feature which was and is present in every territory of the Muslim world, in every layer of the community and in every branch of art. We can visit palaces or museums, or walk in the poorest districts of any town or village of the Muslim countries,

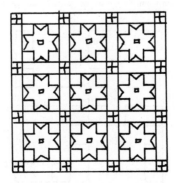

Fig. 51. A Persian rug from Shiraz.

E. RÓZSA

Fig. 52. Inscription decorated tiles.

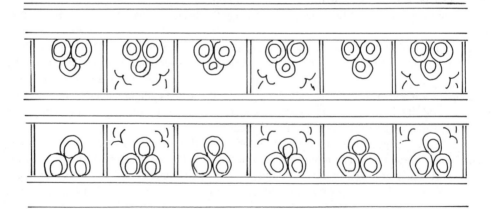

Fig. 53. Painted decoration from a copy of the Quran, from Qetta village, Fezzan, Lybia.

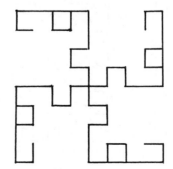

Fig. 54. Ornamental Kufic inscription from the mausoleum of the astronomer Ghazi-Zade Rumi in the necropolis of Shah Zinde, Samarkand, USSR, 1437.

Fig. 55. Painting saying "Praise be to Allah."

Fig. 56. Five-fold "square" Kufic inscription from the mausoleum of Öljeytu at Sultaniye, Iran (1310–16) with the name of the Prophet Muhammad.

Fig. 57. The ṭughrā of Sultan Mahmud II.

and we will always find manifestations of symmetry everywhere. This article presenting such a consciously geometric and symmetrically designed art, perhaps will help us realise symmetry in our own life and surroundings as well.

REFERENCES

1. Sūrat al-Iḫlāṣ, 112, ā.4.: *wa lam yakun lahu kufūan aḥad, Qurān Mujīd,* Libya (1977).
2. Sūrat al-Ḥashar, 59, ā.24.: *al-muṣawwir, Qurān Mujīd,* Libya (1977).
3. "*ṣawwara*" implies giving definite form or colour so as to make a thing exactly suited to a given end or object: hence the title *Muṣawwir,* the Bestower of Forms or Colours, for this shows the completion of the visible stage of Creation." p. 1529, para 5406, *Qurān Mujīd* (1977).
4. Old Testament, *Moses II.*20/4; III.26/1; V.5/8, etc.
5. M. S. Dimand: *A Handbook of Muhammadan* Art, p. 14. 3rd ed. The Metropolitan Museum of Art, New York (1958).
6. ibid., p. 7.
7. U. Scerrato: *Monuments of Civilization: Islam,* p. 21. London (1972).
8. S. H. Nasr: *Al-ʿUlūm fī-l-Islām,* p. 83. Libya-Tunisia (1978).
9. S. H. Nasr: Foreword to K. Critchlow: *Islamic Patterns. An Analytical and Cosmological Approach.* p. 6. New York (1976).
10. K. Critchlow: *Islamic Patterns. An Analytical and Cosmological Approach,* p. 74. New York (1976).

Comp. & Maths. with Appls. Vol. 12B, Nos. 3/4, pp. 751–762, 1986
Printed in Great Britain.

0886–9561/86 $3.00 + .00
© 1986 Pergamon Press Ltd.

NON-SPACE-GROUP SYMMETRY IN CRYSTALLOGRAPHY

Konrad Fichtner

Central Institute of Physical Chemistry, Academy of Sciences of the G.D.R., DDR–1199,
Berlin-Aldershof, G.D.R.

Abstract—After presenting the basic ideas of the characterization of symmetries of a "classical crystal" with its three-dimensional periodicity, approaches are discussed to describing relevant features of certain crystal structures by additional symmetries. Application and usefulness of some conceptions of non-space-group symmetries are illustrated by examples. Polytype structures or OD structures in the sense of Dornberger–Schiff are discussed in more detail.

1. INTRODUCTION: THE THREE-DIMENSIONALLY PERIODIC CRYSTAL

The most impressive feature of a crystal is its regular external form: the plane faces, symmetrically arranged. This regular shape of a crystal is caused by its regular inner structure. The arrangement of atoms, ions or molecules within a small part of a crystal is repeated again and again in all three dimensions of space. The periodicity of the atomic structure is considered as the characteristic property of a crystal. The corners and edges of a crystal may have been completely smoothed, for instance by mechanical treatment. The resulting ball-like body is still regarded as a crystal, if the regularity of the arrangement of atoms is preserved.

Usually the content of a so-called unit cell, i.e. a cube, a cuboid or generally a parallelepiped (Fig. 1) is taken as representative for a crystal structure. Figure 2 shows a very simple example. The unit cell of potassium chloride is a cube with chlorine ions at the corners and a potassium ion in the centre. Unit cells may be much more complex. An example of an organic crystal structure is given in Fig. 3. In Sec. 3 an ytterbium complex is discussed, in which one molecule consists of 160 atoms, and any unit cell contains four molecules. Crystals of biological macromolecular compounds, e.g. globular proteins, have thousands of atoms per unit cell.

The size of a unit cell is very small. In the example of KCL, it is necessary to put side by side about two million unit cells in order to obtain a length of 1 mm. Thus, if one is not especially interested in surface phenomena of a crystal, it is justified to assume that the crystal structure is infinitely extended. Symmetry of real things always includes certain idealizations. Small local deviations from three-dimensional periodicity and the thermal vibrations present in all solid matter are also disregarded, if we speak of (ideal) crystal structures.

In order to express the regularity of a crystal structure in a strict mathematical way, symmetry operations are to be considered. These symmetries are motions, i.e. transformations of the three-dimensional Euclidean space $E^{(3)}$, leaving invariant all distances and angles. The set of all symmetries of a crystal structure consists of all motions of space bringing the structure into coincidence with itself. This set forms a group which is infinite because of the existence of translations among the symmetry operations. Other possible symmetry operations are rotations by certain angles, reflections in a plane and combinations of these motions with each other and with certain translations. Crystallographic space groups are defined as groups of motions of the three-dimensional Euclidean space $E^{(3)}$ with the condition that the subgroup of translations is generated by three linearly independent translations.

There are infinitely many crystallographic space groups. In order to work with them, they have to be sorted into a finite number of classes. The division into space-group types is most important among several useful classifications. Crystallographers often speak of space groups, if they have space-group types in mind. Two groups belong to the same space-group type, if they differ only with respect to the lattice parameters, i.e. with respect to the six metric parameters describing the size of the unit cell. Three parameters give the length of the edges and three the angles between the edges (Fig. 1). The importance of the classification of crystallographic space groups into space-group types follows from the fact that the symmetry of a crystal structure is uniquely characterized by its space-group type and its lattice parameters. The different space-group types are distinguished by appropriate symbols or by numbers.

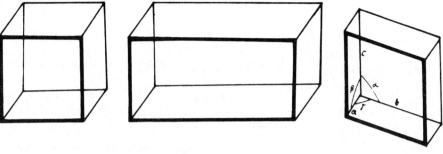

Fig. 1. Possible unit cells of crystal structures.

Up to now the crystal structures of more than 50,000 compounds have been deciphered, mainly by X-ray diffraction methods. More than 200 of the 230 theoretically possible space-group types actually have been observed. The symmetry data (space-group type, lattice constants) have been of great importance in crystal structure determination, for the characterization and identification of crystalline material as well as the investigation of the physical properties of crystals.

For most of the crystal structures, the description of symmetry by space-group type and lattice constants is adequate. Nevertheless, in recent years a rising number of structures have been found for which a refinement of the symmetry description has turned out to be useful or even necessary. In the present paper, the notion of nonspace-group symmetry is used in a very general sense: it covers all regularities in crystal structures not described by lattice parameters and space-group type.

2. GENERALIZATIONS OF SPACE GROUPS AND EXAMPLES OF NON-SPACE-GROUP SYMMETRIES

In recent decades three generalizing conceptions of crystallographic symmetry theory have become of increasing interest, because it became possible to apply them to unusual crystal

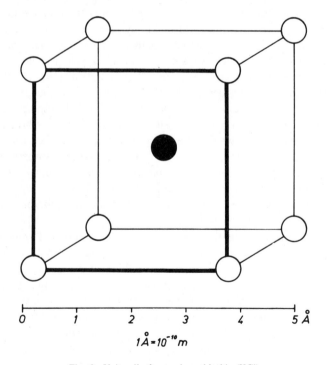

Fig. 2. Unit cell of potassium chloride (KCl).

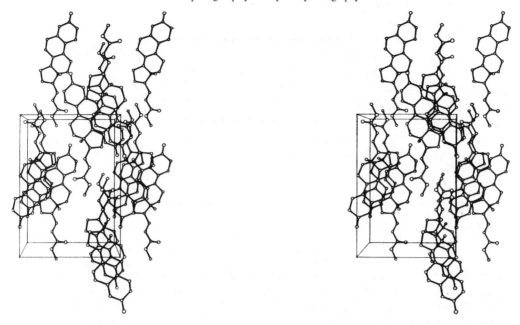

Fig. 3. Stereo plot of the packing of molecules $C_{22}H_{28}O_3$. Only the nonhydrogen atoms are given (after [1]).

structures. In this section, the main ideas of these conceptions are given and it is shown how they can be used for a refinement of the symmetry description of certain crystal structures.

2.1 Crystallographic symmetry in spaces of higher dimension

The classical crystallographic space groups refer to the three-dimensional Euclidean space $E^{(3)}$. The translations form an abelian subgroup of rank 3. Two possibilities for a generalization are obvious: spaces $E^{(n)}$ of dimension n different from 3 may be considered; over this the rank m of the translational subgroup may be less than the dimension of the space. The latter symmetry groups are known as partially periodic crystallographic groups.

For $m = n = 2$, there are 17 essentially different two-dimensional crystallographic symmetry groups describing the possible symmetries of wallpapers. The 80 layer group types ($n = 3$, $m = 2$) were first described in 1930[2,3] and are of great importance for the study of polytypic crystal structures as discussed below. The crystallographic symmetry groups of the four-dimensional Euclidean space ($n = m = 4$) were investigated in the 1960s and 1970s. The derivation of the 4895 space-group types was the result of joint efforts of crystallographers and mathematicians[4].

Higher-dimensional groups ($n > 3$) have been applied to so-called composite crystals. Their atomic structure consists of subsystems with differing three-dimensional periodicities. Figure 4 shows a simple two-dimensional example. In direction **a** the two systems have equal periodicity, but not in direction **b**. Let $\mathbf{b}_2 = r \cdot \mathbf{b}_1$ be the relation between the periodicities in the two subsystems. Then commensurate and incommensurate cases are distinguished in dependence on the rationality of the parameter r. For a rational parameter r the structure is periodic in two dimensions with a special regularity of the distribution of atoms in the cell. This cell has been named supercell, because it describes the periodicity of both subsystems. If r is not rational, no supercell exists; no two-dimensional group describes the symmetry of the composite structure. It is, however, possible to treat the two-dimensional incommensurate structure in a three-dimensional space, the so-called superspace. In this space, a "supercrystal" is periodic in three dimensions and thus has a crystallographic symmetry group. The composite crystal itself is a two-dimensional section of the supercrystal (compare [5]).

"Modulated structures" form another example of the application of higher-dimensional

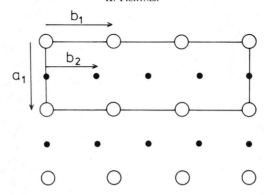

Fig. 4. Schematic example of a composite crystal with supercell.

space groups to special crystal structures. Small deviations from the ideal periodic structure are disregarded by crystallographic space groups. If the deviations are randomly distributed, it is justified to ignore them. The situation changes if the deviations are regularly distributed. The term "modulated structure" has been introduced for quite a number of different wave-like deviations, e.g. displacive modulations of an undistorted crystal structure, modulations of the occupation of certain atomic sites, of the charge density or of the spin density. As in the case of composite crystals, the modulated structure is imbedded in a space of dimension higher than 3. In addition to the normal positional space of three dimensions, the existence of an internal space of d freedom dimensions is assumed ($d = 1, 2$ or 3). Then a clear mathematical description of the symmetry properties of the modulated structure as well as of its X-ray diffraction pattern becomes possible. The $(n + d)$-dimensional extended pattern (the supercrystal) is periodic in $(3 + d)$ dimensions, i.e. it has a $(3 + d)$-dimensional space group. The modulated structure itself is a three-dimensional section of the supercrystal. The observed X-ray diffraction pattern may be understood as a projection of a $(3 + d)$-dimensional pattern onto a three-dimensional space. Figure 5 illustrates a one-dimensional model of a modulated structure. Some observed modulated crystal structures have been quoted and discussed by Janner and Janssen[5].

2.2 Color symmetry, black and white groups

A normal symmetry operation transforms any point of space into a fully equivalent point. In the generalization under discussion equivalent points may differ with respect to a certain property. Colors are a convenient property for illustration and easy perception. Any generalized symmetry operation is connected with a certain permutation of the colors.

In the simplest case, there are only two colors, black and white. Figure 6 shows a black-and-white body. Half of the points are white and the other half are black. Normal symmetries,

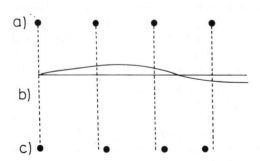

Fig. 5. One-dimensional model of displacive modulation of a structure. (a) Undistorted structure; (b) modulation wave; (c) modulated structure.

Fig. 6. Body with black-and-white symmetry.

e.g. rotations by an angle of 120°, transform the body as a whole into itself and any point of the body into a point of the same color. In addition to the symmetries, "antisymmetries" may be considered, which also transform the body into itself, but which interchange the colors. A rotation by an angle of 60° about the central axis of the body obviously has this property. It is interesting that the number of symmetries of the body equals the number of antisymmetries. For the body of Fig. 6, the symmetry group has the order 12. The symmetries are as follows.

Symmetries:
  Identical mapping;
  Rotations about the central axis by angles of 120° and 240°, respectively;
  Inversion in the centre of gravity;
  Roto-inversions with rotation angles of 60° and 300°, respectively;
  Three rotations by 180° about axes through the midpoints of the side edges;
  Three reflections in planes containing the central axis.
Antisymmetries:
  Rotations about the central axis by angles of 60°, 180° and 300°, respectively;
  Reflection in a plane perpendicular to the central axis;
  Roto-inversions with rotation of 120° and 240°, respectively;
  Three rotations by 180° about axes through the centre of the side faces;
  Three reflections in planes containing the central axis and side edges.

Symmetries and antisymmetries together form a group with an order of 24. The product of two antisymmetries is again a symmetry.

All possible black-and-white space-group types of the three-dimensional Euclidean space have been derived by Zamorzaev[6] and Belov *et al.*[7] in the 1950s. The 1651 different types have been tabulated by Koptsik[8] and are named Shubnikov groups to honor A. V. Shubnikov, who introduced the idea of antisymmetry into crystallography.

Black-and-white symmetry has been used to characterize crystal structures with special electric or magnetic properties. Manganous fluoride, for instance, has an antiferromagnetic structure (Fig. 7). Disregarding the magnetic property, the space group (type) is $P4_2/mnm$. The magnetic moments of the manganese atoms at 0, 0, 0 are antiparallel to the moments at $\frac{1}{2}$, $\frac{1}{2}$, $\frac{1}{2}$. The fluorine atoms have no magnetic moment. The structure of MnF_2, including the magnetic properties, may be described by a black-and-white group. Half of the coincidence operations of $P4_2/mnm$ is connected with a change of the direction of the magnetic moments. If antisym-

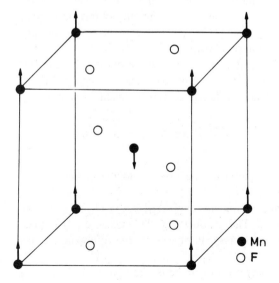

● Mn
○ F

Fig. 7. Magnetic structure of MnF₂.

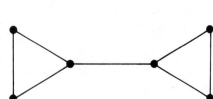

Fig. 8. A fictitious molecule with partial symmetries.

metry operations are marked by a prime, the symbol of the Shubnikov group is $P4_2'/mnm'$ (compare [8,9]).

2.3 Partial symmetries, groupoids

Classical symmetries transform a body or an infinitely extended crystal structure as a whole into itself. The generalization now to be discussed also includes coincidence operations restricted to a part of the body or structure. The simple fictitious molecule of Fig. 8 is used to discuss the basic idea of this generalization. In Fig. 8 the regularity of the distribution of atoms is only partly reflected by the symmetry group of the molecule. Obviously, the molecule consists of two parts (triangles) and three kinds of coincidence operations may be considered.

- Total symmetry operations transforming the whole molecule into itself, e.g. the inversion in the centre of gravity of the molecule.
- Local symmetry operations transforming one of the two triangles into itself, e.g. rotations by an angle of 120°.
- Partial coincidence operations transforming one of the parts into the other one. There is, for instance, a rotation by an angle of 60° transforming the left triangle into the right one (this transformation, however, is not a coincidence operation for the right triangle, because it would be transformed into a region outside the molecule).

The term ''symmetry operation'' of an object is used for the (total) coincidence operations of this object. Accordingly, for a partial coincidence operation, the term ''partial symmetry operation'' may be used. Strictly taken, this notion refers to the coincidence operations of the second and third type, but if we speak of all partial symmetry operations transforming the left triangle into the right one, total symmetry operations are also included.

The additional coincidence operations referring only to parts of an object under consideration have been used in crystallography in different situations. The parts of the crystal structure may be substructures, each of them infinitely extended over the whole space, similar to the case of composite crystals discussed in Section 2.1. The parts may be single molecules, parts of molecules, chains of molecules or layers. In all these cases we will use the term partial symmetry, although in the literature several other names may be found, e.g. ''noncrystallographic symmetry''[10] or ''supersymmetry''[11].

The mathematical notion of a group is adequate not only in the classical case of crystallographic symmetry groups, but also in the two generalizations discussed in Sec. 2.1. and 2.2. If partial coincidence operations are included, a more general notion becomes necessary. Independently of applications in crystallography, a number of appropriate notions have been defined in mathematics: grupoid[12,13], mischgruppe[14], category with invertible morphisms[15], inductive groupoid[13] and inverse semigroup[16].

From the mathematical point of view, these notions are equivalent or closely related. The definition preferred here is a mixture between the notions ''groupoid'' and ''category.'' From its content a groupoid is a special category. Thus the language can be taken from category theory (compare [17]).

A groupoid **G** consists of

(i) A set of objects.
(ii) Sets $[i, j]$ of morphisms or mappings leading from the object i to the object j. (The morphisms are the partial coincidence operations.)
(iii) A partial composition of morphisms. The morphisms $g_1 \epsilon [i_1, j_1]$ and $g_2 \epsilon [i_2, j_2]$ may be composed then and only then, if $j_1 = i_2$. The product $g_2 g_1$ belongs to $[i_1, j_2]$. (The composition is the usual composition of mappings). Furthermore, the following axioms have to be fulfilled.
   (1) *Existence of identical morphisms.* For any object i, there is a morphism $1_i \epsilon [i, i]$ with

$$1_i \cdot f = f, \quad g \cdot 1_i = g,$$

whenever the left sides are defined (1_i: identical map of the object i).

(2) *Existence of inverse morphisms*. For any $g \epsilon [i, j]$ there is a $g^{-1} \epsilon [j, i]$, and

$$g \cdot g^{-1} = 1_j \quad g^{-1}g = 1_i.$$

(As the partial symmetry operations are 1:1 mappings the inverse mapping is again a partial symmetry operation).

(3) *Associative law*. $g_3(g_2 g_1) = (g_3 g_2)g_1$, if the products are defined. (For mappings, the associative law is always fulfilled).

In the example of the fictitious molecule of Fig. 8, there are two objects, the two triangles. Obviously, the set of all partial symmetry operations transforming the two parts of the fictitious molecule into each other forms a groupoid. The following three sections of this paper deal with applications of partial symmetries to real problems in crystallography.

3. PARTIAL SYMMETRIES IN MOLECULAR CRYSTALS

Crystal structure determination consists of finding the spatial distribution of atoms in crystals. Usually, this starts from the determination of the symmetry of the crystal. If lattice constants and space-group type are known, the positions of the atoms in a so-called asymmetric unit have to be determined. Particularly in biological macromolecules, the asymmetric unit contains a great number of atoms and the determination of the structure is very difficult. Therefore, any additional information which simplifies this task is used. This refers especially to partial symmetries. Proteins, for instance, frequently crystallize with several identical molecules in the asymmetric unit. Some very large biological macromolecules, such as viruses, exhibit a high degree of partial symmetry, i.e. contain multiple copies of a certain protein subunit[18]. Satellite tobacco necrosis virus, for instance, has 60 identical subunits in its protective coat. Southern bean mosaic virus contains 30 protein subunits in the crystallographic asymmetric unit. The occurrence of partial symmetries in biological macromolecules is not accidental. It follows from evolutionary influences which favour energetically stable structures.

Partial symmetries may occur not only in case of proteins and viruses, but also in less complicated structures. Zorkii[19], for instance, has quoted 20 examples of structures with partial symmetries. Most of them contain two identical molecules in the asymmetric unit. Other examples are known with a symmetry of the molecule which is not incorporated into the crystal lattice, i.e. does not belong to the space group of the crystal. In the following, partial symmetries of molecular crystals will be demonstrated with an example of moderate complexity.

Kulpe *et al.*[20] have determined the structure of an ytterbium complex, $C_{72}H_{60}N_3O_{18}P_6Yb$. The unit cell contains four molecules; one of them is given in Fig. 9. Each molecule consists of 160 atoms and is asymmetric, i.e. there is no nontrivial operation transforming the molecule as a whole into itself. There are, however, quite a number of partial symmetries. The molecule may be considered as consisting of 13 parts: a central part consisting of the 28 noncarbon and nonhydrogen atoms of the molecule and the 12 phenyl rings C_6H_5 (Fig. 10). The central part has in good approximation the symmetry D_3 (32), i.e. there are rotations by angles of $\pm 120°$ and three rotations by angles of $180°$. The order of the symmetry group of the central part is 6. Each phenyl ring has the symmetry C_{2v} (*mm*2). The consideration of partial symmetries has two advantages: the complexity of the fairly large molecule decreases, and the number of parameters necessary for the description of the structure drops considerably.

4. POLYTYPIC STRUCTURES, THE OD THEORY OF DORNBERGER–SCHIFF

The notion polytypism was coined by Baumhauer[21] in 1915 in connection with the investigation of different modifications of silicon carbide. Other prominent polytypic substances are zinc sulfide, cadmium iodide and micas. The main feature of polytypic substances is that they consist of layers periodic in two dimensions. The layers may be stacked in different ways. For any next layer two or more different positions are possible. Thus an infinite number of stacking variants is theoretically possible. For the substances mentioned above, many different ordered arrangements of layers have been observed. For some hundreds of substances (elements,

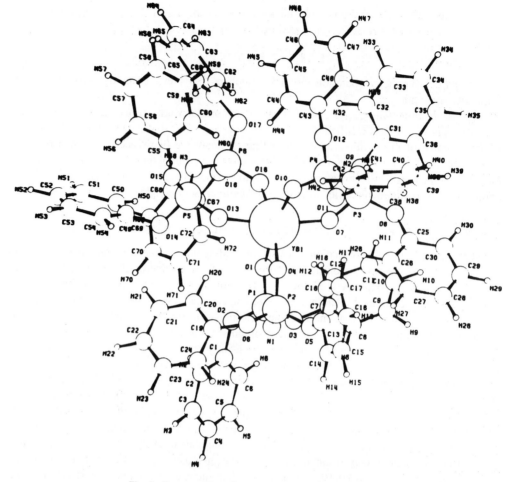

Fig. 9. Representation of a molecule of $C_{72}H_{60}N_3O_{18}P_6Yb$.

alloys, inorganic and organic compounds) several polytypic modifications have been found. Nonperiodic, i.e. disordered, stacking variants do also exit, often in coexistence with ordered stackings. Thus, it is justified to denote polytypic substances as order–disorder structures (OD structures).

The OD theory of Dornberger–Schiff[22–24] explains the phenomenon of polytypism by the presence of partial symmetry operations and uses these partial symmetry operations for an adequate description of symmetry and as a powerful tool in the determination of the atomic structure of such substances.

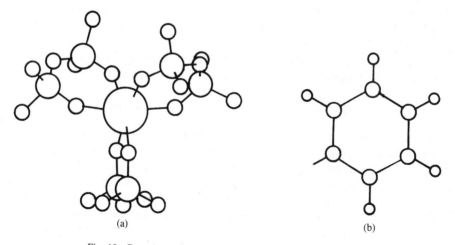

(a)   (b)

Fig. 10. Central part and one of the phenyl rings of the molecule of Fig. 9.

To illustrate an OD structure, let us regard the most typical polytypic structure, silicon carbide. In SiC any layer of the structure consists of the close-packed layer of C atoms with one Si atom below each C atoms at a distance of 1.89 Å (Fig. 11). If one layer, say L_0, is fixed, then the atoms of the next layer may occupy one of two possible positions. Any layer is transformed into the next one by a translation, either to the left or to the right. The symmetry relations of neighboring layers may be seen in Fig. 12. Any layer has the symmetry $P(6)mm$. The threefold axes and one of the mirror planes are common to all layers. But the other mirror plane is valid only for one particular layer. It is a partial symmetry operation. This partial reflection transforms the two possible positions of L_1 into each other. Thus, the existence of the partial reflection may be considered as the reason for the fact that the position of the next layer is not uniquely determined.

Generally speaking, the phenomenon of polytypism is connected with three conditions.

• The influence of the next-but-one layer on the position of a layer may be neglected.
• There are two or more geometrically equivalent ways in which neighboring layers may be arranged with respect to one another.
• For two different layers of the same kind, say L_i and L_j, the relative position of L_i to its adjacent layers is the same as the relative position of the layer L_j to its adjacent layers.

In detail: L_i is of the same kind as L_j means that there is a coincidence operation transforming L_i into L_j. Let us suppose [Fig. 13(a)] that the upper side of L_i is transformed into the upper side of L_j. Then there exists a motion that brings the pair (L_{i-1}, L_i) into coincidence with (L_{j-1}, L_j). Moreover, there exists a motion that brings the pair (L_i, L_{i+1}) into the pair (L_j, L_{j+1}). If, on the other hand, a partial symmetry operation transforms the upper side of L_i into the lower side of L_j [Fig. 13(b)], then there exist motions which transform (L_{i-1}, L_i) into (L_j, L_{j+1}) and (L_i, L_{i+1}) into (L_{j-1}, L_j), respectively. From these considerations we may draw the following conclusions:

• For a polytypic structure, (strictly) partial symmetry operations always exist.
• The symmetry of a polytypic structure (OD structure) may be characterized by the groupoid formed by all partial symmetry operations transforming layers into each other.

5. DESCRIPTION OF THE SYMMETRY OF AN OD STRUCTURE

We know that the symmetry of an OD structure may be described by a groupoid. But this groupoid is infinite and it is necessary to characterize it by a finite number of data. To illustrate how this can be done, we will return to the example of SiC.

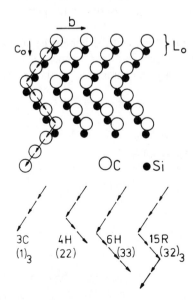

Fig. 11. Stacking possibilities in silicon carbide.

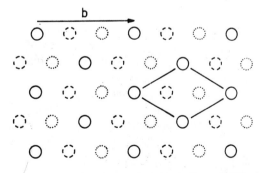

Fig. 12. Possible positions of neighboring layers in SiC. The thick circles represent the atoms of the layer L_0. The atoms of the layer L_1 are either in the position of the broken or of the dotted circles.

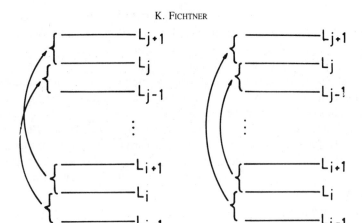

Fig. 13. Relation between layer pairs in an OD structure.

The description of the symmetry of any SiC polytype may be split into two parts. One part describes the stacking of the layers. Different notations for characterizing the stacking have been developed. Hägg's notation describes the two stacking possibilities by $+$ and $-$, so that for instance the polytype *6H* is characterized by $+ + + - - -$, or more succinctly by the Zhdanov symbol (33) (see [25]). The second part describes the symmetry data common to all SiC polytypes. This second part is formed by the layer group and the possible translational vectors leading from one layer to the next one.

For any polytypic substance, the symmetry may be described similarly to the case of SiC. It is possible to separate the stacking. The idea of stacking is finding out which of the possible positions of a layer is really occupied. The first notations for the stacking have been developed for SiC and may be used for quite a number of other polytypic substances. Symbols for more complex polytypic substances have been proposed, e.g. by Zvyagin[26] and Fichtner[27]. General rules for the construction of polytype symbols have been developed by Dornberger–Schiff, Ďurovič and Zvyagin[28] and have been recommended for use by the International Union of Crystallography[29].

The second part has been named the symmetry principle, because it describes the features common to all structures of a polytypic substance, ordered as well as disordered. To this second part belongs the symmetry of a single layer. The layer group may be described by the layer-group type and the net constants a, b, γ. In addition, to the second part belong all partial symmetry operations transforming adjacent layers one into another. Any partial symmetry operation may be split into a homogeneous part and into parameters describing the translational part.

The notion OD groupoid family covers the layer-group type and the homogeneous parts of the partial symmetry operations transforming different layers one into another. Thus, the symmetry principle consists of two parts, OD groupoid family and parameters as shown in the following scheme:

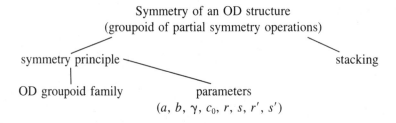

The notion OD groupoid family may be compared with the notion space-group type. As in the case of space-group types, the number of OD groupoid families is finite. For OD structures of one kind of layers, there are 400 OD groupoid families[30].

6. GENERALIZATIONS OF THE DEFINITION OF A CRYSTAL

For a very long period, crystals had been understood as homogeneous and anisotropic solid bodies bounded by plane faces. The investigation of the atomic structure of crystals resulted in a revision of this definition. The three-dimensional periodicity of the inner structure became the outstanding and defining property of crystals. The outer form played a minor role.

Disordered structures of polytypic substances as well as the incommensurate modulated structures are not covered by this definition and suggest a further revision. Dornberger–Schiff[24] writes as a result of the study of OD structures "that the classical definition as a body with three-dimensional periodicity (disregarding dislocations and other local defects) may have its shortcomings; properly speaking, it is not truly atomistic, although it refers to atoms: periodicity, by its very nature, makes statements on parts of the body which are far apart, whereas the forces between atoms are short range forces, falling off rapidly with distance. A truly atomistic crystal definition should, therefore, refer only to neighboring parts of the structure, and contain conditions understandable from the point of view of atomic theory" (compare also [31]).

Janner and Janssen[5] see the problem of defining a crystal from another point of view:

The Euclidean symmetry of an incommensurate crystal phase is fairly low (e.g. no three-dimensional lattice periodicity is possible) and does not explain the regularities of the diffraction pattern: main reflections, satellites, systematic extinctions, a.s.o. Therefore the three-dimensional Euclidean symmetry is a bad characterization of the crystal in question. But then the question arises: Which symmetry is a "good" one? In order to give an answer to this question the concept of internal dimensions is introduced.

. . . A crystal is defined by requiring for its density function $\rho(r)$ a Fourier decomposition of the form

$$\rho(r) = \sum_{k \in M^*} \hat{\rho}(k) \, e^{ikr}$$

with $M^* \simeq Z^{3+d}$, a so-called Z-module of rank $3 + d$ and dimension 3.
The classical case is included with $d = 0$.

At present, a generally accepted new definition of the notion of a crystal is still lacking and certainly far away. There is, however, no doubt that OD structures, modulated structures and magnetic structures are special crystal structures. Therefore, OD groupoids and black-and-white groups, as well as certain symmetry groups in higher-dimensional spaces, have become part of modern crystallographic symmetry theory.

REFERENCES

1. G. Reck and G. Schubert, Crystal and molecular structure of (E)-21-methoxycarbonyl-19-nor-pregna-4, 8(14), 20-trien-3-one. *Cryst. Res. Technol.* **19**, 1101–1105 (1984).
2. L. Weber, Die Symmetrie homogener ebener Punktsysteme. *Z. Kristallogr.* **70**, 309–327 (1930).
3. E. Alexander and K. Hermann, Die 80 zweidimensionalen Raumgruppen. *Z. Kristallogr.* **70**, 328–345 (1930).
4. H. Brown, R. Bülow, J. Neubüser, H. Wondratschek and H. Zassenhaus, *Crystallographic Groups of Four-dimensional Space.* Wiley-Interscience, New York (1978).
5. A. Janner and T. Janssen, Symmetry of incommensurate crystal phases in the superspace group approach. *Comm. Math. Chem.* **10**, 5–26 (1981).
6. A. M. Zamorzaev, *Generalizations of Federov Groups.* Dissertation, Leningrad University (1953).
7. N. V. Belov, N. N. Neronova and T. S. Smirnova, 1651 Shubnikov groups. *Kristallografiya* **2**, 315–325 (1957).
8. V. A. Koptsik, *Shubnikov Groups.* Moscow Univ. Press, Moscow (1966).
9. R. A. Erickson, Neutron diffraction studies of antiferromagnetism in manganous fluoride and some isomorphous compounds. *Phys. Rev.* **90**, 779–785 (1953).
10. D. M. Blow, Non-crystallographic symmetry, in *Crystallographic Computing Techniques* (Edited by F. R. Ahmed), pp. 229–238. Munksgaard, Copenhagen (1976).
11. P. M. Zorkii, The super symmetry of molecular crystal structures. *Acta Crystallogr.* **A31**, S1 (1975).
12. H. Brandt, Über eine Verallgemeinerung des Gruppenbegriffs. *Math. Ann.* **96**, 360–366 (1926).
13. C. Ehresmann, Gattungen von lokalen Strukturen. *Jahresber. DMV* **60**, 49–77 (1957).
14. A. Loewy, Über abstrakt definierte Transmutationssysteme oder Mischgruppen. *J. reine angew. Math.* **157**, 239–254 (1927).
15. S. Eilenberg and S. MacLane, General theory of natural equivalences. *Trans. Am. Math. Soc.* **58**, 231–294 (1945).
16. A. H. Clifford and G. B. Preston, *The Algebraic Theory of Semigroups.* American Mathematical Society, Providence (1964).
17. H. Schubert, *Kategorien I/II.* Akademie-Verlag, Berlin (1970).

18. T. A. Jones and L. Liljas, Crystallographic refinement of macromolecules having non-crystallographic symmetry. *Acta Crystallogr.* **A40**, 50–57 (1984).

19. E. E. Lavut and P. M. Zorkii, Super symmetry in the structural classes of picric acid and pirasole. *Zh. Strukt. Khim.* **20**, 54–58 (1983).

20. S. Kulpe, I. Seidel, K. Szulzewsky and G. Kretschmer, The structure of tris(tetraphenyl imidodiphosphato)ytterbium(III). *Acta Crystallogr.* **B38**, 2813–2817 (1982).

21. H. Baumhauer, Über die verschiedenen Modifikationen des Carborundums und die Erscheinung der Polytypie. *Z. Kristallogr.* **55**, 249–259 (1915).

22. Dornberger–Schiff, On order-disorder structures (OD structures). *Acta Crystallogr.* **9**, 593–601 (1956).

23. K. Dornberger–Schiff, *Grundzüge einer Theorie der OD-Strukturen.* Akademie-Verlag, Berlin (1964).

24. K. Dornberger–Schiff, OD Structures—a Game and a Bit More. *Krist. Tech.* **14**, 1027–1045 (1979).

25. R. A. Verma and P. Krishna, *Polymorphism and Polytypism in Crystals.* Wiley, New York (1966).

26. B. B. Zvyagin, *Electron Diffraction Analysis of Clay Mineral Structures.* Plenum, New York, (1967).

27. K. Fichtner, Generalizations of the *hc* Notation. *Z. Kristallogr.* **167**, 261–273 (1984).

28. K. Dornberger–Schiff, S. Ďurovič and B. B. Zvyagin, Proposal for general principles for the construction of fully descriptive polytype symbols. *Cryst. Res. Technol.* **17**, 1449–1457 (1982).

29. A. Guinier, G. B. Bokij, K. Boll–Dornberger, J. M. Cowley, S. Ďurovič, H. Jagodzinski, P. Krishna, P. M. de Wolff, B. B. Zvyagin, D. E. Cox, P. Goodman, Th. Hahn, K. Kuchitsu and S. C. Abrahams, Nomenclature of polytype structures. *Acta Crystallogr.* **A40**, 399–404 (1984).

30. K. Fichtner, A new deduction of a complete list of OD groupoid families for OD structures consisting of equivalent layers. *Krist. Tech.* **12**, 1263–1267 (1977).

31. K. Dornberger–Schiff and H. Grell, On the Notions: Crystal, OD crystal and MDO crystal. *Kristallografiya* **27**, 126–133 (1982).

Comp. & Maths. with Appls. Vol. 12B, Nos. 3/4, pp. 763–766, 1986
Printed in Great Britain.

0886–9561/86 $3.00 + .00

VORONOI POLYHEDRA AND THE DOCKING PROBLEM

CARL W. DAVID

Department of Chemistry, University of Connecticut, U-60 Rm. 161, 34 Glenbrook Road, Storrs, CT
06268, U.S.A.

(*Received October* 1984)

Abstract—Voronoi polyhedra, used in past work to mensurate volumes and areas of molecules, are here applied to the docking problem in biology, indicating a future use for these polyhedra in studying biological interactions.

Voronoi polyhedra[1] have been used to assign volumes to atoms and ions. Since the volume of atoms or ions is undefined in principle, the assignment of such volumes through the artifice of the Voronoi polyhedra represents one way of inflicting an essentially human point of view onto geometric aspects of structural chemistry. O'Keefe[2] suggested the use of Voronoi polyhedra to obtain coordination numbers in crystals. Bernal[3] used Voronoi polyhedra to characterize the adjacency patterns in hard-sphere fluids. Richards[4] used a variant of Voronoi polyhedra to assign volumes to atoms and groups of atoms in biologically interesting molecules.

In the context of this paper, the Voronoi polyhedra are closed polyhedra surrounding each individual atom in an arbitrary assembly of atoms or molecules, which includes all points closer to the individual atom than any other atom in the system. The faces of the polyhedra are perpendicular to the line joining the centers of nearest neighbors. Two atoms which are adjacent to each other each contribute a face to the other's Voronoi polyhedron, although the areas of the two faces are not equal. The edges of a Voronoi face all intersect in vertices, each labelled by the atoms which are contributing faces to the edges which are forming the vertices. We[5] reintroduced Voronoi polyhedra to the study of liquids and liquid mixtures. Our interest originally was sparked by questions pertinent to the meaning of the partial molar volumes. In speculating about the meaning of partial molar quantities, one starts with the idea of how one measures the volume of a molecule, and ends with the idea that the volume of a molecule is a function of the molecules doing the measuring. This means that there is no single number which can be assigned to the molecular volume, and instead implies that the volume of a molecule is, at least in a theoretical sense, an experimental quantity. Given this central thought, the experiment clearly is one of the statistical mechanical simulation techniques, either molecular dynamics or Monte Carlo simulation. Since the Voronoi polyhedra are the correct measure of molecular size in monatomic regular solids it would appear reasonable to enlarge this concept, and decide to arbitrarily introduce the Voronoi polyhedra per atom in nonmonatomic fluids, i.e. in fluids with polyatomic interacting species.

It is apparent that per solute molecule, each atom will have an instantaneous Voronoi polyhedron, and that the "sum" of all these Voronoi polyhedra for all of the atoms of the solute molecule is a Voronoi polyhedron assignable to the entire solute molecule. Furthermore, it is apparent that if the solute molecule has conformation lability, then the Voronoi polyhedra will be conformational labile even if the solvent molecules were rigid (as in a glass).

Since our simulation work relative to the polarization model[6] for water and its ionic dissociation products has centered on clusters, we applied our original Voronoi (oxygen based, *vide infra*) concept to these clusters, and rapidly found that Voronoi polyhedra had a diagnostic use in discovering which atoms of a cluster were "on the surface"[7].

In the next paper of this series[8] we noted the fact that one could generate two sets of Voronoi polyhedra when dealing with water as solvent, one based on oxygens and one based on both oxygens and protons. The distinction allowed us to design a computer recognition algorithm for recognizing which Voronoi polyhedra of a solute molecule are proton acceptors based on the change in faces experienced in going from the former to the latter method of generating Voronoi polyhedra.

Finally, we noticed[9] that the Voronoi polyhedra of solute molecules identifies the molecules which were adjacent to them so that the meaning of the "first coordination sphere," when there is no sphere (complex solute) and no other definition of "first coordination," becomes quantifiable. The Voronoi polyhedron offers an instantaneous measure of whether or not a given solvent molecule is in the "first coordination sphere."

We have now uncovered two other uses of Voronoi polyhedra in solution studies. First, we have noted that Voronoi polyhedra, or rather the lack of Voronoi polyhedra, may be used as a diagnostic tool for searching for clefts in large molecules, i.e. surface regions where the surface of the molecule is dimpled so that, in possible lock and key fashion, substrates and substrate acceptors might meet and interact (with the elimination of the intervening solvent)[10]. Next, we here report the next step in understanding this process, going one step beyond mere recognition, to measuring in some sense the actual process in which two molecules recognize each other.

The Voronoi polyhedra are completely known when the atomic coordinates of all the atoms in a system are known. In solids, this implies that the Voronoi polyhedra are fixed, stable entities. In liquids and gases, on the other hand, the Voronoi polyhedra are changing constantly as nuclei move. Therefore, only average properties of Voronoi polyhedra are of interest, and the Voronoi polyhedra are uniquely associated with simulations (molecular dynamics or Monte Carlo). In the work on Voronoi polyhedra coming from this laboratory, we have concentrated on the algorithms to be used, rather than on the results which Voronoi polyhedra give. Before proceeding, it is well to note that Voronoi polygons (the two-dimensional analog of Voronoi polyhedra) have been used in a multitude of ways. They have been used in geology[11], biology[12] and in other areas, as indicated by Boots and Murdoch[13].

To illustrate the construction of Voronoi polyhedra in the cases under study here, we use the two-dimensional Voronoi polygons. As can be seen in Fig. 1, connecting the centers of atoms, and erecting the perpendicular bisectors of these centers, results in a pattern of intersecting lines which define areas enclosing each atom center. All points inside a Voronoi polyhedron are closer to the center of the "central" atom than to any other atom in the system.

In three dimensions, the connecting of atom centers is the same, but instead of a perpendicular bisector, one must erect a perpendicular bisector plane: a plane which is itself perpendicular to the line joining atomic centers. Two such planes intersect in a line, and three lines

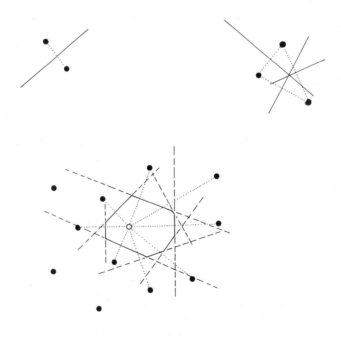

Fig. 1. Constructing Voronoi polygons.

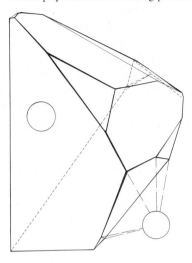

Fig. 2. A typical Voronoi polyhedron, taken from a high-temperature simulation of a water droplet. Only the oxygen in the center, and one which is contributing a face, are shown.

intersecting together generate a corner of a Voronoi polyhedron. Figure 2 shows a typical Voronoi polyhedron, this one taken for a central water molecule (oxygen shown) surrounded by others. The coordinates were taken from a high-temperature simulation of water droplets.

Here, we show how Voronoi polyhedra may be employed to algorithmically detect the "touching" of two molecules in solution, the so-called "docking problem" of biology[14], and we show that Voronoi polyhedra can provide a measure of overlapping areas between juxtaposed docking molecules, which might be used to quantitatively follow a "degree of docking" parameter. Consider two large molecules whose chemical function depends on one of them finding the appropriate region on the other, and attaching itself to this appropriate region in an special fashion. Nature carries out this task all the time. As chemists simulate these systems, an algorithm for detecting when and how efficiently docking has occurred will become important. Since the Voronoi polyhedra of the atoms of solvated solute molecules contain faces which are labelled by either which solute or which solvent molecule "causes" that face to exist, it is apparent that the docked molecules, i.e. those which are adjacent or have fit into

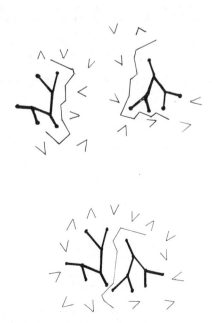

Fig. 3. The approach of two complex solute molecules which were originally solvated. Upon docking, the solvent sheath disappears, and the solute molecules "see" each other.

each other, will share Voronoi polyhedron faces rather than sharing faces with the Voronoi polyhedron of an intervening solvent molecule. As two solute molecules approach each other, the last solvent molecules to leave the docking area, as they leave, will take with them the Voronoi faces which carry that solvent label. Consequently, once adjacency or docking is achieved, certain solute Voronoi polyhedra on both solute molecules will contain faces due to the other solute molecule, where before there were faces due to solvent. This means that these atoms are "touching," i.e. they are reasonably close in space, and no solvent intervenes between them. As docking proceeds, the areas of each solute molecule's faces due to the other solute molecule (and *vice versa*) will grow, thereby providing a semiquantitative measure of the progress of the docking.

Figure 3 shows a sequence of drawings in which the two-dimensional analog of the Voronoi polyhedra for complex solute molecules are shown coming together. It is clear that the Voronoi polyhedra of the atoms of two solute molecules will indicate when no solvent molecules intervene between them. The Voronoi polyhedra offer a multifaceted tool for studying docking under simulation conditions.

REFERENCES

1. G. Voronoi, Nouvelles application des parametres continus a la theorie des formes quadratiques. *J. reine u. angew. Math.* **134**, 199–287 (1908).
2. M. O. O'Keefe, A proposed rigorous definition of coordination numbers. *Acta Cryst.* **A35**, 772–775 (1979).
3. J. D. Bernal, The Bakerian Lecture, 1962—The structure of liquids. *Proc. R. Soc.* **A280**, 299–322 (1964).
4. F. M. Richards, The interpretation of protein structures: Total volume, group volume distributions and packing density. *J. Mol. Biol.* **82**, 1–14 (1974).
5. E. E. David and C. W. David, Voronoi polyhedra as a tool in studying solvation structure. *J. Chem. Phys.* **76**, 4611–4614 (1982). We note here that the algorithm used for obtaining the vertices of Voronoi polyhedra is given in this paper. An alternative algorithm known to us is due to W. Brostow and J-P. Dusasault, Construction of Voronoi polyhedra, *Computat. Phys.* **29**, 81–92 (1978).
6. F. H. Stillinger and C. W. David, The polarization model for water and its ionic dissociation products. *J. Chem. Phys.* **69**, 1473–1484 (1978).
7. E. E. David and C. W. David, Voronoi polyhedra and cluster recognition. *J. Chem. Phys.* **77**, 3288–3289 (1982).
8. E. E. David and C. W. David, Voronoi polyhedra and solvent structure for aqueous solutions (III). *J. Chem. Phys.* **77**, 6251–6254 (1982).
9. E. E. David and C. W. David, Voronoi polyhedra for studying solvation structure (IV). *J. Chem. Phys.* **78**, 1459–1464 (1983).
10. C. W. David, *Comput. Chem.* **8**, 225–226 (1984).
11. D. F. Watson, Computing the n-dimensional Delaunay tesellation with application to Voronoi polytopes. *Comput. J.* **24**, 167–172 (1981); B. N. Boots and D. J. Murdoch, The spatial arrangement of random Voronoi polygons. *Comput. Geosci.* **9**, 351–358 (1983).
12. H. Honda, Description of cellular patterns by Dirchelet domains: the two dimensional case. *J. Theor. Biol.* **72**, 523–543 (1978); H. Honda, T. Morita, and A. Tanabe, Establishment of epidermal cell columns in mammalian skin: computer simulation. *J. Theor. Biol.* **81**, 745–759 (1979); N. Saito, Asymptotic regular pattern of epidermal cells in mammalian skin. *J. Theor. Biol.* **95**, 591–599 (1982); B. J. Gellatly and J. L. Finney, Calculation of protein volumes: an alternative to the Voronoi procedure. *J. Mol. Biol.* **161**, 305–322 (1982).
13. *Loc. cit.*
14. I. D. Kuntz, J. M. Blakely, S. J. Oatley, R. Langridge and T. E. Ferrin, A geometric approach to macromolecule–ligand interactions. *J. Mol. Biol.* **161**, 269–288 (1982).

Comp. & Maths. with Appls. Vol. 12B, Nos. 3/4, pp. 767–781, 1986
Printed in Great Britain.

0886-9561/86 $3.00 + .00
© 1986 Pergamon Press Ltd.

PATTERN SYMMETRY AND COLORED REPETITION IN CULTURAL CONTEXTS

Dorothy K. Washburn

Department of Anthropology, University of Rochester, Rochester, NY, U.S.A.

Abstract—The application of a mathematical principle, symmetry, to an area so highly variable as human culture seems remarkable until one discovers that the principle uncovers hitherto unknown consistencies in human behavior. In Sec. 2, the types of analyses which have shown how symmetry is manifested in culture are discussed. However, not all patterns created by cultural groups maintain structural symmetry when combined with other stylistic aspects. In Sec. 3, colored patterns on pre-Columbian textiles from Peru, which have colorings inconsistent with their underlying structural symmetry are discussed.

1. INTRODUCTION

Symmetry is a type of order with specific geometric parameters. As a mathematical measure it has proved useful for the classification and comparison of patterns on cultural materials. Investigators have been able to study more systematically consistencies and changes in temporal and spatial aspects of design styles and to relate these shifts to other patterns of activity in a given culture.

In general it has been observed that (1) most designs produced by cultures throughout the world are symmetric, and (2) the designs in any given culture are organized by just a few symmetries rather than by all classes of the plane pattern symmetries (7 one-dimensional classes and 17 two-dimensional classes). We do not yet understand why there is preferred use of several symmetries, nor how these preferences relate to other activities in culture.

This paper will focus, however, on designs which, when colored, do not have symmetries identical with their original structural symmetries. That is, the colored design does not preserve the symmetries of the original uncolored design. We have described such colorings as "not consistent" with the structural symmetry[1]. Such colorings may be repeated by another symmetry or they may have no symmetrical repeat scheme at all.

2. SYMMETRY IN CULTURAL ANALYSES

For many years anthropologists have utilized the concept of symmetry to describe both general aspects of the organization of cultural behavior and specific aspects of the structural organization of designs on material culture objects. Two types of treatment of symmetry are typical. One treatment implicitly infers the presence of symmetry by use of terms such as "balance" and "harmony" when describing the structure or association of activities, people, or things in culture. The other treatment explicitly uses the term "symmetry" but does not go beyond the popular conception of symmetry as bilateral reflection or rotation. In none of these latter studies except [2] are the specific plane pattern motion classes used.†

Perhaps the most famous and certainly one of the earliest important considerations of the symmetry in art forms is found in Franz Boas' *Primitive Art*[3]. Boas saw symmetry as a universal "formal" property in "the art of all times and all peoples"[3, p. 32], citing examples from body painting among the Andaman Islanders and designs on wood shields of the Australian aborigines, Pueblo pottery, and Peruvian textiles. Boas observed that "symmetrical arrangements to the right and left of a vertical axis"[3, p. 33] are most frequent because this is the orientation of the human body and most things in the natural world. But it is significant that even at this early date investigators like Boas noticed not only the symmetry but also the deviations from symmetry in design. Boas goes into some detail on the unusual repetitions of form and color on Peruvian textiles, a topic to which we will return later in this paper.

---

†This paper will treat only the classes of finite, one-dimensional, and two-dimensional infinite symmetries since these classes fully characterize designs found on flat or curved objects whose designs can be "unrolled" (i.e. ceramic vessels, carved wood drums, pipes, etc.).

A number of other investigators have focused on the symmetry in art forms of a particular cultural group. Holm's[4] classic work on bilaterally split figures in North West Coast art, Critchlow's[5] relation of symmetries in Islamic architecture to the structure of the Islamic cosmos, Lévi–Strauss'[6] analysis of the symmetries in Caduveo body art and Glassie's[7] discussion of the bilateral tripartite structure of folk architecture in Virginia are several outstanding examples.

Anthropological studies which enumerated the discrete plane pattern symmetries found on decorated material did not appear until the 1970s, even though the classes had been defined much earlier[eg. 8]. There were two notable exceptions. Brainerd's short paper[9] is pivotal because it was the first mention in the anthropological literature of how this type of classification might enable systematic comparative studies. Anna Shepard's seminal monograph, *The Symmetry of Abstract Design with Reference to Ceramic Decoration*[2], contained an explicit discussion of the symmetry classes, description of the classification procedure, and a detailed analysis of patterns from three cultural traditions.

During the 1970s a number of investigators in different disciplines coincidentally "rediscovered" symmetry analysis, both as a classificatory tool and as an aid in the discovery of consistencies and differences in cultural activities. Detailed descriptions of the symmetry classes were presented in Washburn[10] and Zaslow[11] and flow charts for their identification were presented in Crowe[12].

Using these procedures a number of researchers produced descriptions of the structural foundation of a culture's design system. Such studies are typified by Crowe's studies of African art[13–15]; Campbell's study of Pueblo pottery[16]; Donnay and Donnay's analysis of Maori rafter patterns[17]; and Hargittai and Lengyel's description of Hungarian folk needlework[18,19].

Other investigators have used the regularities described by symmetry analysis to aid in the clarification and interpretation of cultural behavior patterns. Ascher and Ascher[20] showed that preferential use of several symmetries was consistent with their theory of "Inca Insistence"; Van Esterik[21] compared the symmetry rules for Ban Chiang pottery designs to language grammars as a form of coded information; Kent[22] correlated continuities and shifts in Anasazi prehistory with continuities and changes in symmetries of textile design; and Washburn applied symmetry analysis to Anasazi ceramics to study site associations[10,23], to Greek Neolithic ceramics in order to study changes in interaction spheres over space and time[24], and to California Indian basketry design to study correlations of structural homogeneity in design with language, marriage practices, and trade networks[25,26].

Still other investigators[27–32] have suggested that these regularities represent the expression of cognitively held organizational principles in the arrangements of forms and parts of forms in decorative design, house architecture, dance, music, village layout, etc.

The above is but a brief sampling of the types of applications of symmetry analysis. A more complete and detailed discussion can be found in [1]. The major thrust of the past research reviewed above has been the application of finite, one-dimensional and two-dimensional plane pattern symmetry classes to classify the structural organization of design motifs. Such classifications have revealed consistencies in use or changes in use of these structures over time and space in cultural contexts.

In contrast, this paper focuses on cultural patterns which are *not* symmetric. Textile designs from sites in pre-Columbian Peru form the basis for a pilot study which analyzes the impact of color on the symmetry of the design structure.

3. SYMMETRY AND COLORED REPETITION

Peruvian textiles have been the object of detailed analyses for many years. These studies have focused on materials and weaving techniques[33–37] or design motifs[38]. Styles have been defined in terms of temporal and spatial changes in these attributes.

While the capacities of the ancient Peruvian textile art remained nearly constant, period expressed itself in style; and style varied according to the favor accorded this or that technical process, as well as its decorative patterns and color combinations[36, p. 31].

In many symmetrically colored patterns the coloring is consistent with the underlying symmetry; that is, the colored symmetry is a coloring of the symmetry of the underlying structure. In other colored patterns, the coloring has no relation to the underlying symmetry of the structure of the pattern. On many Peruvian textiles the superimposition of color reduces the symmetry of the colored pattern to translation. This paper focuses on this type of colored symmetry using eight examples from the Ica Valley, Nazca, Paracas, and the highlands. However, before the patterns are described, it is necessary to discuss four features of layout used by the Peruvian weavers because these seem to be integrally involved in the design organization process.

First, although the design parts are structured along horizontally and vertically oriented axes, frequently the colorings make the pattern appear to be oriented on the diagonal.‡ Sequences of a single predominant color or colors of similar hues move diagonally across the textile. Repetition of these diagonally oriented colors follows certain rhythmic sequences which are regular although they are not consistent with the isometries of the underlying pattern.§

Second, to the Western observer accustomed to patterns arranged in an up/down orientation in a design field surrounded by a border, these patterns seem to be incomplete. There are a number of ways this aspect can be manifested.

(1) Frequently, although the textile itself is a complete piece of cloth, the pattern is repeated to the edges, leaving no border. This layout gives the appearance that the pattern continues beyond the edges of the cloth. For example, in Fig. 14 no motif has been truncated; the motif shapes and the design symmetry used to arrange them allow them to appear as complete motifs on the textile.

(2) There are patterns where the repeated motifs are truncated by the edges of the cloth. For example, in Fig. 11 both the asymmetric shape of the motif, and the symmetries used (glide reflection arranges motifs in offset layouts and thus creates patterns with "ragged edges"), combine to emphasize the appearance of pattern continuation beyond the cloth edges.

We begin with a short history of anthropologists who have studied the unusual color repeat systems used on Peruvian textiles.

As mentioned earlier, the unusual kinds of repeats in Peruvian textiles were first noted around the turn of the century. Charles Mead wrote a short article describing what he called the "six-unit design"[43].

> The six units of the design are of the same size and pattern, but each varies from the preceding and following one in the color or colors employed. This is the common form; but in some cases we find the design composed of three units of one color, followed by three of another, or four of one and two of another, etc. In every case the six-unit design is retained[43,p. 194].

Had Mead examined a larger sample of cloths he would have noticed that not all Peruvian textiles are limited to this repeat cycle, as we shall see from the examples in this paper. Furthermore, although Mead is not explicit about this, the textile in his Plate XI shows both color and motif changing in series. He does note that the colors repeat along the diagonal, rather than along the vertical or horizontal axes of the textile.

Some years later Gladys Reichard investigated whether these "rhythm-units" of repeated colored motifs were present in the decoration of beadwork and embroidery from several subarctic Indian groups. She defined a rhythm-unit as ". . . a combination of motives, either of color or design, which is regularly or symmetrically repeated"[44, p. 187]. For example, despite the failure for the colors on a colored beaded necklace to be systematically alternated, there was mirror symmetry between the right and left halves of the necklace[44, p. 192, schematic No. 6]. In addition, certain predominant colors were repeated less frequently and were positioned so as to fall on one of the repeats in the center of the pattern.

(3) Finally, truncated patterns can be created by certain weaving techniques. Crawford gives an example[34, Figs. 11 and 12] where the order of shedding the warps produces a design

---

‡For this discussion the rectangular textiles are oriented with their longest edge horizontal. In this orientation the bands of identical color are positioned on the diagonal.

§Certain technological processes can, however, produce diagonal stripes. In tye dying diagonal stripes can occur if the fabric is folded on the bias before being rolled, tied, and dyed[34, p. 154].

Fig. 1. Garment fragment, brocade. Catalogue #B-4334, American Museum of Natural History, New York.

within a motif. The decorative wefts pick up the warps of the plain weave background and create another pattern inside the motifs. Although this pattern appears to "run off" the edges of the fret and step motif in Fig. 1, it does not actually continue into the white background which is a simple plain weave. The white outline pattern is created only inside the motifs because the decorative dark wefts have picked up the background white warps.

Third, a problem associated with the incomplete aspect of these patterns arises when they are colored. Frequently, not enough repeats are shown to allow determination of whether there is symmetric repetition of colors. For example, the intact lower section of a fragmentary garment from the Ica Valley (Textile Museum #1966.7.37a) (Fig. 2) has three rows of units in $p2$ symmetry.† Although some of the colors are unclear in this black and white photograph, the dark diagonal and the light diagonal series of units predominate, so that we might think that there is some kind of color alternation. However, the pattern ends at the textile edges before the dark or light white diagonals are repeated again. Although it is clear that motifs of the same color are arranged diagonally, not enough repeats are present for the analyst to identify the color repetition scheme. One could surmise that the black diagonal might begin in the next

Fig. 2. Garment fragment, Ullujaya, group 3, Ica Valley, Peru. Catalogue #1966.7.37a, The Textile Museum, Washington, D.C.

†The different symmetry classes are named by the crystallographic plane pattern nomenclature for one color symmetries as listed in [39], for two color symmetries in [40], for three color symmetries in [41], and for four and high color symmetries in [42].

bifold unit in the upper right-hand column if the cloth were extended beyond the present right edge. However, there are other examples, such as the Paracas mantle described in this paper (Fig. 15), which do not have this black/white alternation, but apparently have, instead, a black/black/white/white alternation scheme.‡

Fourth, often details of the pattern, such as motifs on the costume of figures, are not precisely identical and so do not allow superposition, even though the general shape of the figures and their relative positions in the design suggest one of the symmetry classes. This is the case for the textiles discussed in Figs. 5 and 7, where details of the costume vary and so technically do not allow superpositions. We shall say that such designs have "apparent symmetry," a concept in keeping with psychological research[49] which indicates that viewers scan for major, fundamental aspects of form, such as symmetry, and "overlook" minor irregularities when making their initial identifications of a form. In this paper these detail variations have been noted and then ignored in order to discuss the interrelationship of the structural symmetry and colored symmetry.

For example, the /Y/ (yellow) in the necklace sequence outlined below[44, p. 192, schematic #12]

Y BR Y BR Y RBRB Y B Y RB/Y/BR Y R Y BRB   Y BR Y BR Y BR

not only falls in the center of the line of beads, but demarcates the middle of a repeat sequence of the R (red) and B (blue) beads. Immediately on either side of the center /Y/ bead the red and blue beads are positioned so that it appears that a mirror reflection line is present across which bead colors are preserved: on the left of the /Y/ the RB reflects into the BR to the right of the /Y/. However, except for a set of beads on the left which are inconsistent in number, the other sets change colors across this center reflection line. This color reversal is in counterpoint to the preservation of the BR color sequence as the bead sets "translate" in a linear array from left to right.

Boas continued Mead and Reichard's work on color and repetition by examining systems of color repetition in very complex two-dimensional patterns. On several examples he showed how to correct "mistakes" where the color repeat sequence was not exactly followed in order to obtain a perfectly colored pattern[3, p. 48]. In light of the recent research of Brett–Smith[45], who has suggested that "irregularities" in Mali textiles were actually purposely included to offset the awesome powers of perfect "speech," and of Morris[46], who has shown that the "irregular" repeats in a Chamula huipil design actually depict the numbers of gods and worlds in the Mayan cosmology, the frequency and consistency in the type of "mistake" in Peruvian textiles should enjoin future researchers to investigate the possible relationship of this stylistic feature to other aspects of the cultural system.

Perhaps the most important work on color repetition, at least on Peruvian textiles, is an analysis by Cora Stafford[47] of one- and two-dimensional color repetition sequences, which she labels "surface patterns," on mantles, shawls, ponchos, loincloths, skirts and coca bags from the site of Paracas.

In the following analyses we shall show how symmetry analysis of the two-dimensional mantle patterns complements and enhances Stafford's studies of color repetitions. Use of symmetry principles to systematically describe the arrangements of the motifs avoids more subjective descriptions, such as Stafford's description of the relationships of pumas in a mantle (#32-30-30/45 Peabody Museum, Harvard): ". . . the large free motifs turn alternately right and left in vertical rows and also alternate up and down position in consecutive rows"[47, p. 40]. We can more simply describe this structure with the symmetry class designation *pgg*. Additionally, we can determine whether the color repetition sequences which Stafford describes fall within one of the classes of color symmetry (see [40–42] for illustrations and lists of these classes).

We begin our discussion of the symmetry of color superposition schemes by describing an

‡The fact that the other colors are not clear from this black and white photograph is a good warning to the reader always to analyze either color photographs or the actual object. Pre-Columbian weavers used a wide palette of colors, many of subtle hue changes which are not clearly distinguishable in black and white photographs. In this paper a schematic color repetition scheme is associated with each textile discussed.

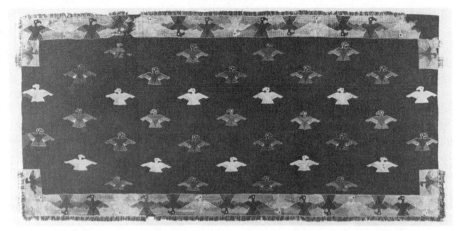

Fig. 3. Mantle, Paracas, Peru. Catalogue #34.1553, Brooklyn Museum, New York.

exception on a mantle from Paracas in the Brooklyn Museum collection (#34.1553, Fig. 3). Light blue/green, dark blue/green and gold birds on a dark green/blue wool background are structured by one color pg symmetry; the addition of colors results in a coloring of the class pg, the three color class $pg[3]_1$. Birds of the same three colors are also found on the bordering strip in a band arranged by one-dimensional glide reflection ($p1a1$).

Because the birds in the center design area change the direction they face every second row instead of every row, it at first appears that this design cannot be a symmetrical pattern. But, because the birds in each succeeding row are offset, vertical glide axes pass through every other bird. Thus the birds change direction as they move vertically along the glide axes to create a pg pattern. In the diagram in Fig. 4, vertical glide axes pass through, for example, q1, p2, q3, and p1. (The rows labeled p represent birds facing right and the rows labeled q represent birds facing left. The glide axis is represented by a dashed line.)

The most unusual aspect of this pattern is that the addition of colors results in a perfectly colored $pg[3]_1$ pattern. Such examples, where the coloring actually results in a colored symmetry consistent with the structural symmetry, are unusual on Paracas textiles; most have colorings which are not symmetrical or which have no relation to the underlying symmetry. In this pattern the three colors (1, 2, 3 in Fig. 4) are preserved along horizontal rows, but are changed along the vertical glide axes. The same coloring results if *pairs* of rows of birds facing the same direction are moved along the glide axes. For example, in Fig. 4, if row pair p3 and p2 is moved up so that p3 is superimposed on q1 and p2 is superimposed on q3, and, likewise, if

```
                    |
                    |
      p1    p1    p1    p1    p1    p1
                    |
         p2 |  p2    p2    p2    p2
                    |
      q3    q3    q3    q3    q3    q3
                    |
         q1 |  q1    q1    q1    q1
                    (
      p2    p2    p2    p2    p2    p2
                    |
         p3 |  p3    p3    p3    p3
                    |
      q1    q1    q1    q1    q1    q1
                    |
                    |
```

Fig. 4. Schematic color scheme for Paracas mantle #34.1553 (Fig. 3)

Fig. 5. Mantle, Paracas, Peru. Catalogue #34.1549, Brooklyn Museum, New York.

the row pair q1 and q3 is moved up so that q1 is superimposed on p2 and q3 is superimposed on p1, the coloring is consistent with *pg[3]₁*.

Glide reflection is a very common symmetry in Paracas textile design. The Paracas mantle in Fig. 5 (#34.1549, Brooklyn Museum) has human figures positioned along the long horizontal axis, each row alternating head-up and feet-up orientations. The snake (?) alternates curled positions on the left and right sides of the figure's feet. For the purposes of this discussion the focus is on the general position of the figures and the snake; other small differences in costume detail are ignored.

First, if the colors and a few irregularities in figure position along the left side of the mantle are not considered, the snake position alternates consistently from the left to the right side of the figure along vertical glide axes which pass through the figures. The figures alternate head-up/feet-up positions in successive vertical columns. There are horizontal glide axes between the figures in rows A and B, C and D, and E and F, but there are only points of rotation at the corners of the figure squares which superimpose the figures in rows B and C, D and E, and F and G (Fig. 6). The basic symmetry of this pattern is *pgg*.

The addition of colors, however, does not result in a colored *pgg* pattern. The figures are of the same color along diagonals running from lower left to upper right, but are of different colors along diagonals running from lower right to upper left.

This pattern is similar to the beaded and embroidered examples described by Reichard in that one color is predominant and is positioned in the center of the design. In this pattern it is color 7, which passes diagonally through the center of the textile. The diagonal emphasis of

Rows

G	1	2	3	4	5	6	/7/	1	2	3	
F		2	3	4	5	6	/7/	1	2	3	4
E		2	3	4	5	6	/7/	1	2	3	4	5
D			3	4	5	6	/7/	1	2	3	4	5
C			3	4	5	6	/7/	1	2	3	4	5	6
B				4	5	6	/7/	1	2	3	4	5	6
A				4	5	6	/7/	1	2	3	4	5	6	7

Fig. 6. Schematic color scheme for Paracas mantle #34.1549 (Fig. 5).

Fig. 7. Mantle, Paracas, Peru. Catalogue #16.32, Museum of Fine Arts, Boston.

the coloring is superimposed upon a *pgg* structure that has a vertical/horizontal orientation of the glide axes.

Another Paracas mantle (#16.32, Boston Museum of Fine Arts) (Fig. 7) has human figures also arranged in alternating feet-up/feet-down position similar to that in Fig. 5. Again we shall ignore the differences in costume detail except for the positions of the branch (?) which alternates position in the left and right hand of the figure. Although this layout appears identical to the pattern in Fig. 5, closer examination reveals that this pattern cannot be a *pgg* pattern because there is also a vertical mirror reflection line (indicated by the solid line in Fig. 8) which divides the whole mantle design into two reflecting halves, but only for some of the rows. That is, for columns E and F, only figures in rows N, P, and R are holding branches in mirror reflection across the central vertical mirror axis. The figures in rows, S, Q (except for figures on the left edge in column A), O (except for figures on the left edge in column A), and M (except for figures on the left edge in column A and figures on the right edge in column J) all are holding their branches in the same hand throughout a given row. In any of the columns the figures move along glide axes which pass vertically through the figures, so that the hand holding the branch is reversed. The successive columns alternate figures feet-up and head-up (except for the figures in column A).

Colors are superimposed upon this pattern so that the diagonals are of alternating colors. Figure 8 displays the color scheme as recorded by Stafford. From lower right to upper left,

Column A	B	C	D	E	F	G	H	H	J	Row
1	2	1	3	4	2	1	3	4	2	M
	4	3	4	2	1	3	4	2	1	N
3	1	2	1	3	4	2	1	3	4	O
	2	4	3	4	2	1	3	4	2	P
4	3	1	2	1	3	4	2	1	3	Q
	1	2	4	3	4	2	1	3	4	R
2	4	3	1	2	1	3	4	2	1	S

Fig. 8. Schematic color scheme for Paracas mantle #16.32 (Fig. 7).

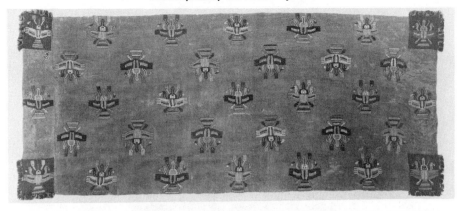

Fig. 9. Mantle, Paracas, Peru. Catalogue #34.1556, Brooklyn Museum, New York.

alternating diagonals alternate colors 2 and 3 and colors 1 and 4. All four colors are present along all diagonals running from lower left to upper right. Although Stafford claimed that the color sequence 1342 characterized this pattern, there are many deviations from this repetition sequence, particularly along the edges. Because many Peruvian textiles have "irregularities" along the edges, such deviations may not be simple "mistakes."

A Paracas mantle (#34.1556, Brooklyn Museum) (Fig. 9) with figures arranged in *pmg* symmetry, is colored with four colors: (1) gold body, (2) red body, (3) dark green body with green and gold head, and (4) dark green body with red and yellow head. Again ignore the small differences in details of the figures' costumes. Figure 10 shows Stafford's color repetition scheme.

The diagonal color arrangement, where diagonals from lower left to upper right alternate two colors but figures in diagonals from lower right to upper left alternate four colors, is not described by any of the 11 four-color *pmg* color classes. There is no systematic change of colors across vertical mirror reflection lines which pass through the figures. The four colors alternate along the horizontal glide axes which run between rows of figures. This is a particularly clear example of how the underlying *pmg* structural arrangement of the figures and the diagonal coloring is not related. It is as if the color repetition and the underlying symmetry of arrangements of figures were thought of as two different and separate design features. Again, it is notable that, especially along the left edge of this textile, there are significant deviations from the 1324 color sequence which Stafford observed.

This analysis of figurative patterns can be extended to geometric patterns. For example, the rather complicated asymmetric geometric motif repeated in a design on a small Nazca bag (Memorial Art Gallery, University of Rochester #74.79) (Fig. 11) has *pg* structure.

This unit motif appears in three different colors: black (B), red (R), and yellow (Y). The entire scheme of the color repeats is shown in Fig. 12. The same color falls along diagonals running from lower left to upper right. The yellow diagonals predominate. This coloring, however, is not one of the two colorings of three-color symmetries of class *pg*. The three colors change along any one vertical glide axis which passes through the columns of units. However, the fact that only some colors change along any glide axis which passes between any two columns of units means that the coloring is not consistent with the glide reflection symmetry.

```
1   2   4   1   3   2   4

  1   3   2   4   1   3

3   4   1   3   2   4   1

  3   2   4   1   3   2

2   1   3   2   4   1   3
```

Fig. 10. Schematic color scheme for Paracas mantle #34.1556 (Fig. 9).

Fig. 11. Bag, Nazca, Peru. Catalogue #74.79, Memorial Art Gallery, University of
Rochester, Rochester, New York.

```
/Y/   B   R   /Y/   B   R   /Y/   B

B   R   /Y/   B   R   /Y/   B   R

R   /Y/   B   R   /Y/   B   R   /Y/

/Y/   B   R   /Y/   B   R   /Y/   B

B   R   /Y/   B   R   /Y/   B   R

R   /Y/   B   R   /Y/   B   R   /Y/

/Y/   B   R   /Y/   B   R   /Y/   B

B   R   /Y/   B   R   /Y/   B   R

R   /Y/   B   R   /Y/   B   R   /Y/
```

Fig. 12. Schematic color scheme for Nazca bag #74.79 (Fig. 11).

Y	B	R	B	O	B	R	B	R	B	O	B	R	B
B	O	B	R	B	Y	B	O	B	R	B	Y	B	R
R	B	Y	B	R	B	O	B	Y	B	R	B	O	B
B	R	B	O	B	R	B	R	B	O	B	R	B	Y
O	B	R	B	Y	B	R	B	R	B	Y	B	R	B
B	Y	B	R	B	O	B	O	B	R	B	O	B	R
R	B	O	B	R	B	Y	B	Y	B	R	B	Y	B
B	R	B	Y	B	R	B	R	B	O	B	R	B	R
Y	B	R	B	O	B	R	B	R	B	Y	B	R	B

Fig. 13. Schematic color scheme, Tunic, Peru. Catalogue #1959.20.12, The Textile Museum, Washington, D.C.

A repeat sequence of four colors incompatible with any of the four color symmetries is superimposed on a simple *pmm* rectangular checkerboard design on a tunic possibly from the highlands (#1959.20.12, Textile Museum, Washington, D.C.). The color scheme of black (B), red (R), orange (O), and yellow (Y) is shown schematically in Fig. 13. (The vertical dashed line marks the center line of the tunic where two pieces of cloth were sewn together.)

The diagonal orientation of the coloring predominates. There is a series of three diagonals ascending from lower right to upper left which alternate black/red/black rectangles. Interspersed between these sets of colors is a diagonal of alternating orange and yellow rectangles which ascend from lower right to upper left. In effect this orange/yellow diagonal replaces one of the red diagonals in the red/black alternation sequence. Thus, there is definite regularity of repetition in this pattern. The coloring, however, only admits translation symmetry.

A very complex *p4m* pattern on a tunic, also possibly from the highlands (#91.341, Textile Museum, Washington, D.C.) (Fig. 14) is colored with many similar hues. Two predominate:

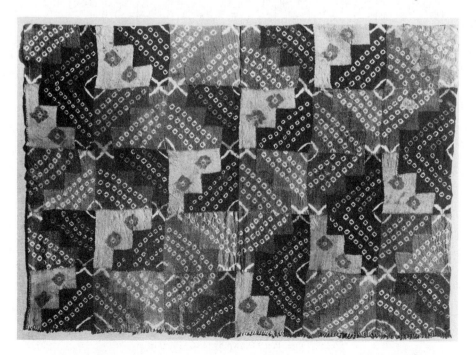

Fig. 14. Tunic, Peru. Catalogue #91.341, The Textile Museum, Washington, D.C.

single white-stepped units moving diagonally from lower left to upper right in two different orientations, and double black-stepped units combined in two different arrangements also moving diagonally from lower left to upper right. The underlying *p4m* structure of this pattern is clearly apparent if the textile is viewed on the diagonal with the tie-dyed white squares as centers of fourfold units. The pattern is not bordered but "runs off" the edges. A sufficient portion of the pattern is repeated to reveal that the coloring is not consistent with the underlying *p4m* symmetry of the pattern.

Finally, the most complex pattern to be examined is on a mantle from Paracas in the collection of the Museo Nacional de Antropologia, Pueblo Libre, Peru (illustrated in [48], Fig. #373). In the orientation shown in the schematic drawing in Fig. 15 the three predominant (and a small section of the fourth) lines of black/white interlocking stepped units ascend diagonally from the lower right to the upper left. The black/white diagonals are separated from each other by four other pairs of colored interlocking stepped units. Since these four pairs are of four different color sets, the repeated sequence is five units (from left to right across the upper edge of the textile: blue/red; dark green/yellow; black/white; light green/pink; brown/tan).

The diagonal orientation of the design makes determination of the "coloring" confusing. Examination of the two full repeats of five sets along the horizontal upper edge (from blue/red to brown/tan and from red/blue to tan/brown) reveals an apparent color reversal between the two repeats. However, the black/white diagonals are arranged so that the dominant features are the two sets of black/white, followed by the two sets of white/black units in translation across the fabric.

The diagonal appearance of this *p2* pattern of interlocking stepped units suggests that there are almost four repeats, but actually only one entire diagonal section of five units is completely visible as it moves across the textile from lower right to upper left. The other three are only

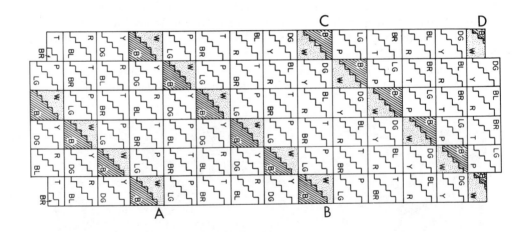

B = BLACK
W = WHITE
DG = DARK
     GREEN
Y = YELLOW
BL = BLUE
R = RED
BR = BROWN
T = TAN
P = PINK
LG = LIGHT
     GREEN

Fig. 15. Schematic color scheme of mantle, Paracas, Peru, Coll. of Museo Nacional de Antropologia, Pueblo Libre, Peru. Drawn by G. C. Bommelje.

partially present because of the diagonal orientation and truncation of the pattern by the textile edges.

Closer examination reveals that although the pattern is composed of units of many colors, the color alternation of the sets of five units is not either of the two color colorings of *p2*. Look first at the black/white stepped units. Both the lower left black/white diagonal (A) and the next middle black/white diagonal (B) have the black section on the bottom and white section on the top—a one-color translation from A to B, not a two-color bifold rotation from A to B. The next two diagonals C and D have the white section on the bottom and the black section on the top—again a one-color translation from C to D.

Viewed in its entirety there are four sets of five units each, albeit some are truncated by the fabric edges. Colors appear to change every two sets: black/white, black/white, white/black, white/black. There are no true color reversals; the pattern remains a one-color *p2* pattern. Viewed in one orientation, the artist has reversed the colors in two successive sets of five units so that it "appears" as if there has been a 180° rotation and color change on every other set. But if the textile was actually rotated, the colors would stay the same. A true *trompe l'oeil*!

Clearly, the Peruvian weavers were experts at manipulating the coloring of complex symmetrical patterns by adding a predominant diagonal orientation and repeating large numbers of colors so that, although the designs are regularly colored, they are not consistent with the underlying two-dimensional symmetry.

It is possible, as Stafford[47] and O'Neale[37] point out, that the large number of colors may be a result of the visual discrimination of hues by 20th century analysts. Colors might have changed or faded so that some of the hues which appear to be different to us may have been ". . . equated in the minds of the workers. Colors occasionally used interchangeably are pink and gold, olive and golden brown, gold and golden brown, dark blue and dark green, red brown and gray. In all of these combinations except the last, the two colors are similar in hue and value"[47, p. 103]. Further studies of colored Peruvian textiles should be coordinated with studies of the discrimination of colors by present-day Peruvian weavers.

SUMMARY

In this paper I have discussed several examples of patterns which are structured by complex two-dimensional symmetries but which are colored so as to reduce the symmetry of the pattern, generally to simple translation. These textiles have structural styles which are not sets of preferred symmetries and their colorings, but involve more complex conjunctions of structural symmetry, "apparent symmetry," and colored repetition. The deviation from the structural order has been noted on decorated material in other cultural contexts, such as on designs on Cuna molas[50], Maori burial chests[51], and Cashinahua hammocks[52], although in these cases it was not caused by coloring. Much remains to be learned about the relationships of design structure with other aspects of what is generally called the design style of a culture, as well as with other beliefs and activities of culture. Here I have simply introduced the problem of the various kinds of interfacing of two aspects of style: structure and color.

Acknowledgments—Grateful thanks are extended to Don Crowe and Kate Kent for critical appraisal of the article. Figure 15 was drawn by Gerald Bommelje. The photographs are reproduced here through the courtesy of the Brooklyn Museum, the Boston Museum of Fine Arts, the American Museum of Natural History, The Textile Museum, and the Memorial Art Gallery.

REFERENCES

1. D. K. Washburn and D. W. Crowe, Symmetries of Culture: A Handbook of Plane Pattern Design (to appear).
2. A. O. Shepard, The symmetry of abstract design with special reference to ceramic decoration, *Contribution to American Anthropology and History* No. 47, pp. 211–293. Carnegie Institution of Washington Publication No. 574 (1948).
3. F. Boas, *Primitive Art*. H. Ascheroug & Co., Oslo (1927).
4. B. Holm, *Northwest Coast Indian Art*. University of Washington Press, Seattle (1965).

5. K. Critchlow, *Islamic Patterns*. Schocken Books, New York (1976).
6. C. Lévi-Strauss, *Tristes Tropiques*. Atheneum, New York (1974).
7. H. Glassie, *Folk Housing in Middle Virginia*. University of Tennessee Press, Knoxville (1975).
8. A. Speiser, *Theorie der Gruppen von endlicher Ordung*, 2nd Ed. Springer-Verlag, Berlin (1927).
9. G. W. Brainerd, Symmetry in primitive conventional design. *Am. Antiq.* **8**, 164–166 (1942).
10. D. K. Washburn, *A Symmetry Analysis of Upper Gila Area Ceramic Design*. Papers of the Peabody Museum of Archaeology and Ethnology, Vol. 68. Harvard University (1977).
11. B. Zaslow, A guide to analyzing prehistoric ceramic decorations by symmetry and pattern mathematics, in *Pattern Mathematics and Archaeology*, Anthropological Research Paper Vol. 2. Arizona State University, Tempe (1977).
12. D. W. Crowe, Symmetry in African art. *Ba Shiru* **11**, 57–71 (1980).
13. D. W. Crowe, The geometry of African art I. Bakuba art. *J. Geometry* **1**, 169–181 (1971).
14. D. W. Crowe, The geometry of African art II. A catalog of Benin patterns. *Historia Mathematica* **2**, 253–271 (1975).
15. D. W. Crowe, The geometry of African art III. The smoking pipes of Begho, in *The Geometric Vein* (Edited by C. Davis, B. Grünbaum and F. A. Sherk), pp. 177–189. Springer-Verlag, New York (1982).
16. P. J. Campbell, The geometry of decoration on prehistoric pueblo pottery from Starkweather Ruin (unpublished).
17. J. D. H. Donnay and G. Donnay, Symmetry and antisymmetry in Maori rafter designs. *Empirical Studies of the Arts* **3**, No. 1, 23–45 (1985).
18. I. Hargittai and G. Lengyel, The seven one-dimensional space-group symmetries by Hungarian folk needlework. *J. Chem. Educ.* **61**, 1033–1034 (1984).
19. I. Hargittai and G. Lengyel, The seventeen two-dimensional space-group symmetries in Hungarian needlework. *J. Chem. Educ.* **62**, 35–36 (1985).
20. M. Ascher and R. Ascher, *Code of the Quipu*. University of Michigan Press, Ann Arbor (1981).
21. P. Van Esterik, *Cognition and design production in Ban Chiang painted pottery*. Paper in International Studies, Southeast Asia Series, No. 58. Ohio University Center for International Studies, Athens (1981).
22. K. P. Kent, Temporal shifts in the structure of traditional southwestern textile design, in *Structure and Cognition in Art* (Edited by D. K. Washburn), pp. 113–137. Cambridge University Press, Cambridge (1983).
23. D. K. Washburn and R. G. Matson, Use of multidimensional scaling to display sensitivity of symmetry analysis of patterned design to spatial and chronological change: examples from Anasazi prehistory, in *Decoding Prehistoric Ceramics* (Edited by B. Nelson). Southern Illinois University Press, Carbondale, pp. 75–101 (1984).
24. D. K. Washburn, Symmetry analysis of ceramic design: two tests of the method on Neolithic material from Greece and the Aegean, in *Structure and Cognition in Art* (Edited by D. K. Washburn), pp. 138–164. Cambridge University Press, Cambridge (1983).
25. D. K. Washburn, The neighbor factor: basket designs in northern and central California (unpublished).
26. D. K. Washburn, Symmetry analysis of Yurok, Karok, and Hupa Indian basket design. *Empirical Studies of the Arts* **4**, No. 1 (1986).
27. M. J. Adams, Structural aspects of a village art. *Am. Anthrop.* **75**, 265–279 (1973).
28. C. E. Cunningham, Order in the Atoni house, in *Right and Left* (Edited by R. Needham), pp. 204–238. University of Chicago Press, Chicago (1973).
29. A. Lomax *et al. Folk Song Style and Culture*. Am. Assoc. for the Adv. of Science Publication No. 88, Washington, D.C. (1968).
30. P. G. Roe, Art and residence among the Shipibo Indians of Peru: a study in microacculturation. *Am. Anthrop.* **82**, 42–71 (1980).
31. D. E. Arnold, Design structure and community organization in Quinua, Peru, in *Structure and Cognition in Art* (Edited by D. K. Washburn), pp. 56–73. Cambridge University Press, Cambridge (1983).
32. F. A. Hanson, When the map is the territory: art in Maori culture, in *Structure and Cognition in Art* (Edited by D. K. Washburn), pp. 74–89. Cambridge University Press, Cambridge (1983).
33. M. D. C. Crawford, Peruvian textiles, in *Anthropological Papers of the American Museum of Natural History, XII*, Part III, New York (1915).
34. M. D. C. Crawford, Peruvian fabrics in, *Anthropological Papers of the American Museum of Natural History, XII*, Part IV, New York (1916).
35. R. D'Harcourt, *Textiles of Ancient Peru and their Techniques*. University of Washington Press, Seattle (1974).
36. L. M. O'Neale and A. L. Kroeber, Textile periods in ancient Peru. *University of California Publications in American Archaeology and Ethnology, XXVIII* (1930).
37. L. M. O'Neale, Textiles periods in ancient Peru: II Paracas Caverns and the Grand Necropolis. *University of California Publications in American Archaeology and Ethnology, XXXIX* (1945).
38. C. W. Mead, Ancient Peruvian cloths. *The American Museum Journal* **XVI**, 6 (1916).
39. N. F. M. Henry and K. Lonsdale, *International Tables for X-Ray Crystallography*, Vol. 1. Kynoch Press for the International Union of Crystallography, Birmingham (1952).
40. N. V. Belov and E. N. Belova, Mosaics for the dichromatic plane groups, in *Colored Symmetry* (Edited by A. V. Shubnikov, *et al.*), pp. 220–221. Pergamon Press, New York (1964).
41. B. Grünbaum and G. C. Shephard, *Tilings and Patterns*. Freeman & Co., New York (1986).
42. T. W. Weiting, *The Mathematical Theory of Chromatic Plane Ornaments*. Marcel Dekker, New York (1982).
43. C. W. Mead, The six-unit design in ancient Peruvian cloth, in *Anthropological Papers*, Boas Anniversary Vol., pp. 193–195. G. E. Stechert & Co., New York (1906).
44. G. A. Reichard, The complexity of rhythm in decorative art. *Am. Anthrop.* **24**, 183–208 (1922).
45. S. C. Brett–Smith, Speech made visible: the irregular as a system of meaning. *Empirical Studies of the Arts* **2**, 127–147 (1984).
46. W. F. Morris, Jr., The ceremonial huipil of magdalenas. *Papers of the New World Archaeological Foundation* (to appear).
47. C. E. Stafford, *Paracas Embroideries*. J. J. Augustin, New York (1941).
48. A. R. Sawyer, *Mastercraftsmen of Ancient Peru*. The Solomon R. Guggenheim Foundation, New York (1968).
49. J. Freyd and B. Tversky, Force of symmetry in form perception. *Am. J. Psychol.* **97**, 109–126 (1984).

50. M. W. Helms, *Cuna Molas and Cocle Art Forms, Working Papers in the Traditional Arts*, (Edited by P. Ben–Amos). Institute for the Study of Human Issues (1981).
51. A. Fox, *Carved Maori Burial Chests*. Bull. No. 13, Auckland Institute and Museum, Auckland (1983).
52. K. M. Kensinger *et al.*, *The Cashinahua of Eastern Peru*, Studies in Anthropology and Material Culture Vol. 1 (Edited by J. P. Dwyer). The Haffenreffer Museum of Anthropology, Brown University (1975).

Comp. & Maths. with Appls. Vol. 12B, Nos. 3/4, pp. 783–787, 1986
Printed in Great Britain.

0886-9561/86 $3.00 + .00
© 1986 Pergamon Press Ltd.

SYMMETRY IN ARCHAEOLOGY†

J. Brandmüller

Sektion Physik der LM Universität München, D-8000 München 40, F.R.G.

and

B. Hrouda and A. V. Wickede

Institut für Vorderasiatische Archäologie der LM Universität München, D-8000 München 40, F.R.G.

Abstract—Often symmetry is simply understood as meaning only mirror-symmetry. But the term "symmetry" has to be thought of in a much wider sense. There are several steps in the definition of symmetry reaching from isometry with the two possibilities of point and translational symmetry, over homoeometry, antisymmetry and colour symmetry to the most general definition of symmetry as "harmony of the different parts of a whole". For all these concepts of symmetry one can find examples in archaeology. Group theoretical methods start playing an important role for the classification and as well for the typological characterization of objects in the field of archaeology. Cooperation of certain scientific branches provides new methods and also new possibilities of perception in archaeology (B. Hrouda, Methoden der Archäologie. Beck, Munich (1978). In analyzing the ornaments of prehistoric Tell Halaf pottery of northern Mesopotamia of the 5th millennium B.C., different symmetry operations in successive phases of development of pottery could be found by employing group theoretical methods which were developed in crystallography and mineralogy.

When talking about this subject, one cannot really isolate it from the much more general idea: "symmetry in art", because the archaeological discoveries, in so far as they are of interest in this context, represent documents of the artistic skill and the conception of art people had at that time and efforts have been undertaken to obtain, on this basis, informations about those people, their way of thinking, their sociology. I can contribute only a few brief thoughts to this general question.

The widely used sense of the term "symmetry" is only "mirror-symmetry", or, in particular in the field of archaeology, "antithetical principle". This restriction is not necessary, and not only for mathematical reasons. The term "symmetry" has to be understood in a more general way.

1. THE PROBLEMS CONCERNING MATHEMATICS AND ART

It often occurs that artists and art historians believe that mathematics and art have nothing in common at all. Sometimes, however, one recognizes that mathematics play at least the role of an auxiliary science. In reality, these relations are much stronger: there does exist an identity of character between mathematics and art. A. Speiser[1] writes: "Mathematics is the science of the things which exist and which do not depend on the I . . . the foundations of mathematics are also those of art". To H. Wölfflin, the art historian[2], the essential thing about art is its form, whereas the essential point concerning history of art is the inner evolution of its form. In the context of our subject one partial aspect of this form is being dealt with. This constitutes finally the justification of the following argument.

2. STEPS IN THE DEFINITION OF "SYMMETRY"

Different kinds of symmetries correspond to different kinds of groups.

2.1. *Isometry*

"Symmetry means the repetition of equal elements in space and/or time"[3]. This repetition can take place in two different ways.

---

†In memoriam Prof. Dr. Hans Sedlmayr, Salzburg.

2.1.1. *Point symmetry*. The form of a configuration can be such that certain rotations and rotoinversions can be carried out which transform the configuration in itself while, at the same time, at least *one* point does not change its position. The rotation of a vector within a fixed coordinate system can be described by the matrix

$$C_n = \begin{pmatrix} \cos(2\pi/n), & -\sin(2\pi/n), & 0 \\ \sin(2\pi/n), & \cos(2\pi/n), & 0 \\ 0, & 0, & 1 \end{pmatrix}.$$

This matrix refers to an *n*-fold rotation. One obtains the matrix for the corresponding rotoinversion by multiplying with the negative unity matrix. The special case of the two-fold rotoinversion means mirror symmetry or, better, bilateral symmetry which plays an important role in all cultures. The perhaps oldest known example of bilateral symmetry in art is the pair of leopards in the second-oldest town on earth, in Çatal Hüyük, which dates back to about 5800 B.C.[4]. Further examples in the field of archaeology are the lions' gate in Mycenae[5] and presentations of man in the archaic but also in Egyptian and Iberic art. If there is only mirror-symmetry, apart from identity, one has got the point group $m \equiv C_s$. Among the altogether 32 classical crystallographic point groups there are ten in particular which play a special role because plane structures can be classified by them, as for example seals from old Crete and Mycenae[6]. There the point groups 2, 3, 4, 6, 8, *m*, *mm*2, *3m* and *4mm* appear. D. Speiser[7] has found the group *6mm* on small plates of gold along with the treasure of Mycenae.

2.1.2. *Translational symmetry*. The repetition of equal elements that is required by the definition of symmetry (isometry) can take place by means of a translation as well. Each one-sided, one-coloured band ornament, in whatever cultural background it has its origin, can be associated with one of the seven band groups with simple rapport. For mathematical reasons the artist is, mostly unconsciously, obliged to that rule. His literally artistic liberty resides only in the fact how he models the object. The art historian H. Sedlmayr[8] proposed that there is a characteristic frequency distribution of the appearance of these seven band groups in the different cultural societies and that there are probably critical groups as well within which changes of style take place or show up. Partial examinations referring to these questions have already been made[3,9–13]. Overlapping lines, such as they appear in ornaments from time to time, lead, so to say, to a double-sidedness. Such ornaments are to be associated with the 31 one-coloured relief band groups[13]. Mrs. E. Müller[14] in her thesis made under A. Speiser, Zurich, has undertaken an examination using group theory and analyzing the structures of the Moorish ornaments of the Alhambra in Granada. The relief band group *p*222, in particular, is found there very often.

When admitting the doubly infinite rapport, one ends up with the areal ornaments the symmetry of which can be described by 17 plane groups. Egyptian areal ornaments were classified by Speiser[15], Moorish areal ornaments by Müller[14] and areal ornaments in African Bakuba and Benin cultures by Crowe[11,16], all applying group theory. In the case of overlapping lines one is forced, because of the double-sidedness of areal ornaments, to classify according to one of the eighty layer groups[18].

Further steps of symmetry are:

2.2. Homoeometry

"Symmetry finds its expression in the repetition in space or time of similar elements, motifs or behaviours"[17]. For this phenomenon one finds examples in archaeology, too[13].

2.3. Antisymmetry

If, together with an operation of symmetry, some quality (e.g. a colour of a configuration or the direction of rotation of a magnetic moment inside a crystal) changes, we speak, according to the proposition made by Shubnikov[18], of *antisymmetry*. There are 122 so-called magnetic point groups[19], 31 one-sided, but two-coloured band ornament groups[18], 179 double-sided and two-coloured relief band groups[20] and in addition 528 two-coloured and double-sided layer groups[21].

2.4. *Colour symmetry*

In an extended sense, colour symmetry is referred to every time one quality cannot be classified any more by just two signs (plus/minus or black/white) but when more than two possibilities or "colours" may appear. The theory of colour groups has been worked out by Belov[22] (see also Loeb[23]).

3. ORNAMENTS OF THE TELL HALAF CERAMICS

Until now, there have been only few studies demonstrating the application of group theory for the classification and characterization of archaeological material. In a recent study of painted Tell Halaf ceramics 400 vessels with approximately 700 patterns were classified according to their symmetrical composition[13]. The Tell Halaf period in the Near East covers the time from about 5200 to 4500 B.C. The following is a summary of the results of the study:

The analysis of the motifs, the patterns, and the composition of the decoration of the entire vessel showed a development of the ornaments in four phases for the painted Tell Halaf pottery.

The bilateral symmetry in the composition of the patterns is characteristic of the pottery of this culture.

The earliest phase of the Tell Halaf pottery includes the band groups *pm*, *pmm2* and *pma2* as well as the plane groups *p1m1*, *p2mg*, *c1m1* and *c2mm*. The patterns in the second phase are characterized by the introduction of ornaments showing point symmetry. In the third and fourth phases antisymmetrical patterns are introduced.

The use of point groups is limited to the interior bottom of the vessel, whereas band groups and plane groups are primarily employed for the decoration of the inner and outer wall surfaces. Present are the point groups 1, *m*, *mm2*, *5m*, *7m*, *4mm*, *6mm*, *8mm*, *12mm*, *16mm*, *32mm* and *64mm*. Of these, the groups *4mm* and *8mm* predominate.

Of the seven possible band groups all but the group *p1a* are present. *pm* and *pmm2* are by far the most frequently used band groups.

Relief band groups on Tell Halaf pottery do not exist. For the first time these groups occur in the second half of the third millennium B.C. in Mesopotamia with the appearance of the guilloche.

The plane groups *c2mm* and *p4mm* predominate over the other existing groups *p1*, *p2*, *p1m1*, *p2mg*, and *c1m1*.

Frequently used are simple linear or plane patterns belonging to the semicontinuous groups $p_0 1$, $p_0 m$, $p_0 mm2$ or the continuous group $p_{00} \infty m/18/$.

With the introduction of polychrome painting in the late Tell Halaf period, bichromatic patterns appear showing the antisymmetrical band groups, *p'mm2*, *pmm'2'*, *p'ma2 and pma'2'* or the plane groups *p2'*, *p'2mg*, *c2'mm'*, *p'4mm*, *p4m'm'*, *p'4mg* and *p6'mm'*. The most widely used bichromatic pattern is the chess-board, i.e. the plane group *p'4mm*.

Comparable early cultures in the Near East show differences with respect to the pottery decoration. Thus ornaments with rotation and glide symmetry were preferred on the Hassuna and Samarra pottery of northern Mesopotamia. Interesting here is the occurrence of the point group 4 and the band groups *p112* and *p1a*. The symmetry group appearing most frequently on the Obed pottery of southern Mesopotamia is the band group *pma2*. This symmetry group can be found on all decorated prehistoric pottery of the Near East.

The symmetrical structure of the patterns on the pottery of the Hassuna, Samarra, and early Obed cultures seems to have a common origin whereas the original symmetry groups of the early Tell Halaf have no predecessors in the Near East.

4. THE MOST GENERAL DEFINITION OF SYMMETRY

A. Speiser[3,15] wrote:

> The same λόγος (we mean the group concept) which generates the outer world of space in the large and the small does also work in art . . . here we find ourselves in the heart of art; we have thus penetrated not only to the inward part of Nature, but also into that of Art.

On the other hand Picasso wrote[24]:

> In my opinion a work of art is the product of calculations, calculations which often are unknown to the artist himself. He behaves himself exactly like the carrier pigeon calculating its way back to the loft. Such calculations precede our intelligence.

Such obviously deep correlations between mathematics and art have led to a very far reaching definition of symmetry. H. Weyl[25] has defined symmetry as "that form of concordance of several parts by means of which they form a whole." Speiser[26] defined symmetry as "harmony of different parts of a whole". An example of this step of symmetry in the domaine of archaeology, is, in my eyes, the parthic agora of Assur[27]. The reconstruction of the archaeological discoveries and its results show the overwhelming harmony of the horizontal and the vertical lines and the curves. It is the concordance of the same elements which in Venice (horizontal line: fronts of the Ducal Palace, of the library of S. Marco and of the "Procuratie"; vertical line: Campanile; curves: the five domes of S. Marco) but also in its predecessor Torcello (horizontal line: nave of the cathedral; vertical line: Campanile; curves: S. Fosca) determines its particular beauty.

5. SCIENCE AND ART

P. Feyerabend, professor of philosophy of science at the ETH in Zurich[28] formulated the thesis:

> Perception without support from art is impossible. If one separates sciences and art not only by words but also in action, i.e. if any kind of communication between the two disciplines is cut, then art will continue to exist, but sciences will not.

I believe, however, that not only is it useful for art to seriously study sciences in particular, but that there is a common root—Speiser would speak of λόγος—which should not be forgotten.

Acknowledgements—Thanks are due to A. Cornelsen and Th. Humer-Hager for the translation into English.

REFERENCES

 1. A. Speiser, *Die geistige Arbeit*. Birkhäuser, Basel und Stuttgart (1955).
 2. See H. Sedlmayr, *Heinrich Wölfflin und die Kunstgeschichte*. University of Munich (1964).
 3. W. V. Engelhardt, Symmetrie. *Studium Generale* **2**, 203–212 (1949).
 4. J. Mellaart, *Çatal Hüyük*. Bergisch-Gladbach (1967).
 5. S. Hiller, Das Löwentor von Mykene, Antike Welt. *Zeitschrift für Archäologie und Urgeschichte* **4**, 21–30 (1973).
 6. K. L. Wolf und R. Wolff, *Symmetrie*. Böhlau, Münster und Köln (1956).
 7. D. Speiser, La Symétrie de l'Ornement sur un bijou du Trésor de Mycènes. *Annali dell'Istituto e Museo di Storia della Scienza di Firenze* **1**, 3–8 (1976).
 8. H. Sedlmayr, private communication.
 9. B. Otto, Symmetrietypen frühminoischer Ornamente. *Jahrestagung der Arbeitskreise für Archäometrie*, Berlin (1979).
10. A. O. Shepard, *Ceramics for the Archaeologist*. Washington (1948).
11. D. W. Crowe, The geometry of african art II. A catalogue of Benin patterns. *Historia Mathematica* **2**, 253–271 (1975).
12. B. Otto und H. Wondratschek, Symmetrietypen antiker Band-Ornamente auf griechischen Vasen. *Berichte der Deutschen Keramischen Gesellschaft* **54**, 296–299 (1977).
13. A. v. Wickede, Die Ornamentik der Tell Halaf-Keramik, Ein Beitrag zu ihrer Typologie, Thesis, University of Munich (1981).
14. E. Müller, Gruppentheoretische und strukturanalytische Untersuchung der maurischen Ornamente aus der Alhambra in Granada, Thesis, Zurich (1944).
15. A. Speiser, *Die mathematische Denkweise*. Birkhäuser, Basel-Stuttgart (1952).
16. D. W. Crowe, The geometry of African art I. Bakuba Art, *J. of Geometry* **1**, 169–182 (1971).
17. K. L. Wolf, Symmetrie und Polarität. *Studium Generale* **2**, 213 (1949).
18. A. V. Shubnikov and V. A. Koptsik, *Symmetry in Science and Art*. New York (1974).
19. H. Heesch, Über die vierdimensionalen Gruppen des dreidimensionalen Raumes. *Zeitschrift für Kristallographie und Kristallgeometrie* **73**, 325–345 (1930).
20. A. Pabst, The 179 two-sided, two-colored band groups and their relations. *Zeitschrift für Kristallographie* **117**, 128–134 (1962).

21. N. N. Neronova and N. V. Belov, A single scheme for the classical and black-and-white crystallographic symmetry groups. *Sov. Phys. Cryst.* **6**, 1–9 (1961), with a correction by B. K. Vainshtein, communication by letter.
22. A. V. Shubnikov and N. V. Belov, *Colored Symmetry*. Oxford-London-New York 1964.
23. A. L. Loeb, *Color and Symmetry*. Krieger, Huntington, New York (1978).
24. Quoted from the catalogue of the exhibition: *Picasso in Vienna* (1981/82).
25. H. Weyl, *Symmetrie*. Birkhäuser, Basel und Stuttgart (1955).
26. A. Speiser, *Die Theorie der Gruppen von endlicher Ordnung*. Birkhäuser, Basel und Stuttgart (1956).
27. W. Andrae, *Das wiedererstandene Assur*, (edited by B. Hrouda). 2. Auflage, Beck, Munich (1977).
28. P. Feyerabend, Keine Erkenntnis ohne Kunst, in *Wissenschaft und Tradition*, (edited by P. Feyerabend and Chr. Thomas). Zurich (1983).

Comp. & Maths. with Appls. Vol. 12B, Nos. 3/4, pp. 789–801, 1986
Printed in Great Britain.

0886–9561/86 $3.00 + .00
© 1986 Pergamon Press Ltd.

STRANGE PATTERNS GENERATED FROM SIMPLE RULES WITH POSSIBLE RELEVANCE TO CRYSTALLIZATION AND OTHER PHENOMENA

REIDAR STØLEVIK, JON BRUNVOLL, BJØRG CYVIN and SVEN CYVIN†
The University of Trondheim, NLHT (N-7000) and NTH (N-7034), Trondheim, Norway

Abstract—Patterns of five integers in arrays are generated from simple rules. The studies are focused upon the strange figures formed by the highest integer. Some effects of breaking the rules by misplacement of single integers are investigated. The procedure may have relevance to different phenomena such as crystallization and crystal growth. However, this work does not propose any actual models for such phenomena.

At the first sight the present subject may seem to be merely a pastime exercise in recreational mathematics. On the other hand, it may provide the inspiration to find useful models for certain chemical phenomena. The present approach has relevance to such a well-known feature as the repulsion of identical particles, and perhaps to the crystallization and crystal growth processes.

BASIC PRINCIPLES

In the present approach a two-dimensional array of integers is built up successively according to definite rules:

(i) Proceed in a spiral starting from an origin.
(ii) Observe the criterion of noncontact between equal numbers. This applies to neighbours in the horizontal, vertical and diagonal directions.
(iii) Give the numbers priority so that a is chosen before b when $a < b$.

The procedure is illustrated in Fig. 1. In this process obviously 4 is the maximum number of neighbours to be accounted for. Hence not more than five different integers in the array are necessary. An example (see Fig. 1) shows that exactly five integers really are necessary.

PRELIMINARY OBSERVATIONS OF SOME PROPERTIES

The first surprising observation concerns the abundancies of the different integers. In spite of the priority criterion (iii) it is realized that the numbers 1, 2, 3 and 4 occur in roughly the same proportion, while number 5 is very rare. Table 1 shows the abundancies in some quadratic arrays of different dimensions.

Figure 2 displays the overall confusing pattern of a moderately sized array. The numbers 5 occur in a sequence 252525 (horizontal) and 5151 (vertical). Otherwise some isolated numbers 5 are scattered around. The given array (Fig. 2) also indicates the existence of rectangles with a regular pattern: 1414 . . . parallel with 3232 . . . in horizontal lines. Consequently, these rectangles have 1313 . . . parallel with 4242 . . . in vertical lines. The upper-right corner of Fig. 2 shows the corresponding regular pattern where the horizontal and vertical lines are reversed. Similar lines of the composition 1212 . . . or 3434 . . . are not found anywhere in Fig. 2. In general, a horizontal or vertical contact between the numbers 1 and 2 on one hand and 3 and 4 on the other are relatively rare; the corresponding diagonal contacts are frequent.

---

† Correspondence address: Prof. S. J. Cyvin, Division of Physical Chemistry, N-7034 Trondheim-NTH, Norway.

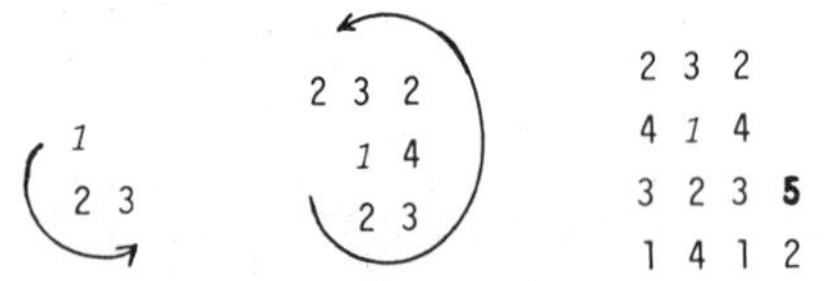

Fig. 1. Illustration of the rules (i)–(iii) in the text. The array is generated until the first occurrence of the numeral 5 (in boldface). The origin (*1*) is in italics.

MATHEMATICAL REPRESENTATION

The array positions may conveniently be identified by indices $i, j = 0, \pm 1, \pm 2, \ldots$. Thus we have introduced the matrix elements $R(i, j)$ in such a way that the first 14 elements along the spiral (see Fig. 1) read

$$R(0, 0) = 1, \quad R(1, 0) = 2, \quad R(1, 1) = 3, \quad R(0, 1) = 4,$$

$$R(-1, 1) = 2, \quad R(-1, 0) = 3, \quad R(-1, -1) = 2, \quad R(0, -1) = 4,$$

$$R(1, -1) = 3, \quad R(2, -1) = 1, \quad R(2, 0) = 4, \quad R(2, 1) = 1.$$

$$R(2, 2) = 2, \quad R(1, 2) = 5,$$

In other words, the number 5 occurs for the first time in the position (1, 2).

COMPUTERIZATION

A computer program was written to find the matrix elements and print out the array. Arrays up to the dimension of 363×363 were produced. The computerization made it feasible to detect more peculiarities about the pattern, which hardly could be foreseen. All the properties described in the following were detected empirically.

ABUNDANCIES OF THE NUMBERS

When the dimension of the array increases the percentage abundancies of numbers 5 decrease (cf. Table 2). For large arrays the abundancies of numbers 1 and 2 are practically equal, and the same is the case for 3 and 4. However, the former abundancies (1 and 2) are slightly, but significantly, larger than the latter (3 and 4). Figure 3 shows the distribution of the numbers 1, 2, 3 and 4 in the 39×39 array. The numbers 1 and 2 give the same overall impression, and the same may be said about 3 and 4. In the latter case some open gates are faintly visible; they correspond to the lines of fives (cf. Fig. 4).

THE PATTERN OF FIVES

In Fig. 4 two branches of lines (one horizontal and one vertical) with the number 5 have emerged in the lower-left corner in addition to the upper-right branches described in the pre-

Table 1. Abundancies of the integers in some small or moderately sized arrays

Absolute number						Percentage				
1	2	3	4	5	Total	1	2	3	4	5
8	5	5	6	1	25	32.0	20.0	20.0	24.0	4.0
10	13	13	11	2	49	20.4	26.5	26.5	22.4	4.1
22	19	17	20	3	81	27.2	23.5	21.0	24.7	3.7
28	30	31	28	4	121	23.1	24.8	25.6	23.1	3.3
71	74	65	70	9	289	24.6	25.6	22.4	24.2	3.1

```
2 3 2│1 4 1 4 1 4 1 4│1│2 4 2 4 2
4 1 4│3 2 3 2 3 2 3 2│5│3 1 3 1 3
2 3 2│1 4 1 4 1 4 1 4│1│2 4 2 4 2
1 4│5│3 2 3 2 3 2 3 2│5│3 1 3 1 3
3 2│1 4 1 4 1 4 1 4│1 4│2 4 2 4 2
1 4│3 2 3 2 3 2 3 2│5 3 1 3 1 3 1
3 2│1 4 1 4 1 4 1 4│1 2 5 2 5 2 5
1 4│3 2 3 2 3 2 3 2│3│4 1 4 1 4 1
2│5│1 4 1 4 1 4 1 4│1│2 3 2 3 2 3
4│3│2 3 2 3 2 3 2 3│5│4 1 4 1 4 1
2│1│4 1 4 1 4 1 4 1│2 3 2 3 2 3 2
4│3│2 3 2 3 2 3 2 3│4 1 4 1 4 1 4
2│1│4 1 4 1 4 1 4 1│2 3 2 3 2 3 2
4│3│2 3 2 3 2 3 2 3│4 1 4 1 4 1 4
2│1│4 1 4 1 4 1 4 1│2 3 2 3 2 3 2
4│3│2 3 2 3 2 3 2 3│4 1 4 1 4 1 4
2 1│4 1 4 1 4 1 4 1│2 3 2 3 2 3 2
```

Fig. 2. The array up to the dimension 17 × 17. Origin in italics; the fives in boldface. Boundaries for rectangular areas with regular patterns are indicated.

liminary observations. A still larger portion of the array displays completely the regularities of the pattern (cf. Fig. 5), where the fives are represented by dots. Each branch consists of parallel subbranches shifted by one unit at certain intervals. The subbranches have increasing lengths outwards; the numbers of dots are 6, 7, 8, . . . when starting with the innermost (right-hand) horizontal subbranch and proceeding in the spiral form. The branches are found on a background of isolated dots with decreasing density outwards; the positions of these dots are easily detected with reference to the shifting points of the subbranches. These regularities were found throughout the 363 × 363 array and are believed to extend to infinity, although a proof has not been given to this effect.

REGULAR PATTERNS

The rectangles of regular patterns (cf. Fig. 2) were found to extend throughout the 363 × 363 array; they increase in size outwards. Details may be studied on Fig. 6. Both of the horizontal branches are found to have the structure 5252 . . . , while the vertical branches consist of 5151 . . . . Sequences of 535 . . . or 545 . . . are not found anywhere. Figure 7 shows the possible regular patterns of the four numbers excluding 5. In Z and Z' ($Z = A, B, C$) only the horizontal and vertical lines are reversed. An inspection of the array (Fig. 6) reveals that the versions A and A' are common, but the others do not exist to any extent. More precisely, the 12 and 34 combinations do not extend to more than single, horizontal or vertical contacts. Fragments of the patterns B' and C' (horizontal contacts), as well as B and C (vertical contacts), are found in terms of two parallel lines in each case. Still, the patterns of A and A' are the dominating ones; A is found "outside" the branches (including the origin), while A' occurs in the "inside" regions, as indicated in Fig. 5.

Table 2. Abundancies of the integers in two large quadratic arrays

	1	2	3	4	5	Total
Absolute number	22934	22908	22784	22776	407	303^2
Percentage	24.98	24.95	24.82	24.81	0.44	100
Absolute number	32899	32910	32741	32727	492	363^2
Percentage	24.97	24.98	24.85	24.84	0.37	100

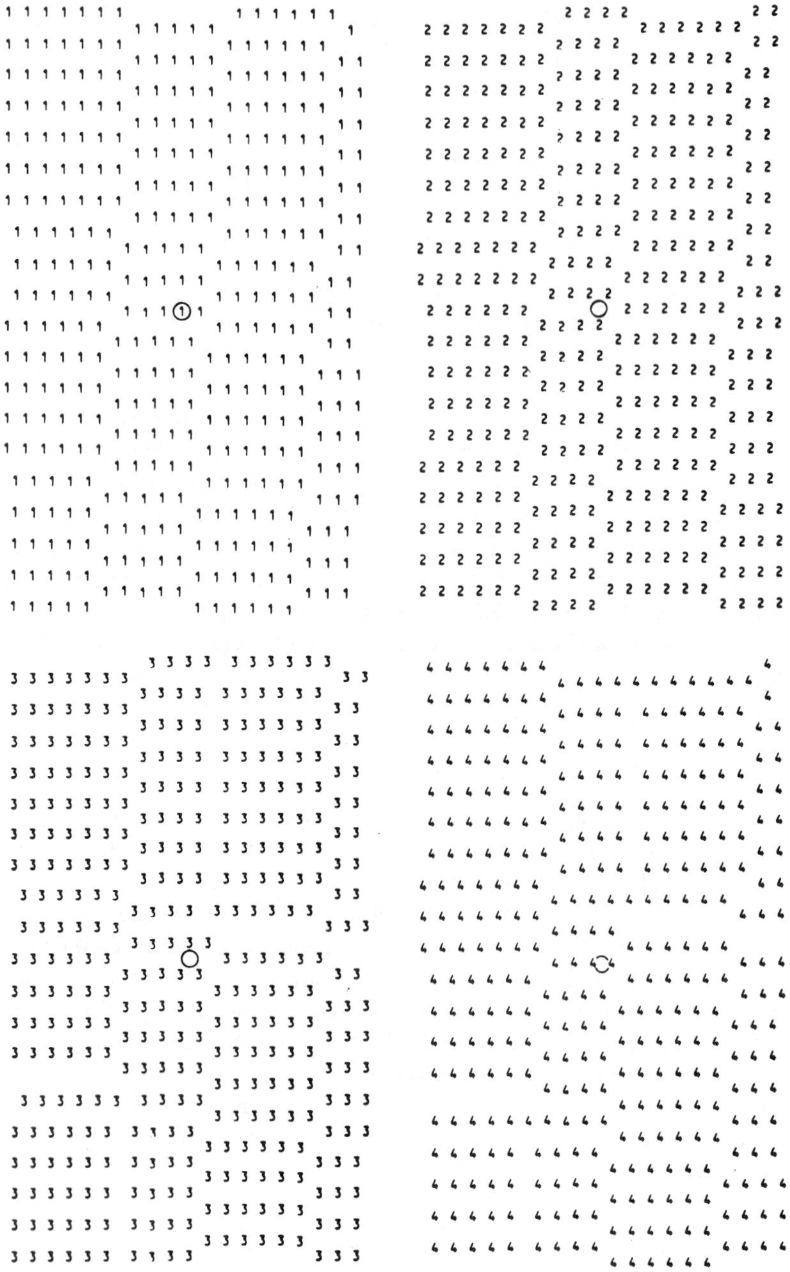

Fig. 3. Distribution of the integers 1, 2, 3 and 4. The origin is marked with a circle.

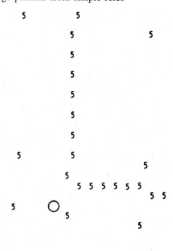

Fig. 4. Distribution of the integer 5. The origin is marked with a circle.

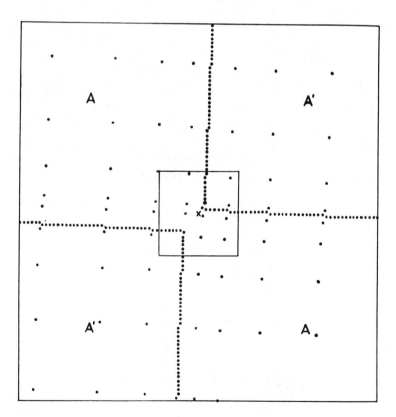

Fig. 5. Pattern of fives (represented by dots) of the 181 × 181 array. The origin is marked with a cross. The inner frame corresponds to Fig. 4.

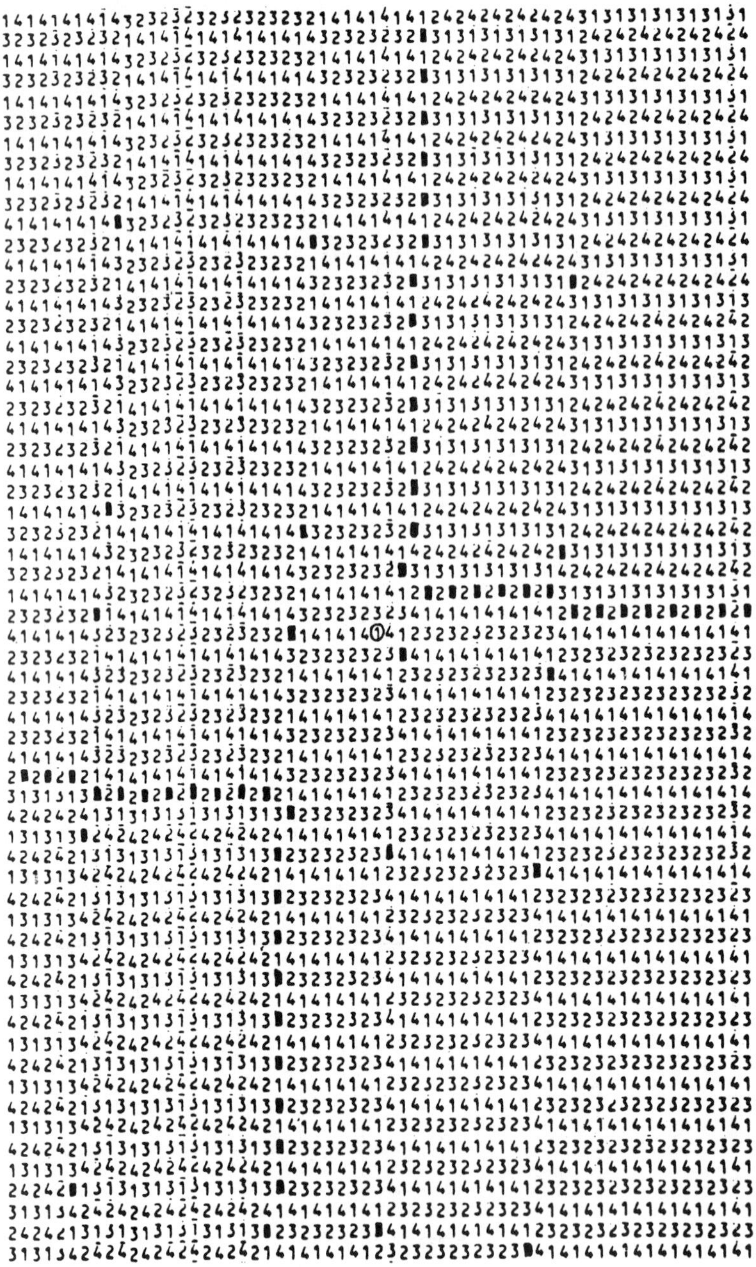

Fig. 6. The complete representation of the array up to 61 × 61. The origin is encircled. The integers 5 are represented as black rectangles.

	A		B		C
	1 4 1 4		1 3 1 3		1 4 1 4
	3 2 3 2		2 4 2 4		2 3 2 3
	1 4 1 4		1 3 1 3		1 4 1 4
	3 2 3 2		2 4 2 4		2 3 2 3

	A'		B'		C'
	1 3 1 3		1 2 1 2		1 2 1 2
	4 2 4 2		3 4 3 4		4 3 4 3
	1 3 1 3		1 2 1 2		1 2 1 2
	4 2 4 2		3 4 3 4		4 3 4 3

Fig. 7. Regular patterns of the integers 1, 2, 3 and 4. Within these patterns (even-number dimensional rectangles) the abundance of each integer is exactly 25%.

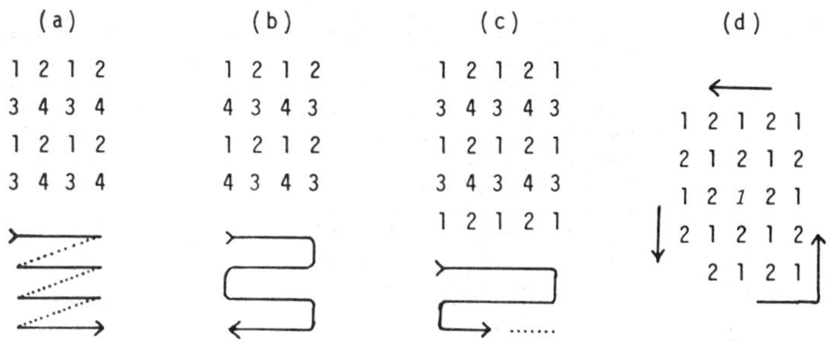

Fig. 8. Regular patterns generated by simplified procedures.

SIMPLIFIED PROCEDURES

Regular patterns may be produced by simplified procedures in different ways. Figure 8 shows some examples. In the cases (a)–(c) the rule (i) of the basic principles was changed. The patterns B' in cases (a) and (c), and C' in case (b) were generated when choosing some nonspiral paths. The simple pattern of (d) was produced by changing the rule (ii); diagonal contacts of equal integers are allowed.

PHYSICAL ANALOGIES

It is tempting to relate the generated arrays to layers of crystal lattices. The simple case (d) of Fig. 8 corresponds to a layer of the NaCl crystal. The large array (Fig. 6) gives intuitively the impression of the face of a natural substance. The numbers 5 could be alien particles (impurities) or holes in a crystal lattice. We believe that the present procedure, with different modifications, may be useful in the studies of physical phenomena. It should be stressed, however, that we have not been aiming at a definite application. Hence we do not claim our array (Fig. 6) to represent a model of a real crystal.

EFFECT OF DIFFERENT INTEGERS AT ORIGIN

It was found to be of interest to investigate the effect of single misplaced integers (analogue to particles) in relation to the basic rules. This has the analogy to crystal defects. Will such a defect be "repaired" during the development? Will the pattern of fives only be affected locally? Or will the defect cause damages throughout the array? Will the pattern of fives be changed substantially or completely destroyed? Will perhaps new patterns emerge? In order to answer some of these questions we decided to try different integers at the origin. From that point on all the basic rules were followed.

Table 3 shows the abundancies of the different integers in the five cases. It was found that number 2 at the origin causes the substitution of all integers 1 with 2 in the normal array (Fig. 6) and *vice versa*. It does not affect the other integers; hence the pattern of fives is unaffected. The regular patterns A and A' occur in reversed positions in relation to the normal array. When numbers 3, 4 or 5 are placed at the origin the arrays become identical after some rounds (cf. Fig. 9). Table 3 shows that the number of fives is substantially larger than normal at the expense of 3 and 4. Numbers 1 and 2 appear with exactly the same abundancies as in the case with 2 at the origin when the array dimension is 363 $\times$ 363. The pattern of fives is very interesting; see Fig. 10, where the regular patterns with reference to Fig. 7 are also indicated. Apart from the common structures A and A' one finds slim rectangles of the structure C', mainly above the origin and right below. One finds some tendencies of "repairing the damage" in this case. Most of the left-hand parts of the array (Fig. 10) is actually equal to the one with integer 2 at the origin. A narrow stripe to the left of the origin must then be excepted. The identical areas include the whole of the lower-left branches of fives with their interior. As the most striking feature of the array with $R(0, 0) = 5$ we note the formation of double branches in the lower-right part and the peculiarity below the origin. A pair of branches (one horizontal and one vertical) have also emerged at the upper-right part, abnormally far from the origin.

Table 3. Abundancies of the integers in the 363×363 arrays with different numbers at the origin, $(0, 0)$

$R(0,0)$		1	2	3	4	5
1	Absolute number	32899	32910	32741	32727	492
	Percentage	24.97	24.98	24.85	24.84	0.37
2	Absolute number	32910	32899	32741	32727	492
	Percentage	24.98	24.97	24.85	24.84	0.37
3	Absolute number	32910	32899	32595	32568	797
	Percentage	24.98	24.97	24.74	24.72	0.60
4	Absolute number	32910	32899	32594	32569	797
	Percentage	24.98	24.97	24.74	24.72	0.60
5	Absolute number	32910	32899	32594	32568	798
	Percentage	24.98	24.97	24.74	24.72	0.61

EFFECTS OF MISPLACED INTEGERS NEAR THE ORIGIN WHILE $R(0, 0) = 1$

Introduction

It was decided to investigate the effects of single misplaced integers in other positions than the origin. Such an error is called here "unessential" if it violates the rule of priority (iii), but not the criterion of noncontact (ii). The error is "essential" when it creates a neighbourhood between identical integers. After introducing an error the spiral is continued strictly according to the postulated principles.

Different integers at $(1, 0)$

The position $(1, 0)$ is right below the origin; it is the first one after the origin in the development of the spiral. In the normal case $R(1, 0) = 2$. In Fig. 11 the normal pattern of fives is compared with the cases when $R(1, 0) = 1\dagger$, 3 and 4; $\dagger$ indicates an "essential" error here and throughout. The case of $R(1, 0) = 5$ is identical with $R(1, 0) = 4$ except for the very position at $(1, 0)$.

Different integers at $(1, 1)$

The position $(1, 1)$ is the second one after the origin. In the normal case $R(1, 1) = 3$. Figure 12 shows the pattern of fives for $R(1, 1) = 1\dagger$ and 4. The case of $R(1, 1) = 2\dagger$ displays a pattern identical to the one of $R(1, 0) = 4$ (cf. Fig. 11), except for a narrow stripe of isolated dots to the left of the origin. $R(1, 1) = 5$ gives a similar pattern to $R(1, 1) = 4$; one has only $R(2, 3) = 5$ instead of $R(1, 2) = 5$.

Different integers at $(1, -1)$

This position is situated to the left of $(1, 0)$. It is number eight after origin in the development of the spiral. Normally $R(1, -1) = 3$. The cases of $R(1, -1) = 1\dagger$ and $4\dagger$ are shown in Figs. 13 and 14, respectively. For $R(1, -1) = 2\dagger$ one obtains a pattern of fives identical to

```
  1 3 1 2 1 5 2          1 3 1 2 1 5 2          1 3 1 2 1 5 2
  4 2 4 3 4 3 1          4 2 4 3 4 3 1          4 2 4 3 4 3 1
  1 3 1 2 1 2 4          1 5 1 2 1 2 4          1 5 1 2 1 2 4
  2 5 4 3 4 3 1          2 4 3 4 3 5 1          2 4 3 5 3 5 1
  3 1 2 1 2 5 2          3 1 2 1 2 4 2          3 1 2 1 2 4 2
↓   4 3 4 3 1 3        ↓   4 3 4 3 1 3        ↓   4 3 4 3 1 3
▼ 1 2 1 2 4 2          ▼ 1 2 1 2 4 2          ▼ 1 2 1 2 4 2
```

Fig. 9. The starting developments of the arrays with integers 3, 4 and 5 at the origin, respectively. Further on, the developments are identical in these three cases.

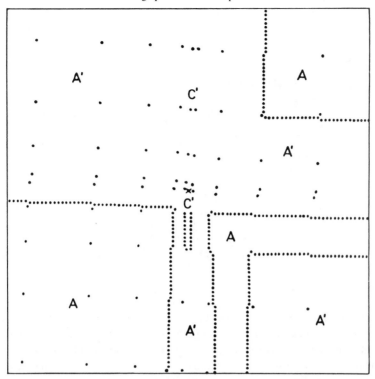

Fig. 10. The pattern of fives (represented by dots) for the 181×181 array with $R(0, 0) = 5$. The origin is marked with a cross. Regular patterns (cf. Fig. 7) are indicated.

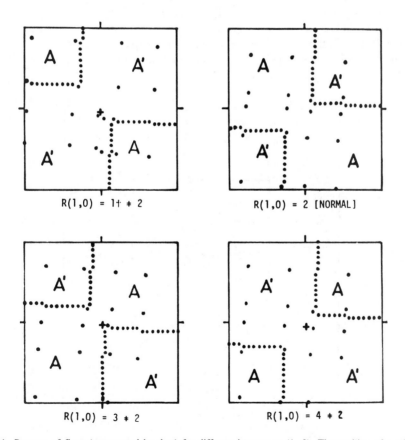

Fig. 11. Patterns of fives (represented by dots) for different integers at $(1, 0)$. The position of a misplaced integer is marked by $+$. "Essential" error is indicated by $\dagger$. Dimension 61×61.

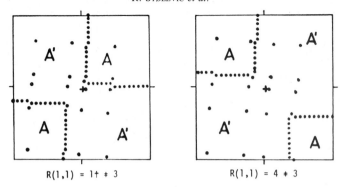

R(1,1) = 1† ≠ 3   R(1,1) = 4 ≠ 3

Fig. 12. See the legend of Fig. 11.

the normal case except for a narrow stripe to the left of the origin. $R(1, -1) = 5$ gives a pattern very similar to the one of Fig. 14; only six dots including the first lower vertical subbranch are slightly shifted.

Discussion

All the patterns developed here display the "normal" features of two right-angled branches on the background of isolated dots. The position of the branches is either upper right and lower left (as normal), but sometimes reversed to upper left and lower right. The density of the dots fluctuates on both sides of the normal case; Table 4 shows the absolute numbers and percentages of fives when referred to the dimension 61×61. There are many variations within the features referred to as "normal." In particular, the origins of the branches are found at different distances from the origin of the array. In the cases of $R(1, -1) = 1†$, 4† and 5 the two branches are abnormally close to each other. Furthermore, in the case of $R(1, -1) = 1†$ the background dots are abnormally dense. In consequence, the abundance of fives is abnormally high in this case (cf. Table 4). In the analogy to a crystal structure one might speak about a tendency of local melting in the neighbourhood of the origin. The distinction between "essential" and "unessential" errors is not reflected in any obvious way in the resulting patterns.

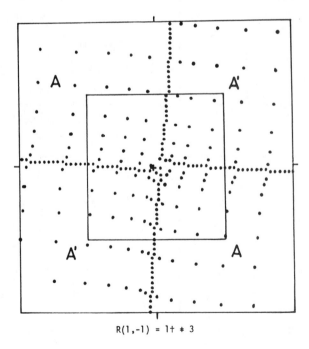

R(1,-1) = 1† ≠ 3

Fig. 13. Dimension 121×121; the inner frame is 61×61. See also the legend of Fig. 11.

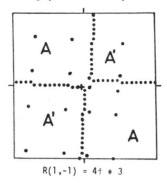

R(1,-1) = 4† ≠ 3

Fig. 14. See the legend of Fig. 11.

EFFECTS OF MISPLACED INTEGERS NEAR ORIGIN WHILE $R(0, 0) = 5$

Starting the spiral with number 5 at the origin instead of 1 proved to have a major effect on the pattern of fives (cf. Fig. 10). What will happen if single misplacements are introduced in addition to the number 5 at origin? The same 12 cases of misplacements as described in the preceding section were investigated systematically. The abundancies of fives referred to the dimension 61 × 61 now fluctuate between 1.45 and 2.28%, while the case of $R(0, 0) = 5$ without further misplacements holds 2.15%. It was quite unexpected to find the ''normal'' features (as defined in the preceding section) restored in many cases. In particular, the pattern for $R(1, 0) = 2$ ($\neq 1$) was found to be very similar to $R(0, 0) = 1$, $R(1, 0) = 3$ (Fig. 11). Furthermore, the cases of $R(1, 1) = 3$, 4 and 5† ($\neq 2$) all displayed similar patterns, which resemble the case of $R(0, 0) = 1$, $R(1, 0) = 4$ (Fig. 11) or $R(1, 1) = 1$† (Fig. 12).

In the cases of $R(1, 0) = 3$, 4 and 5† new features emerge. The abundance of fives is abnormally low (1.45–1.48% referred to 61 × 61). The pattern of fives for $R(1, 0) = 5$† is illustrated in Fig. 15. It shows one single vertical branch on the usual background of isolated dots. Within the dimension of 61 × 61 (the inner frame of Fig. 15) hardly any more features would be detected. On increasing the dimension two right-angled branches appear in the lower-left and lower-right regions, respectively. Notice that the middle and right-hand vertical branches approach each other; they are supposed to meet eventually. This event is realized in Fig. 16, which illustrates the case of $R(1, 1) = 1$†. Qualitatively these two cases are alike.

Table 4. Absolute numbers (N) and percentages (%) of fives in 61 × 61 arrays with one misplaced integer at (i, j). The bracketed figures pertain to the normal case

	$(i, j) = (1, 0)$		$(i, j) = (1, 1)$		$(i, j) = (1, -1)$	
	N	%	N	%	N	%
$R(i, j) = 1$	64	1.72	69	1.85	114	3.06
$R(i, j) = 2$	[66]	[1.77]	64	1.72	66	1.77
$R(i, j) = 3$	69	1.85	[66]	[1.77]	[66]	[1.77]
$R(i, j) = 4$	64	1.72	62	1.67	73	1.96
$R(i, j) = 5$	65	1.75	62	1.67	73	1.96

R. STØLEVIC *et al.*

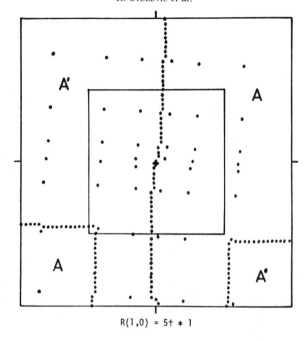

R(1,0) = 5† ≠ 1

Fig. 15. Pattern of fives (represented by dots) for one misplacement in a position marked by +, while $R(0, 0) = 5$. Dimension 121 × 121; the inner frame is 61 × 61.

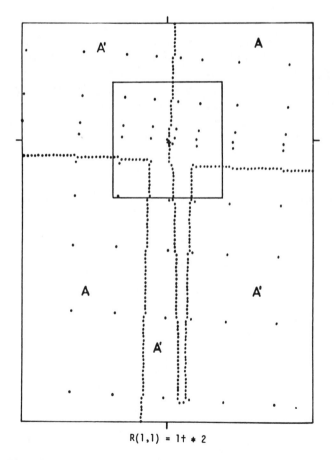

R(1,1) = 1† ≠ 2

Fig. 16. Dimension 161 × 206; the inner frame is 61 × 61. See also the legend of Fig. 15.

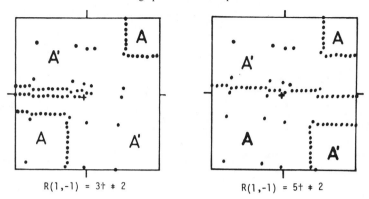

$$R(1,-1) = 3\dagger \neq 2 \qquad\qquad R(1,-1) = 5\dagger \neq 2$$

Fig. 17. Dimension 61 × 61; see also the legend of Fig. 15.

For $R(1, -1) = 1\dagger$, 4 and $5\dagger$ ($\neq 2$) the patterns of fives are identical, except for the additional 5 in the misplaced position for $R(1, -1) = 5\dagger$; this case is shown in Fig. 17. Here the three horizontal branches extend in a parallel way. Figure 17 also illustrates the case of $R(1, -1) = 3\dagger$. It shows a new variant of a pattern where three horizontal branches extend in a parallel way.

CONCLUSION

This work shows the generation of a strange pattern from a set of simple rules. The effects of breaking the rules were studied systematically in a limited number of instances. This empirical material already shows several interesting and unexpected features. There are, of course, infinitely many other possibilities of introducing "errors" in the normal pattern.

Comp. & Maths. with Appls. Vol. 12B, Nos. 3/4, pp. 803–824, 1986
Printed in Great Britain.

0886–9561/86 $3.00 + .00

TOWARDS A GRAMMAR OF INORGANIC STRUCTURE

A. L. MACKAY

Department of Crystallography, Birkbeck College, University of London, Malet Street, London
WC1E 7HX, England

and

J. KLINOWSKI

Department of Physical Chemistry, University of Cambridge, Lensfield Road, Cambridge
CB2 1EP, England

Abstract—The description of inorganic crystals has moved on from the early concentration on a picture of linked symmetrical polyhedra, and we outline several further systems for visualising inorganic structures, discuss their relative usefulness and draw attention to the value of the concepts of curvature in understanding the "grammar" underlying inorganic and biological structure.

The "letters" are the atoms and the "words" are the smallest clusters of bonded atoms. "Grammar" is the way in which the words are combined into larger units and "syntax" is the next level of organisation. We will not here touch on what "meaning" represents, but the metaphor of language is immensely productive and suggestive. Since the discovery of the role of DNA, this metaphor has become firmly established in molecular biology. We suggest now that it is useful also in the inorganic field.

One important element of this "grammar" is *intrinsic curvature*, which can be considered as a measure of strain in structures caused by a misfit of neighbouring components. We concentrate particularly on this element of structuration at the expense of many others which would require lengthy development. For example, strain in flat sheets (for instance silicate sheet structures), besides making them cylindrical, may make them curve either elliptically, producing closed domains (such as spherulites or icosahedral virus particles) or hyperbolically into the third dimension, yielding periodic minimal surfaces.

Periodic minimal surfaces occur as a dominant secondary structure in lipids, lyotropic colloids ("soaps"), framework silicates (zeolite molecular sieves) and many other systems. We describe a number of minimal surfaces and show how their profiles and areas can be calculated numerically. Going a step further, the curvature of 3-D structures into 4-D is difficult to imagine, but this concept is nevertheless important. The mapping of a structure of one curvature onto a structure of an incompatible curvature may be seen as the basis for possible reactions or transformations.

It will be found that everything depends on the composition of the forces with which the particles of matter act upon one another; and from these forces, as a matter of fact, all phenomena of Nature take their origin. [Roger Joseph Boscovich (1711–1787), *Theoria Philosophiae Naturalis*, Venice, 1763 (Sect. I.5)]

INTRODUCTION

How are inorganic structures best to be described? Pauling's Rules[1,2], which account for many structural features of complex inorganic compounds in terms of simple electrostatics, are merely the first level of a hierarchic grammar of structure—building words from letters. As formulated by Pauling himself, the rules follow.

1. *The nature of the coordinated polyhedra*

A coordinated polyhedron of anions is formed about each cation, the cation–anion distance being determined by the radius sum and the ligancy of the cation by the radius ratio.

2. *The electrostatic valency rule*

In a stable ionic structure the valence of each anion is closely equal and opposite to the sum of the strengths of the electrostatic bonds (Maxwell's "lines of force") to it from the adjacent cations. [That is, the lines of force are of minimum length under the action of their mutual repulsions and of their elastic "tensions" when linking positive charges to negative charges.]

The sharing of polyhedron corners, edges and faces

3. The presence of shared edges and especially of shared faces in a coordinated structure decreases its stability; this effect is large for cations with large valence and small ligancy.

4. In a crystal containing different cations, those with large valence and small coordination number tend not to share polyhedron elements with each other.

These rules, which give rise to the concept of symmetrical "coordination polyhedra," have been very successful in describing the immediate environments of individual atoms, but much less so when applied for describing and predicting larger-scale structures. They work well for highly symmetrical coordination polyhedra, which cover a wide range of crystal structures, but are not so easy to extend to the quantitative treatment of less symmetrical polyhedra. Furthermore, many structures, for example the mineral opal which is formed from silica spherulites, are hierarchic in nature and cannot be adequately described by Pauling's Rules alone, which operate on only one level. High-resolution electron microscopy (HRTEM), scanning electron microscopy (SEM) and X-ray diffraction (XRD) have recently provided much new information on the "*syntax*" governing these higher levels of structure. It has been shown by H. A. Simon[3] that hierarchy minimises complexity and is intrinsic to the process of structure building: hierarchic structures have higher probabilities of formation. Modern techniques for the imaging of the crystal structures of highly complex proteins are now seen to be based on the "maximisation of entropy"[4] and exploit the foreknowledge that these structures are hierarchic. Many inorganic structures are also recognisably hierarchic[4A].

It has become customary to describe three-dimensionally periodic structures using the 14 Bravais unit cells (Fig. 1), but space-filling zonohedra, of which there are 5 kinds and 24 aspects (Fig. 2) can also be used for this purpose, thus providing an alternative formal geometrical system for classifying structures. If we wish to describe the crystal structure in terms of linked coordination polyhedra[57], then the space-filling notation is the more convenient. The so-called Voronoi dissection[5–8] (or better, the radical-plane dissection), which furnishes an exact definition for space-filling polyhedra, can profitably be used also in describing amorphous materials, thus enabling both states of matter to be handled with the same language.

Delone *et al.*[9] have shown how an alternative formalism for classical symmetry theory can be constructed using only local order (as in the formalism of cellular automata) and eschewing infinite translations. They demonstrated that, if all lattice points in a structure are to have identical surroundings (without absolute orientation) out to a distance R from each lattice point, then if R is greater than a certain critical value, between $2r$ and $6r$, where r is the radius of the largest empty ball about a lattice point, the whole structure will be constrained to belong to one of the usual Bravais lattices and to be an infinite crystal. Thus, crystallinity results naturally from local order of sufficient range (Fig. 3).

Another feature in the description of complex solids, at a level which is one step up from Pauling's Rules, involves the concept of minimal surfaces, and we will here devote more attention to this, since it is a less familiar concept than that of coordination polyhedra. At a level of resolution insufficient to distinguish atoms, many complex silicates, lyotropic colloids and other systems can be modelled as *membranes* ["of force"] rather than by *lines of force* or by *polyhedra*. We will see, however, that there are important relationships between these descriptions and that minimal surfaces are closely similar to the equipotentials which can be drawn perpendicular to systems of lines of force. This naturally involves the concepts of curvature in three-dimensional space and of minimal surfaces which, like soap films, are surfaces of the lowest free energy and which, given defined boundaries, take up a configuration of minimum area. Even the earliest workers studying minimal surfaces pointed out that, if a closed circuit is placed in a minimal surface and is then removed, a soap film over this circuit is identical with the soap film which it enclosed before removal. The area of every element of the surface is a minimum.

As we shall see, the concept of curvature can be extended to the fourth dimension into which ordinary 3-D space can be considered to be curved. Just as curvature of 2-D space (a planar structure) into 3-D (a curved soap film) can be seen as a consequence of the minimisation of 2-D area, so this concept is also useful in discussing the minimisation of the strain energy in a 3-D structure. This may be applicable, for example, in understanding Loewenstein's Rule for aluminosilicates, which forbids two A1-centred tetrahedra to be neighbours in the framework,

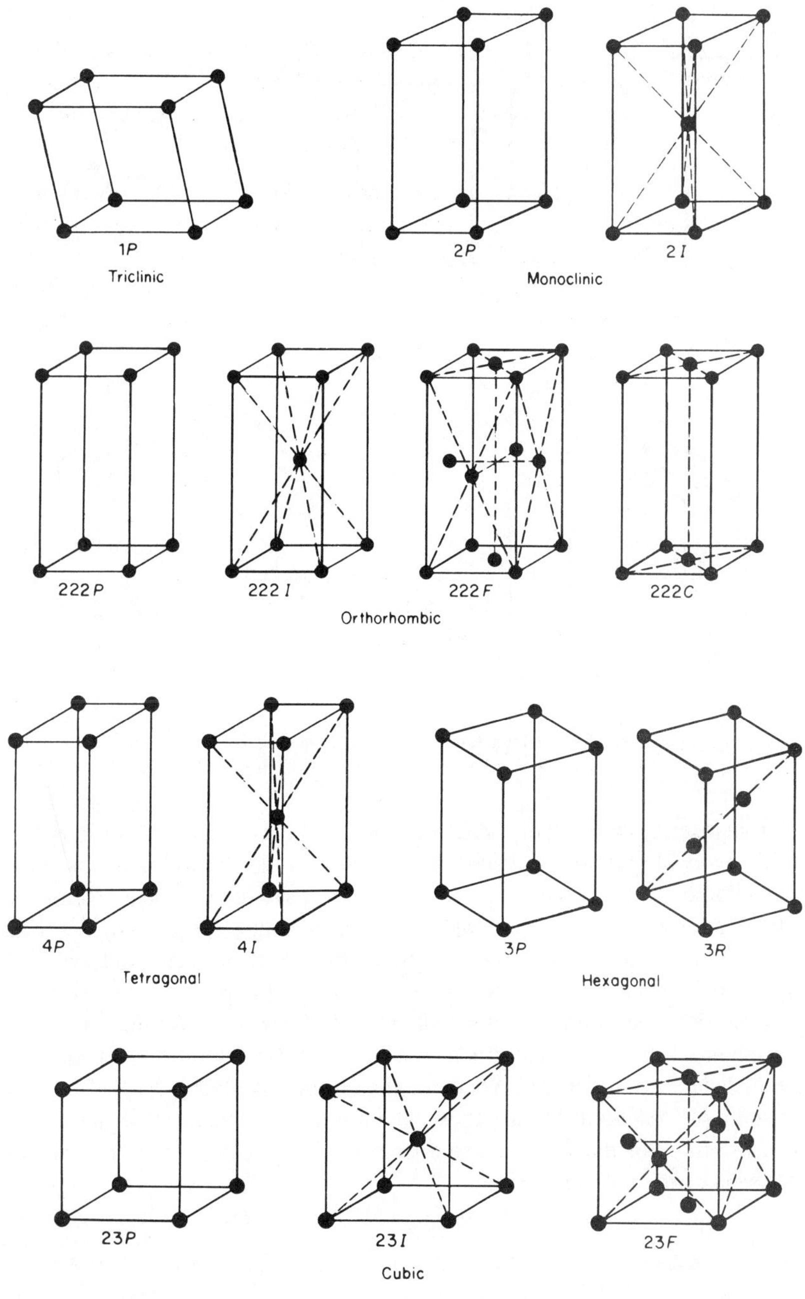

Fig. 1. The 14 Bravais lattices. Each represents the unit cell of an infinite lattice where every lattice point has the same symmetry and surroundings as every other. The 14 lattices are distinguished by their different symmetries.

the general correctness of which has recently been confirmed by high-resolution solid-state NMR[10]. Finally, a number of chemical processes leading to solid-state structures (such as the transformation of the sheet aluminosilicate kaolin to the framework aluminosilicate ultramarine) may be expressed in catastrophe theory terms, following the catastrophic transformations ("jump-like" processes with hysteresis) in assemblies of soap bubbles.

A GRAMMAR OF STRUCTURE

We take the term "grammar" in the extended meaning as encompassing not only the description of components but also the several levels of rules which generate the whole by the combination of the components. This follows the word "grammar" itself, which has similarly

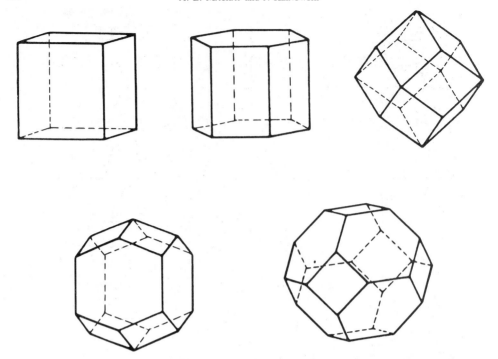

Fig. 2. Five of the paralleohedra which pack to fill space and which generate the lattices by their association, the lattice points being imagined at their centres.

progressed from its early meaning of "lettering." Ideas of the appearance of a complex structure by the *self-assembly* of individual components as a result of their local interactions, rather than of the intervention of some designer outside the system, have now become extremely important under the designation of *cellular automata*[11,12]. It is no longer adequate to describe a structure in an arbitrary way[33,33A]. Some ways are less arbitrary than others and the least arbitrary is the way in which Nature describes it, that is, in terms of the generative rules in the way that a DNA sequence describes a protein. We think it fruitful to ask of inorganic structures the same question we might ask of a biological structure. Thus, taking a mineral comparable in complexity to a protein molecule we might ask, Where are the genes in paulingite?[13] We will anticipate the answer and reply that there are no genes, the information is distributed and is not localised. The rules of assembly of the whole emerge from the local rules, which in turn emerge from the energy levels of the individual atoms.

THE CONCEPT OF CURVATURE: THE GENERALISED SPHEROMETER

Ptolemy's Theorem states that for a cyclic quadrilateral (with vertices numbered 1234 and with a distance d_{12} between the points 1 and 2),

$$(d_{12} + d_{34})(d_{14} + d_{23}) = d_{13}d_{24}, \tag{1}$$

and conversely, this is the condition that four points should be concyclic and coplanar (Fig. 4). The theorem can be generalised to N dimensions with the definition by Blumenthal[14,14A] of a quantity P_N which he named the "Ptolemaicity." This is a determinant of the squares of the distances between the N points. Thus

$$P_4 = \begin{vmatrix} 0 & d_{12}^2 & d_{13}^2 & d_{14}^2 \\ d_{21}^2 & 0 & d_{23}^2 & d_{24}^2 \\ d_{31}^2 & d_{32}^2 & 0 & d_{34}^2 \\ d_{41}^2 & d_{42}^2 & d_{43}^2 & 0 \end{vmatrix}. \tag{2}$$

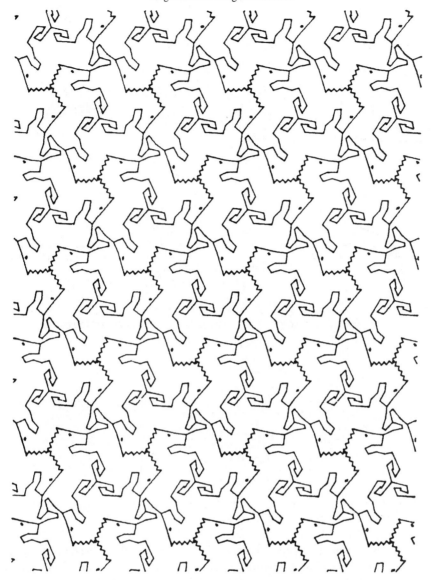

Fig. 3. A crystal lattice results automatically from the local order produced by the interlocking of neighbouring parts like a jigsaw puzzle (motif by M. Nakamura).

The generalisation of Ptolemy's Theorem is simply that $P_N = 0$; that is, if four points lie on a two-dimensional circle, then $P_4 = 0$; if five points lie on a sphere then $P_5 = 0$, etc.

The same determinant, when bordered by ones, gives the volume [the $(N - 1)$ dimensional content] of an $(N - 1)$-dimensional simplex:

$$V_N^2 = A \begin{vmatrix} 0 & d_{12}^2 & d_{13}^2 & \dots & d_{1N}^2 & 1 \\ d_{21}^2 & 0 & d_{23}^2 & \dots & d_{2N}^2 & 1 \\ d_{31}^2 & d_{32}^2 & 0 & \dots & d_{3N}^2 & 1 \\ \dots & \dots & \dots & 0 & d_{4N}^2 & 1 \\ d_{N1}^2 & d_{N2}^2 & d_{N3}^2 & d_{N4}^2 & 0 & 1 \\ 1 & 1 & 1 & 1 & 1 & 0 \end{vmatrix}. \tag{3}$$

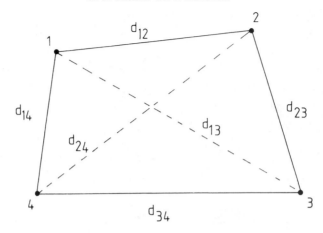

Fig. 4. Ptolemy's Theorem states that for a cyclic quadrilateral (with vertices numbered 1234 and with a distance d_{12} between the points 1 and 2), $(d_{12} + d_{34})(d_{14} + d_{23}) = d_{13}d_{24}$, and conversely, this is the condition that four points should be concyclic and coplanar.

The coefficient $A = (-1)^{(N+1)}/(2^N(1/N!)^2)$ takes, for $N = 2, 3, 4, \ldots$, the values -16, $+288$, -9216, $+460800$, etc. respectively. The general expression for the *radius of curvature* R of the circle, sphere, etc. on which the points lie is then

$$2R^2 \begin{vmatrix} 0 & d_{12^2} & d_{13^2} & d_{14^2} & d_{15^2} & 1 \\ d_{21^2} & 0 & d_{23^2} & d_{24^2} & d_{25^2} & 1 \\ d_{31^2} & d_{32^2} & 0 & d_{34^2} & d_{35^2} & 1 \\ d_{41^2} & d_{42^2} & d_{43^2} & 0 & d_{45^2} & 1 \\ d_{51^2} & d_{52^2} & d_{53^2} & d_{54^2} & 0 & 1 \\ 1 & 1 & 1 & 1 & 1 & 0 \end{vmatrix} = P_5. \tag{4}$$

This expression acts as a generalised spherometer giving the curvature of the space of $N - 2$ dimensions in which N points lie. In fact the radius of curvature is only the simplest parameter and an N-dimensional manifold has $N^2(N^2 - 1)/12$ independent components of curvature. For $N = 1, 2, 3, 4$ this number is respectively 0, 1, 6 and 20.

For testing a tellurometer (a microwave interferometer now widely used by surveyors for measuring distances of up to 45 km) the 10 distances between 5 points are measured as accurately as possible. Five points require only 9 parameters and there is thus one relationship between the distances. The 4-D content of the simplex defined by the 5 points should be zero if the space is Euclidean 3-D. If there is a discrepancy then this can be regarded as a curvature of the 3-D space, although it would usually just be ascribed to error. Thus, incipient distortion of a chemical group can be also described as a tendency to curvature. This incipient curvature is what Sadoc[15] and Kerner[6] discern in random networks. The sign of the curvature can be positive or negative and there may be local cancellations produced by the local attraction between positive and negative regions. If SiO_4 and AlO_4 tetrahedra had opposite curvatures, then we would expect the Loewenstein Rule to result, and Al and Si would tend to alternate in a framework.

Regular tetrahedra do not fill space (see Fig. 2, which shows the space-filling regular solids). Twenty tetrahedra do not quite give an icosahedron (for unit edges the distances from the centre to the vertices are 0.951056. . . instead of 1, and the angles between the edges of the tetrahedra are 60° instead of 63.43°). This distortion could be expressed by changing the metric of the space or in terms of curvature in the fourth dimension. If the lengths of all the edges were 1 but d_{12} were $\tau = 1.618. \ldots$, then using the expressions given in eqns (2)–(4)

the radius of curvature of three-dimensional space in the fourth dimension is $1/\tau = 0.618. . .$, and we obtain the 4-D regular polytope called the "120-cell."

Curvature is an expression of a metric and can be thought of as follows. Suppose that there is a regular hexagonal packing of atoms in a planar structure with nearest neighbours at a distance a. The second nearest neighbours should then be at a distance $a3^{1/2}$ away. Imagine that, perhaps as a result of interatomic interactions more complex than pairwise, the *preferred* second nearest neighbour distance is slightly different from $a3^{1/2}$. The metric then becomes inconsistent with planarity and the space which the packing occupies becomes curved. This is observed in the structure of the poliovirus and other spherical viruses which often form shell structures with positive Gaussian curvature and icosahedral symmetry (Fig. 9).

Extending this argument to a three-dimensional structure, consider the tetrahedral SiO_4^{4-} group. Suppose that the preferred Si–O and O–O distances which, for an unstrained tetrahedron, are in the ratio of $3^{1/2}$ to $8^{1/2}$ (i.e. $0.6124. . .$), differ somewhat from it. The tetrahedron will then be strained and, if this results in deformation, the curvature is in the fourth dimension[17].

MINIMAL SURFACES

Consider the unit vector normal to the surface at the point P (Fig. 5) and all planes containing it. These planes intersect the surface in curves which, in general, have different curvatures at P. Now consider the curves of maximum and minimum curvature. These are the principal curvatures at P and will be in planes at right angles to each other. The mean curvature H of the surface at P is defined as

$$H = (K_1 + K_2)/2, \qquad (5)$$

where K_1 and K_2 are the maximum and minimum values of the curvature at P. For a *minimal surface $H = 0$* at all points in the surface.

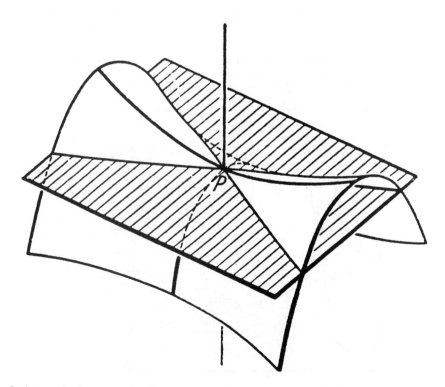

Fig. 5. At any point P on a curved surface there are two principal curvatures where the curvature is a maximum or a minimum. If these lines lie on opposite sides of the tangent plane at P (drawn shaded), then the surface has a hyperbolic (anticlastic) curvature at that point.

Thus, minimal surfaces are surfaces in three-dimensional space which have everywhere a mean curvature (H) of zero. They are exemplified by soap films, where the pressure on the two sides of the film is the same. The Laplace–Young partial differential equation, if $z = f(x, y)$, is (where f_x is the first partial derivative with respect to x, etc.)

$$H = \frac{(1 + f_y f_y)f_{xx} - 2f_x f_y f_{xy} + (1 + f_x f_x)f_{yy}}{2(1 + f_x f_x + f_y f_y)^{3/2}}. \tag{5}$$

The Gaussian curvature K (the product of the principal curvatures $K_1 K_2$) is given at each point by

$$K = (f_{xx} f_{yy} - f_{xy}^2)/(1 + f_x^2 + f_x^2)^2. \tag{6}$$

Since $H = 0$, K is everywhere negative or zero, the surface is hyperbolic and the average coordination number locally in the surface can thus be greater than 6. If K is positive, as on a sphere, the surface has an elliptic curvature and the local coordination number can be less than 6. This is the case for the spherical shells of icosahedral virus particles. The circumference of a small circle on a surface of Gaussian curvature K is given by

$$s(r) = 2\pi r - \tfrac{1}{3}\pi K r^3 + \text{terms in } r^5.$$

The sum of the angles of a triangle bounded by geodesics is $\alpha + \beta + \gamma = \pi + \int K \, dS$.

Periodic minimal surfaces are infinite, three-dimensionally repeating structures, where the unit of pattern is a segment of surface fitting into one of the fundamental regions of a particular space group, so that it is usually repeated by rotation about a twofold axis along its perimeter and thus giving rise to an infinite surface.

These surfaces are geometrical invariants (Figs. 6 and 7), just as the 5 Platonic regular solids (Fig. 8) and the 14 Archimedean semiregular solids are invariants. Indeed, the periodic minimal surfaces are related to the infinite semiregular polyhedra, the first two of which were discovered by Coxeter[18] and Petrie. Each type of surface divides all space into two congruent

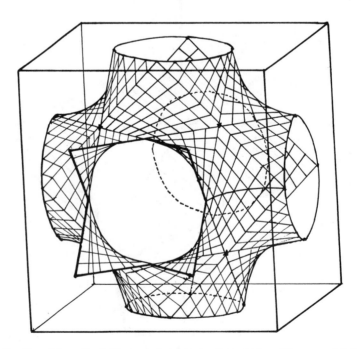

Fig. 6. The *P* surface. Schwarz's primitive periodic minimal surface which partitions space into two congruent domains. Each cavity is connected to the next by six tubes. It has been found, for example, in etioplasts and in sea urchin spines[54].

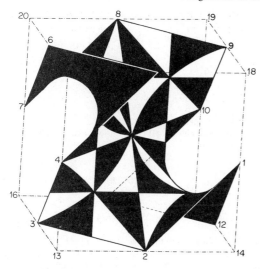

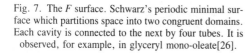

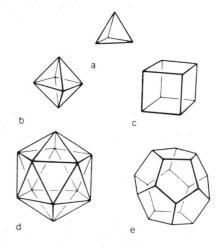

Fig. 7. The *F* surface. Schwarz's periodic minimal surface which partitions space into two congruent domains. Each cavity is connected to the next by four tubes. It is observed, for example, in glyceryl mono-oleate[26].

Fig. 8. The five Platonic regular solids. Each face is a regular polygon and every vertex is symmetrically equivalent to every other. (a) The tetrahedron; (b) the octahedron; (c) the cube; (d) the icosahedron; (e) the dodecahedron.

regions, but in the case of the gyroid surface the two regions are not identical but enantiomorphous. Minimal surfaces were known to the great 18th century mathematicians and the equation [eqn (6)] now known as the Laplace–Young equation[19] was known earlier to Legendre. H. A. Schwarz and his student E. R. Neovius wrote extensively on their discovery of 5 infinite periodic minimal surfaces. These were studied more recently by Schoen[20], who added 13 more, but much of his work remains, apparently, unpublished.

MATHEMATICAL DESCRIPTION

If the equation for the surface is expressed (in the Monge form) as $z = f(x, y)$, then the condition for zero mean curvature is given by eqn (1) with $H = 0$.

In vector notation the mean curvature (which Weatherburn prefers to designate as the first curvature) takes the particularly simple form

$$\text{div } \mathbf{n} = -H. \tag{8}$$

In this notation the Gaussian curvature K (the second curvature of Weatherburn) is given by

$$2K = \mathbf{n} \cdot \text{div grad } \mathbf{n} + (\text{div } \mathbf{n})^2. \tag{9}$$

The corresponding expression for the area of the element $dx\, dy$ is

$$dA = [1 + f_x f_x + f_y f_y]^{1/2}\, dx\, dy. \tag{10}$$

The particular boundary conditions for which a solution had been found by H. A. Schwarz in 1890 are for a quadrilateral lying in two planes in space, but neither Schwarz nor Nitsche[22] gave computations of coordinates or areas but expressed the results as a complex integral. Numerical calculations of the forms of various surfaces have been made and are summarised in Table 1.

The symmetry of the periodic minimal surface is that of a space group with the corresponding fundamental region and symmetry. The fundamental regions which were described for the 230 space groups by Schoenflies are those motifs of pattern which, when repeated by the symmetry elements of the group, fill all space. If it is possible to place a segment of "soap

Table 1. Summary of surface areas

In each case the surface area is that found in a cubic unit cell. If this is not so, then an index $I = A/V^{2/3}$ can be calculated.

(1) Primitive cubic packing of spheres (of radii 0.5): $\pi = 3.14159$.
(2) Body-centred cubic packing of spheres (of radii $3^{1/2}/4$): $1.5\pi = 4.71239$.
(3) Face-centred cubic packing of spheres (of radii $8^{-1/2}$: $2\pi = 6.28318$.
(4) Simple cube with plane faces: 3.
(5) Rhombic dodecahedron (the Voronoi polyhedron for cubic close packing): $2.12132 = 3/2^{1/2}$.
(6) Thomson cube-tetrakaidecahedron (the Voronoi polyhedron for body-centred cubic close packing): Isolated surfaces: $(6 + 12\ 3^{1/2})/8 = 3.34808$.
   Connected: $(3 + 12 \times 3^{1/2})/8 = 2.973076$.
   If only hexagonal faces are counted: 2.598076.
(7) Polyhedral version of P surface = inflated primitive cubic = $(9 + 3 \times 3^{1/2})/(2 + 2 \times 2^{1/2}) = 2.43567$ ([58], p. 255, due to W. Meier).
(8) Surface of three right circular cylinders (radius .5) intersecting at right angles = 2.617074.
(9) Cycloid of revolution approximation to the P surface = 2.34619. [For $k = .25$, $\int_{z=0}^{z=k} (\pi/4 - \arctan(3/k))(k^2 + 2z^2)^{.5}\ dz$].
(10) Hyperbolic paraboloid fitted into a tetrahedral frame = approximation to the F surface: 1.921181, only 1.00105 times greater than the exact value. Area from $A = \int_{x=0}^{x=.5} dx \int_{y=0}^{y=.5-x} (1 + 8x^2 + 8y^2)^{.5}\ dy$.
(11) Hyperbolic paraboloid fitted to the P surface: 2.36589, which is 1.009 times the exact value (this gives a z height of .25 instead of .198).
(12) The area of the film slung on a rhombic frame (of unit edge and unit hinge length, folded by raising both ends through dihedral angles theta). The height of the saddle point is z and A is the area of one-quarter (the asymmetric unit), two asymmetric units in the case of the F surface.
   $0°$, $Z = 0$, $A = 0.216506$ Planar;
   $30°$, $Z = 0.1677$, $A = 0.201607$;
   $35.26°$, $Z = 0.2007$, $A = .195440$ P surface;
   $45°$, $Z = 0.2591$, $A = 0.18059$;
   $54.73°$, $Z = 0.3535$, $A = 0.159930$ F surface;
   $60°$, $Z = 0.3860$, $A = 0.14706$.
(13) P surface. Area in unit cube containing 96 asymmetric units ($Im3m$) = $0.19544 \times 12 = 2.3453$ (in good agreement with Schoen, who finds 2.3451).
(14) F surface area in unit cube containing 48 asymmetric units ($Pn3m$) = $0.159930 \times 12 = 1.91916$ (there is here some difference because Schoen finds 2.4177).
(15) Neovius surface ($Im3m$). $1.17035 \times$ area projected on cube faces, which gives 3.51105 per unit cube (in good agreement with A. H. Schoen, who uses an analytic method and obtains 3.51048).
(16) Tetragonal saddle ($P4_2/nmc$: point symmetry at saddle points $\overline{4}2m$): 2.46784 (Schwarz[21] gives 2.46741).
(17) The gyroid (Schoen gives the area as 2.4533).

film'' across the fundamental region, so that it is mechanically stable and fulfills the symmetry requirements (such as that of being perpendicular to a mirror plane) and the boundary conditions, then an infinite periodic minimal surface results.

However, in view of the relationship to equipotentials (discussed below), it is convenient to colour the two distinct sides of the surface differently and to consider them to be related by an antisymmetry operation. Accordingly, the space groups are more appropriately the black and white groups. The simplest P surface is thus not $Pm3m$ nor $Im3m$ but $P_I m3m$ (the Neovius surface also has this symmetry) and the diamond surface F is not $Pn3m$ but $Pn'3m$.

THE OCCURRENCE OF MINIMAL SURFACES

A surface has everywhere two curvatures (called by Weatherburn[23] the first and second curvatures, but also known as mean and Gaussian curvatures and here abbreviated to H and K). The prescription that the mean curvature H should be zero (defining a minimal surface) corresponds to requiring a minimum area. The second or Gaussian curvature K remains to be prescribed. The mean surface coordination number of equal discs on a curved surface is given by $6 - 3SK/\pi$, where K is the Gaussian curvature and S is the area per unit. This condition requires, for example, that a triangulated net on a sphere, like a Buckminster Fuller geodesic dome or a virus shell, should have 12 pentagonal nodes among the remaining hexagonal nodes (Fig. 9). Since for a minimal surface $K_1 = -K_2$, the Gaussian curvature $K = K_1 K_2$ is intrinsically negative and the radius of curvature is thus imaginary. The mean surface coordination number must then be greater than 6. We suggest that this is the simple explanation for the structure of certain crinkly seaweeds (like *Fucus letuca*) which cannot be flattened on to a plane surface. If we insist on flattening it, tears will appear.

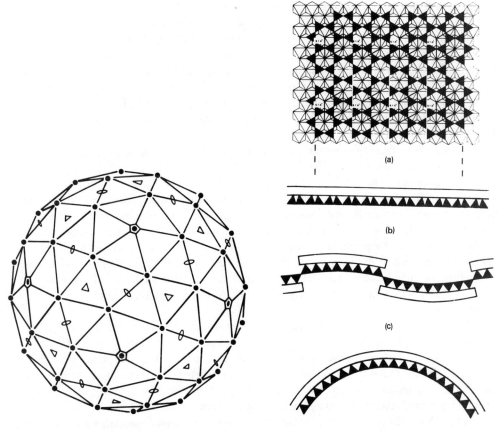

Fig. 9. The T = 7 tessellation of the icosahedron[53] showing that, in order to map a hexagonal net around a sphere, 12 disclinations, where 5 edges meet instead of 6, are necessary.

Fig. 10. A silicate sheet where the two surfaces are different may curl up like a bimetallic strip. In certain silicates the disturbance due to this curvature may be reduced by regular alternation of the direction.

Liebau[24] studied the geometry of cylindrical rolls in silicates, where asymmetric surfaces roll up to relieve the mismatch or strain (Fig. 10). One may compare this to a bimetallic thermometer strip in which the two metals have different coefficients of thermal expansion. This is by way of being a one-dimensional example of the catalytic effects to which we may refer later since, if the movement towards curvature cannot take place, one of the strips (usually the one in tension) will break. Cylinders are not, however, minimal surfaces and have $K = 0$ and H nonzero. They are developable and can be unrolled to give planar sheets.

Minimal surfaces then all have the same zero mean curvature, but the extent and localisation of the negative Gaussian curvature distinguishes them. A periodic minimal surface provides an infinite space of negative Gaussian curvature. This Gaussian curvature determines the dislocation or disclination structure of the network mapped onto the surface.

Minimal surfaces are found in soap films, and also in assemblies of lipid layers in contact with water, lyotropic colloids and in some biological assemblies such as etioplasts [Figs. 11(a) and 11(b)] and lung surfactants. They can also be seen in inorganic crystal structures[33] on an atomic scale, and on a much larger scale in calcified deposits. The first and most notable of such occurrences was found in the spines of sea urchins[25]. The best example of a molecular structure taking the form of the F surface is that of glyceryl mono-oleate with water[26]. The structure is cubic with the space group $Pn3m$ with $a = 105$ Å. The corresponding F surface would have an area per molecule of 37 Å^2, which is in good agreement with that observed elsewhere[27]. A dislocation structure can be suggested.

It seems probable that the gyroid surface G would explain the results of Luzzati and Spegt[28], who postulate a cubic structure of space group $Ia3d$ for a soap (Strontium myristate at 223°C). The structure consists of two separate but interlocked three-connected networks, similar in principle to the structure of cuprite where the minimal surface F separates the two interlocked 4-connected networks. By Babinet's principle (to which Luzzati himself alludes) the diffraction from a structure of layers between the rods would be indistinguishable from that due to the rods. (Schoen[20] gives the full space group symbol for his gyroid surface G as $I4_1/$ a 3 2/d, but in error attaches the space group number of 214 instead of 230 which is Ia3d).

Taylor[29A] has shown that the surface of beta-sheet segments in proteins has a hyperbolic curvature, and Fig. 13 gives his example of Concanavalin A. The curvature of the component beta-sheets is even expressed in the form of the crystalline lamellae of silk fibroin forming helical surfaces. The major and minor grooves of the DNA double helix are also regions of negative Gaussian curvature (although, since the two curvatures are not equal and opposite, not of zero mean curvature).

Some of the surfaces of constant energy in reciprocal space, used in solid-state physics as the "Fermi surface," are very similar to certain of the periodic minimal surfaces (Fig. 12)[30].

Periodic minimal surfaces provide a geometry for the arrangement of two-dimensional components in very-large-scale integrated devices in three dimensions, necessary for massively parallel processes expected in the computational realisation of cellular automata[31]. It is obvious, for example, that with microcircuits arranged on the P surface, the number of chips increases proportionately to the cube of the linear dimensions of the device, as it should do if 3-D phenomena such as the weather are to be modelled, and that the channels provide access to the chips by three bus bars parallel to the X,Y,Z axes on each side. The channels are also clearly available for cooling. This clearly would be a structure on a scale much larger than the atomic.

The correspondence between the periodic minimal surfaces and the surfaces discerned in silicates and lyotropic colloids has been pointed out by Mackay[32] and by Andersson et al.[33]. In zeolites (molecular sieves) the form of the silicate surfaces is isomorphous with that of minimal soap films found in regular arrays of bubbles (where the pressure is the same on both sides)[34,35]. A minimisation of the surface energy of silicate sheets during their synthesis under hydrothermal conditions and also at high temperatures (of the order of 600°) may "guide" the structure. Since the rearrangement of soap films can be represented by catastrophe theory[36], so can the transformations of silicates. Work in this fascinating field is in hand. It is evident that, like the five Platonic solids, periodic minimal surfaces are omnipresent in the mineral and in the living world and are also structural invariants.

MINIMAL SURFACES AND ELECTROSTATIC POTENTIAL

The minimal surface follows the Laplace–Young differential equation [eqn (6)] and if the slope of the surface is small this equation reduces to the Laplace equation in two dimensions:

$$f_{xx} + f_{yy} = 0. \tag{12}$$

The Laplace equation, which in its general form is

$$\text{div grad } V = 0,$$

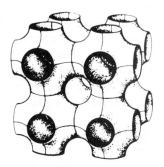

Fig. 11. (a) Etioplasts from Avena (Gunning); in the absence of light chloroplasts fail to develop properly. (b) The P surface of Schwarz[54].

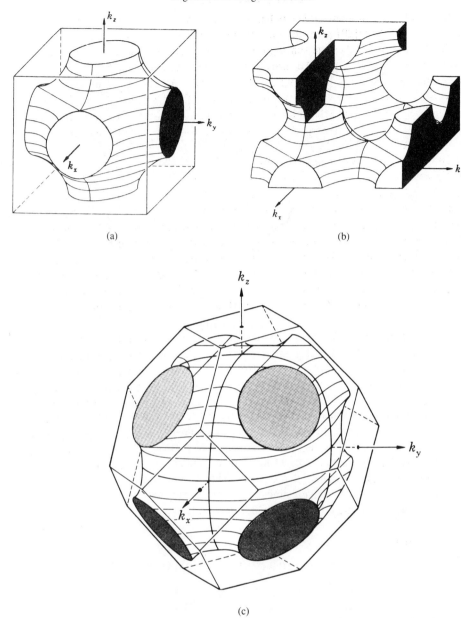

(a)   (b)

(c)

Fig. 12. Some of the Fermi surfaces, which are surfaces in reciprocal space representing constant energy of electron waves, take forms very similar to those of the periodic minimal surfaces. (a) and (b) show the two symmetrical regions of the P surface and (c) shows an eight-tube surface which is close to one reported by Schoen[20]. (Figure courtesy of John Wiley, Inc.)

where V is the electric potential, describes equipotential surfaces in regions where electric charges are absent. Thus, locally, a minimal surface may be identical with the zero equipotential surface. However, the electric field

$$\mathbf{E} = -|E|\mathbf{n} \tag{14}$$

is normal to the equipotential surface at every point, and therefore

$$\operatorname{div} \mathbf{n} = -\frac{1}{|E|} \operatorname{grad} |E| \cdot \mathbf{n} + \operatorname{div} \operatorname{grad} V. \tag{15}$$

By comparison with the Laplace equation [eqn (13)] we see that even though the second term on the right may be zero, the first is not zero. Thus, an equipotential surface is not, in general, a minimal surface.

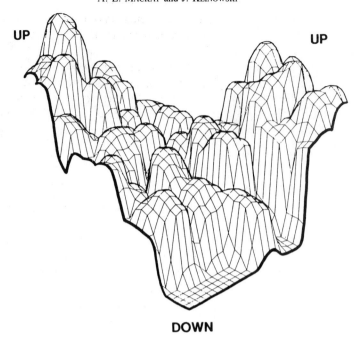

Fig. 13. Part of the surface of the protein molecule Concanavalin A showing that locally it is hyperbolically curved[29].

Since an equipotential surface is uniquely defined by a point charge distribution and, with certain reservations *vice versa*, the electrostatic equipotential surface in, for example, CsCl must be similar to the periodic P surface. To each minimal surface should correspond a charge distribution for which it would represent the zero equipotential. Thus for the P surface, the charge distribution is approximately that of CsCl. The Neovius surface (Fig. 14) may be more like $0.8CsCl \cdot 0.2NaCl$. It will be interesting to look for ionic crystal structures which correspond, at least approximately, to the other periodic minimal surfaces.

NUMERICAL CALCULATION OF MINIMAL SURFACES

As was mentioned earlier, minimal surfaces cannot in general be given in analytic form. However, they may often be defined as soap films with certain boundary conditions. The profile of the surface can be obtained numerically. We have done this for various of the periodic minimal surfaces using an iterative procedure which resembles the natural way in which each element of area in a soap film adjusts itself into conformity with the neighbouring elements. The soap film is an example of a natural cellular automaton, and this procedure is mimicked numerically. Working usually with the Monge form $z = f(x, y)$ of the surface we begin with an approximate surface. The hyperbolic paraboloid

$$z = (x/a)^2 - (y/b)^2 \tag{16}$$

is a very close approximation to the film over a skew rhombus frame—a frame made of two equilateral triangles sharing an edge at a dihedral angle. In this case the area of the hyperbolic paraboloid differs by only about 0.1% from that of the true minimal surface, so that it is a very convenient form to use as an analytic representation of the P and F surfaces which are built of such segments.

In the numerical calculation the z coordinate of each point is adjusted in turn to fulfill the Laplace–Young condition [eqn (6)]. We find that it does not matter whether the cells are updated sequentially or simultaneously. This requires the calculation of the first and second derivatives from the z coordinates of the eight immediate neighbours of each point (see Fig. 15). The boundary conditions have to be satisfied.

The calculated areas (of the area of film in a unit cube) agree well with those found by Schoen, who used a much more complex method involving complex integrals (2.351 for the P surface and 3.5104 for the Neovius surface, the $C(P)$ surface in Schoen's notation) but for the F surface we find 1.9193 where Schoen has 2.4177. Some matter of definition may be involved here. Other results are given in Table 1.

THE RELATIVE STABILITY OF SURFACES

It can be seen from Table 1 that a primitive cubic packing of spheres, such as might occur in the stacking of liposomes or the unit spheres of opal, has a specific surface area of 3.14159, while the corresponding P surface has an area of 2.3460. The P surface has zero mean curvature, while the spheres have a positive curvature. Thus matter may be removed from convex areas and redeposited in concave areas, if the system is driven by surface tension until the P surface results. Iler[55] gives a figure (Fig. 16) where he postulates this process in silica gel.

The face-centred cubic packing of spheres has the high specific surface of 6.283 and may be expected to transform, if driven by the surface tension, to the surface which Schoen[20] designates as F-RD.

The body-centred cubic packing of spheres (which is in fact a packing of minimum density, in that any mutual movement increases the density) has a specific area of 4.712 and its surface is thus liable to transform to the Neovius surface (also space group $Im3m$) with specific surface 3.511.

In each case there will be an adjusting factor because the volumes of material and void space are not initially equal (although they are nearly equal in the primitive cubic packing: packing fraction = 0.5236). We could recalculate, on the basis of area per unit mass, the volume of the solid phase permeated by a liquid phase where the initial lattice had expanded. Surfaces of constant mean curvature can thus be produced by erosion or equilibration of vapour or solution pressure.

We can show that the infinite periodic minimal P surface is, by itself as a soap film, in fact unstable, in the sense that a spherical soap bubble requires an excess internal pressure for its existence. Is this excess pressure stabilising or destabilising? "Small bubbles (with a higher pressure) blow up larger ones" and we may imagine a distortion in which one of the tubes of the P surface becomes smaller while its length, the repeat distance of the structure, remains the same. The principal radii of curvature then can no longer remain the same and the mean curvature must rise, and this could only be countered by a rise in the internal pressure. However, since the channels are connected together, local distortions would appear to be unstable.

We may recall Earnshaw's Theorem, which states that no system of positive and negative electric charges can be stable. Other forces are necessary, such as the repulsive overlap forces which permit NaCl to be stable with electrostatic forces between the ions. We have shown that Laplace's equation, which describes equipotential surfaces in a system of charges, is similar to the Laplace–Young equation which describes the shape of soap films of minimal curvature. If the gradients are small, as in a drumhead, they become identical. It seems reasonable to expect, then, that a theorem similar to Earnshaw's should apply to systems of minimal surfaces. Foam might appear to be stable in spite of this but it is compartmentalised, with different pressures in separate cells. We suggest that a foam with permeable film should be unstable.

However, as experiments with Langmuir films showed, membrane surfaces do not contract indefinitely, but when they reach a position where the molcules are close packed, achieve some incompressibility. Thus, vesicles and similar structures are mechanically more complicated than simple bubbles and have more possibilities of stability. The shape of red blood cells shows the complexity arising.

The 18 periodic minimal surface due to Schoen show regions of many different coordination numbers, 4, 6, 8, 12, 14, etc. The question remains open as to whether any reasonable but irregular structure can be represented as a minimal surface.

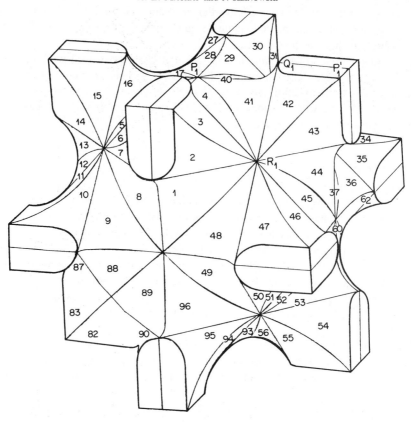

Fig. 14. The "Neovius" periodic minimal surface. One cell is connected to the next by 12 tubes in the [110] directions so that the spaces on each side of the film membrane are congruent but not in communication. The figure is from the thesis of E. R. Neovius[56].

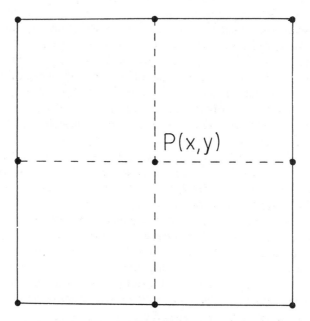

Fig. 15. A surface can be described as $z = f(x,y)$. The gradients and curvatures (the first and second derivatives) at a point P are found from the z coordinates of eight neighbouring points. Each middle point is continually adjusted to conformity with its neighbours, as happens automatically in a physical soap film.

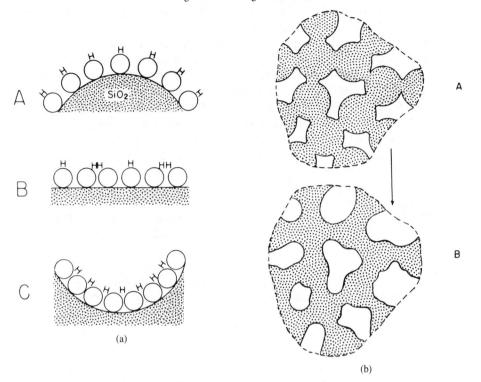

Fig. 16. (a) The vapour pressure of water over a convex surface is greater than that over a plane surface, and that over a concave surface is less. Consequently, (b) a connected network of silica gel particles tends to become a surface of constant mean curvature[55] (figures by courtesy of John Wiley & Sons).

PENTAGONAL SYMMETRY: SHORT-RANGE *vs* LONG-RANGE ORDER

We have seen that the five regular Platonic solids are still the foundation of the grammar of structure. However, although the tetrahedron, the cube and the octahedron figured extensively in the description of crystals, the dodecahedron and the icosahedron were long outlawed because it was believed that crystals could not permit the occurrence of fivefold symmetry. Pentagonal axes are not compatible with an infinite lattice where every lattice point is identically situated. Groups containing fivefold axes did not figure at all in the Bible of crystallography, the *International Tables for X-ray Crystallography* (Vol. I, Symmetry Groups, 1952).

From time to time, however, pentagonal crystals have been reported[37–39]. Local pentagonal symmetry does, of course, occur. For example, the individual molecule crystals of ferrocene and of cyclopentane have fivefold symmetry. Icosahedral clusters of atoms were recognised in crystals of poliomyelitis virus and in some metals such as alpha-manganese, boron (Fig. 19), tungsten and certain alloys, and in some organic molecules such as dodecahedrane[40], the last a tour de force of organic synthesis.

Examples were even known where the icosahedral symmetry persisted for several atomic diameters as in $Mg_{32}(Al,Zn)_{49}$[41], but eventually crystalline order set in. It has been known at least since the time of Newton that 12 spheres packed around a central sphere were furthest from each other in an icosahedral arrangement. If we have one atom 0.90 of the diameter of 12 others, then the icosahedral arrangement of the 12 around the smaller is strongly promoted. This gives a nearly spherical cluster, and in 1962 it was proposed that 13 such large spheres could be packed together to give a still bigger quasisphere and that this process could be continued recursively[42].

However, it has been reported most recently that splat-cooled Al_6Mn (thin sheets of a "metallic glass" produced by extremely rapid cooling) shows icosahedral symmetry over distances of some 100 atomic diameters[43]. The diffraction patterns are exactly like those obtained from computer drawings of the Penrose pattern[44]. Levine and Steinhardt[45] have indeed identified this occurrence of icosahedral order as a materialisation of the Penrose pattern.

Nevertheless more proof is required because strongly icosahedral groups, such as that of $Mg_{32}(Al,Zn)_{49}$, must reflect this symmetry in their diffraction patterns.

The Penrose pattern[11,44,46–48] in three dimensions consists of a packing of acute and obtuse rhombohedra, with angles 63.43° and 116.57° respectively, in a composition ratio of the golden number (1.618. . .) to each other (Figs. 17, 18 and 19). The proportions of the acute rhombodedron are exactly those found for boron and, for example, the carbide $B_{12}C_3$. The pattern has beautiful properties too numerous to mention, but in essence it is a quasicrystalline arrangement where every component is in a prescribed position, where the structure is self-similar and recursive in the sense of Mandelbrot[49], where there is local fivefold (icosahedral) symmetry and where components are not in exactly equivalent positions, as demanded in classical crystallography but only in quasiequivalent positions. The structure consists of rhombic triacontahedra which sometimes touch and sometimes overlap to fill all space. A theory of quasiequivalence with regard to packings on the surface of a sphere was developed by Caspar and Klug[50], but quasiequivalence in the solid is a more recent development. The late N. V. Belov, doyen of Soviet crystallographers, graphically put the situation revealed on the discovery of poliovirus particles as follows:

> The exclusion of a fivefold axis for crystals, as well recognised, results from the impossibility of reconciling it (and axes of order greater than 6) with the "lattice state" of crystalline

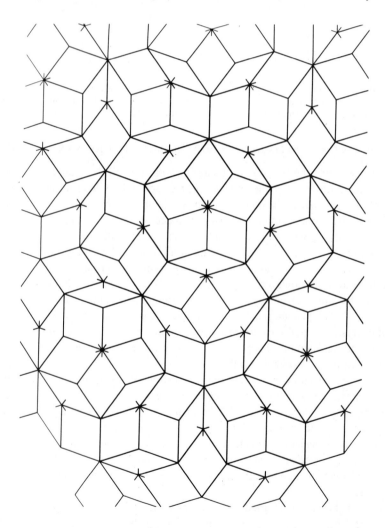

Fig. 17. The Penrose pattern in two dimensions. Two kinds of tile (rhombuses with angles 36° and 72°) cover the plane without exact repetition. If the edges of the pieces are tongued as in a jigsaw puzzle the long-range pattern automatically results from the interaction of neighbours, but there are several other algorithms for its generation. The ratio of the numbers of the two pieces is 1.618. . . , the golden number [46,48].

matter. It would appear, then, that for small organisms the fivefold axis represents a distinctive instrument in their struggle for existence, acting as insurance against petrifaction, against crystallisation, in which the first step would be their "capture" by the lattice[51].

Much earlier, Belov's predecessor Vernadskii had coined the aphorism "Crystallisation is death."

The significance of these observations of the peaceful (if metastable) coexistence of icosahedral symmetry with long-range order is that they witness to the struggle between the local short-range order and the extended, crystalline translational long-range order for the minimisation of the free energy of the system, bearing in mind that the minimisation need not be towards the absolute minimum but towards trapping, by reason of the historical development of the specimen, in a local metastable minimum. If local ordering tends sufficiently strongly towards being icosahedral, then the material cannot properly crystallise but may remain a glass. But as in the compromise case found by Shechtman *et al.*[45], crystallisation spreads, in that each icosahedral cluster influences the orientation of the neighbouring icosahedral cluster, and the interspace is filled up by a definite arrangement of atoms, to give what has been termed the quasilattice structure of the Penrose pattern predicted earlier by Mackay[48]. Here we see how the balance of forces at one level determines the next higher level of structure.

CONCLUSIONS

The purpose of this paper has been to draw attention to four concepts which can be used to understand the organisation of inorganic and other structures.

Cellular structure. This is already well known in the form of dissections of structures into Voronoi and radical polyhedra which have flat faces. The next extension is to subdivision into cells with curved faces.

Periodic minimal surfaces. Sheets or surfaces occur as dominant secondary structure, not only in lipids and lyotropic colloids, where they are to be seen in the electron microscope, but also in silicates, particularly cage or framework silicates and in spherulites and cylindrical rolls.

Cellular automata, the patterns formed in space and time by the interaction of adjacent elements, give a general formalism which allows us to escape from the rigidity of the classical space groups and to treat, eventually, not only classical symmetry, but the phenomena of biological morphogenesis.

Intrinsic curvature provides a measure for the strain in structures, due to partial misfit of neighbouring components, which may express itself in the appearance of curved sheets. It is a

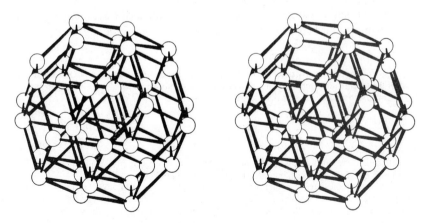

Fig. 18. In three dimensions the Penrose pattern consists of two kinds of rhombohedra (with angles 63.43° and its supplement). The whole pattern is seen to be built of rhombic triacontahedra (each made of 10 acute and 10 obtuse rhombohedra) which pack in contact or with some sharing of rhombohedra. Again the ratio of the two components is 1.618. . . , the golden number. The figure shows one such rhombic triacontahedral unit and is designed for viewing with the "crossed-eyes" technique with the left view on the right and *vice versa*[48].

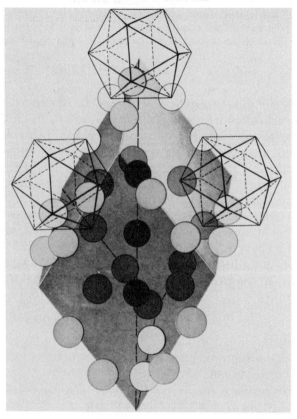

Fig. 19. The rhombohedral cell of $B_{12}C_3$, which has $\alpha = 65°$ (close to the acute rhombohedron of the Penrose pattern) and is built of B_{12} icosahedra with interstitial carbon atoms. (The drawing is reproduced by courtesy of *Scientific American* and W. H. Freeman).

development of ideas, first expressed by Coxeter (1961), that some configurations found in random structures (dodecahedra) approximate to close packing in a higher dimensionality. Strain in a structure can be expressed as curvature in a higher dimension. The forced juxtaposition of two structures of different curvatures (of different metrics) will produce disruption of the weaker.

Strain in sheets is expressed in their crinkling into the third dimension, which makes them into elliptically or hyperbolically curved structures. Elliptic curvature gives closed domains (like icosahedral virus particles) and hyperbolic curvature gives periodic minimal sheets. Strain in three-dimensional structures expresses itself in a more complicated way. Many structural features arise from the competition between short-range and long-range ordering.

Acknowledgments—We are particularly indebted to R. H. Mackay for the provision of a NASCOM II microcomputer on which most of the calculations were carried out, and we are also most grateful also to Professor Sten Andersson of Lund University for his hospitality and for discussions on this subject.

REFERENCES

1. M. O'Keefe and A. Navrotsky (Eds.), *Structure and Bonding in Crystals*. Academic Press (1981).
2. R. C. Evans, *Crystal Chemistry*. Cambridge University Press (1939).
3. H. A. Simon, The architecture of complexity. *Proc. Amer. Phil. Soc.* **106**, 467–482 (1962).
4. G. Bricogne, Maximum entropy and the foundations of direct methods. *Acta Cryst.* **A40**, 410–445 (1984).
4A. R. Sadanaga, Y. Takeuchi and N. Morimoto, Complex structures of minerals. *Recent Progress of Nat. Sci. in Japan* **3**, 141–206 (1978).
5. J. L. Finney, Random packings and the structure of simple liquids. *Proc. Roy. Soc.* **A319**, 479–493 (1970).
6. W. Fischer, E. Koch and E. Hellner, Zur Berechnung von Wirkungsbereichen in Strukturen anorganischer Verbindungen. *Neues Jahrbuch der Mineralogie Monatsheft* **5**, 227–237 (1971).
7. A. L. Mackay, Stereological characteristics of atomic arrangements in crystals. *J. Microscopy* **95**, 217–227 (1972).

8. W. Brostow, J.-P. Dussault and B. L. Fox, Construction of Voronoi polyhedra. *J. Comp. Physics* **29**, 81–92 (1978).

9. B. N. Delone, N. P. Dobilin, M. I. Shtogrin and R. V. Galiulin, A local criterion for regularity of a system of points. *Sov. Math. Dokl.* **17**, 319–322 (1976).

10. J. Klinowski, Nuclear magnetic resonance studies of zeolites. *Progress in NMR Spectroscopy* **16**, 237–309 (1984).

11. A. L. Mackay, Crystal symmetry. *Physics Bulletin* 495–497 (Nov. 1976).

12. S. Wolfram (Ed.), Universality and complexity in cellular automata. *Physica* **D10**, 1–35 (Jan. 1984).

13. E. K. Gordon, S. Samson and W. B. Kamb, Crystal structure of the zeolite, paulingite. *Science* **154**, 1004–1006 (1966).

14. L. Blumenthal, *Distance Geometry*. Oxford (1953).

14A. L. Blumenthal and K. Menger, *Studies in Geometry*. Freeman, San Francisco (1970).

15. J.-F. Sadoc, *J. Non-cryst. Solids* **38** (1981).

16. R. Kerner, Remarks on the curved-space description of amorphous solids. *Phil. Mag.* **B47**, 151–162 (1982).

17. D. R. Nelson, Order, frustration, and defects in liquids and glasses. *Phys. Rev.* **B28**, 5515 (1983).

18. H. S. M. Coxeter, *Twelve Geometric Essays*, S. Illinois Press (1968); *Proc. Lond. Math. Soc. Ser. 2* **43**, 33–62 (1937).

19. C. Isenberg, *The Science of Soap Films and Soap Bubbles*. Tieto, Clevedon (1978).

20. A. H. Schoen, Infinite periodic minimal surfaces without self-intersection, NASA Technical Note C-98 (NASA Electronics Research Center, Cambridge, Mass., Sept. 1969); NASA Technical Note D-5541, (1970); *Notices of the Amer. Math. Soc.* **16**, 519 (1969); **15**, 929 (1968); **15**, 727 (1968); **14**, 661 (1967).

21. H. A. Schwarz, "*Gesammelte mathematische Abhandlungen* (2 vols.). Springer-Verlag, Berlin (1890).

22. J. C. C. Nitsche, *Vorlesungen ueber Minimalflaechen*. Springer-Verlag, Berlin (1975).

23. C. E. Weatherburn, *Differential Geometry of Three Dimensions*, I & II, Cambridge Univ. Press (1939).

24. F. Liebau, *Acta Cryst.* **B24**, 690 (1968).

25. G. Donnay and D. L. Pawson, X-ray studies of echinoderm plates. *Science* **166**, 1147–1150 (1969); H.-U. Nissen, *Science* **166**, 1150–1152 (1969).

26. W. Longley and T. J. MacIntosh, A bicontinuous tetrahedral structure in a liquid-crystalline lipid. *Nature* **303**, 612–614 (1983).

27. S. H. White, Formation of solvent-free black lipid bilayer membranes from glyceryl mono-oleate dispersed in squalene. *Biophysical.* **23**, 337–347 (1978).

28. V. Luzzati and P. A. Spegt, Polymorphism of lipids. *Nature* **215**, 701–704 (1967).

29. B. Lotz, A. Gonthier-Vassal, A. Brack and J. Magoshi, Twisted single crystals of *Bombyx mori* silk fibroin and related model polypeptides with beta-structure, *J. Mol. Biol.* **156**, 345–357 (1981).

29A. W. R. Taylor, Ph.D. Thesis, Oxford (1981).

30. M. Ya. Azbel, M. I. Kaganov and I. M. Lifshits, Conduction electrons in metals. *Sci. Amer.* 88–98 (Jan. 1973).

31. V. Zakharov, Parallel processing. *IEEE Trans. Computers* **C33**(1), 45–78 (Jan. 1984).

32. A. L. Mackay, Silicate membranes as minimal surfaces, (poster and abstract 73-P1-5a), Abstracts, IUC Regional Conference, Copenhagen (1979).

33. S. Andersson, On the description of complex inorganic crystal structures. *Angew. Chem. Int. Ed. Engl.* **22**, 69–81 (1983).

33A. S. Andersson and B. G. Hyde, An attempted exact, systematic, geometrical description of crystal structures. *Zeit. Krist.* **158**, 119–131 (1982).

33B. S. Andersson, B. G. Hyde and H. G. von Schnering, The intrinsic curvature of solids. *Zeit. f. Krist.* **168**, 1–17 (1984).

34. W. Thomson, On the division of space with minimum partitional area. *Phil. Mag.* **4**, 503–574 (1877).

35. J. Lifschitz, On the cellular division of space with minimum area. *Science* **97**, 268 (1943).

36. R. Thom, *Structural Stability and Morphogenesis*. Benjamin, Reading, MA (1972).

37. R. L. Schwoebel, *J. Appl. Phys.* **37**(6), 2515 (1966).

38. G. Milman, S. G. Uzman, A. Mitchell and R. Langridge, Pentagonal aggregation of virus particles. *Science* **152**, 1381–1383 (1966).

39. K. Kobayashi and L. M. Hogan, Fivefold twinned silicon crystals grown in an Al-16 wt.% Si melt. *Phil. Mag.* **40**, 399–407 (1979).

40. L. A. Paquette, D. W. Balogh, R. Usha, D. Kountz and G. G. Christoph, Crystal and molecular structure of a pentagonal dodecahedrane. *Science* **211**, 575–576 (1981).

41. G. Bergmann, J. L. T. Waugh and L. Pauling, The crystal structure of the metallic phase $Mg_{32}(Al, Zn)_{49}$. *Acta Cryst.* **10**, 254–259 (1957).

42. A. L. Mackay, A dense noncrystallographic packing of equal spheres. *Acta Cryst.* **15**, 916–919 (1962).

43. D. Shechtman, I. Blech, D. Gratias and J. W. Cahn, Metallic phase with long-range orientational order and no translational symmetry. *Phys. Rev. Letters* **53**(20), 1951–1953.

44. A. L. Mackay, Crystallography and the Penrose pattern. *Physica* **114A**, 609–613 (1982).

45. D. Levine and P. J. Steinhardt, Quasi-crystals: a new class of ordered structures. *Phys. Rev. Letters* **53**(26), 2477–2480 (1984).

46. R. Penrose, *Bull. Inst. Maths. Appl.* **10**(7/8), 266–271 (1974).

47. M. Gardner, *Scientific American* **236**, 110 (1977).

48. A. L. Mackay, De nive quinquangula: On the pentagonal snowflake, *Kristallografiya* **26**(5), 910–919 (1981); *Sov. Phys. Cryst.* **26**, 517–522 (1981).

49. B. Mandelbrot, *Fractals*. Freeman, San Francisco (1977).

50. D. L. D. Casper and A. Klug, *Cold Spring Harbor Symp. Quant. Biol.* **37**, 1–24 (1962).

51. N. V. Belov, Essays on structural crystallography, XIII, Minerals. *Sbornik Lvov. Geol. Obshch.* **16**, 41 (1962).

52. H. S. M. Coxeter, *Introduction to Geometry*, P. 411. Wiley, New York (1961).

53. I. Rayment, T. S. Baker, D. L. D. Casper and W. T. Murakami, Polyoma virus capsid structure at 22.5 Å resolution. *Nature* **295**, 110–115 (1982).

54.   K. Fontell, Liquid crystallinity in lipid–water systems. *Molec. Cryst. Lig. Cryst.* **63**, 59–82 (1981).
55.   R. K. Iler, *The Chemistry of Silica.* Wiley-Interscience, New York (1979).
56.   E. R. Neovius, *Bestimmung zweier speziellen periodischen Minimalflaechen. Akad. Abhandlung*, Helsinki (1883).
57.   F. C. Frank and J. S. Kasper, *Acta Cryst.* **11**, 184–190 (1958).
58.   A. F. Wells, *Three-Dimensional Nets and Polyhedra.* Wiley, New York (1977).

Comp. & Maths. with Appls. Vol. 12B, Nos. 3/4, pp. 825–834, 1986
Printed in Great Britain.

0886–9561/86 $3.00 + .00
© 1986 Pergamon Press Ltd.

SYMMETRIES OF SOAP FILMS†

A. T. Fomenko

Department of Higher Geometry and Topology, Faculty of Mechanics and Mathematics, Moscow
State University, Moscow 119899, U.S.S.R.

Abstract—The structure of soap films in the neighborhood of a singularity is studied by observing networks of minimal curves formed by the intersection of the film cone with a sphere. Questions of stability of the film cones formed by these networks are discussed. The existence of film cones other than portions of simple discs is shown for minimal surfaces embedded in spaces of dimension greater than 7. Uniqueness of soap films is studied by carrying out continuous transformations of a Douglas contour and obscuring the evolution of the spanning films under these transformations. Finally, the author introduces a method of studying the evolution of soap films by investigating graphs of their areas.

Consider the geometry of soap films spanning a wire contour without closed bubble cycles. Of particular interest is the behavior of soap films near their singular points. Using the well-known Plateau principle, we can assume that in a sufficiently small neighborhood of each singular point, a minimal surface, i.e. a soap film, consists of a few smooth segments of a two-dimensional surface which can be considered to be approximately flat. As we will see, this property can be invalid for minimal surfaces of dimension greater than 2. There exist "multi-dimensional" singular points for which no neighborhood exists, consisting of flat segments. Let us go back to the two-dimensional case. Consider a sphere of small radius with center at the singular point and let us study its intersection of a finite number of smooth arcs lying on the surface of the sphere and forming a net on it.

It is easy to become convinced that each segment of the arc must be a segment of some equator, i.e. a circumference obtained as the result of the intersection of a sphere and a plane through its center. In fact, since we consider that the film near a singular point consists of flat segments, they converge at the singular point, and the portion of the film inside the sphere is a cone with vertex at the singular point. The cone has as its "base" the system of smooth arcs forming a net on the sphere. A soap film lying inside the sphere is formed by all possible radii emerging from the center of the sphere and ending on the one-dimensional net. Some of these radii could be singular edges of a film where a few of its faces meet. A film is composed of a few segments which are portions of a regular plane disc. There exists a very simple relation between the lengths, l, of a one-dimensional net and the area, s, of the film lying in a sphere. It is clear that

$$s = \tfrac{1}{2} r_0 l,$$

where r_0 is the radius of the sphere. It follows from this that the net consists of arcs of equators, i.e. geodesics of a sphere. If we assume that part of an arc is not a geodesic, then there exists a slightly deformed arc between the same end points but with a smaller length. Therefore, there exists a deformation which reduces the area of the film cone since a reduction of l implies a reduction of s. It follows that if a few smooth arcs of a net meet at a point on a sphere, their number is equal to three and the angles between them are $2\pi/3$.

Let us pose the following question: What can be said about one-dimensional nets which are intersections of soap films and a small sphere with center at a singular point of a film? The answer to this question will yield a description of the local structure of a film in a small neighborhood of its singular point not lying on the bounding contour of the film.

Summarizing: (1) each smooth arc of a net is a segment of an equator; (2) at each singular point of a net, its node, only three arcs can meet and they form angles $2\pi/3$. This enables us to list all possible configurations[2], i.e. nets on a sphere. There are exactly ten such config-

---

†Translated from Russian by V. V. Goldberg and J. Kappraff, Department of Mathematics, New Jersey Institute of Technology, Newark, NJ 07102.

urations, and they are illustrated in Fig. 1. In order to convince ourselves that there are no other nets, it is sufficient to use some elementary considerations of spherical trigonometry.

Let us now consider the following question. Can all these configurations be realized as intersections of soap films with a small sphere centered at a singular point? The answer is that most of the indicated configurations cannot. Of course, if we consider cones over these nets with vertexes at the center of a sphere, they then have zero mean curvature at all of their regular points and therefore are minimal surfaces almost everywhere. These cones are shown in Fig. 1. In other words, if we neglect some set of measure zero, then we can say that they are two-dimensional minimal surfaces. However, singular points of a film cannot be ignored, and the conditions of stability and minimality impose stringent restrictions on the structure of singularities of soap films, especially if we are interested in stable films. The main interest for us is in films which are minimal in the following sense: any small disturbance does not reduce the area of the film. Most of the cones in Fig. 1 are not minimal according to this definition. First of all, it is clear that the first three surfaces, which include, respectively, a plane circumference, three arcs meeting at equal angles at two vertices, and a regular tetrahedron, are stable minimal films. Second, it turns out that the other cones are not minimal. For each of them, there exist reducing deformations (disturbances that reduce area). In addition, the vertex of the cone is exploded into more complicated topological configurations having singular points satisfying the Plateau principle. Figure 2 illustrates actual soap films that are obtained in an attempt to realize minimal surfaces with the boundaries shown in Fig. 1 (for details see [2]). The structure of these surfaces are quite complicated, but all their singular points are located either on threefold singular edges or are fourfold singularities at which four edges of a film meet at equal angles. An interested reader, using patience and care, can obtain soap films of these described topological types from the wire contours shown in Fig. 1. These films [of types (a), (b) and (c)] are stable, and any small disturbance does not decrease their area. The films shown in Figs. 2(d)–(j) can no longer be considered "small" because, as we previously discovered, they do not model the structure of an infinitesimal neighborhood of a stable singular point on a soap film. By contrast, the first three films, shown in Figs. 2(a)–(c), can be transformed by similarity (contraction with the center at the origin) into themselves, reducing their size. These films are models of the structure of a film in an infinitesimal neighborhood of a singularity. This discussion touches on the interesting question of minimal cones in $\mathbf{R}^n$ with vertex at the origin having as their "base" (boundary) a minimal submanifold, A^{n-2}, in the standard sphere S^{n-1}.

Consider, in $\mathbf{R}^n$, the sphere S^{n-1} of radius R with center at the origin, and let A^{n-2} be an $(n-2)$-dimensional compact submanifold in the sphere which is almost everywhere (with the exception of a set of points of measure zero) a smooth $(n-2)$-dimensional submanifold in the sphere. Consider the cone CA with vertex at point O and base A. The cone is generated

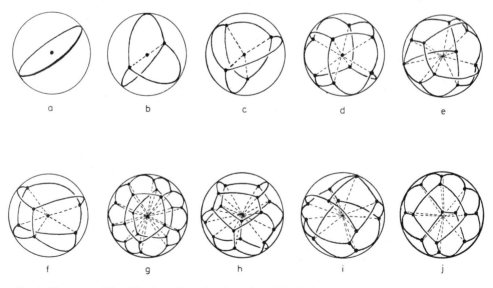

Fig. 1. The ten possible minimal one-dimensional nets formed by the intersection of a film cone by a sphere. Only 1(a), (b) and (c) form stable film cones.

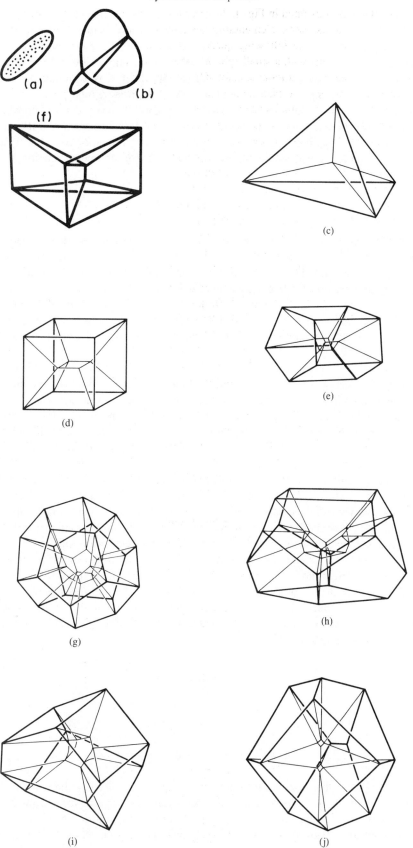

Fig. 2. Actual soap films having boundaries shown in Fig. 1. Figures 2(a)–(c) correspond to Figs. 1(a)–(c) and are models of film cones in an infinitesimal neighborhood of a singular point. Figures 2(d)–(h) correspond to Figs. 1(d)–(h) and cannot be considered as models of films in a small neighborhood of a singular point.

by radii from the center to the points of the set A. Let us fix the boundary A of cone CA and assume that the cone is a minimal surface in the sense that any disturbance of it, not altering the boundary, increases (does not decrease) the volume or area in the two-dimensional case. The disturbance of the cone is understood to take place in the above sense. It is quite clear that for any choice of boundary A, cone CA is minimal. If cone CA is a minimal surface in $\mathbf{R}^n$, then its boundary A is a minimal surface in sphere S^{n-1} at all its regular points.

In particular, with the exception of the set of points of measure zero, boundary A is, locally, a minimal submanifold of the sphere, i.e. its mean curvature vanishes. It follows from elementary integral properties that the $(n-1)$-dimensional volume of cone CA is connected with the $(n-2)$-dimensional volume of its boundary A by the following relation:

$$\text{Vol}_{n-1} \, CA \, = \, \frac{R}{n-1} \, \text{Vol}_{n-2} \, A,$$

where R is the radius of the sphere. Therefore, any small disturbance of the boundary reducing its volume induces (along radii) a corresponding disturbance of the whole cone, reducing its volume. Thus, if we allow the existence of reducing deformations for the boundary, then there exist reducing deformations for the whole cone, which contradicts its assumed minimality and proves the statement formulated above. It follows from this that the intersection of a two-dimensional soap film with a sphere of infinitesimal radius, with center at a singular point of a film, can be considered as consisting of arcs of equators meeting each other at threefold points at equal angles.

For simplicity, we restrict ourselves to the case where the boundary of the cone is a smooth, closed submanifold on a sphere. The question is, does there exist, in $\mathbf{R}^n$, cones with vertex at the origin and minimal in the sense that any small disturbance of them increases their volume? The simplest such example is the standard plane $(n-1)$-dimensional euclidean disc passing through the origin and intersecting the sphere along an equator. Because of this example, we will sharpen our question. Do there exist nontrivial minimal cones distinct from the standard plane disc? The answer is so unexpected that we will discuss it in a little more detail. It is clear that in a three-dimensional euclidean space $\mathbf{R}^3$, there do not exist any minimal cones (with the exception of plane discs) having as the boundary a smooth one-dimensional submanifold (a curve). This follows from the facts that the boundary must be a curve made up of locally minimal lengths and all geodesics on a sphere must be great circles. Therefore a cone is a plane disc. With increasing dimension of the sphere S^{n-1}, we will get locally minimal submanifolds A^{n-2} not coinciding with equatorial hyperspheres. For example, in the sphere S^3, in addition to regular equators there is the locally minimal two-dimensional torus embedded as follows. Let $S^3 \subset R^4 = C^2(z, w)$, then torus T^2 can be given as the intersection of this sphere, $S^3 = \{|z|^2 + |w|^2 = 1\}$, with the surface, $M^3 = \{|z| = |w|\}$. A sphere is a union of two half-spaces: $|z| \leq |w|$ and $|z| \geq |w|$. A torus is their common boundary. If ϕ and ψ are angle coordinates, then points of the torus are given as follows:

$$(z, w) \, = \, \frac{1}{\sqrt{2}} \, (e^{i\phi}, e^{i\psi}).$$

The metric induced on the turns is a flat euclidean metric since

$$dz \, d\bar{z} \, + \, dw \, d\bar{w} \, = \, \tfrac{1}{2}(d\phi^2 \, + \, d\psi^2).$$

It is easy to check that this torus is a locally minimal surface (i.e. a soap film without boundary) in a sphere. However, one can be convinced that a minimal surface, located inside a sphere and having a torus as its boundary, is not a cone. It turns out that in $\mathbf{R}^4$, as it was in $\mathbf{R}^3$, there are no nontrivial cones with center at the origin. If the dimension increases, the picture changes. If the dimension is sufficiently high, there exist minimal nontrivial cones[6]. In order to obtain an intuitive notion of this interesting fact, we will consider, for definiteness, a boundary A^{n-2} of some fixed topological type. The simplest manifolds, after spheres, allowing locally minimal embeddings in a sphere are products of spheres, $S^p \times S^q$, where $p + q = n - 2$.

Let $n = 2$, and consider a circumference S^1 and $A = S^0 \times S^0 \subset S^1$ shown in Fig. 3. In this case, a cone with boundary A is a union of two parallel segments, a "one-dimensional cylinder" as shown in Fig. 3. Now let $n = 3$. Let us take $S^1 \times S^0$ as the boundary A in the sphere S^2. It is clear that a two-dimensional cone with boundary A is not minimal as before since there exists a reducing deformation in a neighborhood of the vertex, decreasing its area. As a result of this variation, the singular point, the vertex of the cone, is decomposed and becomes a circumference with the neck of a catenoid, an actual minimal film with boundary $S^1 \times S^0$ (pairs of circumferences). Comparing a two-dimensional minimal film (a catenoid) with a one-dimensional film (two parallel segments), we note that the two-dimensional film sags in the direction of the origin, unlike the one-dimensional film. It turns out that this sagging of a minimal film in the direction of the origin increases with increasing dimension. Consider S^3 and the torus T^2 in it described above as the boundary of the cone. One can determine that a three-dimensional minimal film has a narrower neck than the two-dimensional catenoid, as shown in Fig. 3. In the first column of Fig. 3 we illustrate hypothetical cones with boundary $S^p \times S^q$, in the second column we illustrate actual minimal surfaces with the same boundary, and in the third column are shown flat sections (generators) of these minimal surfaces by two-dimensional half-planes passing through the axes of symmetry of the surface. A symmetric cross section takes the form of a curve that, with an increase of dimension, sags more and more towards the origin O. Therefore, in this monotonic process, there will be a point where the symmetric minimal surface sags so much that its neck contracts to a point, and the symmetric film becomes a symmetric cone. This happens starting with dimension 8. For dimension 7 or less, there do not exist minimal nontrivial cones distinct from plane discs[1,2,14].

We make the following important statement[14]. Let A^{n-2} be a closed smooth locally minimal submanifold (i.e. multidimensional soap film) in the sphere S^{n-1} standardly embedded in $\mathbf{R}^n$. Let A not be an equator. Then, for $n \leq 7$, the cone CA [i.e. $(n-1)$-dimensional surface generated by all radii from the origin to the points of the manifold A] is not minimal; in other words, there exists a reducing deformation of the cone decreasing its volume.

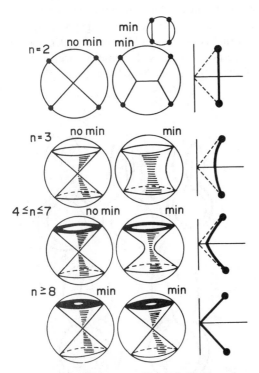

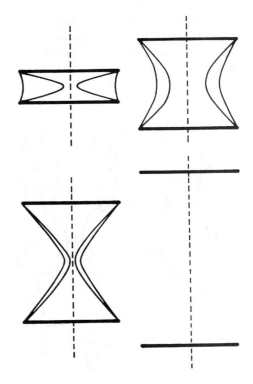

Fig. 3. Relation between soap films embedded in a sphere and dimension. In the first column are hypothetical cones with boundaries in $S^p \times S^q$. The second column illustrates actual soap films with the same boundary. The third column shows flat sections of the minimal surfaces by half-planes passing through axes of symmetry of the surface.

Fig. 4. Two catenoids and a pair of plane discs spanning two coaxial parallel circumferences as an illustration of the nonuniqueness of soap films spanning a given contour.

Now let us consider the question of the change of topological type of minimal surfaces depending on their stability or instability. Generally speaking, several different minimal surfaces (soap films) can span the same contour. We are already familiar with such cases. For example, we can span three distinct soap films on two coaxial parallel circumferences: two catenoids and a pair of plane parallel discs, as shown in Fig. 4. The uniqueness of a soap film can be guaranteed for the case of plane closed non-self-intersecting contours, i.e. contours realized in the form of a system of curves embedded in a plane. It is clear that any film different from a portion of a plane and bounded by these curves has a greater area than a planar film. In addition, we can consider two variational problems: finding films of the smallest area given a contour, and finding minimal films of the minimum topological type (i.e. simple type). We have not excluded the case of a film having minimal area not being of minimum topological type. Conversely, a minimal film of the minimum type could be a film not having minimal area in the class of all films with a given boundary. In other words, the absolute minimum of the area can be reached on films of nonminimum topological type, as shown in Fig. 5. If pairs of circumferences near symmetric contours are close enough, then the area of the first film is less than the area of the second one. At the same time, the first film is homeomorphic to the torus with one hole as shown in Fig. 5, i.e. it has a topological genus equal to 1. The second film is homeomorphic to a disc (also shown in Fig. 5), and because of this it has topological genus 0. The genus is the number of handles on a film.

Consider the symmetric contour in Fig. 5 (the so-called Douglas contour). It is quite suitable for the purpose of studying nontrivial evolutions of minimal surfaces and changes of genus depending on stability or instability of soap films. It is clear that a symmetric contour is homeomorphic to a circumference. We will assume that a contour is realized in $\mathbf{R}^3$ in the form of a closed curvilinear wire. Let u and v be the distances between the upper and lower rings, as shown in Fig. 5. For each fixed condition of the contour, there exist, generally speaking, several types of minimal surfaces. Our goal is to investigate their evolutions under deformation of the boundary contour. This problem was considered by T. Poston in [4]. The author considered this problem from another point of view. He studied the behavior of areas of soap films under deformations of a contour and constructed graphs of the areas which are multivalued functions of some interesting topological structure. In addition to an analytical investigation based, in particular, on [5], the author, together with A. A. Tuzhilin (a student of Moscow State University) made a series of physical experiments with symmetric soap films. In these experiments, the following soap films were obtained: (a)–(e) of Figs. 6, 7, and 8 and the films (f)–(h) of Fig. 9.

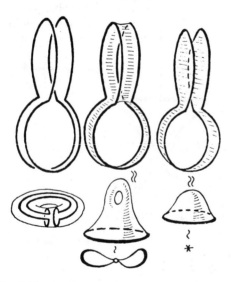

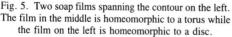

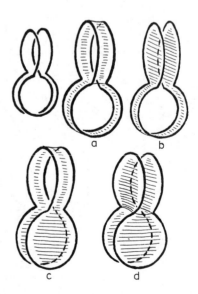

Fig. 5. Two soap films spanning the contour on the left. The film in the middle is homeomorphic to a torus while the film on the left is homeomorphic to a disc.

Fig. 6. The evolution of soap films as a result of continuous transformations of a symmetric contour (the Douglas contour).

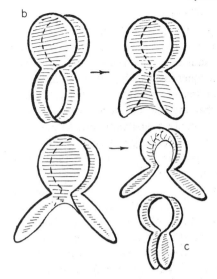

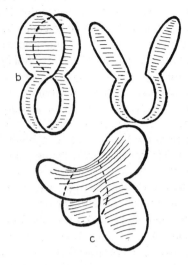

Fig. 7. See caption to Fig. 6.   Fig. 8. See caption to Fig. 6.

Films of types (b)–(e) are homeomorphic to a disc. Films of type (f) are not manifolds that can be contracted along themselves to a point like the films of types (b)–(e). We will restrict ourselves to the study of soap films (b)–(e), i.e. of films of the disc type. T. Poston in [4] indicated the following evolution of these films generated by a continuous deformation of the contour.

Take film (b) and deform the boundary contour (at the beginning symmetrical), implying a deformation of the soap film spanning this contour as shown in Fig. 7. Let us move the two lower circumferences apart step by step. We can assume that this process approximately coincides with the dilatation of a catenoid when two of its sides are moved away from each other. Here we use the fact that if the initial distance of the circumferences from each other is small enough, then a soap film (a strip spanning each pair of circumferences) is well approximated by a catenoid. During an elongation of the catenoid, there will be a critical moment when the catenoid is reconstituted into a pair of discs spanning two circumferences moving away from each other (symmetric contours). However, in our case, these two discs are still connected by a narrow strip stretched over the two upper circumferences, which are influenced to a lesser degree by the described deformation. As a result, we obtain the film shown in Fig. 7. Now moving the two lower circumferences in the opposite direction and returning them to the initial position, we do not change the topological type or genus of the film and arrive at film (c). Therefore, we have illustrated a continuous deformation of a contour, resulting in a step change of the soap film. It changed its genus from position (b) to position (c).

Using the symmetry of the contour, we can, similarly, make a transition in the opposite direction from soap film (c) to film (b). For this, it is enough to perform the operation indicated above on the two upper circumferences: first moving them away from each other, and finally

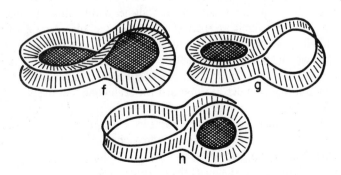

Fig. 9. See caption to Fig. 6.

returning them to the initial position. In the process of deformation, the film is reconstituted stepwise and soap film (c) transforms to film (b). This "pendulum" can be repeated many times if the soap film is formed from a sufficiently elastic liquid. Along with the described stepwise evolution of a film, there is another remarkable deformation of the boundary contour, resulting in a *smooth* change of a film with the same final result(!) In fact, consider soap film (b), and unlike the previous case, start to move the upper circumference apart, as shown in Fig. 8. As a result, we obtain a symmetric film of the saddle type; its center is a saddle point. The film goes up in one direction and down in another. This deformation is smooth without jumps.

Since film (e) is symmetric, applying the same operation to it, we can smoothly transfer it to film (c). Similarly, by repeating all these steps in the opposite direction we can smoothly transform film (c) to film (b) without jumps or change in topology. Therefore, in the space of all surfaces with homeomorphic boundary contours, we obtain two essentially different ways of joining positions (b) and (c). The first way was realized by the smooth deformation of a contour, in the process of which the film was subject to a step change. The second way is realized by another smooth deformation, in the process of which the film evolves without jumps, but exactly as in the first case, the film transforms from position (b) to (c).

As the first approximation, we can assume that the described deformations of contours can be expressed by a variation of two parameters: u, the distance between the two upper circumferences, and v, the distance between the two lower circumferences. In other words, u and v are the widths of upper and lower ring strips on surface (a) as shown in Fig. 6. We can assume that each position of the boundary contour is determined by a pair of real numbers which, for purposes of visualization, we can consider to be a point in a two-dimensional plane.

T. Poston[4] suggested a way to describe this process with the help of the (so-called) Whitney assembling. However, we decided to approach this problem from another point of view: to investigate properties of graphs of the areas of soap films. It turns out that the picture, so obtained, is described by the "swallow tail." Let us state this problem more precisely.

For each point (u, v) we have the corresponding contour which is, generally, spanned by several soap films. Calculating their areas, we get a graph of a multivalued function specifying the areas of the soap films with the given boundary. There are two kinds of films, stable and unstable. Among the films listed above, film (d) is unstable. All others are stable. Consider the unstable films in greater detail. We will use one of Plateau's physical principles, according to which a part of a minimal film bounded by a closed contour, drawn on the film, is also minimal with respect to this contour. In actual physical experiments, unstable films can be obtained as follows: The instability of a film means that there exist arbitrarily small disturbances that destroy the film or force it to change its position with respect to the boundary contour. At the same time, if one is able to locate, on the film, sufficiently many filaments with ends fixed to the contour, then one can achieve stability if the net of filaments is sufficiently dense.

In other words, we can attempt to keep an unstable soap film in a stable position by increasing the contour, adding a net of filaments to the initial contour. Take, for example, unstable film (d) shown in Fig. 6, spanning a contour to which a filament is attached [as shown in Fig. 10(a)]. Let this filament lie freely on the film, i.e. no forces act on it. Let us slightly deform the contour by moving the right circumferences apart. We find that a portion of the film sags to the left, and the filament stretches, becoming a singular edge of the film and keeping it in the equilibrium position, not allowing further deformation. Similar events happen if one moves the left circumferences apart (in this case the filament sags to the right and stretches) to the middle position, i.e. on film (d) the filament is not stretched. If, at this critical position of the film, we remove the filament from one end (instead of fixing the second end of the filament to the contour, we keep it in our hand), giving it a small initial impulse, then the film jumps to condition (b). In order to show that for each condition of the contour (for small u and v) there exists an unstable film of the type (c), we proceed as follows. Take film (b), fix a filament to one point of the contour as shown in Fig. 10(b), and start to pull the filament, as shown in Fig. 10(c). Very soon, we obtain a critical position of the film in which the film becomes symmetric. Here symmetries of soap films play a big role.

If we continue to stretch the filament, there will be a jump of the film and the filament

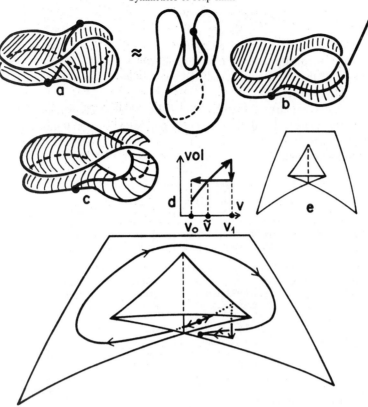

Fig. 10. (a)–(c) Stabilizing soap films of types (a)–(c) by the addition of filaments. (d) Change of area of the film as a result of transformation of the boundary contour. (e) Graph of the multivalued function describing areas of soap films of a given topological type. The surface obtained and the corresponding singularity is called a swallow tail.

sags to the left. If, while we are in a critical symmetric position, we release the filament giving it a large impulse, then the film returns to the previous position (b).

Now consider the position of our system $[u_0, v_0, (b)]$ where parameters u_0 and v_0 are not very large and let us smoothly increase v. Film (b) sags strongly to the center, and finally there will be a condition in which, under further small increases of v, the film begins to move independently under the forces of surface tension of the liquid, changing to condition (b) by a jump and with an almost unchanged boundary contour. If we now decrease parameter v to its initial value, the film will have no jumps. It keeps its type (b), which was obtained as the result of the jump. Because of the brevity of this paper, we will skip the details and given only the final graph of the change of area of the film under these deformations [Fig. 10(d)]. The details can be found in the author's book[8]. The exact proof is based on the results of [5].

The final result of our analysis is shown in Fig. 10(e). The graph of a multivalued function, describing the areas of soap films of a given topological type, is shown in this figure. The surface obtained and the correesponding singularity is called a swallow tail.

It is known that any smooth mapping of a three-dimensional manifold is approximated by mappings having only Whitney singularities of the following forms: pleat, assembling and swallow tail[15]. A swallow tail can be represented as a surface in three-dimensional space of polynomials of the form

$$x^4 + ax^2 + bx + c,$$

consisting of points (a, b, c), corresponding to polynomials with multiple roots.

Let us return to the graph of the area of soap films. In Fig. 10(f) we find the interesting possibility of a transition from condition $[u, v, (b)]$ to condition $[u, v, (c)]$ with the help of jumps as well as smoothly (continuously). The first way is the rise to a singular edge [the path to the right in Fig. 10(f)] and afterwards "falling" from a point on the edge to a point of type (c). The second way, we go smoothly down from point (b) along a leaf of the graph, around a singularity of the graph, and we come to point (c).

Thus, unstable soap films of type (d) fill (from the point of view of the graph of its area) the upper triangle of the swallow tail, and the descending wings of the graph correspond to stable soap films (b) and (c).

As we see, the symmetry of soap films and the algebraic properties at extremals of the variational problem connected with this symmetry play special roles in this topological picture. For more details concerning the role of symmetry in multidimensional variational problems, see the author's books[7–9].

REFERENCES

1. F. J. Almgren, Existence and regularity almost everywhere of solutions to elliptic variational problem among surfaces of varying topological type and singularity structure. *Ann. Math. Ser.* 2 **87**(2), 321–391.
2. F. J. Almgren and J. E. Taylor, The geometry of soap films and soap bubbles. *Scientific American*, 82–93 (July 1976).
3. H. Federer, Geometric measure theorie. Springer, Berlin (1969).
4. T. Poston, The Plateau problem. An invitation to the whole of mathematics. Summer College on global analysis and its applications, 4 July–25 August 1972. International Centre for Theoretical Physics, Trieste, Italy.
5. M. J. Beeson and A. J. Tromba, The cusp catastrophe of Tome in the bifurcation of minimal surfaces. *Manuscripta Matematica* **46**, 273–308 (1984).
6. E. Bombieri, De Georgi E. Giusty, Minimal cones and the Bernstein problem. *Invent. Math.* **7**(3), 243–268 (1969).
7. A. T. Fomenko, *Variational Methods in Topology* (in Russian). Nauka, Moscow (1982).
8. A. T. Fomenko, *Topological Variational Problems* (in Russian). Moscow State University, Moscow (1984).
9. A. T. Fomenko, Multidimensional variational problems in the topology of extrema (in Russian). *Uspekhi matem. nauk* **36**(6), 105–135 (1981).
10. J. C. Nitsche, Vorlesungen über Minimalflächen. Die Grundlehren der Matn. Wissenschaften in Einzeidarstellungen, Bd. 199. Springer-Verlag (1975).
11. H. B. Lawson, The equavariant Plateau problem and interior regularity. *Trans. Amer. Math. Soc.* **173**, 231–249 (1972).
12. S. Hildebrandt and J. C. C. Nitsche, Optimal boundary regularity for minimal surfaces with a free boundary. *Manuscripta Math.* **33**(3/4), 357–364 (1981).
13. S. S.-T. Yau, Kohn-Rossi Cohomology and its application to the complex Plateau problem. I. *Ann. Math.* v. **113**(1), 67–110 (1981).
14. J. Simons, Minimal varieties in Riemannian manifolds. *Ann. Math.* **88**(1), 62–105 (1968).
15. V. I. Arnol'd, A. N. Varchenko and S. M. Gusei-Zade, *Peculiarities of Differential Reflections* (in Russian). Nauka, Moscow, Vol. 1 (1982), Vol. 2 (1984).

Comp. & Maths. with Appls. Vol. 12B, Nos. 3/4, pp. 835–848, 1986
Printed in Great Britain.

0886-9561/86 $3.00 + .00
© 1986 Pergamon Press Ltd.

NOTES ON A VISUAL PHILOSOPHY

AGNES DENES†
595 Broadway, New York, NY 10012, U.S.A.

Abstract—A visual artist takes a look at her own work in which symmetries embody an anatomy of substructure. Essences of ideas are presented in visual form to serve as a new language of communication. These notes consist of three parts. Thoughts on symmetry related to art followed by comments on the artist's own work, and are concluded by 18 of her images.

The symmetries operating in my work are subtle and complex. Some are more easily discernible than others, some operate on the surface of perceptions on the visual level, others are deeply hidden in the ideas and philosophies underlying my work. Whether obvious or subtle, visible or elusive, these symmetries are inherent and real. The fact that they were never consciously sought, since the work was not created with this goal in mind, makes their presence even more exciting.

In mathematics symmetries are precise, well-defined operations such as rotation, translation and inversion. In the sciences these operations are applied to an idealized physical world where they abound, ranging from bilateralism in people to celestial motion to time-reversible laws. Even broken symmetry, so popular among physicists today, has precise meaning. Logic, which manipulates concepts mathematically, has exact symmetries. In art and music they appear on many levels serving esthetic functions and have been the subject of numerous essays, including papers in this journal.

In my work everything, including symmetry, is created through a conscious use of instinct, intellect and intuition. When I visualize (give form to) processes such as math and logic, or when I apply X-ray technology and electron microscopy to organic and crystal structures, one might say I reveal well-defined symmetries and antisymmetries. When I deal with abstract concepts definitions blur and the symmetries go beyond ordinary mathematical confines. Some examples follow.

- Mapping the loss that occurs in communication, i.e. between viewer and artist, between giver and receiver, between specific meaning and symbol, between nations, epochs, systems and universes.
- Mapping human parameters within the changing aspects of reality, within the transformations and interactions of phenomena.
- Working with the paradox, the contradictions of human existence such as our illusions of freedom and the inescapability of the system; our alienation in togetherness; the individual human dilemma, struggle and pride versus the whole human predicament; our importance or insignificance in the universe.
- Trying to give form to invisible processes such as evolution, changing human values, thought processes and time aspects (pinpointing the moment growth becomes decay in an organism; penetrating the "folds of time" to record its "instants").
- Finding contradictions and balances, pitting art against existence, illusions versus reality, imagination versus fact, chaos versus order, the moment versus eternity, universals versus the self.

These symmetries are less available and definable but they are there nevertheless, working on mysterious levels in the interactions of phenomena and ideas. I may even venture to say that

†Agnes Denes combines art with science, mathematics, languages and philosophy. She has been exhibiting in the U.S. and Europe since 1965 and her work is in major museums and private collections around the world. In 1982, the author was chosen "Master of Drawing" in Nuremberg Germany, representing the U.S. Published works: *Paradox and Essence, Sculptures of the Mind, Isometric Systems in Isotropic Space: Map Projections*, and *Book of Dust (The Beginning and the End of Time and Thereafter)*.

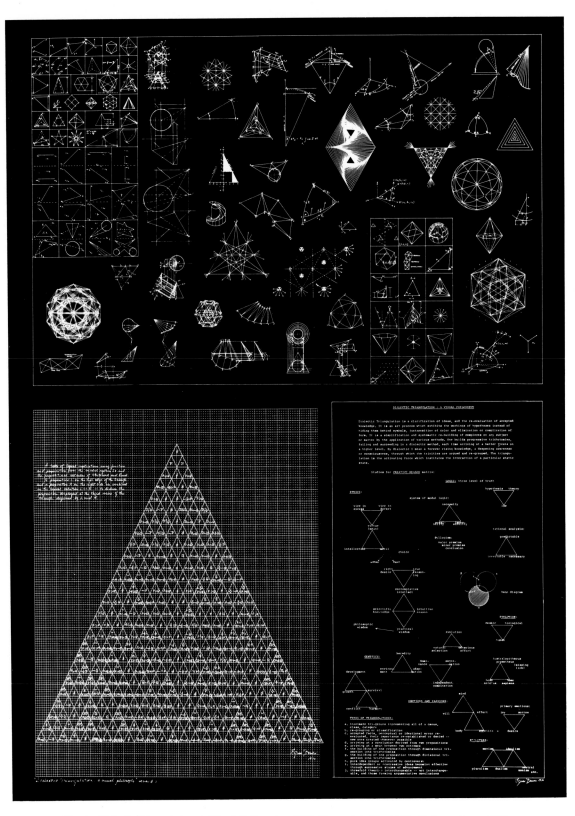

Fig. 1. Dialectic Triangulation: A Visual Philosophy. 1970. Monoprint, 37 × 28″. © Agnes Denes 1970.

Fig. 2. "4000 Years." 1975. Ink on graph paper, 29 × 22". "If the mind possesses universal validity—art reveals a universal truth. I want that truth." (transliterated into Middle Egyptian by the artist). © Agnes Denes 1975.

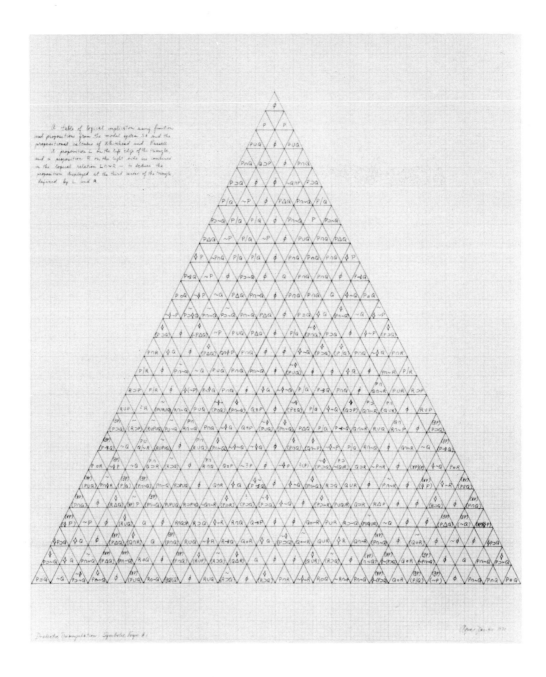

Fig. 3. The Human Argument. (A table of logical implications, using functions and propositions from the modal system S8 and the propositional calculus of Whitehead and Russell.) 1970. Ink on graph paper, 29 × 22″. © Agnes Denes 1970.

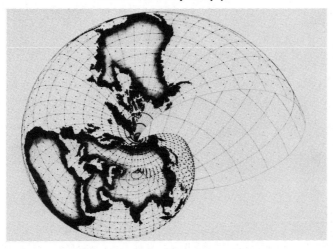

Fig. 4. Isometric Systems in Isotropic Space—Map Projections. The Snail. 1974. Ink and charcoal on graph paper and Mylar, 24 × 30″. © Agnes Denes 1974.

these unnamed, unmeasured symmetries operating in the network of concepts are the anatomy or substructure of invisible underlying patterns of existence that make new associations and analogies possible.

As the world is becoming more complex and knowledge and ideas are coming in faster than they can be assimilated, disciplines become progressively more alienated from each other through specialization, words lose their precision as they take on multi-meanings and communication breaks down. A new type of analytical attitude is called for, a clear overview or a summing up. Not losing sight of abstract reasoning, but using induction and deduction in the discovery of the real and concrete structures, the substances of things and ideas emerge. In this sense they represent the primary being of things to act as universal forms or ideas when brought to the surface to interact with each other. When "things" are pared down to their core or essence, superfluous data fall away and new associations and insights become possible.

My concern is with the creation of a language of perceptions that allows the flow of information among alien systems and disciplines, in which essences carry pure meaning into pure form and all things can be considered once more simultaneously. From specializations to essences, from patterns to symmetries, to form, seeking ultimates in the elemental nature of things and vice versa. Thus analytical propositions are presented in visual form where both the proposition and its deductive reasoning achieve their own essence and communicate visually. The resulting art is a dramatization or "visualization" of these forms, entities or summations. They are words, sentences or paragraphs in a language of seeing. They are the universal concepts or substances I often refer to in my writings.

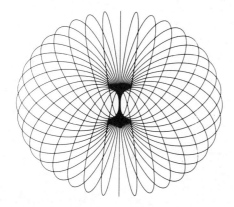

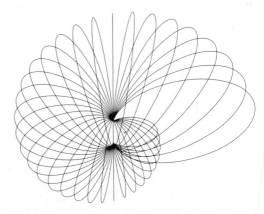

Fig. 5. Isometric Systems in Isotropic Space—Map Projections. The Doughnut—Fragmentation. 1974. Ink on vellum, 24 × 30″. © Agnes Denes 1974.

Fig. 6. Isometric Systems in Isotropic Space—Map Projections. The Snail—Fragmentation. 1976. Ink on vellum, 24 × 30″. © Agnes Denes 1976.

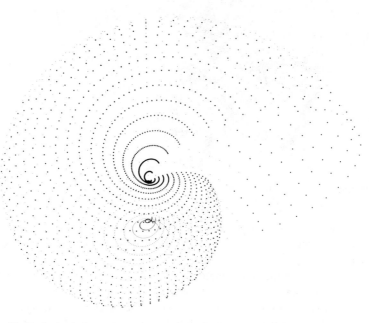

Fig. 7. Isometric Systems in Isotropic Space—Map Projections. The Snail—Fragmentation. 1976. Ink on vellum, 24 × 30″. © Agnes Denes 1976.

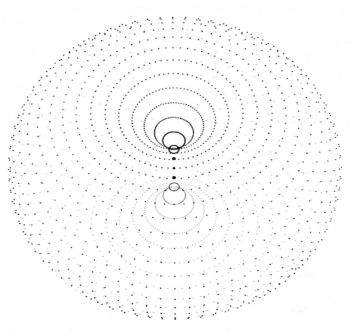

Fig. 8. The Doughnut—Fragmentation. 1976. Ink on vellum, 24 × 30″. © Agnes Denes 1974.